CATECHOLAMINE RESEARCH

From Molecular Insights to Clinical Medicine

ADVANCES IN BEHAVIORAL BIOLOGY

Recent Volumes in This Series

Volume 38B BASIC, CLINICAL, AND THERAPEUTIC ASPECTS OF ALZHEIMER'S AND PARKINSON'S DISEASES, Volume 2
Edited by Toshiharu Nagatsu, Abraham Fisher, and Mitsuo Yoshida

Volume 39 THE BASAL GANGLIA III
Edited by Giorgio Bernardi, Malcolm B. Carpenter, Gaetano Di Chiara, Micaela Morelli, and Paolo Stanzione

Volume 40 TREATMENT OF DEMENTIAS: A New Generation of Progress
Edited by Edwin M. Meyer, James W. Simpkins, Jyunji Yamamoto, and Fulton T. Crews

Volume 41 THE BASAL GANGLIA IV: New Ideas and Data on Structure and Function
Edited by Gérard Percheron, John S. McKenzie, and Jean Féger

Volume 42 CALLOSAL AGENESIS: A Natural Split Brain?
Edited by Maryse Lassonde and Malcolm A. Jeeves

Volume 43 NEUROTRANSMITTERS IN THE HUMAN BRAIN
Edited by David J. Tracey, George Paxinos, and Jonathan Stone

Volume 44 ALZHEIMER'S AND PARKINSON'S DISEASES: Recent Developments
Edited by Israel Hanin, Mitsuo Yoshia, and Abraham Fisher

Volume 45 EPILEPSY AND THE CORPUS CALLOSUM 2
Edited by Alexander G. Reeves and David W. Roberts

Volume 46 BIOLOGY AND PHYSIOLOGY OF THE BLOOD–BRAIN BARRIER: Transport, Cellular Interactions, and Brain Pathologies
Edited by Pierre-Olivier Couraud and Daniel Scherman

Volume 47 THE BASAL GANGLIA V
Edited by Chihoto Ohye, Minoru Kimura, and John S. McKenzie

Volume 48 KINDLING 5
Edited by Michael E. Corcoran and Solomon Moshé

Volume 49 PROGRESS IN ALZHEIMER'S AND PARKINSON'S DISEASES
Edited by Abraham Fisher, Israel Hanin, and Mitsuo Yoshida

Volume 50 NEUROPSYCHOLOGY OF CHILDHOOD EPILEPSY
Edited by Isabelle Jambaqué, Maryse Lassonde, and Olivier Dulac

Volume 51 MAPPING THE PROGRESS OF ALZHEIMER'S AND PARKINSON'S DISEASE
Edited by Yoshikuni Mizuno, Abraham Fisher, and Israel Hanin

Volume 52 THE BASAL GANGLIA VII
Edited by Louise F. B. Nicholson and Richard L. M. Faull

Volume 53 CATECHOLAMINE RESEARCH: From Molecular Insights to Clinical Medicine
Edited by Toshiharu Nagatsu, Toshitaka Nabeshima, Richard McCarty, and David S. Goldstein

CATECHOLAMINE RESEARCH

From Molecular Insights to Clinical Medicine

Edited by

Toshiharu Nagatsu
Fujita Health University
Toyoake, Aichi, Japan

Toshitaka Nabeshima
Nagoya University
Nagoya, Japan

Richard McCarty
Vanderbilt University
Nashville, Tennessee

and

David S. Goldstein
National Institute of Neurological Disorders and Stroke
Bethesda, Maryland

Springer Science+Business Media, LLC

Catecholamine research: from molecular insights to clinical medicine/edited by Toshiharu Nagatsu ... [et al.].
p. cm. — (Advances in behavioral biology; 53)
Includes bibliographical references and index.

1. Catecholamines—Congresses. I. Nagatsu, Toshiharu. II. Series.

QP801.C33 C383 2002
612.8'042—dc21

2002028277

Proceedings of the 9th International Catecholamine Symposium, held from March 31st to April 5th, 2001, in Kyoto, Japan

ISBN 978-1-4419-3388-1 ISBN 978-1-4757-3538-3 (eBook)
DOI 10.1007/978-1-4757-3538-3

Originally published by Kluwer Academic / Plenum Publishers, New York in 2002.
Softcover reprint of the hardcover 1st edition 2002

http://www.wkap.nl/

10 9 8 7 6 5 4 3 2 1

A C.I.P. record for this book is available from the Library of Congress

ACKNOWLEDGEMENTS

The Organizing Committee of the 9th International Catecholamine Symposium gratefully acknowledges the support of the following organizations:

Eli Lily Japan K.K.

Eisai Co., Ltd. & Pfizer Pharmaceuticals Inc.
Pharmacia Corporation
The Japanese Owners of Race Horses Association

Osaka Pharmaceutical Manufactures' Association
The Pharmaceutical Manufacturers' Association of Tokyo

Dr. Tairiku Amakusa
Dr. Banri Amakusa
Fujimoto Pharmaceutical Corp.
Kissei Pharmaceutical Co.
Kyowa Hakko Kogyo Co.
Medtronic Japan Co., Ltd.
National Parkinson Foundation
Nippon Boehringer Ingelheim Co.
Novartis
Sumitomo Pharmaceuticals

Dr. Taro Masaoka
Dr. Masahiro Nomoto
Dr. Masatoshi Morooka
Dr. Mitustoshi Yamamoto
Daiichi Radioisotope Labs., Ltd.
GlaxoSmithKline K.K.
Janssen-Kyowa Co., Ltd.
Meiji Seika Kaisha, Ltd.
Foundation for Catecholamine Research, U.S.A.
(I. J. Kopin Travelling Fellowship Award)
Nippon Zoki Pharmaceutical Co., Ltd.
Smoking Research Foundation

Solvay SA
Takeda Chemical Industries, Ltd.
Uehara Memorial Life Science Foundation

Electronic Industries Association of Japan
The Federation of Electric Power Companies
The Federation of Steel Manufacturing Association of Japan
Tokyo Bankers Association, Inc.
Toyota Motor Corporation

Bearing Industry Association of Japan
Chemical Textile Association of Japan
Communication Device Manufacturing Association of Japan
Electronics Manufacturing Association of Japan
Japan Federation of Construction Contractors
Japan Foreign Trade Council, Inc.
Japan Securities Dealers Association
Mining Industry Association of Japan
Paper Manufacturing Industry Association of Japan
Petroleum Association of Japan
Photoindustry Association of Japan
Regional Banks Association of Japan
The Japan Gas Association
The Japan Warehousing Association Inc.
The Japanese Shipowner's Association
The Real Estate Companies Association in Japan
The Life Insurance Association of Japan
Trust and Banking Association of Japan
The second Regional Banking Association of Japan

Organizing Committee

T. Nagatsu, Japan (President)
D.S. Goldstein, U.S.A.
R. McCarty, U.S.A.

Local Organizing Committee

H. Ichinose, I. Kanazawa, M. Kimura, K. Kobayashi, Y. Mizuno, S. Nakamura, I. Nagatsu, M. Sasa, M. Satoh, M. Sawada, S. Tsuji, Y. Watanabe, T. Nabeshima (Secretary General)

International Scientific Advisory Board

I. Armando, Argentina
J. Axelrod, U.S.A. (Honorary Chairperson)
M. R. Bennett, Australia
R. Kvetnansky, Slovakia
J. Lenders, Netherlands
J. Mallet, France

H. Bönisch, Germany
A. A. Boulton, Canada
A. Carlsson, Sweden
M. W. H. Coughtrie, UK
C. R. Creveling, U.S.A.
W. P. De Potter, Belgium
G. Eisenhofer, U.S.A.
T. Flatmark, Norway
D. S. Goldstein, U.S.A.
A. Graybiel, U.S.A.
T. Hökfelt, Sweden
T. H. Joh, U.S.A.
I. J. Kopin, U.S.A. (Honorary Chairperson)
R. McCarty, U.S.A.
O. A. Nedergaard, Denmark
L. Oreland, Sweden
M. Palkovits, Hungary
P. F. Riederer, Germany
E. L. Sabban, U.S.A.
M. Scheinin, Finland
J. C. Shih, U.S.A.
P. Soares-da-Silva, Portugal
L. Stjärne, Sweden
H. Winkler, Austria
M. B. H. Youdim, Israel

PREFACE

This book is based on invited presentations at the Ninth International Catecholamine Symposium. Over several decades, each International Catecholamine Symposium (ICS) has provided a uniquely important forum for updating basic as well as clinical research on the catecholamines, dopamine, norepinephrine, and epinephrine. The first ICS took place in Bethesda, Maryland, in the USA in 1958; the 2nd in Milan, Italy in 1965; the 3rd in Strasbourg, France in 1973; the 4th in Asilomar, California, USA in 1978; the 5th in Goteborg, Sweden in 1983; the 6th in Jerusalem, Israel in 1987; the 7th in Amsterdam, Netherlands in 1992; and the 8th in Asilomar, California, USA in 1996.

The 9th International Catecholamine Symposium (9ICS) was held in Kyoto, Japan, March 31–April 5, 2001. This ICS was held concurrently with the 5th International Conference on Progress in Alzheimer's and Parkinson's Disease (5th ADPD). These international meetings were coordinated to be a joint international congress. Catecholamines and related neurotransmitters and neuromediators play important roles in the pathogenesis of symptoms and neurodegenerative processes of Alzheimer's disease and Parkinson's disease. Therefore, the Joint Congress provided an opportunity for investigators and clinicians working in these fields to exchange their most recent investigational results and clinical experiences.

This Joint International Congress turned out to be an enormous success, attended by 1258 participants from 38 countries, in an enthusiastic and pleasant atmosphere.

Following along the lines of the previous highly successful 8th ICS, the 9th ICS maintained a focus on bridging genetic, molecular biological, physiological, neurochemical, behavioral, and clinical interests. Catecholamines are neurotransmitters related to higher brain functions such as movement emotion, reward, cognition, memory, and learning, and more "automatic" lower brain functions such as autonomic regulation of the cardiovascular system. The interactions of various neurotransmitters and the related compounds in complex neuronal circuits are considered to be extremely important in understanding functions of central and peripheral catecholamine systems in health and disease. In the program of the 9th ICS, the roles of catecholamines were considered in relation to other neurotransmitters within neuronal networks, such as glutamate, gamma-aminobutyric acid, serotonin, histamine, acetylcholine, neuropeptides, and nitric oxide, as well as in relation to other neuromediators such as neurotrophins and cytokines in neuron-glial cells interactions.

We thank the members of the International Scientific Advisory Board, and the Local Organizing Committee, especially Prof. Yoshikuni Mizuno, the President of 5th ADPD, for their constructive input and their excellent service as chairpersons and speakers in the Symposium.

We also deeply appreciated the financial support of several contributors that ensured the success of the meeting noted in the preceeding acknowldgements section that lists sources of support of this Joint Congress of the 9th ICS with the 5th ADPD.

It is hoped that this book about catecholamine research will be a valuable addition to the libraries of students and scientists wordlwide who are interested in catecholamines specifically and neurotransmitters in general and will foster development of further collaborative efforts in basic and clinical science in several disciplines.

Toshiharu Nagatsu
Toshitaka Nabeshima
Richard McCarty
David S. Goldstein

CONTENTS

THE DOPAMINERGIC SYSTEM

DEVELOPMENTAL NEUROBIOLOGY OF CATECHOLAMINERGIC SYSTEMS

PHYSIOLOGICAL EFFECTS OF CATECHOLAMINES

IN MEMORY OF DR. SIDNEY UDENFRIEND

Toshiharu (Toshi) NAGATSU*

Dr. Sidney Udenfriend
(1918~1999)

I was deeply shocked and extremely saddened, when I received the news that Dr. Sidney Udenfriend had passed away on December 29, 1999, after his cardiac surgery. I had had a long talk by telephone with Dr. Sidney Udenfriend at his home in New Jersey on August 6, 1999, when I was at the NIH as a Fogarty Scholar. I wanted to see him, but his health condition did not permit my visit. When I talked with Dr. Udenfriend, he was very kind and encouraged me to continue my research, as he always did whenever I talked with him.

* Toshiharu Nagatsu, Institute for Comprehensive Medical Science, Fujita Health University, Toyoake, Aichi 470-1192 Japan.

Catecholamine Research, Edited by Nagatsu *et al.*
Kluwer Academic/Plenum Publishers, 2002

I worked with Dr. Udenfriend twice, first at the NIH as an NIH International Postdoctoral Fellow from 1962 to 1964, and then at the Roche Institute of Molecular Biology as a Visiting Scientist from 1972 to 1973. My wife, Ikuko, also worked with Dr. Udenfriend at the Roche Institute of Molecular Biology from 1972 to 1973.

Dr. Sidney Udenfriend was one of the great pioneers in catecholamine (CA) research, and opened the way for the development of CAs as the key molecules bridging basic science with clinical science. In 1950 Dr. Udenfriend by isotope methods had completely elucidated the biosynthetic pathway of CAs: tyrosine ⟶ 3,4-dihydroxyphenylalanine(DOPA) ⟶ dopamine(DA)⟶norepinephrine/noradrenaline (NE/NA)⟶ epinephrine/adrenaline (EN/AD) (Udenfriend et al., 1953). In Dr. Udenfriend's laboratory, the second enzyme of CA biosynthesis, "DOPA decarboxylase," was isolated, characterized, and termed "aromatic L-amino acid decarboxylase (AADC)", as it was found to decarboxylate L-DOPA to dopamine (DA) in CA cells and also L-5-hydroxytryptophan to 5-hydroxytryptamine (serotonin) in serotonin cells (Lovenberg et al., 1962).

During my work with Dr. Udenfriend at the NIH from 1962 to 1964, we discovered tyrosine hydroxylase (tyrosine 3-monooxygenase), the first-step, rate-limiting, and pteridine-requiring monooxygenase for CA biosynthesis, which converts tyrosine to L-DOPA, the substrate for aromatic L-amino acid decarboxylase (Nagatsu et al., 1964). I recall distinctly how delighted we were to see the enzyme activity of isolated tyrosine hydroxylase being recorded by that liquid scintillation counter at midnight so long ago.

In 1958 Dr. Sidney Udenfriend, along with Dr. Sydney Spector and a number of others who were at the NIH at that time, planned the 1st International Catecholamine Symposium (1ICS), which was followed by the 2nd International Catecholamine Symposium (2ICS) in Milan, Italy, in 1965. The 1ICS at the NIH in 1958 included such notable scientists as J. Axelrod, S.M. Bacq, H. Blaschko, B.B. Brodie, C. Cori, V.Z. Gorkin, P. Greengard, O. Hayaishi, S. Kaufman, I.J. Kopin, E.W. Sutherland, S. Udenfriend, J.R. Vane, U.S. von Euler, M. Vogt, and E.A. Zeller. As we thought that it would be nice for the young people in the field, who had never had the opportunity to have known him, to appreciate the impact he had had in organizing these Symposia, we planned the Dr. Sidney Udenfriend Memorial Lectures for this 9ICS.

Dr. Sidney Udenfriend was born in Brooklyn, New York, on April 5, 1918. He went to the City of College of New York where he received his B.S. and then to New York University to receive his M.S. and Ph.D. (Biochemistry). During the period Dr. Udenfriend was at New York University he also worked at the Goldwater Memorial Hospital in New York, where in this rich intellectual environment he interacted with such outstanding scientists as Drs. James Shannon (who later became director of the NIH), Bernard B. Brodie, Robert Berliner, and Julius Axelrod. Following his stay at Goldwater Memorial Hospital he worked at Washington University in St. Louis in Dr. Carl Cori's department of biochemistry, where his reputation as a biochemist grew. In 1950 he went to the NIH, where three years later he headed up the Section of Cellular Pharmacology in Dr. Bernard Brodie's laboratory. Then, from 1956 to 1968, he held the position of Laboratory Chief in his own Laboratory of Clinical Biochemistry. In 1968, he organized a basic research institute at Hoffman LaRoche and became Director of the Roche Institute of Molecular Biology. Under Dr. Sidney Udenfriend's direction of the Roche Institute of Molecular Biology developed into a truly internationally recognized center of excellence. Over the years this institute has been responsible for training more than 1100 young scientists, mostly post doctoral fellows, from all over the world. Many of the senior staff at Roche Institute of Molecular Biology built their scientific careers there, and thereafter assumed prestigious positions in academia and industry.

Dr. Udenfriend was Director and Laboratory Chief of Molecular Neurobiology at Roche from 1968 to 1983. In 1983, he stepped down as Director of the Roche Institute, and was replaced by Dr. Herbert Weissbach; however, he still remained as active as ever as the Laboratory Chief of Molecular Neurobiology.

In 1997, after he had retired from the Roche Institute at the end of 1996, he became Director of the Charles A. Dana Research Institute for Scientists Emeriti (RISE) at Drew University in Madison, N.J. The purpose of this institution is to afford retired scientists, who are still productive and who could be inspirational to students, the opportunity to mentor young scientists. The scientists at the Institute are retirees who are still at the top of their respective fields.

Dr. Sidney Udenfriend was one of the most fertile and original of thinkers. As he expressed to Dr. Sydney Spector, "if others are thinking of doing the same experiment that I am, it is time to get out of the field" (Spector, 2000). He made great contributions not only in the field of catecholamines but also in that of serotonin. He also conducted many original experiments in the fields of collagen, opioid peptides (especially enkephalins), glycosylphosphatidylinositol (GPI)-anchored proteins such as alkaline phosphatase, and lastly amino acid transporters. For starting new fields, he always developed and introduced new methods, and took advantage of state-of-the-art methods in fields, such as radiochemistry (radioactive and nonradioactive isotope methods), spectrofluorimetry, immunochemistry, immunohistocytochemistry, and molecular biology.

Over his long distinguished career Dr. Sidney Udenfriend was awarded many honors. He was the only foreign scientist who ever became an honorary member of both the Japanese Biochemical Society and the Japanese Pharmacological Society. He was also a member of the National Academy of Sciences USA and the American Academy of Arts and Sciences.

Dr. Udenfriend was not only an outstanding scientist of great genius but also an extremely warm-hearted teacher who kindly guided young scientists. Many Japanese scientists worked with Dr. Udenfriend at the NIH and at the Roche Institute of Molecular Biology, and they all greatly admired him. Dr. Udenfriend visited Japan 13 times, including 4 visits to our Institute for Comprehensive Medicine Science at Fujita Health University.

We have suffered the loss of a great teacher and friend. Dr. Udenfriend told me that a leader in science is like an orchestra conductor who brings out the maximum from each member of the group and is also like a father or mother who loves his or her students but who is strict with respect to research. He also told me that a scientific leader should have both an excellent brain and a warm heart. He was really outstanding in having had that combination. I had great respect for this man of science and am so grateful for having known him.

In closing I want to thank Dr. Paul Greengard (Vincent Astor Professor and Head, Laboratory of Molecular and Cellular Neuroscience, The Rockefeller University, New York, USA), who had worked with Dr. Udenfriend at the NIH from 1958 to 1959 and was awarded a Nobel Prize in 2000, for kindly presenting his lecture in memory of Dr. Sidney Udenfriend at this 9ICS. Likewise, I am also grateful to Dr. Osamu Hayaishi (Director Emeritus, Osaka Bioscience Institute, Osaka, Japan), who was a good friend of Dr. Sidney Udenfriend, for serving as the chairperson of my lecture in memory of Dr.

Udenfriend, and to Dr. Herbert Weissbach (Director, Center for Molecular Biology and Biotechnology, Florida Atlantic University, Boca Raton, Florida, USA) and Dr. Sydney Spector (Professor, Department of Psychiatry, Vanderbilt University, Nashville, Tenesee, USA) for their encouragements in organizing these Dr. Sidney Udenfriend Memorial Lectures.

REFERENCES

Lovenberg, W., Weissbach, H., and Udenfriend, S., 1962, Aromatic L-amino acid decarboxylase, *J. Biol. Chem.* 237: 89-93.

Nagatsu, T., Levitt, M., and Udenfriend, S., 1964, Tyrosine hydroxylase: the initial step in norepinephrine biosynthesis, *J. Biol. Chem.* 239: 2910-2917.

Spector, S., 2000, In Memorium: Sidney Udenfriend 1918-1999, *Neuropsychopharmacology* 22: 447-448.

Udenfriend, S., Cooper, J. R., Clark, C. T., and Baer, J. E., 1953, Rate of turnover of epinephrine in the adrenal medulla, *Science* 117: 663-665.

DR. SIDNEY UDENFRIEND MEMORIAL LECTURE

MOLECULAR GENETICS OF CATECHOLAMINES: KEY MOLECULES BRIDGING BASIC SCIENCE WITH CLINICAL SCIENCE

Toshiharu (Toshi) Nagatsu*

1. HISTORICAL OVERVIEW OF CATECHOLAMINE RESEARCH

Sidney Udenfriend (1918-1999) was one of the great pioneers in catecholamine (CA) research, and opened the way for the development of CA's as the key molecules bridging basic science with clinical science. The history of CA research is summarized in Table 1. CA's in vivo are composed of dopamine (DA), norepinephrine (NE) [noradrenaline (NA)], and epinephrine (EN) [adrenaline (AD)]. Historically, in 1898 John Jacob Abel first isolated an active CA compound in the adrenal medulla and named it EPINEPHRINE. Then, two years later, Jokiti Takamine isolated and crystallized the same active compound from the adrenal medulla and named it ADRENALIN. Later, ADRENALINE became the generic name. Thus EN/AD was discovered as a hormone in the adrenal medulla. Ulf S. von Euler (1946) discovered a CA neurotransmitter in the peripheral sympathetic nerves and named it NORADRENALINE, which is also called NOREPINEPHRINE. Then NE/NA and EN/AD were also found in the brain as neurotransmitters. This is the historical reason for having two names, EPINEPHRINE (EN) / ADRENALINE (AD) and NOREPINEPHRINE (NE) / NORADRENALINE (NA), for the same two molecules. Arvid Carlsson (1958) discovered DOPAMINE (DA), which is also called 3-HYDROXYTYRAMINE, as yet another CA neurotransmitter in the brain.

Although the three CA's, i.e., DA, NE/NA, and EN/AD, are similar in their structures, their physiological and pathological functions in the central and peripheral nervous systems (CNS, PNS) are quite different and versatile.

* Toshiharu Nagatsu, Institute for Comprehensive Medical Science, Fujita Health University, Toyoake, Aichi 470-1192 Japan.

Table 1. History of catecholamine research

CatecholamineResearch in Basic and Clinical Science

1900 | 40 | 50 | 60 | 70 | 80 | 90 | 2000

Bioassay | Histofluorescence Fluorimetry | Immunohistochemistry Immunoassay | Molecular Cloning | Mouse Genetics | Human Genetics

EN/AD NE/NA DA BH4
DDC/AADC PAH DBH PNMT TH
Biosynthetic pathway

MAO COMT
Metabolic pathway

Reuptake

TH gene
AADC gene
DBH gene
PNMTgene
MAO gene
COMT gene

hTH Tg mice
hDBH Tg mice
hPNMT Tg mice
TH KO mice
DBH KO mice
DA(-) mice

β-R
α-R
DA-R
DA-T
NE/NA-T
VMAT1
VMAT2
DARPP-32

PD (DA deficiency)
L-DOPA therapy
Schizophrenia (DA hypothesis)
α-,β-,DA-R agonists & antagonists
DBH deficiency (DOPS therapy)
AD-DRD/HPD (GCH1 deficiency)
AR-DRD (TH deficiency)

Studies on the biosynthesis and metabolism of CA's and their related enzymes began around the 1930's. L-3,4-Dihydroxyphenylalanine (DOPA) decarboxylase (DDC), which decarboxylates L-DOPA to DA, was discovered among the CA-synthesizing enzymes by Holtz (1937), and the following biosynthetic pathway was proposed by Blaschko (1938): tyrosine ⟶ DOPA ⟶ DA ⟶ NE ⟶ EN. This hypothetical pathway was later confirmed by isotope experiments by Udenfriend et al. (1953, 1956), and the CA-synthesizing enzymes were found at the laboratory of Sidney Udenfriend at the NIH. DDC was isolated and characterized to be aromatic L-amino acid decarboxylase (AADC), which produced both L-DOPA in CA cells and L-5-hydroxytryptophan (serotonin) in serotonin cells (Lovenberg et al., 1962). The last enzyme to be discovered was tyrosine 3-monooxygenase (tyrosine hydroxylase, TH), a new pteridine-dependent monooxygenase (Nagatsu et al., 1964). Phenylalanine hydroxylase (PAH), which converts phenylalanine to tyrosine in the liver, an essential step for complete degradation of phenylalanine, was discovered by Udenfriend & Cooper (1952), and its previously unrecognized cofactor was discovered and determined to be reduced biopterin by Kaufman (1963).

CA's were found to be mainly inactivated by both monoamine oxidase (MAO), by Blaschko et al. (1937), and by catechol O-methyltransferase (COMT), by Axelrod (1957). Thus all of the enzymes related to CA biosynthesis and metabolism were discovered from 1930 to 1964. The metabolism of CA's in vivo was elucidated by Irwin J. Kopin. A very important discovery by Axelrod was the reuptake of CA neurotransmitters, which is now established as a general principle for neurotransmitter termination (Axelrod et al., 1959,

Wolfe et al., 1962). Later, plasma membrane DA and NE transporters (DA-T, NE-T) and vesicular monoamine transporters (VMAT 1, VMAT 2) for reuptake of CA's were isolated, and their structures were determined by molecular cloning around 1990.

The physiology and pharmacology of alpha and beta-adrenergic receptors (alpha-R, beta-R) and DA receptors (DA-R) were studied in the 1960's and 1970's by binding assays using isotope-labeled ligands, and their structures were determined by molecular cloning in the1980's and 1990's. After 1980, the structures of the genes and proteins of CA-synthesizing and –metabolizing enzymes were also determined by molecular cloning.

New analytical methods such as radioisotopic, fluorescence, and immuno assays, toward the development of which Dr. Udenfriend greatly contributed, have been promoting the biochemical and molecular biological studies on CA's. Up to the 1950's, the bioassay was the only analytical method for CA's. Then, around 1960 the fluorescence assay was introduced by Udenfriend for the chemical analysis of CA's (Udenfriend, 1962, 1969). Electrochemical or fluorescence detection with high-performance liquid chromatography (HPLC) further increased the sensitivity of CA assays up to the pmol~fmol order (for reviews, see Nagatsu et al., 1988; 1991).

The signal transduction pathway engaged after CA receptor stimulation in the postsynaptic neurons was also elucidated; and DARPP-32 (dopamine- and cyclic AMP-related phosphoprotein of 32 kDa), which was discovered by Paul Greengard and his coworkers (Wallas & Greengard, 1984), was shown to be a key molecule for DA signal transduction. The complex DA signal transduction pathways were elegantly explained by Paul Greengard in his Dr. Sidney Udenfriend Memorial Lecture.

Since 1990, mouse genetics has been directed toward elucidating the functions of CA systems and creating animal models of diseases. On the other hand, progress in basic science on CA's has contributed greatly to the progress in clinical science. A deficiency of CA's in the striatum in Parkinson's disease (PD) was predicted by Carlsson (1959), based on animal experiments in which depletion of DA in the striatum produced PD-like movement disorders.

DA deficiency in the striatum in PD was discovered by Ehringer & Hornykiewicz (1960). In Japan, although it was only one case, Sano also reported a reduced level of DA in the brain of a PD patient. DA supplementation therapy by L-DOPA was also predicted by Carlsson from the results of animal experiments (1959); and after many clinical trials by several groups in the world, L-DOPA therapy with a peripherally acting DDC inhibitor to increase the transport of DOPA into the brain was established as the most effective therapy for PD by 1970. Antagonists and agonists of alpha- and beta-adrenergic receptors were found to be valuable drugs for cardiovascular diseases, hypertension, asthma, and various diseases. DA-receptor antagonists were found to be effective for the treatment of schizophrenia. NE/serotonin transporter inhibitors, i.e., NE and serotonin re-uptake inhibitors (NSRI), are potent for the treatment of depression.

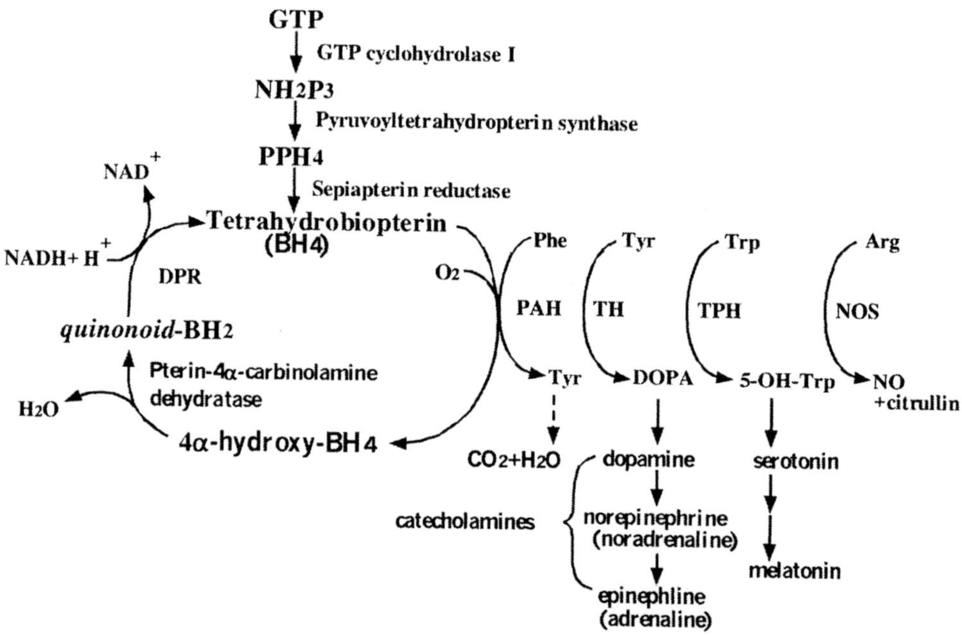

Figure 1. Relation between the biosynthesis of catecholamines / other peridine-dependent monooxygenase-related metabolites and that of tetrahydrobiopterin. TPH, tryptophan hydroxylase; NOS, nitric oxide synthase.

2. GENES OF HUMAN CATECHOLAMINE-SYNTHESIZING ENZYMES

The structures of the genes and proteins of all enzymes related to CA biosynthesis in humans were determined between 1985 and 1995 by molecular cloning (Tables 2, 3). CAs are synthesized from tyrosine by 4 enzymes, i.e., (1) TH, (2) AADC (DDC), (3) dopamine beta-monooxygenase (dopamine beta-hydroxylase, DBH), and (4) noradrenaline N-methyltransferase (phenylethanolamine N-methyltransferase, PNMT): tyrosine $\xrightarrow{(1)}$ DOPA $\xrightarrow{(2)}$ DA $\xrightarrow{(3)}$ NE $\xrightarrow{(4)}$ EN (for a review, see Nagatsu, 1991). TH requires tetrahydrobiopterin [(6R)-L-erythro-5,6,7,8-tetrahydrobiopterin, BH4] as a cofactor, which is synthesized from guanosine triphosphate (GTP): GTP $\xrightarrow{(1')}$ D-erythro-7,8-dihydroneopterin triphosphate (NH2P3) $\xrightarrow{(2')}$ 6-pyruvoyltetrahydropterin (PPH4) $\xrightarrow{(3')}$ BH4, with these steps being catalyzed by (1')GTP cyclohydrolase I (GCH), (2') 6-pyruvoyltetrahydropterin synthase (PTPS), and (3') sepiapterin reductase (SPR). In the hydroxylation reaction of tyrosine to DOPA by TH, the cofactor BH4 is oxidized and recycled back to BH4 by 2 enzymes: BH $\rightarrow$ BH4-4 alpha-carbinolamine $\xrightarrow{(4')}$ quinonoid dihydrobiopterin $\xrightarrow{(5')}$ BH4, by (4') pterin-4 alpha-carbinolamine dehydratase (PCD) and (5') dihydropteridine reductase (DPR) (for reviews, see Nichol et al., 1985; Nagatsu & Ichinose, 1999; Thöny et al., 2000). Therefore, four CA-synthesizing enzymes and five BH4-synthesizing and recycling enzymes that supply the BH4 cofactor to TH, are related to CA biosynthesis.

Table 2. Genes and proteins of catecholamine-synthesizing enzymes

Human Catecholamine-Synthesizing Enzymes

Enzyme (EC number)	Gene		cDNA	Base pairs (coding region)	Alternative mRNA splicing	Protein	
	Chromosome	Exons				Mr (kDa× subunit numbers)	Amino acid residues
Tyrosine 3-monooxygenase (Tyrosine hydroxylase) (TH) (1.14.16.2)	11p15	14	type1	1491	Internal donor sites and cassette	5553×4	497
			type2	1503		5597×4	501
			type3	1572		5808×4	524
			type4	1584		5852×4	528
Aromatic L-amino acid decarboxylase (Dopa decarboxylase) (AADC) (4.1.1.28)	7q12-12	16	typeN, typeL	1440	Alternative promoter	5389×2	480
Dopamine β-monooxygenase (Dopamine β-hydroxylase) (DBH) (1.14.17.1)	9q34	12	typeA, typeB	1809	Alternative polyadenylation	6486×4	578 (603)*
Noradrenaline N-methyltransferase (Phenylethanolamine N-methyltransferase) (PNMT) (2.1.1.28)	17q21-22	3	typeA, typeB	846	Alternative promoter	3085×1	282

*The numbers in parentheses of amino acid residues of DBH include the signal sequences.

The human genes and proteins of CA-synthesizing enzymes are shown in Table 2. Isoforms of mRNA are produced by alternative mRNA splicing from a single gene in the case of TH, AADC, DBH, and PNMT. However, except for human TH, which produces at least 4 protein isoforms (hTH1-hTH4), the other enzymes are composed of a single protein. Sugimoto et al. (2001) recently found (6R)-5,6,7,8-tetrahydro-D-monapterin as a new reduced pteridine in *Tetrahymena pyriformis,* which has GTP cyclohydrolase I but almost no BH4 (Nomura et al., 1999). Tetrahydromonapterin may also exist in humans (Ogiwara et al., 1992), and the physiological function of it remains as an interesting problem.

Humans produce four isoforms of TH (hTH1-hTH4) from a single gene (Grima et al., 1987; Kaneda et al., 1987; O'Malley et al., 1987; Kobayashi et al., 1988; and for reviews, see Nagatsu & Ichinose, 1991; Nagatsu, 1995). New- and Old-World monkeys and the gorilla produce only two of the TH isoforms corresponding to hTH1 and hTH2 (Ichikawa et al., 1990; Ichinose et al., 1993). Comparison among the genomic DNA sequences of various primates revealed that mutations that had accumulated in the genomic DNA created a new exon 2, resulting in the appearance of two new TH isoforms in humans. Non primate animals such as rats (Grima et al., 1985) and mice (Ichikawa et al., 1991; Iwata et al., 1992) produce a single protein corresponding to the shortest hTH1 from a single gene, in which the exon 2 of the human TH gene does not exist. The cyclic AMP-response element (CRE) and AP1 binding site are present in the 5'-flanking region of human and mouse TH genes. We cloned the human AADC gene, and assigned it to chromosome 7p12.1-p12.3. The gene, being approximately 100 kb in size and composed of 15 exons, is expressed in both CA and serotonin neurons as well as in non-neuronal tissues such as the liver. An alternative usage of the non-neuronal (L1) and neuronal (N1) first exons in the 5'-untranslated regions produces neuronal (catecholaminergic and serotonergic) and non-neuronal (liver type) mRNAs encoding a single protein (Sumi-Ichinose, et al., 1992; Ichinose, et al., 1992).

We cloned the human DBH gene and found it to be approximately 23 kb in size, to comprise 12 exons, and to exist as a single gene in the genome (Kobayashi et al., 1989). Two kinds of cDNA (types A and B) different only in their 3'-untranslated region encode a single human DBH protein. The DBH promoter contains a glucocorticoid-response element (GRE) and a cyclic AMP-response element (CRE) and also a novel phorbol ester-responsive DNA sequence in the 5'-flanking region of the gene (Ishiguro et al., 1998).

Also, we cloned human PNMT gene and assigned it to chromosome 17 (Kaneda et al., 1988; Sasaoka et al., 1989). This gene consists of 3 exons and contains a glucocorticoid-response element (CRE) and an Sp1 binding site in its 5'-flanking region. Two types of hPNMT mRNA (types A and B) are produced by a single gene through the use of two alternative promoters.

The structures of the genes and enzyme proteins for BH4-synthesizing enzymes have also been elucidated, as shown in Table 3 (for reviews, see Nagatsu & Ichinose, 1999; Thöny et al., 2000). The human GTP cyclohydrolase I (GCH) gene is encoded by a single-copy gene composed of six exons spanning about 30 kb (Ichinose et al., 1995), and has been assigned to chromosome 14q22.1-22.2 (Ichinose et al., 1994a).

Table 3. Genes and proteins of tetrahydrobiopterin-synthesizing enzymes

Human Tetrahydrobiopterin-Synthesizing Enzymes

Enzyme (EC number)	Gene		Protein	
	Chromosome	Exons	Mr (kDa × subunit numbers)	Amino acid residues
Guanosine triphosphate (GTP) cyclohydrolase I (GCH1) (3.5.4.16)	14q22.1-q22.2	6	30×10	250
Pyruvoyl-tetrahydropterin synthase (PTS) (4.6.1.10)	11q22.3-q23.3	6	17×6	145
Sepiapterin reductase (SPR) (1.1.1.153)	2p13	3	25×2	261
Pterin-4α-carbinolamine dehydratase (PCBD) (4.2.1.96)	10q22	4	11×2	103
Dihydropteridine reductase (QDPR) (1.6.99.7)	4q15.3	7	25×2	244

Multiple mRNAs are produced by an alternative usage of the splicing acceptor site at the sixth exon (Togari et al., 1992; Ichinose et al., 1995), but only the longest mRNA produces active protein. The elucidation of the structures of CA-related genes and proteins, especially those of humans and mice, is important for their application to mouse genetics and gene therapy, as well as for determining the mechanism of human diseases.

3. ANALYSIS OF CATECHOLAMINE SYSTEMS BY MOUSE GENETICS.

Mouse genetics is a powerful approach to elucidate the function of CA systems and to create animal models of diseases. We (Kaneda et al., 1991) created hTH transgenic mice tissue-specifically overexpressing hTH at high levels. For the tissue-specific expression of the human TH gene, the entire gene containing 2.5 kb of the 5' upstream region, the entire-exon-intron sequence, and 0.5 kb of the 3'-flanking region is required. The level of human TH mRNA in the brain was about 50-fold higher than that of endogenous mouse TH mRNA in this organ. TH protein levels and the in vitro enzyme activities were increased about 5-fold, but in vivo enzyme activity measured by microdialysis and the tissue CA levels were increased only slightly and not significantly. These results suggest that regulatory mechanisms exist for human TH expression and for CA levels. Transgenic mice harboring only the 4 kb 5' flanking region of the human TH gene did not express the reporter gene in a tissue-specific manner (Sasaoka et al., 1992). This finding also suggests that there exist tissue-specific enhancer elements somewhere other than in the 5' flanking region, perhaps in introns or in the 3' downstream region.

Similar regulatory mechanisms for CA levels were also found in human DBH transgenic mice (Kobayashi et al., 1994). The 4-kb upstream promoter region of human DBH can express the minigene tissue-specifically in NE and EN systems and is useful in transgenic approach (Morita et al., 1993). Transgenic mice generated by transfection with a human DBH minigene construct containing the full-length cDNA connected downstream of the 4-kb upstream promoter region overexpressed human DBH tissue-specifically in the brain and adrenal gland. We detected a marked increase in DBH activity in various CA-containing tissues; but in these transgenics the steady-state levels of NE and EN were normally maintained, suggesting that there are some regulatory mechanisms to preserve a constant rate of NE synthesis in spite of the increased amount of DBH protein. These transgenic mice with the minigene construct provide one approach to study the mechanisms underlying the biogenesis of the DBH polypeptide and regulation of NE synthesis.

Genetic alternation of CA specificity from NE cells to EN cells was achieved in transgenic mice (Kobayashi et al., 1992). We generated a line of transgenic mice carrying a chimeric gene containing human PNMT cDNA fused to the 4-kb fragment of the human DBH promoter to switch NE phenotype to EN one in brain and sympathetic ganglia as well as in the adrenal medulla. The results indicate that the additional PNMT expression in NE-producing cells can convert these cells to EN-producing ones and that the former actually possess the basic machinary required for the synthesis of EN except for PNMT. Thus, it appears that the only major difference between NE- and EN-producing cells is the expression of PNMT. These transgenic mice provide an experimental model to investigate the functional differences between NE and EN. In these transgenic mice, we showed that the conversion of NE to EN causes down-regulation of beta2-adrenergic receptors, which suggests a compensatory or adaptive mechanism and that alteration of CA specificity leads to regulated expression of the beta-receptor subtypes in the target tissues of such mice (Kobayashi et al., 1995b).

We also (Kobayashi et al., 1995a) produced TH gene knockout mice. The homozygous TH(-/-) mutant mice died at a late stage of embryonic development or shortly after birth. Both TH mRNA and enzyme activity were lacking in the homozygous mutants, which have a severe depletion of CA's. Analysis of electrocardiograms showed bradycardia and arrhythmia, which may account for the lethality of this mutation. Transfer of a human TH transgene into the homozygous mice corrected the mutant phenotype, showing recovery of TH activity by expression of the human enzyme. These results indicate that TH is essential for survival of the animals during late gestational development and after birth, as it would be required by sympathetic NE neurons to support cardiac function. This was confirmed by conditional disruption of the TH gene in peripheral sympathetic neurons and in the adrenal medulla by our newly developed immunotoxin-mediated cell targeting (IMCT) method (Kobayashi et al., 1995c; 1997). Targeting of the peripheral CA cells by IMCT resulted in severe and progressive phenotypic abnormalities mainly characterized by cardiac dysfunction (Sawada et al., 1998). Zhou et al. (1995) also produced TH gene knockout mice, and showed that the mice died as embryos due to cardiovascular failure and that administration of L-DOPA to pregnant females resulted in complete rescue of the mutant mice in utero. Their results agree with ours in that TH is essential for cardiovascular function during development. Nakanishi's group has successfully applied IMCT for ablation and elucidation of the neuronal function of Golgi cells in the cerebellum (Watanabe et al., 1998) and of the acethylcholine interneurons in the striatum (Kaneko et al., 2000).

Kobayashi et al. (2000) found that heterozygous (+/-) TH mice showed about 40% of the wild-type TH activity, 70~80% of the NE level in the brain, and modest neuropsychological deficits, for example, impairment in the water finding task for assessment of latent learning performance. These animals also exhibited mild impairment in long-term memory formation in three distinct forms of associative learning, i.e., active avoidance, cued fear conditioning, and conditioned taste avoidance. These deficits were restored by the drug-induced stimulation of noradrenergic activity. In contrast, the spacial learning and hippocampal long-term potentiation were normal in the mutants. These results provide genetic evidence that the central NE system plays an important role in memory formation, particularly in the long-term memory of conditioned learning.

DBH gene knockout mice were produced by Thomas et al. (1995), and the deficiency in NE and EN was proved to be lethal due to failure of sympathetic neurotransmission. These mice could be rescued by L-threo-DOPS (Narabayashi et al., 1984), which produces NE by AADC action. These results also confirm that sympathetic nerve NE is essential for mouse fetal development and postnatal survival.

DA-deficient mice were generated by a transgenic rescue approach to clarify the role of DA signaling during postnatal development. Introduction of the TH transgene directed by the DBH gene promoter into TH knockout mice restored NE and EN synthesis, preventing cardiac dysfunction in the knockout mice. The lack of TH expression in the cells that normally express the DA phenotype resulted in a marked reduction in DA accumulation in the tissues, which led to multiple behavioral abnormalities at the juvenile stage. These abnormalities were characterized by a reduction in spontaneous locomotor activity, blockade of methamphetamine-induced hyperactivity, cataleptic behavior, and defects in active avoidance learning. In contrast, development of the pituitary gland as well as production and secretion of the pituitary peptide hormones which depend on hypothalamic DA control were normally maintained, despite defective DA synthesis. These results demonstrate that DA neurotransmission is essential for controlling spontaneous and voluntary movement and associative learning during postnatal development through the nigrostriatal (A9) and mesocortico limbic (A10) pathways (Nishii et al., 1998; Kobayashi & Nagatsu, 2000).

4. MOLECULAR GENETICS OF CATECHOLAMINE-RELATED DISEASES

CA's have a wide range of functions in the brain, sympathetic nervous system, and the adrenal medulla as neurotransmitters and hormones, and therefore their dysfunction is related to various diseases (for review, see Goldstein et al., 1998). In the CNS, CA's are slow neurotransmitters modulating the action of rapid neurotransmitters such as glutamate and gamma-aminobutyric acid (GABA), and are closely related to movement, emotion, reward, and memory.

DA deficiency in the nigrostriatal DA (A9) system causes movement disorder as in PD. We (Ichinose et al., 1994a) discovered autosomal dominant, hereditary progressive dystonia with marked diurnal fluctuation [HPD (Segawa's disease)/ DOPA responsive dystonia (DRD)] is caused by mutations of GTP cyclohydrolase I (GCH) and the resultant partial decrease in the enzyme activity, deficiencies in BH4 cofactor level, TH activity, and DA level in the nigrostriatal DA neurons. Supplementation of DA by oral L-DOPA administration at a small dose completely attenuates the symptoms without any side effects for many years, and thus HPD/DRD is a controllable disease.

Autosomal recessive DOPA-responsive dystonia is caused by TH deficiency (a point mutation, Q381K, in hTH1) and a partial reduction (~15% of wild-type hTH1 activity expressed in E. coli) in the DA level in the nigrostriatal DA neurons (Knappskog et al., 1995). It is interesting that in TH deficiency with a very low homospecific activity (~1.5% of the wild-type hTH1 expressed in E. coli) the clinical phenotype is DOPA-responsive parkinsonism (Lüdecke et al., 1996). Considering the fact that TH(-/-) mice are lethal (Kobayashi et al., 1995a; Zhou et al., 1995), complete loss of TH activity in humans would also be lethal. The symptoms of dystonia or parkinsonism depending upon the degree of DA decrease due to TH deficiency are completely normalized by L-DOPA administration without side effects. This is different from idiopathic Parkinson's disease, in which DA supplementation therapy by L-DOPA is effective in the early phase, but in which the effect is gradually reduced and dyskinesia occurs as a side effect. The difference between DOPA-responsive dystonia and idiopathic Parkinson's disease is the absence in the former and the presence in the latter of nigrostriatal DA neuronal cell death.

Idiopathic PD may be caused by the genetic background and some environmental factors such as neurotoxins. 1-Methyl-4-phenyl-1,2,3,6-tetrahydropyridine (MPTP), which produces parkinsonism, has contributed to the molecular understanding of PD. MPTP-like neurotoxins such as isoquinolines have been found in the PD brain, but their significance is not clear yet (Nagatsu, 1997). Mitochondrial complex I dysfunction was found in MPTP-induced PD models. The recent finding that long-term administration of rotenone to rats produces parkinsonism with Lewy bodies as seen in PD indicates that inhibition of complex I may be important (Betarbet et al., 2001). In collaboration studies with Narabayashi et al. on postmortem brains, we showed decreases in GCH activity, BH4 level, TH activity, TH mRNA (hTH1, hTH2, hTH3, and hTH4), DBH activity and

PNMT activity in catecholaminergic regions in the PD brain, specifically in the nigrostriatum and locus coeruleus (Nagatsu 1993; Ichinose et al., 1994b). These decreases were caused by neuronal death. We recently found increased levels of proinflammatory cytokines and apoptosis-related proteins, and decreased levels of neurotrophins, in the striatum of the PD brain and of PD model animals, and suggested that the cell death in PD is caused by glial activation and subsequent apoptotic neuronal cell death (Nagatsu et al., 2000). DBH deficiency causes hypotention, which can be controlled by supplementation of NE in the sympathetic nerves by L-DOPS (Robertson & Hale, 1998).

Supplementation of DA by L-DOPA is effective in PD, but it has limitations in reduced effects and the side effect of dyskinesia. Ozawa's group succeeded in triple transduction with adeno-associated virus vectors expressing TH, AADC, and GCH in the striatum for gene therapy of PD in rats with 6-hydroxy-DA-induced hemiparkinsonism. DA was produced, resulting in behavioral recovery that persisted for at least 12 months (Shen et al., 2000). The three enzymes were expressed in the same neurons without any toxic side effects or immune response. The same efficacy was also observed in MPTP-induced PD monkeys. Therefore, gene therapy for DA supplementation appears to be promising for the treatment of PD.

5. CONCLUSION

The initial great contributions of Dr. Sidney Udenfriend to CA research has been highly appreciated by all. Recent molecular biological and genetic studies have elucidated complex and important physiological and pathological functions of CA's in the CNS or PNS. CA research will continue to contribute to elucidation of the molecular mechanisms of neurological and psychiatrical disorders in CNS and PNS, and of those of endocrinological, cardiovascular, and metabolic disorders in the periphery.

ACKNOWLEDGEMENTS

This review is mainly based on our findings, but there are many important contributors to the field whom we have cited. I thank the Ministry of Education and Science and the Ministry of Health and Labor of Japan for their support of our studies.

REFERENCES

Axelrod, J. 1957, O-Methylation of epinephrine and other catecholamines in vitro and in vivo, *Science* **126**: 400-401.

Axelrod J., Weil-Malherbe, H., and Tomchick, R., 1959, The physiological disposition of H^3-epinephrine and its metabolite metanephrine, *J. Pharmacol. Exp. Therap.* **127**: 251-256.

Betarbet, R., Sherer, T. B., Mackenzie, G., Garcia-Osuna, M., Panov, A. V., Greenamyre, J. T., 2001, Chronic systemic pesticide exposure reproduces features of Parkinson's disease, *Nature Neurosci.* **3**: 1301-1306.

Blaschko, H., Richtei, D., and Schlossmann, H., 1937, The oxidation of adrenaline and other amines, *Biochem. J.* **31**: 2187-2196.

Blaschko, H., 1939, The specific action of L-dopa decarboxylase, *J. Physiol.* **96**: 509.

Carlsson A., Lindquist, M., Magnusson, T., and Waldeck, B., 1958, On the presence of 3-hydroxytyramine in brain, *Science* **127**: 471.

Carlsson, A., 1959, The occurrence, distribution and physiological role of dopamine in the nervous system, *Pharmacol. Rev.* **11**: 490-493.

Ehringer, H., and Hornykiewicz, O., 1960, Verteilung von Noradrenalin und Dopamin (3-Hydroxytyramin) im Gehirn des Menschen und ihr Verhalten bei Erkrankungen des extrapyramidalen Systems, *Klin. Wschr.* **38**: 1236-1239.

Euler, U. S. von, 1946, A specific sympathomimetic ergone in adrenergic nerve fibers (Sympathin) and its relation to adrenaline and noradrenaline, *Acta Physiol. Scand.* **12**:73-97.

Goldstein, D. S., Eisenhofer, G., and McCarty, R., 1998, Catecholamines: Bridging Basic Science with Clinical Medicine, *Adv. Pharmacol.* **42**: 1-1069.

Grima, B. Lamouroux, A., Blanot, F., Biguet, N. F., and Mallet, J., 1985, Complete coding sequence of rat tyrosine hydroxylase mRNA, *Proc. Natl. Acad. Sci. USA* **82**: 617-621.

Grima, B., Lamouroux, A., Boni, C., Julien, J. -F., Javoy-Agid, F., and Mallet, J., 1987, A single human gene encoding multiple tyrosine hydroxylase with different predicted functional characteristics, *Nature* **326**: 707-711.

Holtz, P., 1939, Dopadecarboxylase, *Naturwissenschaften*, **27**: 724-725.

Ichikawa, S., Ichinose, H., and Nagatsu, T., 1990, Multiple mRNAs of monkey tyrosine hydroxylase, *Biochem. Biophys. Res. Commun.* **173**: 1331-1336.

Ichikawa, S., Sasaoka, T., and Nagatsu, T., 1991, Primary structure of mouse tyrosine hydroxylase deduced from its cDNA, *Biochem. Biophys. Res. Commun.* **176**: 1610-1616.

Ichinose, H., Sumi-Ichinose, C., Ohye, T., Hagino, Y., Fujita, K., and Nagatsu, T., 1992, Tissue-specific alternative splicing of the first exon generates two types of mRNAs in human aromatic L-amino acid decarboxylase, *Biochemistry* **31**: 11546-11550.

Ichinose, H., Ohye, T., Fujita, K., Yoshida, M., Ueda, S., and Nagatsu, T., 1993, Increased heterogeneity of tyrosine hydroxylase in humans, *Biochem. Biophys. Res. Commun.* **195**: 158-165.

Ichinose, H., Ohye, T., Takahashi, E., Seki, N., Hori, T., Segawa, M., Nomura, Y., Endo, K., Tanaka, H., Tsuji, S., Fujita, K., and Nagatsu, T., 1994a, Hereditary progressive dystonia with marked diurnal fluctuation caused by mutations of the GTP cyclohydrolase I gene, *Nature Genet.* **8**: 236-242.

Ichinose, H., Ohye, T., Fujita, K., Pantucek, K., Lange, K., Riederer, P., and Nagatsu, T. 1994b, Quantification of mRNA of tyrosine hydroxylase and aromatic L-amino acid decarboxylase in the substantia nigra in Parkinson's disease and schizophrenia, *J. Neural Transm. [P-D Sect]* **8**: 149-158.

Ichinose, H., Ohye, T., Matsuda, Y., Hori, T., Blau, N., Burlina, A., Rouse, B., Matalon, R., Fujita, K., and Nagatsu, T., 1995, Characterization of mouse and human GTP cyclohydrolase I genes: Mutations in patients with GTP cyclohydrolase I deficiency, *J. Biol. Chem.* **270**: 10062-10071.

Ishiguro, H., Yamada, K., Ichino, N., and Nagatsu, T., 1998 Identification and characterization of a novel phorbol ester-responsive DNA sequence in the 5'-flanking region of the human dopamine beta-hydroxylase gene, *J. Biol. Chem.* **273**: 21941-21949.

Iwata, N., Kobayashi, K., Sasaoka, T., Hidaka, H., and Nagatsu, T., 1992, Structure of the mouse tyrosine hydroxylase gene, *Biochem. Biophys. Res. Commun.* **182**: 348-354.

Kaneda, N., Kobayashi, K., Ichinose, H., Kishi, F., Nakazawa, A., Kurosawa, Y., Fujita, K., and Nagatsu, T., 1987, Isolation of a novel cDNA clone for human tyrosine hydroxylase: alternative mRNA splicing produces four kinds of mRNA from a single gene, *Biochem. Biophys. Res. Commun.* 146: 971-975.

Kaneda, N., Ichinose, H., Kobayashi, K., Oka, K., Kishi, F., Nakazawa, A., Kurosawa, Y., Fujita, K., and Nagatsu, T., 1988, Molecular cloning of cDNA and chromosomal assignment of the gene for human phenylethanolamine N-methyltransferase, the enzyme for epinephrine biosynthesis, *J. Biol. Chem.* **263**: 7672-7677.

Kaneda, N., Sasaoka, T., Kobayashi, K., Kiuchi, K., Nagatsu, I., Kurosawa, Y., Fujita, K., Yokoyama, M., Nomura, T., Katsuki, M., and Nagatsu, T., 1991, Tissue-specific and high-level expression of the human tyrosine hydroxylase gene in transgenic mice, *Neuron* **6**: 583-594.

Kaneko S., Hikida T., Watanabe D., Ichinose H., Nagatsu T., Kreitman R. J., Pastan I., and Nakanishi S., 2000, Synaptic integration mediated by striatal cholinergic interneurons in basal ganglia function, *Science* **289**: 633-637.

Kaufman, S., 1963, The structure of the phenylalanine-hydroxylation cofactor, *Proc. Natl. Acad. Sci. USA* **50**: 1085-1093.

Knappskog, P. M., Flatmark, T., Mallet, J., Lüdecke, B., and Bartholomé, K., 1995, Recessively inherited L-DOPA-responsive dystonia caused by a point mutation (Q381k) in the tyrosine hydroxylase gene, *Human Mol. Genet.* **4**: 1209-1212.

Kobayashi, K., Kaneda, N., Ichinose, H., Kishi, F., Nakazawa, A., Kurosawa, Y., Fujita, K., and Nagatsu, T., 1988, Structures of the human tyrosine hydroxylase gene: alternative splicing from a single gene accounts for generation of four mRNA types, *J. Biochem.* **103**: 907-912.

Kobayashi, K., Kurosawa, Y., Fujita, K., and Nagatsu, T., 1989, Human dopamine beta-hydroxylase gene: two mRNA types having different 3'-terminal regions are produced through alternative polyadenylation, *Nucleic Acids Res.* **17**: 1089-1102.

Kobayashi, K., Sasaoka, T., Morita, S., Nagatsu, I., Iguchi, A., Kurosawa, Y., Fujita, K., Nomura, T., Kimura, M., Katsuki, M., and Nagatsu, T., 1992, Genetic alteration of catecholamine specificity in transgenic mice, *Proc. Natl. Acad. Sci. USA* **89**: 1631-1635.

Kobayashi, K., Morita, S., Mizuguchi, T., Sawada, H., Yamada, K., Nagatsu, I., Fujita, K., and Nagatsu, T., 1994, Functional and high-level expression of human dopamine beta-hydroxylase in transgenic mice, *J. Biol. Chem.* **269**: 29725-29731.

Kobayashi, K., Morita, S., Sawada, H., Mizuguchi, T., Yamada, K., Nagatsu, I, Hata, T., Watanabe, Y., Fujita, K., and Nagatsu, T., 1995a, Targeted disruption of the tyrosine hydroxylase locus results in severe catecholamine depletion and perinatal lethality in mice, *J. Biol. Chem.* **270**: 27235-27243.

Kobayashi, K., Ota, A., Togari, A., Morita, S., Mizuguchi, T., Sawada, M., Yamada, K., Nagatsu, I., Matsumoto, S., Fujta, K., and Nagatsu, T., 1995b, Alteration of catecholamine phenotype in transgenic mice influences expression of adrenergic receptor subtypes, *J. Neurochem.* **65**: 492-501.

Kobayashi, K., Morita, S., Sawada, H., Mizuguchi, T., Yamada, K., Nagatsu, I., Fujita, K., Kreitman, R. J., Pastan, I., and Nagatsu, T., 1995c, Immunotoxin-mediated conditional disruption of specific neurons in transgenic mice, *Proc. Natl. Acad. Sci.* USA **92**: 1132-1136.

Kobayashi, K., Pastan, I., Nagatsu, T., 1997, Controlled genetic ablation by immunotoxin-mediated cell targeting, in: *Transgenic Animals* : L. M. Houdebine ed., Harwood Academic Publishers, Amsterdam, pp.331-336.

Kobayashi, K., and Nagatsu, T., 2000, Transgenic rescue of tyrosine hydroxylase-deficient mice: application for generating animal models with catecholamine dysfunction, in: *Progress in Gene Therapy: Basic and Clinical Frontiers.*, R. Bertolotti, S. H. Parvez, and T. Nagatsu, eds., VSP, Utrecht, pp. 267-288.

Kobayashi K., Noda Y., Matsushita N., Nishii K., Sawada H., Nagatsu T., Nakahara D., Fukabori R., Yasoshima Y., Yamamoto T., Miura M., Kano M., Miyama T., Miyamoto Y., and Nabeshima T., 2000, Modest neuropsychological deficits caused by reduced noradrenaline metabolism in mice heterozygous for a mutated tyrosine hydroxylase gene, *J. Neurosci.* **20**: 2418-2426.

Lovenberg, W., Weissbach, H., and Udenfriend, S., 1962, Aromatic L-amino acid decarboxylase, *J. Biol. Chem.* **237**: 89-93.

Lüdecke, B., Knappskog, P. M., Clayton, P. T., Surtees, R. A. H., Clelland, J. D., Heales, S. J. R., Brand, M. P., Bartholomé, K., and Flatmark, T., 1996, Recessively inherited L-DOPA-responsive parkinsonism in infancy caused by a point mutation (L205P) in the tyrosine hydroxylase gene, *Human Mol. Genet.* **5**: 1023-1028.

Morita, S., Kobayashi, K., Muzuguchi, T., Yamada, K., Nagatsu, I., Titani, K., Fujita, K., Hidaka, H., and Nagatsu, T., 1993, The 5'-flanking region of human dopamine beta-hydroxylase gene promotes the neuron subtype-specific gene expression in the central nervous system of transgenic mice, *Mol. Brain Res.* **17**: 239-244.

Nagatsu, T., Levitt, M., and Udenfriend, S., 1964, Tyrosine hydroxylase-the initial step in norepinephrine biosynthesis, *J. Biol. Chem.* **239**, 2910-2917.

Nagatsu, T., and Kojima, K., 1988, Application of electrochemical detection in HPLC to the assay of biologically active compounds, *Trends Anal. Chem.* **7**: 21-27.

Nagatsu, T., 1991, Application of high-performance liquid chromatography to the study of biogenic amine-related enzymes, *J. Chromatogr. Biomed. Appl.* **566**: 287-307.

Nagatsu, T., 1991, Genes for human catecholamine-synthesizing enzymes, *Neurosci. Res.* **12**: 315-345.

Nagatsu, T., and Ichinose, H., 1991, Comparative studies on the structure of human tyrosine hydroxylase with those of the enzyme of various mammals, Comp. Biochem. Physiol., 98C: 203-210.

Nagatsu, T., 1993, Biochemical aspects of Parkinson's disease, in: *Advances in Neurology*, Vol. 60. Parkinson's Disease: From Basic Research To Treatment, H. Narabayashi, T. Nagatsu, N. Yanagisawa, Y. Mizuno, eds., Raven Press, New York, pp. 165-174.

Nagatsu, T., 1995, Tyrosine hydroxylase: human isoforms, structure and regulation in physiolosy and pathology, in: *Essays in Biochemistry* Vol.30, D. K. Apps, K. F. Tipton, eds., Portland Press, London, pp.15-35.

Nagatsu, T., 1997, Isoquinoline neurotoxins in the brain and Parkinson's disease, *Neurosci. Res.* **29**: 99-111.

Nagatsu T., and Ichinose H., 1999, Regulation of pteridine-requiring enzymes by the cofactor tetrahydrobiopterin, *Mol. Neurobiol.* **19**: 79-96.

Nagatsu T., Mogi M., Ichinose H., and Togari A., 2000, Changes in cytokines and neurotrophins in Parkinson's disease, *J. Neural Transm. [Suppl]* **60**: 277-290.

Narabayashi, H., Kondo, T., Nagatsu, T., Hayashi, A., and Suzuki, T., 1984, DL-Threo-3,4-dihydroxyphenylserine for freezing symptom in parkinsonism, in: *Advances in Neurology*, Vol. 40, R. G. Hassler, J. F. Christ, eds., Raven Press, New York, pp. 497-502.

Nichol, C. A., Smith, G. K., and Duch, D. S., 1985, Biosynthesis and metabolism of tetrahydrobiopterin and molybdopterin, *Ann. Rev. Biochem.* **54**: 729-762.

Nishii, K., Matsushita, N., Sawada, H., Sano, H., Noda, Y., Mamiya, T., Nabeshima, T., Nagatsu, I., Hata, T., Kiuchi, K., Yoshizato, H., Nakashima, K., Nagatsu, T., and Kobayashi, K., 1998, Motor and learning dysfunction during postnatal development in mice defective in DA neuronal transmission, *J. Neurosci. Res.*, **54**: 450-464.

Nomura, T., Tazawa, M., Ohtsuki, M., Ichinose, C. S., Hagino, Y., Ota, A., Nakashima, A., Mori, K., Sugimoto, T., Ueno, O., Yazawa, Y., Ichinose, H., and Nagatsu, T., 1998, Enzymes related to catecholamine biosynthesis in tetrahymena pyriformis. Presence of GTP cyclohydroxylase I, *Comp. Biochem. Physiol. B* **120**: 753-760.

Ogiwara, S., Nagatsu, T., Teradaira, R., Fujita, K., and Sugimoto, T., 1992, Diastereomers of neopterin and biopterin in human urine, *Biol. Chem. Hoppe-Seyler*, **373**: 1061-1065.

O'Malley, K. L., Anhalt, M. J., Martin, B. M., Kalsoe, J. R., Winfield, S. L., and Ginns, E. I., 1987, Isolation and characterization of the human tyrosine hydroxylase gene: identification of 5' alternative splice sites responsible for multiple mRNAs, *Biochemistry* **26**: 6910-6914.

Robertson, D., and Hale, N., 1998, Genetic disease of hypotension, *Adv. Pharmacol.* **42**: 61-65.

Sasaoka, T., Kaneda, N., Kurosawa, Y., Fujita, K., and Nagatsu, T., 1989, Structure of human phenylethanolamine N-methyltransferase gene: existence of two types of mRNA with different transcription initiation sites, *Neurochem. Int.* **15**: 555-565.

Sasaoka, T., Kobayashi, K., Nagatsu, I., Takahashi, R., Kimura, M., Yokoyama, M., Nomura, T., Katsuki, M., and Nagatsu, T., 1992, Analysis of the human tyrosine hydroxylase promoter-chloramphenicol acetyltransferase chimeric gene expression in transgenic mice, *Mol. Brain Res.* **16**: 274-286.

Sawada, H., Nishii, K., Suzuki T., Hasegawa, K., Hata, T., Nagatsu, I., Kreitman, R. J., Pastan, I., Nagatsu, T., and Kobayashi, K., 1998, Autonomic neuropathy in transgenic mice caused by immunotoxin targeting of the peripheral nervous system, *J. Neurosci. Res.* **51**: 162-173.

Shen, Y., Muramatsu, S., Ikeguchi, K., Fujimoto, K., Fan, D. -S., Ogawa, M., Mizukami, H., Urabe, M., Kume, A., Nagatsu, I., Urano, F., Suzuki, T., Ichinose, H., Nagatsu, T., Monahan, J., Nakano, I., and Ozawa, K., 2000, Triple transduction with adeno-associated virus vectors expressing tyrosine hydroxylase, aromatic-L-amino-acid decarboxylase, and GTP cyclohydrolase I for gene therapy of Parkinson's disease, *Human Gene Therapy* **11**: 1509-1519.

Sugimoto, T., Ikemoto, K., Murata, S., Tazawa, M., Nomura, T., Hagino, Y., Ichinose, H., and Nagatsu, T., 2001, Identification of (6R)-5,6,7,8-tetrahydro-D-monapterin as the native pteridine in *Tetrahymena pyriformis*, *Helv. Chim. Acta* **84**: 918-927.

Sumi-Ichinose, C., Ichinose, H., Takahashi, E., Hori, T., and Nagatsu, T., 1992, Molecular cloning of genomic DNA and chromosomal assignment of the gene for human aromatic L-amino acid decarboxylase, the enzyme for catecholamine and serotonin biosynthesis, *Biochemistry* **31**: 2229-2238.

Thomas, S. A., Matsumoto, A. M., and Palmiter, R. D., 1995, Noradrenaline is essential for mouse fetal development, *Nature* **374**: 643-646.

Thöny, B., Auerbach, G., and Blau, N., 2000, Tetrahydrobiopterin biosynthesis, regeneration and functions, *Biochem. J.* **347**: 1-16.

Togari, A., Ichinose, H., Matsumoto, S., Fujita,K., and Nagatsu, T., 1992, Multiple mRNA forms of human GTP cyclohydrolase I, *Biochem. Biophys. Res. Commun.* **187**: 359-365.

Udenfriend, S., and Cooper, J. R., 1952, The enzymatic conversion of phenylalanine to tyrosine, *J. Biol. Chem.* **194**: 503-511.

Udenfriend, S., Cooper, J. R., Clark, C. T., and Baer, J. E., 1953, Rate of turnover of epinephrine in the adrenal medulla, *Science* **117**: 663-665.

Udenfriend, S., and Wyngaarden, J. B., 1956, Precursors of adrenal epinephrine and norepinephrine in vivo, *Biochim. Biophys. Acta* **20**: 48-52.

Udenfriend, S., 1962, *Fluorescence Assay in Biology and Medicine Vol. 1*, Academic Press, New York.

Udenfriend, S., 1969, *Fluorescence Assay in Biology and Medicine, Vol.* 2, Academic Press, New York.

Watanabe, D., Inokawa, H., Hashimoto, K., Suzuki, N., Kano, M., Shigemoto, R., Hirano, T., Toyama, K., Kaneko, S., Yokoi, M., Moriyoshi, K., Suzuki, M., Kobayashi, K., Nagatsu, T., Kreitman, R. J., Pastan, I., and Nakanishi, S., 1998, Ablation of cerebellar Golgi cells disrupts synaptic integration involving GABA inhibition and NMDA receptor activation in motor coordination, *Cell* **95**: 17-27.

Wallas, S. I., and Greengard, P., 1984, DARPP-32, a dopamine and adenosine 3': 5"-monophosphate regulated phosphoprotein enriched in dopamine-innervated brain regions. I. Regional and cellular distribution in the rat brain. *J. Neurosci.* **4**: 84-88.

Wolfe, D. E., Potter, L. T., Richardson, K. C., and Axelrod, J., 1962, Localizing tritiated norepinephrine in sympathetic axons by electron microscopic autoradiography, *Science* **138**: 440-442.

Zhou, Q. -Y., Quaife, C. J., and Palmiter, R. D., 1995, Targeted disruption of the tyrosine hydroxylase gene reveals that catecholamines are required for mouse fetal development, *Nature* **374**: 640-643.

CHROMOGRANINS / CURRENT CONCEPT

Dominique Aunis and Marie-Hélène Metz-Boutigue[1]

1. INTRODUCTION

The large dense core secretory granules of adrenal medulla chromaffin cells synthesize, store and release catecholamines. Upon cell stimulation intragranular stored material is liberated into circulation to activate target organs. In addition to catecholamines, numerous peptides and proteins are present in chromaffin secretory granules and are exocytotically co-secreted with the catecholamines (Aunis, 1997).

The proteins chromogranin A (CGA) and B (CGB) are the major water-soluble constituents of chromaffin granules; they are also widely distributed in many neuropeptide- and hormone-containing secretory granules and in large dense core vesicles of neuronal cells. CGA was the first discovered member of this large chromogranin/ secretogranin (CG) family followed by CGB (Simon and Aunis, 1989). In addition, the group of CGs comprises secretogranin II (SgII) and III, and the proteins VGF and NESP55.

Since their discovery many groups have looked for the function of these proteins. They have been assumed to be prime candidates in the sorting mechanism of secretory proteins, to play a role in granule matrix condensation and organisation, to act as inhibitors of granule proteases or to regulate Ca^{2+} homeostasis (Helle and Aunis, 2000).

The unique amino acid sequences of CGA, CGB and SgII are characterized by a ramdom structure, the unusual high content in acidic residues and the presence of multiple pairs of basic residues representing putative cleavage sites for endoproteases. Therefore since large dense core secretory granules contain a large variety of trypsin-like endopeptidases and carboxypeptidases, CG processing generates multiple peptides that may play a role as hormones or neuropeptides. CGs may thus be considered as the precursors for small peptides, for which the physiological function should be established (Dillen et al., 1993).

---

[1] Unité Inserm U.338, Centre de Neurochimie, 5 Rue Blaise Pascal, F-67084 – Strasbourg Cedex, France

2. CHROMOGRANIN-DERIVED PEPTIDES

The use of antibodies raised against native CGA, CBG and SgII has shown that many low molecular mass immunoreactive components are present in secretory granules, indicative of an active processing of these proteins within granule matrix. A very careful study on CGA and CGB processing in chromaffin granules has revealed that these two proteins are degraded on both their N- and C-terminal sides, the central core being protected by the presence of carbohydrate moieties hampering the proteolytic attack (Metz-Boutigue et al., 1993; Strub et al., 1995, 1997; Gadroy et al., 1998). This processing gives rise to a multiplicity of low molecular weight CG-derived peptides. The generation of derived peptides proceeds further in the extracellular space after liberation, producing peptides with lower molecular mass; the situation is however rendered complex because the generation of peptides has been shown to be cell- and tissue-specific (Cetin et al., 1993). Therefore the degradation of CGA, CGB and SgII generates a multitude of peptides that are released and recovered as circulating in fluids.

The presence of numerous, well conserved dibasic sites along CGA, CGB and SgII sequences and the generation of peptides form the basis of the pro-hormone concept (Eiden, 1987). The first discovered CG-derived peptide with biological activity was pancreastatin, the $CGA_{248-296}$-derived peptide able to inhibit the release of insulin from glucose-stimulated pancreatic beta-cells (Tatemoto et al., 1987). Since then, other CGA-derived peptides have been described: vasostatins I (CGA_{1-76}) and II (CGA_{1-110}) are peptides located in the N-terminal sequence of CGA that have been implicated in the control of the human vascular system suggesting a role in blood flow regulation (Aardal et al., 1993), parastatin ($CGA_{348-420}$) located in the C-terminal domain of CGA shown to control parathyroid release of PTH (Fasciotto et al., 1993) and catestatin ($CGA_{344-364}$) involved in the adrenal medullary release of catecholamines and in the release of histamine from mast cells (Mahata et al., 1997).

3. ANTIMICROBIAL ACTIVITY OF CG-DERIVED PEPTIDES

When using a plasmid designed to express the human protein in *E. coli*, a complete disappearance of bacteria was noticed after overnight incubation concomittant with CGA accumulation in bacteria. The systematic examination of antibacterial activity in every fraction of the water soluble peptide material from bovine chromaffin granules has permitted the discovery of a large number of CGA- and CGB-derived peptides but also of pro-enkephalin A-derived peptides able to kill a large variety of $Gram^+$ and $Gram^-$ bacteria (Metz-Boutigue et al., 1998 ; Goumon et al., 1998).

The major active fraction in bovine chromaffin granule was found to correspond to vasostatin I; this peptide with a molecular mass of 8545 displays a sequence highly conserved amongst species, kills not only bacteria but also filamentous fungi and yeasts such as *S. cerevisiae* and *C. albicans*. This property is also shared by recombinant human peptide and synthetic rat vasostatin I peptide. Shorter vasostatin I-derived peptides able to inhibit fungi and yeasts were synthetized and studied; the shortest peptide CGA_{47-66} named chromofungin kills fungi and yeasts at micromolar concentration and is capable of penetrating into microorganisms (Lugardon et al., 2000). This peptide is totally inactive towards erythrocytes, fibroblasts, and neurons showing its specificity of action.

The presence of vasostatin I was immunologically detected in human polymorphonuclear neutrophils (PMN) with a CGA pattern similar to that in chromaffin granule and in secretions from PMN suggesting its presence in infectious fluids

(Lugardon et al., 2000). The presence of CGA and CGA-derived peptides in immunocytes is a new finding that opens a new way in the neuroimmunity field: both antibacterial and antifungal activities present in secretory granules asociated with catecholamines represent highly beneficial survival strategy particularly in stress situations. These peptides represent a shield protecting organisms against bacterial and fungal invasion. This shield that is triggered in early defense reactions forms the innate immunity working on until acquired immunity becomes activated.

4. CGA-INDUCED TOXICITY TO BRAIN MICROGLIAL CELLS

Recent investigation in our laboratory has shown that microglial cells isolated from newborn rat brain and maintained in primary culture become activated when nM concentration of CGA is added to culture medium: from a quiescent state these cells adopt the activated macrophage phenotype (Taupenot et al., 1996; Ciesielski-Treska & Aunis, 2000). Once activated by CGA, microglial cells secrete factors that were shown to be highly toxic to neurons since culture medium conditioned by CGA-activated microglia provokes neuronal injury with massive degeneration shown to proceed via apoptotic pathway. The identity of factors responsible for neuronal injury remain to be established although NO, peroxides and TNF-alpha have been excluded. The domain of CGA responsible for microglia activation and release of neuron toxic factors is the CGA_{1-76} region that corresponds to vasostatin I (Ciesielski-Treska et al., 1998, 2001).

In the central nervous system CGA is present in neurons either as an intact molecule or as smaller fragments. The processing of CGA differs depending on brain regions and the significance of these CGA-derived peptides in brain remains to be investigated. However CGA has been found in senile and pre-amyloid plaques (Munoz, 1991) and in large dystrophic neurites containing the amyloid precursor protein. Microglial cells are the resident macrophages of the nervous system where they play an important role in supporting neuronal survival; in addition, they also function as scavenger cells due to their phagocytic properties. They may also cause damage to neighbouring cells particularly to neurons known to be extremely sensitive to extracellular factors when microglia have to respond to sustained activation (Chao et al., 1995). It is now well accepted that the inflammatory process initiated by microglia is implicated in the pathogenesis of chronic degenerative disorders, i.e. Alzheimer's disease, Parkinson's disease and multiple sclerosis (Giulan et al., 1995). The observations that CGA accumulates in senile and pre-amyloid plaques and in dystrophic neurites and that CGA and vasostatin I both cause neuron apoptosis suggest that these molecules could be involved as neurotoxic signaling molecules in central nervous system diseases.

The occurrence of N-terminal fragments of CGA (and perhaps of other CG-derived peptides) in neurodegenerative brain lesions represents an interesting challenge for future investigations.

REFERENCES

Aardal, S., Helle, K.B., Elsayed, S., Reed, R.K., and Serck-Hanssen, G., 1993, Vasostatins, comprising the N-terminal domain of chromogranin A, suppress tension in isolated human blood vessel segments. *J. Neuroendocrinol.* **5**: 405-412.

Aunis, D., 1997, Exocytosis in chromaffin cells of the adrenal medulla. *Int. Rev. Cytol.* **181**: 213-320.

Cetin, Y., Aunis, D., Bader, M.F., Galindo, E., Jörns, A., Bargsten, G., and Grübe, D., 1993, Chromostatin, a chromogranin A-derived bioactive peptide, is present in human pancreatic insulin (ß) cells. *Proc. Natl. Acad. Sci. USA* **90**: 2360-2364.

Chao, C.C., Hu, S. and Peterson, P.K., 1995, Glia, cytokines and neurotoxicity. *Critic. Rev. Neurobiol.* **9:** 189-205.

Ciesielski-Treska, J. and Aunis, D., 2000, Chromogranin A induces a neurotoxic phenotype in brain microglial cells. *Adv. Exptl. Med. Biol.* **482:** 291-298.

Ciesielski-Treska, J., Ulrich, G., Taupenot, L., Chasserot-Golaz, S., Corti, A., Aunis, D., and Bader, M. F., 1998, Chromogranin A induces a neurotoxic phenotype in brain microglial cells. *J. Biol. Chem.* **273**: 14339-14346.

Ciesielski-Treska, J., Ulrich, G., Chasserot-Golaz, S., Zwiller, J., Revel, M.O., Aunis, D., and Bader, M. F., 2001, Chromogranin A induces a neurotoxic phenotype in brain microglial cells. *J. Biol. Chem.* **276**: 13113-13120.

Eiden, L.E., 1987, Is chromogranin a prohormone? *Nature (London)* **325**: 301.

Dillen, L., Miserez, B., Claeys, M., Aunis, D., and De Potter, W., 1993, Posttranslational processing of proenkephalins and chromogranins/secretogranins. *Neurochem. Int.* **22**: 315-352.

Fasciotto, B.H., Trauss, C.A., Greeley, G.H., and Cohn, D.V., 1993, Parastatin (porcine chromogranin A347-419), a novel chromogranin A-derived peptide, inhibits parathyroid cell secretion. *Endocrinology* **133**: 461-466.

Giulian, D., Haverkamp, L.J., Li, J., Karshin, W.L., Yu, J., Tom, D., Li, X., and Kirkpatrick, J.B., 1995, Senile plaques stimulate microglia to release a neurotoxin found in Alzheimer disease. *Neurochem. Int.* **27:** 119-137.

Goumon, Y., Lugardon, K., Kieffer, B., Lefèvre, J. F., Van Dorsselaer, A., Aunis, D., and Metz-Boutigue, M.H., 1998, Characterization of antibacterial COOH-terminal proenkephalin-A-derived peptides (PEAP) in infectious fluids. Importance of enkelytin, the antibacterial PEAP209-237 secreted by stimulated chromaffin cells. *J. Biol. Chem.* **273**: 29847-29856.

Helle, K.B, and Aunis, D., 2000, *Chromogranins : from Fundamental to Clinics*, Kluwer Academic/Plenum Publishing, New York.

Lugardon, K., Raffner, R., Goumon, Y., Corti, Y., Delmas, A., Bulet, P., Aunis, D., and Metz-Boutigue, M.H., 2000, Antibacterial and antifungal activities of vasostatin-1, the N-terminal fragment of chromogranin A. *J. Biol. Chem.* **275**: 19745-19753.

Mahata, S.K., O'Connor, D.T., Mahata, M., Yoo, S.H., Taupenot, L., Wu, H., Gill, B.M., and Parmer, R.J., 1997, Novel autocrine feedback control of catecholamine release. A discrete chromogranin A fragment is a noncompetitive nicotinic cholinergic antagonist. *J. Clin. Invest.* **100**: 1623-1633.

Metz-Boutigue, M. H., Garcia-Sablone, P., Hogue-Angeletti, R., and Aunis, D., 1993, Intracellular and extracellular processing of chromogranin A determination of cleavage sites. *Eur. J. Biochem.* **217**: 247-257.

Metz-Boutigue, M.H., Goumon, Y., Lugardon, K., Strub, J.M., and Aunis, D., 1998, Antibacterial peptides are present in chromaffin cell secretory granules. *Cell. Mol. Neurobiol.* **18:** 249-266.

Munoz, D.G., 1991, Chromogranin A-like immunoreactivity neurites are major constituents of senile plaques. *Lab. Invest.* **64:** 826-832.

Simon, J.P., and Aunis, D., 1989, Biochemistry of the chromogranin A protein family. *Biochem. J.* 262: 1-13.

Strub, J.M., Garcia-Sablone, P., Lonning, K., Taupenot, L., Hubert, P., Van Doorsselaer, A., Aunis, D., and Metz-Boutigue, M.H., 1995, Processing of chromogranin B in bovine adrenal medulla : identification of secretolytin, the endogenous C-terminal fragment of residues 614-626 with antibacterial activity. *Eur. J. Biochem.* **229**: 356-368.

Strub, J.M., Sorokine, O. , Van Doorsselaer, A., Aunis, D., and Metz-Boutigue, M.H., 1997, Phosphorylation and *O*-glycosylation of bovine chromogranin A from adrenal medullary chromaffin granules and their relationship with biological activities. *J. Biol. Chem.* **272**: 11928-11936.

Tatemoto, K., Efendic, S., Mutt, V., Makk, G., Feistner, G. J., and Barchas, J. D., 1986, Pancreastatin, a novel pancreatic peptide that inhibits insulin secretion. *Nature (London)* **324**: 476-478.

Taupenot, L., Ciesielski-Treska, J., Ulrich, G., Chasserot-Golaz, S., Aunis, D. and Bader, M.F., 1996, Chromogranin A triggers a phenotypic transformation and the generation of NO in brain microglial cells. *Neuroscience* **72**: 377-389.

THE VESICULAR MONOAMINE TRANSPORTERS (VMATs): ROLE IN THE CHEMICAL CODING OF NEURONAL TRANSMISSION AND MONOAMINE STORAGE IN AMINE-HANDLING IMMUNE AND INFLAMMATORY CELLS

L. E. Eiden,* B. Schütz,** M. Anlauf,** C. Depboylu,** M.K.-H. Schäfer, ** E. Weihe*

1. INTRODUCTION

Monoamines can act as neurotransmitters, hormones, autocrine and paracrine factors, or autacoids. How they function depends on the locations of the cells that synthesize and store them, and the stimuli that release them. All amine transmitters are first sequestered in a storage vesicle or granule, from which they are secreted from the cell. This requires specific transporters that reside on the vesicle. All of the vesicular transporters for classical neurotransmitters inferred to exist as individual proteins based on functional studies, have been cloned and characterized in a detailed molecular way over the last ten years (see Table 1). As a result, an understanding has developed that the role of these transporters in the chemical coding of neurotransmission is dynamic, and a novel view of what constitutes a neurotransmitter phenotype for a given neuron has emerged. The purpose of this contribution is to highlight recent progress from our laboratories and others in understanding the evolution of vesicular transporter structure, transport properties and cell-specific expression, as these relate to the physiological and regulatory functions of mammalian monoamine-containing*** neuronal, endocrine, and hematopoietic cells.

---

*Section on Molecular Neuroscience, National Institute of Mental Health, NIH, Bethesda MD 20892, USA
**Institute of Anatomy and Cell Biology, Philipps-Universität, Marburg, Germany

*** 'Monoamines' are understood here to include DA, NE, Epi, HIS and related invertebrate or vestigial mammalian transmitters like tyramine and octopamine. Thus, monoamines are all the substrates of the VMATs. The term 'amine transmitter' is used to include the monoamines plus acetylcholine. Thus, the amines are all the substrates of the VATs (VMATs plus VAChT). This usage is not universal, but conveniently distinguishes the positively charged VAT substrates from the classical neurotransmitters that are substrates for VIAAT/VGAT or VEAAT/VGluT/BNPI (see Table 1).

2. VMAT'S BASIC STRUCTURE AND FUNCTION

Despite their widely different substrate specificities, the vesicular transporters work in similar ways. A proton-intruding vacuolar ATPase causes a proton gradient to be generated in a proton-impermeant vesicle. Neurotransmitters are then accumulated in the vesicle by a transport system sensitive to, and powered by, the electrochemical gradient that the proton-pumping ATPase has created.[1-4] The vesicular amine transporters (VATs) include the vesicular acetylcholine transporter (VAChT) and the vesicular monoamine transporters (VMATs).[5] The VATs have twelve predicted transmembrane domains based on Kyte-Doolittle analysis, and ten based on PHD.[6, 7] The localization of the C-terminal tail to the cytoplasm is their only empirically established structural feature.[8, 9] They are members of the TEXAN (toxin-extruding antiporter) family of transporters, which employ proton antiport for either vesicular accumulation of amines or cellular discharge of toxins/antibiotics [10]. The vesicular GABA and glutamate transporters are related to bacterial amino acid permeases and inorganic phosphate transporters, respectively,[11-14] with ten predicted transmembrane domains for VIAAT/VGAT, and nine for VEAAT/VGluT/BNPI.[6, 7]

Table 1. Vesicular Transporters and Chemicallly Coded Phenotypes in Neuronal and Non-neuronal Amine- and Amino Acid-Secreting Cells

Transmitter	Biosynthetic Enzyme(s)	Vesicle Type(s)	Vesicular Transporter	Nominal Function(s)
DA	TH; AADC	LDCV/SSV; LDCV	VMAT2; VMAT1	neurotransmitter; paracrine factor
NE	TH; AADC; DßH	LDCV; SDCV/SSV	VMAT2; VMAT1	neurotransmitter; hormone
Epi	TH; AADC; DßH; PNMT	LDCV; SDCV/SSV	VMAT2; VMAT1	neurotransmitter; hormone
5-HT	TrH; AADC	SSV; LDCV/SG	VMAT2; VMAT1	neurotransmitter; autacoid/hormone
HIS	HDC	LDCV; SG	VMAT2	neurotransmitter; autacoid; hormone
ACh	ChAT	SSV	VAChT	neurotransmitter
GABA	GAD	SSV	VIAAT	neurotransmitter
Gly	--	SSV	VIAAT	neurotransmitter
Glu	--	SSV	VEAAT	neurotransmitter

DA, dopamine; NE, norepinephrine; Epi, epinephrine; 5-HT, serotonin; HIS, histamine; ACh, acetylcholine; GABA, gamma-aminobutyric acid; Gly, glycine; Glu, glutamate; TH, tyrosine hydroxylase; AADC, aromatic amino acid decarboxylase; DßH, dopamine ß-hydroxylase; PNMT, phenylethanolamine N-methyltransferasae; TrH, tryptophan hydroxylase; HDC, histidine decarboxylase; ChAT, choline acetyltransferase; GAD, glutamic acid decarboxlyase; LDCV, large dense-core vesicle; SSV, small synaptic vesicle; SDCV, small dense-core vesicle; SG, secretory granule; VMAT, vesicular monoamine transporter; VAChT, vesicular acetylcholine transporter; VIAAT/VGAT, vesicular inhibitory amino acid or GABA transporter; VEAAT/VGluT/BNPI, vesicular excitatory amino acid or glutamate transporter, also called brain-specific Na+-dependent inorganic phosphate transporter. References to molecular characterization of vesicular transporters: VMAT1/2[15, 16]; VAChT[17, 18];VIAAT[11, 12]; VEAAT.[13, 14]

Despite the overall mechanistic similarity among the vesicular neurotransmitter transporters, there is significant variation in dependence on transvesicular charge versus protonmotive force. The VATs display a greater dependence on protonmotive force, while the GABA and glutamate transporters display a greater dependence on the charge gradient across the vesicular membrane.[19] All monoamines are transported as singly-positively charged species, and two protons are released from the vesicle for each amine transported to the vesicle interior.[5] However, even among monoamine-containing cells, transporter-containing secretory vesicles exhibit variation in size, cellular localization, and protein composition. These may be critically important in determining quantal size, kinetics of exocytosis, latency and refractory properties of secretion, and other features that specialize amine-handling cells for their specific physiological roles in the immunoinflammatory and neuroendocrine axes.

3. VESICULAR TRANSPORTERS AND THE CHEMICAL CODING OF NEUROTRANSMISSION

Specific vesicular neurotransmitter transporters are required for the chemical coding* of inhibitory and excitatory amino acid, cholinergic and monoaminergic neurotransmission (VIAAT, VEAAT, VAChT and VMAT respectively, see Table 1). Figure 1 depicts diagramatically specific biosynthetic enzyme and vesicular transporter combinations** required for expression of the four monoaminergic phenotypes, as well as the cholinergic and excitatory and inhibitory amino acid phenotypes. Genes encoding the vesicular neurotransmitter transporters have become the focus of investigation into the transcriptional regulatory details of how chemical coding is enforced during nervous system development.[20-23] Trans-acting factors have been identified that control neurotransmitter-specific phenotypes by specifically interacting with the promoters of the genes encoding biosynthetic enzymes, vesicular transporters, or both. Transcription of ChAT and VAChT is apparently regulated by transactivation at the ß cis-acting element of the cholinergic gene locus (CGL) in C. elegans.[24] Unc-30 is a homeodomain protein which controls GAD (unc-47) and VIAAT/VGAT (unc-25) expression in C. elegans by direct transcriptional activation of both genes,[25] and whose mammalian ortholog Pitx2 has recently been identified.[26] VMAT2, in contrast to VAChT and VIAAT/VGAT, is associated with at least four specific neuronal phenotypes with distinct patterns of biosynthetic enzyme gene expression (see Table 1). and is also expressed in amine-handling cells of hematopoietic origin, and in gut enterochromaffin-like (ECL) cells,[21] that arise from different stem cell populations than neuronal progenitors. The VMAT2 gene is therefore likely to contain a mosaic of cis-regulatory domains that allow it to access multiple neuronal and non-neuronal lineage-specific transactivators. For example, Phox2A/Arix and Phox2b, together with the bHLH protein MASH-1, control the development of noradrenergic neurons and Phox2 interacts directly with the DßH gene to regulate its transcription in noradrenergic neurons.[27-31] However, lineage-determining factors do not appear to control the genes encoding all the proteins that confer traits for a given type of chemically coded neuron. The gene encoding the plasma membrane

* 'Chemical coding of neurotransmission' refers here to the expression of sets of proteins that allow neuroendocrine cells to employ specific neurotransmitters for intercellular communication. Neurons that express TH, AADC, VMAT2, and DAT are 'chemically coded' as dopaminergic, whereas neurons that express TH without any other dopamine-associated traits are not 'chemically coded' as dopaminergic neurons.

transporter NET, for example, is apparently not regulated by Phox2,[32] and VMAT2 is expressed in dopaminergic and serotonergic neurons of MASH knock-out mice (Weihe, Damadzic and Eiden, in preparation), although VMAT2-containing noradrenergic neurons are absent.

4. TWO ISOFORMS OF VMAT--FUNCTION AND EVOLUTION

Besides the substrate specificity that exists between the transporters for acetylcholine, monoamines, GABA/glycine, and glutamate, there are differences in substrate recognition between VMAT1, the 'neuroendocrine' isoform, and VMAT2, 'neuronal' isoform, of mammalian VMAT. Neurons in both central and peripheral nervous system express only VMAT2. Paracrine SIF cells of sympathetic ganglia express predominantly VMAT1, while adrenal chromaffin cells express both VMAT1 and VMAT2, with VMAT1 the major isoform in rodents and VMAT1 and VMAT2 about equally abundant in human adrenal medulla. In the gut, VMAT2 is expressed in the histamine-containing ECL cells of the stomach that control acid secretion from the stomach, while serotonin-containing EC cells of the intestine express VMAT1.[33]

The evolutionary emergence of the two isoforms of VMAT in vertebrates can be considerd from the standpoint of the structure and function of the single VMAT protein present in nematodes. Transport studies have revealed that VMAT2 has a higher affinity than VMAT1 for non-catechol substrates and histamine.[33-37] The C. elegans VMAT sequence is equally similar to VMATs 1 and 2, but CelVMAT is more like VMAT2 in substrate preference[38](Table 2). VMAT1 thus seems to have emerged as a 'specialized' or 'restricted' VMAT, rather than as a second isoform with broader substrate specificity than the primordial VMAT transporter. The high affinity of CelVMAT for tyramine and octopamine may reflect the use of parahydroxyphenylethylamines, as well as catecholamines, as neurotransmitters in nematodes. Arthropods also employ non-catechol monoamines as transmitters. Drosophila with mutations in the tyrosine decarboxylase gene, required for generation of the trace monoamine tyramine, show altered sensitivity to cocaine;[39] and a functional tyramine ß-hydroxylase, responsible for octopamine biosynthesis, is required for egg-laying.[40]

Table 2. Substrate specificities of human VMAT1 and VMAT2 compared to the single-copy VMAT of C. elegans (CelVMAT)

	DA	Tyr	5-HT	NE	Oct	His
CelVMAT	0.07	0.06	0.91	2.8	3.4	120
hVMAT1	3.8	4.5	1.4	13.7	95	4696
hVMAT2	1.4	0.6	0.9	3.4	5.0	143

IC50 values (μM) for inhibition of 5-HT uptake by monoamines in permeabilized CV-1 cells expressing CelVMAT, hVMAT1, or hVMAT2. Data from Duerr et al., 1999 and Eiden, unpublished observations [octopamine (Oct) IC50 values for hVMAT1 and hVMAT2].

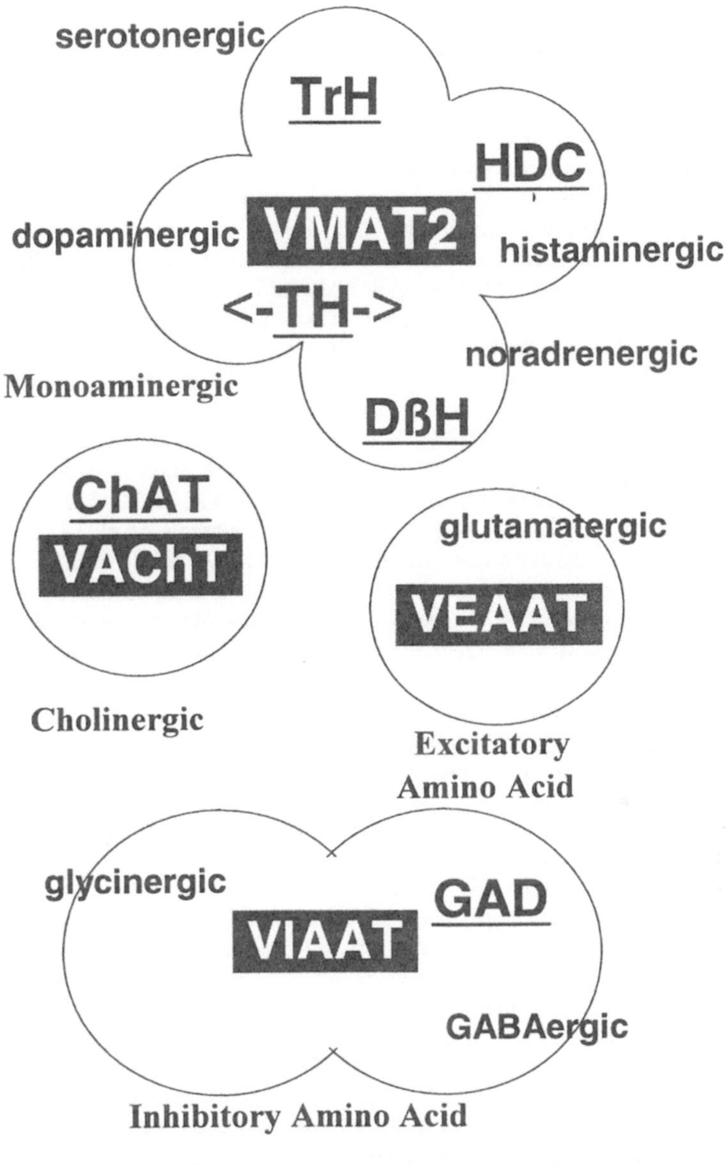

Figure 1. Vesicular transporter expression in the four major types of chemically coded neurons. Vesicular transporters are boxed, and biosynthetic enzymes are underlined. The bounded figures depict minimal protein sets whose expression is required to produce each chemically coded phenotype. Plasma membrane transporters that also contribute to amine and amino acid chemical coding are not shown here.

The role of VMAT in fly tyraminergic and octopaminergic neurotransmission has not been examined biochemically or genetically. The rodent nervous system contains trace amounts of stored tyramine, and thus this system may be vestigial in mammals.[41, 42] Phenylethylamine, a preferred substrate for VMAT2 compared to VMAT1, has been reported to be excreted in humans under conditions of extreme stress.[43] Lineage-specific expression of VMAT1 and VMAT2 may therefore serve to limit the vesicular storage of substrates besides the classical monoamines to specific types of neuroendocrine and hematopoietic/inflammatory cells.

5. VMAT2 AND CHEMICAL CODING IN THE GASTROINTESTINAL TRACT

The coordinate expression of the vesicular amine transporters with other proteins that contribute to chemical coding has been examined by double-label immunohistochemistry throughout the neuroendocrine axis ([44] and references therein). This is of particular interest in the enteric nervous system. Uncertainty about the functional neurotransmitter phenotypes of intrinsic enteric neurons with catecholaminergic or serotonergic traits, has hindered understanding of how monoamines modulate gastrointestinal function. Nominally serotonergic neurons in the rodent gut, thought to be critical for bowel function, have been identified as transiently TH- and DßH-positive neurons, derived from both vagal and truncal neural crest, which depend on the bHLH transcription factor MASH-1 for their development *in vivo*, and can accumulate exogenously applied tritiated serotonin.[45] In human gut, we have used double fluorescence and confocal microscopy to reveal presumably dopaminergic intrinsic neurons in submucous and myenteric plexus, with highest proportions in the stomach (Anlauf, Eiden and Weihe, in preparation). These neurons co-express TH and VMAT2, but not DßH, 5-HT, VMAT1, or VAChT. Their targets are longitudinal, circular, and submucosal smooth muscle cells. In contrast, the extrinsic noradrenergic innervation of the human gut (TH-, DßH-, and VMAT2-positive fibers lacking 5-HT or VAChT) targets mainly blood vessels of all layers, and intrinsic ganglionic cells of Auerbach's and Meissner's plexus. We have been unable to identify VMAT2-positive, TH-negative neurons in human gut, corresponding to rodent putative serotonergic enteric neurons. It is important to note that direct evidence for the presence of dopamine in the 10-20% of intrinsic neurons of the human gut that are VMAT2- and TH-positive is still lacking. Likewise, the putative serotonergic neurons of the intrinsic nervous system of the rodent gastrointestinal tract have not yet been convincingly demonstrated to possess tryptophan hydroxylase. These may be 'orphan' neurons [44] that express only VMAT2, and the serotonergic plasma membrane transporter 5-HTT/SERT, and accumulate serotonin from outside the cell, as postulated for developing thalamocortical neurons of the rodent central nervous system.[46] Likewise, the transient catecholaminergic neurons which become 'serotonergic' in rodent may remain catecholaminergic throughout life in the human gastrointestinal tract. This would be comparable to the persistent expression of TH and VMAT2 in adult human cholinergic sympathetic eccrine sweat gland innervation and the lack of any expression of VMAT2, and only transient expression of TH, in corresponding neurons in the rat (see below). The possibility exists, therefore, that monoaminergic enteric neurons may be neither truly 'dopaminergic' or 'serotonergic' in either rodent or human, but rather posses a complex mixture of monoaminergic traits, including TH, TrH, 5-HTT/SERT, and VMAT2, perhaps allowing either dopaminergic or serotonergic modes of neurotransmission to dominate in different regions of gut, at different times of development, or in response to different physiological stimuli. Both human and rodent enteric neuronal phenotypes must be re-examined using all available markers including VATs, plasma membrane transporters, and antibodies that visualize the endogenous transmitters themselves, to establish their actual phenotype *in vivo*, which may be more plastic, and more various, than previously thought.

6. THE DISTRIBUTED MONOAMINERGIC PHENOTYPE IN THE CENTRAL NERVOUS SYSTEM

As described above (Section 5), some neurons may have an incomplete neurotransmitter phenotype, i.e. expression of some monoaminergic traits with insufficient complementation by other monamine-specific proteins to permit functional neurotransmission. Cells of the primate locus coeruleus, for example, express TH and DßH as expected for a noradrenergic phenotype, and neurons of the nucleus tractus solitarious (NTS) express TH but not DßH, and have been assigned a dopaminergic phenotype based on this evidence.[47] However, NTS neurons do not express VMAT2, and are therefore 'orphan neurons'--expected to synthesize DA but not store or release it as a neurotransmitter from synaptic vesicles.[48] Ugrumov and co-workers[49] have identified similar neurons within the 'dopaminergic' projections from the arcuate nucleus to the median eminence of the hypothalamus, that contain TH but not dopa decarboxylase (AADC), and thus would be capable of synthesizing L-dopa, but not dopamine. Since neighboring neurons have been identified that express AADC but not TH, Ugrumov and colleagues have hypothesized that intercellular complementation may occur between AADC-containing neurons that can take up L-dopa and synthesize dopamine, and TH-containing neurons that can synthesize and secrete L-dopa but cannot decarboxylate L-dopa to form dopamine. It remains to be determined which of these neuronal populations expresses VMAT2, although the fibers innervating the median eminence itself are fully positive for VMAT2.[9]

7. VMAT2 EXPRESSION IN CHOLINERGIC SYMPATHETIC NEURONS

It has been claimed, with considerable supporting evidence, that the small number of principal (post-ganglionic) cells of the sympathetic ganglia that are cholinergic become so by switching from a noradrenergic to a cholinergic phenotype only after establishing synaptic connections with their targets, such as the eccrine sweat glands and periosteum ([22, 50]' and references therein). It has also been observed that cholinergic principal ganglion cells are found in sympathetic ganglia much earlier than these synaptic connections occur, suggesting target-independent establishment of the cholinergic sympathetic phenotype.[20] Simultaneous visualization of TH, VMAT2 and VAChT in the sympathetic nervous system of mouse, rat and human has allowed us to determine that important differences exist in the monoaminergic traits exhibited by cholinergic neurons during development and in the adult. Adult rat cholinergic projections to the eccrine sweat glands express neither TH- nor VMAT2; adult mouse projections are TH-and VMAT2-negative, and adult human projections express both TH and VMAT2, suggesting cholinergic/noradrenergic co-transmission in adult human sweat glands, and perhaps relevant to human 'adrenergic sweating'.[51] During early post-natal development in both mouse and rat, eccrine sweat gland projections are TH-positive, VAChT-positive, and VMAT2-negative. In both species, the sparse innervation of the sweat gland is weakly VAChT-positive, and becomes steadily more strongly VAChT-positive during the first three weeks of life. The patterns of expression of VAChT, TH and VMAT2 in the rodent stellate ganglion and eccrine projection fields could be interpreted as follows: the cholinergic phenotype and some noradrenergic traits exist in some sympathetic post-ganglionic neurons prior to birth, and the projections of these neurons to the sweat glands are essentially cholinergic as early as such innervation exists. The cholinergic phenotype appears to become more robust at the nerve terminal throughout early postnatal development because synaptic vesicle density increases with synaptic stabilization during this period. According to this view, target-dependent cholinergic phenotypic 'switching' at sweat gland terminals for example, is more likely to reflect synaptic maturation than changes in transcription from the cholinergic gene locus.

8. VMAT2 IN HEMATOPOIETIC AMINE-HANDLING CELLS

Monoamines are stored in, and secreted from, secretory granules of non-neuroendocrine hematopoietic cells, including mast cells, basophils and platelets, that respond to inflammation and tissue injury. VMAT2 protein and mRNA are present in rodent tryptase- and histidine decarboxylase-positive mast cells[48, 52] of skin, lymph nodes, thymus, and in paravascular areas of most tissues. Stomach mucosal mast cells stain relatively weakly for VMAT2, perhaps indicating down-regulation of mast cell VMAT2 in oxyntic gut, where high levels of extracellular histamine contributed by ECL cells might otherwise allow adventitious refilling of previously histamine-depleted tissue mast cells. VMAT2-positive mast cells are prominent in autonomic and sensory nerves and ganglia, around cerebral arteries and meninges in the brain. Embryonic and adult rodent bone marrow cells are frequently VMAT2-positive. VMAT2 is present in mast cells of rhesus monkey and human in all tissues examined. Human tonsils are rich in VMAT2. In human mastocytosis, cells of the tumor mass co-stain for tryptase, histidine decarboxylase, and VMAT2, and contain VMAT2 mRNA. VMAT2-positive splenic megakaryocytes are found in rodents, but not primates. Human platelets, CD1-positive skin Langerhans cells, and tissue and bone marrow basophils all express VMAT2. Monoamine storage in mammalian basophils, platelets, and mast cells depends on levels of VMAT2 expression as well as VMAT2 affinity for histamine, serotonin and other substrates, which differs among mammalian species, and may account for different ratios of histamine and serotonin in platelets, for example, of various species.[53-55] Therapeutic intervention in injury and inflammatory responses involving amine-handling cells could conceivably be focused on inhibition of monoamine storage at these loci.

9. VAT EXPRESSION IN SECRETORY VESICLE SUBTYPES

Several types of secretory vesicles are known to exist. Monoamines, are stored in both large and small dense-core vesicles (LDCVs and SDCVs), and in some cases in small synaptic vesicles (SSVs), in which most small-molecule classical neurotransmitters are stored prior to regulated secretion. It has been assumed that the small dense-core vesicles are essentially identical to the small synaptic vesicles of non-noradrenergic neurons, since without the dense core visualized by permanganate fixation, the vesicles are of the same size and appearance as vesicles from other types of neurons, and sediment similarly on sucrose and velocity gradients when depleted. However, recent evidence suggests that noradrenergic small dense-core vesicles (SDCVs) are distinct from other small synaptic vesicles (SSVs) in their ontogenesis and in their cell lineage restriction. In PC12 cells that co-express VMAT1 and VAChT, VMAT1 is found almost wholly associated with LDCVs, and VAChT with SSVs, in nerve terminals after NGF-differentiation.[56] VAChT may be transported to the nerve terminal in LDCVs, and then transferred to SSVs however, since VAChT is observed by immune electronmicroscopy to be associated with LDCVs rather than constitutive vesicles in PC12 cell neurites.

Introduction of VMAT2 protein via retroviral expression vector transduction in PC12 cells was expected to show VMAT2 distribution to both LDCVs and SSVs, given the distribution of VMAT2 to both LDCVs and SSV/SDCVs (SDCVs appear as SSVs if electron microscopy is performed in the absence of permanganate or chromate salts) in noradrenergic neurons *in vivo*. Instead, VMAT2 was excluded from SSVs containing VAChT in PC12 cells, implying the existence of distinct 'cholinergic' and 'noradrenergic' SSVs. The possibility that each of the vesicular neurotransmitter transporters are targeted to distinct synaptic vesicle types specific to monoaminergic, cholinergic, GABAergic and glutamatergic neurotransmission provides an intriguing new direction for future research. It is already apparent, for example, that phosphorylation sites on the vesicular transporters themselves can affect targeting to specific vesicle sub-types (reviewed in [57]).

10. ACKNOWLEDGEMENTS

We thank Jim Rand and Janet Duerr, and members of the Eiden and Weihe laboratories, for comments and suggestions on the manuscript. Work described here was supported in part by the Volkswagen-Stiftung and by German Research Foundation Grant SFB 297.

REFERENCES

1. B. I. Kanner and S. Schuldiner, Mechanism of transport and storage of neurotransmitters, *CRC Crit. Rev. Biochem.* **22,** 1-38 (1987).
2. R. G. Johnson Jr., Accumulation of biological amines into chromaffin granules: a model for hormone and neurotransmitter transport, *Physiol. Revs.* **68,** 232-307 (1988).
3. P. R. Maycox, T. Deckwoerth, J. W. Hell, and R. Jahn, Glutamate uptake by brain synaptic vesicles, *J. Biol. Chem.* **263,** 15423-15428 (1988).
4. P. M. Burger, J. Hell, E. Mehl, C. Krasel, F. Lottspeich, and R. Jahn, GABA and glycine in synaptic vesicles: Storage and transport characteristics, *Neuron* **7,** 287-293 (1991).
5. S. M. Parsons, Transport mechanisms in acetylcholine and monoamine storage, *FASEB J.* **14,** 2423-2434 (2000).
6. B. Rost, PHD: predicting one-dimensional protein structure by profile based neural networks, *Meth. Enzymol.* **266,** 525-539 (1996).
7. B. Rost, P. Fariselli, and R. Casadio, Topology prediction for helical transmembrane proteins at 86% accuracy, *Protein Science* **7,** 1704-1718 (1996).
8. M. L. Gilmor, N. R. Nash, A. Roghani, R. H. Edwards, H. Yi, S. M. Hersch, and A. I. Levey, Expression of the putative vesicular acetylcholine transporter in rat brain and localization in cholinergic synaptic vesicles, *J. Neurosci.* **16,** 2179-2190 (1996).
9. E. Weihe, J.-H. Tao-Cheng, M. K.-H. Schäfer, J. D. Erickson, and L. E. Eiden, Visualization of the vesicular acetylcholine transporter in cholinergic nerve terminals and its targeting to a specific population of small synaptic vesicles, *Proc. Natl. Acad. Sci. USA* **93,** 3547-3552 (1996).
10. S. Schuldiner, A. Shirvan, and M. Linial, Vesicular neurotransmitter transporters: From bacteria to humans, *Physiol. Rev.* **75,** 369-392 (1995).
11. C. Sagné, S. E. Mestikaway, M.-F. Isambert, M. Hamon, J.-P. Nenry, B. Giros, and B. Gasnier, Cloning of a functional vesicular GABA and glycine transporter by screening of genome databases, *FEBS Letts.* **417,** 177-183 (1997).
12. S. L. McIntire, R. J. Reimer, K. Schuske, R. H. Edwards, and E. M. Jorgensen, Identification and characterization of the vesicular GABA transporter, *Nature* **389,** 870-876 (1997).
13. E. E. Bellochio, R. J. Reimer, R. T. Fremeau Jr., and R. H. Edwards, Uptake of glutamate into synaptic vesicles by an inorganic phosphate transporter, *Science* **289,** 957-960 (2000).
14. S. Takamori, J. S. Rhee, C. Rosenmund, and R. Jahn, Identification of a vesicular glutamate transporter that defines a glutamatergic phenotype in neurons, *Nature* **407,** 189-194 (2000).

15. J. D. Erickson, L. E. Eiden, and B. Hoffman, Expression cloning of a reserpine-sensitive vesicular monoamine transporter, *Proc. Natl. Acad. Sci. USA* **89,** 10993-10997 (1992).
16. Y. Liu, D. Peter, A. Roghani, S. Schuldiner, G. G. Prive, D. Eisenberg, N. Brecha, and R. H. Edwards, A cDNA that suppresses MPP+ toxicity encodes a vesicular amine transporter, *Cell* **70,** 539-551 (1992).
17. A. Alfonso, K. Grundahl, J. R. McManus, and J. B. Rand, Cloning and characterization of the choline acetyltransferase structural gene (cha-1) from C. elegans, *J. Neurosci.* **14,** 2290-3300 (1994).
18. J. D. Erickson, H. Varoqui, M. Schäfer, M.-F. Diebler, E. Weihe, W. Modi, J. Rand, L. E. Eiden, T. I. Bonner, and T. Usdin, Functional characterization of the mammalian vesicular acetylcholine transporter and its expression from a 'cholinergic' gene locus, *J. Biol. Chem.* **269,** 21929-21932 (1994).
19. S. Schuldiner, A molecular glimpse of vesicular transporters, *J. Neurochem.* **62,** 2067-2078 (1994).
20. M. K.-H. Schäfer, B. Schütz, E. Weihe, and L. E. Eiden, Target-independent cholinergic differentiation in the rat sympathetic nervous system, *Proc. Natl. Acad. Sci. USA* **94,** 4149-4154 (1997).
21. B. Schütz, M. K.-H. Schäfer, L. E. Eiden, and E. Weihe, Vesicular amine transporter expression and isoform selection in developing brain, peripheral nervous system and gut, *Dev. Brain Res.* **106,** 181-204 (1998).
22. S. E. Asmus, S. Parsons, and S. C. Landis, Developmental changes in the transmitter properties of sympathetic neurons that innervate the periosteum, *J. Neurosci.* **20,** 1495-1504 (2000).
23. C. Goridis and J.-F. Brunet, Transcriptional control of neurotransmitter phenotype, *Curr. Opin. Neurobiol.* **9,** 47-53 (1999).
24. D. Frisby, J. McManus, J. Duerr, and J. Rand, Regulation of cholinergic gene expression in C. elegans, *Soc. Neurosci. Abstr.* **22,** 1032 (1996).
25. C. Eastman, H. R. Horvitz, and Y. S. Jin, Coordinated transcriptional regulation of the unc-25 glutamic acid decarboxylase and the unc-47 GABA vesicular transporter by the Caenorhabditis elegans UNC-30 homeodomain protein, *J. Neurosci.* **9,** 6225-6234 (1999).
26. J. J. Westmoreland, J. McEwen, B. A. Moore, Y. Jin, and B. G. Condie, Conserved function of C. elegans UNC-30 and mouse Pitx2 in controlling GABAergic neuron differentiation, *J. Neurosci.* **in press,** (2001).
27. M.-R. Hirsch, M.-C. Tiveron, F. Guillemot, J.-F. Brunet, and C. Goridis, Control of noradrenergic differentiation and Phox2a expression by MASH1 in the central and peripheral nervous system, *Development* **125,** 599-608 (1998).
28. L. Lo, M.-C. Tiveron, and D. J. Anderson, MASH1 activates expression of the paired homeodomain transcription factor Phox2a, and couples pan-neuronal and subtype-specific components of autonomic neuronal identity, *Development* **125,** 609-620 (1998).
29. J.-F. Brunet and A. Ghysen, Deconstructing cell determination: proneural genes and neuronal identity, *BioEssays* **21,** 313-318 (1999).
30. C. Yang, H.-S. Kim, H. Seo, C.-H. Kim, J.-F. Brunet, and K.-S. Kim, Paired-like homeodomain proteins, Phox2a and Phox2b, are responsible for noradrenergic cell-specific transcription of the dopamine ß-hydroxylase gene, *J. Neurochem.* **71,** 1813-1826 (1998).
31. M. Adachi, D. Browne, and E. J. Lewis, Paired-like homeodomain proteins Phox2a/Arix and Phox2b/NBPhox have similar genetic organization and independently regulate dopamine ß-hydroxylase gene transcription, *DNA Cell Biol.* **19,** 539-554 (2000).
32. C.-H. Kim, H.-S. Kim, J. F. Cubells, and K.-S. Kim, A previously undescribed intron and extensive 5' upstreamm sequence, but not Phox2a-mediated transactivation, are necessary for high level cell type-specific expression of the human norepinephrine transporter gene, *J. Biol. Chem.* **274,** 6507-6518 (1999).
33. J. D. Erickson, M. K.-H. Schäfer, T. I. Bonner, L. E. Eiden, and E. Weihe, Distinct pharmacological properties and distribution in neurons and endocrine cells of two isoforms of the human vesicular monoamine transporter, *Proc. Natl. Acad. Sci. USA* **93,** 5166-5171 (1996).
34. D. Peter, J. Jimenez, Y. Liu, J. Kim, and R. H. Edwards, The chromaffin granule and synaptic vesicle amine transporters differ in substrate recognition and sensitivity to inhibitors, *J. Biol. Chem.* **269,** 7231-7237 (1994).
35. E. Weihe, M. K.-H. Schäfer, J. D. Erickson, and L. E. Eiden, Localization of vesicular monoamine transporter isoforms (VMAT1 and VMAT2) to endocrine cells and neurons in rat, *J. Mol. Neurosci.* **5,** 149-164 (1994).
36. J. D. Erickson, L. E. Eiden, M. K.-H. Schäfer, and E. Weihe, Reserpine- and tetrabenazine-sensitive transport of 3H-histamine by the neuronal isoform of the vesicular monoamine transporter, *J. Mol. Neurosci.* **6,** 277-287 (1995).
37. A. Merickel and R. H. Edwards, Transport of histamine by vesicular monoamine transporter-2, *Neuropharmacol.* **34,** 1543-1547 (1995).

38. J. S. Duerr, D. L. Frisby, J. Gaskin, A. Duke, K. Asermely, D. Huddleston, L. E. Eiden, and J. B. Rand, The cat-1 gene of Caenorhabditis elegans encodes a vesicular monoamine transporter required for specific monoamine-dependent behaviors, *J. Neurosci.* **19,** 72-84 (1999).
39. C. McClung and J. Hirsh, The trace amine tyramine is essential for sensitization to cocaine in Drosophila, *Curr. Biol.* **9,** 853-860 (1999).
40. M. Monastirioti, C. E. Linn, and K. White, Characterization of Drosophila tyramine beta-hydroxylase gene and isolation of mutant flies lacking octopamine, *J. Neurosci.* **16,** 3900-3911 (1996).
41. J. F. Tallman, J. M. Saavedra, and J. Axelrod, Biosynthesis and metabolism of endogenous tyramine and its normal presence in sympathetic nerves, *J. Pharmacol. Exp. Ther.* **199,** 216-221 (1976).
42. J. B. Rand, J. S. Duerr, and D. L. Frisby, Neurogenetics of vesicular transporters in C. elegans, *FASEB J.* **14,** 2414-2422 (2000).
43. M. A. Paulos and R. E. Tessel, Excretion of beta-phenethylamine is elevated in humans after profound stress, *Science* **215,** 1127-1129 (1982).
44. E. Weihe and L. E. Eiden, Vesicular amine transporter expression in amine-handling cells of the nervous, endocrine and inflammatory systems, *FASEB J.* **14,** 2435-2449 (2000).
45. E. Blaugrund, T. D. Pham, V. M. Tennyson, L. Lo, L. Sommer, D. J. Anderson, and M. D. Gershon, Distinct subpopulations of enteric neuronal progenitors defined by time of development, sympathoadrenal lineage markers and Mash-1 dependence, *Development* **122,** 309-320 (1996).
46. C. Lebrand, O. Cases, C. Adelbrecht, A. Doye, C. Alvarez, S. El Mestikawy, I. Seif, and P. Gaspar, Transient uptake and storage of serotonin in developing thalamic neurons, *Neuron* **17,** 823-835 (1996).
47. K. Kitahama, N. Sakamoto, A. Jouvet, I. Nagatsu, and J. Pearson, Dopamine-beta-hydroxylase and tyrosine hydroxylase immunoreactive neurons in the human brainstem, *J. Chem. Neuroanat.* **10,** 137-146 (1996).
48. E. Weihe and L. E. Eiden, Chemical neuroanatomy of the vesicular transporters, *FASEB J.* **14,** 2435-2449 (2000).
49. I. S. Balan, M. V. Ugrumov, A. Calas, P. Mailly, M. Kreiger, and J. Thibault, Tyrosine hydroxylase-expressing and/or aromatic L-amino acid decarboxylase-expressing neurons in the mediobasal hypothalamus of perinatal rats: differentiation and sexual dimorphism, *J. Comp. Neurol.* **425,** 167-176 (2000).
50. G. Guidry and S. C. Landis, Target-dependent development of the vesicular acetylcholine transporter in rodent sweat gland innervation, *Dev. Biol.* **199,** 175-184 (1998).
51. S. A. Shields, K. A. MacDowell, S. B. Fairchild, and M. L. Campbell, Is mediation of sweating cholinergic, adrenergic, or both-A comment on the literature, *Psychophysiology* **24,** 312-319 (1987).
52. E. Weihe, M. Anlauf, M.-K. H. Schäfer, W. Hartschuh, and L. E. Eiden, VMAT2 is the transporter mediating sequestration of monoamines in rat and human platelets, mast cells, and cutaneous dendritic cells, *Soc. Neurosci. Abstr.* **Nov. 7-12,** #301.301 (1998).
53. M. da Prada, A. Pletscher, J. P. Tranzer, and H. Knuchel, Subcellular localization of 5-hydroxytryptamine and histamine in blood platelets, *Nature* **216,** 1315-1317 (1967).
54. M. H. Fukami, H. Holmsen, and K. Ugurbil, Histamine uptake in pig platelets and isolated dense granules, *Biochem. Pharmacol.* **33,** 3869-3874 (1984).
55. A. Pletscher, M. Da Prada, K. H. Berneis, H. Steffen, B. Lütold, and H. G. Weder, *Molecular organization of amine storage organelles of blood platelets and adrenal medulla*, in *Advances in Cytopharmacology*, B. Ceccarelli, F. Clementi, and J. Meldolesi, Editors. 1974, Raven Press: New York. p. 257-264.
56. S. Tao-Cheng and L. E. Eiden, The vesicular monoamine transporter VMAT2 is targeted to large dense-core vesicles, and the vesicular acetylcholine transporter VAChT to small synaptic vesicles, in PC12 cells, *Adv. Pharmacol.* **42,** 250-253 (1998).
57. J. D. Erickson, D. Yao, H. Zhu, H. Ming, and H. Varoqui, Domains of vesicular amine transporters important for substrate recognition and targeting to secretory organelles, *FASEB J.* **14,** 2450-2458 (2000).

MUTANTS AND VARIANTS OF THE HUMAN ANTIDEPRESSANT-SENSITIVE NOREPINEPHRINE TRANSPORTER

Heinz Bönisch[1], Philipp Wiedemann[1], Fabian Runkel[1], Bruno Giros[2], Christine Roubert[2], and Michael Brüss[1]

1. INTRODUCTION

The norepinephrine transporter (NET) is responsible for the rapid re-uptake of synaptically released NE. Substrate transport by neurotransmitter transporters (NTTs) is principially posible in both directions. In Na^+- dependent transporters the normal transport direction is given by the Na^+ gradient mainted by Na^+/K^+-ATPase. The ions Na^+ and Cl^- are cosubstrates of the NET and both ions are also needed for substrate (e.g., NE) transport and binding of NET inhibitors such as nisoxetine or the tricyclic antidepressant desipramine (DMI). In transfected cells overexpressing the NET, a channel-like mode of the NET has also been described.[1] The NET transports the catecholamines NE, DA and epinephrine and amines like amphetamine, and the NET is a primary target of cocaine and of clinically important antidepressants (such as DMI, doxepine and reboxetine). The NET is not only expressed in noradrenergic neurons of the central and peripheral nervous system and in adrenal medullary cells but also in the placenta and in endothelial cells of small blood vessels of the lung where the NET is involved in the inactivation of circulating catecholamines.[2]

After the rat GABA transporter (rGAT) the human NET (hNET) was the second cloned NTT[3]; thereafter further neurotransmitter transporters (NTTs) such as the dopamine transporter (DAT), the serotonin transporter (SERT), glycine transporters (GLYTs) or vesicular monoamine transporters (VMATs) have been cloned.[2, 4] On the basis of sequence homology, ion dependence, and predicted topology, the cloned NTTs have been subdivided into at least three large families. The NET belongs to the family of Na^+ and Cl^--dependent plasma membrane transporters. This family includes beside the NET the monoamine transporters DAT, SERT and fET (the frog epinephrine transporter)

[1] Heinz Bönisch (corresponding author; e-mail: boenisch@uni-bonn.de); Institute of Pharmacology & Toxicology, University of Bonn, Reuterstr. 2 b, D-53113 Bonn, Germany.
[2] INSERM U-513, Neurobiology & Psychiatry, Faculty of Medicine, F-94000 Creteil, France

which are most closely related to the NET, the amino acid transporters GAT and GLYT as well as the transporters for proline, taurine and betaine. Transporters of this family are characterized by 12 transmembrane domains (TMs) and a large extracellular loop between TM3 and TM4, and intracellularly located N- and C-termini. At least for the hNET the topology has been confirmed.[5] The hNET protein is a 617 amino acid (aa) polypeptide with three potential N-glycosylation sites in the loop between TM3 and TM4. The hNET has four potential casein kinase II and one potential protein kinase C (PKC) pophosphorylation sites; the latter is obviously not responsible for PKC-mediated internalization of the hNET.[6]

The hNET gene (SLC6A2) has been mapped to chromosome 16q12.2[7] and the coding region consists of 14 exons.[8] In the 3′region an alternatively spliced exon has been identified which is responsible for an alternative C-terminus of the hNET.[9] Recently, in healthy controls and in patients suffering from schizophrenia or bipolar affective disorder, the coding region of the hNET gene has been systematically analysed for naturally occurring mutations and five rare missense mutations causing aa substitutions (Val69Ile, Thr99Ile, Val245Ile, Val449Ile and Gly478Ser) have been detected which, however, were not associated with one of the above disorders.[10]

We have now examined the functional and pharmacological properties of the five naturally occurring hNET mutants and of the C-terminal splice variant, and we generated and studied a series of hNET mutants in which aa potentially involved in binding and/or transport of substrates or of the cosubstrates Na^+ and Cl^- or in binding of inhibitors such as the tricyclic antidepressant desipramine (DMI) have been exchanged. Parts of these studies have been published.[11-13]

2. RESULTS AND INTERPRETATIONS

2.1 Naturally occurring hNET variants

All naturally occurring hNET variations are located in transmembrane domains: Val69Ile in TM1, Thr99Ile in TM2, Val245Ile in TM4, Val449Ile in TM9 and Gly478Ser in TM 10. Alignment of the aa sequence of the hNET with bovine, mouse and rat NET and with human, bovine, rat and mouse DAT and SERT shows that Val245 is not conserved, whereas Val69 and Gly478 are absolutely conserved in these transporters, and Thr99 as well as Val449 are conserved only among the NETs and SERTs.

Transfected HEK293 cells expressing these hNET variants or the wild-type hNET (WT-hNET) were examined for the kinetics of ^{3}H-NE uptake and for inhibition of NE uptake by DMI. DMI showed about the same inhibitory potency (mean K_i: 2 nM) at the variant hNETs and at the WT-hNET. Also none of the hNET variants, except the Gly478Ser variant, exhibited differences in the kinetics of NE transport as compared to the WT-hNET (mean K_m: 0.3 μM, mean V_{max}:53 pmol/mg protein). Only the Gly478Ser variant showed an about 4-fold lower apparent affinity (higher K_m) with no change in the V_{max}.[12] Thus, this glycine residue may play a role in the binding and translocation of NE.

2.2 Residues involved in binding/translocation of substrate or cosubstrate

Trp80 and Arg81 (in TM1 of hNET) and Glu113 (in TM2 of hNET) are absolutely conserved in all Na^+- and Cl^--dependent NTTs. Replacement of these residues by other aa have been shown in the rat GAT not to affect plasma membrane expression but to cause an almost complete loss of transport function.[14-16] We now show that the hNET variants Trp80Ser, Arg81His and Glu113Asp exhibit no transport of ^{3}H-NE; this may indicate that Arg (as a positively charged aa) could be involved in binding and/or translocation of chloride and Trp and the negatively charged Glu could be involved in binding/translocation of the sodium ion. The latter was supported by the finding that much lower concentrations of Na^+ were needed to cause half-maximum stimulation of ^{3}H-DMI binding at the Trp80Ser and the Glu113Asp variants than at the WT-hNET. The higher affinity (lower EC_{50}) for Na^+ must result in a slower dissociation of Na^+ from the carrier and therefore should cause a considerable reduction in the number of transport cycles per unit of time, explaining the pronounced effect on NE uptake.

In adrenoceptors, the amino group of catecholamines interacts with a conserved aspartate residue (in TM1) and the phenolic hydroxyl groups with two conserved and neighbouring serine residues in TM5. Since an Asp residue in TM1 (Asp75 in hNET) and two serine residues in TM 7 (Ser354 and 357 in hNET) are conserved in the monoamine transporters, replacement of these residues on ^{3}H-NE uptake and ^{3}H-DMI binding were studied. The Asp75Ala mutant showed neither uptake of NE nor binding of DMI, indicating that Asp75 might be involved in binding of the charged amino group of NE and DMI. NE transport and DMI binding was either marginally, or strongly or almost completely suppressed in the hNET mutants Ser357Ala, Ser354Ala, and the double mutant Ser354Ala/Ser357Ala, respectively, indicating that the serine residues are necessary for the transporter function but not specifically for the recognition of phenolic hydroxyl groups of catecholamines.

2.3 Residues involved in desipramine binding

From studies with chimeric NET/DAT proteins it is known that the region between TM5 to TM8 of the NET is important for the recognition of the tricyclic antidepressant DMI.[17] We therefore studied NET mutants in which nonconservative aa of the NET were replaced against those of the DMI-insensitive DAT. From the various mutants tested the double mutant Ser399Pro/Gly400Leu (in TM8) caused a reduction of the DMI affinity (IC_{50}) by a factor of about 3000 but it showed no change in the affinities for NE, DA and cocaine, indicating that these two residues are critically and specifically involved in binding of the tricyclic antidepressant desipramine.[13] This knowledge may be helpful in the development of new antidepressants.

2.4 Functional properties of the C-terminal splice variant

The C-terminal hNET splice variant in which the alternatively spliced exon 15 is connected to the coding exon 13 (skipping exon 14) was studied in transfected HEK293 cells. Compared to the WT-hNET this splice variant showed about 40% less uptake of ^{3}H-NE and binding of ^{3}H-nisoxetine, indicating a reduced plasma membrane expression. Co-expression of this splice variant together with the WT-hNET caused a reduction in the function of the WT-hNET which indicates that this C-terminal splice variant has a dominant negative effect on the expression of the WT-hNET. This splice variant was expressed in the human neuroblastoma cell line SKN-SH-SY5Y and in human placenta. Expression of this splice variant may regulate hNET surface expression by a protein-protein interaction.

REFERENCES

1. M.S. Sonders and S.G. Amara, Channels in transporters, *Neurobiol.* **6**, 294-302 (1996)
2. H. Bönisch and M. Brüss, Catecholamine transporter of the plasma membrane, *Ann. New York Acad. Sci.* **733**, 193-202 (1994).
3. T. Pacholczyk, R.D. Blakely, and S.G. Amara, Expression cloning of a cocaine- and antidepressant-sensitive human noradrenaline transporter. *Nature* **350**, 350-354 (1991).
4. S.L. Povlock and S.G. Amara, The structure and function of norepinephrine, dopamine, and serotonin transporters, in: *Neurotransmitter Transporters: Structure, function, and regulation*, edited by M.E.A. Reith (Humana Press, Totowa, New Jersey 1997), pp. 1-28
5. M. Brüss, R. Hammermann, and H. Bönisch, Antipeptide antibodies confirm the topology of the human norepinephrine transporter, *J. Biol. Chem.* **270**, 9197-9201 (1995)
6. H. Bönisch, R. Hammermann, and M. Brüss, Role of protein kinase C and second messengers in regulation of the norepinephrine transporter, in: *Catecholamines. Bridging basic science with clinical medicine*, edited by D.S. Goldstein, G. Eisenhofer, and R. McCarty (Academic Press, San Diego, 1998) *Adv. Pharmacol.* **42**, 149-164
7. M. Brüss, J. Kunz, B. Lingen, and H. Bönisch, Chromosomal mapping of the human gene for the tricyclic antidepressant-sensitive noradrenaline transporter, *Hum. Genet.* **91**, 278 280 (1993)
8. P. Pörzgen,, H. Bönisch, and M. Brüss, Molecular cloning and organization of the coding region of the human norepinephrine transporter gene, *Biochem. Biophys. Res. Commun.* **215**, 1145-1150 (1995).
9. P. Pörzgen, R. Hammermann, H. Bönisch, and M. Brüss, The human norepinephrine transporter gene contains multiple polyadenylation sites and two alternatively spliced C-terminal exons, *Biochim. Biophys. Acta* **1398**, 365-370 (1998).
10. G. Stöber, M. Nöthen, P. Pörzgen, M. Brüss, H. Bönisch, M. Knapp, H. Beckmann, and P. Propping, Systematic search for variation in the human norepinephrine transporter gene: Identification of five naturally occurring missense mutations and study of association with major psychiatric disorders. *Am. J. Med. Genet.* **67**, 523-532 (1996).
11. H. Bönisch, F. Runkel, C. Roubert, B. Giros, and M. Brüss, The human desipramine-sensitive noradrenaline transporter and the importance of defined amino acids for its function. *J. Auton. Pharmacol.* **19**, 327-333 (2000)
12. F. Runkel, M. Brüss, M.M. Nöthen, G. Stöber, P. Propping, and H. Bönisch, Pharmacological properties of naturally occurring variants of the human norepinephrine transporter. *Pharmacogenetics* **10**, 1-9 (2000).
13. C. Roubert, P. Cox, M. Brüss, M. Hamon, H. Bönisch, and B. Giros, Determination of residues in the norepinephrine transporter that are critical for tricyclic antidepressant affinity. *J. Biol. Chem.* **276**, 8254-8260 (2001).
14. S. Pantanowitz, A. Bendahan, and B.I. Kanner, Only one of the charged amino acids located in the transmembrane α-helices of the γ-aminobutyric acid transporter (subtype A) is essential for its activity. *J. Biol. Chem.* **268**, 3222-3225 (1993).
15. N. Kleinberger-Doron, and B.I. Kanner, Identification of tryptophan residues critical for the function and targeting of the γ-aminobutyric acid transporter (subtype A). *J. Biol. Chem.* **269**, 3063-3067 (1994).
16. G.I. Keshet, A. Bendahan, H. Su, S. Mager, H.A. Lester, and B.I. Kanner, Glutamate 101 is critical for the function of the sodium and chloride-coupled GABA transporter GAT-1. *FEBS Lett.* **371**, 39-42 (1995).
17. B. Giros, Y.-M. Wang, S. Suter, S.B. McLeskey, C. Pifl, and M.G. Caron, Delineation of discrete domains for substrate, cocaine, and tricyclic antidepressants using chimeric dopamine-norepinephrine transporters. *J. Biol. Chem.* **269**, 15985-15988 (1994).

D_2 RECEPTOR-MEDIATED REGULATION OF VESICULAR DOPAMINE UPTAKE

Annette E. Fleckenstein, Jeffrey M. Brown, Verónica Sandoval, Evan L. Riddle, J. Paul Hansen, Yvette V. Ugarte, James W. Gibb and Glen R. Hanson*

1. INTRODUCTION

The vesicular monoamine transporter-2 (VMAT-2) sequesters cytoplasmic monoamines, including dopamine (DA), into synaptic vesicles. It has been suggested that VMAT-2 is resistant to regulation by drug treatment. However, recent data indicate that agents which alter DA disposition also rapidly modify VMAT-2 function. For example, our laboratory reported that multiple injections of the DA-releasing agent, methamphetamine (METH), rapidly (within 1 h) *decrease* vesicular [^{3}H]DA uptake and binding of the VMAT-2 ligand, [^{3}H]dihydrotetrabenazine ([^{3}H]DHTBZ), as assessed in purified striatal vesicles obtained from treated rats; an effect unrelated to residual METH introduced by the original drug injections.[1] Similarly, Hogan et al. have demonstrated decreases in [^{3}H]DA uptake and [^{3}H]DHTBZ binding 24 h after high-dose METH treatment.[2] Accordingly, our laboratory investigated whether other stimulants, including cocaine and methylenedioxymethamphetamine (MDMA), rapidly affect VMAT-2. Our results indicate that these psychostimulants rapidly and differentially regulate vesicular DA uptake. In addition, D_2 receptors mediate the effects of cocaine on vesicular DA uptake. These data may explicate mechanisms contributing to psychostimulant-induced neurotoxicity, and provide insight into the physiological regulation of DA neurons.

2. COCAINE ADMINISTRATION INCREASES VESICULAR DA UPTAKE VIA A D_2 RECEPTOR-MEDIATED MECHANISM

It is well established that cocaine increases extracellular DA levels principally by inhibiting plasmalemmal DA transporter (DAT) function. In addition, recent reports

* Department of Pharmacology and Toxicology, University of Utah, Salt Lake City, Utah 84112, USA.

Catecholamine Research, Edited by Nagatsu *et al.*
Kluwer Academic/Plenum Publishers, 2002

demonstrate that cocaine treatment rapidly alters vesicular DA transport. Specifically, cocaine rapidly (within 1 h) and reversibly (within 6 h) increases [^{3}H]DA uptake, as assessed *ex vivo* in purified vesicles obtained from the striatum of treated rats.[3] This effect is not due to cocaine introduced by the original drug injection. It can be speculated that this increase in vesicular DA sequestration represents a redistribution of vesicles so that more vesicles containing VMAT-2 are detached from the plasmalemmal membranes and thus included in the purified vesicular preparations under study. Although this hypothesis is yet to be tested, such an alteration in "trafficking" of vesicles is suggested by findings that [^{3}H]DHTBZ binding is increased concurrently with DA uptake in the vesicular preparations. Interestingly, selective inhibitors of the DAT (amfonelic acid and GBR12935), but not the plasmalemmal serotonin transporter (fluoxetine), also increase vesicular DA uptake.[3] Pretreatment with the D_2 antagonist, eticlopride, but not the D_1 antagonist, SCH23390, prevents the cocaine-induced increases in [^{3}H]DA uptake and [^{3}H]DHTBZ binding. In addition, administration of the D_2 agonist, quinpirole, but not the D_1 agonist, SKF81297, increases vesicular DA uptake in these preparations.[4]

3. PSYCHOSTIMULANTS DIFFERENTIALLY ALTER VESICULAR DA UPTAKE: IMPLICATIONS FOR NEUROTOXICITY

It is well established that high-dose METH administration can cause long-term damage (i.e., persisting months) to DA neurons in rodents, non-human primates, and presumably humans (for review, see[5,6]). Administration of a neurotoxic regimen of METH rapidly decreases plasmalemmal[7] and vesicular[1,2] DA uptake as well; as assessed in synaptosomes and purified vesicles, respectively, prepared from the striata of treated rats or mice. We hypothesize that these concurrent acute deficits may contribute to the persistent DA deficits caused by METH treatment by promoting intraneuronal DA accumulation and reactive oxygen species generation. This hypothesis is consistent with numerous studies wherein a role for DA[8,9,10] and reactive oxygen species[11,12,13,14] in promoting METH-induced toxicity has been suggested (for review, see[5,6]). Accordingly, we hypothesize that stimulants that do not effect acute deficits in both plasmalemmal and vesicular DA transport would predictably be less toxic to DA neurons than agents that exert this effect. Consistent with this hypothesis, high-dose cocaine administration (i.e., a regimen that *increases* vesicular DA uptake) causes little, if any, persistent DA deficits.[15,16,17] Moreover, multiple high-dose administrations of MDMA, a stimulant with lesser DA neurotoxic potential than METH,[18,19] cause *decreases* in vesicular[20] and plasmalemmal[6] DA uptake that are lesser in magnitude and/or duration than those observed after METH treatment. Hence, there are important differences among the effects of cocaine, METH and MDMA on DAT and VMAT-2 function; effects that may underlie the dissimilar DA neurotoxic potential of these agents.

4. SUMMARY

In conclusion, the results demonstrate that psychostimulants differentially alter vesicular [^{3}H]DA uptake and [^{3}H]DHTBZ binding. Specifically, METH and MDMA *decrease*, whereas cocaine *increases*, vesicular DA uptake in purified vesicular preparations. The increase caused by cocaine is mediated via D_2 receptor activation.

Whether the presently reported alterations in vesicular DA sequestration after stimulant or D_2 agonist treatment represent: 1) a redistribution of vesicles so that a greater or lesser number of vesicles containing VMAT-2 are detached from the plasmalemmal membranes and are included in the purified vesicular preparations; or 2) enhanced or diminished VMAT-2 activity *per se* remains to be determined. Still, these data may help to explain mechanisms contributing to psychostimulant-induced neurotoxicity, as well as provide insight into the physiological regulation of DA neurons.

5. ACKNOWLEDGEMENTS

The authors acknowledge the excellent technical assistance provided by Jarom E. Hanson, Raul K. Weston and Michael J. Crosby. This work was supported by PHS grants DA11389, DA00869 and DA04222. Presentation of this work was supported by an Irwin J. Kopin Traveling Fellowship from the Foundation for Catecholamine Research.

REFERENCES

1. J. M. Brown, G.R. Hanson, and A.E. Fleckenstein, Methamphetamine rapidly decreases vesicular dopamine uptake, *J. Neurochem.* **74**, 2221-2223 (2000).
2. K.A. Hogan, R.G. Staal, P.K. Sonsalla, Analysis of VMAT2 binding after methamphetamine or MPTP treatment: disparity between homogenates and vesicle preparations, *J. Neurochem.* **74**, 2217-2220 (2000).
3. J.B. Brown, G.R. Hanson, A.E. Fleckenstein, Regulation of the vesicular monoamine transporter-2: A novel mechanism for cocaine and other psychostimulants, *J Pharmacol Exp Ther* **296:** 762-767 (2001).
4. J.M. Brown, G.R. Hanson, and A.E. Fleckenstein, Cocaine-induced increase in vesicular dopamine uptake: Role of dopamine receptors *J Pharmacol Exp Ther* , 1150-1153 (2001).
5. J.W. Gibb, M. Johnson, I. Elayan, H.K. Lim, L. Matsuda, and G.R. Hanson, Neurotoxicity of amphetamines and their metabolites, *NIDA Res Monogr.* **173**, 128-145 (1997)
6. A.E. Fleckenstein, J.W. Gibb, and G.R. Hanson, Differential effects of stimulants on monoaminergic transporters: Pharmacological consequences and implications for neurotoxicity, *Eur J Pharmacol.* **406**, 1-13 (2000).
7. J.M. Kokoshka, R.A. Vaughan, G.R. Hanson GR and A.E. Fleckenstein, Nature of methamphetamine-induced rapid and reversible changes in dopamine transporters, Eur. J. Pharmacol. **361**, 269-75 (1998).
8. A.J. Hotchkiss and J.W. Gibb, J.W., Long-term effects of multiple doses of methamphetamine on tryptophan hydroxylase and tyrosine hydroxylase in rat brain, *J. Pharmacol. Exp. Ther.* **214**, 257-262 (1980).
9. F. Fumagalli, R.R. Gainetdinov, K.J. Valenzano, and M.G. Caron, Role of dopamine transporter in methamphetamine-induced neurotoxicity: evidence from mice lacking the transporter, *J Neurosci.* **18**, 4861-9 (1998).
10. M.J. LaVoie and T.G. Hastings, Dopamine quinone formation and protein modification associated with the striatal neurotoxicity of methamphetamine: evidence against a role for extracellular dopamine, *J Neurosci.* **19**:1484-1491 (1999).
11. J.L. Cadet, P. Sheng, S. Ali, R. Rothman, E. Carlson, and C. Epstein, Attenuation of methamphetamine-induced neurotoxicity in copper/zinc superoxide dismutase transgenic mice, *J. Neurochem.* **62**,380-3 (1994).
12. J.F. Cubells, S. Rayport, G. Rajendran, and D. Sulzer, Methamphetamine neurotoxicity involves vacuolation of endocytic organelles and dopamine-dependent intracellular oxidative stress. *J. Neurosci.* **14**, 2260-2271 (1994).
13. B.K. Yamamoto and W. Zhu, The effects of methamphetamine on the production of free radicals and oxidative stress, *J. Pharmacol. Exp. Ther.* **287**,107-14 (1998).
14. G.D. Cappon, H.W. Broening, C. Pu, L. Morford, C.V. Vorhees, alpha-Phenyl-N-tert-butyl nitrone attenuates methamphetamine-induced depletion of striatal dopamine without altering hyperthermia, *Synapse* **24**, 173-181 (1996).
15. M.S. Kleven, W.L. Woolverton, and L.S. Seiden, Lack of long-term monoamine depletions following repeated or continuous exposure to cocaine, *Brain Res. Bull.* **21**, 233-237 (1988).

16. S. Benmansour, S.M. Tejani-Butt, M. Hauptmann, and D.J. Brunswick, D.J., Lack of effect of high-dose cocaine on monoamine uptake sites in rat brain measured by quantitative autoradiography, *Psychopharmacology (Berl)* **106**, 459-462 (1992).
17. G.D. Cappon, L.L. Morford and C.V. Vorhees, C.V. Enhancement of cocaine-induced hyperthermia fails to elicit neurotoxicity. *Neurotoxicol. Teratol.* **20**, 531-535 (1998).
18. M. Johnson, A.A. Letter, K. Merchant, G.R. Hanson, and J.W. Gibb, J.W., Effects of 3,4-methylenedioxyamphetamine and 3,4-methylenedioxymethamphetamine isomers on central serotonergic, dopaminergic and nigral neurotensin systems of the rat, *J. Pharmacol. Exp. Ther.* **244**, 977-982 (1988).
19. T.R. Insel, G. Battaglia, J.N. Johannessen, S. Marra, and E.B. De Souza, 3,4-Methylenedioxymethamphetamine ("ecstasy") selectively destroys brain serotonin terminals in rhesus monkeys. *J. Pharmacol. Exp. Ther.* **249**, 713-720 (1989).
20. E.L. Riddle. J.P. Hansen, V. Sandoval, J.M. Brown, G.R. Hanson, J.W. Gibb and A.E. Fleckenstein. Methylenedioxymethamphetamine decreases plasmalemmal and vesicular dopamine transport: Mechanisms and implications for neurotoxicity. Society for Neuroscience abstract (submitted; 2001).

A VARICOSITY ENSEMBLE MODEL OF SYMPATHETIC NEUROMUSCULAR TRANSMISSION

Lennart Stjärne*

1. INTRODUCTION

Varicosities within sympathetic nerve terminals appear to differ widely in function, *e.g.*, in secretory responsiveness to action potentials.[1] By recording excitatory junction currents (EJCs) at visualised nerve terminals in mouse vas deferens, the probability of release of a quantum of ATP from single varicosities (p_{ATP}) was found to be highly non-uniform, in a few 0.1-0.5 but in most 0.01 or less.[2] ATP-mediated responses to nerve impulse trains may thus be evoked by repeated release of ATP quanta from 'strong' varicosities while 'weak' varicosities remain 'silent' (hotspot transmission). The question is if this applies as well to transmission mediated by co-released noradrenaline (NA).[3]

2. RANDOM OR HOTSPOT TRANSMISSION BY NORADRENALINE?

Conclusive proof is lacking but indirect evidence offers clues. One, based on study of NA overflow,[3,4,5,6] is that each nerve action potential may release the NA content of a single small synaptic vesicle (a NA quantum[5]) from 1-3% of the varicosities, *i.e.*, that average p_{NA} (0.01-0.03) is similar to average p_{ATP}. Further clues have been obtained in studies of the NA-mediated neurogenic contractile responses (ΔC) in rat tail artery (RTA), with and without cocaine to block neuronal reuptake of released NA; an example is shown in Fig. 1A. Here $\Delta[NA]_{CF}$, the nerve stimulation-evoked rise in the concentration of released endogenous NA near a carbon fibre (CF) electrode applied to the densely innervated adventitial surface,[7] was monitored amperometrically.[8,9] Single stimulus pulses triggered a small, mainly α_2-mediated contraction (ΔC), trains (*e.g*, 15 pulses) at 20 Hz a much (500-fold) larger, mainly α_1-mediated ΔC (traces *a,b*). This effect was not due to facilitated per pulse NA release,[6] but to cooperative local interaction between NA quanta released by successive pulses, probably from the same or nearby sites. Another clue: cocaine amplified the single pulse ΔC by 20-fold by recruiting a large α_1-component; its relative effect on ΔC responses at 20 Hz declined steeply with the train

* Lennart Stjärne, Dpt of Physiology and Pharmacology, Karolinska Institutet, S-17177 Stockholm, Sweden

Catecholamine Research, Edited by Nagatsu *et al.*
Kluwer Academic/Plenum Publishers, 2002

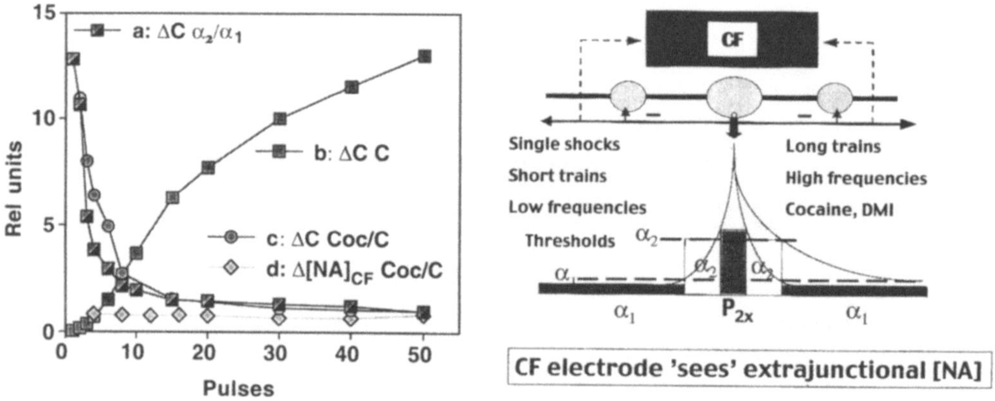

Figure 1A. Roles of postjunctional α_1- and α_2-adrenoceptors and of neuronal NA reuptake for the NA-mediated neurogenic contractions (ΔC) of RTA in response to 1-50 pulses at 20 Hz.[5,6] Trace **a** (ΔC α_2/α_1):the ratio of α_2- and α_1-adrenoceptor-mediated components of ΔC in controls, trace **b** (ΔC C): the amplitude of ΔC in controls, traces **c** (ΔC Coc/C) and **d** ($\Delta[NA]_{CF}$ Coc/C): the ratios of ΔC or NA oxidation current at the carbon fibre (CF) electrode, respectively, with and without 3 μM cocaine. Traces **a,c,d** are normalised to responses to 50 pulse trains. **B**. Hypothetical hotspot model, showing the CF electrode (diameter: 8 μm) near a string of varicosities at the surface of the artery,[4] exocytosis of NA (and ATP?) in a single vesicle,[1,3,7] the clearance of released NA by diffusion and neuronal reuptake, (-) the autoinhibition of release from nearby sites,the (log) NA concentration gradient from the exit point[6] (≤100 mM) towards perijunctional high-treshold α_2- and extrajunctional low-threshold α_1-adrenoceptors[5] (also shown: intrajunctional P_{2x}-purinoceptors[10]), and the increased lateral diffusion of released NA towards, and recruitment of, extrajunctional α_1-adrenoceptors, caused by block (cocaine, DMI) or local saturation (high frequency trains) of neuronal NA reuptake.[8,9] For further comments see the text.

length (trace *c*).[8,9] The α_2-adrenoceptors were thus few and close to the release sites, the (20-fold?) more numerous α_1-adrenoceptors mainly located beyond a 'barrier' posed by local NA reuptake transporters with capacity to remove NA in single quanta, but not in quanta repeatedly released at ≥ 2.5 Hz (turnover time: 0.4 sec) from the same (or nearby) sites.[3] One more clue: cocaine amplified $\Delta[NA]_{CF}$ to single pulses or 2-50 pulse trains at 20 Hz to the same extent, by about two-fold (trace *d*).[8,9] High frequency trains did thus not alter *overall* fractional removal of released NA by neuronal reuptake, only its *'geometry'*.[3,8,9] In the working hypothesis in Fig. 1B, the CF electrode cannot 'see' intrajunctional events: the NA concentration at the exit point, or its steep gradient towards pre- and postjunctional adrenoreptors and neuronal reuptake transporters near the release site, all of which compete for the released NA before it escapes into the effluent.[3,8,9] At low frequency, local neuronal reuptake maintains the steepness of the slope and curtails ΔC by restricting access to α_1-adrenoceptors; block (by cocaine or DMI), or local saturation of reuptake transporters near the release sites by repeated release from the same (or small groups of closely related) 'strong' varicosities, force released NA to diffuse further laterally to recruit more α_1-adrenoceptors before being taken up into varicosities at the margin of the growing 'hotspot', or escape into the effluent.[3,8,9]

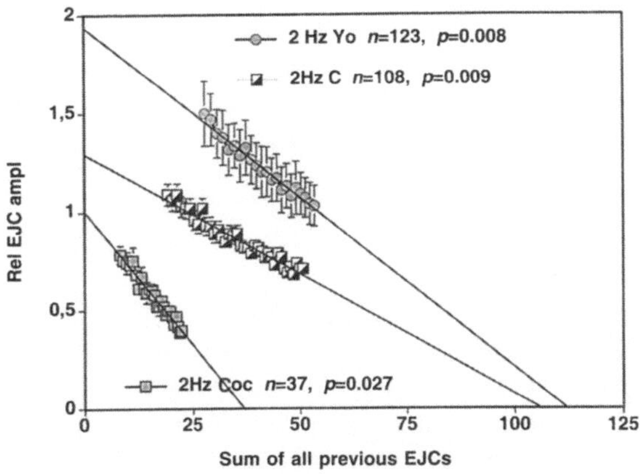

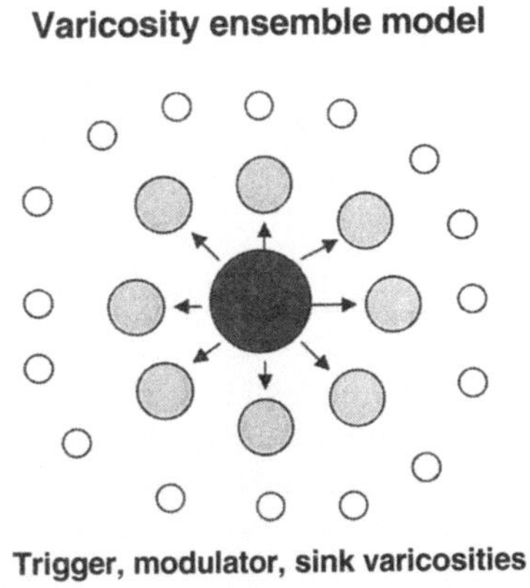

Figure 2. Left panel: The amplitude of each successive EJC (normalized to the first EJC in controls) during (parts of) 100 pulse trains at 2 Hz in the absence of drugs (2Hz C), with 3 uM cocaine (2 Hz Coc), and with cocaine plus 1 μM yohimbine (2 Hz Yo), plotted against the sum of all previous EJCs.[3,9] The intercepts of the regression lines with the abscissa are assumed to represent the *relative* numbers (n) of activated monoquantal release sites within the area sensed by the electrode, and their reciprocal the *average* release probabilities (p).[3,9] Right panel: In the proposed 'ensemble' model varicosities differ in p as well as sensitivity to α_2-adrenoceptor mediated autoinhibition. 'Trigger' varicosities (p=0.1-0.5) are 'immune' to the NA they release, which instead turns off neighbouring 'modulator' varicosities (p=0.01-0.1). The large majority (p<0.01) hardly participate in release, but serve as local 'NA sinks', i.e. take up NA released from 'trigger' and 'modulator' varicosities.

A corollary of the hotspot model is that 'strong' varicosities are not themselves autoinhibited by the NA they release.[3,4,11,12] Direct evidence for this has been reported from studies in MVD. While activation of prejunctional α_2-adrenoceptors by released NA depressed the Ca^{2+} transient[13] and ATP release in most varicosities, some appeared to lack α_2-autoreceptors and were therefore 'immune' to autoinhibition.[14] Indirect functional evidence pointing in the same direction has been obtained by analysis of the α_2-autoinhibition of EJCs in RTA during 100 pulses at 2 Hz[11] (when ATP release apparently parallels that of NA[6,11]). In controls, the first few EJCs grew in amplitude ('facilitation'), later EJCs declined progressively ('depression'). Block of NA reuptake by addition of cocaine, to enhance α_2-autoinhibition, almost totally eliminated the early facilitation and strongly enhanced the subsequent depression of EJCs. Further addition of yohimbine to block α_2-autoreceptors more than reversed the effects on the facilitation, but did not prevent the subsequent relative depression of EJCs.[11] In the analysis by a modification of the method of Martin[15] shown in Fig. 2A, the amplitudes of successive EJCs under the three conditions were plotted against the summed amplitudes of all previous EJCs.[3,15] The question asked was if activation of α_2-autoreceptors by released NA acted mainly by reducing the number (n) of active varicosities, or their probability (p_{ATP}) of releasing single ATP quanta. It can be seen that cocaine reduced n but enhanced p, and that these effects were reversed by yohimbine. The results strongly suggest that the concentration of released NA at α_2-autoreceptors tonically depressed the number (n) of monoquantal release sites activated, but enhanced their *average* release probabilities (p_{ATP}). These results also support a hotspot model, in this case one in which NA released from 'strong' varicosities turns off release from the 'weak'. One physiological function of α_2-autoinhibition[3,11] may thus be to sharpen the signal emitted from 'strong' varicosities by reducing the noise from surrounding 'weak' varicosities ('lateral inhibition').[4,11,12]

3. CONCLUSIONS

Sympathetic varicosities are not equal; an 'ensemble model' is required (Fig. 2B), in which they belong to three different classes whose roles vary with the frequency and number of nerve impulses. Neurotransmission is driven by preferential release of quanta from a minority of 'strong' ('trigger') varicosities immune to autoinhibition by the transmitters they release. NA accumulating around them turns off 'weak' ('modulator') varicosities in their surround. The more numerous 'silent' varicosities serve to take up excess released NA, *i.e.*, influence transmission by acting as a powerful NA 'sink'.

REFERENCES

1. M.R. Bennett, Autonomic neuromuscular transmission at a varicosity, *Prog. Neurobiol.*, **50**, 505-532 (1996).
2. N.A. Lavidis, and M.R. Bennett, Probabilistic secretion of quanta from visualised varicosities along single sympathetic nerve terminals. *J. Auton. Nerv. Syst.*, **43**, 41-50 (1993).
3. L. Stjärne, and E. Stjärne, Geometry, kinetics and plasticity of release and clearance of ATP and noradrenaline as sympathetic cotransmitters: roles for the neurogenic contraction, *Prog. Neurobiol.*, **47**, 45-94 (1995).
4. P. Alberts, T. Bartfai, and L. Stjärne, Site(s) and ionic basis of α-autoinhibition and facilitation of [^{3}H]-noradrenaline secretion in guinea-pig vas deferens, *J. Physiol.*, **312**, 297-334 (1981)
5. L. Stjärne, Do sympathetic nerves release noradrenaline in 'quanta'? *J. Auton. Nerv. Syst.*, **81**, 236-243 (2000).
6. M. Msghina, F. Gonon, and L. Stjärne, Facilitation and depression of ATP and noradrenaline release from sympathetic nerves of rat tail artery, *J. Physiol.*, **515**(2), 523-531 (1999).
7. S.E. Luff, and E.A. McLachlan, Frequency of neuromuscular junctions on arteries of different dimensions in the rabbit, guinea-pig and rat, *Blood Vessels*, **26**, 95-106 (1989).
8. J.X. Bao, F. Gonon, and L. Stjärne, Frequency- and train length-dependent variation in the roles of postjunctional α_1- and α_2-adrenoceptors for the field stimulation-induced neurogenic contraction of rat tail artery, *Naunyn-Schmiedeberg's Arch. Pharmacol.*, **347**, 601-616 (1993).
9. L. Stjärne, J.X. Bao, F. Gonon, and M. Msghina, Nerve activity-dependent variations in clearance of released noradrenaline: regulatory roles for sympathetic neuromuscular transmission in rat tail artery. *Neuroscience*, **60**(4), 1021-1038 (1994).
10. M.A. Hansen, J.L. Dutton, V.J. Valcar, J.A. Barden, and M.R. Bennett, P_{2x} (purinergic) receptor distribution in rat blood vessels. *J. Auton. Nerv. Syst.*, **75**, 147-155 (1999).
11. M. Msghina, F. Gonon, and L. Stjärne, A novel electrophysiological approach to monitor pulse by pulse the concentration of released NA at the presynaptic α_2-adrenoceptors of sympathetic nerves in rat tail artery, *Naunyn-Schmiedeberg's Arch. Pharmacol.*, **351**, 173-185 (1995).
12. J.A. Brock, and T.C. Cunnane, Local application of drugs to sympathetic nerve terminals: an electrophysiological analysis of the role of prejunctional α-adrenoceptors in the guinea-pig vas deferens, *Br. J. Pharmacol.*, **102**, 595-600 (1991).
13. K.L. Brain, and M.R. Bennett, Calcium in sympathetic varicosities of mouse vas deferens during facilitation, augmentation and autoinhibition, *J. Physiol.*, **502**(3), 521-536 (1997).
14. S.C. O'Connor, K.L. Brain, and M.R. Bennett, Individual sympathetic varicosities possess different sensitivities to alpha 2 and P2 receptor agonists and antagonsits in mouse vas deferens. *Br. J. Pharmacol.*, **128**, 1739-1753 (1999).
15. A.R. Martin, Junctional transmission II. Presynaptic mechanisms. In *Handbook of Physiology*, pp. 329-355 (1977).

SINGLE SYMPATHETIC VARICOSITIES AND THEIR RECEPTOR CLUSTERS

Regulation of transmitter release and cluster formation

Max R. Bennett*

1. INTRODUCTION: THE AUTONOMIC GROUND PLEXUS OF HILLARP AND A VARICOSITY

The autonomic ground plexus of Hillarp[1] consists of postganglionic nerves which terminate in systems of collateral branches each of which possesses in the order of one hundred varicosities.[2] Single strings of varicosities of these collateral systems can be found on occasion at the surface of smooth muscles after their labelling by orthograde transport of rhodamine coupled to dextran into the varicosities and their subsequent identification with confocal microscopy.[3] Labelling the synaptic vesicles in these varicosities in the vas deferens with fluorescent coupled antibodies to the vesicle proteoglycan SV2 shows that the vesicles are concentrated in a volume about one-third that of the varicosity and in close apposition to one of its membranes.[4] The clump of synaptic vesicles apposes a site on the vesicle membrane that possesses a high concentration of syntaxin, necessary for vesicle docking.[4] Each varicosity in a single string possesses vesicles, together with their associated proteins synaptobrevin and SNAP25, accumulated at a site which is defined by the high concentration of syntaxin, whether the string of varicosities occurs at the surface of the muscle or in its media.[5] The observations described below are our most recent attempts to determine the properties of individual varicosites in Hillarp's ground plexus, principally using the vas deferens as the experimental preparation.

* The Neurobiology Laboratory, Dept. of Physiology and the Institute for Biomedical Research, University of Sydney, NSW 2006, Australia

2. THE DISTRIBUTION AND PROPERTIES OF P2X RECEPTOR CLUSTERS AT INDIVIDUAL VARICOSITIES

The P2X1 subunit class of receptors is found in two different size clusters, one about 1 μm diameter and the other about 0.4 μm diameter in all muscle containing effectors of the autonomic nervous system so far studied.[5, 6, 7] The large P2X1 receptor clusters are uniquely associated with varicosities of both sympathetic[5] and parasympathetic[6] nerve terminals, whereas the smaller clusters are found in much greater numbers over the surface of individual smooth muscle cells. About 90% to 95% of the varicosities that are clearly delineated as belonging to a single string possess large P2X1 receptor clusters.

Although nearly all the varicosities that belong to a string possess a large cluster of P2X1 receptors on the apposing smooth muscle cells in visceral and vascular organs, other P2X subunits can also exist in these clusters. Subunits of P2X receptors ranging through P2X1, P2X3, and P2X5 are found in junctional clusters, although few P2X4 and P2X6 subunits are evident in such clusters in either the bladder detrusor or the vas deferens muscle.[5, 8] It is not unusual to find junctional clusters that possess both P2X1 and P2X3 subtypes, and more rarely both P2X1 and P2X5 subunits, suggesting hetero-polymerization. In many cases varicosities on a string may possess P2X1 without P2X3 subunits and on rare occasions P2X3 without P2X1 subunits (see Fig. 6 in 8).

Junctional P2X receptor clusters have been observed to form from the small 0.4 μm diameter clusters of different subunits found in large numbers around nerves in the developing rat detrusor muscle. It appears that small clusters of different subunit types migrate beneath the varicosities where it is possible that a heteropolymerization takes place.[8]

These observations on the heterogeneity of the subunit composition of the junctional clusters and their plasticity suggest that these clusters are labile. In order to test this proposition P2X1 and a chimera of the P2X1 purinoceptor and green fluorescent protein (P2X1-GFP) were delivered into cells using either microinjection, transient transfection or infection with a replication deficient adenovirus.[9] P2X1-GFP chimera was then used to study the time course of P2X1 receptor clustering in plasma membranes. Both P2X1 and P2X1-GFP chimera were found to cluster in the plasma membranes of *Xenopus* oocytes from an initial diffuse distribution, forming junctional-size receptor clusters. These results indicate that junctional cluster formation is independent of nerve varicosities and suggests that the failure of such clusters to form outside of varicosities on muscle cells *in situ* is due to an action of the varicosities to suppress junction clustering except at the varicosity.

HEK 293 cells also form junctional-size clusters from an initial diffuse distribution after infection with the adenovirus, Ad-CMV-P2X1-GFP[9]. Exposure of these cells to alpha,beta-methylene ATP for a few minutes leads to the disappearance of P2X1-GFP fluorescence from the surface of the cells as a consequence of the endocytosis of the chimera into endosomes.

A junctional varicosity can now be identified for the first time since the description by Hillarp[1] of the autonomic ground plexus of varicosities. It is a varicosity that possesses a highly localized concentration of syntaxin on its membrane, apposed by synaptic vesicles that possess vesicle-associated proteins. This varicosity is juxtaposed against a large 1 μm diameter cluster of P2X1 receptors on the underlying smooth muscle cell membrane.

3. SECRETION OF TRANSMITTER AT INDIVIDUAL JUNCTIONAL VARICOSITIES

Placing a loose-patch recording electrode over a single varicosity in relative isolation from the rest on a string, ascertained by previously visualising the varicosity, allows recording of the electrical signs of transmitter release from the varicosity.[10] The interpretation of such recordings requires a quantitative description of how current flows around a synapse located on a muscle cell in a volume conductor. The particular case of a synapse on a muscle cell coupled into an electrical syncytium presents its own special problems. It requires consideration of the currents generated at the surface of the muscle in the vicinity of the loose - patch electrode as a consequence of transmitter release from varicosities deeper in the muscle generating currents at the surface.[11, 12] We have now solved this problem which provides a description of all sources of current arriving at the recording electrode, including that from the visualised varicosity immediately beneath the electrode.[13] The analysis shows that currents arising from varicosities at the surface have a rise time that is typically less than 3 ms whereas those from deeper in the tissue have slightly longer rise times and are of such a small amplitude as to be barely detectable.

Graphs which plot the rise times against amplitude of excitatory junctional currents -(EJCs), recorded experimentally with a small diameter loose-patch electrode, show that these consist of fast (rise times less than 3 ms) potential transients with variable amplitudes and slow (rise times greater than 3 ms and up to 12 ms) potential transients that are of small amplitude.[13] The fast rise time EJCs appear to have a different origin from those of the slow rise time EJCs. Proof of this distinction is provided by experiments in which classical exocytotic secretion is blocked with peptides that interact with the core complex of proteins that are responsible for regulated exocytosis. A loose-patch electrode is placed over visualised varicosities and perfused with a compound consisting of a complex of Penetratin and a domain of the vesicle-associated protein alpha SNAP, which is known to block vesicle exocytosis and transmitter release.[14] After perfusion, all the fast rise-time EJCs with rise times less than about 3 ms are eliminated, indicating that it is these that are generated by a classical exocytotic mechanism. Amplitude-frequency histograms of the EJCs with rise times less than about 3 ms show that the amplitudes of these EJCs are consistent with occasional multiquantal release of transmitter from a varicosity, as has previously been argued to be the case using statistical criteria.[15, 16]

The mechanism that gives rise to the slow EJCs that are not blocked by the alpha SNAP peptide is not established. According to the syncytial analysis of current flow after the release of a quantum of ATP they do not arise as a consequence of any electrotonic conduction in the syncytium to the point of recording.[13] This being the case, the slow and small EJCs seem to arise from a non-classical form of exocytosis.

Most importantly, heptanol, which is known to block the major slow component of the intracellularly recorded EJP[17], also selectively blocks the slow and small EJCs (Bennett et al., personal observations). This points to the slow and small EJCs providing a major component of the EJP and so emphasizes the importance of elucidating the mechanism of secretion that gives rise to these EJCs.

What then is the origin of the small and slow EJCs that give rise to the slow component of the EJP? One possibility is that ATP is released from axons found in bundles which are only partially ensheathed by Schwann cells, in much the same way as ATP release from sensory axons occurs[18]. This source of ATP could then act on the very large number of small clusters of P2X receptors that have now been discovered throughout smooth muscles and in close proximity to nerve trunks. These may then be responsible for the deterministic changes in conductance found in smooth muscle cells. Thus ATP released from axons onto small P2X receptor clusters in their vicinity could supply a deterministic increase in conductance of smooth muscle cells that gives rise to the current responsible for the principal component of the EJP. Alternatively, transmitter from a junctional varicosity from vesicles that do not use a classical exocytotic secretory mechanism may be involved. It is interesting in this regard that Stjarne[19] has recently presented evidence for two different classes of small synaptic vesicles in sympathetic varicosities, one with a high content of ATP molecules (about 1000) and the other with a low content of such molecules (about 20). If this is the case, then the former type of vesicle would give rise to the quantal EJC and the fast component of the EJP and the latter type of vesicle would be responsible for the small and slow EJC and the dominant slow component of the EJP. Direct experimental tests of these hypotheses are now in progress.

REFERENCES

1. N. A. Hillarp, Structure of the synapse and the peripheral innervation apparatus of the autonomic nervous system, *Acta anat.* **Suppl. iv**, 1-153 (1946).
2. T. Malmfors, Studies on adrenergic nerves, *Acta physiol. scand.* **64, Suppl. 248**, 1-93 (1965).
3. M. R Bennett, A. Cheung, and K. L Brain, Sympathetic neuromuscular transmission at a varicosity in a syncytium, *Micros. Res. Tech.* **42**, 433-450 (1998).
4. K. L. Brain and M. R. Bennett, Calcium in sympathetic varicosities of mouse vas deferens during facilitation, augmentation and autoinhibition., *J. Physiol.* **502**(Pt 3), 521-536 (1997).
5. J. A. Barden, L. J. Cottee, and M. R..Bennett, Vesicle-associated proteins and P2x receptor clusters at single sympathetic varicosities in mouse vas deferens,*J. Neurocytol.* **28**, 469-480 (1999).
6. M. A. Hansen, V. J. Balcar, J. A. Barden, and M. R. Bennett, The distribution of single P2x1 receptor clusters on smooth muscle cells in relation to nerve varicosities in the rat urinary bladder, *J. Neurocytol.* **27**, 529-539 (1998).
7. M. A. Hansen, M. R. Bennett, and J. A. Barden, Distribution of purinergic P2x receptors in the rat heart, *J. Autonom. Nerv. Syst.* **78**, 1-9 (1999).
8. J. L. Dutton, M. A. Hansen, V. J. Balcar, J. A. Barden, and M. R. Bennett, Development of P2x receptor clusters on smooth muscle cells in relation to nerve varicosities in the rat urinary bladder, *J. Neurocytol.* **28**, 3-16 (1999).
9. J. L. Dutton, P. Poronnik, G. H. Li, C. A. Holding, R. A. Worthington., R. J. Vandenberg, D. I. Cook, J. A. Barden, and M. R. Bennett, P2x1 receptor membrane redistribution and down-regulation visualized using receptor coupled green fluorescent protein chimeras, *Neuropharmacology* **39**, 2054-2066 (2000).

10. G. T. Macleod, N. A. Lavidis, and M. R. Bennett,. Calcium dependence of quantal secretion from visualized sympathetic nerve varicosities on the mouse vas deferens, *J. Physiol.* **480**(Pt 1), 61-70 (1994).
11. M. R. Bennett, W. G. Gibson, and R. R. Poznanski, Extracellular current flow and potential during quantal transmission from varicosities in a smooth muscle syncytium, *Phil. Trans. Roy. Soc. Lond. Ser. B: Biol Sci.* **342**, 89-99 (1993).
12. M. R. Bennett, and W. G. Gibson, On the contribution of quantal secretion from close-contact and loose-contact varicosities to the synaptic potentials in the vas deferens., *Phil. Trans. Roy. Soc. Lond. Ser. B: Biol. Sci.* **347**, 187-204 (1995).
13. M. R. Bennett, L. Farnell, W. Gibson, Y. Q. Lin, and D. H. Blair, Quantal and non-quantal current and potential fields around individual sympathetic varicosities on release of ATP, *Biophys. J.* **80**, 1311-1328 (2001).
14. D. Blair, S. Robson, G. King, and M. R. Bennett,. Vesicle associated proteins and transmitter release from sympathetic ganglionic boutons, *NeuroReport* **12**, 607-610 (2001).
15. M. R. Bennett, J. Robinson, M. C. Phipps, S. Karunanithi, Y. Q. Lin, and L. Cottee, Quantal components of spontaneous excitatory junction potentials at visualised varicosities, *J. Autonom. Nerv. Syst.* **56**, 161-174 (1996).
16. S. Karunanithi, M. C. Phipps, J. Robinson, and M. R. Bennett, Statistics of quantal secretion during long trains of sympathetic nerve impulses in mouse vas deferens, J. *Physiol.* **489**(Pt 1), 171-181 (1995).
17. R. Manchanda, and K. Venkateswarlu, Quantal evoked depolarizations underlying the excitatory junction potential of the guinea - pig isolated vas deferens., *J. Physiol.* **520**(Pt 2), 527-537 (1999).
18. B. Stevens, and R. D. Fields, Response of Schwann cells to action potentials in development, *Science* **287**, 2267-2271 (2000).
19. L. Stjarne, Novel dual 'small' vesicle model of ATP- and noradrenaline-mediated sympathetic neuromuscular transmission, *Autonom. Neurosci.* **87**, 16-36 (2001).

ENHANCEMENT OF NORADRENERGIC PHENOTYPE EXPRESSION IN TRANSGENIC MICE OVEREXPRESSING V-1, A CYTOPLASMIC ANKYRIN REPEAT PROTEIN

Tohru Yamakuni, Toshifumi Yamamoto, Hideko Yamamoto, Si-Young Song, Toshiharu Nagatsu, Kazuto Kobayashi, Minesuke Yokoyama, Asako Nakano, Rika Suzuki, Nobuhide Suzuki, Shintaro Iwashita, Akira Omori, Yukiyo Ichinose, Chieko Kato, Masumi Kobayashi, and Yukisato Ishida*

Catecholaminergic cells exhibit three types of transmitter phenotypes: dopaminergic, noradrenergic and adrenergic. The catecholaminergic phenotypes are thought to be determined by a phenotype-specific coordinate regulation of gene expression of the four catecholamine synthesizing enzymes, *i.e.* tyrosine hydroxylase (TH), aromatic L-amino acid decarboxylase (AADC), dopamine beta-hydroxylase (DBH) and phenylethanolamine *N*-methyltransferase (PNMT). However, molecular mechanisms that control the coordinate transcriptional regulation of these genes remain to be fully elucidated, although the second messenger systems and transcription factors involved in catecholaminergic cell-specific gene expression for catecholamine synthesizing enzymes have been so far identified and extensively investigated (Zetterström et al., 1997; Morin et al., 1997; Lo et al., 1999). Recently, an ankyrin repeat protein designated V-1 has been suggested to act as a possible coordinate regulator for the expression of genes encoding catecholamine synthesizing enzymes, since it has been demonstrated that stable overexpression of V-1 protein in the catecholaminergic cell line, PC12D cell, augments

* Tohru Yamakuni, Si-Young Song, Minesuke Yokoyama, Asako Nakano, Rika Suzuki, Nobuhide Suzuki, Shintaro Iwashita, Akira Omori, Yukiyo Ichinose, Chieko Kato, Masumi Kobayashi, and Yukisato Ishida, Mitsubishi Kasei Institute of Life Sciences, Machida, Tokyo 194-8511, Japan, Toshifumi Yamamoto, Laboratory of Molecular Recognition, Graduate School of Integrated Sciences, Yokohama City University, Yokohama 236-0027, Japan, Hideko Yamamoto, Department of Molecular Psychopharmacology, Tokyo Institute of Psychiatry, Tokyo 156-0057, Japan, Toshiharu Nagatsu, Institute for Comprehensive Medical Science, School of Medicine, Fujita Health University, Toyoake 470-1192, Japan. Kazuto Kobayashi, Department of Molecular Genetics, Institute of Biomedical Sciences, Fukushima Medical University, Fukushima 960-1295, Japan.

the expression of transcripts encoding TH, AADC and DBH, and as a result, catecholamine production is upregulated (Yamakuni et al., 1998). V-1 is a 12 kDa protein containing 2.5 copies of the ankyrin repeat, which has been originally isolated as a developmentally regulated protein from murine cerebellum (Taoka et al., 1992). Consistent with the function of V-1 as a promoter of catecholamine biosynthesis demonstrated by the V-1-overexpression experiment using the cell culture system, the expression of V-1 mRNA occurs in all catecholaminergic cell populations, including adrenal medulla chromaffin cells, sympathetic neurons, and dopaminergic and noradrenergic neurons of the central nervous system, and is especially abundant in noradrenergic cell subpopulations. In addition, as previously reported (Yamakuni et al., 1998), V-1 augments mRNA expression for TH, AADC and DBH in PC12D cells. Therefore, these findings suggest the involvement of V-1 in noradrenergic phenotype expression *in vivo*. In the present study, we generated transgenic mice that selectively overexpress V-1 in noradrenergic or adrenergic cell subpopulations to further analyze the functions of V-1 in control of catecholaminergic phenotype expression *in vivo*.

Rat V-1 cDNA fragment derived from pRV17 clone (Taoka et. al, 1994) was inserted into an expression vector carrying 4-kb human DBH gene promoter region, pHDBH (Kobayashi et al., 1994) excised human DBH cDNA with EcoRI, which results in the V-1 transgene designated pDV1. Following digestion of pDV1 with XbaI and XhoI and the subsequent purification by low melting agarose gel electrophoresis, the transgene was microinjected into fertilized mouse eggs, and the two-cell stage eggs were transferred into the oviducts of psuedopregnant foster females as previously described (Kaneda et al., 1991). Genomic DNA from the tail tissues of animals was analyzed by Southern blotting, and consequently three lines of mice carrying the V-1 transgene (DV1-4, DV1-9 and DV1-11) were established.

We first analyzed the protein levels of V-1 in adrenal glands of V-1 transgenic mice using Western blotting and immunohistochemisry as previously described (Yamakuni et al., 1998) to determine whether the transgenic animals show an increase in the expression of V-1. When assayed extracts from the adrenal glands by Western blot analysis using anti-V-1 antibody, the V-1 protein level in the adrenal glands was 2 to 3-fold higher in the three lines of V-1 transgenic mice than in non-transgenic mice. The expression of TH, an enzyme for the first and rate-limiting step of catecholamine biosynthesis was increased 2-fold in V-1 transgenic mice compared with non-transgenic mice, while no changes in the expression of 14-3-3 beta protein were detected between the transgenic and non-transgenic mice. In addition, the expression of V-1 was observed to be remarkably increased in noradrenergic chromaffin cells but not adrenergic chromaffin cells in the adrenal glands of the transgenic mice, as examined by immunohistochemistry. These findings suggest that V-1 functions at least as a positive regulator for the expression of the TH gene *in vivo*.

To examine the effect of the overexpression of V-1 on catecholaminergic cell function, we next assayed catecholamine contents in adrenal glands and vas deferens of transgenic and non-transgenic mice, because vas deferens is well-known to be rich in the nerve terminals of sympathetic neurons (Holman, 1975). Age-matched male individuals of the DV1-11 were subjected to catecholamine determination. As shown in Table 1, the norepinephrine level in the vas derefens of V-1 transgenic mice was 1.6-fold higher compared with those in non-transgenic mice. On the other hand, in the adrenal glands,

Table 1. Catecholamine content in the V-1-overexpressing mice (V-1 Tg) and non-transgenic mice (non Tg)

Tissue	Norepinephrine		Epinephrine	
	non Tg	V-1 Tg	non Tg	V-1 Tg
Adrenal gland[a]	8.955	8.534	17.40	14.52
Vas deferens[b]	0.1489	0.2369	ND	ND

Each value is the mean of data obtained from five animals. Tissues were extracted with 0.4 N perchloric acid and catecholamine content assayed by HPLC-ECD. [a]nmol/adrenal gland; [b]nmol/mg protein; ND, not detected.

the epinephrine level in V-1 transgenic mice was reduced to 83% of the non-transgenic mice, whereas there were no changes detected in the norepinephrine level between transgenic and non-transgenic mice. These results demonstrate that V-1 enhances expression of the noradrenergic phenotype of sympathetic neurons typically in the vas deferens, but inhibits expression of the adrenergic phenotype in the adrenal glands.

Furthermore, we recently mapped the V-1 gene to the 7q33 locus of human chromosome, in which a gene responsible for obesity has been shown to be localized by linkage analysis (Bouchard and Pérusse, 1996). V-1 transgenic mice exhibited a small but significant reduction of body mass as compared with the non-transgenic mice. These observations could be explained by the enhancement of sympathetic nervous system activity, since lipolysis in adipose tissues is stimulated by the activation of beta 3 adrenergic receptor (Susulic et al., 1995).

It is of great importance to elucidate how V-1 coordinately controls gene expression for three catecholamine synthesizing enzymes, although V-1 has been proposed to act as a cytoplasmic protein-protein adapter. V-1 is a small protein and 73% of the amino acid sequence is composed of the 2.5 tandem ankyrin repeats. The ankyrin repeat has been identified in a variety of proteins with multiple functions in eukaryotic organisms and demonstrated to serve as an interface for protein-protein interactions with their specific partner proteins (Blank et al., 1992). As the first step to disclose the mechanism for the action of V-1, we searched for proteins that interact with V-1 by metabolic labelling of proteins expressed in V-1-overexpressing PC12D cells (Yamakuni et al., 1998) followed by immunoprecipitation with anti-V-1 antibodies. At least two 33-35 kDa proteins was specifically co-immunoprecipitated with V-1 with three kinds of anti-V-1 antibodies which recognize the N- or C-terminal sequences of V-1, whereas neither V-1 nor these proteins of 33-35 kDa were immunoprecipitated with the preimmune serum. The data given here indicate that V-1 physically associates with at least these proteins.

We reported here the generation of transgenic mice that overexpress the V-1 gene under control by the 4-kb human DBH gene promoter. In the present study, we provided the first evidence that V-1 controls noradrenergic function of sympathetic neurons *in vivo*. It is reasonable to interpret that an increase in norepinephrine production observed in the vas deferens of the V-1 transgenic mice results from the upregulation of gene expression for catecholamine synthesizing enzymes by V-1, because it was observed that the expression of TH protein is augmented in the adrenal glands of the transgenic mice. This increase in the expression of the TH gene is consistent with the results from our stable V-

1-overexpression experiment in PC12D cells as reported elsewhere (Yamakuni et. al., 1998). Therefore, it is possible that the upregulated expression of the TH gene is directed by V-1 *in vivo* via the same regulatory mechanism mediated by protein-protein interaction as *in vitro*. On the other hand, the adrenal glands of V-1 transgenic mice exhibited a reduction in epinephrine levels and no change in norepinephrine levels despite a rise in TH protein levels. These complicated results raise the possibility that this decrease in epinephrine production may be due to suppression of the expression of the PNMT gene by V-1, and V-1 may modulate the activity of catecholamine secretion from the adrenal chromaffin cells.

The results obtained from the present study strengthen our working hypothesis that V-1 may couple with cytoplasmic signaling pathways which direct coordinate regulation of catecholamine synthesizing enzyme genes via protein-protein interaction between specific partner proteins to control catecholaminergic cell specification. Furthermore, the present study suggests that V-1 possibly participates in lipolysis and blood pressure regulation through the modulation of sympathetic functions *in vivo*.

REFERENCES

Blank, V., Kourilsky, P., and Israël, A., 1992, NF-kappa B and related proteins: Rel/dorsal homologies meet ankyrin-like repeats, *Trends Biochem.* Sci. **17**: 135-140.

Bouchard, C., and Pérusse, L., 1996, Current status of the human obesity gene map, *Obesity Res.* **4**: 81-90.

Holman, M., 1975, Nerve-muscle preparations of vas deferens, in: *Methods in Pharmacology Vol. 3. Smooth Muscle*, E. E. Daniel and D. M. Paton, ed., Plenum Press, New York, pp. 403-417.

Kaneda, N., Sasaoka, T., Kobayashi, K., Kikuchi, K., Nagatsu, I., Kurosawa, Y., Fujita, K., Yokoyama, M., Nomura, T., Katsuki, M., and Nagatsu, T., 1991, Tissue-specific and high level expression of the human tyrosine hydroxylase gene in transgenic mice, *Neuron* **6**: 583-594.

Kobayashi, K., Morita, S., Mizuguchi, T., Sawada, H., Yamada, K., Nagatsu, I., Fujita, K., and Nagatsu, T., 1994, Functional and high level expression of human dopamine beta-hydroxylase in transgenic mice, *J. Biol. Chem.* **269**: 29725-29731.

Lo, L., Morin, X., Brunet, J.-F., and Anderson, D. J., 1997, Specification of neurotransmitter identity by Phox2 proteins in neural crest stem cells, *Neuron* **22**: 693-705.

Morin, X., Cremer, H., Hirsch, M.-R., Kapur, R. P., Goridis, C., and Brunet, J.-F., 1997, Defects in sensory and autonomic ganglia and absence of locus coeruleus in mice deficient for the homeobox gene *Phox2a*, *Neuron* **18**: 411-423.

Suslic, V. S., Frederich, R. C., Lawitts, J., Tozzo, E., Kahn, B. B., Harper, M.-E., Himms-Hagen, J., Flier, J. S., and Lowell, B. B., 1995, Targeted disruption of the beta3-adrenergic receptor gene, *J. Biol. Chem.* **270**: 29483-29492.

Taoka, M., Yamakuni, T., Song, S.-Y., Yamakawa, Y., Seta, K., Okuyama, T., and Isobe, T., 1992, A rat cerebellar protein containing the cdc10/SWI6 motif, *Eur. J. Biol.* **207**: 615-620.

Taoka, M., Isobe, T., Okuyama, T., Watanabe, M., Kondo, H., Yamakawa, Y., Ozawa, H., Hishinuma, F., Kubota, M., Minegishi, A., Song, S.-Y., and Yamakuni, T., 1994, Murine cerebellar neurons express a novel gene encoding a protein related to cell cycle control and cell fate determination protein, *J. Biol. Chem.* **269**: 9946-9951.

Yamakuni, T., Yamamoto, T., Hoshino, M., Song, S.-Y., Yamamoto, H., Kunikata-Sumitomo, M., Minegishi, A., Kubota, M., Ito, M., and Konishi, S., 1998, A novel protein containing cdc10/SWI6 motifs regulates expression of mRNA encoding catecholamine biosynthesizing enzymes, *J. Biol. Chem.* **273**: 27051-27054.

Zetterström, R. H., Solomin, L., Jansson, L., Hoffer, B. J., Olson, L., and Perlmann, T., 1997, Dopamine neuron agenesis in Nurr 1-deficient mice, *Science* **276**: 248-250.

MULTIPLE ROLES OF ACTIN CYTOSKELETON IN CATECHOLAMINE RELEASE FROM CHROMAFFIN CELL

Konosuke Kumakura, Mica Ohara-Imaizumi, Fiorenzo Battaini,* Nobuyuki Sasakawa, Shingo Ohkubo

1. INTRODUCTION

In adrenal chromaffin cells, it is generally accepted that sub-plasmalemmal actin network acts as a barrier to secretory vesicles.[1] On the other hand, we have demonstrated previously that myosin light chain kinase which regulates actin-myosin interaction are required for ATP-dependent priming or for step before the priming of exocytosis in adrenal chromaffin cells,[2] and suggested that actin-myosin interaction is required for the vesicle recruitment in these cells. In order to clarify the implication of actin-myosin interaction and of actin cytoskeleton in the mechanism of CA release, we studied the effects of Mycalolide B (MLB), Cytochalasin D (CD), and Wortmannin (WT) on the vesicle movement and kinetics of exocytotic events. Mycalolide B depolymerizes F-actin to G-actin and therefore inhibits actin-myosin interaction.[3] Cytochalasin D severs F-actin into short filaments, but does not inhibit actin-myosin interaction.[3] Wortmannin inhibit the phosphorylation of myosin light chain thus inhibit actin-myosin interaction, but does not depolymeraize F-actin.[4]

2. METHODS

To study mechanism of granule recruitment to the releasable pool before Ca^{2+}-triggered fusion, we examined the track motions of LysoTracker Green DND26-labeled secretory granules after stimulation using fluorescence microscopy and cooled CCD system. Position of the granules were recorded in normal buffer every 2 s for 2 min after stimulation with high K^+ for 2 min. For analysis of the verocity of the vesicle movement,

* Konosuke Kumakura, Mica Ohara-Imaizumi, Nobuyuki Sasakawa, Shingo Ohkubo, Life Science Institute, Sophia University, Tokyo 102-8554, Japan. *Firoenzo Battaini, Uiversity of Rome Tor Vergata, 00133 Rome, Italy.

Catecholamine Research, Edited by Nagatsu *et al.*
Kluwer Academic/Plenum Publishers, 2002

30 to 40 vesicles were randomly selected and the verocity of each vesicle were calculated from their trac motion by MetaMorph 4.5 (Universal Imaging).

Amperometric Detection of CA released from a single chromaffin cell was conducted at 25 ℃ using a carbon fiber microelectrode (5 μm diameter). The holding voltage applied to the microelectrode was +700 mV versus Ag/AgCl reference electrode in the recording chamber. The oxidative current was detected by a voltammetric amperometric amplifier and the signal was digitized at 4KHz by a MacLab/4s system coupled to a Macintosh computor.

3. RESULTS AND DISCUSSION

Among the examined three substances, MLB and WT inhibited the evoked release of CA from digitonin permeabilized chromaffin cells. These two are the substances that inhibit actin-myosin interaction. The inhibition by both substances are on the secretion rate of the second phase in the time course, suggesting that actin-myosin interaction is important for the priming of exocytosis or the vesicle recruitment. Inhibition of the vesicle recruitment is reflected in the kinetics of CA release in intact cells, as we examined by a real-time monitoring of the secretory response.[5] A persistent stimulation with ACh resulted a transient secretory response in control cells, and WT clearly reduced the duration of the response without any effects on the initial burst. Mycalolide B had the same effect, while CD did not affect the kinetics. Repetitive stimulation produced repetitive responses with rather constant magnitude. The magnitudes of the initial response was not affected by WT nor by MLB, but the magnitude of the following response was dramatically decreased by both treatments .

The magnitude of the repetitive responses were clearly decreased by two inhibitors of actin-myosin interaction, while CD did not affect it at all. This decrease in the magnitudes is due to the decrease in the frequency of exocytotic events, as was shown by the amperometric detection of the

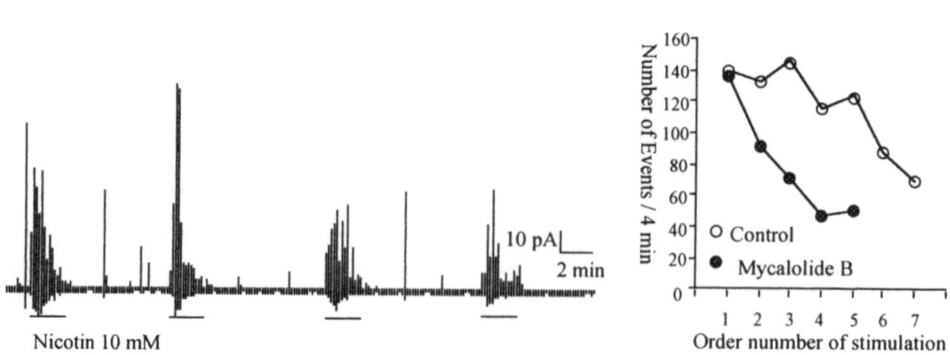

Figure 1. Effect of mycalolide B (1 mM x 30 min) on nicotin (10 μM)-evoked catecholamine release from individual cells. Exocytotic events were detected amperometoricaly.

response of a single chromaffin cell to the repetitive stimulation. In the amperometric detection with a single carbon fiber electrode, the frequency of the events in the response to the second and further stimulation decreased by MLB (Figure 1). These results strongly suggest that actin-myosin interaction is essential for the vesicle recruitment for CA release.

To clarify a possible implication of the vesicle movement for the recruitment, we examined track motion of the vesiscles in living chromaffin cells during 2min after deporalizing stimulation for 2min. Under this condition, about 25% of granules moved rapidly (more than 0.5 μm/s) migrating a long distances (more than 2.5 μm in 2 min) in control cells. The population of granules with such a rapid movement was remarkably decreased in MLB-treated cells and also in WT-treated cells in a dose dependent manner, and in turn a large population of vesicles stopped movement. Again CD failed to affect it at all. These results suggest that this type of granule movement is mediated through the actin-myosin interaction in chromaffin cells and regulate the granule recruitment to the release sites.

Protein Kinase C (PKC) has been suggested to increase the size of release-ready pool of vesicles by different authors, and we also have suggested that PKC accelerate the priming of exocytosis in adrenal chromaffin cells.[6] PKC activation by phorbol ester TPA significantly increased the population of the vesicles with the rapid movement to about 35%, suggesting that such a rapid movement plays a role for the vesicle recruitment.

Activation of PKC by TPA increased the secretory response by facilitating ATP-dependent priming, and MLB abolished the effect of TPA, while CD had no effect[6]. Upon the PKC activation, PKC α and β translocated to the sub-plasmalemma where F-actin and the receptor for activated PKC (RACK1)[7] are colocalizing. After MLB that abolishes the effect of TPA, the cortical F-actin was completely disrupted, colocalizing RACK1 disappeared and the translocation of PKC was abolished. In contrast, CD had almost no effects. Disruption of the cortical F-actin upon exocytosis is well documented in chromaffin cell,[1] and RACK1 seems to change its localization together. Indeed, RACK1 was detected in Triton insoluble fraction from chromaffin cells where F-actin is recovered. When the suspension of TritonX 100 insoluble pellet was treated with MLB and recentrifuged. both actin and RACK1 were detected in the supernatant, but not in the pellet. This suggest that RACK1 tightly binds F-actin, but dissociates if F-actin is depolymerized to G-actin. In the immunoprecipitation with anti-RACK1 antibody, PKCα and β were co-precipitated with RACK1 and actin from the extract of TPA-treated cells but not from control cells, suggesting their tight interaction after PKC translocation. Therefore, the cortical F-actin seems to have an important role also for the regulation of CA release by PKC. All of these results indicate that actin cytoskeleton has multiple roles in the mechanism of CA release in adrenal chromaffin cells.

REFERENCES

1. J – M. Trifaró, M. L. Vitale, Cytoskeleton dynamics during neurotransmitter release, Trends Neurosci. **16,** 466-472 (1993).
2. K. Kumakura, K. Sasai, T. Sakurai, M. Ohara-Imaaizumi, H. Misonou. S. Nakamura, and Y. Nonomura, Essential role of myosin light chain kinase in the mechanism for MgATP-dependent priming of exocytosis in adrenal chromaffin cells, J. Neurosci. **14,** 7695-7703 (1994).
3. S. Saito, S. Watanabe, H. Ozaki, N. Fusetani, and H. Karaki, Mycalolide B, a novel actin depolymerizing agent, J. Biol. Chem. **269,** 29710-29714 (1994).

4. S. Nakanishi, S. Kakita, I. Takahashi, K. Kawahara, E. Tsukada, T. Sano, K. Yamada, M. Yoshida, H. Kase, Y. Matsuda, Y. Hashimoto, and Y. Nonomura, Wortmanni, a microbial product inhibitor of myosin light chain kinase, J. Biol. Chem. **267,** 2157-2163 (1992).
5. K. Kumakura, M. Ohara, and G. P. Satô, Real-time monitoring of the secretory function of cultured adrenal chromaffin cells, J. Neurochem. **46,** 1851-1858 (1986).
6. H. Misonou, M. Ohara-Imaizumi, T. Murakami, M. Kawasaki, K. Ikeda, T. Wakai, and K. Kumakura, Protein kinase C controls the priming step of regulated exocytosis in adrenal chromaffin cells, Cell. Mil. Neurobiol. **18,** 379-390 (1998).
7. D. Ron, C.-H. Chen, J. Caldwell, L. Jamieson, E. Orr, and D. Mochly-Rozen, Cloning of an intracellular receptor for protein kinase C: A homolog of the b subunit of G proteins, Proc. Natl. Acad. Sci. USA, **91,** 839-843 (1994).

INTERACTION OF TYROSINE HYDROXYLASE AND 14-3-3 PROTEINS IN PC12 CELLS

Tohru Ichimura, Chiharu Itagaki, Akiko Wakamiya, and Toshiaki Isobe*

1. INTRODUCTION

The 14-3-3 protein is a family of acidic, dimeric proteins distributed widely among eukaryotic cells. This protein family binds to a variety of proteins in a phosphorylation-dependent manner and participates in the regulation of cell proliferation, differentiation and function. To date, more than 70 proteins, localized in the cell membrane to nuclei, are reported to bind with the 14-3-3 family. These proteins include at least 19 membrane components (receptors, ion channels and effectors), 22 signal transduction kinases and phosphatases, 5 cytoskelton proteins, 18 nuclear proteins including transcription factors, 10 metabolic enzymes, and others. We previously showed, using PC12nnr5 cells transfected with myc-tagged 14-3-3β and η isoforms, that 14-3-3s form a complex with tyrosine hydroxylase (TH) particularly when the hydroxylase is phosphorylated at Ser-19 by endogeneous Ca^{2+}, calmodulin-dependent protein kinase II[1]. In this study we have employed proteomic techniques for systematic characterization of PC12 proteins and identified the TH protein as a major constituent of the proteins that are co-immunoprecipitated with 14-3-3. We have also shown that the TH protein is predominantly associated with 14-3-3ε and η isoforms, but not with ζ and θ isoforms. These results indicate that the activity of TH is regulated through phosphorylation-dependent interaction with limited members of the 14-3-3 family.

2. METHODS

PC12- and PCnnr5-derived cells were cultured and passaged as described previously[1]. Retroviral infections were carried out as described[1]. Infected cells were cultured for 1 week and then subjected to selection with G418 for at least 1 month. Surviving cells were polyclonally expanded and cultured. The cDNAs for bovine 14-3-3η and rat 14-3-

* Tohru Ichimura, and Toshiaki Isobe, Tokyo Metropolitan University, Hachioji, Tokyo 192-0397, Japan. Chiharu Itagaki, Akiko Wakamiya, and Toshiaki Isobe, MEXT, Chiyoda-ku, Tokyo 100-8959, Japan. E-mail: ichimura@mail.comp.metro-u.ac.jp.

Catecholamine Research, Edited by Nagatsu *et al.*
Kluwer Academic/Plenum Publishers, 2002

3β, γ, ε, ζ and θ were synthesized by PCR using oligonucleotides according to the procedures described previously[1]. The PCR fragments were inserted into the *MluI* and *BamHI* sites of myc-tag cloning vector, pMUM1. The myc-tagged 14-3-3 cDNAs were then cloned into retroviral expression vector pLNC5 as described[1].

Myc-14-3-3 was immunoprecipitated from PC12 cell extracts expressed myc-14-3-3η with myc antibody-conjugated Sepharose beads. After washing the beads, bound proteins to immobilized myc-14-3-3 were dissociated with 1 mM synthetic phosphopeptides (LSQRQRSTpSTPNVHA, based on Raf-1 sequence), and were visualized by silver staining following SDS-PAGE. Protein bands were cut out, in-gel digested with trypsin, and the resulting peptides were identified by nano-spray tandem MS/MS (Q-Tof, Micromass) technique and database search with a Mascot software (Matrix Science).

Western blotting was performed as described[1]. The cross-reacting bands were visualized with ECL Plus reagent (Amersham Pharmacia Biotech) and Storm 860 Scanning Image analyzer (Molecular Dynamics). The intensities of the visualized proteins were quantified with ImageQuant software (Molecular Dynamics). The linearity range was established with a calibration curve.

3. RESULTS

3.1. Identification of TH as a Major Ligand for 14-3-3 in PC12 Cells

Figure 1 (lane 1) shows SDS-PAGE photograph of 14-3-3-binding proteins dissociated from the myc-14-3-3-immunoprecipitates with Raf-1 phosphopeptides (see Methods). About 40 proteins were reproducibly detected by this procedure. However, only a few proteins were recovered from the myc-14-3-3-immunoprecipitates with Raf-1 non-phosphopeptides (lane 2), suggesting that most of the detected proteins were specifically associated with 14-3-3 through the phosphoSer recognition pocket in the 14-3-3 molecule.

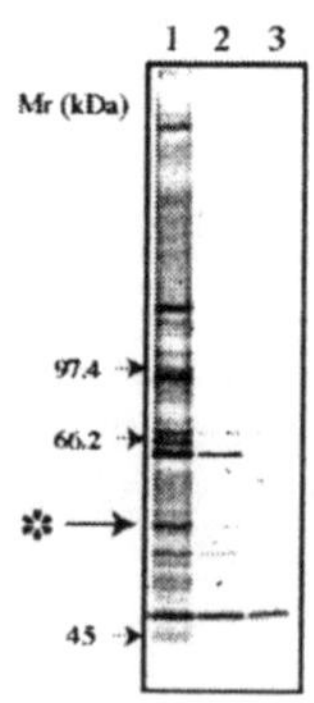

Figure 1: SDS-PAGE analysis of proteins coimmunoprecipitated with myc-14-3-3η. PC12 cell extracts stably expressed myc-14-3-3η were subjected to immunoprecipitation with myc-antibody-conjugated Sepharose beads. Bound proteins to immobilized myc-14-3-3 were dissociated with the synthetic Raf-1 phosphopeptide (lane 1) or Raf-1 non-phosphopeptide (lane 2). The experiment was also carried out with the phosphopeptide and control PC12 cell extracts (lane 3). Dissociated proteins were analyzed by SDS-PAGE (7.5 % PAGE gel) and silver staining. ✲ indicates TH identified by MS (see text).

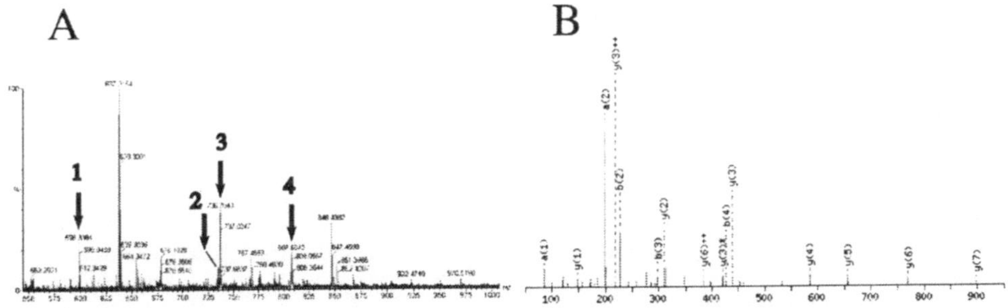

Figure 2: Mass spectrometry analysis of the ✲ protein. (A) NanoESI/MS spectrum of total tryptic peptides. Arrows indicate the peptides that were selected to generate peptide sequence information. (B) Averaged MS/MS spectrum obtained from the peptide at mass to charge ration (m/z) 598.3 (arrow #1 in A). The fragment ions that originate either from the N terminus (b type ions) or the C terminus (y type ions) correspond to an amino acid sequence that was identified using the Mascot database search program.

Among the ~40 proteins, 30 major protein bands were cut out, digested with trypsin, and the resulting peptides were subjected to MS/MS analysis (see Methods). Figure 2A shows a total ion chromatogram of tryptic peptides from one of the bands, shown by ✲ in Figure 1. Among numbers of peptides, four (marked in Figure 2A as #1-4) were selected for MS/MS analysis. Figure 2B shows the MS/MS spectrum of the peptide #1. Database search with this fragment ions identified the peptide sequence LIAEIAFQYK which is found in rat TH at residues 206-215 (accession number, gi:3319077). Similar analysis indicated that the peptides #2-4 were also assigned to rat TH (#2, residues 460-476; #3, residues 216-234; #4, residues 184-204). This observation, together with the strong density of the ✲ protein in the dissociated fraction (see Fig. 1, lane 1), suggest that TH is a major ligand for 14-3-3 in PC12 cells.

3.2. Isoform Specificity for TH Binding

Previous studies suggested that multiple isoforms of the 14-3-3 family were co-expressed in PC12 cell lines. Consistent with this observation, six (β, γ, ε, ζ, η and θ) out of seven mammalian 14-3-3 isoforms were found in PC12nnr5 cells as analyzed by RT-PCR and Western blotting with isoform-specific antibodies (data not shown). We prepared six PC12nnr5-derived cells, each of which stably expresses one of myc-tagged 14-3-3 isoforms (see Methods). The expressed isoforms were immunoprecipitated in parallel with a myc-antibody, and the immunoprecipitates were analyzed by Western blotting with TH antibodies. As shown in Figure 3 (left, upper panel), significant signal of TH was observed with myc-ε. A much lower, but detectable signal was found with myc-η, -γ, and -β, and no signal was detected with myc-ζ and -θ.

The same blot was also probed with antibodies against C-Raf-1 and Cdc25C to compare the binding specificity to the 14-3-3 isoforms. As shown in Fig.3 (right panels), each protein shows a characteristic spectrum in isoform specificity; Cdc25C binds preferentially to γ and ε, while Raf-1 kinase binds to all isoforms to a similar extent. This suggests that the observed binding of TH with the 14-3-3ε, η, γ and β isoforms is not due to non-specific interaction and that 14-3-3 isoforms have different specificity in target recognition.

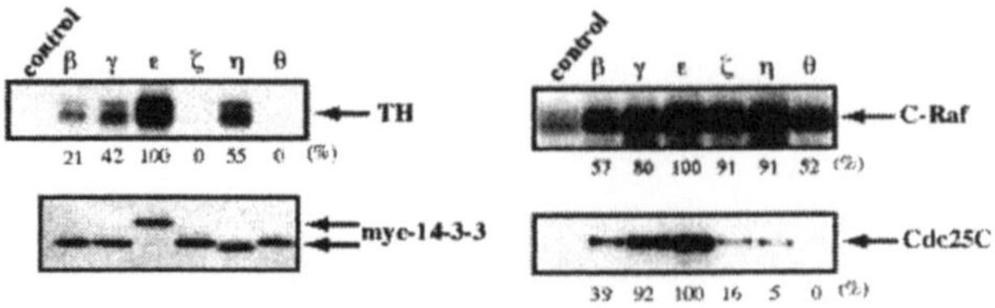

Figure 3: Distinct specificity of 14-3-3 isoforms for the association with TH. PC12nnr5 cell extracts stably expressed the myc-14-3-3 isoforms were subjected to immunoprecipitation with a myc antibody. The immunoprecipitants were analyzed by Western blotting with a TH, a C-Raf or a Cdc25C antibody. A control experiment was performed with nontransfected PC12nnr5 cells. An aliquot (1/80) of each immunoprecipitate was also analyzed by Western blotting with a myc antibody (left, lower panel) to ensure the same amount of inputs of 14-3-3 isoforms. Values represent intensity of each band measured by a scanning image analyzer.

4. DISCUSSION

We have shown here that TH is a major ligand for 14-3-3 in PC12 cells. This observation supports our previous report[1] that 14-3-3 is a regulatory factor of catecholamine synthesis *in vivo*. We note, however, that this protein forms complexes with many additional targets in PC12 cells (Figure 1, lane 1). This finding strongly suggests that other binding partners for 14-3-3 are also present in the cells and are regulated by Ser/Thr phosphorylation and interaction with the 14-3-3 protein.

Previous studies using synthetic phosphoSer peptides suggested that the 14-3-3 family recognizes a consensus sequence, RxxpSxP or RxY/FxpSxP (where pS is phosphoSer), which is found in common with many 14-3-3 target proteins including TH, Cdc25C and Raf-1 kinase. No isoform preferance was also found in these studies. However, the present study demonstrates that TH and Cdc25C bind preferentially to ε and η isoforms in PC12nnr5 cells, while Raf-1 kinase has broader binding specificity (Figure 3). Consistent with this observation, a recent study has shown that A20, an inhibitor of tumor necrosis factor-induced apoptosis, binds preferentially to the η isoform. It has also been reported that η and ζ isoforms bind preferentially to Cdc25B. These findings suggest that additional sequence(s) or motif(s) may also be essential for target recognition of the 14-3-3 protein. This would be important for the elucidation of the physiological roles of the 14-3-3 family, since different 14-3-3 heterodimers may facilitate interaction of different signaling molecules.

ACKNOWLEDGMENTS

This work was supported by Grants-in-Aid for Scientific Research, and by Grants for Integrated Proteomics Project, Pioneer Research on Genome the Frontier from MEXT of Japan.

REFERENCE

1. C. Itagaki, T. Isobe, M. Taoka, T. Natsume, N. Nomura, T. Horigome, S. Omata, H. Ichinose, T. Nagatsu, L. A. Greene, and T. Ichimura, Stimulus-coupled interaction of tyrosine hydroxylase with 14-3-3 proteins, *Biochemistry* **38**, 15673-15680 (1999).

ACETYLCHOLINESTERASE INHIBITOR ELICITS MUSCARINIC RECEPTOR-MEDIATED CHOLINERGIC TRANSMISSION IN THE RAT ADRENAL MEDULLA

Tsuyoshi Akiyama, Toji Yamazaki, and Hidezo Mori*

1. INTRODUCTION

Both nicotinic and muscarinic receptors exist in the adrenal gland and their receptor agonists evoke catecholamine (CA) release from adrenal gland. The CA release induced by nerve stimulation is, however, predominantly suppressed by nicotinic antagonists and resistant to muscarinic antagonists.[1, 2] From these findings, it has been suggested that nicotinic receptors are primarily concentrated in the synaptic regions, with muscarinic receptors primarily localized on the extra-synaptic regions of chromaffin cells. We are interested in whether the differential anatomical localization can entirely explain the predominance of nicotinic receptors in cholinergic synaptic transmission and what conditions are relevant to muscarinic receptor-mediated synaptic transmission. We focused on the function of acetylcholinesterase (AChE), and hypothesized that the inhibition of AChE expanded the regions involved in cholinergic transmission to the extra-synaptic regions. To verify this hypothesis, we applied the microdialysis technique to the adrenal gland of anesthetized rats and investigated the influence of AChE inhibitor on cholinergic transmission in the adrenal gland.

2. MICRODIALYSIS TECHNIQUE

We made a transverse type dialysis probe.[3, 4] The length of the exposed dialysis fiber was 3 mm (50,000 MW cutoff). The dialysis probe was implanted in the medulla of the left adrenal gland of anesthetized rats. Using a microinjection pump, we perfused a dialysis probe with Ringer's solution or Ringer's solution containing pharmacological agents at 10 μl/min. We chose a sampling period of 1 min. Dialysate epinephrine (Epi)

* Tsuyoshi Akiyama, Toji Yamazaki, and Hidezo Mori, Department of Cardiac Physiology, National Cardiovascular Center Research Institute, Fujishiro-dai 5-7-1, Suita, Osaka, 565-8565 Japan.

Catecholamine Research, Edited by Nagatsu *et al.*
Kluwer Academic/Plenum Publishers, 2002

and norepinephrine (NE) concentrations were measured as indexes of Epi and NE release from adrenal gland by high-performance liquid chromatography with electrochemical detection. The left splanchnic nerve was transected in all protocols. In protocols requiring splanchnic nerve stimulation, electrodes were applied to the distal end of the nerve, which was then stimulated with a digital stimulator. We started the protocols after basal dialysate Epi and NE concentrations had reached steady levels. All pharmacological agents tested in the protocols were locally administered by perfusion through the dialysis probe.

We used two types of cholinergic stimulation to evoke CA release: exogenous acetylcholine (ACh) by local administration of ACh and endogenous ACh by electrical stimulation of nerves. Exogenous ACh acts on the cholinergic receptors in both synaptic and extra-synaptic regions. Endogenous ACh acts predominantly on the cholinergic receptors within synaptic regions. To compare the two types of stimulation, we chose an almost equipotent stimulation so that the CA responses were identical between these stimulations. Dialysate Epi and NE levels rapidly increased by 1 min-nerve stimulation or 1 min-local administration of ACh, and declined immediately after ceasing stimulation or administration. Dialysate Epi levels were almost identical between local administration of ACh (1 mM, 1 min) and nerve stimulation (4Hz, 1 min). But dialysate NE levels induced by nerve stimulation were about three fourths of those induced by local administration of ACh. Furthermore, the increase in dialysate CA concentration correlated with the frequency of nerve stimulation from 2 to 20 Hz when we electrically stimulated the left splanchnic nerve consecutively at different frequencies: 2, 4, 10, 20, and 40 Hz for 1 min.

3. CONTRIBUTION OF NICOTINIC AND MUSCARINIC RECEPTORS

We observed the peak of CA release during the local administration of ACh (1 mM, 1 min) or the electrical stimulation of splanchnic nerve (2 or 4 Hz, 1min). We locally administered the nicotinic antagonist, hexamethonium (1 mM), or the muscarinic antagonist, atropine (10 μM), and investigated the influence on this peak of CA release.

3.1. Exogenous ACh-induced CA Release

Hexamethonium significantly suppressed the release of both Epi and NE induced by local administration of ACh. Atropine did not affect NE release, but significantly suppressed Epi release. There was a differential contribution of nicotinic and muscarinic receptors between NE and Epi release from rat adrenal gland. Exogenous ACh evokes nicotinic receptor-mediated Epi and NE release and muscarinic receptor-mediated Epi release.

3.2. Endogenous ACh-induced CA Release

Hexamethonium significantly suppressed the release of both Epi and NE induced by nerve stimulation at 4 Hz, whereas atropine did not affect the Epi or NE release. Epi and NE release induced by nerve stimulation at 2 Hz was almost one-half of that induced by stimulation at 4 Hz. But the manner of suppression by hexamethonium or atropine was identical between the 2 frequencies of nerve stimulation. Nerve stimulation predominantly evokes nicotinic receptor-mediated Epi and NE release.

In the rat adrenal gland, NE release induced by either exogenous or endogenous ACh is mediated through nicotinic receptors but not muscarinic receptors. Muscarinic receptors may be nearly absent on the NE-storing chromaffin cells. Epi release induced by endogenous ACh is predominantly mediated through nicotinic receptors, but Epi release induced by exogenous ACh is mediated through both nicotinic and muscarinic receptors. Muscarinic receptors localize on the extra-synaptic regions of the Epi-storing chromaffin cells. Exogenous ACh acts on both synaptic and extra-synaptic regions of chromaffin cells. On the other hand, endogenous ACh mainly acts on synaptic regions. Therefore, exogenous ACh evokes Epi release through extra-synaptic muscarinic receptors of Epi-storing chromaffin cells, while endogenous ACh does not.

4. EFFECT OF ACETYLCHOLINESTERASE INHIBITOR

In the *in vivo* state, AChE rapidly decomposes exogenous and endogenous ACh. AChE restricts the diffusion of released ACh from the synaptic region to the extra-synaptic region. Thus, the inhibition of AChE could not only elevate tissue ACh concentration but also extend the site of action of ACh. We investigated the role of AChE in cholinergic transmission induced by exogenous and endogenous ACh. We locally administrated neostigmine (10 μM) or eserine (100 μM) through the dialysis probe.

4.1. Exogenous ACh-induced CA Release

Neostigmine did not affect the basal CA release, but significantly enhanced the exogenous ACh-induced CA release. The CA responses induced by ACh of 10 μM in the presence of neostigmine were almost identical with those induced by ACh of 1 mM in the absence of neostigmine. In the presence of neostigmine, hexamethonium significantly suppressed the Epi and NE release. Atropine did not affect the NE release but significantly suppressed the Epi release. Thus neostigmine enhanced the exogenous ACh-induced Epi and NE release, but did not affect the manner of suppression by hexamethonium or atropine. These data suggest that the inhibition of AChE did not affect the qualitative contribution of nicotinic and muscarinic receptors in exogenous ACh-induced CA release.

4.2. Endogenous ACh-induced CA Release

Neostigmine significantly enhanced the nerve stimulation-induced CA release. The CA responses induced by stimulation at 2 Hz in the presence of neostigmine were almost identical with that induced by stimulation at 4 Hz in the absence of neostigmine. In the presence of neostigmine, hexamethonium significantly suppressed Epi and NE release. Atropine did not affect NE release but significantly suppressed Epi release. Another AChE inhibitor, eserine also enhanced the nerve stimulation-induced CA release. This enhancement was less than that induced by neostigmine. But the manner of suppression by hexamethonium or atropine was identical with that in the presence of neostigmine. In the presence of AChE inhibitors, the contribution of nicotinic and muscarinic receptors in the endogenous ACh-induced CA release was identical with that in the exogenous ACh-induced CA release. These data suggest that AChE inhibitors extend the sites of action of endogenous ACh to the extra-synaptic regions of chromaffin cells.

In the exogenous ACh-induced CA release, AChE inhibitors enhanced the nicotinic receptor-mediated Epi and NE release and the muscarinic receptor-mediated Epi release, but did not affect the qualitative contribution of nicotinic or muscarinic receptors. On the other hand, in the endogenous ACh-induced CA release, AChE inhibitors not only enhanced the nicotinic receptor-mediated Epi and NE release, but also elicited the muscarinic receptor-mediated Epi release. These results indicate that AChE inhibitor not only elevates tissue ACh concentration by blocking decomposition of exogenous and endogenous ACh, but also extends the site of action of endogenous ACh to the extra-synaptic regions of chromaffin cells. This extension elicits the cholinergic transmission mediated through extra-synaptic muscarinic receptors of Epi-storing chromaffin cells.

5. CONCLUSION

Microdialysis technique with high-performance liquid chromatography makes it possible to continuously monitor Epi and NE release from *in vivo* rat adrenal medulla at 1min-intervals. In the rat adrenal gland, muscarinic receptors are present on the extra-synaptic region of the Epi-storing chromaffin cells, and AChE restricts this muscarinic receptor-mediated cholinergic transmission. Thus, a decrease in cholinesterase activity could not only enhance the nicotinic receptor-mediated Epi and NE release but also elicit the muscarinic receptor-mediated Epi release.

ACKNOWLEDGEMENT

This study was supported by grants in aid for scientific research from the Ministry of Education, Science, and Culture of Japan and by a grant from the Technology Agency, Encourage System of COE.

REFERENCES

1. T. Kimura, T. Shimamura, and S. Satoh, Effects of pirenzepine and hexamethonium on adrenal catecholamine release in responses to endogenous and exogenous acetylcholine in anesthetized dogs, *J. Cardiovasc. Pharmacol.* 20, 870-874 (1992).
2. A. R. Wakade, and T. D. Wakade, Contribution of nicotinic and muscarinic receptors in the secretion of catecholamines evoked by endogenous and exogenous acetylcholine, *Neuroscience* 10, 973-978 (1983).
3. T. Akiyama, and T. Yamazaki, Adrenergic inhibition of endogenous acetylcholine release on postganglionic cardiac vagal nerve terminals, *Cardiovasc. Res.* 46, 531-538 (2000).
4. T. Akiyama, and T. Yamazaki, Myocardial interstitial norepinephrine and dihydroxyphenylglycol levels during ischemia and reperfusion, *Cardiovasc. Res.* 49, 78-85 (2001).

LEPTIN STIMULATES CATECHOLAMINE SYNTHESIS VIA TYROSINE HYDROXYLASE PHOSPHORYLATION-DEPENDENT AND - INDEPENDENT PATHWAYS IN CULTURED BOVINE ADRENAL MEDULLARY CELLS

Nobuyuki Yanagihara, Kensuke Utsunomiya, Yumiko Toyohira, Susumu Ueno, Masato Tsutsui, Tat Beng Cheah and Futoshi Izumi*

Obesity occurs as a consequence of an increased ratio of caloric intake to energy expenditure, and is often associated with cardiovascular and metabolic disorders such as hyperglycemia, hyperlipidemia and hypertension. Although the molecular factors regulating food intake and energy expenditure are not well understood, energy expenditure and body weight are influenced by exercise, physical activity and thermogenesis through the sympathetic nervous system. Leptin, the obese gene product, has been recognized as another mediator of energy expenditure (Zhang et al., 1994). Leptin is secreted from adipose tissues and plays an important role in regulating body weight and food intake through its receptors in the hypothalamus, the center of energy homeostasis (Zhang et al., 1994 ; Tartaglia et al., 1995). Recent studies have shown that leptin acts on peripheral tissues, such as hematopoietic cells, pancreatic islets and the reproductive system, suggesting that leptin also has extrahypothalamic actions. Furthermore, leptin receptors (ObRs) are reported to occur not only in the hypothalamus, but also in various peripheral tissues (Tartaglia et al., 1995; Lee et al., 1996). ObRs are assumed to activate several signal transduction pathways, including the mitogen-activated protein kinase (MAPK) cascade and the Janus protein-tyrosine kinase (JAK)-signal transducers and activators of transcription (STAT) cascade. However, the relation between leptin-induced signal transduction pathway and responses caused by leptin in these tissues remains largely unknown. Catecholamines are neurotransmitters or hormones that can cause weight loss via a number of mechanisms, including lipolysis, increased blood glucose and

* Nobuyuki Yanagihara,[1] Kensuke Utsunomiya, Yumiko Toyohira, Susumu Ueno, Masato Tsutsui, Department of Pharmacology, University of Occupational & Environmental Health, School of Medicine, Kitakyusyu, 807-8555, Japan. Tat Beng Cheah, Discipline of Medical Biochemistry, Faculty of Medicine and Health Sciences, University of Newcastle, Newcastle, Australia. Futoshi Izumi, University of Occupational & Environmental Health, Kitakyushu, 807-8555, Japan.

Catecholamine Research, Edited by Nagatsu *et al.*
Kluwer Academic/Plenum Publishers, 2002

thermogenesis. Recent study reported that leptin influences catecholamine secretion from the adrenal medulla through the central sympathetic nervous system (Satoh et al., 1999). On the other hand, the presence of ObRs has been reported in the rat, murine, human and porcine adrenal medulla. From these data, it is tempting to speculate that leptin controls the metabolic homeostasis via modulating catecholamine synthesis or secretion in adrenal medulla. In our recent study, we identified and characterized the ObR and its mRNA in the bovine adrenal medulla by ^{125}I-leptin binding and the reverse transcriptase-polymerase chain reaction (RT-PCR), respectively (Yanagihara et al., 2000). In the present study, we studied the effect of leptin on catecholamine synthesis in cultured bovine adrenal medullary cells.

Bovine adrenal medullary cells were isolated and maintained as previously reported (Yanagihara et al., 1996). ^{14}C-Catecholamine synthesis from [^{14}C] tyrosine or [^{14}C] DOPA (Yanagihara et al., 2000), phosphorylation and activity of tyrosine hydroxylase (Yanagihara et al., 1984), in situ activity of tyrosine hydroxylase and Western blot analysis of phosphorylated MAPKs were measured (Utsunomiya et al., 2001).

Leptin (10 nM) caused a significant increase in ^{14}C-catecholamine synthesis from [^{14}C] tyrosine by about 20-30% over the control during incubation for 10-30 min. Leptin at 3-30 nM induced a concentration-dependent increase in ^{14}C-catecholamine synthesis. When [^{14}C] DOPA was used as a substrate instead of [^{14}C] tyrosine, leptin failed to increase ^{14}C-catecholamine synthesis, indicating that stimulation of catecholamine synthesis induced by leptin must occur mainly before the DOPA decarboxylase step, and probably step of tyrosine hydroxylase, the rate-limiting enzyme of catecholamine biosynthesis (Nagatsu et al., 1964 ; Weiner, 1970). Indeed, the activation (24% increase over the control) of tyrosine hydroxylase in the soluble fraction derived from the cells was observed after stimulation of cells with leptin (10 nM) for 20 min. ^{32}P-Labeled adrenal medullary cells were stimulated with or without leptin (10 nM), and the supernatants derived from the cells were immunoprecipitated with antityrosine hydroxylase antiserum. The immunoprecipitates were then analyzed by sodium dodecyl sulfate-polyacrylamide gel electrophoresis. Leptin stimulated the phosphorylation of tyrosine hydroxylase by 29% over the control. A number of in vitro and in situ experiments have shown that tyrosine hydroxylase can be phosphorylated by multiple protein kinases, including cyclic AMP-dependent protein kinase (protein kinase A), protein kinase C, and Ca^{2+}/calmodulin-dependent protein kinases. To study which kinase participates in leptin-induced phosphorylation of tyrosine hydroxylase, forskolin, 4β-phorbol-12β-myristate-13α-acetate (PMA) and ionomycin were used to activate protein kinase A, protein kinase C and Ca^{2+}/calmodulin-dependent protein kinases, respectively. At concentrations producing maximal responses, all the activators caused an additive or more than additive increase of in situ activity of tyrosin hydroxylase in leptin-treated cells. This result suggests that leptin phosphorylates and activates tyrosine hydroxylase by a mechanism distinct from that of these activators for protein kinase A, protein kinase C and Ca^{2+}/calmodulin-dependent protein kinases. To investigate another signal transduction pathway of tyrosine hydroxylase phosphorylation induced by leptin, we measured the phosphorylation of p44/42 MAPK in the cells. Incubation of cells with leptin (30 nM) for 5 min resulted in an increase of phospho-p44/42 MAPK by 35% over the control. To study an involvement of MAPK in the stimulation of catechlamine synthesis by leptin, U0126 was used as a

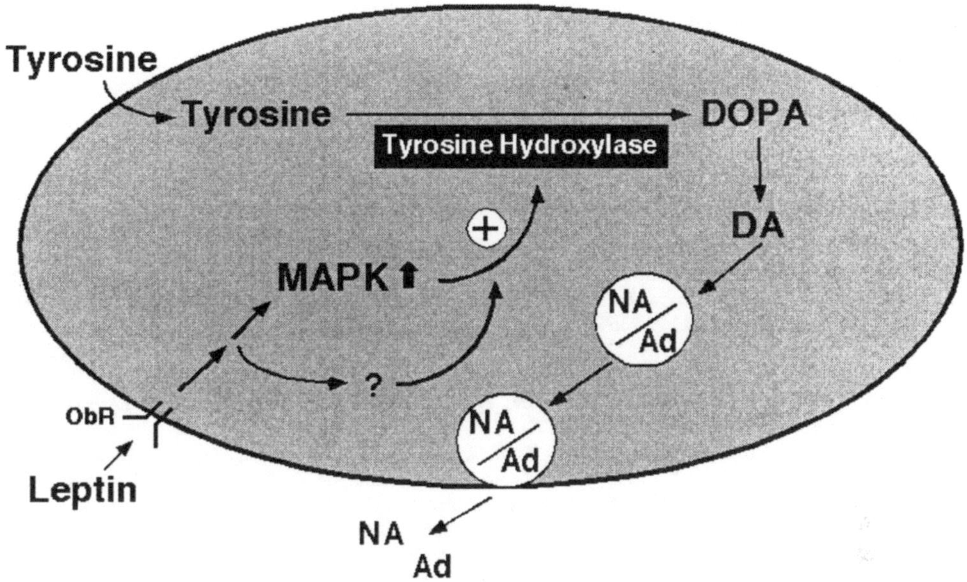

Figure 1. Signal transduction of catecholamine synthesis in bovine adrenal medullary cells by leptin

selective inhibitor for MAPK kinase. U0126 (1-10 μM) completely nullified the stimulatory effect of leptin of ^{14}C-catecholamine synthesis. Since tyrosine hydroxylase can be phosphorylated and activated by MAPK (Haycock et al., 1992), the present findings suggest that leptin phosphorylates and activates tyrosine hydroxylase through activation of the MAPK pathway. Concurrent treatment of cells with leptin (10 nM) and acetylcholine (0.3 mM) greatly potentiated the stimulatory effect of acetylcholine on ^{14}C-catecholamine synthesis by 220% over the control while acetylcholine (0.3 mM) alone increased it by 130%. The potentiation by leptin of acetylcholine-induced synthesis of ^{14}C-catecholamines was evident at 1 nM and the maximal effect was observed at 10-30 nM. On the other hand, leptin (10 nM) did not potentiate the effect of acetylcholine (0.3 mM) on phosphorylation and activity of tyrosine hydroxylase. These results suggest that leptin enhances the effect of acetylcholine on catecholamine synthesis by a mechanism other than tyrosine hydroxylase phosphorylation. This may involve an increase in availability of the cofactor, tetrahydrobiopterine, a decrease in intracellular pH or other unknown mechanisms (Zigmond et al., 1989). The leptin level in human serum is reported to be around 0.2-0.7 nM. It is noteworthy that leptin (1-10 nM) strongly enhances ^{14}C-catecholamine synthesis induced by acetylcholine. This finding leads us to hypothesize that leptin accelerates catecholamine synthesis produced by some stress or emotional excitation which induces stimulation of the splanchnic nerve and subsequently the adrenal medulla.

In the signal transduction of adrenal medullary cells, leptin stimulates catecholamine synthesis via the activation of tyrosine hydroxylase, which is dependent on tyrosine hydroxylase phosphorylation through the MAPK pathway and is also independent of enzyme phosphorylation (Figure 1). This information would be helpful for the better understanding of the mechanism by which leptin acts on metabolic homeostasis.

ACKNOWLEDGEMENTS

This research was supported by a Grant-in-Aid (11670109) for Scientific Research from the Japan Society for the Promotion of Science to FI and NY.

REFERENCES

Haycock, J. W., Ahn, N. G., Cobb, M. H., and Krebs, E. G., 1992, ERK1 and ERK2, two microtubule-associated protein 2 kinases, mediate the phosphorylation of tyrosine hydroxylase at serine-31 in situ, Proc. Natl. Acad. Sci. USA. **89**:2365-2369.

Lee, G.-H., Proenca, R., Montez, J. M., Carroll, K. M., Darvishzadeh, J. G., Lee, J. I., and Friedman, J. M., 1996, Abnormal Splicing of the leptin receptor in diabetic mice, Nature **379**:632-635.

Nagatsu, T., Levitt, M., and Udenfriend, S., 1964, Tyrosine hydroxylase:The initial step in norepinephrine biosynthesis, J. Biol. Chem. **239**:2910-2917.

Satoh, N., Ogawa, Y., Katsuura, G., Numata, Y., Tsuji, T., Hayase, M., Ebihara, K., Masuzaki, H., Hosoda, K., Yoshimasa, Y., and Nakao, K., 1999, Sympathetic activation of leptin via the ventromedial hypothalamus: Leptin-induced increase in catecholamine secretion, Diabetes **48**:1787-1793.

Tartaglia, L. A., Dembski, M., Weng, X., Deng, N, Culpepper, J., Devos, R., Richards, G. J., Campfield, L. A., Clark, F. T., Deeds, J., Muir, C., Sanker, S., Moriarty, A., Moore, K. J., Smutko, J. S., Mays, G. G., Woolf, E. A., Monroe, C. A., and Tepper, R. I., 1995, Identification and expression cloning of a leptin receptor, OB-R, Cell **83**:1263-1271.

Utsunomiya, K., Yanagihara, N., Tachikawa, E., Cheah, T. B., Kajiwara, K., Toyohira, Y., Ueno, S., and Izumi, F., 2001, Stimulation of catecholamine synthesis in cultured bovine adrenal medullary cells by leptin, J. Neurochem. **76**:926-934.

Weiner, N., 1970, Regulation of norepinephrine biosynthesis, Annu. Rev. Pharmacol. **10**:273-290.

Yanagihara, N., Tank, A. W., and Weiner, N., 1984, Relationship between activation and phosphorylation of tyrosine hydroxylase by 56 mM K^+ in PC12 cells in culture, Mol. Pharmacol. **26**:141-147.

Yanagihara, N., Oishi, Y., Yamamoto, H., Tsutsui, M., Kondoh, J., Sugiura, T., Miyamoto, E., and Izumi, F. , 1996, Phosphorylation of chromogranin A and catecholamine secretion stimulated by elevation of intracellular Ca^{2+} in cultured bovine adrenal medullary cells, J. Biol. Chem. **271**:17463-17468.

Yanagihara, N., Utsunomiya, K., Cheah, T. B., Hirano, H., Kajiwara, K., Hara, K., Nakamura, E., Toyohira, Y., Uezono, Y., Ueno, S., and Izumi, F., 2000, Characterization and functional role of leptin receptor in bovine adrenal medullary cells, Biochem. Pharmacol. **59**:1141-1145.

Zhang, Y., Proenca, R., Maffei, M., Barone, M., Leopold, L., and Friedman, J. M., 1994, Positional cloning of the mouse obese gene and its human homologue, Nature **372**:425-432.

Zigmond, R. E., Schwarzschild, M. A., and Rittenhouse, A. R., 1989, Acute regulation of tyrosine hydroxylase by nerve activity and by neurotransmitters via phosphorylation, Annu. Rev. Neurosci. **12**:415-461.

SYMPATHETIC POTENTIATION OF CYCLIC ADP-RIBOSE FORMATION IN RAT CARDIAC MYOCYTES

Haruhiro Higashida, Jia-Sheng Zhang, Shigeru Yokoyama, Mami Noda, Zhen-Guo Zhong, Sumiko Mochida and Alla Egorova*

1. INTRODUCTION

Sympathetic nerve excitation stimulates β-adrenergic receptors on cardiac myocytes by release of noradrenaline, leading to an increase in the contractility. This cardiostimulant effect is mediated by an increase in Ca^{2+} permeability resulted from cyclic AMP-dependent phosphorylation of voltage-gated ion channels (Figure 1). Opening of phosphorylated L-type Ca^{2+} channels and tetrodotoxin-sensitive Na^{+} channels results in a transient intracellular Ca^{2+} concentration increase ($[Ca^{2+}]_i$ transient) which is greater than without sympathetic stimulation. The increased $[Ca^{2+}]_i$ is further amplified by Ca^{2+}-induced Ca^{2+}-release (CICR) from the sarcoplasmic reticulum ryanodine receptor Ca^{2+} release channels, leading to strengthened contraction. In CICR in the heart, both cyclic ADP-ribose (cADP-ribose) and Ca^{2+} cooperatively activate type-II ryanodine receptors to release Ca^{2+}. However, no information on the concentration of cADP-ribose after β-adrenoceptor stimulation has yet been reported (Figure 2).

Membrane-bound and cytosolic ADP-ribosyl cyclases constitutively synthesize cADP-ribose from β-NAD^{+}. Formation of cADP-ribose is increased or decreased by stimulation of muscarinic acetylcholine receptors (mAChRs) in a subtype-specific manner, and this is mimicked by GTP and blocked by bacterial toxins in NG108-15 neuronal cells (Higashida et al., 1997). ADP-ribosyl cyclase thus seems to be coupled directly with neurotransmitter or hormone receptors via different G proteins in the surface membrane of these cells. The same control of cADP-ribose formation could be carried out by ventricular adrenergic receptors. To address this question, we measured ADP-ribosyl cyclase activity in crude membrane fractions of rat ventricular myocytes in the presence or absence of an adrenergic agonist (Higashida et al., 1999).

---

* Department of Biophysical Genetics, Kanazawa University Graduate School of Medicine, 13-1 Takara-machi, Kanazawa 920-8640, Japan.

2. ACTIVATION OF ADP-RIBOSYL CYCLASE BY ADRENOCEPTOR STIMULATION

[^{3}H]cADP-ribose was produced by the crude membrane preparation of ventricular myocytes. The activity increased at a constant rate for at least the first 1 min of incubation with 2.11 μM NAD^+ as substrate. The average specific activity of ventricular myocytes was 3.70 ± 0.65 nmol/min/mg of protein (mean ± S.E.M., n=7). Addition at zero time of 2 μM isoproterenol, a β-adrenergic agonist, increased the rate of [^{3}H]cADP-ribose production, with a mean value of 5.62 ± 0.97 nmol/min/mg of protein (n=7). The maximum activation found was 165.8 ± 14.3% (n=8) of the control activity at 1 μM isoproterenol ($p<0.002$). Propranolol (10 μM), the β-adrenergic receptor antagonist, inhibited the stimulation of ADP-ribosyl cyclase by isoproterenol.

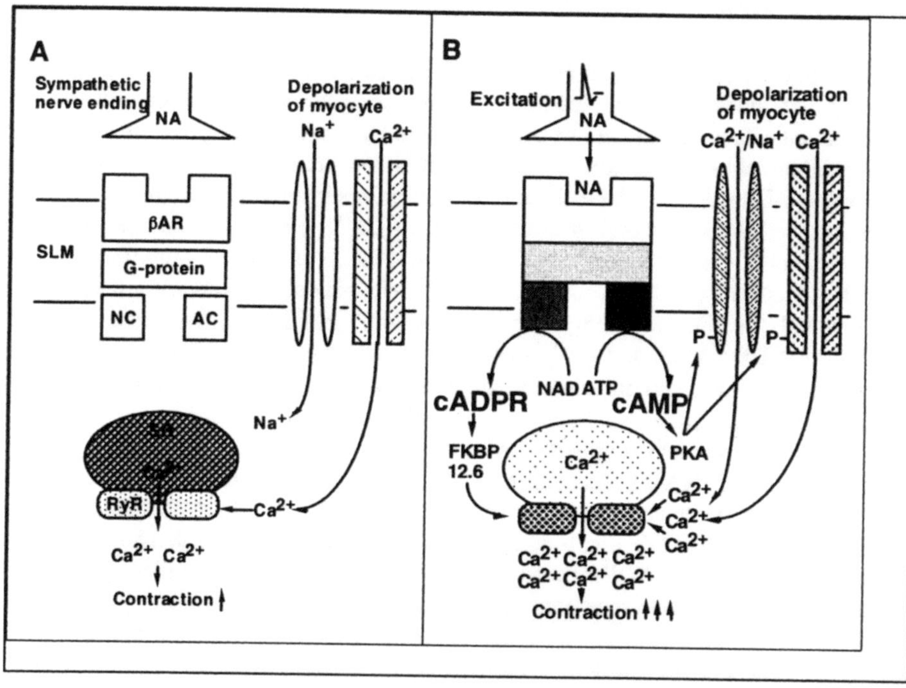

Figure 1. Model depicting signal pathways for cardiac contraction in the control state (A) and in response to β-adrenergic receptor stimulation due to sympathetic nerve excitation (**B**). Noradrenaline (NA) released from sympathetic nerve ending by depolarization binds to β-adrenergic receptor (βAR), producing a G protein-dependent activation of both ADP-ribosyl cyclase (NC) and adenylyl cyclase (AC) in the cell surface sarcolemma (SLM) membrane in ventricle muscle. ADP-ribosyl cyclase activation produces cADP-ribose (cADPR) from NAD^+ that binds to cADP-ribose binding protein (FKBP12.6). Ca^{2+} permeation is higher in the stimulated state (**B**) than the unstimulated state (**A**), due to formation of cyclic AMP (cAMP) and resulting protein kinase A (PKA)-dependent phosphorylation (P) of both Ca^{2+} channels (square) and Na^+ channels (oval). The latter Ca^{2+} permeation is mediated by so-called slip-mode conductance. Increased cADP-ribose and cytosolic Ca^{2+} function co-operatively at ryanodine receptors (RyRs) in sarcoplasmic reticulum (SR) to activate Ca^{2+}-induced Ca^{2+}-release, which results in an amplified cytosolic Ca^{2+} concentration that leads to increased cardiac contraction.

3. EFFECT OF GTP AND BACTERIAL TOXIN

We next determined whether or not β-adrenergic receptor-mediated activation of ventricular ADP-ribosyl cyclase requires GTP. Addition of 10 nM GTP resulted in a detectable increase in enzyme activity at each time tested between 0.5-4 min, while 10 μM GTP inhibited it. The stimulatory effect in ventricular muscle required as little as 1 to 100 nM GTP. The maximum stimulation was obtained at 10 nM GTP with an average increase of 168.3 ± 10.7% (n=15) of the control value ($p<0.001$). This low concentration is the same dose at which GTP effectively binds on G_s-types of G proteins.

When rats were pretreated with cholera toxin (CTx) at 100 ng/g of body weight for 16-24 h, the activation of ADP-ribosyl cyclase by 1 μM isoproterenol was markedly inhibited to 119.3 ± 7.7% of the control level (n=6) from 171.5 ± 11.7% (n=11) in non-treated rats ($p<0.05$). This toxin action is similar to the finding that the m1 and m3 mAChR-induced activation of ADP-ribosyl cyclase is sensitive to CTx in NG108-15 cells (Higashida et al., 1997). Altogether, the effects of GTP and CTx suggest that G proteins mediate coupling between β-adrenergic receptors and ADP-ribosyl cyclase.

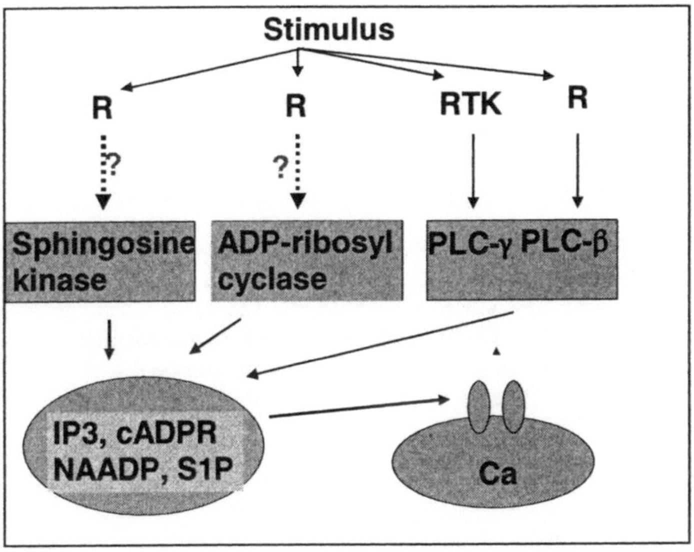

Figure 2.
A scheme depicting receptor-stimulated second messenger production

4. MODEL FOR NORADRENALINE-INDUCED POTENTIATION OF MYOCARDIAC CONTRACTION

Our results provide the first evidence for the role of cADP-ribose as a second messenger downstream of β-adrenoceptors in heart cells after sympathetic excitation (Figures 1 and 3; Higashida et al., 1999).

Recently, a new physiological effect of β-adrenergic receptor stimulation has been reported, namely the regulation of $[Ca^{2+}]_i$ transient by activation of Na^+ channels phosphorylated by cyclic AMP-dependent protein kinase, which is referred to as slip-mode conductance. The Ca^{2+} flux through phosphorylated Na^{+} channels (30% of the total content) as well as through the classical pathways of phosphorylated Ca^{2+} channels (70%) can increase subcellular Ca^{2+} and thereby activate SR Ca^{2+} release, Ca^{2+} sparks, and the $[Ca^{2+}]_i$ transient. On the cADP-ribose action site, it has been shown that FKBP12.6 is its binding protein and that FKBP12.6 bound with cADP-ribose is removed from ryanodine receptors and delays inactivation of Ca^{2+} release channels, which results in a larger $[Ca^{2+}]_i$ transient. Thus, we hypothesize that Ca^{2+} and cADP-ribose levels, both of which are increased by adrenergic stimulation in the ventricle by sympathetic nerve excitation, function cooperatively at the ryanodine receptor level, as depicted in Figure 1B. These changes may contribute significantly to local and global cardiac Ca^{2+} signaling, which controls the force of contraction.

Subtype	β_1	β_2	β_3	β_4
G protein	Gs	Gs	Gi/o	Gs
cAMP	↑	↑	↓	↑
cADPR		↑		

Figure 3.
Signal transduction of β-adrenergic receptor subtypes.

In conclusion, our results suggest that stimulation of β-adrenergic receptors in the ventricle can upregulate cardiac function through cADP-ribose as a second messenger, in concert with cytosolic Ca^{2+} promoting CICR via ryanodine receptors. This finding is further extended to cortical astrocytes (Hotta et al., 2000) and Angiotensin II receptors (Higashida et al., 2000). And not only in cardiac cells but also in neuronal Ca^{2+} signaling, cADP-ribose plays an important role (Hashii, et al., 2000, Higashida, et al., 2001).

REFERENCES

Hashii, M., Minabe, Y. and Higashida, H., 2000, Cyclic ADP-ribose potentiates Ca^{2+} elevation and Ca^{2+} entry via L-type voltage-activated Ca^{2+} channels in NG108-15 neuronal cells. *Biochem. J.* **345,** 207-215.

Higashida, H., Egorova, A., Higashida, C., Zhong, Z.-G., Yokoyama, S., Noda, M. and Zhang, J.-S., 1999, Sympathetic potentiation of cyclic ADP-ribose formation in rat cardiac myocytes. *J. Biol. Chem.* **274,** 33348-33354.

Higashida, H., Hashii, M., Yokoyama, S., Hoshi, N., Asai, K. and Kato, T., 2001, Cyclic ADP-ribose as a potential second messenger for neuronal Ca^{2+} signaling. *J. Neurochem.* **76,** 321-331.

Higashida, H., Yokoyama, S., Hashii, M., Taketo, M., Higashida, M., Takayasu, T., Ohshima, T., Takasawa S., Okamoto, H. and Noda, M., 1997, Muscarinic receptor-mediated dual regulation of ADP-ribosyl cyclase in NG108-15 neuronal cell membranes. *J. Biol. Chem.* **272,** 31272-31277.

Higashida, H., Zhang, J., Hashii, M., Shintaku, M., Higashida, C., Takeda, Y., 2000, Angiotensin II stimulates cyclic ADP-ribose formation in neonatal rat cardiac myocytes. *Biochem. J.* **352,** 197-202.

Hotta, T., Asai, K., Fujita, K., Kato, T. and Higashida, H., 2000, Membrane-bound form of ADP-ribosyl cyclase in rat cortical astrocytes in culture. *J. Neurochem.* **74,** 669-675.

RAB3A SMALL G PROTEIN AND ITS REGULATORS IN NEUROTRANSMITTER RELEASE AND SYNAPTIC PLASTICITY

Hiroyuki Nakanishi, Takuya Sasaki, Jun Miyoshi, and Yoshimi Takai*

1. INTRODUCTION

Rab3A is a member of the Rab3 small GTP-binding protein (G protein) subfamily along with Rab3B, -3C, and –3D, and the Rab3 proteins belong to the Rab family (Takai et al., 2001). The Rab family members are involved in vesicle trafficking, such as exocytosis and endocytosis (Takai et al., 2001). Evidence is accumulating that Rab3A is involved in Ca^{2+}-dependent exocytosis, particularly neurotransmitter release from nerve terminals (Takai et al., 2001). The Ca^{2+}-dependent neurotransmitter release is regulated by several steps: translocation of synaptic vesicles from the reserve pool to the active zone of the presynaptic plasm membrane where Ca^{2+} channels localize, docking of the vesicles to the active zone, transition from the docking to the priming of the vesicles in the readily releasable pool, and fusion of the vesicles with the plasma membrane induced by Ca^{2+} influx (**Fig. 1**) (Takai et al., 2001). Studies on Rab3A knockout mice have revealed that Rab3A is not essential for basal neurotransmitter release but modulates synaptic plasticity (Geppert et al.,1994; 1997; Castillo et al., 1997). In Rab3A knockout mice, synaptic depression is increased during repetitive stimulation in the CA1 region of the hippocampus (Geppert et al., 1994), and mossy fiber long-term potentiation (LTP) in the CA3 region is abolished (Castillo et al., 1997). In cultured hippocampal neurons derived from Rab3A knockout mice, release probability is increased, while the size of the readily releasable pool measured by hypertonic solution is normal (Geppert et al., 1997). Rab3A appears to up-regulate the steps of the translocation and docking as well as to down-regulate the step of the fusion. To understand the precise role and mode of action of Rab3A in neurotransmitter release, it is important to clarify the physiological roles of the Rab3A regulators.

* Hiroyuki Nakanishi and Yoshimi Takai, Department of Molecular Biology and Biochemistry, Osaka University Graduate School of Medicine/Faculty of Medicine, Suita 565-0871. Takuya Sasaki, Department of Biochemistry, Tokushima University, 770-8503. Jun Miyoshi, Department of Molecular Biology, Osaka Medical Center for Cancer and Cardiovascular Diseases, Osaka 535-8511, Japan.

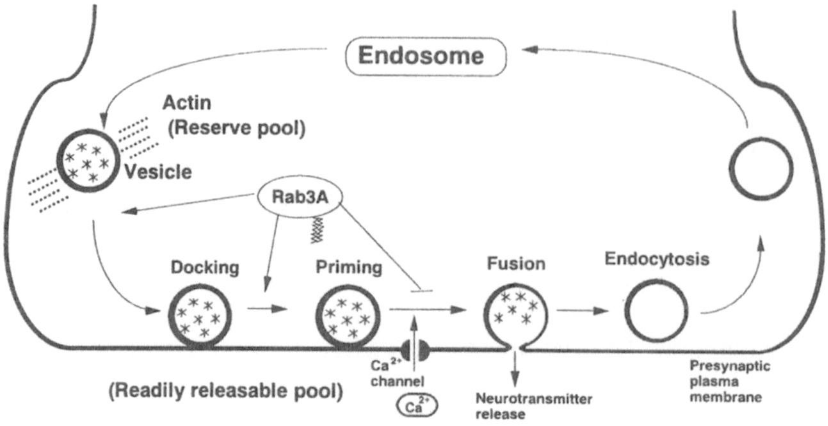

Fig. 1. Role of Rab3A in neurotransmitter release

2. MODE OF ACTION OF RAB3A IN NEUROTRANSMITTER RELEASE

Like other small G proteins, Rab3A has two interconvertible forms: GDP-bound inactive and GTP-bound active forms (Takai et al, 1996; 2001). Activation is regulated by a GDP/GTP exchange protein specific for the Rab3 subfamily members (Rab3 GEP). This activation is inhibited by Rab GDP dissociation inhibitor (Rab GDI) that is active on all the Rab family members including Rab3A (Takai et al., 1996; 2001). Inactivation of Rab3A from the GTP-bound form to the GDP-bound form is regulated by a GTPase-activating protein specific for the Rab3 subfamily members (Rab3 GAP). The precise mode of action of Rab3A in neurotransmitter release remains to be clarified, but its current model is as follows (Takai et al., 1996): GDP-Rab3A forms an inactive complex with Rab GDI and stays in the cytosol of nerve terminals (**Fig. 2**). GDP-Rab3A is released from Rab GDI by an unknown mechanism and converted to GTP-Rab3A by the action of Rab3 GEP. Once GTP-Rab3A is formed, it binds two downstream effector molecules, rabphilin-3 and Rim which localize at synaptic vesicles and the active zone, respectively, causing the translocation and docking of synaptic vesicles to the active zone. GTP-Rab3A is converted to GDP-Rab3A by the action of Rab3 GAP when the vesicles fuse with the plasma membrane, although its precise timing (i.e., before, during, or after the fusion) is unknown. Once GDP-Rab3A is formed, it is associated with Rab GDI and retrieved from the membrane to the cytosol. Thus, the Rab3A recycling of activation and inactivation closely correlates with its cyclical translocation between the cytosol and the plasma membrane, which is essential for the mode of action of Rab3A in neurotransmitter release.

3. PHYSIOLOGICAL ROLE OF RAB GDIα

Rab GDI is a general regulator of all the Rab family small G proteins that are

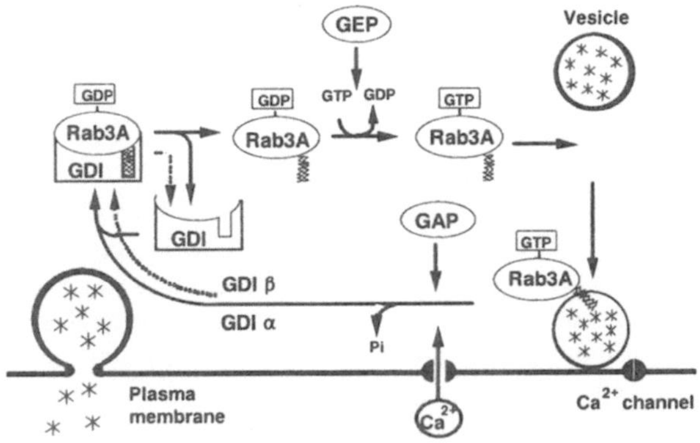

Fig. 2. A model for the mode of action of Rab3A in neurotransmitter release

implicated in intracellular vesicle trafficking (Takai et al., 1996; 2001). Rab GDI have at least three isoforms: Rab GDIα, -β, and -γ. Rab GDIα is specifically expressed in neuronal tissues. To clarify the neuron-specific role of Rab GDIα, we have generated Rab GDIα knockout mice. These mice are viable and fertile, and show no morphological abnormalities in brain. In the knockout mice, however, synaptic potentials in the CA1 region of the hippocampus display larger enhancement during repetitive stimulation. This phenotype is apparently opposite to that of Rab3A knockout mice. Furthermore, Rab GDIα knockout mice show hypersensitivity to bicuculine, an inducer of epileptic seizures. Consistently, mutations of the Rab GDIα gene cause human X-linked mental retardation associated with epileptic seizures (D'Adamo et al., 1998). Because Rab GDIβ as well as Rab GDIα is expressed in brain, Rab GDIβ apparently fulfills essential GDI functions in Rab GDIα knockout mice (**Fig. 2**). Such basic functions are required for survival because yeasts expressing only one isoform of Rab GDI are inviable when the gene is deleted (Takai et al., 1996; 2001). Thus, Rab GDIα is not essential for basal neurotransmitter release but plays a specialized role in the Rab3A recycling to suppress hyperexcitability through modulation of presynaptic forms of plasticity. In addition, our studies on Rab GDIα knockout mice have provided an additional line of evidence that Rab3A up-regulates the steps of the translocation and docking in neurotransmitter release.

4. PHYSIOLOGICAL ROLE OF RAB3 GEP

To determine the physiological role of Rab3 GEP in neurotransmitter release, we have generated Rab3 GEP knockout mice. The knockout embryos develop normally but die immediately after birth. The embryos at embryonic day 18.5 show no evoked action potentials of the diaphragm and gastrocunemius muscles in response to electrical stimulation of the phrenic and sciatic nerves, respectively. In contrast, axonal conduction of the spinal cord and the phrenic nerve is not impaired. Total number of synaptic vesicles, especially those docked at the presynaptic plasma membrane, are reduced at neuromuscular junctions approximately 10-fold compared to controls, while postsynaptic structures and functions appear normal. Thus, loss of Rab3 GEP induces more severe phenotypes than does loss of Rab3A (Geppart et al., 1994) or Rab GDIα, and Rab3 GEP is essential for neurotransmitter release and probably for formation and trafficking of synaptic vesicles at least in the peripheral nervous system. Consistently, mutations of Aex-3, a Rab3 GEP homolog of C. elegans, shows reduced synaptic transmission (Iwasaki et al., 1997). However, the results obtained with Rab3 GEP knockout mice do not provide any evidence for the mode of action of Rab3A in neurotransmitter release and furthermore suggest that a novel role of Rab3 GEP which is mediated not only through Rab3A but also through other substrate small G proteins, such as Rab3C and –3D, and an unidentified Rab3 GEP-interacting protein(s) involved in the formation of synaptic vesicles.

5. DISCUSSION

Our studies on Rab GDIα knockout mice have provided the evidence for the specialized role of Rab GDIα in the Rab3A recycling. However, our studies on Rab3 GEP knockout mice have not provided any evidence for the mode of action of Rab3A. The role of Rab3 GEP in neurotransmitter release in the central nervous system also remains to be clarified, because it is technically difficult to keep Rab3 GEP knockout mice alive up to 3 weeks while central synapses are formed and matured. Cultured hippocampal neurons derived from Rab3 GEP knockout mice will be useful not only to obtain another line of evidence for the mode of action of Rab3A but also to clarify the role and mode of action of Rab3 GEP in the central nervous system. Such studies may clarify that in addition to the Rab3 subfamily-dependent mechanism, Rab3 GEP regulates neurotransmitter release by a novel mechanism. We are now generating Rab3 GAP knockout mice. Studies on Rab3 GAP knockout mice will also lead us to a better understanding of the molecular mechanisms of neurotransmitter release and synaptic plasticity.

REFERENCES

Castillo, P.E., Janz, R., Südhof, T.C., Tzounopoulos, T., Malenka, R.C., and Nicoll, R.A, 1997, Rab3A is essential for mossy fiber long-term potentiation in the hippocampus, Nature **388**:590-593.

D'Adamo, P., Menegon, A., Lo Nigro, C., Grasso, M., Gulisano, M., Tamanini, F., Bienvenu, T., Gedeon, A.K., Oostra, B., Wu, S.K., Tandon, A., Valtorta, F., Balch, W.E., Chelly, J., Toniolo, D., 1998, Mutations in GDI1 are responsible for X-linked non-specific mental retardation. Nature Genet. **19**:134-139.

Geppert, M., Bolshakov, V.Y., Siegelbaum, S.A., Takei, K., De Camilli, P., Hammer, R.E., and Südhof, T.C., 1994, The role of Rab3A in neurotransmitter release, Nature **369**:493-497.

Geppert, M., Goda, Y., Stevens, C.F., and Südhof, T.C., 1997, The small GTP-binding protein Rab3A regulates a late step in synaptic vesicle fusion, Nature **387**:810-814.

Iwasaki, K., Staunton, J., Saifee, O., Nonet, M., and Thomas, J.H., 1997, *aex-3* encodes a novel regulator of presynaptic activity in C. elegans. Neuron **18**:613-622.

Takai, Y., Sasaki, T., Shirataki, H., and Nakanishi, H., 1996, Rab3A small GTP-binding protein in Ca^{2+}-dependent exocytosis, Genes Cells **1**: 615-632.

Takai, Y., Sasaki, T., and Matozaki, 2001, Small GTP-binding proteins, Physiol. Rev. **81**: 153-208.

ASSESSING SYMPATHETIC FUNCTION FROM DYNAMICS OF NOREPINEPHRINE METABOLISM AND FALSE TRANSMITTERS

Irwin J. Kopin*

1. INTRODUCTION

About 40 years ago, tritiated norepinephrine (^{3}H-NE) and epinephrine (^{3}H-EPI) were introduced as a means for examining the fate of catecholamines in experimental animals and in humans. By examining the urinary products of ^{3}H-EPI and ^{14}C-metanephrine, the O-methylated product of EPI, it was shown that O-methylation of the administered catecholamine was the major means of metabolic termination of its action.[1] When ^{3}H-NE was administered, a major portion of the administered radioactivity appeared to be sequestered in the tissues, such as the heart, that were richly innervated by sympathetic nerves.[2] After chronic denervation, however, this uptake of ^{3}H-NE was abolished, indicating that intravenously administered catecholamines are taken up into sympathetic nerve terminals.[3] The supersensitivity to NE released from sympathetic nerves that developed after such denervation indicated that reuptake was the major means of terminating the action of released NE.

On the basis of pharmacological evidence, Trendelenburg[4] was the first to suggest that the stores of NE in the sympathetic nerve terminal were not uniformly available for release. The presence of multiple pools of tissue NE was confirmed using isotopic studies that showed the multiphasic decline in the specific activity (SA) of stored ^{3}H-NE, *in vitro* as well as *in vivo*. Two days after administration of ^{3}H-NE endogenous NE was found to be preferentially released by tyramine[5]. Finally, it was demonstrated that there was preferential release of newly synthesized ^{14}C-NE from isotopically labeled tyrosine[6]. Combined with fluorometric estimation of tissue levels of NE, estimates of NE turnover of the catecholamine became possible, but these were subject to underestimation because only the slower phase was used in the calculations.

The pattern of urinary metabolites excreted shortly after intravenous administration of ^{3}H-NE differs from that found after long intervals. There are similar differences in the urinary metabolites of NE depleted by reserpine and NE released in an active form by

*National Institute of Neurological Disorders and Stroke, National Institutes of Health, Bethesda, MD 20892.

tyramine. These differences indicated that MAO (type A) within the nerve deaminates NE, whereas NE metabolized outside the nerve is primarily O-methylated (by COMT). Thus, there emerged a concept of the sympathetic nerve terminal releasing NE by exocytosis from the vesicles. The released NE acts on receptors to elicit responses and is inactivated mainly by recapture via the neuronal transporter and largely sequestered in the synaptic vesicles for reuse. A small portion of the recaptured NE is destroyed by intraneuronal MAO, to form an aldehyde that is mainly reduced, resulting in the production of 3,4–dihydroxyphenylglycol, the major metabolite of NE sympathetic neurons.[7] Most of DHPG is derived from NE that leaks from the vesicular storage sites into the cytoplasm. When uptake of NE is blocked by desipramine, DHPG formation from leaked NE remains constant regardless of exogenously elevated plasma NE levels.

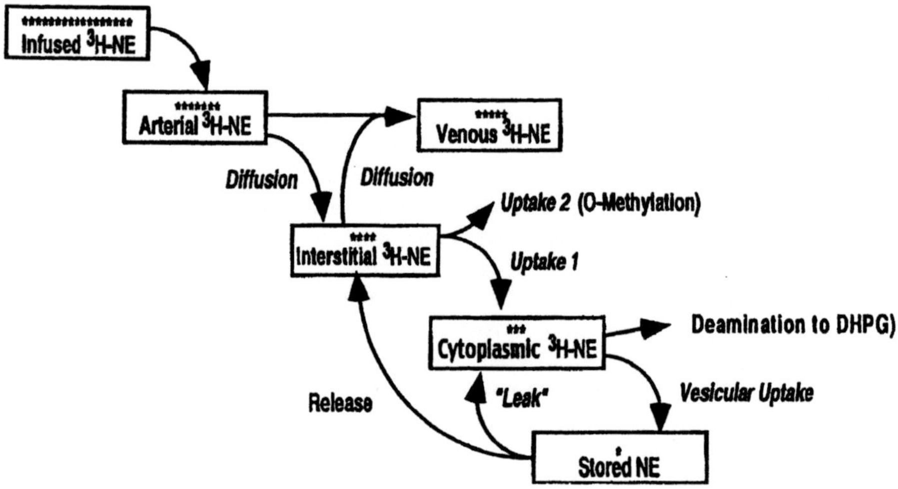

Figure 1. Sequential isotopic dilution of infused ^{3}H-NE in the various sites of its compartmentalization. Specific activities (represented by asterisks) of the NE decrease as a greater proportion of the catecholamine is derived from unlabeled endogenous NE. Infused>arterial>venous>interstitial> cytoplasmic>stored NE.

2. ISOTOPE DILUTION STUDIES OF NE METABOLISM

When ^{3}H-NE is intravenously infused, the isotopically labeled catecholamine is diluted by the endogenous NE that is being released into the plasma. The rate of spillover of endogenous NE into plasma can be determined by dividing the rate of infusion of tritiated ^{3}H-NE (dpm/min), which contributes negligible amounts of unlabeled NE, by the specific activity of arterial NE (dpm/pmole) to obtain pmole/min entering the plasma. In each tissue, the arterial ^{3}H-NE enters the capillaries from which a portion diffuses into the interstitial tissue. The endogenous NE released into the interstitial fluid further dilutes the ^{3}H-NE to reduce its specific activity. This low-specific activity NE enters the capillary where it mixes with the arterial ^{3}H-NE which has not entered the interstitial fluid. Thus, the ^{3}H-NE that leaves the capillary via the venous outflow has a specific activity lower than arterial and higher than interstitial ^{3}H-NE.

Regional spillover of endogenous NE into the venous outflow is calculated from the blood flow, F, the arterial concentration of NE, NE_a, the extraction fraction of arterial ^3H-NE and the specific activities of arterial and venous NE, SA_a and SA_v:

$$SO = (1-E)\cdot F \cdot NE_a \cdot \left(\frac{SA_a}{SA_v} - 1 \right)$$

Spillover of NE into the circulation represents only a fraction of the released catecholamine because a major portion of the NE is actively transported back into the neuron. There are several means of estimating the relative amounts of released NE that are recaptured or spill over into the circulation. One method is by use of the DHPG that is formed from recaptured NE, which can be determined by the difference in plasma levels of DHPG with and without desipramine-induced inhibition of NE uptake. After desipramine, there was a 36% decrease in plasma DHPG levels, indicating the fraction of plasma DHPG derived from NE taken up into the sympathetic nerves. The decrease in ^3H-DHPG spillover, however was only about 10% of the decrease in neuronal uptake of ^3H-NE . Thus it can be concluded that only about 10% of NE released from the sympathetic nerve terminals spills over into the circulation and about 90% is recaptured.

Another means of estimating the relative proportion of released NE that is recaptured is based on the same principle as measurement of NE spillover. The equation that is used to determine the rate of NE reuptake is similar to that used to estimate spillover. The blood flow, F, the arterial concentration of NE, NE_a, the extraction fraction of arterial ^3H-NE and the specific activities of arterial and interstitial fluid NE, SA_a and SA_I:

$$U_r = E \cdot F \cdot NE_a \left(\frac{SA_a}{SA_i} - 1 \right)$$

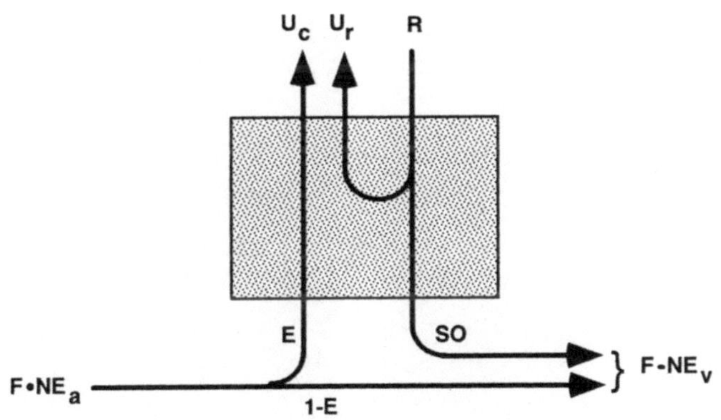

$$R = SO + U_r$$

Figure 2. Diagrammatic representation of NE release (R), Spillover (SO) and reuptake (U_r) . F represents blood flow, NE_a and NE_v the arterial and venous concentrations of NE, and E the extraction fraction of ^3H-NE.

As represented in figure 2, the sum of the rates of endogenous NE reuptake (U_r) and spillover (SO) provides an estimate of the rate of NE release from the sympathetic nerve terminals into the interstitial fluid. There are several means for estimating the SA of interstitial NE. The SA of NE in microdialysates or in lymph flowing out of an organ can be determined directly. Since normetanephrine (NM) is derived from NE in the interstitial fluid, the SA of NM that spills over into the venous effluent may be assumed to be similar to that of the NE from which it was formed. We used the latter method to estimate the SA of interstitial NE[9] and from this determined the rates of NE reuptake and SO and release of NE into the interstitial fluid of the kidneys, skeletal muscle and heart. In the kidneys, NE release averaged three times NE spillover, in skeletal muscle, 12 times NE spillover, and in the heart, >20 times NE spillover.[9]

3. IMAGING FALSE SYMPATHETIC NEUROTRANSMITTERS

Many compounds structurally related to NE can replace the catecholamine in its storage sites and be released in response to nerve stimulation, although they are relatively inactive at the receptors. Such "false transmitters" could reduce the efficacy of sympathetic nervous activity.[10] Use of false transmitters labeled with positron-emitting isotopes (^{11}C or ^{18}F) that are stored and metabolized similarly to NE allow imaging by positron emission tomography (PET). ^{18}F-6-fluorodopamine is such a compound.[11] It has the advantage of being converted to ^{18}F-NE and, like NE, metabolized by both MAO and COMT. Imaging of ^{18}F-fluorodopamine-derived radioactivity over time provides information about the rates of NE turnover as well as the integrity of sympathetic nerve terminals in more densely innervated tissues.

One of the first clinical applications of PET imaging of the sympathetic nerves was in a study of patients with orthostatic hypotension. About 25 years ago, Ziegler et al,[12]

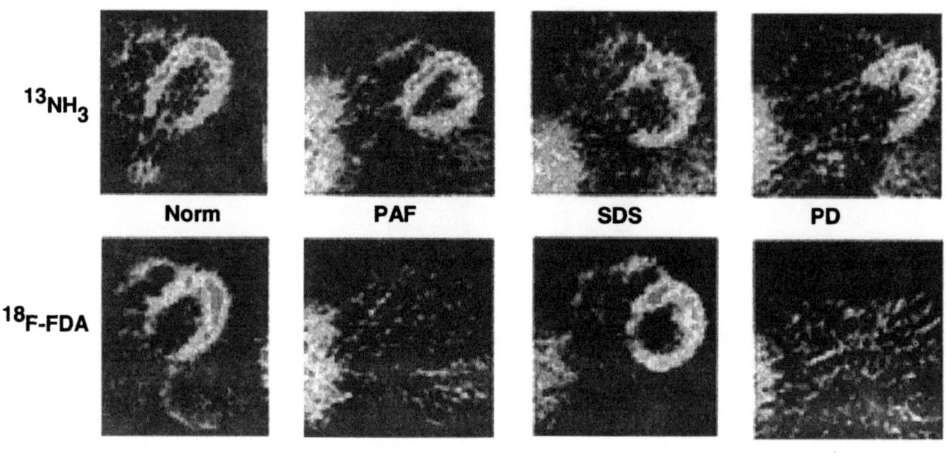

Figure 3. PET images of the heart of a normal subject compared with that of patients with PAF, SDS or Parkinson's disease with autonomic failure (PD)

demonstrated a difference between the plasma catecholamine levels in patients with orthostatic hypotension who did not have any clinical indication of involvement of the central nervous system (Primary Autonomic Failure, PAF) and those that presented with evidence of cerebellar, basal ganglia, or cortical dysfunction (Shy-Drager Syndrome, SDS). Plasma NE levels were normal in SDS patients, whereas NE levels were markedly lower in PAF patients. Neither SDS nor PAF had the normal increments in plasma NE with standing or during exercise. It was concluded that in SDS the peripheral sympathetic nerves were intact, but could not be appropriately activated because of a deficit in the central nervous system regulation of these nerves. In PAF, the deficit was degeneration of the peripheral sympathetic nerves; the central nervous system was intact, but the sympathetic effector system was absent. PET studies with ^{18}F-6-fluorodopamine confirmed this conclusion (figure 2). Perfusion of the heart was imaged using $^{13}NH_3$ and revealed that the delivery of the isotopic tracer was normal in all patients. ^{18}F-6-fluorodopamine-derived radioactivity was absent in the myocardium of patients with PAF, whereas SDS patients appeared to have normal levels of cardiac ^{18}F-6-fluorodopamine.[12] Patients with autonomic failure who had parkinsonian symptoms were generally regarded as having the SDS or hypotension as a result of dopamine formed

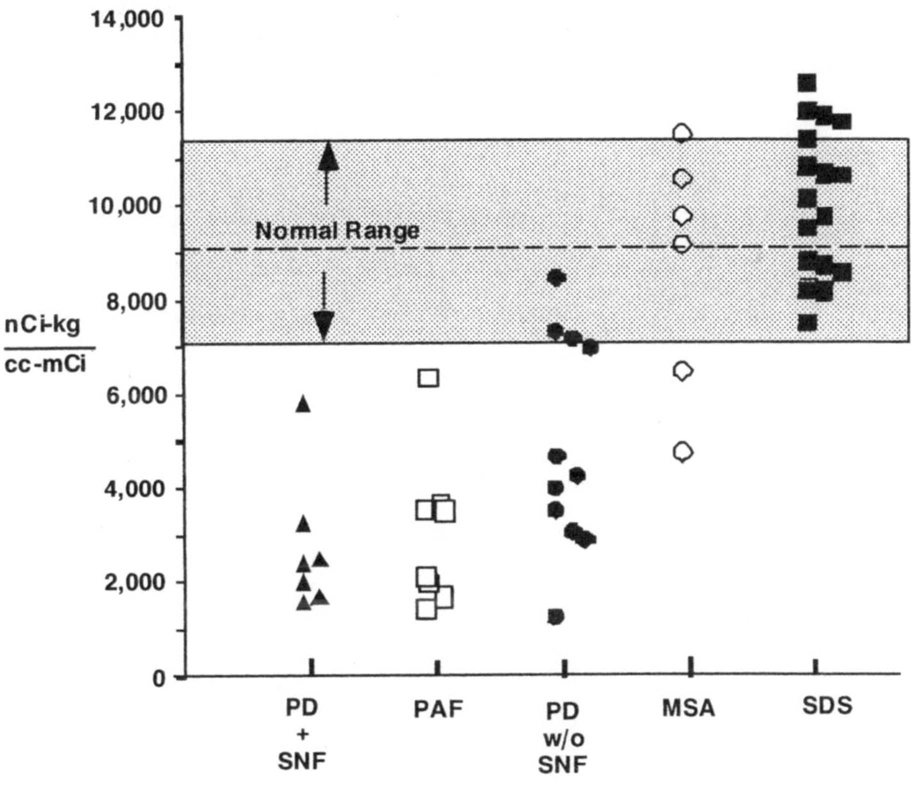

Figure 4. Myocardial levels of ^{18}F-6-fluorodopamine-derived radioactivity in normal subjects (shaded area) and in patients with PAF, Parkinson's disease with or without autonomic failure, SDS, or MSA (see text).

from the DOPA with which they were being treated. However, PET scanning of these patients with ^{18}F-6-fluorodopamine demonstrated that they had sympathetically denervated hearts with as markedly diminished ^{18}F-6-fluorodopamine-derived radioactivity as patients with PAF (figure 4).[12] When patients with Parkinson's disease without orthostatic hypotension were examined, they were also found to have at least partial cardiac sympathetic denervation (figure 4).

Cardiac sympathetic denervation in parkinsonian patients appears to be progressive (Goldstein et al., unpublished observation). The occurrence of sympathetic cardiac denervation in one patient with the α-synuclein mutation that was the basis for his Parkinson's disease also showed cardiac sympathetic denervation. This indicates that the degenerative process in Parkinson's disease resulting from different causes involves peripheral ssympathetic as well as central dopaminergic neurons.

4. CONCLUSION

The development of methods to measure precisely the levels of endogenous as well as isotopically labelled catecholamines and their metabolites and the use of PET imaging to visualize sympatheic nerves have provided means for assessing sympathetic nerve integrity and function. This has led to discovery of involvement of sympathetic nerves in the degenerative process of Parkinson's disease and other disorders.

REFERENCES

1. Kopin, I.J., Axelrod, J. and Gordon, E.: The metabolic fate of H^3-epinephrine and C^{14}-metanephrine in the rat. J. Biol Chem **236**:3072-3075,1961.
2. Whitby, L.G., Axelrod J., and Weil-Malherbe, H.: The fate of ^3H-norepinephrine in animals J. Pharmacology **132**: 193-201,1961
3. Hertting, G., Axelrod, J., Kopin, I.J., and Whitby, L.G.: Lack of uptake of catecholamines after chronic denervation of sympathetic nerves. Nature **189**: 66,1961
4. Trendelenburg U. Modification of the effect of tyramne by various agents and procedures. J. Pharmacol. Exp. Ther. **134**: 8-17,1961.
5. Potter L., Axelrod J. and Kopin I.J. Differential binding and release of norepinephrine and tachyphylaxis. Biochem Pharm **11**:254-256,1962
6. Kopin, I.J., Breese, G.R., Krauss, K.R. and Weise, V.K.: Selective release of newly synthesized norepinephrine from the cat spleen during sympathetic nerve stimulation. J Pharmacol Exp Ther **161**: 271-278, 1968
7. Kopin, I.J. Catecholamine Metabolism: Basic and Clinical Significance. Pharmacological Reviews **37**:333-364,1985.
8. Esler M, Jennings G, Lambert G, Meredith I, Horne M, Eisenhofer G. Overflow of catecholamine neurotransmitters to the circulation: source, fate, and functions. Physiol Rev. **70**:963-85,1990.
9. Kopin IJ, Rundqvist B, Friberg P, Lenders J, Goldstein DS, Eisenhofer G. Different relationships of spillover to release of norepinephrine in human heart, kidneys, and forearm. Am J Physiol. 275:R165-73, 1998.
10. Kopin, I.J.: False adrenergic transmitters. Annu Rev Pharmacol **8**: 377-394, 1968
11. Eisenhofer G, Hovevey-Sion D, Kopin IJ, Miletich R, Kirk KL, Finn R, Goldstein DS. Neuronal uptake and metabolism of 2- and 6-fluorodopamine: false neurotransmitters for positron emission tomographic imaging of sympathetically innervated tissues. J Pharmacol Exp Ther. 1989 Jan;**248**:419-427.
12. Ziegler, M.G., Lake, C.R. and Kopin, I.J.: The sympathetic nervous system defect in primary orthostatic hypotension. N Engl J Med **296**: 293-297, 1977
13. Goldstein DS, Holmes C, Cannon RO 3rd, Eisenhofer G, Kopin IJ. Sympathetic cardioneuropathy in dysautonomias. Adv Pharmacol. **42**:615-9.1998

ADENYLYL CYCLASE: A MOLECULE OF DIVERSITY

Yoshihiro Ishikawa*

1. INTRODUCTION

It is now well known that adenylyl cyclase is made of multiple subtypes that are divergent in tissue distribution and biochemical characteristics. So far nine subtypes have been cloned and characterized (Hanoune, et al., 2001, Ishikawa, 1997). These subtypes share the common property, i.e., the conversion of ATP to cAMP upon stimulation by Gsα. Other properties such as regulation by Gβγ or calcineurin, however, are different among the subtypes. Further, the tissue distribution of these subtypes is also different. Types I and VIII are localized to neuronal tissues and type V is found in the corpus striatum and the heart. Type IV is found elsewhere. It is thus speculated that each tissue or cell type expresses an unique combination of multiple adenylyl cyclase subtypes. In this article, I will summarize the divergent regulation of each adenylyl cyclase subtypes from various aspects including that by newly developed pharmacological reagents.

2. WHAT IS COMMON AMONG THE SUBTYPES?

Besides the production of cAMP, the transmembrane topology of adenylyl cyclase is shared by all the subtypes (Greenberger, et al., 1994). This structure is reminiscent of that of ion channels or transporters; it has a tandem repetition of a six-transmembrane domain followed by a large cytoplasmic catalytic domain. Interestingly, the catalytic domain within the cytoplasmic domain has relatively high sequence homology, which also shows homology to the C-terminal, catalytic domain of mammalian guanylyl cyclases. A dendrogram based upon the homology score map has demonstrated that the adenylyl cyclase types IX and III most diverged from the others, thereafter types I/VIII, II/IV/VII and V/VI diverged forming subgroups within the family.

* Department of Physiology, Yokohama City University School of Medicine, Yokohama 236-0004, Japan

3. MECHANISMS OF CATALYTIC ACTIVATION

Soluble forms of guanylyl cyclase have two distinct subunits, GC-A and GC-B (Nakane, et al., 1990). Although each subunit contains an apparent catalytic domain, they do not function independently; heterodimer formation is necessary. This finding is consistent with observations on the soluble forms of adenylyl cyclase that were artificially generated by overexpressing only the cytoplasmic domain. Such subunits, constituting either the first or the second half of the molecule, must be expressed together to generate catalytic activity (Patel, et al., 2001). This is in agreement with recent crystalographic studies that there is an ATP and a Gsα binding site within a pocket that must be created by the association of the two catalytic domains within adenylyl cyclase (Hurley, 1999).

4. WHAT IS DIFFERENT AMONG THE SUBTYPES?

Although the structure of these subtypes is relatively well conserved, their biochemical properties are quite different. Indeed, no two isoforms are regulated in the same manner, even within the same subgroup. There are several intracellular molecules that have been known to regulate the catalytic activity of adenylyl cyclase. Potential regulators vary from simple cation (Ca, Mn, and Mg) concentration to various kinases (PKA and PKC) or calmodulin and, of course, G protein subunits (Giα and Gβγ). The pattern of regulation also differs, even for the same isoform, when expressed in different cell types or between splice variants. Inhibition of activity by Giα is not universal among the subtypes; variable degrees of inhibition have been demonstrated. It is even possible that the same receptor stimulation may lead to increased or decreased cAMP production depending upon the subtype of adenylyl cyclase dominantly expressed in the cell. Alpha2-adrenergic stimulation, for example, leads to the activation of Giα protein and thus inhibition of adenylyl cyclase. Such inhibition may occur when the cell expresses Gi-sensitive subtype of adenylyl cyclase, such as type VI. If the cell expresses a subtype that is Gi-insensitive but sensitive to Gβγ stimulation, such as type II, alpha2-adrenergic stimulation may lead to increased cAMP production.

A simple question relates to why there is such a wide diversity among the adenylyl cyclase subtypes. Apparently, this diversity seems to be evolutionally conserved. We and others have demonstrated that all the adenylyl cyclase subtypes are located on different chromosomes. It was rather surprising when type V and type VI, the most well-conserved subgroup family members within the adenylyl cyclase family are located different chromosomes, i.e., 3q13.2-21 and 12q12-13, respectively. These findings indicate that the evolutionary development of different adenylyl cyclase subtypes makes a difference in chromosomal organization from other components within the same signaling pathway, such as catecholamine receptors and G proteins (Magovcevic, et al., 1992).

5. SUBCELLULAR LOCALIZATION OF ADENYLYL CYCLASE

Catecholamine release from the synaptic terminal leads to the sequential activation of the receptor, G protein, and adenylyl cyclase at the plasma membrane, and the production of a diffusible second messenger cAMP in the cytoplasm. cAMP activates cAMP-dependent kinase or protein kinase A and initiates a reaction cascade of phosphorylation of multiple proteins within the cells. Importantly, these changes occur on a second-to-second basis *in vivo*. Despite the rapid physiological regulation *in vivo*, it has been proposed that the β-adrenergic receptor/G-protein/effector interaction within the cell membrane is based upon "the random collision mechanism," in which signaling proteins are randomly drifting within the lipid membrane and contact with the specificity at the sites of protein-protein interaction. In this mechanism, however, a rapid signaling cascade involving multiple proteins may not be efficiently processed.

Caveolae, flask-shaped plasma membrane invaginations, are subcellular compartments of the plasma membrane that exist in most cell types (Schlegel, et al., 2001). Caveolin is a principal component of caveola *in vivo*. As recent evidence suggests that caveolae participate in G protein-coupled signalling, we also demonstrated the interaction of adenylyl cyclase with caveolin (Toya, et al., 1998). The type III, type IV, and type V adenylyl cyclase subtypes were localized in caveolin-enriched membrane fractions, suggesting that they co-exist with caveolin in caveolae. Similarly, beta-adrenergic receptors and Gsα were also found in the same fraction (Schwencke, et al., 1999). Thus, caveolae may act as a plasma membrane-bound site to compartmentalize these molecules and that the interaction of adenylyl cyclase with G proteins may be more highly organized *in vivo* than previously appreciated. Importantly, the above mentioned method has enabled the detection of adenylyl cyclase protein subtypes by immunoblotting (Ishikawa, et al., 2000).

6. OTHER REGULATORS OF ADENYLYL CYCLASE

Besides classic regulators of adenylyl cyclase described above, multiple regulators have been recently identified. Caveolin interacts with adenylyl cyclase in an adenylyl cyclase subtype- and a caveolin-subtype dependent manner. A scaffolding domain peptide derived from caveolin-1 inhibited the catalytic activity of type V and type III while the same caveolin had little effect on type II. A similar peptide derived from the corresponding domain of caveolin-3 showed similar inhibition of type III and type V while that derived from caveolin-2 did not show any inhibition (Toya, et al., 1998). Similar subtype-specific regulation of adenylyl cyclase has been demonstrated by n-alkanols (Ebina, et al., 1997), oxidized catecholamine (Ebina, et al., 1997), or proteolysis (Ebina, et al., 1997) or membrane solubilization (Ebina, et al., 1997).

7. SUBTYPE TARGETED REGULATION OF ADENYLYL CYCLASE

It is known that forskolin, a natural diterpene compound, interacts directly with adenylyl cyclase. We have developed forskolin derivatives with enhanced adenylyl cyclase subtype-selectivity (Toya, et al., 1998). 6-Dimethylaminopropionyl-forskolin, a water-soluble forskolin derivative, stimulated type V more potently (~2-fold) than type II (1-fold) and type III (0.8-fold) relative to forskolin. Similarly, 6-dimethylaminopropionyl-14,15-dihydro-forskolin stimulated type V even more potently

(1.4-fold) than types II (0.6-fold) and type III (0.3-fold). This selectivity was maintained under different assay conditions (membrane preparations versus intracellular cAMP accumulation assays). Accordingly, cardiac adenylyl cyclase stimulation by such derivatives was greater than lung or brain adenylyl cyclase stimulation.

More recently, we have developed forskolin derivatives that selectively stimulate other subtypes as well, such as type II and type III. Further, we have found a P-site inhibitor and a non-nucleic acid compound that inhibit the type V adenylyl cyclase in a subtype-specific manner. These findings suggest that it is now feasible to develop pharmacological compounds that regulate adenylyl cyclase in a subtype-specific manner and therefore a tissue-specific manner.

REFERENCES

1. J. Hanoune and N. Defer, Regulation and role of adenylyl cyclase isoforms, *Annu. Rev. Pharmacol. Toxicol.* **41**, 145-174 (2001).
2. L. Greenberger and Y. Ishikawa, ATP-binding cassette proteins: common denominators between ion channels, transporters and enzymes., *TCM* **4**, 193-198 (1994).
3. M. Nakane, K. Arai, S. Saheki, T. Kuno, W. Buechler and F. Murad, Molecular cloning of cDNA coding for 82 kD and 70 kD subunits of soluble guanylate cyclase., *J. Biol. Chem.* **265**, 16841-16845 (1990).
4. T. Patel, Z. Du, S. Pierre, L. Cartin and K. Scholich, Molecular biological approaches to unravel adenylylcyclase signaling and function, *Gene* **16**, 13-25 (2001).
5. J. Hurley, Structure, mechanism, and regulation of mammalian adenylyl cyclase, *J. Biol. Chem.* **274**, 7599-7602 (1999).
6. I. Magovcevic, S. L. Ang, J. G. Seidman, C. J. Tolman, E. J. Neer and C. C. Morton, Regional localization of the human G protein alpha i2 (GNAI2) gene: assignment to 3p21 and a related sequence (GNAI2L) to 12p12-p13, *Genomics* **12**, 125-9 (1992).
7. A. Schlegel and M. P. Lisanti, Caveolae and their coat proteins, the caveolins: from electron microscopic novelty to biological launching pad, *J Cell Physiol* **186**, 329-37. (2001).
8. Y. Toya, C. Schwencke, J. Couet, M. P. Lisanti and Y. Ishikawa, Inhibition of adenylyl cyclase by caveolin peptides, *Endocrinology* **139**, 2025-31. (1998).
9. C. Schwencke, M. Yamamoto, S. Okumura, Y. Toya, S. J. Kim and Y. Ishikawa, Compartmentation of cyclic adenosine 3',5'-monophosphate signaling in caveolae, *Mol Endocrinol* **13**, 1061-70. (1999).
10. Y. Ishikawa, B. S. Grant, S. Okumura, C. Schwencke and M. Yamamoto, Immunodetection of adenylyl cyclase protein in tissues, *Mol Cell Endocrinol* **162**, 107-12. (2000).
11. T. Ebina, Y. Toya, J. Kawabe and Y. Ishikawa, Regulation of adenylyl cyclase isoforms by N-alkanols, *J Cell Biochem* **66**, 450-6. (1997).
12. T. Ebina, Y. Toya, N. Oka, C. Schwencke, J. Kawabe and Y. Ishikawa, Isoform-specific regulation of adenylyl cyclase by oxidized catecholamines, *J Mol Cell Cardiol* **29**, 1247-54. (1997).
13. T. Ebina, Y. Toya, N. Oka, J. Kawabe, C. Schwencke and Y. Ishikawa, Isoform-dependent activation of adenylyl cyclase by proteolysis, *FEBS Lett* **401**, 223-6. (1997).
14. T. Ebina, J. Kawabe, T. Katada, S. Ohno, C. J. Homcy and Y. Ishikawa, Conformation-dependent activation of type II adenylyl cyclase by protein kinase C, *J Cell Biochem* **64**, 492-8. (1997).
15. Y. Toya, C. Schwencke and Y. Ishikawa, Forskolin derivatives with increased selectivity for cardiac adenylyl cyclase, *J Mol Cell Cardiol* **30**, 97-108. (1998).

L-DOPA UPTAKE AND DECARBOXYLATION IN THE BRAIN

Kunio Kitahama[1*], Ryohachi Arai[2], Michel Geffard[3], Nobuyuki Karasawa[4], and Ikuko Nagatsu[4]

1. INTRODUCTION

L-dihydroxyphenylalanine (L-DOPA) therapy has been well accepted in order to improve extrapyramidal motor disturbance in Parkinson disease, and the uptake and metabolism of loaded DOPA as a precursor of dopamine (DA) in the brain has been studied, using autoradiography and high-resolution positron emission tomography (PET). However, no detailed anatomical information has been provided. It is known that L-DOPA is transformed by a second step catecholamine-synthesizing enzyme, aromatic L-amino acid decarboxylase (AADC). In the present study, we examined the localization of this enzyme, and elucidated in what neuronal structures L-DOPA is converted to DA, using immunohistochemistry in conjunction with the antisera, anti-DA as well as anti-AADC.

2. MATERIALS AND METHODS

Ten cats weighing 3-5 kg each were used. Animals were treated with L-DOPA (100 mg/kg) or saline, with or without an inhibitor of monoamine oxidase (MAO) (Pargyline, 10 mg/kg) 2 hours before sacrifice. Under deep anesthesia, each animal was perfused through the ascending aorta with 1 liter of 0.01M phosphate-buffered saline (PBS) followed by 1 liter of fixative containing 2% glutaraldehyde, 1% sodium metabisulfite, and 0.25% picric acid in 0.1 M PB. The brains were removed and cut into several blocks at a thickness of 5 mm. Tissue blocks were postfixed in the same fixative for 8 hours, followed by a rinse for 3 days in PBS containing 15% sucrose and 1% sodium metabisulfite. Sections were cut using a cryostat (25μm) in coronal planes.

The sections of the cat brain were incubated in antibodies against DA or against AADC diluted 1: 30,000 in PBS containing 0.3% Triton X-100 at 4°C for 1 week. Then,

---

[1*] Kunio Kitahama, INSERM U480, Neurobiologie des états de sommeils et d'éveil, Faculté de Médecine, 8, Avenue Rockefeller, 69373 LYON CEDEX 08 France, [2]Department of Anatomy, Shiga University of Medical Science, Japan, [3]INSERM, Talence, France, [4]Department of Anatomy, Fujita Health University, Japan.

they were incubated in biotinylated rabbit IgG (Vector Laboratory, 1:1,000) for 12 hours at 4°C, and finally avidin-biotin-peroxidase complex (Vector Laboratory, 1:1,000) for 1 hour at room temperature. Peroxidase activity was then revealed in 50 mM Tris-HCl buffer (pH 7.6) containing 0.0003% H_2O_2, 0.01% 3,3'-diaminobenzidine-4HCl (DAB) and 1% nickel ammonium sulfate. For DA-immunochemistry, a monoclonal antibody against DA was used. Details of the production, characterization, and specificity of the AADC and DA antisera have been described elsewhere (Kitahama et al. 1988; Chagnaud et al., 1987). AADC-immunoreactive (ir) structures in the postmortem human brain were also examined with an immunohistochemical method using AADC antisera.

3. RESULTS

AADC-immunoreaction could be seen in many discrete regions of the human and feline brains. In the human brain (Fig.1), the staining is intense in serotonin (5-HT) cells, moderate in norepinephrine (NE) and DA ones, and weak in nonaminergic ones such as those in the preoptic area. But, as has been already reported (Kitahama et al., 1988), such difference in stainability is not distinct in the feline brain (Fig.2). A large number of intensely stained AADC-cell bodies are present in the feline midbrain and hypothalamus.

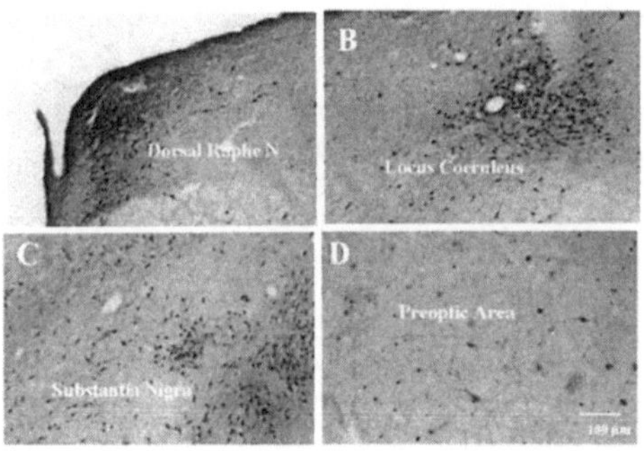

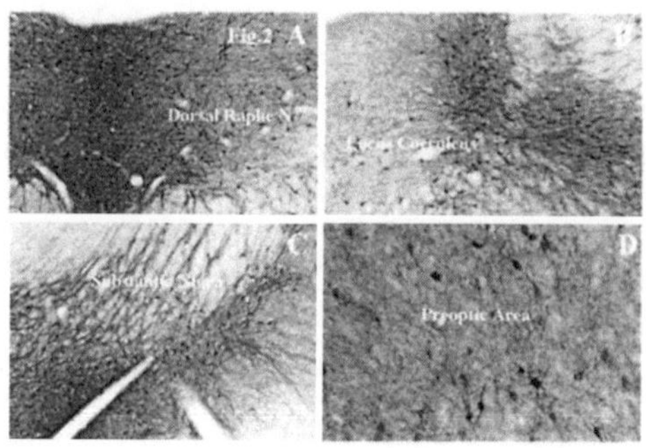

Figures 1, 2: AADC-ir structures in the 5-HT (A), NE (B), DA (C) and nonaminergic (D) structures in human (Fig.1) and cat (Fig.2) brains.

In saline-treated cats, DA-immunostaining was intense enough in DA cell groups, as well as in the nigrostriatal pathway and striatum. It is, however, very weak in NE cells and fibers, and completely absent in 5-HT ones even after MAO inhibitor (IMAO) treatment. After L-DOPA administration, DA is densely accumulated in 5-HT cells and fibers (Figs.3,4) and NE ones. DA is also synthesized in nonaminergic cells and fibers, i.e., in the area postrema, ventrolateral posterior hypothalamus, pontine and suprachiasmatic nuclei (Fig.5). For many other small AADC-ir cells which did not show DA-immunoreactivity by L-DOPA administration, such as those in the habenular and central tegmental ones, it could be visualized in conjunction with IMAO administration (Fig.6). Finally, in the thalamic AADC-ir cell region, no or a few cells displayed DA-immunoreactivity even after L-DOPA and IMAO treatments.

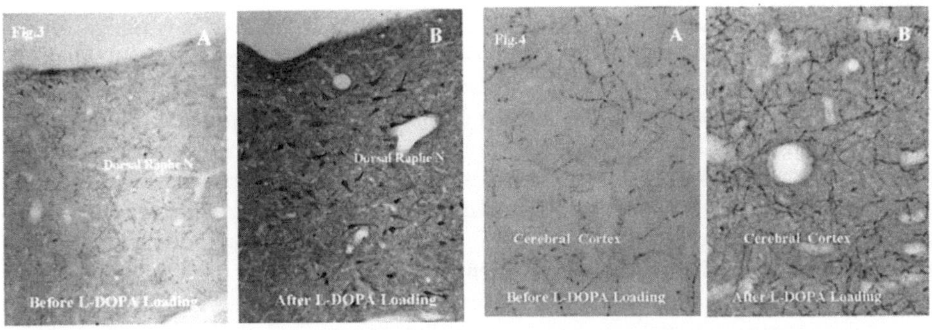

Figures 3, 4: DA synthesis in 5-HT structures containing AADC after L-DOPA loading in the dorsal raphe nucleus and cerebral cortex.

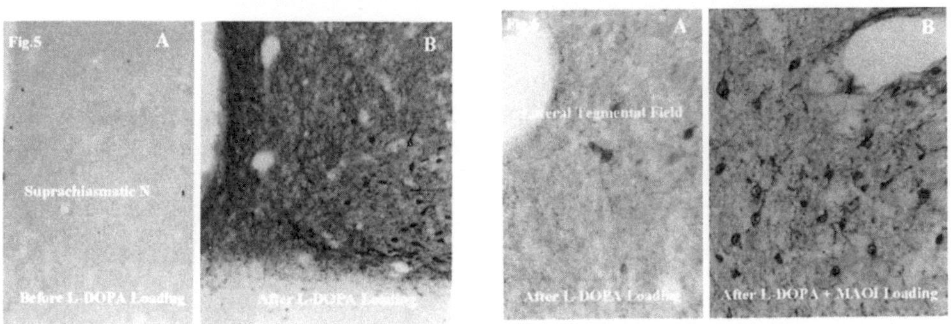

Figures 5, 6: DA synthesis in nonaminergic structures containing AADC after L-DOPA loading with or without IMAO.

4. DISCUSSION

We elucidated the localization of AADC-ir structures in the human and feline brains, and examined their chemical nature in normal and L-DOPA-injected cats, with or without IMAO. AADC-ir structures could be visualized in these two species more distinctly than in the rat (Jaeger et al., 1984). Regarding L-DOPA uptake and decarboxylation, as reported in the rat (Arai et al., 1994), we confirmed in the cat, that DA is accumulated after L-DOPA loading in 5-HT cells and fibers, which contain higher amounts of AADC and consequently transform L-DOPA to DA much more than in the other AADC-ir ones. This accumulation may also be due to the absence or low concentration of 5-HT degrading enzyme MAO-A (type A MAO) in 5-HT structures. DA-immunoreactivity (DA-ir) was less intense in NE cells, since part of synthesized DA may be oxidized by MAO-A.

Many AADC-ir nonaminergic cell bodies can take up L-DOPA and decarboxylate it to DA. But, there are also AADC-ir ones which could display DA-ir only after MAO inhibition. That is, in such cells, synthesized DA might be rapidly oxidized by MAO-A. All of these cells correspond to 'L-DOPA-positive cells' which can be revealed by a histofluorescence method after treatment with L-DOPA and IMAO (Butcher et al., 1972;

Lidbrink et al., 1974). Thus, AADC is, in fact, enzymatically active in virtually all the AADC-ir cells and fibers in the brain. This enzyme also transforms loaded 5-hydroxytryptophan (5-HTP) to 5-HT (Denoyer et al., 1989; Karasawa et al., 1994).

However, several problems should be pointed out. In DA terminals, DA once released into synaptic cleft could be taken up again into pre-terminals to be oxidized by MAO-A. However, at present, DA transport mechanism is not known out of DA system, such as 5-HT terminals. It is also puzzling that MAO-A has not been demonstrated histochemically in most of AADC-ir nonaminergic cells and fibers, although it should be present being based upon pharmacological evidence.

In conclusion, L-DOPA loading increases, in fact, DA concentration in virtually all the AADC structures, especially in 5-HT terminal structures, including those extending in the cerebral cortex, hypothalamus and limbic system, in which has been demonstrated a decrease in physiological action of 5-HT. This may be due to a competitive effect of L-DOPA for AADC enzymatic activity against 5-HTP, and the consequent decrease in presynaptic 5-HT concentration (Trouvin et al., 1991; Hartvig et al., 1993).

ACKNOWLEDGMENTS

This work was supported by INSERM U480 and CNRS (French Government) for K.K., and a grant from the program grants-in-aid for Scientific Research on Priority Areas, Ministry of Education, Science, Sports and Culture of Japan to I.N.

REFERENCES

Arai, R., Karasawa, N., Geffard, M., Nagatsu, T. and Nagatsu, I., 1994, Immunohistochemical evidence that central serotonin neurons produce dopamine from exogenous L-DOPA in the rat, with reference to the involvement of aromatic L-amino acid decarboxylase, *Brain Res.* 667 295-299.

Butcher, L.L., Engel, J. and Fuxe, K., 1972, Behavioral, biochemical histochemical analysis of the central effects of monoamine precursors after peripheral decarboxylase inhibition, *Brain Res.* 41 387-411.

Chagnaud, J.L., Mons, N., Tuffet, S., Grandier-Vazelles, X. and Geffard, M., 1987, Monoclonal antibodies against glutaraldehyde-conjugated dopamine, *J.Neurochem.* 49 487-494.

Denoyer, M., Kitahama, K., Sallanon, M., Touret, M. and Jouvet, M., 1989, 5-hydroxytryptophan uptake and decarboxylating neurons in the cat hypothalamus, *Neuroscience*, 31 203-211.

Hartvig, P., Tedroff,J., Lindner, K.J., Bjurling, P., Chang, C.W., Tsukada, H., Watanabe, Y., and Langstrom, B. , 1993, Positron emission tomographic studies on aromatic L-amino acid decarboxylase activity in vivo for L-dopa and 5-hydroxy-L-tryptophan in the monkey brain. J Neural Transm Gen Sect. 94 127-135.

Jaeger, C.B., Ruggiero, D.A., Albert, V.R., Park, D.H., Joh, T.H., Reis, D.J., 1984 Aromatic L-amino acid decarboxylase in the rat brain: immunocytochemical localization in neurons of the brain stem, Neuroscience.11 691-713.

Karasawa, N., Arai, R., Isomura, G., Yamada, K., Sakai, K., Sakai, M., Nagatsu, T. and Nagatsu, I., 1994, Phenotypic changes of AADC-only immunopositive premammillary neurons in the brain of laboratory shrew *Suncus murinus* by systemic administration of monoamine precursors, *Neurosci.Lett.* 179 65-70.

Kitahama, K., Denoyer, M., Raynaud, B., Borri-Voltattorni, C., Weber, M. and Jouvet, M., 1988, Immunohistochemistry of aromatic L-amino acid decarboxylase in the cat forebrain, *J.Comp.Neurol.* 270 337-353.

Lidbrink, P., Jonsson, G. and Fuxe, K., 1974, Selective reserpine- resistant accumulation of catecholamines in central dopamine neurons after DOPA administration, *Brain Res.* 67 439-456.

Trouvin, J.H., Maubrey, M.C, Raynal, H, Jacquot, C. 1991, Effect of L-dopa loading on 5-HTP decarboxylation in rat brain areas, *Fundam. Clin. Pharmacol.* 5 497-505.

DOPAMINE SYNTHESIS FROM L-TYROSINE BY NON-DOPAMINERGIC NEURONS IN CO-OPERATION

Michael V. Ugrumov, Petr V. Ershov and Irina S. Balan*

I. INTRODUCTION

Since the development of the histofluorescence technique for detection of catecholamines (Björklund and Lindvall, 1984) and then of immunostaining for tyrosine hydroxylase (TH) (Hökfelt at al., 1984), the neurons showing fluorescence or possessing TH were considered as catecholaminergic neurons. Later, neurons possessing only one of the enzymes of dopamine (DA) synthesis, TH or aromatic L-amino acid decarboxylase (AADC), were discovered (Meister et al., 1988). Although the number of monoenzymatic neurons greatly exceeds that of bienzymatic DAergic neurons (Björklund and Lindvall, 1984; Van den Pol et al., 1984) and they are widely distributed in the brain (Meister et al., 1988; Ikemoto et al., 1998), the functional significance of the former remains uncertain. Arcuate nucleus (AN) is considered as a promising model to approach this problem since it contains neurons expressing only TH (Meister et al., 1988; Okamura et al., 1988a) and most probably the neurons possessing only aromatic L-amino acid decarboxylase (AADC) (Krajnak and Nunez, 1996). On the other hand, the AN is the functionally important DAergic system, and DA produced in this region inhibits prolactin secretion (McCann et al., 1984). This paper summarizes our recent data on the development and functional significance of monoenzymatic TH- or AADC-expressing neurons in the AN of male rats during ontogenesis.

2. NEURONS EXPRESSING ENZYMES OF DOPAMINE SYNTHESIS

Neurons expressing TH and AADC were identified in the AN of rats at the 17th embryonic day (E17), E18, E20, E21, the 9th postnatal day (P9), and in adulthood by using the double immunofluorescent technique and fluorescent or confocal microscopy. Three neuron populations have been distinguished according to the spectrum of the

* Michael V. Ugrumov, Institute of Developmental Biology, Russian Academy of Sciences, Moscow, Russia, 117808; Institute of Normal Physiology, Academy of Medical Sciences, Moscow, Russia, 125315. Petr V. Ershov and Irina S. Balan, Institute of Normal Physiology, Academy of Medical Sciences, Moscow, Russia, 125315.

Catecholamine Research, Edited by Nagatsu *et al.*
Kluwer Academic/Plenum Publishers, 2002

enzymes. The 1st population is represented by TH-IR / AADC-immunonegative (IN) neurons, which were first detected at E18 (Balan et al., 2000). So far, TH-IR neurons lacking AADC were demonstrated only in adults (Meister et al., 1988; Okamura et al., 1988b). The 2nd population of TH-IN / AADC-IR neurons and the 3rd population of both TH-IR and AADC-IR neurons were initially observed in the AN at E20. At E20, 21, 99% of all the neurons expressing enzymes of DA synthesis were represented by the 1st and 2nd populations, and the fraction of the 3rd population was about 1%. Postnatally to adulthood, the 1st population decreased gradually, the 3rd population proportionally increased, and the 2nd population did not change (Table 1).

Table 1. Total numbers (N) and fractions (%) of the neurons immunoreactive (IR) for tyrosine hydroxylase (TH) but immunonegative (IN) for aromatic L-amino acid decarboxylase (AADC), as well as TH-IN / AADC-IR and TH-IR / AADC-IR neurons in the arcuate nucleus in male rats at the 21st embryonic day (E21), the 9th postnatal day (P9), and in adulthood (Ad) (n=4 for each age group)

Age	TH-IR / AADC-IN		TH-IN / AADC-IR		TH-IR / AADC-IR	
	N	%	N	%	N	%
E21	862.3	49.4	867.5	49.7	14.3	0.9
P9	1150.5	39.4	957.1	32.7	814	27.9
Ad	447.8	17.9	1057.2	42.2	998.9	39.9

3. FUNCTIONAL SIGNIFICANCE OF MONOENZYMATIC NEURONS

As mentioned above, the total numbers and fractions of the neurons expressing individual complementary enzymes of DA synthesis highly prevailed that of DAergic neurons in the AN during the perinatal period, and even in adulthood their total fraction is about 60%. This fact raised the question of whether the monoenzymatic neurons contribute to the synthesis of either DA or L-DOPA. In order to approach this problem, the DA and L-DOPA contents were measured by HPLC-ED in the dissected mediobasal hypothalamus (MBH = AN + median eminence) and in its primary dissociated cell culture (Melnikova et al., 1999).

At E21, the DA concentration in the MBH (Melnikova et al., 1999) was essentially higher than that in the whole brain and diencephalon (Coyle and Henry, 1973). The concentration of L-DOPA in the fetal MBH by 3-fold exceeded that of DA (Melnikova et al., 1999). This is in contrast to other brain regions enriched with DAergic neurons because of the high activity of AADC (Moore et al., 1985). It means that L-DOPA in the fetal MBH is synthesized in the neurons expressing TH but lacking AADC. A surprisingly high level of DA detected in the fetal MBH almost lacking DAergic neurons was the main reason to hypothesize that the neurons expressing individual complementary enzymes of DA synthesis produce DA in co-operation. It means that L-DOPA is synthesized from L-tyrosine in the monoenzymatic TH neurons that is followed by L-DOPA release and uptake by the monoenzymatic AADC-containing neurons for its final conversion to DA (Fig. 1). L-DOPA is taken up by different cells due to the membrane transporter for large neutral amino acids (Yoshida et al., 1963). In particular, the L-DOPA uptake by DAergic neurons may serve to increase the efficiency of DA synthesis (Fig. 1).

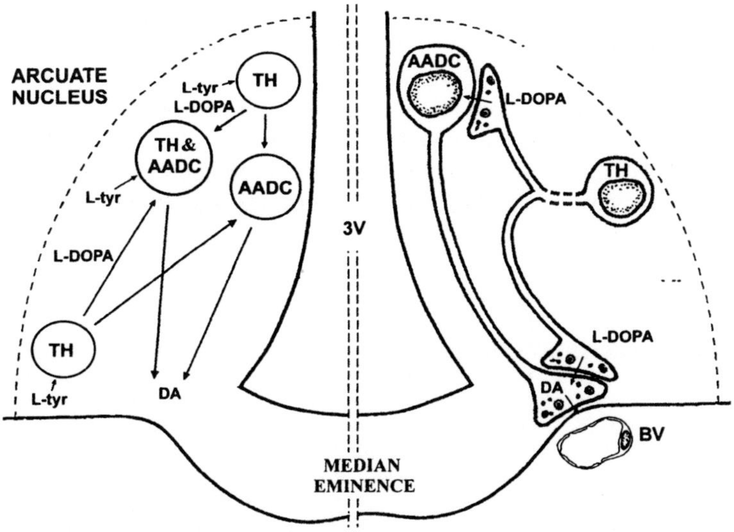

Figure 1. Schematic representation of dopamine synthetic pathways in the arcuate nucleus of rats (left side) and topographic relations of monoenzymatic tyrosine hydroxylase (TH)-expressing neurons with monoenzymatic aromatic L-amino acid decarboxylase (AADC)-expressing neurons and portal blood vessels (BV) (right side). DA, dopamine; 3V, the third ventricle.

If the hypothesis about the DA synthesis by non-catecholaminergic neurons in co-operation is correct, one can expect to find close topographic relations between the monoenzymatic TH and AADC neurons facilitating the L-DOPA transport from the former to the latter. In fact, this was observed at any studied age (E21, P9, adults) by using double-immunolabeling with confocal microscopy (Fig. 1).

Although serious arguments were obtained in the *in vivo* and *ex vivo* studies in favour of the co-operative synthesis of DA, the shortcoming of this approach lies in failure to differentiate the DA synthesized in cell bodies of the AN and that coming to the MBH via catecholaminergic afferents or from the fetal circulation (Peleg et al., 1985; Everitt et al., 1986). A cell culture study was undertaken to overcome this problem.

A primary cell culture of the MBH explanted at E17 and maintained for 7 days was used to evaluate the metabolism of DA and its K^+-stimulated release in the neurons of the AN (Melnikova et al., 1999). There was a rather high level of DA and L-DOPA in cell extracts, and DA release increased in response to an excess of K^+ (Melnikova et al., 1999). Moreover, the ratio of the concentrations of L-DOPA to DA in cell extracts exceeded significantly this index observed *ex vivo*, suggesting a great contribution of the monoenzymatic TH-expressing neurons to L-DOPA synthesis. These data confirm our hypothesis about the co-operative synthesis of DA by non-DAergic neurons expressing individual enzymes, TH and AADC (Fig. 1).

Although there was a relatively high level of DA synthesis in the AN of fetal rats, it remained uncertain whether DA production was sufficient to provide inhibitory control of adenohypophysial prolactin secretion. According to our data, DA began to inhibit prolactin release from the pituitary between E20 and E22 that was proved by the haloperidol-provoked increase in prolactin concentrations in plasma and decrease in the pituitary (Melnikova et al., 1998). If DA is synthesized by monoenzymatic non-DAergic neurons, it should be released from axon terminals of the monoenzymatic AADC neurons in the median eminence toward the pituitary (Fig. 1). Indeed, many axon terminals contained AADC but lacked serotonin, TH, and DA-β-hydroxylase were observed with double-immunolabeling in the median eminence at any studied age (E21, P9, adults).

Thus, the rat AN contains numerous neurons expressing individual complementary enzymes of the DA synthetic pathway, which appear to synthesize DA in co-operation.

REFERENCES

Balan, I. S., Ugrumov, M. V., Calas, A., Mailly, P., Krieger, M., and Thibault, J., 2000, Tyrosine hydroxylase-expressing and/or aromatic L-amino acid decarboxylase-expressing neurons in the mediobasal hypothalamus of perinatal rats: differentiation and sexual dimorphism, *J. Comp. Neurol.* **425**:167-176.

Björklund, A., and Lindvall, O., 1984, Dopamine-containing system in the CNS, in: *Handbook of Chemical Neuroanatomy*, A. Björklund, and T. Hökfelt, eds., Elsevier, Amsterdam, pp. 55-122.

Coyle, J. T., and Henry, D., 1973, Catecholamines in fetal and newborn rat brain, *J. Neurochem.* **21**:61-67.

Everitt, B., Meister, B., Hökfelt, T., Melander, T., Terenins, L., Rokaeus, A., Theodorsson-Norheim, E., Dockray, G., Edwardson, J., Cuello, C., Elde, R., Goldstein, M., Hemmings, H., Ouimet, C., Walaas, I., Greengard, P., Vale, W., Weber, E., Wu, J., and Chang, K., 1986, The hypothalamic arcuate nucleus-median eminence complex: Immunocytochemistry of transmitters, peptides and DARPP-32 with special reference to coexistence in dopamine neurons, *Brain Res. Rev.* **11**:97-155.

Hökfelt, T., Johansson, O., and Goldstein, M., 1984, Central catecholamine neurons as revealed by immunocytochemistry with special reference to adrenaline neurons, in: *Handbook of Chemical Neuroanatomy. Classical Neurotransmitters in the CNS*, A. Björklund, and T. Hökfelt, eds., Elsevier, Amsterdam, pp. 157-276.

Ikemoto, K., Nagatsu, I., Nishimura, A., Nishi, K., Arai, R., 1998, Do all of human midbrain tyrosine hydroxylase neurons synthesize dopamine?, *Brain Res.* **805**:255-258.

Krajnak, K., and Nunez, A., 1996, Short-photoperiod exposure reduces L-aromatic-amino-acid decarboxylase immunostaining in the arcuate nucleus and median eminence of male Syrian hamsters, *Brain Res.* **712**:95-101.

McCann, S. M., Lumpkin, M. D., Mizunuma, H., Khorram, O., Ottlecz, J., and Samson, W. K., 1984, Peptidergic and dopaminergic control of prolactin release, *Trends in Neurosci.* **7**:127-131.

Meister, B., Hökfelt, T., Steinbusch, H. W. M., Skagerberg, G., Lindvall, O., Geffard, M., Joh, T., Cuello, A. C., and Goldstein, M., 1988, Do tyrosine hydroxylase-immunoreactive neurons in the ventro-lateral arcuate nucleus produce dopamine or only L-DOPA?, *J. Chem. Neuroanat.* **1**:59-64.

Melnikova, V., Orosco, M., Calas, A., Sapronova, A., Gainetdinov, R., Delhaye-Bouchaud, N., Nicolaidïs, S., Raevsky, K., and Ugrumov, M., 1999, Dopamine turnover in the mediobasal hypothalamus in rat fetuses, *Neuroscience.* **88**:235-241.

Melnikova, V., Orosco, M., Rouch, C., Calas, A., Sapronova, A., Gainetdinov, R., Delhaye-Bouchaud, N., Nicolaidïs, S., Proshlyakova, E. V., Sapronova, A. Y., and Ugrumov, M., 1998, Prolactin secretion and its dopamine inhibitory control in rat fetuses, *Eur. J. Endocrinol.* **139**:337-342.

Moore, K. E., Riegle, G. D., and Demarest, K. T., 1985, Regulation of tuberoinfundibular dopaminergic neurons: Prolactin and inhibitory neuronal influences, in: *Catecholamines as Hormone Regulators*, N. Ben-Jonathan, J. M. Bahr, and R. I. Weiner, eds., Raven Press, New York, pp. 31-48.

Okamura, H., Kitahama, K., and Nagatsu, I., 1988a, Comparative topography of dopamine and tyrosine hydroxylase immunoreactive neurons in the rat arcuate nucleus, *Neurosci. Lett.* **95**:347-353.

Okamura, H., Kitahama, K., Raynaud, B., Nagatsu, I., Borri-Volatorni, C., and Weber, M., 1988b, L-amino acid decarboxylase (AADC)-immunoreactive cells in the tuberal region of the rat hypothalamus, *Biomed. Res.* **9**:261-267.

Peleg, D., Arbogast, L., Peleg, E., and Ben-Jonathan, N., 1985, Catecholamines in maternal and fetal plasma and the amniotic fluid during late gestation in the rat, in: *Catecholamines as Hormone Regulators*, N. Ben-Jonathan, J. M. Bahr, and R. I. Weiner, eds., Raven Press, New York, pp. 350.

Van den Pol, A. N., Herbst, R. S., and Powell, J. F., 1984, Tyrosine hydroxylase-immunoreactive neurons of the hypothalamus: a light and electron microscopic study, *Neuroscience,* **13**:1117-1156.

Yoshida, H., Kaniike, K., and Namba, J., 1963, Properties of a carrier system to transport L-DOPA into brain slices, *Nature,* **198**:191-192.

DOPAMINE NEURONS ARE NOT AFFECTED BY EXPANDED POLYGLUTAMINE STRETCHES IN HD GENE KNOCK-IN MICE

Hiroshi Ishiguro*, Kouji Yamada, Hirohide Sawada, Kazuhiro Nishii, Makoto Sawada, Jun Goto, Ichiro Kanazawa, and Toshiharu Nagatsu

1. INTRODUCTION

Huntington's Disease (HD) is an autosomal dominant neurodegenerative disorder characterized by involuntary movement, personality change and dementia. The HD gene was identified using gene linkage analysis.[1] The HD gene contains a triplet (CAG) repeat expansion (>36 CAG repeats) in exon 1, something that is also a feature of polyglutamine diseases. A clear relation exists between the CAG repeat length of the HD gene and age of disease onset. HD's pathological abnormalities include loss of medium spiny neurons in the striatum, especially the caudate nucleus and putamen, which can begin at an early stage of the disease. HD model mice, in the form of transgenic (Tg) or gene knock-in animals, have been produced by several groups. Several HD gene knock-in mice were also produced by replacing the mouse exon 1 with mutated forms of the human HD gene exon 1.[2-6] In contrast to Tg mice, HD gene knock-in mice showed relatively mild abnormalities, including aggregation in nuclei of neurons, neuronal cell death and abnormal behavior.

2. CREATION OF HUNTINGTON' S DISEASE GENE KNOCK-IN MICE AND THEIR GENERAL CHARACTERISTICS

HD gene knock-in mice have been produced by replacing the normal mouse exon 1 in embryonic stem (ES) cells with a human IT-15 (HD) gene exon 1 containing an expansion of 77 CAG repeats. As a result, chimeric proteins coded for by the mutated human exon 1 and the remaining normal exons of the mouse HD gene are expressed ubiquitously in heterozygous mice, and have been found in the brain, lung, heart, liver,

* Hiroshi Ishiguro, Kouji Yamada, Hirohide Sawada, Kazuhiro Nishii, Makoto Sawada, Toshiharu Nagatsu, Institute for Comprehensive Medical Science, Fujita Health University, Toyoake, Aichi 470-1192, Jun Goto, Ichiro Kanazawa, Graduate School of Medicine, The University of Tokyo, Hongo, Tokyo 113-8655 JAPAN

Catecholamine Research, Edited by Nagatsu *et al.*
Kluwer Academic/Plenum Publishers, 2002

spleen, kidney and testis. On electrophoresis, the faster band was wild-type mouse huntingtin (approximately 350 kDa), and the slower band was the protein translated from the mutant allele of the HD gene (approximately 360 kDa), which contained expanded polyglutamine stretches within its N-terminal. These results indicate that the structure and distribution of huntingtin in these knock-in mice may be similar to what is found in HD patients. This suggests that these mice may be useful for investigations of HD abnormalities such as inflammation of the brain and impairment of neuron survival mechanisms and neuroregeneration.

Astrocytosis is typically observed in the brain tissue of HD patients, and is indicative of inflammation.[7] Levels of Glial fibrillary acidic protein (GFAP)-positive cells are elevated in the substantia nigra (SN) and globus pallidus (GP) of the brains of 40-week-old heterozygous mice. These results indicate that biological responses in these HD gene knock-in mice are the same as those found in HD patients.

In order to clarify the details of tissue specificity of somatic CAG repeat instability, we examined CAG repeat instability in 9 different brain areas and peripheral tissues. In the brain, somatic CAG repeat instability was particularly pronounced in the striatum, midbrain and pons/medulla, but was not detected in the cerebellum. Other brain areas (cortex, olfactory bulbs, hippocampus, thalamus and hypothalamus) and peripheral tissues showed moderate or low levels of CAG repeat instability.[6] Instability of 77-CAG repeats occurred in somatic DNA in the brain in a time-dependent and tissue-specific manner. Polyglutamine aggregations translated from 77-CAG repeat expansions were first detected in striatum of HD gene knock-in mice when they were 60 weeks old. The formation of polyglutamine aggregations progressed at least until the mice were 100 weeks old. Polyglutamine fragments within the huntingtin protein may continuously accumulate in the cytoplasm or nucleus. These aggregations were classified into 4 types of polyglutamine formations: (1) diffuse formations in the nucleus, (2) mixture of diffuse and dense formations in the nucleus, (3) dense formations in the nucleus, and (4) aggregations in both the nucleus and the cytoplasm. These aggregations are localized in striatum, and are not found in other brain areas.

3. MECHANISM OF POLYGLUTAMINE DISEASE

There are many hypotheses regarding mechanisms of neuronal cell death from polyglutamine aggregation or related processes. One possible mechanism of neuronal cell death from polyglutamine disease is decreases in the amounts of transcriptional factors such as TAFII130 (in the nucleus)[8] or cAMP response element-binding protein (CBP) (in the cytoplasm).[9] Decreased levels of gene transcription factors could interfere with gene expression, reducing levels of translated proteins. Persistence of these conditions in neuronal cells could disrupt the supply of energy from the mitochondria or reduce levels of proteins that are needed for survival of the cells.

4. DOPAMINE (DA) LEVELS WERE NOT ALTERED IN STRIATUM OF HD GENE KNOCK-IN MOUSE

DA neurons are generally considered candidate targets of HD neurotoxicity.[10] To clarify the changes in DA levels in the striatum of HD gene knock-in mice, we isolated cytosol proteins from the mouse striatum (60 and 100-week-old heterozygous mice). DA levels were not significantly different between the normal mice and HD gene knock-in mice, even though polyglutamine aggregates had formed in the nuclei of striatum neurons in the knock-in mice. DA neurons project to the striatum from the substantia nigra. We also confirmed the presence of healthy DA neurons in the substantia nigra by immunohistochemical analysis using anti-tyrosine hydroxylase antibody (Dia Sorin). There was no evidence of neuronal cell death in the striatum (no TUNEL-positive cells or anti-single strand DNA antibody-positive cells), even though there was astrogliosis, CAG repeat expansion of somatic DNA, and polyglutamine aggregation in the striatum. In conclusion, DA neurons did not affect neuronal cells of the striatum (which are generally considered candidate targets of HD neurotoxicity), as evidenced by the normal DA levels observed in the striatum. Normal DA concentrations did not affect the degree of neurotoxicity in the striatum of the HD gene knock-in mice, which had expanded 77-CAG repeats.

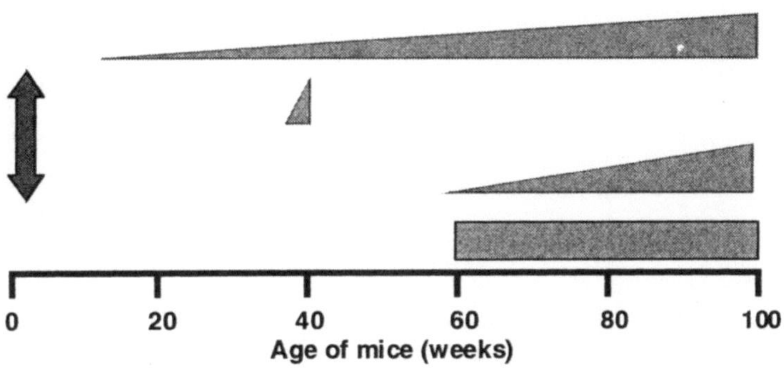

Figure 1. Abnormalities of HD gene knock-in mice. Ages of mice are shown on horizontal bar, and vertical arrows indicate level of each abnormal condition. Somatic CAG repeat instability, astrogliosis and neuronal intranuclear inclusions (NIIs) indicate polyglutamine aggregates in nuclei of neurons. DA levels in striatum are shown by triangular and square boxes.

REFERENCES

1. The Huntongton's Disease Collaborative Research Group, A novel gene containing a trinucleotide repeat that is expanded and unstable on Huntington's disease chromosome, *Cell* **72**, 971-983 (1993).
2. V.C. Wheeler, W. Auerbach, J.K. White, J. Srinidhi, A. Auerbach, A. Ryan, M.P. Duyao, V. Vrbanac, M. Weaver, J.F. Gusella, A.L. Joyner, M.E. MacDonald, Length-dependent gametic CAG repeat instability in the Huntington's disease knock-in mouse. *Hum. Mol. Genetics* **8**, 115-122 (1999).
3. P.F. Shelbourne, N, Killeen, R, F, Hevner, H, M, Johnston, L, Tecott, M, Lewandoski, M, Ennis, L, Ramirez, Z, Li, C, Iannicola, D, R, Littman,R,M, Myers, A Huntington's disease CAG expansion at the murine Hdh locus is unstable and associated with behavioral abnormalities in mice. *Hum. Mol. Genetics* **8**, 763-774 (1999).
4. M.S. Levine, G,J, Klapstein, A, Koppel, E, Gruen, C, Cepeda, M,E, Vargas, E,S, Jokel, E,M, Carpenter, H, Zanjani, R,S, Hurst, A, Efstratiadis, S, Zeitlin, M, Chesselet, Enhanced sensitivity to N-methyl-D-aspartate receptor activation in transgenic and knockin mouse models of Huntington's disease. *J. Neurosci. Res.* **58**, 515-532 (1999).
5. C.H. Lin, S. Tallaksen-Greene, W.M. Chien, J.A. Cearley, W.S. Jackson, A.B. Crouse, S. Ren, X.J. Li, R.L. Albin, and P.J. Detloff, Neurological abnormalities in a knock-in mouse model of Huntington's disease, *Hum. Mol. Genetics* **10 (2)**, 137-144 (2001).

6. H. Ishiguro, K. Yamada, H. Sawada, K. Nishii, N. Ichino, M. Sawada, Y. Kurosawa, N. Mastushita, K. Kobayashi, J. Goto, H. Hashida, N. Masuda, I. Kanazawa, and T. Nagatsu, Age-dependent and tissue-specific CAG repeat instability occurs in mouse knock-in for a mutant HD gene. *J. Neurosci. Res.* (*in-press*) (2001).
7. J.C. Hedreen, and S.E. Folstein, Early loss of neostriatal striosome neurons in Huntington's disease, *J. Neuropathol. Exp. Neurology* **54**, 105-120 (1995).
8. T. Shimohata, T. Nakajima, M. Yamada, C. Uchida, O. Onodera, S. Naruse, T. Kimura, R. Koide, K. Nozaki,Y. Sano, H. Ishiguro, K. Sakoe, T. Oohshima, A. Sato, T. Ikeuchi, M. Oyake, T. Sato, Y. Aoyagi, I. Hozumi, T. Nagatsu, Y. Takiyama, M. Nishizawa, J. Goto, I. Kanazawa, I. Davidson, N. Tanase, H. Takahashi, S. Tsuji, Expanded polyglutamine stretches interact with TAFII130, interfering with CREB-dependent transcription. *Nature Genetics* **26**, 29-36 (2000).
9. F.C. Nucifora Jr., M. Sasaki, M.F. Peters, H. Huang, J.K. Cooper, M. Yamada, H. Takahashi, S, Tsuji, J. Troncoso, V. L. Dawson, T.M. Dawson, and C.A. Ress, Interference by Huntingtin and atrophi-1 with CBP-mediated transcription leading to cellular toxicity, *Science* **291**, 2423-2428 (2001).
10. R.J. Jakel and W.F. Maragos, Neuronal cell death in Huntington's disease: a potential role for dopamine, *TINS* **23 (6)**, 239-245 (2000).

CATECHOLAMINE - NEUROPEPTIDE INTERACTIONS

Tomas Hökfelt, Zhi-Qing D. Xu, Xiasong Ma,Yong-Guang Tong and Marc Landry*

1. INTRODUCTION

It is now well established that neuropeptides act as messenger molecules in the nervous system, and they often coexist with classical transmitters such as noradrenaline (NA), acetylcholine, and amino acids (see 1). In fact, the first example in the mammalian nervous system showed that peripheral sympathetic noradrenergic neurons express the tetradecapeptide somatostatin, a study based on immunohistochemistry using antibodies to the catecholamine synthesizing enzyme dopamine beta-hydroxylase (DBH).[2] This was possible thanks to our collaboration with the late Professor Menek Goldstein at New York University School of Medicine, since he isolated all four enzymes in the catecholamine (CA) synthesis pathway and raised antibodies to these enzymes.[3] This was the basis for Kjell Fuxe's and my interaction with Menek over a period of three decades starting in 1969. It is an honour to acknowledge Menek Goldstein's contributions to our work on CA-peptide interactions.

2. NEUROPEPTIDES IN THE ADRENAL GLAND

The four CA synthesizing enzymes, including DBH, were purified from the adrenal gland, a gland that has not only served as a source of enzymes and other markers, but also has represented a model organ for studies of, for example, CA storage and release mechanisms. It turned out early that the chromaffin cells in the adrenal medulla express neuropeptides, the first example being enkephalin.[4] Peptides are present in efferent and afferent nerves to the adrenal gland, and are in this way involved in the regulation of the CA synthesis and secretion (see 5). Also peptide expression in the chromaffin cells is markedly regulated by the neuronal input, whereby the type of manipulation plays an

* Tomas Hökfelt, Department of Neuroscience, Karolinska Institutet, Stockholm, Sweden and Zhi-Qing D. Xu, Xiasong Ma and Yong-Guang Tong, same address; Marc Landry, Institute Francois Magendie, University of Bordeaux, Bordeaux, France

Catecholamine Research, Edited by Nagatsu *et al.*
Kluwer Academic/Plenum Publishers, 2002

important role. For example, whereas enkephalin is upregulated after denervation, pharmacological blockade by cholinergic antagonists and immunosympathectomy, only the latter treatment, which specifically destroys the preganglionic sympathetic fibers, causes a strong increase in galanin expression.[6] Taken together the adrenal gland has served as a model organ also for studies of interactions between neuropeptides and CAs, both through localization of these messenger molecules with CAs in chromaffin cells as well as via neuropeptide-containing afferent and efferent nerves.

3. NEUROPEPTIDES IN THE BRAIN - FOCUS ON GALANIN

Subsequently many examples of CA-peptide coexistence were observed in the brain, including galanin in many noradrenergic locus coeruleus neurons.[7] Galanin is a 29 amino acid peptide discovered by Tatemoto, Mutt and collaborators,[8] and it has a wide distribution in the central nervous system, where it not only coexists with NA but also with a number of other classical transmitters as well as peptides.[7] Here we would like to describe in some detail our work on galanin in the locus coeruleus system as an example of CA-neuropeptide interactions.

Locus coeruleus neurons have a robust galanin synthesis.[7, 9] It is well known that locus coeruleus neurons project to wide spread areas of the brain including cortex and hippocampus, but only with recent sensitive immunohistochemical techniques has it been possible to clearly demonstrate galanin in NA fibers also in cortical/hippocampal areas.[9] Moreover, after reserpine treatment, a drug known to deplete central and peripheral monoamine stores,[10] galanin disappears from these cortical and hippocampal fibers, presumably via reflex activation of LC.[9] These findings indirectly show that endogenous galanin can be released from cortical NA nerve terminals.

The functional role of galanin in locus coeruleus has been elucidated using electrophysiological approaches demonstrating that galanin reduces firing rate and induces hyperpolarization of LC neurons, presumably via an increase in K^+ conductance.[11, 12, 13] The receptor mediating this effect seems to be located on the postsynaptic membrane of LC neurons, since the effect persisted under conditions in which synaptic input was probably blocked (TTX or/and plus low CA^{2+} medium). Recently a further effect of galanin has been recorded. Thus, studying interactions between NA and galanin, it was found that the well-known outward current induced by NA[14, 15] is enhanced and prolonged by preincubation with low concentrations of galanin (0.05-0.1 nM), that is at concentrations that have no detectable effects on the membrane current by themselves.[16] Such an indirect, sensitizing modulation could be of physiological importance in view of the low galanin concentrations needed.

The NA effect is mediated via α2-adrenoceptors and has been considered to represent autoinhibition by NA released from soma and dendrites of the locus coeruleus neurons,[14, 15] but it is also possible that the α2-adrenoceptors are activated by adrenaline released from nerve terminals originating from medullary adrenaline neurons.[17, 18] Also galanin may originate from the LC neurons, since around 80% of all LC neurons produce galanin[19] and since galanin mRNA and galanin have been demonstrated not only in soma but also in dendrites of LC neurons (Landry, Hökfelt, unpublished observations). Moreover, there are fairly few galanin-positive nerve endings in the locus coeruleus.[13] It is therefore possible that galanin can be released from dendrites/soma of the noradrenergic LC neurons,[9, 13] as has been shown for oxytocin and vasopressin in the hypothalamic magnocellular neurons.[20]

Three galanin receptors have recently been cloned, the galanin-R1, -R2 and -R3 receptors (see 21). In situ hybridization studies have demonstrated the presence of both GALR1 and -R2 in LC neurons.[22] The receptor(s) involved in hyperpolarization has not been definitely identified, but we have shown that a mixed GALR1/R2 agonist, but not a selective GALR2 agonist, causes the same effect as galanin (1-29) on LC neurons.[23] Instead, the GALR2 receptor may be a presynaptic receptor transported centrifugally to the nerve terminals in cortical and hippocampal areas. The mechanism(s) underlying the above mentioned sensitizing, modulating effect of galanin at low concentrations is not known.

4. CONCLUDING REMARKS

Many drugs used to treat mental disorders, such as depression, target the ascending NA and serotonin systems. The fact that galanin (and other peptides) are coexpressed in these neurons raises the possibility that also peptidergic mechanisms are of interest in relation to such diseases. The work described in the present brief article aims at forming a conceptual basis for understanding the role of neuropeptides under physiological and pathophysiological conditions and, ultimately, for drug development.

ACKNOWLEDGEMENTS

This work was supported by Marianne and Marcus Wallenbergs Stiftelse, an Unrestricted Bristol-Myers Squibb Neuroscience Grant and the Swedish MRC (04X-2887).

REFERENCES

1. T. Hökfelt, C. Broberger, Z.-Q. Xu, V. Sergeyev, R. Ubink and M. Diez, Neuropeptides - an overview. *Neuropharmacology* **39**, 1337-1356 (2000).
2. T. Hökfelt, L. G. Elfvin, R. Elde, M. Schultzberg, M. Goldstein and R. Luft, Occurrence of somatostatin-like immunoreactivity in some peripheral sympathetic noradrenergic neurons. *Proc. Natl. Acad. Sci. USA* **74**, 3587-3591 (1977).
3. M. Goldstein, B. Anagnoste, L. S. Freedman, M. Roffman, R. P. Ebstein, D. H. Park, K. Fuxe and T. Hökfelt. (1973) Characterization, localization and regulation of catecholamine synthesizing enzymes. In *Frontiers in Catecholamine Research.* (editors, E. Usdin and E. Snyder) pp. 69-78. Elsevier/Pergamon Press, New York.
4. M. Schultzberg, J. M. Lundberg, T. Hökfelt, L. Terenius, J. Brandt, R. P. Elde and M. Goldstein, Enkephalin-like immunoreactivity in gland cells and nerve terminals of the adrenal medulla. *Neuroscience* **3**, 1169-1186 (1978).
5. H. Holgert, Å. Dagerlind and T. Hökfelt, Immunohistochemical characterization of the peptidergic innervation of the rat adrenal gland. *Horm. Metabol. Res.* **30**, 315-322 (1998).
6. Å. Dagerlind, M. Pelto-Huikko, J. Lundberg, R. Ubink, A. Verhofstad, S. Brimijoin and T. Hökfelt, Immunologically induced sympathectomy of preganglionic nerves by antibodies against acetylcholinesterase: increased levels of peptides and their mRNAs in rat adrenal cells. *Neuroscience* **62**, 217-239 (1994).

7. T. Melander, T. Hökfelt, Å. Rökaeus, A. C. Cuello, W. H. Oertel, A. Verhofstad and M. Goldstein, Coexistence of galanin-like immunoreactivity with catecholamines, 5-hydroxytryptamine, GABA and neuropeptides in the rat CNS. *J. Neurosci.* **6,** 3640-3654 (1986).
8. K. Tatemoto, Å. Rökaeus, H. Jörnvall, T. J. McDonald and V. Mutt, Galanin- a novel biologically active peptide from porcine intestine. *FEBS* **164,** 124-128 (1983).
9. Z.-Q. D. Xu, T.-J. S. Shi and T. Hökfelt, Galanin/GMAP- and NPY-like immunoreactivities in locus coeruleus and noradrenergic nerve terminals in the hippocampal formation and cortex with notes on the galanin-R1 and - R2 receptors. *J. Comp. Neurol.* **392,** 227-251 (1998).
10. A. Carlsson. (1965) Drugs which block the storage of 5-hydroxytryptamine and related amines. In *Handbuch Exp. Pharmacol.* (ed, V. Erspamer) pp. 529-592. Springer, Berlin.
11. V. Seutin, P. Verbanck, L. Massotte and A. Dresse, Galanin decreases the activity of locus coeruleus neurons in vitro. *Eur. J. Pharmacol.* **164,** 373 (1989).
12. J. Sevcik, E. P. Finta and P. Illes, Galanin receptors inhibit the spontaneous firing of locus coeruleus neurones and interact with μ-opioid receptors. *Eur. J. Pharmacol.* **230,** 223-230 (1993).
13. V. Pieribone, Z.-Q. Xu, X. Zhang, S. Grillner, T. Bartfai and T. Hökfelt, Galanin induces a hyperpolarization of norepinephrine-containing locus coeruleus neurons in the brainstem slice. *Neuroscience* **64,** 861-874 (1995).
14. G. K. Aghajanian, J. M. Cedarbaum and R. Y. Wang, Evidence for norepinephrine-mediated collateral inhibition of locus coeruleus neurons. *Brain Res.* **136,** 570-577 (1977).
15. J. T. Williams, R. A. North, S. A. Shefner, S. Nishi and T. M. Egan, Membrane properties of rat locus coeruleus neurones. *Neuroscience* **13,** 137-56 (1984).
16. Z.-Q. D. Xu, Y.-G. Tong and T. Hökfelt, Galanin enhances noradrenaline-induced outward current on locus coeruleus noradrenergic neurons. *NeuroReport,***12,**1779-1782 (2001).
17. T. Hökfelt, K. Fuxe, M. Goldstein and O. Johansson, Immunohistochemical evidence for the existence of adrenaline neurons in the rat brain. *Brain Res.* **66,** 235-251 (1974).
18. G. Aston-Jones, B. Astier and M. Ennis, Inhibition of locus coeruleus noradrenergic neurons by C1 adrenergic cells in the rostral ventral medulla. *Neuroscience* **48,** 371-382 (1992).
19. V. R. Holets, T. Hökfelt, A. Rökaeus, L. Terenius and M. Goldstein, Locus coeruleus neurons in the rat containing neuropeptide Y, tyrosine hydroxylase or galanin and their efferent projections to the spinal cord, cerebral cortex and hypothalamus. *Neuroscience* **24,** 893-906 (1988).
20. J. F. Morris, D. V. Pow, H. W. Sokol and A. Ward. (1993) Dendritic release of peptides from magnocellular neurons in normal rats, Brattleboro rats and mice with hereditary nephrogenic diabetes insipidus. In *Vasopressin* (ds, P. Gross, D. Richter and G. L. Robertson) pp. 171-182. John Libbey Eurotext, Paris.
21. T. A. Branchek, K. E. Smith, C. Gerald and M. W. Walker, Galanin receptor subtypes. *TiPS* **21,** 109-116 (2000).
22. D. O'Donnell, S. Ahmad, C. Wahlestedt and P. Walker, Expression of the novel galanin receptor subtype GALR2 in the adult rat CNS: Distinct distribution from GALR1. *J. Comp. Neurol.* **409,** 469-481 (1999).
23. X. Ma, Y.-G. Tong, R. Schmidt, W. Brown, K. Payza, L. Hodzic, C. Pou, C. Godbout, T. Hökfelt and Z.-Q.D. Xu, Effect of galanin receptor agonists on locus coeruleus neurons. *Brain Res., in press* (2001).

THE NORADRENERGIC NEURON, A MULTIPEPTIDE SECRETORY CELL

Werner De Potter*, Zesheng Wang, Jan Depreitere, Anne-Marie Lambeir, Christine Durinx and Etienne Nouwen

1. INTRODUCTION

It is generally accepted that each neuron can secrete a classical transmitter and one or more peptides via small synaptic vesicles (SSVs) and large dense-cored vesicles (LDCVs), respectively. According to the current concept both SSVs and LDCVs undergo regulated exocytosis but their release process is differentially regulated. However, this general concept of neurotransmitter release seemed not to hold for sympathetic neurons since the latter store their classical transmitter noradrenaline (NA) not in SSVs but in small dense-cored vesicles (SDCVs) and moreover also in LDCVs together with proteins and peptides. In addition, NA in sympathetic neurons was shown to be released, at least in part, by exocytosis from LDCVs, a finding long disputed but now supported by many recent physiological, pharmacological, electron microscopical and confocal immunomicroscopical studies (De Potter et al., 1997; Annaert et al., 1997). At present the debate merely concerns the part of NA released from SDCV and LDCV.

2. NA IS EXCLUSIVELY RELEASED FROM LDCVs

Using the in situ perfused porcine spleen, it was demonstrated that high-frequency stimulation of the splenic nerve caused a preferential release of neuropeptide Y (NPY). Since NPY is an established component of LDCVs and NA is stored in both SDCVs and LDCVs, it was implied that LDCVs release their content only at high frequency stimulation (Lundberg et al., 1994). These findings, seemingly in accord with the vesicle hypothesis and differential regulation of SSVs and LDCVs, however, were not compatible with earlier observations on the co-release of NA, dopamine-β hydroxylase and chromogranins; nevertheless they directly supported the exocytotic release from LDCV

* Werner De Potter, Zesheng Wang, Jan Depreitere and Etienne Nouwen, Laboratory of Neuropharmacology, University of Antwerp, 2610 Wilrijk, Belgium. Anne-Marie Lambeir, Christine Durinx, Laboratory for Medical Biochemistry, University of Antwerp, 2610 Wilrijk, Belgium

Catecholamine Research, Edited by Nagatsu *et al.*
Kluwer Academic/Plenum Publishers, 2002

and further demonstrated that NPY is a convenient marker to evaluate the release from these vesicles (De Potter et al., 1997).

To clarify these conflicting results we used the perfused spleen preparation (dog, sheep or porcine) and examined the release of NA and NPY, under various conditions of stimulation. These included stimulation in the absence and presence of phentolamine (an α-blocking agent well known to potentiate the exocytotic release of NA from sympathetic neurons), stimulation at various frequencies, continuous stimulation versus stimulation in bursts and repeated exhaustive stimulation. In all those conditions the ratio of NA to NPY remained essentially constant. This also applied for the superfused rat vas deferens, a tissue which, in contrast to the dog spleen e.g., contains predominantly SDCVs (more than 90%) and which is believed to release NA exclusively from SDCVs (Stjarne, 2001). Thus also for this tissue NPY (a marker for LDCVs) was released from the nerves and in addition and similarly as for the spleen, exhaustive stimulation also resulted in a gradual and parallel decline in the amounts of NA and NPY released. Moreover for this tissue ω-conotoxine GVIA, a drug which selectively blocks N-type calcium channels, as expected, inhibited the release of NA (IC_{50} =15 nM) yet was equally effective in blocking the release of NPY (IC_{50}=20 nM). On the other hand nifedipine, an L-type calcium channel blocker, had no effect, neither on the release of NA nor of NPY. Similar results have been reported for the release of NA and NPY from the guinea pig heart (Haass et al., 1990). Accordingly, since under these various conditions of stimulation no significant differences were found between the release patterns of NA and NPY, neither for the spleen nor the vas deferens, it was concluded that both the classical transmitter NA and the peptide NPY are released from one pool or secretory vesicle, the LDCV (De Potter et al., 1997). Interestingly, these results are in agreement with the findings that synaptobrevin, a membrane protein which plays a key role in exocytosis, and synaptotagmin, which has a regulatory role in this process, are not only present on SSVs but also on many secretory vesicles, including chromaffin granules and LDCVs of sympathetic neurons , but not on SDCVs (Strecker et al.,2001) .

3. PEPTIDES STORED IN LDCVs AND RELEASED FROM SYMPATHETIC NEURONS

Because of the original demonstration of release of chromogranin A (CgA) from splenic nerves and the increasing evidence for a role of chromogranins as precursor of peptides, we studied the processing of CgA, CgB, SgII and NESP55 during axonal transport using the pig splenic nerve as a model. Therefore use was made of RIA's for peptides earlier identified in bovine chromaffin granules such as GE-25, PE-11, secretoneurin, and GAIPIRRH derived from CgA, CgB, SgII and NESP55, respectively. High-performance gel filtration of extracts of splenic nerve and spleen or spleen perfusate and analysis by RIA revealed that the chromogranins are proteolytically processed to varying degrees during axonal transport. The main finding however was that for the spleen perfusate, CgB, SgII and NESP55 are essentially fully processed to the corresponding free peptides.These data indicate that in the nerve terminals, from which these peptides are released, processing must also be nearly complete. For CgA however, in addition to the free peptide also a significant part of intermediate-sized peptide was still present (Liang et al., 1999; Lovisetti-scamihorn et al., 1999). These results de facto also reveal that peptides very similar or identical to GE-25, PE-11, secretoneurin and GAIPIRRH are secreted from the stimulated porcine splenic nerve.

Subsequently we examined directly the release from the splenic nerve of chromogranin-derived peptides and first focused on pancreastatin (PST).a 49 amino acid peptide derived from CgA.The peptide , first characterized from the porcine pancreas,was shown to inhibit insulin and parathyroid hormone release. Indeed, upon stimulation of the splenic nerve, PST-IR in the spleen perfusate significantly increased and, as expected, was further enhanced in the presence of the α-blocking agent phentolamine. RP-HPLC of the perfusate however revealed that PST-IR was due to both PST(1-49) and PST(33-49), a C-terminal fragment (unpublished results).

In a recent study the cDNA and deduced amino acid sequence of porcine CgB was reported and it was demonstrated that the C-terminal region of CgB (CgB 586-646) can be further processed in the secretory granules of the adrenal medulla to three smaller peptide fragments, i.e. S586-R602 (SR-17), H603-Q636 (HQ-34) and a secretolytin-like peptide fragment KR-11 (Wang et al., 2001). In addition also a N-terminal fragment of porcine CgB, CgB(1-27) (MK-27) has been identified in the adrenal medulla. In further search for the possible release of CgB derived peptides we have focused on MK-27, SR-17 and KR-11.

Using antisera against CgB (bovine) and NPY (porcine), light microscope immunohistochemical studies on the porcine spleen demonstrated a total overlap for the two staining patterns. Moreover, CgB immunoreactivity was only found in nerve fibres and varicosities, indicating that CgB and its derived peptides are exclusively present in sympathetic neurons. Interestingly, CgB positive nerve varicosities were found mainly associated with smooth muscle, yet also occured in T-lymphocyte zones (PALS).

Size fractionation by high-performance gel filtration of spleen extracts and analysis for KR-11, SR-17 and MK-27 IR revealed a single peak eluting at the position of the standards for the first two peptides and somewhat earlier for the MK-27 IR. In the spleen perfusate however a large immunoreactive peak was found to co-elute with MK-27. Stimulation experiments and analysis of samples by ELISA showed a frequency dependent release for the 3 peptides and a further enhancement, as for NPY, when stimulation was performed in the presence of the α-blocking agent phentolamine.

4. INTERACTION OF CHROMOGRANIN DERIVED PEPTIDES WITH CD26

It has been proposed that dipeptidyl-peptidase IV (DPP IV), an exopeptidase earlier identified as the leukocyte differentiation marker CD26, regulates receptor specificity of NPY by removing the N-terminal dipeptide (Zukowska-Grojec et al., 1998). Inspection of the novel molecules revealed that MK-27 and SR-17 have, respectively, a proline and an alanine at the penultimate position and might therefore be substrates for DPP IV/CD26. DPP IV/CD26 is known to play an important co-stimulatory role in T cell activation (De Meester et al., 1999) and is localised on T cells in the spleen (Mentlein, 1999). To assess their ability as substrates, MK-27 and SR-17 were incubated in the presence of DPP IV and truncation was followed by RP-HPLC. After incubation, MK-27(1-27) was totally converted to the VK-25 (3-27) peptide as confirmed by mass spectrometry. The cleavage rate of MK-27 was somewhat slower than for NPY, the best reported peptide substrate for DPP IV. Although even slower, the conversion rate of SR-17 was comparable to other bioactive peptides.

5. CONCLUSIONS

Evidence is presented that LDCVs are the sole secretory vesicles of the sympathetic system directly involved in the release of NA and peptides by exocytosis. Evidence is also provided that chromogranins, similar to pro-enkephalins (Dillen et al., 1993), are proteolytically processed during their axonal transport, a process which is essentially complete at the nerve terminals. Moreover the resulting peptides are co-released with NPY and opioid peptides from sympathetic neurons and presumably also immediately adjacent to lymphoid cells. Therefore and since two of the novel peptides, MK-27 and SR-17, are substrates for DPP IV/CD26, it would be interesting to further study the involvement of these peptides in immune regulation.

REFERENCES

Annaert, W. G., Partoens, P., Slembrouck, D., Bakker, A., Jacob, W., and De Potter, W. P., 1997, Rab3 dissociation and clathrin-mediated endocytosis, two key steps in the exo-endocytotic pathway of large dense cored vesicles.in primary cultures of superior cervical ganglia, *Eur. J. Cell Biol.* **74**:217-229.

De Meester, I., Korom, S., Van Damme, J., and Scharpé, S., 1999, CD26, let it cut or cut it down, *Immunol. Today* **20**:367-375.

De Potter, W. P., Partoens, P., and Strecker, S., 1997, Noradrenaline storing vesicles in sympathetic neurons and their putative role in neurotransmitter release: an historical overview of controversial issues, *Neurochem.Res.* **22**:911-919.

Dillen, L., Miserez, B., Claeys, M., Aunis, D., and De Potter, W. P., 1993, Post-translational processing of proenkephalins and chromogranins/secretogranins, *Neurochem. Int..* **22**:315-352.

Haass, M., Förster, C., Richardt, G., Kranzhöfer, R., and Schömig, A., 1990, Role of calcium channels and protein kinase C for release of norepinephrine and neuropeptide Y, *Am. J. Physiol.* **259**:925-930.

Liang, F., Li, J.-Y., Lovisetti-Scamihorn, P., Coen, E., Depreitere, J., Claeys, M., Wechsung, E., Dahlstrom, A., Winkler, H., and De Potter, W. P., 1999, The processing of secretogranin II in the peripheral nervous system :release of secretoneurin from porcine sympathetic nerve terminals, *Brain Res.* **818**:459-467.

Lovisetti-Scamihorn, P., Liang, F., Leitner, B., De Potter, W. P., and Winkler, H., 1999, Pig splenic nerve :peptides derived from chromogranins by proteolytic processing during axonal transport, *Regul. Pept.* **79**: 63-67.

Lundberg, J. M., Franco-Cereceda, A., Lou, Y. P., Modin, A., and Pernow, J., 1994, Differential release of classical transmitters and peptides, in : *Molecular and Cellular Mechanisms of Transmitter Release*, L. Stjarne, P. Greengard, S. Grillner, T. Hokfelt , and D. Ottoson, ed., Raven Press, Ltd, New York, pp. 223-234.

Mentlein, R., 1999, Dipeptidyl-peptidase IV (CD26)-role in the inactivation of regulatory peptides, *Regul. Pept.* **85**:9-24.

Stjarne, L., 2001, Novel dual 'small' vesicle model of ATP- and noradrenaline-mediated sympathetic neuromuscular transmission, *Auton. Neurosci..* **87**:16-29.

Strecker, S., Partoens, P., Slembrouck, D., Wang, J-M., Courtoy, P.J., Nouwen, E., and De Potter, W.P., 2001, Small dense -cored vesicles are depleted of the classical synaptic vesicle membrane proteins synaptobrevin,synaptotagmin and Rab 3, *Neurochem Res* **28** :173-179.

Wang, Z., Vandenberghe, I. , Depreitere, J., Devreese, B., Clerens, S., Nouwen, E. J., Van Beeumen, J., and De Potter, W. P., 2001, Identification and characterization of novel chromogranin B-derived peptides from porcine chromaffin granules by liquid chromatography/electrospray tandem MS, *Eur. J. Biochem.* **268**:235-242.

Zukowska-Grojec, Z., Karwatowska-Prokopczuk, E., Rose, W., Rone, J., Movafagh, S., Ji, H., Yeh, Y., Chen, W. T., Kleinman, H. K., Grouzmann, E., and Grant, D. S., 1998, Neuropeptide Y: a novel angiogenic factor from the sympathetic nerves and endothelium, *Circ. Res.* **83**:187-195.

TRANSIENT TYROSINE HYDROXYLASE EXPRESSION IN THE NON-CATECHOLAMINERGIC NEURONS OF PRE- AND POST-NATAL MICE

Ikuko Nagatsu,[1*] Kunio Kitahama,[3] Nobuyuki Karasawa,[1] and Keiki Yamada[2]

1. INTRODUCTION

In the 8th ICAS, we reported "Immunocytochemical evidence of novel catecholamine- or biopterin-related neurons of mammalian brain" (Nagatsu et al., 1998). Here we report transient appearance of tyrosine hydroxylase-immunoreactive (TH-ir) cells in the non-catecholaminergic regions in mice (anterior olfactory n., striatum, medial geniculate n, paraventricular n., dorsomedial hypothalamic n., spinal sensory trigeminal n.) simultaneously with expression of GTP cyclohydrolase I (GCH), aromatic L-amino acid decarboxylase (AADC), and dopamine (DA).

2. MATERIALS AND METHODS

C57BL/6J mice and ddy mice were employed from embryonic day 12 (E12) to adult. Antisera were raised in rabbits and well characterized in our laboratory. Peroxidase antiperoxidase method, avidin-biotin peroxidase complex method, and confocal laser scanning microscopy were used (Nagatsu et al., 1997). Using anti-GCH antibody, we could distinguish TH-only neurons from L-DOPA neurons or DA neurons.

3. RESULTS

Transiently expressed TH-ir cells were found in the mouse spinal dorsal root ganglion(DRG) at E14. These neurons were GCH(-) / AADC(-), thus might be TH-only

[1*] Ikuko Nagatsu; Departments of Anatomy, [1]School of Medicine, [2]School of Health Sciences, Fujita Health University, Toyoake 470-1192, Japan, [3] INSERM U480, Neurobiologie des états de sommeils et d'éveil, Faculté de Médecine, 8, Avenue Rockefeller, 9373 LYON CEDEX 08 France.

neurons. In contrast, advanced appearance of TH and delayed expression of GCH and AADC were observed in the noradrenergic neurons of the mouse sympathetic ganglion (SyG) at E14 (Fig.1). We also observed TH-ir / GCH(-) / AADC(-) cells in the neocortex, supraoptic n., amygdala n., and dorsal motor n. of vagus.

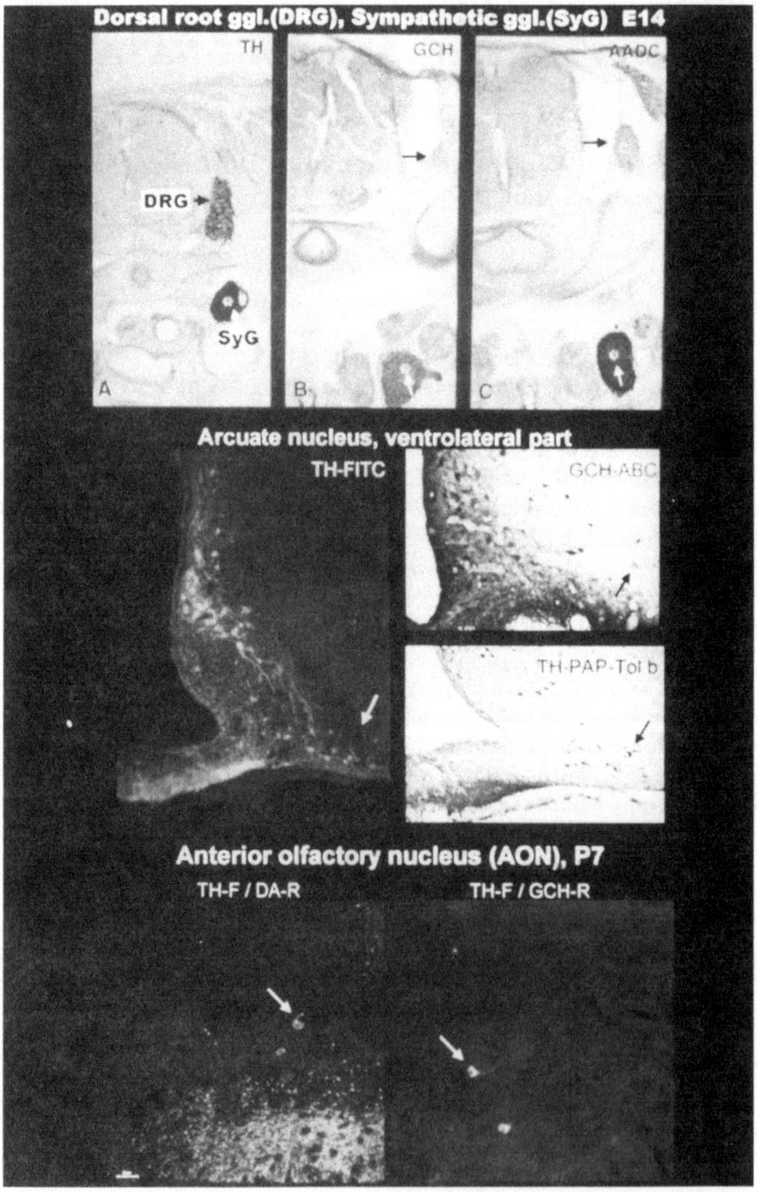

Fig. 1. Transient expression of TH(+)/GCH(-)/AADC(-) cells in the mouse DRG at E14 (top). TH(+)/GCH(+) L-DOPA neurons in the ventrolateral arcuate n (middle), and in the AON at P7 (bottom).

We found TH(+) / GCH(+) / AADC(-) L-DOPA-only neurons in the ventrolateral part of the hypothalamic arcuate nucleus (Fig. 1). L-DOPA neurons were reported in the human ventral tegmental area (Komori et al., 1993). At postnatal day 7 (P7), TH(+) / GCH(+) / DA(-) L-DOPA neurons were present in the anterior olfactory nucleus (AON) (Fig.1).

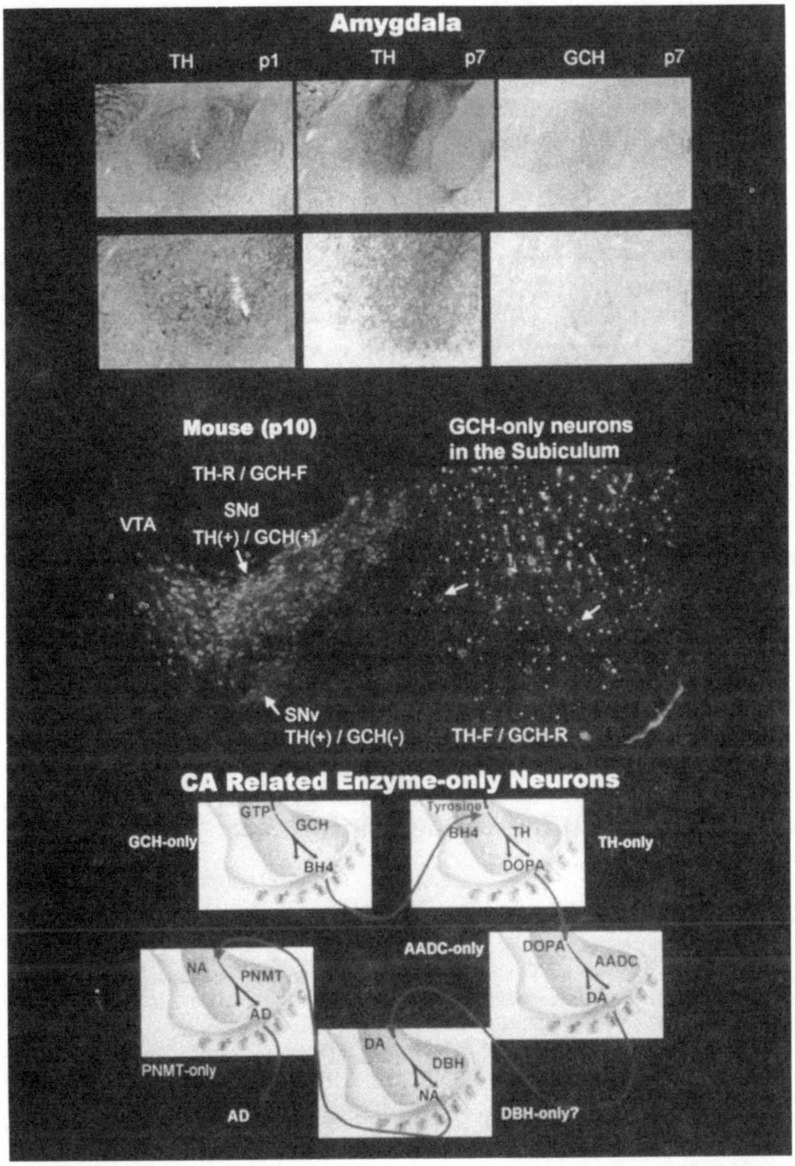

Fig. 2. TH(+)/GCH(-) TH-only neurons in the central n of amygdala from P1~P7 (top), and in the ventral tier of the SN and VTA at P10 (middle left). Transient GCH(+)/TH(-) GCH-only neurons in the subiculum (middle right). Hypothesis for functions of CA related enzyme-only neurons (bottom).

We distinguished TH(+) / GCH(-) TH-only neurons in the central n. of amygdala from P1~P7 (Fig.2). Even in catecholaminergic regions such as the ventral tier of the substantia nigra (SN) and ventral tegmental area (VTA), TH-ir / GCH(-) TH-only cells were distributed among DA neurons at P10 by confocal laser scanning microscopy (Fig.2).

4. DISCUSSION

We previously found transiently appeared GCH(+) / TH(-) GCH-only neurons in the intergeniculate leaflet and ventral lateral geniculate n. at P14 (Nagatsu et al., 1996). Such GCH-only neurons were scattered around many TH(+) / GCH(+) L-DOPA neurons in the subiculum (Fig. 2).

For proving the synthesis of catecholamines in the TH(+) neurons, we tried to determine the coexistence of GCH, a rate-limiting enzyme of tetrahydrobiopterin (BH4) synthesis, which is a cofactor of TH. Based on our present results, BH4 produced in GCH-only neurons may be released from GCH neuron terminals, and may be taken up into TH-only neurons to synthesize L-DOPA. BH4-containing nerve terminals of GCH-only neurons may donate this cofactor not only to the TH-only neurons but also to the neuronal nitric oxide synthase-containing cells.

Furthermore, TH-only neurons might be converted to either DOPA- or DA-forming neurons under developmental changes in order to add new, fully functional neurons into existing brain circuitry (Fig.2).

CONCLUSIONS: Although GCH and AADC are probably present below the level of immunocytochemical detectability, the transiently appeared TH-only cells in the neonatal mouse may synthesize neither DOPA nor DA, and are supposed to be functionally different from TH-ir catecholaminergic neurons. The function of the TH-only neurons remains to be determined.

ACKNOWLEDGMENTS

This work was supported by a grant from the program grants-in-aid for Scientific Research on Priority Areas, Ministry of Education, Science, Sports and Culture of Japan to I.N. and INSERM U480 and CNRS (French Government) for K.K.

REFERENCES

Komori K, Uesaka S, Yamaoka H, Fujita K, Yamaoka K, Naitoh H, Kuroda M, Karasawa N, Ito T, Kasahara Y, Nagatsu I (1993) Identification of L-DOPA immunoreactivity in some neurons in the human mesencephalic region: a novel DOPA neuron group? Neurosci. Lett. **157** 13-16

Nagatsu I, Arai R, Sakai M, Yamawaki Y, Takeuchi T, Karasawa N, Nagatsu T (1997) Immunohistochemical colocalization of GTP cyclohydrolase I in the nigrostriatal system with tyrosine hydroxylase. Neurosci Lett **224** 185-188

Nagatsu I, Sakai M, Karasawa N, Takeuchi T, Arai R, Yamada K, Nagatsu T (1998) Immunohistochemical evidence of novel catecholamine- or biopterin-related neurons of mammalian brain. Adv Pharmacol. **42** 859-862

Nagatsu I, Takeuchi T, Sakai M, Arai R, Karasawa N, Nagatsu T (1996) Transient appearance of GTP cyclohydrolase I-positive non-monoaminergic neurons in the ventral lateral geniculate nucleus of postnatal mice. Neurosci Lett **215** 79-82

GDNF-INDUCED EXPRESSION OF TYROSINE HYDROXYLASE

K. Kiuchi[1, 2] and H. Xiao[1]

1. INTRODUCTION

Glial cell line-derived neurotrophic factor (GDNF) was first identified as a protein with specific neurotrophic activity affecting the dopaminergic neurons of the substantia nigra.[1] Since then, several studies have shown that GDNF plays an important role in mediating the survival of dopaminergic neurons.[2, 3] The signal transduction pathway through which GDNF exerts this effect is thought to be GFRα-1 and Ret-dependent,[4] and to involve MAP kinase and PI-3 kinase.[5, 6] The tyrosine hydroxylase (TH) gene is considered to be a likely target for GDNF signaling, since intrastriatal injection of GDNF enhances nigral TH activity and dopamine content (in postnatal rats[10]). First discovered by Nagatsu *et al.* in 1964,[7] TH catalyzes the rate-limiting step in catecholamine biosynthesis and since then has been the subject of numerous studies investigating the regulation of its expression and activity.[8, 9] The principle processes controlling the number and intrinsic catalytic activity of this enzyme include: (i) modulation of its transcription; (ii) modulation of its translation; (iii) mRNA splicing, (iv) modulation of mRNA stability; (v) phosphorylation; and (vi) feedback inhibition. Here we investigate in more detail the relationship between GDNF stimulation and TH expression.

2. GDNF-INDUCED EXPRESSION OF THE TH GENE

Using RT-PCR and Western blot analyses, we determined that over a time period of 30 h, stimulation with GDNF up-regulates expression of both TH mRNA and protein in TGW cells known to possess Ret[11] (Fig. 1). The level of TH mRNA was increased about 13-fold over that seen in untreated controls, and a concomitant accumulation of TH protein was noted within several hours thereafter. The effect of GDNF seemed to be mainly on TH transcription, since a 15-min pretreatment with Actinomycin D, which inhibits the *de novo* synthesis of RNA, almost completely blocked this induction.

[1] Laboratory for Genes of Motor Systems, Bio-Mimetic Control Research Center, RIKEN, Nagoya 463-0003,
[2] Department of Biomolecular Science, Faculty of Engineering, Gifu University, Gifu 501-1193, Japan

Catecholamine Research, Edited by Nagatsu *et al.*
Kluwer Academic/Plenum Publishers, 2002

3. FROM GDNF STIMULATION TO TH mRNA EXPRESSION

A novel Ret-independent signaling pathway was recently reported to be activated *via* the GDNF/GFRα-1 complex.[12] Nevertheless, in SK-N-MC human neuroblastoma cells, which express only the GFRα-1 co-receptor,[13] GDNF failed to induce expression of TH mRNA under the same culture conditions used for TGW cells (Fig. 2). However, transcription of SK-N-MC cells with exogenous human ret cDNA rendered the resultant SK-N-MC (Ret) cells susceptible to TH gene induction by GDNF. This apparent Ret-dependency was confirmed using SH-SY5Y human neuroblastoma cells, in which intrinsic expression of TH and Ret proteins are up-regulated by stimulation with retinoic acid. In this case, GDNF-induced TH gene expression was correlated with the intrinsic expression of Ret in the SY5Y cells (Fig. 2).

To determine the kinase cascade involved in GDNF-induced TH gene expression *via* Ret in TGW cells, we tested the effects of a variety of protein kinase inhibitors. Two MEK inhibitors, PD098059 and U0126, attenuated TH mRNA expression, whereas a PI-3 kinase inhibitor, LY290042, only weakly affected the induction. Other kinase inhibitors, including SB203580 (p38 kinase) and KT5720 (PKA), also showed little effect. It thus appears that the MAP kinase cascade is the one essential for GDNF-induced up-regulation of TH.

A number of regulatory elements are known to participate in the transcriptional activation of the TH gene.[8, 9] Judging from gel shift assays using nuclear extracts from TGW cells, two AP-1 sites and one CRE site, situated within the 2-kb promoter region upstream of the transcription start site, are involved in the induction of TH mRNA by GDNF. Protein binding to the two AP-1 sites was found to be enhanced following 24 h of

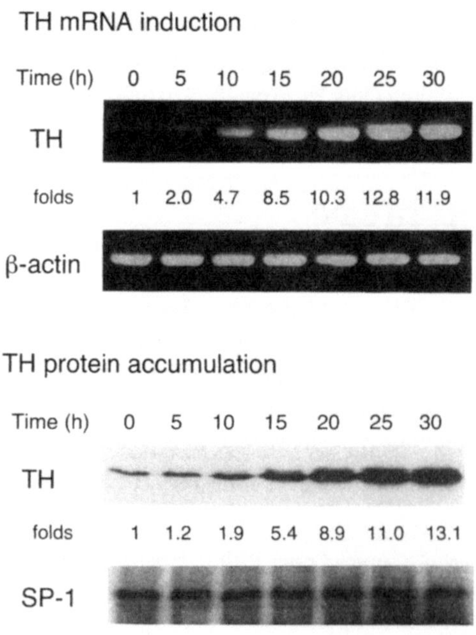

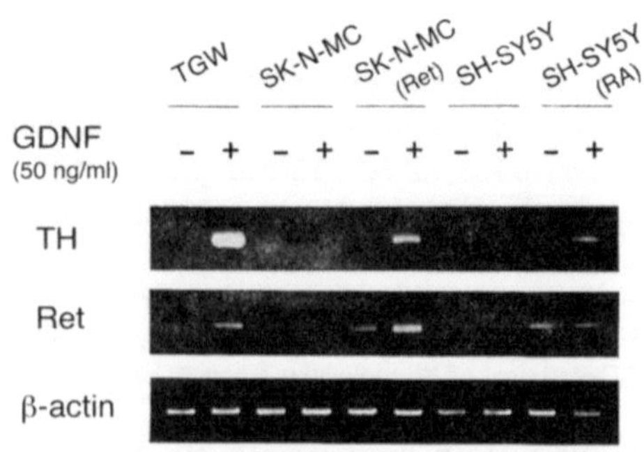

Fig. 1 Time-course of GDNF-induced expression of TH mRNA and protein accumulation in TGW cells. (*left*)
Fig. 2 Correlation of TH mRNA induction by GDNF and Ret mRNA expression in a human neuroblastoma cell-line. (*right*)

exposure to GDNF, whereas protein binding to the CRE site peaked within 2 h. Apparently, the MAP kinase cascade was responsible for the observed accumulation of transcription factors bound to the AP-1 sites, as the MEK inhibitor PD098059 markedly diminished levels of the AP-1 complex. Supershift assays carried out using specific antibodies against members of the Fos/Jun superfamily confirmed the transcription factors primarily complexing with AP-I were c-Fos and c-Jun.

Deletion mutation analysis carried out using a luciferase assay system revealed that the two AP-1 sites and the CRE-site are all necessary for GDNF-induced stimulation of TH gene promoter activity in SK-N-MC (Ret) cells, though the fact that site-directed mutation of the CRE-motif did not severely diminish promoter activity indicated the AP-I motifs to be the more important of the regulatory sites. As shown in Fig. 3, Ret-dependent signals transduced *via* the MAP kinase cascade lead to induction of the immediate early genes c-fos and c-jun, and probably to activation of CREB kinase; the AP-1 complex and phosphorylated CREB would then enhance TH promoter activity.

4. GDNF ENHANCES TH mRNA STABILITY

The rate TH mRNA degradation following Actinomycin D treatment was slowed in GDNF-stimulated cells: the half-lives ($T_{1/2}$) of TH mRNA expressed in GDNF-stimulated and untreated control cells were 24 h and 15 h, respectively (Fig. 4a). The stabilization of TH mRNA by GDNF was confirmed using RNA-EMSA analysis (Fig 4b), which assessed

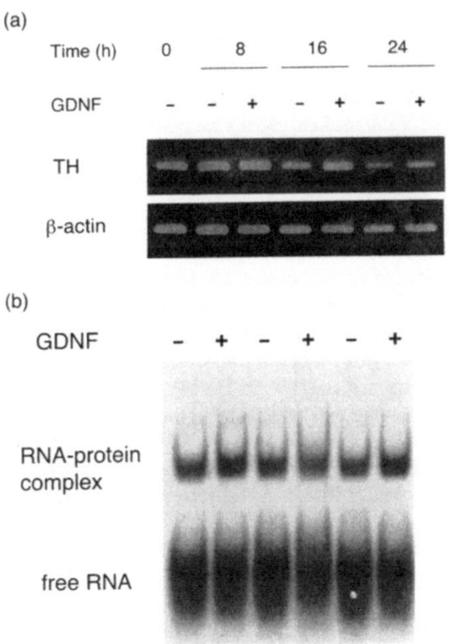

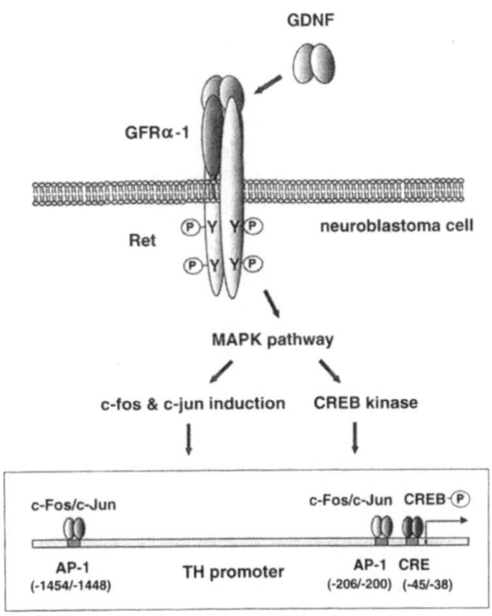

Fig. 3 Signaling pathways leading from GDNF stimulation to TH promoter activation. (*left*)

Fig. 4 (a) Enhanced stability of TH mRNA in GDNF-stimulated TGW cells. (b) Effect of GDNF stimulation on the formation of an RNA-protein complex via the 3'-UTR of TH mRNA. (*right*)

the binding stringency between the 3'-untranslated region (3'-UTR) of TH mRNA and a protein molecule as previously observed.[14] In that regard, it is known that the pyrimidine-rich motif, (U/C)(C/U)CCCU, present in four eukaryotic 3'-UTRs, is responsible for mRNA stability, which is modulated by an RNA-protein interaction.[15]

5. CONCLUSION

GDNF up-regulates TH expression in human neuroblastoma cells primarily by promoting TH gene transcription, but also by increasing the stability of the transcript. Both Ret receptor tyrosine kinase and a MAP kinase cascade play key roles in the signaling pathway leading from GDNF stimulation to TH mRNA expression.

REFERENCES

1. L.F. Lin, D.H. Doherty, J.D. Lile, S. Bektesh, and F. Collins, GDNF: a glial cell line-derived neurotrophic factor for midbrain dopaminergic neurons, *Science* **260**, 1130-1132 (1993).
2. D.M. Gash, Z. Zhang, A. Ovadia, W.A. Cass, A. Yi, L. Simmerman, D. Russell, D. Martin, P.A. Lapchak, F. Collins, B.J. Hoffer, and G.A. Gerhardt, Functional recovery in parkinsonian monkeys treated with GDNF, *Nature* **380**, 252-255 (1996).
3. C. Winkler, H. Sauer, C.S. Lee, and A. Bjorklund, Short-term GDNF treatment provides long-term rescue of lesioned nigral dopaminergic neurons in a rat model of Parkinson's disease, *J. Neurosci.* **16**, 7206-7215 (1996).
4. S. Jing, D. Wen, Y. Yu, P.L. Holst, Y. Luo, M. Fang, R. Tamir, L. Antonio, Z. Hu, R. Cupples, J.C. Louis, S. Hu, B.W. Altrock, and G.M. Fox, GDNF-induced activation of the ret protein tyrosine kinase is mediated by GDNFR-alpha, a novel receptor for GDNF, *Cell* **85**, 1113-1124 (1996).
5. C.A. Worby, Q.C. Vega, Y. Zhao, H.H. Chao, A.F. Seasholtz, and J.E. Dixon, Glial cell line-derived neurotrophic factor signals through the RET receptor and activates mitogen-activated protein kinase. *J. Biol. Chem.* **271**, 23619-23622 (1996).
6. H. Hayashi, M. Ichihara, T. Iwashita, H. Murakami, Y. Shimono, K. Kawai, K. Kurokawa, Y. Murakumo, T. Imai, H. Funahashi, A. Nakao, and M. Takahashi, Characterization of intracellular signals via tyrosine 1062 in RET activated by glial cell line-derived neurotrophic factor, *Oncogene* **19**, 4469-4475 (2000).
7. T. Nagatsu, M. Levitt, and S. Udenfriend, Tyrosine hydroxylase: the initial step in norepinephrine biosynthesis, *J. Biol. Chem.* **239**, 2910-2917 (1964).
8. S.C. Kumer, and K.E. Vrana, Intricate regulation of tyrosine hydroxylase activity and gene expression, J. *Neurochem.* **67**, 443-462 (1996).
9. T. Nagatsu, and L. Stjarne, Catecholamine synthesis and release, *Adv Pharmacol.* **42**, 1-14 (1998).
10. K.D. Beck, I. Irwin, J. Valverde, T.J. Brennan, J.W. Langston, and F. Hefti, GDNF induces a dystonia-like state in neonatal rats and stimulates dopamine and serotonin synthesis, *Neuron* **16**, 665-673 (1996).
11. I. Ikeda, Y. Ishizaka, T. Tahira, T. Suzuki, M. Onda, T. Sugimura, and M. Nagao, Specific expression of the ret proto-oncogene in human neuroblastoma cell lines, *Oncogene* **5**, 1291-1296 (1990).
12. M. Trupp, R. Scott, S.R. Whittemore, and C.F. Ibanez, Ret-dependent and -independent mechanisms of glial cell line-derived neurotrophic factor signaling in neuronal cells, *J. Biol. Chem.* **274**, 20885-20894 (1999).
13. K. Miyazaki, N. Asai, T. Iwashita, M. Taniguchi, T. Isomura, H. Funahashi, H. Takagi, M. Matsuyama, and M. Takahashi, Tyrosine kinase activity of the ret proto-oncogene products in vitro, *Biochem Biophys Res Commun.* **193**, 565-570 (1993).
14. M.F. Czyzyk-Krzeska, and J.E. Beresh, Characterization of the hypoxia-inducible protein binding site within the pyrimidine-rich tract in the 3'-untranslated region of the tyrosine hydroxylase mRNA, *J. Biol. Chem.* **271**, 3293-3299 (1996).
15. M. Holcik, and S.A. Liebhaber, Four highly stable eukaryotic mRNAs assemble 3' untranslated region RNA-protein complexes sharing cis and trans components, *Proc. Natl. Acad. Sci. USA* **94**, 2410-2414 (1997).

CELL-TYPE SPECIFIC GENE REGULATION OF TYROSINE HYDROXYLASE IN THE CENTRAL NERVOUS SYSTEM

Yoshinori Shimizu, Shuei Sugama, Byung Pil Cho and Tong H. Joh*

1. INTRODUCTION

Since 1964 when Nagatsu, Levitt and Udenfriend[1] identified and partially characterized tyrosine hydroxylase (TH) from the rat adrenal gland as the rate limiting enzyme in the catecholamine (CA) biosynthetic pathway, TH has become one of the most studied enzymes in biochemistry, pharmacology and neuroscience. It has been also widely used as a marker protein for catecholamine neurons in the studies on human brain diseases that include behavioral (Schizophrenia, depression, drug abuse, etc.) and neurodegenerative disorders (Parkinson's disease).

Studies on TH regulation have a long history starting in 1960s and various laboratories have contributed to the understanding of how catecholamine neurotransmitters are synthesized and regulated through TH. Joh et al.[2] for the first time demonstrated that increased TH activity by stress or reserpine administration is due to the increased amount of TH protein, and defined TH regulation by two ways, "induction" and "activation". The term "induction" is specified as increased amount of the enzyme that has the same specific activity (activity/protein), and "activation" is termed as the increased enzymatic activity without increasing amount of the enzyme protein. Later, Joh et al.[3] demonstrated TH activation by direct phosphorylation of TH protein through protein kinase A and ATP. In 1980s and 90s, many laboratories include ours have shown TH gene regulation and molecular mechanisms governing TH gene regulation. Since the TH gene is highly regulated in noradrenergic neurons of the locus ceruleus (LC NE), the sympathetic nervous system, and the adrenal medulla by stressful conditions as well as various drugs and chemicals, these tissues are used for most of the in vivo studies on TH regulation. For TH gene regulation studies, TH-expressing clonal cell lines, such as neuroblastoma/

* Yoshinori Shimizu, Shuei Sugama, Byung Pil Cho, Tong H. Joh, The Burke Medical Research Institute, Department of Neurology and Neuroscience, Weill Medical College of Cornell University, White Plains, New York 10605, USA. Correspondence to: Tong H. Joh, Ph. D., Tel.: 914-948-0500 (Ext. 2151); Fax: 914-948-9541; E-mail: tjoh@burke.org

Catecholamine Research, Edited by Nagatsu *et al.*
Kluwer Academic/Plenum Publishers, 2002

pheochromocytoma cell lines are used. At the last International Catecholamine Symposium held in Asiloma (1997), Joh et al.[4] demonstrated cell-type specific gene regulation of TH in the CNS; TH along with c-fos are only induced in LC NE, but not dopaminergic neurons of the substantia nigra pars compacta (SN DA). It has been demonstrated in the authors' laboratory that c-fos is never found to be induced in SN DA,[5, 6] but found in LC NE, the sympathetic neurons and the adrenal medulla where the TH gene is readily up-regulated by many means. The results are indicative for TH gene regulation coupling with c-fos induction. Although many results have shown the critical involvement of the cyclic-AMP response element (CRE) for the basal as well as inducible gene regulatory mechanisms of the TH gene,[7-9] the AP-1 motif in the regulatory gene of TH also plays an important role in TH induction.[10-12] Thus, the lack of c-fos induction in SN DA may be linked to the lack of TH induction. Therefore, in order to find the signal transduction pathway for c-fos induction in LC NE but not in SN DA, we examined expression and activation of various PKs, MAPKs (pERK1/2, pJNK and p-p38MAPK) in LC NE and SN DA by a common stimulus, restraint stress. We demonstrate here highly specific, differential activation of MAP kinases and other PKs in LC NE and SN DA, and propose cell-type specific TH induction through c-fos/c-jun binding to AP-1 as well as CRE/CBP as the common regulatory mechanism for TH gene regulation.

2. METHODS

Male Sprague-Dawley rats were used for these experiments. 2-hour restraint stress was given by placing a rat in a plastic restrainer. Animals were sacrificed immediately after restraint stress or returned to their own cage followed by sacrificed 24 h later. Control-unstressed animals were sacrificed immediately after removal from the animal room. Using immunohistochemical analysis, we examined the availability of known transcription factors, specific protein kinases and activation of the signal transduction system in the control/stressed rat SN DA and LC NE.

3. RESULTS AND DISCUSSION

TH immunoreactivity increased in LC NE 24-h after from the stress (Fig. 1b). No immunohistochemical alteration was observed in SN DA after the stress. pCREB, CBP, PKA, PKC and c-Jun were expressed in control-nonstimulated rat SN DA and LC NE. CaMKIV was expressed only in LC NE, but not in SN DA. CaMII was expressed only in SN DA. Intensive induction of c-Fos was observed only in LC NE (Fig. 2b), but not in SN DA of stressed rats. There is also differential activation of the MAP Kinase signal transduction pathway. Expression of phospho-ERK1/2 was observed only in LC NE (Fig. 2d) and drastically up regulated after the stress (Fig. 2e, f). Immunoreactivity of phospho-p38 MAPK was observed only in SN DA (Fig. 2g). Expression of phospho-p38 MAPK in SN decreased after 2h stress (Fig. 2h) then recovered 24h later (Fig. 2i). The results suggest that the TH gene is differentially and cell-type specifically regulated in the rodent brain.

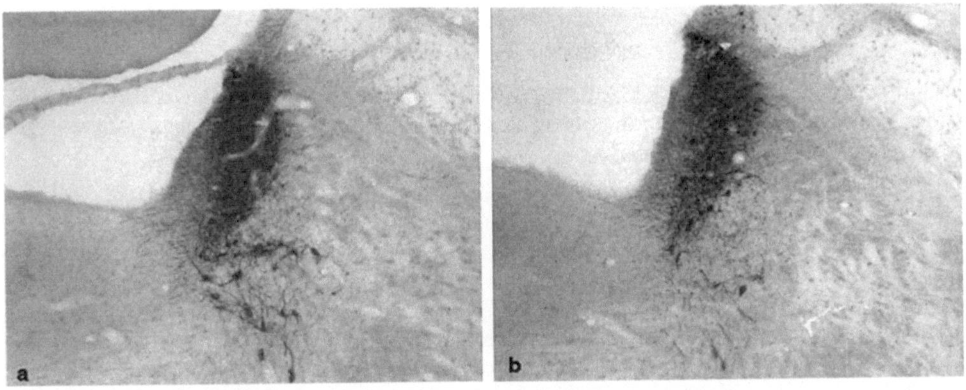

Figure 1. TH immunoreactivity in LC NE in control animals (a) and stressed animals (b). The stress induced expression of TH in LC NE (b).

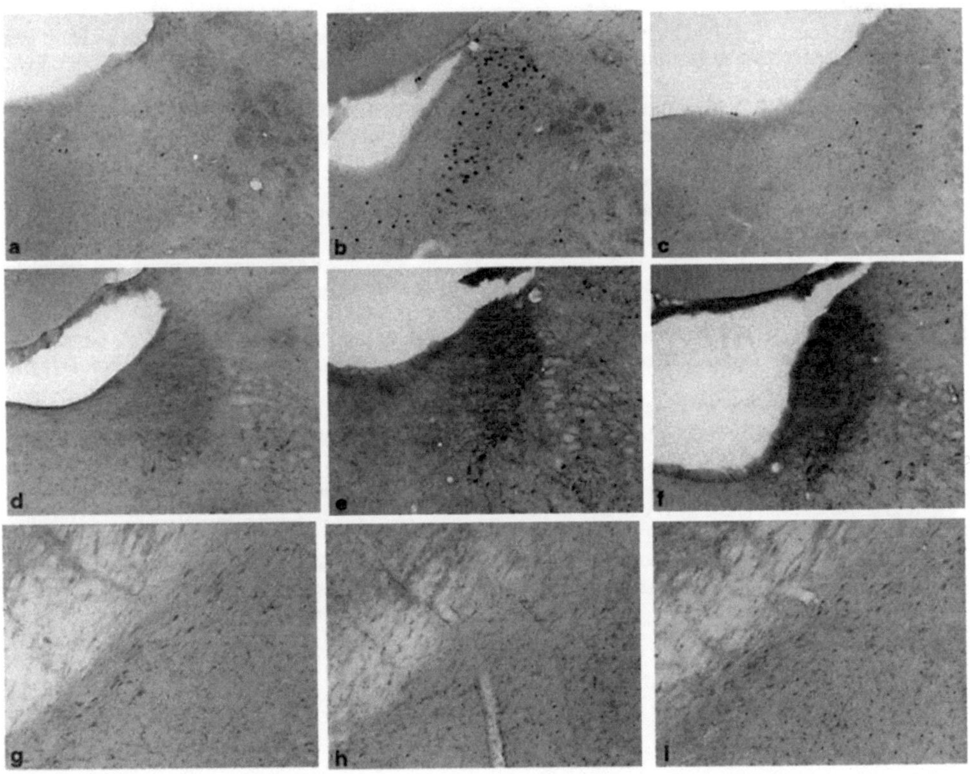

Figure 2. Immunohistochemical analysis of c-Fos (a-c), phospho-ERK1/2 (d-f) and phospho-p38 MAPK (g-i) in LC NE (a-f) and SN DA (g-i). These figures are classified into three groups; control-nonstimulated (left panels), 2h restraint stress (center panels) and 2h restraint followed by sacrificed 24h later (right panels).

4. CONCLUSION

The TH gene in SN DA is not regulated by any known stimuli. Induction of c-Fos, activation of both CaMKs and the MAPK signal transduction pathway may play important roles in cell-type specific TH induction.

REFERENCES

1. T. Nagatsu, M. Levitt, and S. Udenfriend, Tyrsosine hydroxylase: The initial step in norepinephrine biosynthesis, J. Biol. Chem., **239**, 2910-2918 (1964).
2. T. H. Joh, C. Geghman, and D. J. Reis, Immunochemical demonstration of increased accumulation of tyrosine hydroxylase protein in sympathetic ganglia and adrenal medulla elicited by reserpine, Proc. Natl. Acad. Sci., USA, **70**, 2767-2771 (1973).
3. T. H. Joh, D. H. Park, and R. J. Reis, Direct phosphorylation of brain tyrosine hydroxylase by cyclic AMP-dependent protein kinase: mechanism of enzyme activation, Proc. Natl. Acad. Sci., USA, **75**, 4744-4748 (1978).
4. T. H. Joh, J. H. Son, C. Tinti, B. Conti, S. J. Kim, and S. Cho, Unique and cell-type specific tyrosine hydroxylase gene expression, in: Catecholamines, Bridging basic science with clinical medicine, *Adv. Pharmacol*, **42**, edited by D. S. Goldstein, G. Eisenhoffer, and R. McCarty, (1998) pp. 33-36.
5. T. C. Wessel, and T. H. Joh, Parallel upregulation of catecholamine-synthesizing enzymes in rat brain and adrenal gland: effects of reserpine and correlation with immediate early gene expression, Mol. Brain Res., **15**, 349-360 (1992).
6. M. Weiser, H. Baker, T. Wessel, and T. H. Joh, Differential, spatial and temporal gene expression in response to axotomy and deafferentation following transection of the medial forebrain bundle, J. Neurosci., **13**, 3472-3484 (1993).
7. K. S. Kim, D. H. Park, T. C. Wessel, B. Song, J. A. Wagner, and T. H. Joh, A dual role for the cAMP-dependent protein kinase in tyrosine hydroxylase gene expression, Proc. Natl. Acad. Sci., USA, **90**, 3471-3475 (1993).
8. C. Tinti, B. Conti, J. F. Cubells, K. S. Kim, and T. H. Joh, Inducible cAMP early repressor can modulate tyrosine hydroxylase gene expression after stimulation of cAMP synthesis, J. Biol. Chem., **271**, 25375-25381 (1996).
9. C. Tinti, C. Yang, H. Seo, B. Conti, C. Kim, T. H. Joh, and K. S. Kim, Structure/function relationship of the c-AMP response element in tyrosine hydroxylase gene transcription, J. Biol. Chem., **272**, 19158-19164 (1997).
10. K. Nagamoto-Combs, K. M. Piech, J. A. Best, B Sun, and A. W. Tank, Tyrosine hydroxylase gene promoter activity is regulated by both cyclic AMP-responsive element and AP1 sites following calcium influx, J. Biol. Chem., **272**, 6051-6058 (1997).
11. K. M. Piech-Dumas, J. A. Best, Y. Chen, K. Nagamoto-Combs, C. A. Osterhout, and A. W. Tank, The cAMP responsive element and CREB partially mediate the response of the tyrosine hydroxylase gene to phorbol ester, J. Neurochem., **76**, 1376-1385 (2001).
12. Z. Guo, X. Du, and L. Iacovitti, Regulation of tyrosine hydroxylase gene expression during transdifferentiation of striatal neurons: Changes in transcription factors binding the AP-1 site, J. Neurosci., **18**, 8163–8174 (1998).

MECHANISMS OF TYROSINE HYDROXYLASE REGULATION WITH AGE

Nihal Tümer[1, 2], Christopher S. Broxson[1], Haydar A. Demirel[2], Lidia Serova[3], Esther L. Sabban[3], Philip J. Scarpace[1, 2]

1. INTRODUCTION

The incidence of cardiovascular disease increases in the elderly. In particular, the elderly do not regulate blood pressure as well as young people.[1] Plasma catecholamines (CAs) and CA biosynthesis in the adrenal medulla increase with age and may be contributing to the prevalence of hypertension in the elderly[1]. Tyrosine hydroxylase (TH) is the rate-limiting enzyme in catecholamine biosynthesis. TH gene expression and TH activity are elevated in the adrenal medulla of senescent rats. An elevated sympathoadrenal activity with age might play an important role in the pathogenesis of genetically determined blood pressure elevation[2]. The sustained elevation of catecholaminergic activity most likely requires increases in TH gene expression to maintain the augmented CAs. The failure of homeostatic regulation may be contributing to this autonomic dysfunction in the elderly. Autonomic dysfunction not only contributes to medical problems in the elderly, but strongly affects the quality of life in relatively healthy individuals.

The CAs, including dopamine, norepinephrine, and epinephrine are formed from their amino acid precursor, tyrosine, in the brain, in the chromaffin cells of the adrenal medulla, and the sympathetic nerves. TH catalyzes the hydroxylation of tyrosine, producing dopamine, whereas dopamine beta hydroxylase (DßH) activity catalyzes the conversion of dopamine to norepinephrine. TH is generally regarded as the rate-limiting step in the biosynthesis of CAs, and the regulation of TH activity is important in the response of the CAs[3] TH plays a central role in the neuronal transmission and hormonal action of CAs. The exact factors contributing to the increase in TH gene expression with age are unknown but may involve both the protein kinase A and protein kinase C signal transduction pathways.

---

[1]GRECC, VA Medical Center; [2]Dept of Pharmacol., Univ of FL, Gainesville FL 32608;3 Dept. Biochem & Mol. Biol., NY Medical College, Valhalla, NY 10595

2. RESPONSE TO COLD AND LONG-TERM EXERCISE

We have examined adaptations to cold and exercise in young and old Fisher 344 rats. The induction of TH activity was impaired in adrenal medulla of senescent rats following cold-exposure, and the down regulation of TH gene expression was impaired in adrenal medulla of old rats following long-term exercise training. These data indicate that there is a loss of plasticity in CA biosynthesis with age. Stimulation of the known signal transduction pathways involved in TH induction, including cAMP pathway by forskolin [4] and cholinergic pathway by carbachol (see below), indicated that TH regulation by either pathway was not impaired with age. This suggests that neither pathway is involved in the impaired regulation of TH gene expression with cold or long term exercise training with age.

The response to training was also different in the hypothalamus of senescent rats, but opposite of what we found in the adrenals. The old rats upregulated CA biosynthesis with exercise whereas the young rats did not. Exercise training had no significant effect on the level of TH activity, TH immunoreactivity, and TH mRNA in the hypothalamus of young rats. In contrast, training increased the same parameters in senescent rats, such that exercise training reversed the age-related decline in CA biosynthesis.

3. INDUCTION OF DßH AND TH mRNA BY CARBACHOL WITH AGE

In addition to TH, aging is associated with increased DßH activity in the rat adrenal medulla. The cholinergic pathway increases TH expression in the adrenal medulla of young rats[5]. We investigated whether the *in vivo* stimulation by carbachol, a mixed nicotinic-muscarinic agonist, increases TH or DßH expression in old compared with young rats[6]. TH activity, TH mRNA, and DßH mRNA were assessed in the adrenal medulla of 4 and 24 mo-old male F-344 rats with or without carbachol (0.5 mg/kg every 12 hr for 3 days, ip). Animals were sacrificed 5 hrs after the last injection.

Table 1. Effects of Carbachol on TH activity, TH mRNA, and DβH mRNA

	TH activity (nm/AM/hr)		TH mRNA (OD units)		DßH mRNA (OD units)	
	Control	Carbachol	Control	Carbachol	Control	Carbachol
Young	12 ± 1	21 ± 1†	40 ± 3	60 ± 7†	44 ± 6	48 ± 5
Senescent	56 ± 8*	114 ± 26†	60 ± 6*	140 ± 21†	64 ± 6*	93 ± 4†

As expected, TH activity, TH mRNA, and DßH mRNA were significantly higher in the adrenal medulla from control old rats compared with young rats (*p<.05, Table 1). Carbachol significantly elevated TH activity and TH mRNA in the adrenal medulla of both young and old rats (†p<.05, Table 1). In contrast, there was no significant change in DßH mRNA in the adrenal medulla of young rats following carbachol[6]. However, carbachol elevated DßH mRNA in adrenal medulla of old rats compared with controls (†p<.05, Table 1). These results indicate that cholinergic induction of TH or DßH are not impaired with age. It is possible that over-stimulation of the cholinergic pathway in the senescent rat may play a role in the increase in TH activity with age.

4. NPY mRNA LEVELS INCREASE WITH AGE AND ARE INDUCED BY CARBACHOL

In addition to CAs, neuropeptide Y (NPY) is synthesized in the adrenal medulla and is co-released with epinephrine and norepinephrine. It was previously reported that NPY gene expression increased with age in Sprague-Dawley rats[7]. We determined that NPY gene expression increased with age concomitantly with TH mRNA levels in the F344 rats († $p < .001$, Fig 1). Physiological stimulators of TH gene expression such as immobilization stress or cold exposure also increase NPY gene expression in the adrenal medulla.[8] We also assessed NPY mRNA following carbachol. *In vivo,* we found that carbachol increased NPY gene expression in both the young and old rat. Furthermore, the increases in the old, similar to the finding with TH mRNA, were greater than the increases in the young rats ($^*p < .001$). These data suggest that, similar to the young animals, there is also a parallel regulation of both NPY and TH gene expression in senescent rats.

5. SHORT-TERM EXERCISE TRAINING AND CATECHOLAMINE BIOSYNTHESIS WITH AGE

As opposed to the long-term or adaptive effects of exercise training that diminish CA biosynthesis, immediately following a bout of exercise, there is a stimulation of sympathetic outflow.[9] This response is similar to an acute stress response. We examined the acute and adaptive effects of exercise training on TH, DßH, and NPY gene expression in the adrenal medulla in young and senescent rats.

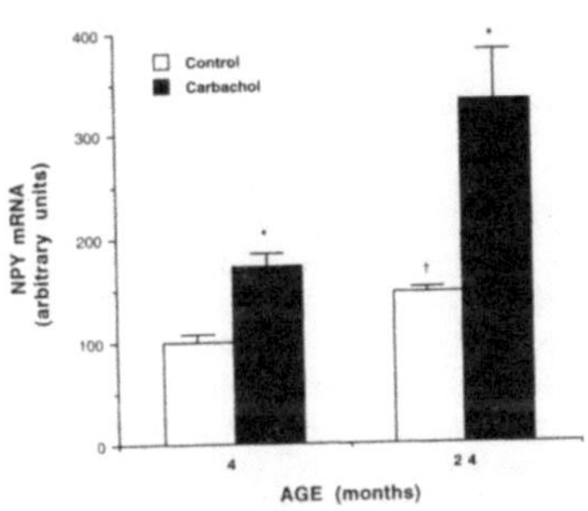

Figure 1. Carbachol increases TH mRNA in young and old rats.

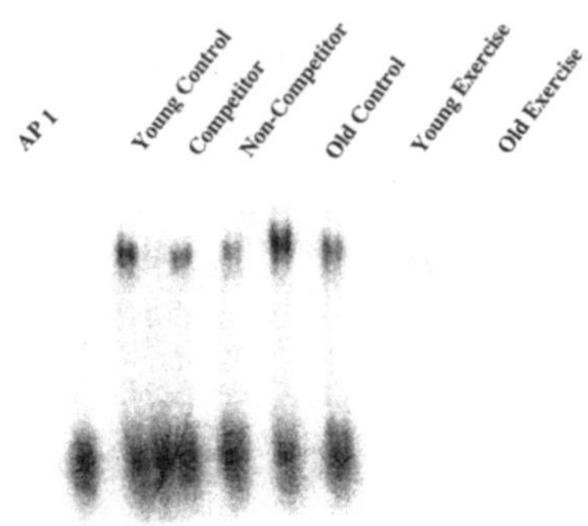

Figure 2. AP-1-DNA binding activity in response to short-term exercise.

In the adrenal medulla, similar to our previous findings, TH mRNA, DßH mRNA, and NPY mRNA levels were significantly elevated with age in control animals.[6] Following exercise training, TH and NPY mRNA were significantly elevated in both the young and old animals, whereas exercise had little effect on DßH mRNA levels in either young or senescent rats.[10]

In addition, we examined whether any exercise-induced changes in TH gene expression are associated with altered complex formation of the transcription factors AP-1 and CREB from the adrenal medulla of young and old rats. Several genomic regulatory elements have been identified in the promoter region of the TH gene, two of the most important being the AP-1 regulatory element and the cAMP response element. We have shown previously that the transcriptional activation of TH in the adrenal medulla of both young and old Fischer-344 rats exposed to cold stimulus, is correlated with increased binding of AP-1 transcription factor to oligonucleotide comprising its AP-1 promoter element.[11] Moreover, many stress-responsive genes, including not only TH but also DßH, contain AP-1-like binding sites in their promoters, and AP-1 factors could be important in establishing stress-induced patterns of gene expression[12]. The exercise protocol used in this study increased the binding activity of AP-1 transcription factor (Fig 2). AP-1 regulatory element binds the products of the immediate early genes, the c-Fos and c-Jun related proteins. AP-1 transcription factor is a dimer of these proteins. However, members of the AP-1 transcription factor family are differentially regulated by single and repeated stress in the rat adrenal medulla, suggesting distinct roles in establishing stress-induced patterns of gene expression in this tissue.[12]

Exercise had no significant effect on CREB binding activity in the adrenals of either young or old animals, whereas exercise evoked similar increases in AP-1 binding activity in the adrenals from both ages. This suggests that the stimulatory pathway that increases binding to the AP-1 regulatory motif is not impaired in the old animals. In line with this, the immobilization stress-mediated induction of mRNA for CA biosynthesis enzymes was also not impaired in the adrenal medulla of senescent rats.[13]

In summary, the induction of TH following cold stimulus and the down-regulation of TH following long-term exercise training are blunted in senescent rats. In contrast, short-term exercise training upregulates TH, DßH and NPY gene expression equally in young and old rats. These data suggest that long-term adaptive responses are impaired in senescent rats and may contribute to the dysregulation of blood pressure with age.

REFERENCES

1. M. L. Tuck, in: Endocrinology of Aging, edited by JA. Armbrecht, R. Coe, and N. Wongsurawat (Springer-Verlag, 1989) pp. 147-160.
2. M. C. Michel, O-E. Brodde, P. A. Insel, Peripheral adrenergic receptors in hypertension. *Hypertension* **16**:107-120, 1990.
3. T. Nagatsu., M. Levitt. and S. Udenfriend, Tyrosine hydroxylase: The initial step in norepinephrine biosynthesis. *J. Biol. Chem.* **238**: 2910-2917 (1964).
4. N. Tumer, C. J. Bowman, J. S. LaRochelle, A. Kelley, and P. J. Scarpace, Induction of tyrosine hydroxylase by forskolin: Modulation with age, *Eur. J. Pharmacol.* **324**:57-62 (1997).
5. M. Stachowiak, E. M. Striker, M. J. Zigmond, and B. B. Kaplan: A cholinergicantagonist blocks cold stress-induced alterations in rat adrenal tyrosinehydroxylase mRNA. *Mol. Brain. Res.* **3**:193-196 (1988).
6. N. Tumer, C. S. Broxson, J. S. LaRochelle and P. J. Scarpace.: Induction of tyrosine hydroxlase and NPY by carbachol: Modulation with age. J *Gerontology: Biological Sciences*, **54A**, B418-B423 (1999).
7. H. Higuchi, K. Yokokawa, A. Iwasa, H. Yoshida, and M. Miki, Age-dependent increase in neuropeptide Y gene expression in rat adrenal gland and specific brain area J. *Neurochem.* **57**:1840-1847 (1991).
8. B. Hiremagalur, R. Kvetnansky, B. Nankova, J. Fleischer, R. Geertman K. Fukuhara. E.Viskupic, and E. L. Sabban Stress elicits trans-synaptic activation of adrenal neuropeptide Y gene expression. *Mol. Brain Res.* **27**, 138-144 (1994).
9. R. S. Mazzeo, Catecholamine responses to acute and chronic exercise *Med. Sci. Sports Exerc.* **23**, 839-844 (1991).
10. N. Tümer, H. A. Demirel, L. Serova, E. L. Sabban, C. S. Broxson, and S.K. Powers: Gene expression of cathecholamine biosynthetic enzymes following exercise: Modulation by age. *Neuroscience.* **103**,703-711, 2001.
11. N. Tümer, P. J. Scarpace, H.V. Baker and J. S. Larochelle, AP-1 transcription factor binding activity in rat adrenal medulla and hypothalamus with age and cold exposure. *Neuropharmacology.* **36**: 1065-1069 (1997).
12. B. B. Nankova, M. Rivkin, M. Kelz, E. J. Nestler, and E. L. Sabban, Fos-related antigen 2: Potential mediator of the transcriptional activation in rat adrenal medulla evoked by repeated immobilization stress. *J. Neurosci.* **20**: 5647-5653 (2000).
13. L. I. Serova, B.B. Nankova, K. Pacak, O. Tjurmina, E. L. Sabban, and R. Kvetnansky Age and stress: differential effect on neurotransmitter gene expression in adrenal medulla and dopaminergic neurons, in *Stress: Neural, Endocrine and Molecular Studies*, edited by R. McCarty, G. Aguilera, E.L. Sabban, and R. Kvetnansky (Harwood Academic Publishers, London, in Press).

TRANSCRIPTIONAL REGULATION OF THE DOPAMINE β-HYDROXYLASE GENE

Kwang-Soo Kim, Hyemyung Seo, Chunhyung Kim and Dong-Youn Hwang*

1. INTRODUCTION

Dopamine β-hydroxylase (DBH) catalyzes conversion of dopamine to norepinephrine and thus is a hallmark protein of noradrenergic (NA) neurons. While DBH is largely co-expressed with tyrosine hydroxylase (TH) in NA and adrenergic neurons, DBH is not expressed in dopaminergic neurons. The highly restricted pattern of DBH expression in the nervous system (typically expressed in <0.0001% of the total neurons) predicts that this gene is subject to neuron-specific as well as to cell type-specific control mechanisms. Therefore, delineation of molecular mechanisms of DBH gene transcription may provide important insight to how NA phenotype is specified during neural development and how its abnormal regulation is related to brain disorders in which NA neurotransmission is dysregulated.

Transgenic mice experiments have shown that the 5' flanking sequences of the human DBH gene can drive expression of the reporter gene in neurons of the locus coeruleus as well as other NA neurons and adrenal chromaffin cells, with some ectopic expression (Mercer et al., 1991; Morita et al., 1993). More detailed analyses of DBH-reporter gene constructs in transgenic animals indicated that the upstream region between -1.1 and -0.6 kb is necessary for expression in adult and fetal NA neurons (Hoyle et al., 1994). In addition, transient transfection assays using cell culture systems have shown that the 5' upstream region of the DBH gene can drive reporter gene expression in a cell-specific manner (Ishiguro et al., 1993; Shaskus et al., 1992). Collectively, both in vivo transgenic mice experiments and in vitro cell culture studies demonstrate that the 5' upstream sequences of the DBH gene have sufficient genetic information for its cell type-specific transcription. It is thus important to identify specific cis-acting elements and their cognate transcription factors for understanding the regulatory mechanism of DBH gene transcription.

---

* Kwang-Soo Kim, Hyemyung Seo, Chunhyung Kim and Dong-Youn Hwang, Molecular Neurobiology Laboratory, McLean Hospital, Harvard Medical School, Belmont, MA 02478

Catecholamine Research, Edited by Nagatsu *et al.*
Kluwer Academic/Plenum Publishers, 2002

2. MULTIPLE CIS-ACTING ELEMENTS RESIDING IN THE 5' PROMOTER CO-OPERATIVELY REGULATE DBH TRANSCRIPTION

Deletional and site-directed mutational analyses indicated that as little as 486 bp of the upstream sequence of the human DBH gene can direct expression of a reporter gene in a cell-specific manner (Ishiguro et al., 1993; Seo et al., 1996). In the 486 bp region of the human DBH gene, the distal part spanning -486 to -263 bp appears to have a cell-specific silencer function which contributed to suppression of the promoter activity in nonneuronal cells (Ishiguro et al., 1993). In contrast, the proximal part spanning -262 to +1 bp was sufficient and essential for the high-level DBH promoter activity in DBH-positive cells (Ishiguro et al., 1993; Seo et al., 1996). In this 262 bp proximal area, four protein-binding regions (domains I to IV) have been identified by DNase I footprinting analysis (Seo et al., 1996). As shown in Figure 1, domain IV is a composite promoter and contains the cAMP response element (CRE), AP1, YY1, and two core motifs of homeodomain (HD) binding sites. Site-directed mutagenesis of each sequence motif has revealed that the CRE is essential for basal promoter activity in every cell line, YY1 is multifunctional, and the AP1-like motif may be transcriptionally inactive (Seo et al., 1996). Interestingly, mutation of either core ATTA motif within domain IV significantly reduced the promoter activity only in DBH-positive cell lines, defining the HD binding sites within domain IV as the NA-specific promoter elements (Yang et al., 1998). In addition, domain II residing at –84 to –66 bp was found to be another NA-specific cis-element and its mutation rendered the DBH promoter to be a non-specific one (Kim et al., 1998). Taken together, multiple cis-acting elements including several cell-specific and general cis-elements synergistically and/or co-operatively regulate the transcriptional activity of the DBH gene (Figure 1).

3. THE DBH GENE IS AN IMMEDIATE TARGET OF PHOX2 PROTEINS

Several homeodomain (HD) proteins have been identified by Brunet and his colleagues (Valarche et al., 1993). One of these transcription factors, termed Phox2 (subsequently renamed Phox2a), is a novel paired-like HD protein found to be expressed restrictively in NA and adrenergic neurons (Valarche et al., 1993; Tiveron et al., 1996). A subsequent gene inactivation experiment showed that Phox2a is critical for development of several NA neuron populations, including the locus coeruleus (Morin et al., 1997). In addition, a closely related Phox2b is widely coexpressed with Phox2a in both the central and peripheral nervous system (Pattyn et al., 1997) and is also critical for development of major NA neurons (Pattyn et al., 2000). A strict correlation between

Phox2a expression and NA phenotype as well as its critical role in development of NA neurons raised the possibility that Phox2a may be involved in specification of neurotransmitter identity. Indeed, analysis of DNA-protein interactions on the DBH promoter demonstrated that both of the NA-specific cis-regulatory elements, i.e., the HD binding sites of domains IV and II, are Phox2a-binding sites (Kim et al., 1998; Yang et al., 1998). Furthermore, using an E.coli-expressed recombinant Phox2a protein, a new Phox2a-binding site has been identified at -101 to -88 bp (Figure 1). Forced expression of Phox2a or Phox2b robustly transactivates the promoter activity of the DBH gene through their interaction with the four binding sites described above (Figure 1). Collectively, these experimental evidences establish that the DBH gene is an immediate downstream target of Phox2 proteins. However, it is at present unclear whether Phox2a/2b directly activate other genes, e.g., TH and norepinephrine transporter, that are critical for NA neurotransmitter phenotype.

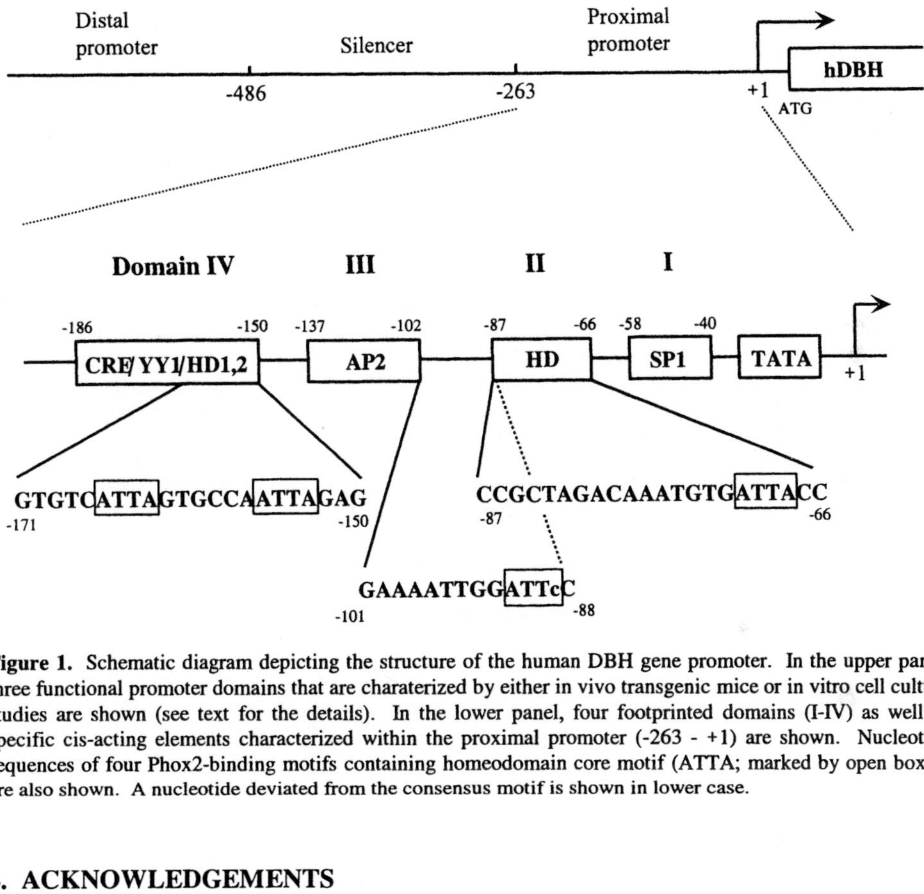

Figure 1. Schematic diagram depicting the structure of the human DBH gene promoter. In the upper panel, three functional promoter domains that are charaterized by either in vivo transgenic mice or in vitro cell culture studies are shown (see text for the details). In the lower panel, four footprinted domains (I-IV) as well as specific cis-acting elements characterized within the proximal promoter (-263 - +1) are shown. Nucleotide sequences of four Phox2-binding motifs containing homeodomain core motif (ATTA; marked by open boxes) are also shown. A nucleotide deviated from the consensus motif is shown in lower case.

4. ACKNOWLEDGEMENTS

This work was supported by NIH grant MH48866 and NARSAD Independent Award (to KSK).

REFERENCES

Hoyle G.W., Mercer E.H., Palmiter R.D., Brinster R.L., 1994, Cell-specific expression from the human dopamine β-hydroxylase promoter in transgenic mice is controlled via a combination of positive and negative regulatory elements, *J. Neurosci.* **14**:2455-2463.

Ishiguro H., Kim K.T., Joh T.H., Kim K.S., 1993, Neuron-specific expression of the human dopamine β-hydroxylase gene requires both cAMP response element and a silencer region, *J. Biol. Chem.* **268**:17987-17994.

Kim H.S., Seo H., Yang C., Brunet J.F., Kim K.S., 1998, Noradrenergic-specific transcription of the dopamine beta-hydroxylase gene requires synergy of multiple cis-acting elements including at least two Phox2a-binding sites, *J. Neurosci.* **18**: 8247-60.

Kobayashi K., Morita S., Mizuguchi T., Sawada H., Yamada K., Nagatsu I., Jujuta K., Nagatsu T., 1994, Functional and high level expression of human dopamine β-hydroxylase in transgenic mice, *J. Biol. Chem.* **269**:29725-29731.

Mercer E.H., Hoyle G.W., Kapur R.P., Brinster R.L., Palmiter R.D., 1991, The dopamine β-hydroxylase gene promoter directs expression of *E. coli* lacZ to sympathetic and other neurons in adult transgenic mice, *Neuron* **7**:703-716.

Morin X., Cremer H., Hirsch M.R., Kapur R.P., Goridis C., Brunet J.F., 1997, Defects in sensory and autonomic ganglia and absence of locus coeruleus in mice deficient for the homeobox gene Phox2a, *Neuron* **18**:411-423.

Morita S., Kobayashi K., Mizuguchi T., Yamada K., Nagatsu I., Titani K., Fujita K., Hidaka H., Nagatsu T., 1993, The flanking region of the human dopamine β-hydroxylase gene promotes neuron subtype-specific gene expression in the central nervous system of transgenic mice, *Mol. Brain Res.* **17**:239-244.

Pattyn A., Morin X., Cremer H., Goridis C., Brunet J.F., 1997, Expression and interactions of the two closely related homeobox genes Phox2a and Phox2b during neurogenesis, *Development* **124**:4065-4075.

Pattyn A., Morin X., Cremer H., Goridis C., Brunet J.F., 1999, The homeobox gene Phox2b is essential for the development of autonomic neural crest derivatives, *Nature* **399**: 366-370.

Seo H., Yang C., Kim H.S., Kim K.S., 1996, Multiple protein factors interact with the *cis*-regulatory elements of the proximal promoter in a cell-specific manner and regulate transcription of the dopamine - hydroxylase gene, *J. Neurosci.* **16**:4102-4112.

Shaskus J., Greco D., Asnani L.P., Lewis E.J., 1992, A bifunctional genetic regulatory element of the rat dopamine β-hydroxylase gene influences cell type specificity and second messenger-mediated transcription, *J. Biol. Chem.* **267**:18821-18830.

Tiveron M.C., Hirsch M.R., Brunet J.F., 1996, The expression pattern of the transcription factor Phox2 delineates synaptic pathways of the autonomous nervous system, *J. Neurosci.* **16**:7649-7660.

Valarche I., Tissier-Seta J.P., Hirsch M.R., Martinez S., Goridis C., Brunet J.F., 1993, The mouse homeodomain protein Phox2 regulates Ncam promoter activity in concert with Cux/CDP and is a putative determinant of neurotransmitter phenotype, *Development* **119**: 881-96.

Yang C., Kim H.S., Seo H., Kim C.H., Brunet J.F., Kim K.S., 1998, *Paired*-like homeodomain proteins, Phox2a and Phox2b, are responsible for cell-specific transcription of the dopamine β-hydroxylase gene, *J. Neurochem.* **71**: 1813-26.

IDENTIFICATION OF ACTIVE SITE AMINO ACID RESIDUES OF PHENYLETHANOLAMINE *N*-METHYLTRANSFERASE

Norio Kaneda, Yoko Itoh, Taeko Fukuo, and Kiyomi Hikita*

1. INTRODUCTION

Phenylethanolamine *N*-methyltransferase (PNMT; EC2.1.1.28) is an enzyme which catalyzes the formation of epinephrine from norepinephrine, the last step of the catecholamine biosynthesis pathway (Axelrod, 1962). PNMT is a monomeric enzyme, and similarly to other mammalian small-molecule *N*-methyltransferase, *S*-adenosyl-L-methionine is required for the enzyme reaction as a methyl donor. In addition to its major role in producing adrenomedullary hormones, PNMT has some important roles in blood pressure and neuroendocrine regulation in the central nervous system. Studies using PNMT inhibitors that act in vivo suggested involvement of epinephrine neurons in blood pressure regulation. However, PNMT inhibitors, more or less, possesses α_2-adrenoceptor blocking activity. Therefore, information regarding the structure of the active site may be useful for development of selective and potent inhibitors for possible use as antihypertensive agents (Fuller, 1982). We have previously isolated a full-length cDNA encoding human PNMT, and revealed the primary structure of the enzyme (Kaneda et al., 1988). Human PNMT consists of 282 amino acid residues with a predicted molecular weight of 30,853 including the initial methionine. Using PCR-based site-directed mutagenesis, we investigated amino acid residues located at the active site of the enzyme.

2. PRODUCTION OF WILD-TYPE AND MUTANT PNMT ENZYMES IN *E. Coli*

Site-directed mutagenesis was performed according to the method of Ito et al. (1991). The entire nucleotide sequences of the mutants were confirmed with a fluorescence DNA sequencer. Wild-type and mutant PNMT cDNAs were subcloned into expression vectors, pKK223-3, and allowed to express the enzyme proteins in *E.*

* Department of Analytical Neurobiology, Faculty of Pharmacy, Meijo University, Nagoya 468-8503, Japan.

Catecholamine Research, Edited by Nagatsu *et al.*
Kluwer Academic/Plenum Publishers, 2002

coli strain JM109 by adding 1 mM isopropyl-β-D-thiogalactopyranoside. Bacterial cells were washed, and homogenized by sonication in a solution of 30 mM TrisHCl (pH 7.5) containing 1 mM dithiothreitol and 1 mM EDTA. After ultracentrifugation of the homogenate, the crude extract was further purified by ammonium sulfate precipitation, followed by anion exchange and gel-filtration chromatographies.

3. ESSENTIAL CYS RESIDUE OF PNMT

Previous studies of chemical modification of the PNMT with reagents specific to Cys such as *p*-chloromercuribenzoate (PCMB) showed that the enzyme contains a Cys residue essential for its activity (Axelrod, 1962; Connett and Kirshner, 1970). By cDNA cloning, we revealed that the human enzyme has six Cys residues at positions 48, 60, 91, 131, 139 and 183 in its primary amino acid sequence. Thus, we changed each Cys residue to Ser to identify the essential Cys residue using PCR-based site directed mutagenesis (Kaneda et al., 1998). The apparent Km value of wild-type enzyme for norepinephrine was 18.4 μM. Those of mutant enzymes, C48S, C60S, C91S, C131S and C139S, ranged from 16 to 20 μM. These value were almost the same as that of the wild-type. These mutants all showed essentially the same Vmax, indicating that these residues had no fundamental effect on enzyme activity. In contrast, C183S showed an apparent Km value of 110 μM for norepinephrine, ca. 6-fold larger than that of the wild-type, and the Vmax value was reduced to only 2.7% of that of the wild-type. Almost the same results were obtained with kinetic parameters for *S*-adenosyl-L-methionine; Only C183S showed a ca. 6-fold larger value of apparent Km, and only 2.6 % of Vmax value compared with the wild-type enzyme. These results strongly suggest that the Cys residue at position 183 is the most important among the six Cys residues for enzyme activity.

4. MUTATIONS IN CONSERVED REGIONS OF MAMMALIAN SMALL-MOLECULE METHYLTRANSFERASES

Overall amino acid sequence homology among several mammalian small-molecule methyltransferases, including PNMT, nicotinamide *N*-methyltransferase (NNMT), indolethylamine *N*-methyltransferase (INMT), thioether *S*-methyltransferase (TEMT), guanidinoacetate *N*-methyltransferase (GAMT), glycine *N*-methyltransferase (GNMT) is relatively low. However, the presence of some amino acid regions conserved through these methyltransferases has been reported (Kagan and Clarke, 1994). Amino acid sequences of regions I and II are aligned as shown in Table 1.

4.1. Mutations in Region I

In region I, two Gly residues are highly conserved, and the GXG motif is reminiscent of GXGXXG, which is known as a nucleotide binding site of NADH in many dehydrogenases (Wierenga and Hol, 1983). When these Gly residues were replaced with Ala to produce G79A and G81A, enzyme activity of G79A was reduced to 1% of the

Table 1. Amino acid sequence of regions I and II conserved among mammalian small-molecule methyltransferases

Enzymes	Region I	Position	Region II	Position	Reference
hPNMT	LIDIGSGPT	75-83	PADALVS	174-180	Kaneda et al., 1988
hNNMT	LIDIGSGPT	59-67	PADCVLS	156-162	Aksoy et al., 1994
hINMT	LIDIGSGPT	59-67	LADCVLT	156-162	Thompson et al., 1999
mTEMT	LIDIGSGPT	60-68	LADCVLT	157-163	Warner et al., 1995
rGAMT	VLEVGFGMA	63-71	LYDTYPL	132-138	Ogawa et al., 1988
rGNMT	VLDVACGTG	60-68	GFDAVIC	131-137	Ogawa et al., 1987

hPNMT, human phenylethanolamine *N*-methyltransferase; hNNMT, human nicotinamide *N*-methyltransferase; hINMT, human indolethylamine *N*-methyltransferase; mTEMT, mouse thioether *S*-methyltransferase; rGAMT, rat guanidinoacetate *N*-methyltransferase; rGNMT, rat glycine *N*-methyltransferase

wild-type enzyme, and G81A showed no enzyme activity. Similar results were reported in GAMT (Hamahata et al., 1996). In GAMT, two conserved Gly residues at position 67 and 69 were shown to be important because replacement of these Gly with Ala resulted in inactivation of the enzyme, probably due to small conformational change as observed with far-ultraviolet CD spectra.

There is a conserved acidic amino acid residue located on the C-terminal side of region I separated by 16-20 residues. It was reported that a conserved acidic residue of Glu89 was important for enzyme activity of GAMT (Hamahata et al., 1996). In PNMT, Asp101 or Glu104 corresponds to this conserved acidic residue. We constructed two mutant enzymes, D101N and E104Q, and assessed the effect of mutation on the enzyme activity. D101N and E104N exhibited 39% and 12 % of the wild-type enzyme activity, respectively. Therefore, these acidic residues are important, but not essential for the enzyme activity.

4.2. Mutations in Region II

In region II, an Asp residue is highly conserved, surrounded by hydrophobic and small neutral amino acid residues. We focused on this region, especially on Asp176, and generated various mutants. In D176A and D176E, the value of Vmax for *S*-adenosyl-L-methionine was about 10% of that of the wild-type, and about 20% in the case of D176N. Almost similar results were obtained regarding Vmax for norepinephrine. In D176E, whereas both D and E have carboxylic side chains, the enzyme activity was reduced to about 10%, and in the case of alteration into Asn (D176N), about 30% activity was remained. From these data, the contribution of Asp176 to enzyme activity seems to be partial, but both size and negative charge of the side chain may be important for the enzyme activity. This situation differs from that reported for GAMT, in which conserved Asp134 in region II was essential (Takata et al., 1994).

4.3. Essential Phe30 in PNMT

We further searched more N-terminal region than region I, and found another conserved region among PNMT, INMT, TEMT and NNMT, which was, however, absent from GAMT, GNMT, catechol *O*-methyltransferase and hydroxyindol *O*-methyltransferase. In order to examine the importance of this region, we constructed deletion mutants of this region. A deletion from positions 26 to 29 of the PNMT had no effect on enzyme activity, but deletion from 30 to 33 resulted in loss of activity. Next, each amino acid from position 30 to 33 was changed into Ala. As a result, only F30A showed no activity. We performed CD spectra measurement and urea denaturation analysis, and found no substantial change in the protein conformation or structural stability between wild-type and F30A mutant enzymes.

REFERENCES

Aksoy, S., Szumlanski, C. L., and Weinshilboum, R. M., 1994, Human liver nicotinamide *N*-methyltransferase: cDNA cloning, expression, and biochemical characterization, *J. Biol. Chem.* **269**: 14835-14840.

Axelrod, J., 1962, Purification and properties of phenylethanolamine *N*-methyltransferase, *J. Biol. Chem.* **237**: 1657-1660.

Connett, R. J., and Kirshner, N., 1970, Purification and properties of bovine phenylethanolamine *N*-methyl transferase, *J. Biol. Chem.* **245**: 329-334.

Fuller, R. W., 1982, Pharmacology of brain epinephrine neurons, *Ann. Rev. Pharmacol. Toxicol.* **22**: 31-55.

Hamahata, A., Takata, Y., Gomi, T., and Fujiolka, M., 1996, Probing the *S*-adenosylmethionine-binding site of rat guanidinoacetate methyltransferase. Effect of site-directed mutagenesis of residues that are conserved across mammalian non-nucleic acid methyltransferases, *Biochem. J.* **317**: 141-145.

Ito, W., Ishiguro, H., and Kurosawa, Y., 1991, A general method for introducing a series of mutations into cloned DNA using the polymerase chain reaction, *Gene* **102**: 67-70.

Kagan, R. M., and Clarke, S., 1994, Widespread occurrence of three sequence motifs in diverse *S*-adenosyl-methionine-dependent methyltransferases suggests a common structure for these enzymes, *Arch. Biochem. Biophys.* **310**: 417-427.

Kaneda, N., Ichinose, H., Kobayashi, K., Oka, K., Kishi, F., Nakazawa, A., Kurosawa, Y., Fujita, K., and Nagatsu, T., 1988, Molecular cloning of cDNA and chromosomal assignment of the gene for human phenylethanolamine *N*-methyltransferase, the enzyme for epinephrine biosynthesis, *J. Biol. Chem.* **263**: 7672-7677.

Kaneda, N., Hikita, K., Naruse, Y., Fukuo, T., Matsubara, K., and Nagatsu, T., 1998, Identification of the essential Cysteinyl residue located in the active site of human phenylethanolamine *N*-methyltransferase, *Biochem. Biophys. Res. Commun.* **249**: 405-409.

Ogawa, H., Date, T., Gomi, T., Konishi, K., Pitot, H. C., Cantoni, G. L., and Fujioka, M., 1988, Molecular cloning, sequence analysis, and expression in *Escherichia coli* of the cDNA for guanidinoacetate methyl transferase from rat liver, *Proc. Natl. Acad. Sci. U. S. A.* **85**: 694-698.

Ogawa, H., Konishi, K., Takata, Y., Nakashima, H., and Fujioka, M., 1987, Rat glycine methyltransferase. Complete amino acid sequence deduced from a cDNA clone and characterization of the genomic DNA, *Eur. J. Biochem.* **168**: 141-151.

Takata, Y., Konishi, K., Gomi, T., and Fujioka, M., 1994, Rat guanidinoacetate methyltransferase. Effect of site-directed alteration of an aspartic acid residue that is conserved across most mammalian *S*-adenosylmethionine-dependent methyltransferases, *J. Biol. Chem.* **269**: 5537-5542.

Thompson, M. A., Moon, E.,. Kim, U. –J., Xu, J., Siciliano, M. J., and Weinshilboum, R. M., 1999, Human indolethylamine *N*-methyltransferase: cDNA cloning and expression, gene cloning, and chromosomal localization, *Genomics* **61**: 285-297.

Warner, D. R., Mozier, N. M., Pearson, J. D., and Hoffman, J. L., 1995, Cloning and base sequence analysis of a cDNA encoding mouse lung thioether *S*-methyltransferase, *Biochim. Biophys. Acta* **1246**: 160-166.

Wierenga, R. K., and Hol, W. G., 1983, Predicted nucleotide-binding properties of p21 protein and its cancer-associated variant, *Nature* **302**: 842-844.

NEURAL MECHANISMS REGULATING PHENYLETHANOLAMINE N-METHYLTRANSFERASE GENE EXPRESSION

Dona L. Wong and T. C. Tai*

1. INTRODUCTION

Stress activates the hypothalamic-pituitary-adrenal (HPA) axis to stimulate hormonal and neural regulatory pathways leading to the release of glucocorticoids and epinephrine into the bloodstream. This response is the classic fight or flight mechanism originally described by Walter B. Cannon in the 1900s.[1] It permits the organism to meet the challenge of a stressor, whether environmental, physiological or psychological. Eventually, however, homeostasis must be restored to ensure the organism's survival. The latter occurs through the activation of inhibitory hormonal and neural feedback pathways impacting each level of the axis to suppress further glucocorticoid and epinephrine production and release.

Patients with psychiatric disorders exhibit apparent HPA axis dysfunction. Basal glucocorticoid levels are highly elevated in these individuals and at best, feedback inhibition of glucocorticoid production is minimal. Adrenal medullary levels of epinephrine are also reportedly elevated, suggesting that neural feedback mechanisms are malfunctioning as well.

The studies which follow focus on identifying neural mechanisms for the genetic control of epinephrine regulated through its biosynthetic enzyme, phenylethanolamine N-methyltransferase (PNMT). PNMT, the final enzyme in the catecholamine biosynthetic pathway, converts norepinephrine to epinephrine.

The adrenal medulla, the primary peripheral source of epinephrine, is innervated by the splanchnic nerve. It has previously been demonstrated that acetylcholine, the classical neurotransmitter released by the splanchnic nerve, activates PNMT gene expression in vivo[2] and in vitro.[3-5] We have recently shown that this neurotransmitter can activate the rat PNMT gene promoter in cells transiently transfected with rat PNMT promoter-luciferase reporter gene constructs and treated with cholinergic agonists. Cholinergic activation appears to occur through stimulation of the cAMP-protein kinase A (PKA) signaling pathway since H89 and PKI, two inhibitors of PKA, can fully block

* Harvard Medical School and McLean Hospital, Belmont, MA 02478.

Catecholamine Research, Edited by Nagatsu *et al.*
Kluwer Academic/Plenum Publishers, 2002

carbachol-induced (100 μM) PNMT promoter activity. However, forskolin (10 μM), which stimulates cAMP production, more markedly induces PNMT promoter driven-luciferase reporter gene expression in transient transfection assays in PC12 cells, suggesting that another neurotransmitter may participate in cAMP-PKA-mediated activation of the PNMT gene promoter.

Many peptidergic transmitters are also released from the splanchnic nerve. While it has previously been suggested that pituitary adenylate cyclase activating polypeptide (PACAP) may not affect PNMT gene expression,[6, 7] the present studies demonstrate that PACAP markedly increases PNMT promoter expression. In part, the activation occurs through recruitment of the cAMP-PKA signal transduction pathway. In addition, both intracellular and extracellular calcium is mobilized as well. These neurally induced changes occur primarily through PAC1 receptor activation and the effects of PACAP occur rapidly and are sustained.

2. PACAP ACTIVATION OF THE RAT PNMT PROMOTER

Figure 1 shows a time course for PACAP (10 nM) activation of the rat PNMT promoter. The PNMT promoter-luciferase reporter gene construct pGL3RP893 was transiently transfected into PC12 cells using polyethylenimine (PEI) and luciferase activity measured after different times of drug treatment. PNMT promoter-driven luciferase reporter gene expression rapidly increases in response to PACAP and continues to rise through 24 h.

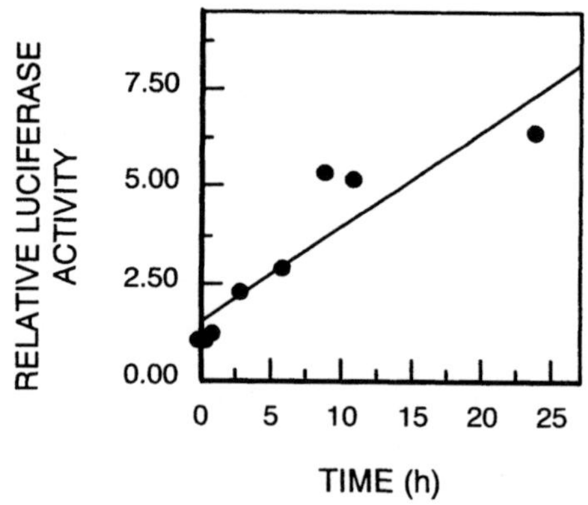

Figure 1. TIME COURSE FOR PACAP ACTIVATION OF THE PNMT PROMOTER. Transfected PC12 cells were treated with 10 nM PACAP for times up to 24 h and luciferase reporter gene expression quantitified. Luciferase activity is expressed relative to the activity of untreated control cells.

It is known that the rat adrenal medulla predominantly expresses PAC1 receptors, with limited expression of VPAC1 receptors.[8] To determine the receptor subtype mediating PACAP-stimulated PNMT promoter activity, PC12 cells transfected with pGL3RP893 were treated with PACAP at concentrations ranging from 0-100 nM and luciferase activity measured after 24 h. PACAP activation of the PNMT promoter increases in a linear manner up to 10 nM where activity levels plateau. In contrast, VIP activation of the PNMT promoter did not reach maximum levels until a concentration of at least 1 μM. These results are consistent with PACAP stimulation of the PNMT promoter occurring through the activation of PAC1 receptors.

Activation of PAC1 receptors has been linked to recruitment of the cAMP-PKA signaling pathway and mobilization of external calcium through L-type Ca^{+2} channels. To investigate the potential role of cAMP and PKA, the transiently transfected cells were pretreated with the PKA inhibitor, H89, prior to PACAP treatment. Alternatively, the PC12 cells were co-transfected with an expression construct for the PKI, a polypeptide inhibitor of PKA along with pGL3RP893 and then exposed to 10 nM PACAP. Both H89 and PKI markedly reduced PACAP stimulation of PNMT promoter activity, although not fully.

The effects of the L-type Ca^{+2} channel blockers, D600 (100 μM) and nifedipine (50 μM), on PACAP driven-PNMT promoter activity were also examined by pretreatment of PC12 cells transfected with the PNMT promoter-luciferase reporter gene construct with D600 or nifedipine, followed by exposure to 10 nM PACAP. While both drugs partially prevented PACAP stimulation of the PNMT promoter, the magnitude was much less than observed with the cAMP-PKA inhibitors (30% versus 70%). In contrast, other Ca^{+2} blockers, e.g. N channel blockers, ω-conotoxin MVIIA and GVIA, the P/Q channel blocker, ω-conotoxin MVIIC, and the T channel blocker, flunarizine, had no effect on PACAP-stimulated PNMT promoter activation. To further investigate the requirement for external Ca^{+2}, the effect of the L-type Ca^{+2} channel agonist, 2,5-dimethyl-4-[2-9phenylmethyl)-benzoyl]-1H-pyrrole-3-carboxlic acid methyl ester (FPL) on PNMT promoter activation was examined. PC12 cells transfected with the rat PNMT promoter-luciferase reporter gene construct pGL3RP893 were exposed to FPL at concentrations ranging from 0-100 μM and changes in luciferase activity determined after 24 h. PNMT promoter driven-luciferase reporter gene expression increased linearly in response to FPL treatment at concentrations between 1-50 μM. However, at 100 μM, FPL stimulation of the PNMT promoter was clearly reduced and approximately equivalent to the stimulation observed with 1 μM FPL. Compared to transfected cells treated with 10 nM PACAP, the maximum increase in PNMT promoter activity (50 μM, 3.5-fold untreated control) was approximately 60% of that observed with the PACAP treated cells. The decrease in PNMT promoter responsiveness observed in the cells treated with 100 μM FPL was most likely due to toxicity of the drug since those cultures looked highly compromised.

The role of intracellular Ca^{+2} in PNMT promoter activation was also examined using a variety of intracellular Ca^{+2} inhibitors. BAPTA-AM, a Ca^{+2} chelator, ryanodine and xestospongin C, which block the release of Ca^{+2} from the endoplasmic reticulum (ER) by inhibiting ryanodine and IP_3 receptor releasable pools, had no significant effect on PACAP-stimulated PNMT promoter activity. However, thapsigargin, which blocks the Ca^{+2}-ATPase pump responsible for replenishing Ca^{+2} in the ER, significantly induced the PNMT promoter beyond levels observed with PACAP alone. Likely, blockade of the Ca^{+2}-ATPase pump sustains elevated cytosolic levels of Ca^{+2} and consequently leads to incremental PNMT promoter stimulation.

In conclusion, PACAP, a polypeptide neurotransmitter released from the splanchnic nerve, can markedly activate the rat PNMT promoter. Activation is rapid and sustained at least through 24 h, and the magnitude of activation by PACAP is much greater than that observed previously in response to cholinergic stimulation. PACAP appears to exert its influence mainly through the cAMP-PKA signaling pathway. In addition, external Ca^{+2} mobilization, apparently through L-type Ca^{+2} channels and sustained elevation of intracellular Ca^{+2} through inhibition of the Ca^{+2}-ATPase pump can contribute to PACAP induction of the PNMT promoter.

3. ACKNOWLEDGEMENTS

This work was supported by a grant from the NIDDK, DK51025, the Spunk Fund, Inc. and the Sobel and Keller Fund.

4. REFERENCES

1. W. G. Cannon and D. De La Paz, Emotional stimulation of adrenal secretion, *Am J. Physiol.*, **28**, 64-70 (1911).
2. R. L. Patrick and N. Kirschner, Acetylcholine-induced stimulation of catecholamine recovery in denervated rat adrenals after reserpine-induced depletion, *Mol. Pharmacol.*, **7**, 389-396 (1971).
3. M. K. Stachowiak, J. S. Hong and O. H. Viveros, Coordinate and differential regulation of phenylethanolamine N-methyltransferase, tyrosine hydroxylase and proenkephalin mRNAs by neural and hormonal mechanisms in cultured bovine adrenal medullary cells, *Brain Res.*, **510**, 277-288 (1990).
4. M. J. Evinger, P. Ernsberger, S. Regunathan, T. H. Joh and D. J. Reis, A single transmitter regulates gene expression through two separate mechanisms: Cholinergic regulation of phenylethanolamine N-methyltransferase mRNA via nicotinic and muscarinic pathways., *J. Neurosci.*, **14**, 2106-2116 (1994).
5. K. Morita and D. L. Wong, Role of Egr-1 in cholinergic stimulation of phenylethanolamine N-methyltransferase promoter, *J. Neurochem.*, **67**, 1344-1351 (1996).
6. C. Tonshoff, L. Hemmick and M. J. Evinger, Pituitary adenylate cyclase activating polypeptide (PACAP) regulates expression of catecholamine biosynthetic enzyme genes in bovine adrenal chromaffin cells, *J. Mol. Neurosci.*, **9**, 127-140 (1997).
7. H. J. Choi, S. Y. Park and O. Hwang, Differential involvement of PKA and PKC in regulation of catecholamine enzyme genes by PACAP, *Peptides*, **20**, 817-822 (1999).
8. D. Vaudry, B. J. Gonzalez, M. Basille, L. Yon, A. Fournier and H. Vaudry, Pituitary adenylate cyclase-activatin polypeptide and its receptors: From structure to functions, *Pharmacological Reviews*, **52**, 269-324 (2000).

REGULATION OF TETRAHYDROBIOPTERIN BIOSYNTHESIS

Catalytic Residues of Sepiapterin Reductase, the Terminal Enzyme of BH_4-Biosynthesis

Setsuko Katoh,[1*] Hirotaka Yamada,[1] Yasumitsu Sakamoto,[2] Takamasa Nonaka,[2] and Kengo Fujimoto[1]

1. INTRODUCTION

Tetrahydrobiopterin (6R-(L-*erythro*-dihdroxypropyl)-5,6,7,8-tetrahydropterin; BH_4) is required as an essential intracellular cofactor of aromatic amino acid hydroxylases in the formation of catecholamines. This compound is a very unstable substance having a very fast turnover *in vivo*, and its tissue level is controlled at least by the rate of *de novo* synthesis. BH_4 is biosynthesized from GTP through two intermediates, which are 7,8-dihydroneopterin triphosphate (H_2NTP) and 6-pyruvoyl-5,6,7,8-tetrahydropterin (PTP) and the three processes are mediated in turn by GTP cyclohydrolase I (GTPCH), 6-pyruvoyl-tetrahydropterin synthase (PTPS), and sepiapterin reductase (SPR) (Fig 1). Crystallographic studies have revealed the subunit conformations of these three enzymes[1-3]; and their corresponding genes (GCH1, PTS, & SPR) have been cloned and each of the genes is located on a different human chromosome (14q22.1-22.2, 11q22.3-23.3, & 2p13, respectively). So far, only the gene expression of GTPCH has been found to be regulated by cytokines[4] or cAMP[5], whereas PTPS and SPR genes have house-keeping functions and are expressed constitutively. Depending on the cell or tissue type, however, all three enzymes are thought to be constitutively expressed, such as in some brain regions and in the liver. And various interesting systems for post-translational modifications have been observed to affect the activities of all enzymes. Feedback regulation is known to occur for GTPCH and SPR (Fig 1). End product BH_4 regulates both GTPCH and SPR. There is a specific feedback regulatory protein for GTPCH (GFRP) in tissues including brain and it works as an effective end-product feedback inhibitor when it binds with BH_4[6]. Excess noradrenaline and adrenaline or serotonin and

* [1]Department of Biochemistry, Meikai University School of Dentistry, Sakado, Saitama 350-0283, Japan. [2]Department of BioEngineering, Nagaoka University of Technology, Nagaoka, Niigata 940-2188, Japan., *To whom correspondence should be addressed. E-mail,katohs@dent.meikai.ac.jp

melatonin directly inhibit SPR activity.[7] Especially N-acetylserotonin (NAS), which is a precursor of melatonin, strongly inhibits SPR competitively against pteridine substrate.[7]

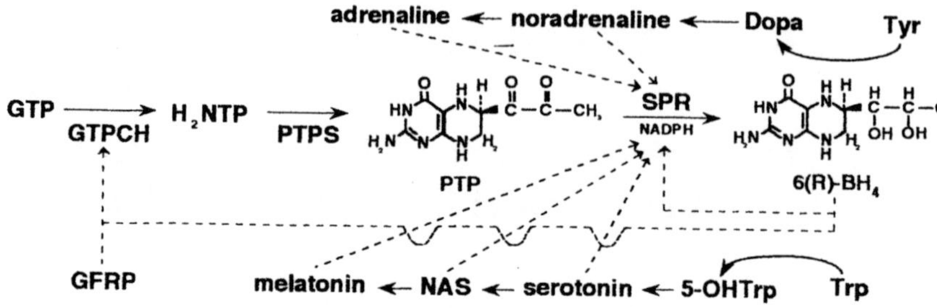

Figure 1. BH_4-Biosynthesis and its feedback inhibition. BH_4, tetrahydrobiopterin; H_2NTP, 7,8-dihydroneopterin triphosphate; PTP, 6-pyruvoyl tetrahydropterin; GTPCH, GTP cyclohydrolase I; PTPS, pyruvoyl tetrahydropterin synthase; SPR, sepiapterin reductase; GFRP, GTP cyclohydrolase feedback regulatory protein; NAS, N-acetylserotonin.

Since PTP is more unstable than BH_4, rapid reduction of it with NADPH by SPR perhaps results in the effective formation of BH_4. To know more clearly the activation mechanism of SPR, we recently estimated the catalytic site and the binding regions for pterin substrate and NADP(H) of SPR by constructing various mutants of the enzyme by site-directed mutagenesis.

2. MATERIALS AND METHODS

SPR belongs to the short-chain dehydrogenase/reductase (SDR) family based on its molecular homology.[8, 9] SPR has the conserved Ser and YXXXK motif, implicated as the catalytic site in the SDR family,[10] as only one set in the C-terminal region (Table 1). Single- or double point mutants against the Ser-YXXXK motif of SPR were constructed from rat SPR cDNA by use of Quick Change™ site-directed mutagenesis kit (Stratagene, CA).[11] Clone encoding rat SPR was amplified by PCR. Plasmids containing wild-type and mutant SPR cDNA were used for transformation of *E. coli* (GI724), and the expressed SPRs were purified by affinity chromatography. Deletion mutants of rat SPR (rSPR13-263, rSPR29-262, rSPR34-262, and rSPR1-227) were also constructed by PCR.[11] SPR activity was measured by a photometric method described previously.[8]

3. RESULTS AND DISCUSSION

A single-point mutation in S158, Y171, or K175 did not change in Km values against a pteridine substrate and NADPH but remarkably decreased the activity[11] (Table 2). Values for kcat/Km of each of these mutants, however, retained about 25% activity of that of wild-type. Then we constructed a double-point mutant in which Tyr and Ser were simultaneously changed to other amino acids. The mutant S158D/Y171V did not show any activity[11] (Table 2). Km values of all single-point mutants for pteridine and NADPH were unaffected by the mutation. This result indicates that all three residues, Ser, Tyr, and Lys, are required for the activity of SPR, and these residues also work as the catalytic site just as the Ser and YXXXK motif of the SDR family. In the catalytic mechanisms of the Ser and YXXXK motif of typical SDR-family members,[10, 12] Lys stabilizes the position and orientation of the nicotinamide nucleotide moiety of the bound NADPH, and both Tyr and Ser are located near the carbonyl group of the substrate. Tyr (or Ser) functions in proton transfer to the carbonyl group of the substrate, and Ser (or Tyr) stabilizes the position of the substrate.[10, 12] In the SDR family, one-point mutants against Tyr or Ser of the motif generally lose almost all activity. However, SPR is a peculiar enzyme of the SDR family because when either Tyr or Ser is mutated to another amino acid, in this study, the remaining residue can work alone for proton transfer to a low degree to express low but significant activity. In this case, Ser or Tyr in SPR probably can act with a water molecule as a proton stabilizer instead of Tyr or Ser, and reduce the carbonyl group of the substrate.[11]

Table 1. The S and YXXXK motifs in sepiapterin reductases

rat	S158	Y171XXXK175
mouse	S158	Y171XXXK175
human	S157	Y170XXXK174
Drosophila	S152	Y166XXXK170

Table 2. Activities of SPR mutants in the S and YXXXK motif and D in C-terminus.

	Sepiapterin	
Mutants	Km(μM)	kcat/Km($s^{-1} \cdot \mu M^{-1}$)
Wild-type	12.6	0.8
S158D	14.4	0.2
Y171V	13.3	0.2
K175I	13.0	0.2
S158D/Y171V	-	0.0
D258A	22.3	0.4
D258S	23.5	0.4

Deletion mutants of rSPR29-262, rSPR34-262, and rSPR1-227 indicated complete loss of activity[11] and one point-mutants against Asp258 show significant increase in Km value for the pteridine substrate (Table 2). SPR has a Rossmann fold (GXXXGXG) of NADP(H) binding in the N-terminal region (G15XXXG19XG21) and a specific Asp residue (Asp258) in the C-terminus has been indicated as an anchoring residue for pteridine moiety by the three-dimensional structure of mouse SPR.[3] The kcat/Km of the rSPR13-262 mutant remarkably decreased.[11] These indicated that the N-terminal region of residues from 13 to 33 is required for NADPH-binding, and residues from 1 to 12 play an important role in supporting the appropriate conformation of the catalytic domain. The C-terminal region of residues from 228 to 262 may be important for pteridine binding and Asp258 is an important residue for pteridine binding.

From these mutagenesis studies, the active site of rat SPR is illustrated by the molecular docking of PTP and NADPH into the crystal structure of SPR (Fig 2). The nicotiamide moiety of NADPH, the Tyr at 171, and Ser at 158 are located within hydrogen bond distance of the carbonyl group of the substrate, PTP. Lys at 175 stabilizes the ribose of NADPH. Asp at 258 is depicted as anchoring N3 and the amino group of the pteridine ring. By X-ray analysis in mouse SPR[3], Asp at 258 was supposed to serve also as an anchoring residue for NAS, a feedback inhibitor of SPR.[7] However, point-mutants for this residue, in this study, were still inhibited by NAS. Thus, the anchoring residue of NAS is still unknown.

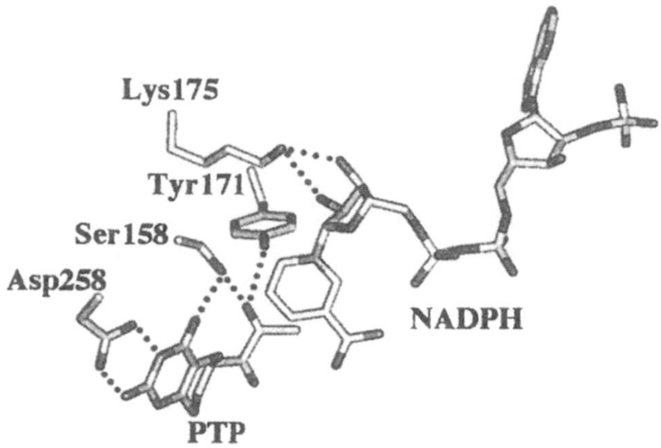

Figure 2. Active site of rat SPR. Localization of Ser158, Tyr171, and Lys175 of the catalytic Ser and YXXXK motif and Asp258 and their hydrogen bondings between 6-pyruvoyl tetrahydropterin and NADPH.

4. ACKNOWLEDGEMENT

This work was supported, in part, by Grant 12771109 from the Ministry of Education, Science, Sports, and Culture of Japan.

REFERENCES

1. H. Nar, R. Huber, W. Meining, C. Schmid, S. Weinkauf, and A. Bacher, *Structure* **3**, 459-466 (1995).
2. T. Ploom, B. Thöny, J. Yim, S. Lee, H. Nar, W. Leimbacher, J. Richardson, R. Huber, and G. Aüerbach, *J. Mol. Biol.* **286**, 851-860 (1999).
3. G. Aüerbach, A. Herrmann, M. Gutlich, M. Fisher, U. Jacob, A. Bacher, and R. Huber, *EMBO J.* **16**, 7219-7230 (1997).
4. E. R. Werner, G. Werner-Felmayer, G. Weiss, and H. Wachter, *Adv. Exp. Med. Biol.* **338**, 203-209 (1993).
5. G. Kapatos, S. L. Stegenga, and K. Hirayama, *J. Biol. Chem.* **275**(8), 5947-5957 (2000).
6. T. Harada, H. Kagamiyama and K. Hatakeyama, *Science* **260**, 1507-1510 (1993).
7. S. Katoh, T. Sueoka, and S. Yamada, *Biochem.Biophys. Res. Commun,* **105**,75-81 (1982).
8. S. Katoh, and T. Sueoka, *Biochem. Biophys. Res. Commun.* **118**, 859-866 (1984).
9. R. Oyama, S. Katoh, T. Sueoka, M. Suzuki, H. Ichinose, T. Nagatsu, and K. Titani, *Biochem. Biophys. Res. Commun.* **173**, 627-631 (1990).
10. H. Jörnvall, B. Persson, M. Krook, S. Atrian, R. Gonzalez-Duarte, J. Jeffery, and D. Ghosh, *Biochem.* **34**, 6003-6013 (1995).
11. K. Fujimoto, H. Ichinose, T. Nagatsu, T. Nonaka, Y. Mitsui, and S. Katoh, *Biochim. Biophys. Acta,* **1431**, 306-314 (1999).
12 .J. M. Jez, M. J. Bennett, B. P. Schlegel, M. Lewis, and T. M. Penning, *Biochem. J.* **326**, 625-636 (1997).

GTP CYCLOHYDROLASE I GENE EXPRESSION AND CATECHOLAMINE SYNTHESIS

Gregory Kapatos* and Kei Hirayama*

1. INTRODUCTION

The reduced pteridine 5,6,7,8-tetrahydrobiopterin (BH4) is the essential cofactor for the family of pterin-dependent monooxygenases that includes tyrosine (TH), tryptophan (TPH) and phenylalanine hydroxylase, the rate-limiting enzymes in the synthesis of the monoamine (MA) neurotransmitters dopamine (DA), norepinephrine and serotonin (5-HT) and the detoxification of L-phenylalanine, respectively (Kaufman, 1974). BH4 serves as an electron donor in these enzyme reactions. GTP cyclohydrolase I (GTPCH) catalyzes the first and rate-limiting step in *de novo* BH4 biosynthesis, the formation of the first pterin intermediate, D-erythro-7,8-dihydroneopterin triphosphate, from GTP (Nichol et al., 1985). Within the brain GTPCH mRNA (Hirayama et al., 1993) and protein (Nagatsu et. al., 1997; Dassesse et. al., 1997) can only be detected within MA neuronal cell bodies. GTPCH protein is transported to the MA nerve terminals, indicating that BH4 is synthesized at its primary site of action (Levine et al., 1981). The rate of BH4 synthesis actually exceeds that of the MA neurotransmitters, with 25% of the intracellular pool being synthesized each hour (Kapatos, 1990). BH4 levels within DA nerve terminals are subsaturating for TH (Kettler et al., 1974) while BH4 levels within 5-HT nerve endings saturate TPH (Wolf et al., 1991). This distinction based upon neurotransmitter phenotype may be important clinically (see below) and presumably is due to the low levels of GTPCH mRNA (Lentz and Kapatos, 1996) and GTPCH protein (Hirayama and Kapatos, 1998) that are expressed within nigrostriatal DA neurons.

2. GTP CYCLOHYDROLASE I WITHIN MONOAMINERGIC NEURONS

After extended periods of stimulation BH4 biosynthetic capacity is increased as part of a coordinated induction of MA biosynthetic enzymes. For example BH4 content and GTPCH enzyme activity (Abou-Donia and Viveros, 1981), GTPCH mRNA (Hirayama et

* Department of Psychiatry and Behavioral Neurosciences and Center for Molecular Medicine and Genetics, Wayne State University School of Medicine, Detroit, Michigan U.S.A. 48201

Catecholamine Research, Edited by Nagatsu *et al.*
Kluwer Academic/Plenum Publishers, 2002

al., 1993) and GTPCH gene transcription rate (Serova et al., 1997) are each elevated in the adrenal gland following extended periods of increased NE turnover. First messengers that effect GTPCH expression include the neurotrophin NGF (Hirayama et al., 1995), neuropoietic cytokines such as LIF and CNTF (Stegenga et al., 1996), glucocorticoids (Serova et al., 1997), vasoactive intestinal peptide (Anastasiatis et al., 1998) and membrane depolarization (Zhu et al., 1994). Increased levels of cAMP in adrenal medullary cells (Abou-Donia et al., 1986), DA neurons (Zhu et al., 1994) and PC12 cells (Anastasiatis et al., 1998) produce a robust increase in GTPCH mRNA and BH4 content that is dependent upon gene transcription. This response to cAMP is not observed in all MA cell types (Kapatos et al., 1981). These studies indicate that GTPCH gene expression is highly dynamic and under the control of a growing number of extracellular signaling molecules and signal transduction pathways that converge upon the GTPCH gene promoter.

3. GTP CYCLOHYDROLASE I GENE TRANSCRIPTION

2627 bp of the GTPCH 5' flanking region of the human (Ichinose et al., 1995; Witter et al., 1996), 788 bp of the mouse (Ichinose et al., 1995; Shimoji et al., 1999) and 5812 bp of the rat (Kapatos et al., 2000) have been cloned and sequenced. Progressive deletion analysis of the rat GTPCH promoter using transient transfection assays of luciferase reporter constructs has defined the proximal promoter for this gene as a highly conserved and GC-rich 142 bp sequence upstream from the cap site (Kapatos et al., 2000). *In vitro* footprint analysis of this region revealed (5'-3') an Sp1-like element, a non-canonical CRE, a CCAAT-box and an E-box. Transcription from the core promoter was enhanced by 8Br-cAMP in PC12 cells, which express GTPCH, but not in C6 or Rat2 cells, which do not. Mutagenesis showed that both the CRE and CCAAT-box independently contribute to basal and cAMP-dependent activity. The combined CRE and CCAAT-box cassette was also found to confer cAMP-sensitivity on a heterologous minimal promoter. Addition of the Sp1-like element to this construct inhibited basal transcription without affecting the response to cAMP. EMSA revealed that nuclear proteins from PC12 but not C6 or Rat2 cells bind the CRE as a complex containing ATF-4 and C/EBPβ while both PC12 and C6 cell nuclear extracts were recruited by the CCAAT-box as a complex containing NF-Y. Over-expression of ATF-4 was also found to transactivate the GTPCH promoter response to cAMP. These studies suggest that the elements required for cell type-specific cAMP-dependent enhancement of gene transcription are located along the GTPCH proximal promoter and include the CRE and adjacent CCAAT-box and the proteins ATF-4, C/EBPβ and NF-Y.

4. CLINICAL SIGNIFICANCE OF GTP CYCLOHYDROLASE I

Humans that are homozygous for GTPCH mutations are detected at birth by hyperphenylalaninemia, have severe depletions of MA neurotransmitters and exhibit mental retardation, developmental delay, convulsions, hypotonia of the trunk and hypertonia of the limbs (Niederwieser et al., 1987). In contrast, individuals with heterozygous mutations in GTPCH (Ichinose et al., 1994) are characterized by a mild dystonia referred to as Segawa's Disease (SD) also known as Hereditary Progressive

Dystonia or Dopa-Responsive Dystonia because of the positive response to low doses of the DA precursor (Segawa et al., 1971). A third class of GTPCH deficiency with a phenotype intermediate between the most severe cases of homozygous GTPCH mutations and SD has recently been reported to result from certain compound heterozygous mutations in the coding region of GTPCH (Furukawa et al., 1998). SD is an autosomal dominant disease and, although the inheritance pattern of SD is not sex linked, the disorder primarily affects females. Genetic studies of maternal versus paternal transmission and gender-related penetrance have shown no evidence for genetic imprinting in SD, indicating that the 4:1 ratio of female to male SD patients and the 2.3 fold higher female penetrance is related to some more subtle aspect of GTPCH gene control (Furukawa et al., 1998). SD is characterized by low concentrations of BH4 and DA in the CSF (LeWitt et al., 1988) and postmortem basal ganglia (Rajput et al., 1994). This loss of DA appears to be selective because levels of 5-HT and NE are within the normal range (DeJong et al., 1989). In contrast to Parkinson's Disease, DA neurons do not appear to degenerate in early SD (Rajput et al., 1993). However, although the neurological deficits associated with SD differ from Parkinson's Disease, many cases of SD eventually develop parkinsonian symptoms, including resting tremor (Segawa et al., 1976).

The neurochemical characteristics of the SD phenotype raise the question of how a universal decline in the capacity to synthesize BH4 can result in a defect in neurotransmitter synthesis that is specific to nigrostriatal DA neurons. To explain this selective loss of DA synthesis we propose a novel form of cell type-specific haplo-insufficiency that is based upon our recent observations that GTPCH gene expression is low within nigrostriatal DA neurons of the rat (Lentz and Kapatos, 1996; Hirayama and Kapatos, 1998) and mouse (Shimoji et al., 1999). Specifically, we hypothesize that a low basal level of GTPCH gene expression is a phenotypic characteristic of nigrostriatal DA neurons and that this property makes DA synthesis by these neurons highly vulnerability to mutations in GTPCH that decrease BH4 content.

REFERENCES

Abou-Donia M.M., and Viveros O.H. 1981, Tetrahydrobiopterin increases in adrenal medulla and cortex: a factor in the regulation of tyrosine hydroxylase *Proc. Natl. Acad. Sci. USA* **78**, 2703-2706.

Abou-Donia M.M., Wilson S.P., Zimmerman T.P., Nichol C.A., and Viveros O.H. 1986, Regulation of GTP cyclohydrolase and tetrahydrobiopterin levels and the role of the cofactor in tyrosine hydroxylation in primary cultures of adrenomedullary chromaffin cells. *J. Neurochem.* **46**, 1190-1199

Anastasiadis P.Z., Bezin L., Gordon L.J., Imerman B., Blitz J., and Levine R.A. 1998, Vasoactive intestinal peptide induces both tyrosine hydroxylase activity and tetrahydrobiopterin biosynthesis in PC12 cells. *Neuroscience* **86**, 179-189.

Dassesse D., Hemmens B., Cuvelier L., and Resibois A. 1997, GTP cyclohydrolase I like immunoreactivity in rat brain. *Brain Res.* **777**, 187-201.

De Jong A.P.J.M., Haan E.A., Manson J.I., Wise G.A., Ouvrier R.A., and Wadman S.K. 1989, Kinetic study of catecholamine metabolism in hereditary progressive dystonia *Neuropediatrics* **20**, 3-11.

Furukawa Y., Kish S.J., Bebin M., Jacobson R.D., Fryburg J., Wilson W.G., Shimadzu M., Hyland K., and Trugman J.M. 1998, Dystonia with motor delay in compound heterozygotes for GTP cyclohydrolase I gene mutations. *Ann. Neurol.* **44**, 10-16.

Furukawa Y., Lang A.E., Trugman J.M., Bird T.D., Hunter A., Sadeh M., Tagawa T., St. George-Syslop P.H., Guttman M., Morris L.W., Hornykiewicz O., Shimadzu M., and Kish S.J. 1998, Gender-related penetrance and de novo GTP-cyclohydrolase I gene mutations in dopa-responsive dystonia. *Neurology* **50**, 1015-1020.

Hirayama, K., Lentz, S.I., and Kapatos, G. 1993, Tetrahydrobiopterin cofactor biosynthesis: GTP cyclohydrolase I mRNA expression in rat brain and superior cervical ganglia. *J. Neurochem.* **61**, 1006-1014.

Hirayama K. and Kapatos G. 1995, Regulation of GTP cyclohydrolase I gene expression and tetrahydrobiopterin content by nerve growth factor in cultures of superior cervical ganglia. *Neurochem. Int.* **27**, 157-162.

Hirayama K., and Kapatos G. 1998, Nigrostriatal dopamine neurons express low levels of GTP cyclohydrolase I protein. *J. Neurochem.* **70**, 164-170.

Ichinose H., Ohye T., Takahashi E., Seki N., Hori T., Segawa M., Nomra Y., Endo K., Tanaka H., Tsuji S., Fujita K., and Nagatsu T. 1994, Hereditary progressive dystonia with marked diurnal fluctuation caused by mutations in GTP cyclohydrolase I gene. *Nature Genetics* **8**, 236-242.

Ichinose H., Ohye T., Matsuda Y., Hori T., Blau N., Burlina A., Rouse B., Matalon R., Fujita K., and Nagatsu T. 1995, Characterization of mouse and human GTP cyclohydrolase genes. Mutations in patients with GTP cyclohydrolase I deficiency. *J. Biol. Chem.* **270**, 10061-10071.

Kapatos, G., Kaufman, S., Weller, J.L., and Klein, D.C. 1981, Biosynthesis of biopterin: Adrenergic cyclic adenosine monophosphate-dependent inhibition in the pineal gland. *Science* **213**, 1129-1131.

Kapatos G. 1990, Tetrahydrobiopterin synthesis rate and turnover time in neuronal cultures from rat mesencephalon and hypothalamus. *J. Neurochem.* **55**, 1995-1201.

Kapatos, G., Stegenga, S.L., and Hirayama, K. 2000, Identification and characterization of basal and cyclic AMP response elements in the promoter of the rat GTP cyclohydrolase I gene. *J. Biol. Chem.* 275:5947-5957.

Kaufman, S. 1974, Properties of the pterin-dependent hydroxylases. In: *Aromatic Amino Acids in the Brain,* CIBA Foundation Symposia, Vol. 22, pp. 85-115. Elsevier, Amsterdam.

Kettler R., Bartholini G., and Pletscher A. 1974, In vivo enhancement of tyrosine hydroxylation in rat striatum by tetrahydrobiopterin. *Nature* **249**, 476-478.

Lentz S.I., and Kapatos G. 1996, Tetrahydrobiopterin biosynthesis in the rat brain: heterogeneity of GTP cyclohydrolase I mRNA expression in monoamine-containing neurons. *Neurochem. Int.* **28**, 569-582.

Levine R.A., Miller L.P., and Lovenberg W. 1981, Tetrahydrobiopterin in striatum: Localization to dopamine nerve terminals and role in catecholamine synthesis. *Science* **214**, 919-921.

LeWitt P.A., Miller L.P., Levine R.A., Lovenberg W., Newman R.P., Papavasiliou A., Rayes A., Eldridge R., and Burns R.S. 1986, Tetrahydrobiopterin in dystonia: identification of abnormal metabolism and therapeutic trials. *Neurol.* **36**, 760-764.

Nagatsu I., Ichinose H., Sakai M., Titani K., Suzuki M., Nagatsu T. 1995, Immunocytochemical localization of GTP cyclohydrolase I in brain, adrenal gland and liver of mice. *J. Neural Transm.* **102**, 175-188.

Nichol C.A., Smith G.K., and Duch D.S. 1985, Biosynthesis and metabolism of tetrahydrobiopterin and molybdopterin. *Ann. Rev. Biochem.* **54**, 729-764.

Niederwieser A. and Curtius H.C. 1987, Tetrahydrobiopterin: biosynthetic pathway and deficiency. *Enzyme* **38**, 302-311.

Rajput, A.H., Gibb, W.R.G., Zhong, X.H., Shannak, K.S., Kish, S., Chang, L.G. and Hornykiewicz, O. 1994, Dopa-responsive dystonia: pathological and biochemical observations in a case. *Ann. Neurol.* **35**, 396-402.

Segawa M., Hosaka A., Miyazawa F., Nomura Y., and Imai H. 1971, Childhood basal ganglia disease with remarkable response to l-Dopa, hereditary basal ganglia disease with marked diurnal fluctuation. *Shinryo* **24**, 667-672.

Segawa M., Hosaka A., Miyagawa F., Nomura Y., Imai H. 1976, Hereditary progressive dystonia with marked diurnal fluctuation. In: *Advances in Neurology*. (Eldridge R., and Fahn S., eds) Vol. **14**, 215-233.

Serova L., Nankova B., Rivkin M., Kvetnansky R., and Sabban E.L. 1997, Glucocorticoids elevate GTP cyclohydrolase I mRNA levels in vivo and in PC12 cells. *Mol. Br. Res.* **48**, 251-258.

Shimoji M., Hirayama K., Hyland K., and Kapatos G. 1999, GTP cyclohydrolase I gene expression in the brains of male and female hph-1 mice. *J. Neurochem.* **72**, 757-764.

Stegenga S.L., Hirayama K., and Kapatos G. 1996, Regulation of GTP cyclohydrolase I gene expression and tetrahydrobiopterin content in cultured sympathetic neurons by leukemia inhibitory factor and ciliary neurotrophic factor. *J. Neurochem.* **66**, 2541-2545.

Togari, A., Ichinose, H., Matsumoto, S., Fujita, K. and Nagatsu, T. 1992, Multiple mRNA forms of human GTP cyclohydrolase I. *Biochem. Biophys. Res. Com.* **187**, 359-365.

Witter K., Werner T., Blusch J.H., Schneider E.M., Riess O., Ziegler I., Rodl W., Bacher A., and Gutlich M. 1996, Cloning, sequencing and functional studies of the gene encoding human GTP cyclohydrolase I. *Gene* **171**, 285-290.

Wolf W.A., Ziaja E., Arthur R.A.Jr., Anastasiadis P.Z., Levine R.A., and Kuhn D.M. 1991, Effect of tetrahydrobiopterin on serotonin synthesis, release, and metabolism in superfused hippocampal slices. *J. Neurochem.* **57**, 1191-1197.

Zhu, M., Hirayama, K., and Kapatos, G. 1994, Regulation of tetrahydrobiopterin biosynthesis in cultured dopamine neurons by depolarization and cAMP. *J. Biol. Chem.* **269**, 11825-11829.

IN VIVO AND IN VITRO FUNCTIONS OF TWO TYPES OF MAO

Jean C. Shih and Kevin Chen*

SUMMARY

Monoamine oxidase (MAO) A and B have selective substrate and inhibitor specificity. MAO A and B knock-out mice exhibit distinct neurotransmitter metabolism and behavior, suggesting that these two isoenzymes have distinct functions in vivo. These two isoenzymes are coded by different cDNAs with 70% amino acid identity. Interestingly, Isoleucine 335 in MAO A and the corresponding amino acid in MAO B, tyrosine 326, determine the substrate and inhibitor selectivity in vitro.

INTRODUCTION

MAO (monoamine; oxygen oxidoreductase, deaminating, EC 1.4.3.4) catalyzes the oxidative deamination of a number of biogenic amines in the brain and peripheral tissues and produces hydrogen peroxide (H_2O_2).[1] On the basis of substrate selectivity and inhibitor sensitivity, two forms of MAO, A and B have been designated. [2,3] MAO A has higher affinity for the substrate serotonin (5-HT), norepinephrine (NE) and the inhibitor clorgyline whereas MAO B has higher affinity for phenylethylamine (PEA), benzylamine and the inhibitor deprenyl (for review, 4).

Both MAO A and B are located on the outer membrane of mitochondria throughout the brain.[5] MAO A inhibitors have been used clinically as antidepressants whereas the MAO B inhibitor deprenyl has been shown to be useful for delaying the progression of Parkinson's and Alzheimer's diseases.[6]

* Jean C. Shih, Department of Molecular Pharmacology and Toxicology, School of Pharmacy and Department of Cell and Neurobiology, Keck School of Medicine, University of Southern California, 1985 Zonal Avenue PSC 528, Los Angeles CA 90089-9121. Kevin Chen, Department of Molecular Pharmacology and Toxicology, School of Pharmacy, University of Southern California, 1985 Zonal Avenue PSC 530, Los Angeles CA 90089-9121.

Cloning of MAO A and B has demonstrated unequivocally that these two isoenzymes are made of different polypeptides.[7] Our understanding of MAO structure, regulation and function has been significantly advanced by studies using their cDNAs. MAO A and B genes are located on the X-chromosome (Xp11.23)[8] and comprise 15 exons with identical intron-exon organization, which suggests that they are derived from the same ancestral gene.[9]

MAO A AND B KNOCK-OUT (KO) MICE EXHIBIT DISTINCT NEUROTRANSMITTER METABOLISM AND BEHAVIORS

MAO A KO mice have elevated brain levels of 5-HT, NE and DA and manifest aggressive behavior[10] similar to men with a deletion of MAO A.[11] This deletion is due to a point mutation in exon 8 of MAO A gene which resulted in a truncated MAO A without activity.[12] In contrast, MAO B KO mice do not exhibit aggression and only have increased levels of PEA. Mice lacking MAO B are resistant to the Parkinsongenic neurotoxin, 1-methyl-4-phenyl-1,2,3,6-tetrahydropyridine (MPTP).[13] Since the substrate specificity of MAO A and B is relative, each form of the enzyme is able to oxidize the other form's preferred substrate . Thus, the *in vivo* function of MAO can be understood only when MAO A and B double KO mice are generated and studied. This work is ongoing.

DIFFERENT AMINO ACID RESIDUES DETERMINE THE SPECIFICITY IN HUMAN AND RAT MAOS

MAO A and B are composed of 527 and 520 amino acids, respectively, and have a 70% amino acid identity.[7] In order to determine the region(s) responsible for the substrate and inhibitor specifities of the two isoenzymes, we and other groups have made point mutations and chimeric MAO A/B enzymes.[14-22] It has been shown that reciprocally interchanging amino acids Phe-208 in MAO A and its corresponding residue in MAO B, Ile-199, was sufficient to partially reverse the substrate and inhibitor specificities of rat MAOs[23] but not human MAOs.[22] This indicated that different amino acid residues may determine specificity in human and rat MAOs.

A SINGLE AMINO ACID CAN SWITCH THE SUBSTRATE AND INHIBITOR SPECIFICITY OF HUMAN MAO A AND B

We have found that when we switched isoleucine (I-335) in MAO A with its corresponding amino acid in MAO B tyrosine (Y-326), we were able to make the MAO A mutant (MAO A-I335Y) more like MAO B.[24] As shown in Fig. 1, the sensitivity to clorgyline (MAO A specific inhibitor) was reduced drastically for MAO A-I335Y ($IC_{50} = 7.1 \times 10^{-6}$M) compared to MAO A WT ($IC_{50} = 1.2 \times 10^{-9}$M), its inhibition profile became more like MAO B. The reciprocal mutant MAO B-Y326I ($IC_{50} = 2.8 \times 10^{-8}$M) showed increased sensitivity to clorgyline compared to MAO B WT ($IC_{50} = 6.7 \times 10^{-7}$M), thus it was more like MAO A.

Similary, the sensitivity of two mutants to deprenyl also switched compared to the wildtype. The sensitivity of mutant MAO A I335Y (I C_{50} = 1.2×10^{-7}M) to deprenyl became 10-fold more sensitive than MAO A wildtype (I C_{50} = 1.3×10^{-6}M), became more like MAO B. The sensitivity of mutant MAO B–Y326I to deprenyl (I C_{50} = 1.8×10^{-7}M) was decreased compared to MAO B wild type (I C_{50} = 4.3×10^{-9}M) suggesting it became more like MAO A.[24]

Taken together, these results suggest that MAO A Ile-335 and its corresponding amino acid, Tyr-326 in MAO B, determine MAO A and MAO B specificity defined by the sensitivity to inhibitors clorgyline and deprenyl.[24]

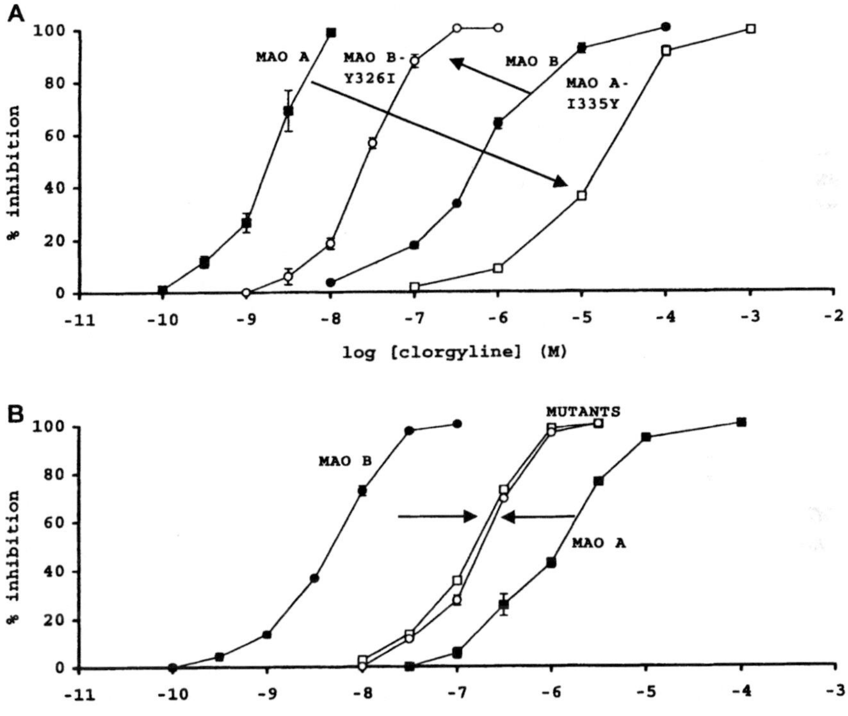

Figure 1. Clorgyline and deprenyl inhibition of wild-type MAOs and the mutants A-I335Y and B-Y326I. The clorgyline (*A*) and deprenyl (*B*) inhibition curves of wild-type MAOs and the MAO mutants that exhibited as change in inhibitor sensitivity, A-I335Y and B-Y326I, are plotted as percent inhibition *versus* log inhibitor concentrations. *Error bars* represent the S.E. of three experiments. The symbols ■, □, ●, and O represent MAO A, A-I335Y, MAO B, and MAO B-Y326I, respectively.

ACKNOWLEDGEMENT

This work was supported by National Institute of Mental Health Grants R01 MH37020, R37 MH39085 (MERIT Award) and by the Welin Professorship.

REFERENCES

1. J.C. Shih, Molecular basis of human MAO A and B. *Neuropsychopharmacology* **4**, 1-7 (1991).
2. J.P. Johnston, Some observations upon a new inhibitor of monoamine oxidase in brain tissue. *Biochem. Pharmacol.* **17**, 1285-1297 (1968).
3. J. Knoll and K. Magyar, Some puzzling pharmacological effects of monoamine oxidase inhibitors. *Adv. Biochem. Psychopharmacol.* **5**,393-408 (1972).
4. J.C. Shih, K. Chen, and M. Ridd, Monoamine oxidases: from gene to behavior. *Ann. Rev. Neuroscience* **22**,197-217 (1999).
5. A.R. Green and M.B.H. Youdim, Effects of monoamine oxidase inhibition by clorgyline, deprenyl or tranycypromine on 5-hydroxy-deprenyl or tranycypromine on 5-hydroxytryptamine concentrations in rat brain and hyperactivity following subsequent tryptophan administration. *Br. J. Pharmacol.* **55**,415-422 (1975).
6. M. Sano, C. Ernesto, R.G. Thomas, M.R. Klauber, K. Schafer, M. Grundman, P. Woodbury, J. Growdon, C.W. Cotman, E. Pfeiffer, L.S. Schneider and L.J. Thal, A controlled trial of selegiline, α-tocopherol, or both as treatment for Alzheimer's disease. The Alzheimer's Disease Cooperative Study. *New. Eng. J.* **336**,1216-1222 (1997).
7. A.W.J. Bach, N.C. Lan, D.L. Johnson, C.W. Abell, M.E. Bemkenek, S.-W. Kwan, P.H. Seeburg and J.C. Shih, cDNA cloning of human liver monoamine oxidase A and B: Molecular basis of differences in enzymatic properties. *Proc. Natl. Acad. Sci.* **85**, 4934-4938 (1988).
8. N.C. Lan, C. Heinzmann, A. Gal, I. Klisak, U. Orth, et. al., Human monoamine oxidase A and B genes map to Xp 11.23 and are deleted in a patient with Norrie disease. *Genomics* **4**,552-559 (1989).
9. J. Grimsby, K. Chen, L.J. Wang, N.C. Lan and J.C. Shih, Human monoamine oxidase A and B gene exhibit identical exon-intron organization. *Proc. Natl. Acad. Sci. USA* **88**,3637-3641 (1991).
10. O. Cases, I. Seif, J. Grimsby, P. Gaspar, K. Chen, et. al., Aggressive behavior and altered amounts of brain serotonin and norepinephrine in mice lacking MAO A. *Science* **268**,1763-1766 (1995).
11. H.G. Brunner, M.R. Nelen, P. Van Zandvoort, N.G.G.M. Abeling, A.H. Van Gennip, et. al., X-linked borderline mental retardation with prominent behavioral disturbance: phenotype, genetic localization, and evidence for disturbed monoamine metabolism. *Am. J. Hum. Genet.* **52**,1032-1039 (1994).
12. H.G. Brunner, M. Nelen, X.O. Breakefield, H.H. Ropers and B.A. Van Oost, Abnormal behavior associated with a point mutation in the structural gene for monoamine oxidase A. *Science* **262**,578-580 (1993).
13. J. Grimsby, M. Toth, K. Chen, T. Kumazawa, L. Klaidman, et. al., Increased stress response and β-phenylethylamine in MAO B-deficient mice. Nature Genet **17**,1-5 (1997).
14. A.M. Cesura, J. Gottowik, H.-W. Lahm, G. Lang, R. Imhof, P. Malherbe, U. Röthlisberger and M. Da Prada, Investigation on the structure of the active site of monoamine oxidase-B by affinity labeling with the selective inhibitor lazabemide and by site-directed mutagenesis. *Eur. J. Biochem.* **236**, 996-1002 (1996).
15. K. Chen, H.-F. Wu, and J.C. Shih, Influence of C Terminus on monoamine oxidase A and B catalytic activity. *J. Neurochem.* **66**, 797-803 (1996).
16. J. Gottowik, A.M. Cesura, P. Malherbe, G. Lang, and M. Da Prada, Characterization of wild-type and mutant forms of human monoamine oxidase A and B expressed in a mammalian cell line. *FEBS Lett.* **317**, 152-156 (1993).
17. J. Gottowik, P. Malherbe, G. Jang, M. Da Prada, and A.M. Cesura, Structure/function relationships of mitochondrial monoamine oxidase A and B chimeric forms. *Eur. J. Biochem.* **230**, 934-942 (1995).
18. Y. Tsugeno, I. Hirashiki, F. Ogata, and A. Ito, Regions of the molecule responsible for substrate specificity of monoamine oxidase A and B: a chimeric enzyme analysis. *J. Biochem. (Tokyo)*, **118**, 974-980 (1995).
19. K. Chen, H.-F. Wu, J. Grimsby and J.C. Shih, Cloning of a novel monoamine oxidase (MAO) cDNA from trout liver. *Mol. Pharm.* **46**:1226-1233 (1994).
20. J. Grimsby, M. Zentner and J.C. Shih: Identification of a region important for human monoamine oxidase B substrate and inhibitor selectivity. *Life Sci.* **58**:777-787 (1996).
21. H.-F. Wu, K. Chen, and J.C. Shih, Site-directed mutagenesis of monoamine oxidase A and B. *Mol. Pharmacol.* **43**, 888-893 (1993).
22. R.M. Geha, K. Chen, and J.C. Shih, Phe^{208} and Ile^{199} in human monoamine oxidase A and B do not determine substrate and inhibitor specificities as in rat. *J. Neurochem.* **75**, 1304-1309 (2000).
23. Y. Tsugeno and A. Ito, A key amino acid responsible for substrate selectivity of monoamine oxidase A and B . *J. Biol. Chem.* **272**, 14033-14036 (1997).
24. R.M. Geha, I. Rebrin, K. Chen and J.C. Shih, Substrate and inhibitor specificities of human monoamine oxidase A and B are influenced by a single amino acid. *J. Biol. Chem.*, **276**,9877-9882 (2001).

CURRENT STATUS OF NEW MAO-A/B INHIBITOR DEVELOPMENT

N. Iwata*

1. MAO INHIBITORS AS ANTIDEPRESSANTS

Although old MAO inhibitors, nonselective and irreversible MAO inhibitors, have antidepressive effects, they had serious side effects, such as hepatotoxicity caused by a hydrazine moiety, the cheese effect and orthostatic hypotension. Due to these adverse events, classical MAO inhibitors with a hydrazine moiety are no longer used clinically.

In order to avoid the severe cheese effect, reversible inhibitors of MAO-A (RIMA) have been developed. Toloxatone (Fig. 1) was launched in 1985, followed by pirlindole and moclobemide (Fig. 1). Although no other RIMA has been launched in the last 10 years, befloxatone (Fig. 1) is expected to be marketed soon. Since approximately 75% of tyramine is metabolized by MAO-A, RIMA cannot avoid causing the cheese effect. However, as the effect of RIMA is not cumulative, the cheese effect is not potentiated by successive administration.

Recently, serotonin reuptake inhibitors (SSRIs), and serotonin-norepinephrine reuptake inhibitors (SNRIs) have been successfully developed and are in use all over the world. Thus, it is increasingly known that depression is associated with deficiencies in norepinephrine and serotonin. However, disruptions in the dopaminergic system are also thought to be involved in depression (Bonhomme and Esposito, 1998). The therapeutic efficacy of RIMAs is as great as tricyclic antidepressants (TCA) and SSRIs, and their adverse events seem to be less severe than TCA and comparable to SSRIs.

2. NEUROPROTECTIVE ACTION OF MAO INHIBITORS.

It is well established that repeated injections of MPTP elicits Parkinson's disease-like behavior in mice and primates. MPTP is metabolized to $MADP^+$ by MAO-B and is further metabolized to a neurotoxin, MPP^+, nonenzymatically.

The mechanisms of the neurotoxic effect of MPP^+ is considered to be due to ① MPP^+ incorporated into the synaptic terminals of DA neurons in the pars compacta of the

* Science Information Co., Ltd. Tokyo 140-8710, Japan.

Catecholamine Research, Edited by Nagatsu *et al.*
Kluwer Academic/Plenum Publishers, 2002

substantia nigra, is displaced by DA in the granules at the terminals. The released DA produces free radicals when oxidized by MAO, and the free radicals attack the cell membranes, ② MPP^+ inhibits complex 1 and -ketoglutarate dehydrogenase activity in mitochondoria and decreases ATP synthesis, and ③ MPP^+ also inhibits tyrosine hydroxylase, and decreases DA.

Some derivatives of MPTP, 2'methyl-MPTP and 2'-ethyl-MPTP, are metabolized by MAO-A to 2'methyl-MPP^+ and 2'ethyl-MPP^+, respectively.

These environmental neurotoxins or endogenous neurotoxins as well as DA bind to MAO and produce hydrogen peroxide. Under normal conditions the hydrogen peroxide is metabolized to water by glutathione peroxidase. Since glutathione peroxidase in the DA neurons of the substantia nigra is deficient in patients with Parkinson's disease, unlike in normal subjects, hydroxyl radicals formed by the Fenton reaction cause lipid peroxidation, which in turn attacks the cell membrane. Therefore, it seems likely that MAO-B inhibitors can protect neurons from oxidative stress. In fact, selegiline (Fig. 1) has neuroprotective activities against not only MPTP but also against ischemia-reperfusion.

Selegiline is the only commercially available MAO-B inhibitor so far. Rasagiline (Fig. 1) is now in phase 3 clinical studies. The neuroprotective actions of selegiline and rasagiline are demonstrated to be due to the propargylamine moiety in their structures and, not due to their MAO-B inhibiting activities (Speiser, et al., 1999). An irreversible MAO-A inhibitor, clorgyline, also has a propargylamine moiety in its structure, while lazabemide, a reversible MAO-B inhibitor, lacks the moiety in its structure. According to Suzuki, et al. (1999), hydrogen peroxide levels increased in the brain of a mouse after reperfusion following ischemia. Either clorgyline or lazabemide suppress the production of hydrogen peroxide. Furthermore, the effects of clorgyline are greater than those of lazabemide. The effects of clorgyline may, in part, be brought about by its chemical structure, independently from its MAO-A inhibiting activity. However, since the DOPAC level in the brain is significantly reduced by clorgyline, the MAO-A inhibiting activity of this compound may possibly be involved. Thus, both inhibitors of MAO-A and MAO-B activities seem to be beneficial in protecting neurons from oxidative stress.

Clinical trials have been carried out in order to investigate whether selegiline has neuroprotective activity or not (DATATOP studies). The conclusion of the study was that selegiline had no neuroprotective effects, although it delayed the time until L-DOPA or DA receptor agonists were required for therapy.

Why were the neuroprotective effects of MAO-B inhibitors not reflected in the clinical studies? One possible reason is a question of timing. In many animal studies, MAO-B inhibitors were administered immediately after reperfusion, or even prior to the ischemia, and were repeatedly administered during and after the ischemia. However, in patients with Parkinson's disease, we do not know when the disease started. When the symptoms started to appear in patients with Parkinson's disease, the DA content in the striatum is already reduced by about 90%, and about 50% of the DA neurons in the substantia nigra are lost. Thus, the compounds have to be administered and their neuroprotective effects in the animal models examined at a timing corresponding closer to the clinical situation, as suggested in studies on glutamate receptor antagonists in ischemia-reperfusion models.

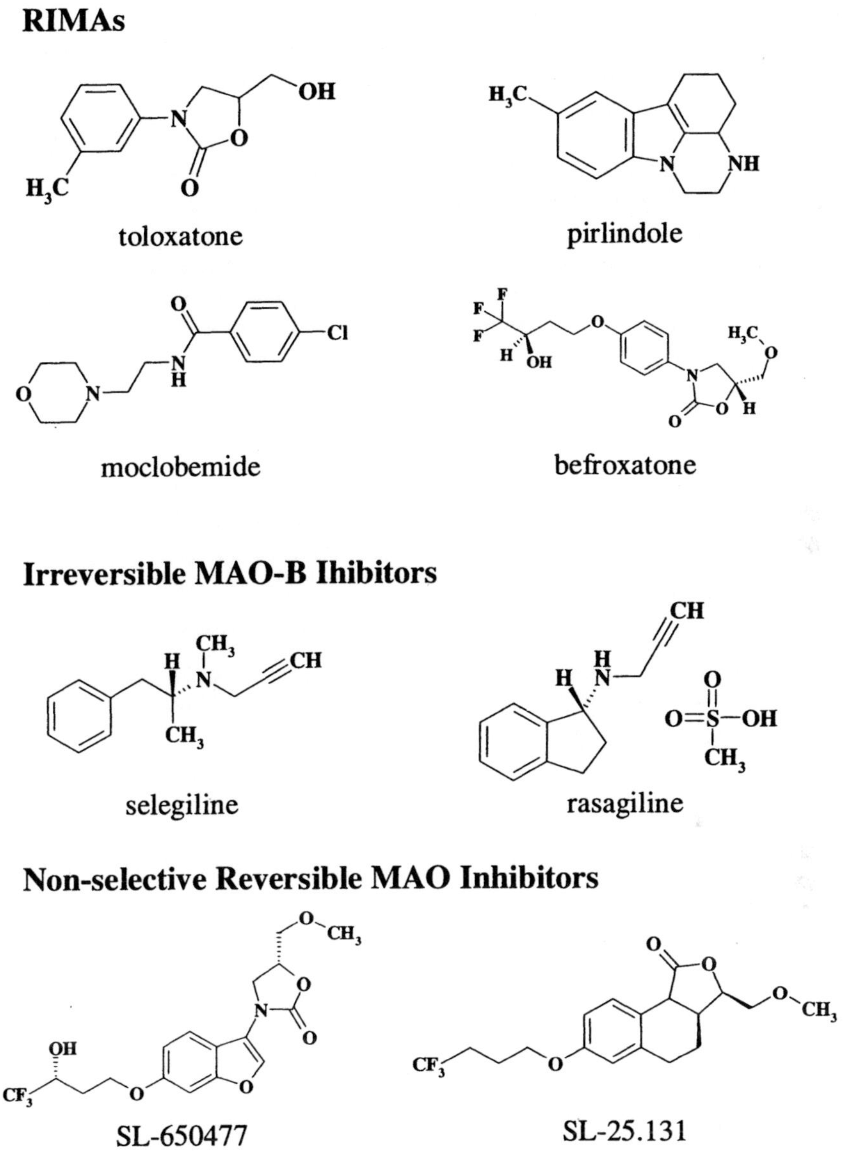

Fig. 1. Chemical structures of new MAO inhibitors.

FUTURE OF MAO INHIBITORS

Tranylcypromine, an old MAO inhibitor, has been demonstrated to slow down the progression of the disease and only 2 out of the 27 patients who completed a 6 months treatment, required L-DOPA or DA agonists in their therapies. During the same period, all patients in the placebo group required dopaminergic agents (Fahn and Chouinard, 1998). This beneficial effect of MAO inhibitors should be confirmed on a larger scale by double blind clinical trials with new reversible and nonselective MAO inhibitors.

Novel potent and non-selective, reversible MAO inhibitors have recently been reported (Curet, et al., 2000). Further evaluation of these non-selective and reversible MAO inhibitors could lead to new therapeutic strategies. If these non-selective and reversible MAO inhibitors exert neuroprotective effects when they are administered at a certain time after reperfusion in ischemia-reperfusion models, or soon after Parkinson's disease-like behavior which appear following repeated injections of MPTP, they could be used as agents not only for depression but also for neurodegenerative diseases. Even though they do not protect neurons from oxidative stress, they may prove to be more potent antidepresants than RIMAs, SSRIs and SNRIs.

REFERENCES

Bonhomme, N. and Esposito, E., 1998, Involvement of serotonin and dopamine in the mechanism of action of novel antidepressant drugs: a review. *J. Clin. Psychopharmacol.*, **18**, 447-54.

Curet, O., Sontag, N., Aubin, N., Marc, C., Jegham, S., Puech, F., Burnier, P., George, P., Scatton, B., 2000, Biochemical profile of SL25.1131 a novel mixed and reversible MAO-A/B inhibitor. 30th Soc. Neurosci.Abs., 666.2, 2000 (New Orleans).

Fahn, S. and Chouinard, S., 1998, Experience with tranylcypromine in early Parkinson's disease, *J Neural Transm* [Suppl]. **52**: 49-61.

Suzuki T, Akaike N, Ueno K, Tanaka Y, Himori, N., 1995, MAO inhibitors, clorgyline and lazabemide, prevent hydroxyl radical generation caused by brain ischemia/reperfusion in mice. *Pharmacology*, **50**: 357-62.

Speiser, Z., Mayk, A., Eliash, S., Cohen, S., 1999, Studies with rasagiline, a MAO-B inhibitor, in experimental focal ischemia in the rat, *J. Neural Transm.*, **106**: 593-606.

MOLECULAR BASIS FOR THE SUBSTRATE SPECIFICITY OF HUMAN CATECHOLAMINE SULFOTRANSFERASE, SULT1A3

Michael W.H. Coughtrie, Jyrki Taskinen and Alan M. Hood*

1. INTRODUCTION

Sulfation is an important component of the body's chemical defence mechanism, and provides a major route for the metabolism, detoxification and elimination of a host of xenobiotics (Coughtrie 1996). In addition to this role, sulfation also functions in the homeostasis of a number of important endogenous compounds, including iodothyronines (Visser 1996), steroid hormones (Roy 1992), oxysterols (Javitt *et al.* 2001) and catecholamines (Goldstein *et al.* 1999). These reactions are catalysed by members of the sulfotransferase (SULT) enzyme superfamily that transfer a sulfuryl moiety from the co-substrate 3'-phopshoadenosine 5'-phosphosulfate (PAPS). In humans, the SULT family numbers at least 11 members derived from at least 10 genes. The human SULTs can be subdivided into 3 different subfamilies based on amino acid sequence similarity and functional activity, and the largest family, SULT1 or phenol SULT, currently comprises 7 members (Coughtrie *et al.* 2001). Humans (and other primates) differ from most animal species in that they have significant levels of circulating catecholamine sulfates; for example up to 98% of circulating dopamine is in its sulfated form (Goldstein *et al.* 1999). It is believed that the gastrointestinal tract is the major source of these circulating catecholamine sulfates (Goldstein *et al.* 1999), and indeed significant increases in plasma dopamine sulfate can be observed after ingestion of bananas (Davidson *et al.* 1981) which are rich in dopamine (Kanazawa *et al.* 2000).

* Michael W.H. Coughtrie & Alan M. Hood, Department of Molecular & Cellular Pathology, University of Dundee, Ninewells Hospital & Medical School, Dundee DD1 9SY, Scotland, UK. Jyrki Taskinen, Division of Pharmaceutical Chemistry, Department of Pharmacy, University of Helsinki, P.O. Box 56, Viikinkaari 5, FIN-00014 Helsinki, Finland.

2. CATECHOLAMINE SULFOTRANSFERASE, SULT1A3

It has been known for some time that humans possess a sulfotransferase with a high degree of specificity towards dopamine (and other catecholamines) (Rein *et al.* 1982). This enzyme, known variously as catecholamine sulfotransferase, thermolabile phenol sulfotransferase, M-PST and, more recently SULT1A3, has a high affinity for dopamine (K_m 3μM) and is characterised by a predominantly extrahepatic expression pattern in adult humans (Rubin *et al.* 1996; Richard *et al.* 2001).

It is interesting that, to date, man is the only species from which a SULT1A3 enzyme has been cloned and characterised. Detailed amino acid sequence analysis of all cloned SULTs shows that SULT1A3 is (so far) unique to humans, and has probably arisen through a series of gene duplication events which have produced 3 highly homologous genes (*SULT1A1*, *SULT1A2*, *SULT1A3*) in close proximity on chromosome 16 (Dooley 1998). *SULT1A1* is likely to be the ancestral gene of this subfamily, since it alone has been found in other species, including rodents.

Physiological evidence for the presence of a gastrointestinal catecholaminergic system producing large amounts of circulating catecholamines (Goldstein *et al.* 1999) is therefore consistent with a) the expression of SULT1A3 at particularly high levels in the upper gastrointestinal tract (Rubin *et al.* 1996) and b) the apparent evolution of a specific SULT enzyme (SULT1A3) with a high degree of specificity towards catecholamines.

3. MOLECULAR BASIS OF SULT1A3 SPECIFICITY

3.1. The Role of Glu^{146}

The 2 human sulfotransferases SULT1A1 and SULT1A3 share 93% amino acid sequence identity (i.e. they by only 20 out of 295 amino acids) yet they have dramatically different substrate specificities. SULT1A3 exhibits high affinity (or low K_m) and reactivity (or high V_{max}) for dopamine and other catecholamines, whereas SULT1A1 is active against a very wide range of phenolic compounds at low (<10μM) concentrations. These two enzymes therefore provide a good model system for studying the factors that govern SULT specificity towards catecholamines and other compounds.

Sequence alignment of SULTs 1A1 and 1A3 demonstrated a short region between amino acids 143 and 148 which contained 4 of the 20 amino acid differences between the two proteins. We hypothesised that this region may therefore contribute to the selectivity of SULT1A3, so we carried out a number of site-directed mutagenesis experiments to determine whether these amino acids were important determinants of selectivity (Dajani *et al.* 1998). SULT1A3 was mutated at positions 143 and 146 to the corresponding amino acids in SULT1A1 (His Tyr and Glu Ala, respectively) and the recombinant proteins expressed and purified from *E. coli*. Mutation of SULT1A3 His^{143} to Tyr as it is in SULT1A1 had no effect on either K_m or V_{max} towards the prototypical isoform-specific substrates dopamine (for SULT1A3) or 4-nitrophenol (for SULT1A1). However, when Glu^{146} in SULT1A3 was mutated to Ala as in SULT1A1, the recombinant SULT1A3 variant displayed the kinetic properties of SULT1A1 towards these 2 substrates (Dajani *et al.* 1998). Thus, Glu^{146} plays a key role in determining the substrate specificity of SULT1A3 towards catecholamines. Further, we showed that this amino acid also contributes significantly for the preference of SULT1A3 for catechols over phenols. The important role of Glu^{146} has been confirmed independently by mutation of SULT1A1 Ala^{146} to Glu (Brix *et al.* 1999)

3.2. Insights from the SULT1A3 X-Ray Crystal Structure and Molecular Models

The first SULT X-Ray crystal structure to be solved was that of a mouse estrogen SULT, in the presence of substrate -estradiol and reaction product PAP (Kakuta *et al.* 1997; Kakuta *et al.* 1998). Since then the structure of SULT1A3 has been solved, in the presence of PAP (Dajani *et al.* 1999; Bidwell *et al.* 1999).

These structural studies have helped elucidate the SULT reaction mechanism (Kakuta *et al.* 1998), which is a nucleophilic S_N2 substitution reaction. An absolutely conserved histidine (His^{108} in SULT1A3) acts as the catalytic base that deprotonates the nucleophile, and Lys^{48} is believed to act as a catalytic acid, protonating the phosphate-sulfate bridge of PAPS thereby enhancing dissociation of the leaving group.

The structural data provide a rational explanation for the influence of Glu^{146} on SULT1A3 specificity towards catecholamines. This amino acid is located at the end of a conserved -helical structure (-6), which sits at the entrance to the active site. Molecular modelling of dopamine and other catecholamine substrates into the SULT1A3 active site (Dajani *et al.* 1999) showed that the basic nitrogen group in the ethylamine side chain is in position to form a salt bridge with the carboxylic acid of Glu^{146}. No other sulfotransferase so far cloned and sequenced from any species has a glutamic acid (or other acidic residue) in this position.

More recently, we have carried out detailed quantitative structure-activity relationship (QSAR) analysis of purified recombinant SULT1A3 and SULT1A1. The data generated from approximately 100 phenols and catechols show clearly that SULT1A1 is a broad specificity enzyme, whereas SULT1A3 exhibits a much greater degree of selectivity. Important features that determine high affinity substrates for SULT1A3 are a) an amine function optimally 2 carbon atoms distant from the aromatic ring, b) lipophilicity, c) the presence of hydrogen bond donors and acceptors, and d) "narrow" molecular dimensions. A major factor influencing reaction rate (i.e. the effectiveness with which the proton on the reacting hydroxyl is abstracted) is the presence of a bulky substituent ortho- to the nucleophile. Thus, compounds such as 2-bromophenol and normetanephrine are exceptionally good substrates. The nucleophilicity of the reacting oxygen on the substrate does not seem to be an important effector of reaction rate, since the electronic properties of substituents in our series of compounds had no effect on V_{max}.

SUMMARY

Considerable advances in our understanding of the molecular basis of SULT specificity have been made recently, considerably aided by the availability of X-ray crystal structure data. These studies have confirmed that SULT1A3 is an enzyme with distinct substrate specificity towards catecholamines and similar compounds. We hypothesise that humans have evolved the SULT1A3 enzyme as part of the gastrointestinal tract catecholaminergic system that produces the large amounts of catecholamine sulfates present in the circulation, and that the specificity of this important enzyme has been determined by introduction of a glutamic acid in position 146 in the protein.

ACKNOWLEDGEMENTS

This work was supported by grants from the Commission of the European Communities (BMH4-97-2621) and the Biotechnology & Biological Sciences Research Council.

REFERENCES

Bidwell L. M., McManus M. E., Gaedigk A., Kakuta Y. *et al.*, 1999, Crystal structure of human catecholamine sulfotransferase. *J Mol Biol,* **293**: 521-530.

Brix L. A., Duggleby R. G., Gaedigk A., and McManus M. E., 1999, Structural characterization of human aryl sulphotransferases. *Biochem J,* **337**: 337-343.

Coughtrie M. W. H., 1996, Sulphation catalysed by the human cytosolic sulphotransferases - chemical defence or molecular terrorism? *Hum Exp Toxicol,* **15**: 547-555.

Coughtrie M. W. H., and Johnston L. E., 2001, Interactions between dietary chemicals and human sulfotransferases – Molecular mechanisms and clinical significance. *Drug Metab Dispos,* **29**: 522-528.

Dajani R., Cleasby A., Neu M., Wonacott A. J. *et al.*, 1999, X-ray crystal structure of human dopamine sulfotransferase, SULT1A3: Molecular modelling and QSAR analysis demonstrate a molecular basis for sulfotransferase substrate specificity. *J Biol Chem,* **274** : 37862-37868.

Dajani R., Hood A. M., and Coughtrie M. W. H., 1998, A single amino acid, Glu_{146}, governs the substrate specificity of a human dopamine sulfotransferase, SULT1A3. *Mol Pharmacol,* **54**: 942-948.

Davidson L., Vandongen R., and Beilin L. J., 1981, Effect of eating bananas on plasma free and sulfate-conjugated catecholamines. *Life Sci,* **29**: 1773-1778.

Dooley T. P., 1998, Molecular biology of the human phenol sulfotransferase gene family. *Journal Of Experimental Zoology,* **282**: 223-230.

Goldstein D. S., Swoboda K. J., Miles J. M., Coppack S. W. *et al.*, 1999, Sources and physiological significance of plasma dopamine sulfate. *J Clin Endocrinol Metab,* **84**: 2523-2531.

Javitt N. B., Lee Y. C., Shimizu C., Fuda H. *et al.*, 2001, Cholesterol and hydroxycholesterol sulfotransferases: Identification, distinction from dehydroepiandrosterone sulfotransferase, and differential tissue expression. *Endocrinology,* **142**: 2978-2984.

Kakuta Y., Pedersen L. G., Carter C. W., Negishi M. *et al.*, 1997, Crystal structure of estrogen sulphotransferase. *Nature Struct Biol,* **4**: 904-908.

Kakuta Y., Petrotchenko E. V., Pedersen L. C., and Negishi M., 1998, The sulfuryl transfer mechanism: crystal structure of a vanadate complex of estrogen sulfotransferase and mutational analysis. *J Biol Chem,* **273**: 27325-27330.

Kanazawa K., and Sakakibara H., 2000, High content of dopamine, a strong antioxidant, in Cavendish banana. *J Agr Food Chem,* **48**: 844-848.

Rein G., Glover V., and Sandler M. , 1982, Mutiple forms of phenolsulphotransferase in human tissues. Selective inhibition by dichloronitrophenol. *Biochem Pharmacol,* **31**: 1893-1897.

Richard K., Hume R., Kaptein E., Stanley E. L. *et al.*, 2001, Sulfation of thyroid hormone and dopamine during human development - ontogeny of phenol sulfotransferases and arylsulfatase in liver, lung and brain. *J Clin Endocrinol Metab,* **86**: 2734-2742.

Roy A. K., 1992, Regulation of steroid hormone action in target cells by specific hormone-inactivating enzymes. *Proc Soc Exp Biol Med,* **199**: 265-272.

Rubin G. L., Sharp S., Jones A. L. , Glatt H. *et al.*, 1996, Design, production and characterisation of antibodies discriminating between the phenol- and monoamine-sulphating forms of human phenol sulphotransferase. *Xenobiotica,* **26**: 1113-1119.

Visser T. J., 1996, Pathways of thyroid hormone metabolism. *Acta Med Austriaca,* **1/2** : 10-16.

STRUCTURAL INSIGHTS INTO D_1 DOPAMINE RECEPTOR PHOSPHORYLATION AND DESENSITIZATION

David R. Sibley, Benjamin Gardner, Jennifer D. Peters, and Ok-Jin Kim*

1. INTRODUCTION

Dopamine receptors belong to the large G-protein coupled receptor superfamily and molecular cloning studies have revealed the existence of five structurally distinct subtypes (Neve and Neve, 1997). These can be divided into two subgroups on the basis of their amino acid sequences as well as their pharmacological and transductional properties. The first subgroup comprises the D_1 and D_5 dopamine receptors and is termed "D_1-like". When expressed in mammalian cells, activated D_1-like receptors stimulate adenylyl cyclase and raise intracellular levels of cAMP. The second dopamine receptor subgroup includes the D_2, D_3 and D_4 receptors and is termed "D_2-like". The D_2-like dopamine receptors are coupled to the inhibition of adenylyl cyclase as well as the modulation of potassium and calcium ion channels. As with other G-protein coupled receptors, dopamine receptors are subject to a wide variety of regulatory mechanisms, which can either positively or negatively modulate their expression and functional activity.

One of the most important forms of regulatory mechanisms that modulate signaling by G-protein coupled receptors is that of agonist-induced desensitization - defined as the tendency of receptor-mediated responses to wane over time despite continued agonist stimulation. Recent studies with β-adrenergic receptor systems have suggested a general paradigm for agonist-induced desensitization (Lefkowitz, 1998). This involves phosphorylation of the receptors by a member of the G-protein coupled receptor kinase (GRK) family leading to the binding of an arrestin-like protein ultimately resulting in uncoupling of the receptor from its cognate G protein and decreased functional activity. The binding of an arrestin molecule also promotes internalization of the receptor through clathrin-coated pits into an endosomal compartment where it may be de-phosphorylated by a specific G-protein coupled receptor phosphatase (GRP) and then recycled to the cell surface or degraded via a lysosomal pathway.

---

* Molecular Neuropharmacology Section, National Institute of Neurological Disorders & Stroke, NIH, Bethesda, MD 20892, USA.

Catecholamine Research, Edited by Nagatsu *et al.*
Kluwer Academic/Plenum Publishers, 2002

Previous investigations of agonist-induced regulation of dopamine receptors have revealed great variability among the subtypes. For instance, agonist-induced desensitization is not always observed with the D_2 dopamine receptors and, in some instances, agonist occupancy of this subtype results in increased receptor expression (Sibley and Neve, 1997). In contrast, the D_1 dopamine receptor has been shown to exhibit agonist-induced refractoriness in both endogenous and recombinant/heterologous cellular expression systems (Sibley and Neve, 1997). Recent data have also provided support for a phosphorylation pathway underlying agonist-induced desensitization of the D_1 dopamine receptors. For instance, we have previously shown that the D_1 dopamine receptor, when expressed in C6 glioma cells, is stoichiometrically phosphorylated in response to agonist activation and that this phosphorylation is both rapid and transient (Gardner et al., 2001). The D_1 dopamine receptor phosphorylation was shown to be pharmacologically specific and correlated with the desensitization response. Our current data show that internalization is not required for D_1 dopamine receptor de-phosphorylation to take place suggesting that the phosphorylated D_1 receptor is processed through a novel recovery pathway. We have also attempted to identify the sites of regulatory phosphorylation within the D_1 dopamine receptor using mutagenesis techniques and transient expression in HEK-293 cells.

2. THE D_1 RECEPTOR IS DE-PHOSPHORYLATED THROUGH A NOVEL PATHWAY

Our initial hypothesis was that once phosphorylated, the D_1 dopamine receptor might be subsequently processed in a manner similar to that of the β_2-adrenergic receptor (see above). Our first approach was to examine the effect of inhibiting internalization of the receptor on the de-phosphorylation process. It has previously been shown that inhibiting the internalization of β_2-adrenergic receptors would inhibit their de-phosphorylation and resensitization. Prior studies have shown that treatment of cells with the plant lectin concanavalin A or hypertonic sucrose will inhibit D_1 receptor internalization. We thus treated the cells with concanavalin A or hypertonic sucrose, to see what effect this would have on the de-phosphorylation of the receptor. The design for these experiments was to treat with dopamine for 10 min, which results in maximal receptor phosphorylation, followed by a 20 min wash-out to allow phosphorylation to return to basal levels. Following a 20 min recovery period, the level of receptor phosphorylation had indeed returned to near basal levels in the control treatment group. Surprisingly, pretreatment of the cells with hypertonic sucrose or concanavalin A did not have any effect on the D_1 receptor de-phosphorylation subsequent to agonist removal despite the fact that receptor internalization was inhibited. These results suggest that receptor internalization is not required for D_1 receptor de-phosphorylation to occur and that the phosphatase responsible for this action may reside in a different cellular compartment than the previously identified G-protein coupled receptor phosphatase (GRP).

To further characterize the de-phosphorylation pathway of the D_1 receptor, we examined the effects of two protein phosphatase inhibitors, okadaic acid and calyculin A which have been shown to inhibit the G-protein coupled receptor phosphatase and/or resensitization of the β_2-adrenergic receptor. However, neither of these inhibitors blocked D_1 receptor de-phosphorylation and, after 20 min of recovery, all treatment groups exhibited receptor phosphorylation states that were near basal levels. These results would seem to suggest that the previously identified "G-protein coupled receptor phosphatase" is not involved in D_1 receptor de-phosphorylation. Further experiments are in progress to identify the phosphatase involved in D_1 dopamine receptor de-phosphorylation and to delineate the role of receptor internalization in the resensitization response.

Several other G-protein coupled receptors have been shown to exhibit similar phenomena as we have observed here for the D_1 dopamine receptor. For instance, both the bradykinin B2 receptor expressed in human fibroblasts and the vasopressin V1a receptor expressed in HEK-293 cells exhibit rapid but transient phosphorylation in response to agonist treatment. Interestingly, as we have observed here with the D_1 dopamine receptor, okadaic acid was reported not to affect the de-phosphorylation of the V1a vasopressin receptor. Similarly, the CCK receptor has been reported to undergo transient agonist-induced phosphorylation as well as a complex internalization process in acinar or CHO cells. It was further suggested that the CCK receptor might be de-phosphorylated in a smooth vesicular compartment adjacent to the plasma membrane. All of these observations suggest that our current findings with the D_1 dopamine receptor may have wide-spread applicability to many G protein coupled receptors and different cell types.

3. MAPPING SITES OF FUNCTIONAL PHOSHORYLATION IN THE D_1 RECEPTOR

In the current study, we have attempted to identify the sites of regulatory phosphorylation within the D_1 dopamine receptor using mutagenesis techniques and transient expression in HEK-293 cells. We created two 3rd cytoplasmic loop mutants - one in which all serine and threonine residues were mutated to alanine or valine residues, respectively (3rdTOT) and another mutant (a subset of the 3rdTOT mutant) in which just serine residues 256, 258 & 259 within the 3rd cytoplasmic loop were mutated to alanines (3rd234). We also prepared a series of mutants where the D_1 dopamine receptor was truncated subsequent to the indicated positions in the COOH terminus: T347, T369, T394 & T404. Transient expression of these mutant and wild-type constructs resulted in similar levels of receptor binding activity (~1 pmol/mg) with the exception of the T347 mutant which was about 50% lower. Dopamine stimulated cAMP accumulation in a similar fashion for all constructs except for the 3rdTOT mutant which showed a 10-fold reduction in dopamine potency. Dopamine-induced desensitization was initially investigated for the 3rdTOT and 3rd234 mutants. Compared to the wild-type receptor, both mutant constructs exhibited an impaired or attenuated desensitization response. When agonist-induced phosphorylation of these constructs was examined, both the 3rdTOT and 3rd234 mutant receptors exhibited less phosphate incorporated - approximately 50% less compared to the wild-type receptor. These results suggest that phosphorylation of residues within the 3rd cytoplasmic loop are important for the desensitization response, but that other residues outside this region are also important since the desensitization was not completely abrogated. Moreover, since the 3rd234 mutant behaved in an identical fashion as the 3rdTOT mutant, this would suggest that one or more of serine residues 256, 258 & 259 within the 3rd cytoplasmic loop are the critical residues involved. Further work involving mutating these three residues individually is currently in progress.

We next examined agonist-induced phosphorylation of the truncated D_1 dopamine receptors. Beginning with a 50% reduction of phosphorylation in the T404 mutant, there was a steady decrease in phosphate content of the receptor as more and more of the COOH terminus was truncated. Surprisingly, we found that the most severely truncated mutant (T347) exhibited no agonist-induced phosphorylation what-so-ever. This must be reconciled with our previous data showing that removal of potential phosphorylation sites in the 3rd cytoplasmic loop resulted in a decrease in receptor phosphorylation. Agonist-induced desensitization of the phosphorylation-defective T347 mutant receptor was next examined. Surprisingly, we found that the T347 receptor construct not only desensitized, but desensitized more rapidly and to a greater extent compared to the wild-type receptor.

In order to explain all of our current findings with the mutated receptors, we are proposing the following hypothesis for how the D_1 receptor undergoes phosphorylation and desensitization. Under basal conditions, the receptor's COOH terminus is in relatively tight association with the 3rd cytoplasmic loop. Agonist activation results in a conformational change in the receptor leading to its phosphorylation by one or more G protein-coupled receptor kinases (GRKs). This phosphorylation takes place in a sequential or ordered fashion – initially residues within the COOH terminus are phosphorylated followed by phosphorylation of serine residues in the 3rd cytoplasmic loop. In this model, phosphorylation of the COOH terminus is actually required for phosphorylation of the 3rd cytoplasmic loop, perhaps by transfer of the GRK from the COOH terminus to the 3rd loop. Phosphorylation of these two receptor regions results in their separation allowing an arrestin protein to associate with the 3rd cytoplasmic loop and quench G protein coupling. It is important to note that the only role of phosphorylation is to remove the steric hindrance of the COOH terminus which then allows arrestin to bind to the activated conformation of the 3rd cytoplasmic loop. With the 3rdTOT and 3rd234 mutants, the COOH terminus can be phosphorylated, but not the 3rd cytoplasmic loop. This results in an incomplete removal of the COOH terminal steric hindrance and thus arrestin binding and desensitization is impaired. With the severely truncated T347 mutant receptor, there is no receptor phosphorylation since the COOH terminus has been removed; however, since the steric hindrance of the COOH terminus has also been removed, an arrestin can still associate with the activated 3rd cytoplasmic loop and desensitize the receptor. Indeed, since the steric hindrance of the COOH terminus has been removed, the arrestin protein can associate more rapidly and readily with the receptor resulting in more rapid and enhanced desensitization. While this hypothesis can explain our current data, certain aspects of it must be rigorously proven. A central feature is the association of an arrestin to the 3rd cytoplasmic loop – that this association is enhanced in a COOH-truncated receptor and impaired in the 3rdTOT and 3rd234 mutants. We are currently setting up experiments to examine the translocation of arrestin-2 and arrestin-3 to the wild-type and mutant D_1 receptors as one test of this hypothesis.

REFERENCES

Gardner, B., Liu, Z.F., Jiang, D. and Sibley, D.R., 2001, The role of phosphorylation/de-phosphorylation in agonist-induced desensitization of D_1 dopamine receptor function: evidence for a novel pathway for receptor de-phosphorylation, *Mol Pharmacol*, **59**: 310-321.

Lefkowitz, R. J., 1998, G protein coupled receptors: New roles for receptor kinases and β-arrestins in receptor signaling and desensitization. *J Biol Chem*, **273**: 18677-18680.

Neve, K. A., and Neve, R. L., 1997 Molecular biology of dopamine receptors, in: *The Dopamine Receptors*, Neve, KA and Neve, RL eds. , Humana Press, Totowa, NJ., pp. 27-76.

Sibley, D. R. and Neve, K. A., 1997, Regulation of dopamine receptor function and expression, in: *The Dopamine Receptors*, Neve, K. A. and Neve, R. L. eds., Humana Press, Totowa, NJ, pp. 383-424.

THE NEUROBIOLOGY OF DOPAMINE SIGNALING

Paul Greengard *

Nerve cells communicate with each other through two mechanisms, referred to as fast and slow synaptic transmission. Fast-acting neurotransmitters, e.g., glutamate (excitatory) and GABA (inhibitory), achieve their effects on their target cells, within one thousandth of a second, by virtue of opening ligand-operated ion channels. Slow-acting neurotransmitters, which include dopamine and serotonin, all peptide neurotransmitters, as well as glutamate and GABA, achieve their effects over seconds to minutes, by an enormously more complicated sequence of biochemical steps involving second messengers, protein kinases, and protein phosphatases. Slow-acting neurotransmitters control the efficacy of synaptic transmission, both by regulating the efficiency of neurotransmitter release from presynaptic terminals and by regulating the efficiency with which both fast- and slow-acting neurotransmitters produce their effects on postsynaptic receptors. Fast synaptic transmission can be considered to represent the hardware of the brain, and slow synaptic transmission its software (see Figure 1).

* Greengard, P., Laboratory of Molecular and Cellular Neuroscience, The Rockefeller University, New York, NY 10021, USA

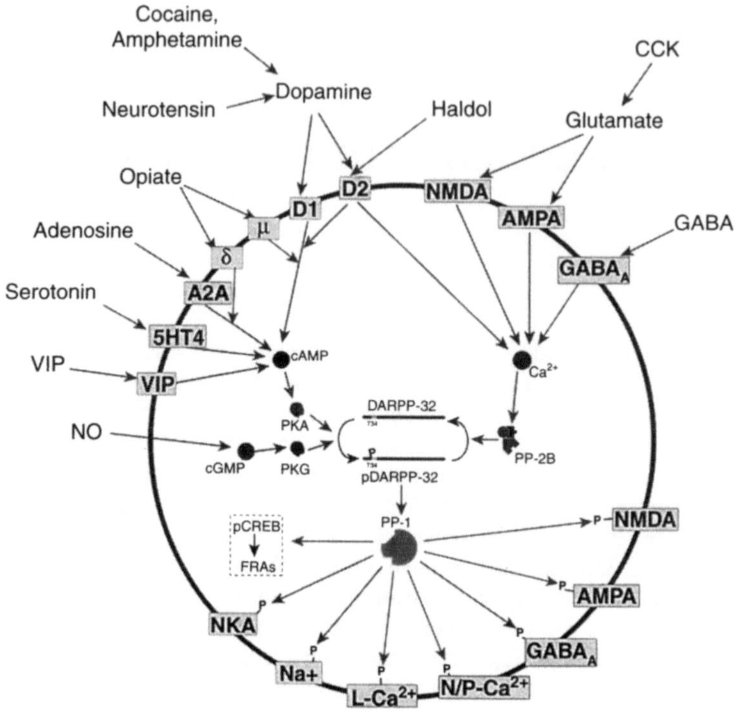
Cocaine,
Amphetamine
Dopamine
Neurotensin
Haldol
CCK
Glutamate
Opiate
D1
D2
NMDA
AMPA
GABA
μ
δ
Adenosine
A2A
GABAA
Serotonin
5HT4
VIP
VIP
cAMP
Ca2+
PKA
DARPP-32
NO
cGMP
PKG
pDARPP-32
PP-2B
PP-1
pCREB
FRAs
NMDA
AMPA
NKA
Na+
L-Ca2+
N/P-Ca2+
GABAA

Figure 1.

MUTANT MICE LACKING DOPAMINE RECEPTOR GENES

Atsu Aiba and Motoya Katsuki*

1. INTRODUCTION

Dopamine is the principle neurotransmitter for several neural systems in the brain. The dopaminergic system is involved in the inhibitory regulation of the secretion of several peptide hormones in the pituitary as well as in the modulation of motor activity via the nigrostriatal dopaminergic pathway. Dopaminergic system is also implicated in motivated behaviors, emotional stability, and certain aspects of learning and memory. So far, five subtypes of dopamine receptor, designated as D1R to D5R, have been found by cloning. These receptors have been divided into two major classes, referred to as D1-like (D1 and D5) and D2-like (D2, D3 and D4) receptors originally based on their pharmacological properties and on their opposite coupling of either up or down regulation to the adenylyl cyclase and cAMP production pathway (Civelli et al., 1993; Gingrich and Caron, 1993). The D1-like receptors are mainly expressed in the caudate putamen, nucleus accumbens, olfactory tubercle, cerebral cortex and hippocampus, while the D2-like receptors are mainly expressed in the caudate putamen, nucleus accumbens, olfactory tubercle, and the pituitary. The D2-like receptors are also expressed on dopaminergic cell bodies within the substantia nigra compacta, ventral tegmental areas, and on the terminals. In order to investigate the selective physiological role of each of dopamine receptor, we generated mutant mice which lack each of five dopamine receptors. Because expression of dopamine receptors overlaps extensively in the brain, and despite the differences in signal transduction mechanisms, the contribution of D1-like receptors and D-2 like receptors to the locomotion or other behaviors is sometimes considered synergistic, we generated multiple mutant mice which lack more than one dopamine receptor simultaneously. We found that D1RD2R double mutant mice showed the sever phenotype that was not observed in the single D1R or D2R mutant mice.

* Atsu Aiba, Motoya Katsuki, Laboratory of DNA Biology and Embryo Engineering, Center for Experimental Medicine and CREST, Institute of Medical Science, University of Tokyo, Tokyo 108-8639, Japan

2. METHODS

To generate mice which lack each of five dopamine receptors (D1R-D5R), we have cloned the genomic DNAs from mouse 129/Sv phage library. Targeting vectors were constructed to generate null allele of each of dopamine receptor genes. The targeting vectors were transfected independently into embryonic stem cells by electroporation, and cells were subjected to G418 selection. The DNA of the resistant clones were analyzed by Southern blotting. Homozygous mutant mice were generated by crossing the heterozygous mutant mice. Double or triple mutant mice were generated by crossing the different mutant mice naturally or by *in vitro* fertilization.

3. RESULTS AND INTERPRETATION

We have generated D1R, D2R, D3R, D4R, and D5R mutant mice (Yamaguchi et al., 1996). In addition, we generated the mice which lack D1-like receptors or D2-like receptors by crossing the single mutant mice. We generated D1RD5R, D2RD4R, and D2RD3RD4R mutant mice. Pharmacological and behavioral analyses of these dopamine receptor mutant mice are under investigation. D1RD2R double mice were born at expected frequency, but phenotypes of D1RD2R double mutant mice are similar to those of dopamine-deficient mice (Zhou & Palmiter, 1995). They became hypoactive, hypothermia and stopped feeding a few weeks after birth, suggesting D1R and D2R play an important role in regulation of movement, body temperature, and feeding.

4. CONCLUSIONS

We have generated five dopamine receptor mutant mice and multiple mutant mice. Because expression of dopamine receptors overlap extensively, it is often difficult to elucidate the role of dopaminergic system *in vivo*. Multiple dopamine receptor mutant mice which lack more than one dopamine receptor simultaneously showed sometimes the phenotypes which were not observed in single mutant mice. Thus, it is critical to generate and analyze multiple mutant mice, in order to understand the role of dopaminergic pathway *in vivo*.

REFERENCES

Civelli, O., Bunzow, J.R., and Grandy, D.K. 1993, Molecular diversity of the dopamine reeptors. *Annu. Rev. Pharmacol. Toxicol.* **32**: 281

Gingrich, J.A., and Caron, M. G. 1993, Recent advances in the molecular biology of dopamine. *Annu. Rev. Neurosci.* **16**:299.

Yamaguchi, H., Aiba, A., Nakamura, K., Nakao, K., Sakagami, H., Goto, K., Kondo, H., and Katsuki, M. 1996, *Genes Cells* **1**:253.

Zhou, Q.Y., and Palmiter R.D., 1995, Dopamine-deficient mice are severely hypoactive, adipsic, and aphagic, *Cell* **83**:1197

INTERACTION BETWEEN GLUTAMATE AND MONOAMINES

Modulation of glutamatergic neurotransmission by dopamine D1 and 5-HT2A receptors

Tomohiro Abekawa*, Minoru Honda, and Tsukasa Koyama

1. INTRODUCTION

Electrophysiological studies revealed that in the prefrontal cortex(PFC), dopamine has an inhibitory action via dopamine D1 receptor,[1] and serotonin has an excitatory action via 5-HT2A receptor,[2] on glutamatergic neurotransmission. The changes in activity of glutamatergic pyramidal neurons play an important role in pathogenesis of phencyclidine(PCP)-induced psychosis[3] or schizophrenia.[4] However, neurochemical basis for the modulation of glutamatergic neurotransmission in the PFC is not well known. In this chapter, we present data on the effects of stimulation or blockade of dopamine D1 receptor, and blockade of 5-HT2A receptor or dopamine D4 receptor on basal levels or PCP-induced increases in extracellular concentrations of glutamate.

2. MATERIALS AND METHODS

Male Wistar-King rats or Sprague-Dawley rats(250-300 g) were used. A dopamine D1 receptor agonist, SKF38393 or a D1 antagonist, SCH23390 was dissolved in artificial CSF, and perfused via membrane of microdialysis probe. PCP, a selective dopamine D4 and 5-HT2A receptors antagonist, NRA0045 (gift from Taisho Pharmaceutical Co., Ltd., Japan) or a selective dopamine D4 receptor antagonist, NRA0160(Taisho Pharmaceutical Co., Ltd., Japan) was dissolved in physiological saline and administered intraperitoneally. Microdialysis probe was implanted in the medial PFC(mPFC)(Coordinate; A: +2.7 mm, L: 0.8 mm, DV: -1.8 mm with respect to the bregma). Perfusion was begun at the speed of

* Tomohiro Abekawa, Department of Psychiatry, Hokkaido University, Graduate School of Medicine, Sapporo 060-8638, Japan.

Catecholamine Research, Edited by Nagatsu *et al.*
Kluwer Academic/Plenum Publishers, 2002

2 µl/min, 24 h after the implantation of the probe, and samples were obtained every 20 or 30 min. Glutamate concentration was assayed by HPLC-ECD or HPLC-FLD(Eicom, Japan) according to the pre-column derivatization method.[5] Data were analyzed by a repeated two-way ANOVA and post-hoc Duncan test. The level of significance was set at $p<0.05$.

3. RESULTS

Perfusion with SKF38393(20 or 200 µM) reduced basal level of extracellular concentration of glutamate, and this effect was blocked by co-perfusion with SCH23390(40 µM). Perfusion with SCH23390 alone had no effect on glutamate level[6] (at least from 0 to 120 min)(Figure was not shown). PCP(7.5 mg/kg, i.p.) increased glutamate level. Perfusion with SKF38393(200 µM) enhanced the PCP-induced increase in glutamate level, and perfusion with SCH23390(40 µM) inhibited this increase in glutamate level(Fig. 1).

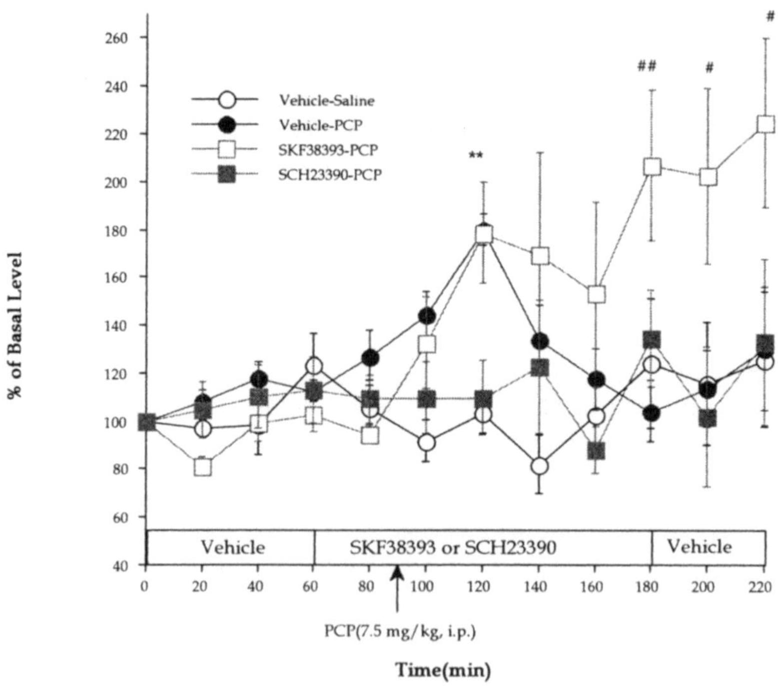

Figure 1. Effect of local perfusion with SKF38393 or SCH23390 into mPFC on PCP-induced increase in glutamate level in this region. SKF38393(200 µM) or SCH23390(40 µM) was perfused from 60 to 180 min. **: $p<0.01$ vs. Vehicle-Saline, #: $p<0.05$, ##: $p<0.01$ Vehicle-PCP vs. SKF38393-PCP.

A selective D4 and 5-HT2A receptors antagonist, NRA0045(3 mg/kg, i.p.) blocked the PCP-induced increase in glutamate level, contrarily, a selective D4 receptor antagonist, NRA0160(3 mg/kg, i.p.) had no effect on this increase in glutamate. Both NRA0045 and NRA1060 had no effect on basal level(after saline injection) of glutamate(Fig. 2).

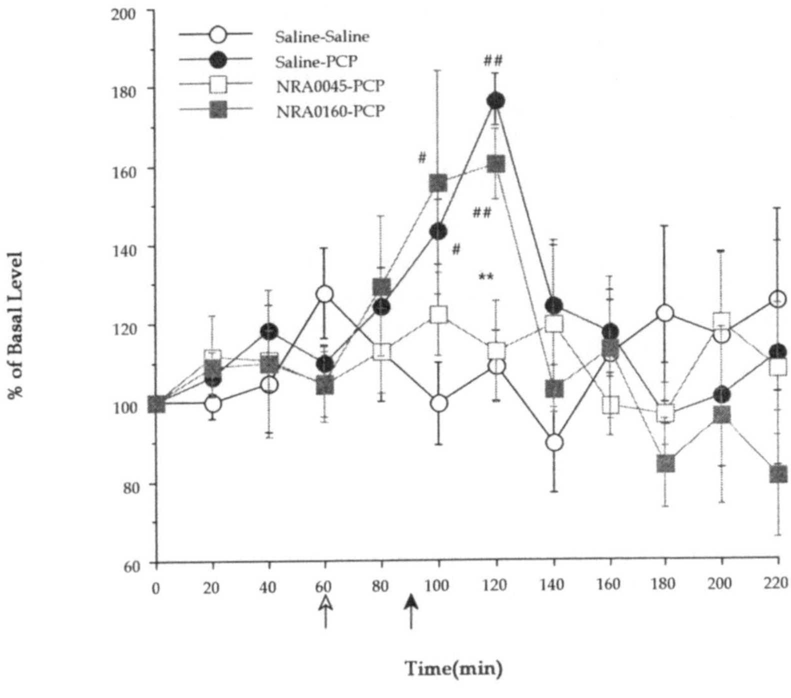

Figure 2. Effect of NRA0045 or NRA0160 on PCP-induced increase in glutamate level in mPFC. NRA0045(3 mg/kg, i.p.), NRA0160(3 mg/kg, i.p.), or saline(1 ml/kg, i.p.) was injected at 60 min(white arrow), afterward, PCP(7.5 mg/kg, i.p.) was injected at 90 min(black arrow). **: $p<0.01$ NRA0045-PCP vs. Saline-PCP, #: $p<0.05$, ##: $p<0.01$ Saline-PCP vs. Saline-Saline, or NRA0160-PCP vs. Saline-Saline

4. DISCUSSION

The stimulation of dopamine D1 receptors reduced basal levels of glutamate in the mPFC, contrarily enhanced the PCP-induced increase in glutamate levels. The blockade of D1 receptor had no effect on basal level of glutamate, however, inhibited the PCP-induced increase in glutamate level. Dopamine D1 receptors are localized to spines of pyramidal neurons in the cerebral cortex.[7] PCP-induced increase in glutamate level in the mPFC is attributed to disinhibition of glutamatergic pyramidal neurons via the blockade of N-methyl-D-aspartate (NMDA) receptors on GABAergic interneurons, therefore, this increase reflects hyperactivity of pyramidal neurons.[3] Taken together, these results suggest that in the absence of hyperactivity of glutamatergic pyramidal neurons, the stimulation of dopamine D1 receptor inhibits excitatory response, and in the presence of PCP-induced hyperactivity of glutamatergic neurons, the stimulation of dopamine D1 receptor enhances excitatory response. In othe words, the modulation of glutamatergic neurotransmission by dopamine D1 receptor is state-dependent. Similar state-dependent modulation by dopamine D1 receptor was reported in the nucleus accumbens.[8]

The selective blockade of dopamine D4 and 5-HT2A receptors inhibited the PCP-induced increase in glutamate level, although it had no effect on basal level of glutamate. Contrarily, the selective blockade of dopamine D4 receptor had no effect on both the PCP-induced increase in glutamate and basal level of glutamate. These results suggest that the blockade of 5-HT2A receptor plays an important role in inhibiting PCP-induced increase in glutamate. Considering the presence of 5-HT2A receptors on the glutamatergic terminals and on the apical dendrites of pyramidal glutamatergic neurons in the mPFC,[9] and the stimulatory modulation of neuronal activity of pyramidal neurons,[3] we can hypothesize that the blockade of 5-HT2A receptor inhibits the PCP-induced hyperactivity of glutamatergic pyramidal neurons which leads to increases in glutamate levels. In summary, these data suggest that glutamatergic neurotransmission in the mPFC is modulated state-dependently by dopamine D1 receptor and modulated stimulatorily by 5-HT2A receptors. These modulations may play some roles in the pathogenesis of PCP-psychosis and schizophrenia.

REFERENCES

1. T.D. Law, J.C. Hirsch, F. Crepel, Dopamine modulation of synaptic transmission in rat prefrontal cortex: an in vivo electrophysiological study, *Neurosci. Res.* **21**, 151-160 (1994).
2. G.K. Aghajanian, G.J. Marek, Serotonin induces excitatory postsynaptic potentials in apical dendrites of neocortical pyramidal cells, *Neuropharmacol.* **36**, 589-599 (1997).
3. B. Adams, B. Moghaddam, Corticolimbic dopamine neurotransmission is temporally dissociated from the cognitive and locomotor effects of phencyclidine, *J. Neurosci.* **18**, 5545-5554 (1998).
4. P.S. Goldman-Rakic, L.D. Selemon, Functional and anatomical aspects of prefrontal pathology in schizophrenia, *Schizophr. Bull.* **23**, 437-445 (1997).
5. P. Lindroth, K. Mopper, High performance liquid chromatographic determination of subpicomole amounts of amino acids by precolumn fluorescence derivatization with o-phthaldialdehyde, *Anal Chem.* **51**, 1667-1674.
6. T. Abekawa, T. Ohmori, T. Koyama, D1 dopamine receptor activation reduces extracellular glutamate and GABA concentrations in the medial prefrontal cortex, *Brain Res.* **867**, 250-254 (2000).
7. J.F. Smiley, A.I. Levey, B.J. Ciliax, P.S. Goldman-Rakic, D1 dopamine receptor immunoreactivity in human and monkey cerebral cortex: predominant and extrasynaptic localization in dendritic spines, *Proc. Natl. Acad. Sci. USA* **91**, 5720-5724 (1994).
8. S. Hernandez-Lopez, J. Bargas, D.J. Surmeier, A. Reyes, E. Galarraga, D1 receptor activation enhances evoked discharge in neostriatal medium spiny neurons by modulating an Ca2+ conductance, *J. Neurosci.* **17**, 3334-3342 (1997).
9. R.L. Jakab, P.S. Goldman-Rakic, 5-hydrotryptamine 2A serotonin receptors in the primate cerebral cortex: possible site of action of hallucinogenic and antipsychotic drugs in pyramidal cell apical dendrites, *Proc. Natl, Acad. Sci. USA* **95**, 735-740 (1998).

ANALYSIS OF DOPAMINE D4 RECEPTOR FUNCTION IN GENE KNOCKOUT MICE: A ROLE IN CORTICAL EXCITABILITY

Malcolm J. Low, David K. Grandy, and Marcelo Rubinstein*

1. INTRODUCTION

The prefrontal cortex (PFC) integrates diverse environmental stimuli and contextual cues to appropriately regulate the temporal organization and execution of motor behaviors. Mesocortical dopamine (DA) projections from the ventral tegmental area play a key role in the modulation of these signals to influence complex processes such as arousal, attention, vigilance, novelty-induced stress and learning. Both DA release and turnover rates in the PFC are increased by stress (Roth et al., 1988), while manipulations that attenuate DA transmission in the PFC impair alertness, behavioral reactions to stress and working memory performance. D1-like receptor agonists promote seizure activity in the PFC and conversely, stimulation of D2-like DA receptors is anticonvulsant and diminishes the electrical excitability of pyramidal neurons in the PFC (Starr, 1996). Among the members of the D2-like receptor family, the D4R has been of particular interest because it is densely expressed in the PFC and it has high affinity for various atypical antipsychotics (Van Tol et al., 1991). In addition, the D4R has been implicated in the pathophysiology of psychiatric diseases that involve PFC malfunction such as attention-deficit with hyperactivity disorder and schizophrenia (Seeman et al., 1993). At the molecular-genetic level some alleles of the highly polymorphic human D4R gene (DRD4) have been associated with human diseases, personality traits including attention deficit hyperactivity disorder, opiate and alcohol use, and novelty-seeking behavior. *In vitro* studies have demonstrated that activated D4Rs couple to inwardly rectifying potassium channels that hyperpolarize neurons (Werner et al., 1996), providing a potential mechanism by which the D4R could play a fundamental role in the overall inhibitory modulation of PFC neuronal activity.

* Malcolm J. Low, Vollum Institute, Oregon Health & Science University, Portland, Oregon, 97201. David K. Grandy, Department of Physiology and Pharmacology, Oregon Health & Science University, Portland, Oregon, 97201. Marcelo Rubinstein, Instituto de Investigaciones en Ingeniería Genética y Biología Molecular (CONICET) and Departamento de Ciencias Biológicas, Facultad de Ciencias Exactas y Naturales, Universidad de Buenos Aires, Vuelta de Obligado 2490, 1428 Buenos Aires, Argentina.

Because of the limited selectivity of DA agonists and antagonists for individual members of the D2-like receptor family, we adopted a genetic approach to characterize the physiological function of D4 receptors in cortex. Mutant mice were generated by homologous recombination in embryonic stem cells and analyzed in a series of multidimensional experimental protocols. Our initial studies with these mice demonstrated normal or somewhat depressed basal activity, but a significant locomotor supersensitivity to psychostimulants including methamphetamine, cocaine, and ethanol (Rubinstein et al., 1997). The mutant mice also exhibited improved performance on the accelerating rotarod, a test of integrated motor skills, suggesting increased vigilance or motor coordination compared to wild-type mice. No abnormalities were demonstrated in the prepulse inhibition of acoustic startle responses in D4 receptor-deficient mice (Ralph et al, 1999). Neurochemical analyses of DA biosynthesis indicated an increased rate of DA turnover in the dorsal striatum, suggesting to us the possibility of an increased excitatory tone from disinhibited cortical-striatal projection neurons. To further test this hypothesis, we conducted additional behavioral, electrophysiological, and morphological analyses of the D4 receptor knockout mice.

2. ALTERATIONS IN APPROACH-AVOIDANCE BEHAVIORS

Several different approach-avoidance paradigms were used to assess the role of the D4 receptor in behavioral responses to novelty and stress. Significant differences were observed between wild type and D4 receptor-deficient mice in each of these tests. The most robust differences were uncovered in a novel object exploration paradigm, a test that maximizes approach behavior and minimizes avoidance behavior because of the familiarity of the environment. In contrast, the weakest genotype differences were observed in the open field test that maximizes avoidance behavior. The open field represents a forced exploration paradigm where escape from the stressful novel environment is impossible. An intermediate genotype effect size was observed in an emergence test from a darkened cylinder. Therefore, these data were most suggestive of a decrease in approach behavior in response to novelty (Dulawa et al., 1999). However, additional studies have demonstrated significant genotype differences on a light-dark preference exploration test, where D4 receptor knockout mice had a longer latency to leave the darkened compartment and spent more time in the darkened compartment overall. Moreover, D4 knockout mice had fewer entries and spent a reduced percentage of time on the open arms of an elevated plus maze (Falzone et al., submitted). These deficits were reversed by the anxiolytic midazolam, a compound that does not increase motivation or novelty seeking, indicating that the absence of D4 receptors increases anxiety-like behavior as well as reducing exploratory behavior in response to novelty.

3. INCREASED EXCITABILITY OF CORTICAL PYRAMIDAL NEURONS

Electrophysiological recordings were performed in brain slice preparations by patch-clamp analysis under both current and voltage clamp conditions. Pyramidal neurons in cortical layers III and V demonstrated that the frequency and duration of spontaneous and induced paroxysmal discharges were increased in D4 receptor-deficient mice (Rubinstein et al., 2001). The activity was dependent on excitatory glutamatergic input because it was blocked by the application of the non-NMDA receptor antagonist CNQX. Following the electrophysiological analyses, biocytin filled neurons were observed microscopically and no morphological abnormalities were found. Enhanced cortical excitability was also observed in brain slices, from wild-type mice, incubated in the presence of the selective D4R antagonist PNU-101387G (a gift from K. Merchant). This drug had no effect in brain slices from the D4 receptor-deficient mice, indicating the specificity of the drug for the D4 receptor subtype. Together these results suggest that the D4R can function as an inhibitory modulator of glutamate activity in the FC. The actions appear to be purely postsynaptic because we found no evidence for altered rates of dopamine synthesis or turnover in the frontal cortex of D4 receptor knockout mice, unlike the increase demonstrated in dorsal striatum. Interestingly, similar experiments performed in striatal brain slices demonstrated increased glutamate-dependent excitability of medium spiny neurons in D2 but not D4 dopamine receptor-deficient mice, suggesting that each D2-like receptor subtype has a specific regional importance in the modulation of neuronal activity by the excitatory transmitter glutamate (Cepeda et al, 2001).

Based on the electrophysiological data we hypothesized that pyramidal neurons in the FC of D4R-deficient mice might become hyperexcitable under proconvulsive conditions. To test this hypothesis wild type and mutant mice were exposed to increasing concentrations of the convulsant bicuculline. At doses up to 3 mg/kg bicuculline produced only mild to medium signs of epileptogenic activity in mice of both genotypes. However, at 4 mg/kg, bicuculline displayed pronounced convulsant effects that differed as a function of genotype. Following the administration of 4 mg/kg bicuculline all of the wild-type mice entered a tonic-clonic seizure phase that was characterized by a loss of postural control, barrel rolling, spontaneous jumping, and convulsions. At this dose 60% of the wild-type animals died by the end of the 15 minute observation period. In contrast the mice lacking D4Rs displayed more pronounced epileptic-like seizure behavior and a significantly earlier onset of tonus, wild running, and clonus. In addition, all of the D4R-deficient mice experienced severe seizures and died within 15 minutes of administration of the bicuculline (Rubinstein et al., 2001).

Recent *in vivo* microdialysis and electron microscopic studies have demonstrated that enhanced extracellular levels of glutamate are correlated with a decrease in the density of terminal glutamate immunolabeling secondary to reduced neurotransmitter accumulation. Given the apparent cortical hyperexcitability that we observed in our mutant mice we hypothesized that glutamate immunolabeling of FC pyramidal neurons might be reduced in the D4R-deficient mice compared to wild-type cells. To test this hypothesis we performed an ultrastructural analysis of glutamate immunolabeling in FC terminals. This study revealed a significant 25% decrease in glutamate-immunoreactivity in FC layer VI of mutant, compared to wild type, mice (Rubinstein et al., 2001).

4. SUMMARY

Based on all these data, we speculate that one of the functions of stress-induced DA release in the PFC is to regulate cortical excitability via D4R stimulation. In the absence of inhibitory modulation by this receptor subtype, environmental stimuli that produce normal arousal in wild-type mice may trigger a hypervigilant state in the D4R-deficient mice that results in exaggerated avoidance behavior and decreased seizure threshold. In humans, anxiety disorders such as agoraphobia and panic attacks are clinically recognized conditions that are normally treated with drugs capable of diminishing cortical excitability, such as benzodiazepines. In addition, it is interesting to note that anxiety is highly comorbid with Parkinson's disease and ADHD, two conditions characterized by compromised midbrain dopaminergic neurotransmission that are treated with drugs that enhance DA receptor stimulation such as the DA precursor L-DOPA and the DA transporter blocker methylphenidate, respectively. The human DRD4 gene is highly polymorphic in its coding region and regulatory regions upstream of the transcription initiation site. Given the putative genetic associations of some of the human DRD4 alleles with the occurrence of psychiatric diseases, together with the heightened anxiety and decreased novelty-seeking behavior demonstrated in mice lacking the D4R receptor, it is tempting to speculate that the D4R gene might be a candidate gene for traits related to emotional reactions to stress or anxiety -like behaviors.

5. ACKNOWLEDGMENTS

The authors wish to thank our collaborators T. Falzone, C. Cepeda, R. Hurst, J Flores-Hernandez, L. Kozell, S. Dulawa, J. Bunzow, T. Phillips, C. Meshul, M. Geyer, and M. Levine for their contributions to work reviewed in this chapter. This work was funded by grants from the National Institutes of Health, National Science Foundation, and an International Research Scholars Grant from the Howard Hughes Medical Institute.

REFERENCES

Cepeda, C., Altemus, K.L., Hurst, R.S., Flores-Hernandez, J., Calvert, C., Jokel, E.S., Grandy, D.K., Low, M.J., Rubinstein, M., Ariano, M.A., Levine. M,S., 2001, Facilitated glutamatergic transmission in the striatum of D2 dopamine receptor-deficient mice, *J. Neurophysiol.* 85:659.

Dulawa, S.C., Grandy, D.K., Larson, J.L., Low, M.J., Paulus, M., Geyer, M.A., 1999, Dopamine D4 receptor-knock-out mice exhibit reduced exploration of novel stimuli, *J. Neurosci.* 19:9550.

Ralph, R.J., Varty, G.B., Kelly, M.A., Wang, Y.-M., Caron. M.G., Rubinstein, M., Grandy, D.K., Low, M.J., Geyer, M.A., 1999, The dopamine D2, but not D3, or D4 receptor subtype is essential for the disruption of prepulse inhibition produced by amphetamine in mice, *J. Neurosci.* 19:4627.

Roth, R.H., Tam, S.-Y., Ida, I., Yang, J.-X., Deutsch, A.Y., 1988, Stress and the mesocorticolimbic dopamine systems, *Ann. N.Y. Acad. Sci.* 537:138.

Rubinstein, M., Cepeda, C., Hurst, R.S., Flores-Hernandez, J., Ariano, M.A., Falzone, T.L., Kozell, L.B., Meshul, C.K., Bunzow, J.R, Low, M.J., Levine, M.S., Grandy, D.K., 2001, Dopamine D4 receptor-deficient mice display cortical hyperexcitability, *J. Neurosci.* 21:3756.

Rubinstein, M., Phillips, T.J., Bunzow, J.R., Falzone, T.L., Dziewczapolski, G., Zhang, G., Fang, Y., Larson, J.L., McDougall, J.A., Chester, J.A., Saez, C., Pugsley, T.A., Gershanik, O., Low, M.J., Grandy, D.K., 1997, Mice lacking dopamine D4 receptors are supersensitive to ethanol, cocaine, and metamphetamine, *Cell* 90:991.

Seeman, P., Guan, H.-C., Van Tol, H.H.M., 1993, Dopamine D4 receptors elevated in schizophrenia, *Nature* 365:441.

Starr, M., 1996, The role of dopamine in epilepsy, *Synapse* 22:159.

Van Tol, H.H.M., Bunzow, J.R., Guan, H.-C., Sunahara, R.K., Seeman, P., Niznik, H.B., Civelli, O., 1991, Cloning of the gene for a human dopamine D4 receptor with high affinity for the antipsychotic clozapine, *Nature* 350:610.

Werner, P., Hussy, N., Buell, G., Jones, K.A., North, R., 1996, D2, D3, and D4 dopamine receptors couple to G protein-regulated potassium channels in Xenopus oocytes, *Mol. Pharmacol.* 49:656.

DOPAMINE D2L RECEPTOR KNOCKOUT MOUSE PROVIDES A UNIQUE MODEL SYSTEM FOR STUDYING THE FUNCTIONS OF D2L AND D2S

Yanyan Wang, Rong Xu, Toshikuni Sasaoka, and Emma-B. Sankoorikal*

Dopamine receptors are classified into two categories: D1-like receptors (including the D1 and D5 subtypes), and D2-like receptors (including the D2, D3, and D4 subtypes). These classifications are based on both gene structure and pharmacological criteria.[1] Among D2-like receptors, D2 is the most abundant form in the brain, with expression levels 10- to 100-fold higher than those of D3 or D4. Two isoforms of the D2 receptor, termed D2L (long form) and D2S (short form), have been identified in the brain. Little is known about the distinct functions of either D2 isoform. No selective pharmacological agents are currently available. We have generated mice lacking the D2L receptor (D2L-/-), but expressing a functional D2S receptor. Thus, the mouse expressing purely D2S (D2L-/-) should allow us to identify a D2 ligand relatively selective for D2S, to assess the effects of drugs on neurons expressing only D2S, and to examine the effects of drugs on behaviors mediated by D2S. Because D2L is the predominant D2 isoform expressed in wild-type (WT) mice, the comparisons made between D2L-/- and WT mice will reveal novel information regarding the functions of D2L. Here, we attempted to investigate the distinct functional roles of D2L and D2S, using D2L-/- mouse as a model system and a combination of approaches.

Dopamine D2L receptor knockout mice (D2L-/-) were generated as previously described in detail.[2] No gross neuroanatomical abnormalities were found on examination of serial brain sections (Fig. 1). D2L-/- mice also had normal fertility and body weight in comparison with WT mice.

Previously, we showed that D2 receptor density was similar in the striatum of D2L-/- and WT mice, due to a compensatory increase in D2S expression in mice lacking D2L.[2] We now performed saturation binding assays of [^{3}H]-YM-09151-2 using hippocampal tissue. The D2 receptor sites (Bmax values) in the hippocampus were also

* Yanyan Wang, Rong Xu, and Emma-B. Sankoorikal, Department of Pharmacology, University of Pennsylvania School of Medicine, M102 John Morgan building, Philadelphia, PA 19104-6084. Toshikuni Sasaoka, National Institute of Neuroscience, Tokyo, Japan. Contact author: Yanyan Wang, yywang@pharm.med.upenn.edu, 215-573-4291 (Tel), 215-573-2236 (Fax).

Catecholamine Research, Edited by Nagatsu *et al.*
Kluwer Academic/Plenum Publishers, 2002

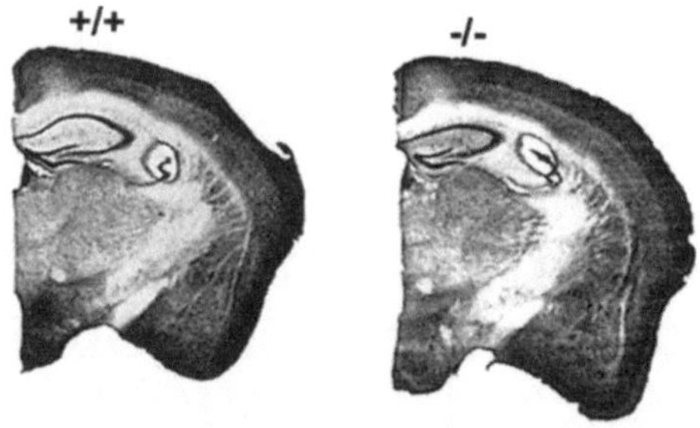

Figure 1. Representative coronal brain section with Nissl staining. The section (on the right) from D2L-deficient (-/-) mouse is a little anterior to the one (on the left) from WT (+/+) mouse, due to the availability of brain sections for Nissl staining.

not significantly different between WT (97.4 ± 6.8 fmol/mg protein, n = 4 independent experiments) and D2L-/- (109.3 ± 9.4 fmol/mg protein, n = 4) mice ($p > 0.34$, Student's t test). In contrast, the affinities (Kd values) of [^{3}H]-YM-09151-2 were significantly different between WT (0.78 ± 0.03 nM) and D2L-/- (1.09 ± 0.06 nM; $p < 0.004$).

Several behavioral tests were conducted to identify the phenotypes of D2L-/- mice. We observed that D2L-/- mice displayed reduced levels of locomotion and rearing activity,[2] but not grooming (not shown) in the open-field test. The mutant mice did not show increased anxiety and may even have reduced anxiety in the elevated zero-maze test. D2L-/- mice initially performed poorly in the rotarod test, but they were able to learn to perform this test in the subsequent trials, suggesting that these animals do not have severe impairment in motor coordination.[2] Interestingly, D2L-/- mice exhibited lower levels of aggression than WT mice in the resident-intruder test.[3]

To assess the role of D2L in the actions of typical antipsychotic drugs, we examined the effect of haloperidol treatment on several behaviors of mutant mice. D2L-/- mice showed reduced sensitivity to haloperidol-induced catalepsy in the bar test (Fig. 2).[2,4] We also observed that haloperidol produced significantly less inhibition of locomotion, rearing, and grooming in D2L-/- mice (Fig. 3).[2] These results suggest that D2L may contribute more than D2S to the induction of catalepsy by haloperidol.

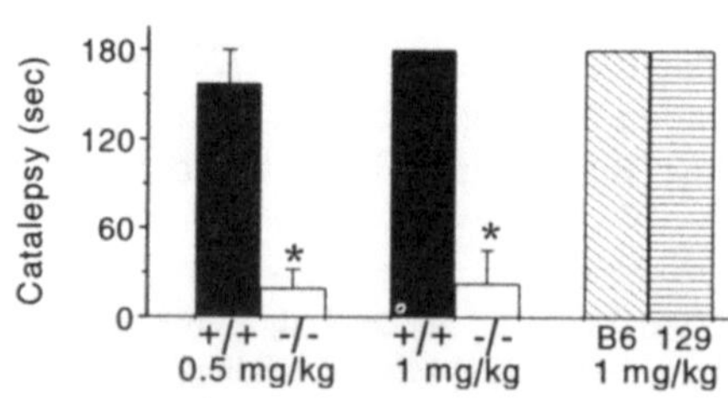

Figure 2. The effect of haloperidol on cataleptic behavior of D2L-/- and WT (+/+) mice on a hybrid background (B6 x 129). The doses of haloperidol used were 0.5 mg/kg (n = 8 for each genotype), and 1 mg/kg (n = 9 and 8 for +/+ and -/-, respectively); *$p < 0.02$ (Mann-Whitney test). The effect of haloperidol (1 mg/kg) was also examined on wild-type mice with a pure C57BL/6 (B6; n = 6) or 129/Sv/J (129; n = 7) background. Data are taken in part from ref. 2 (Copyright 2000 by the Society for Neuroscience).

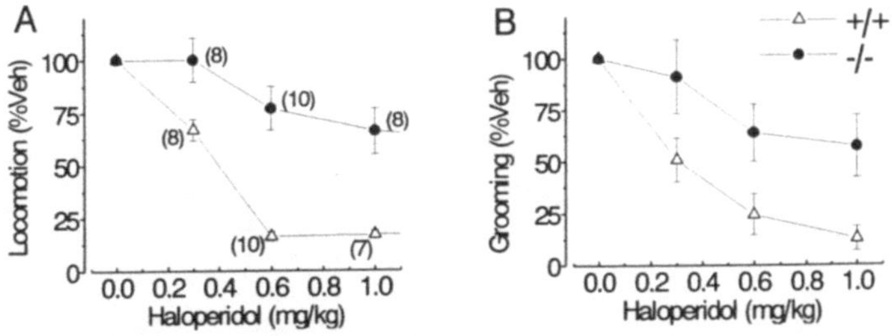

Figure 3. Effects of haloperidol on behaviors of mice in the open-field test, measured within 10 min. *(A)* Inhibitory effect of haloperidol on locomotion. *(B)* Inhibitory effect of haloperidol on grooming behavior. Responses to vehicle injection (Veh) were normalized as 100%. All other data were expressed as percentage of the respective vehicle injection. Haloperidol produced significantly less inhibition in locomotion [$F(1,63) = 60.96$; $p < 0.0001$; two-way ANOVA] and grooming [$F(1,63) = 17.16$; $p = 0.0001$] in D2L-/- than in WT (+/+) mice. Data shown are mean ± SEM. Numbers in parentheses represent numbers of mice. +/+ (open triangles) and -/- (black circles). Part of the data are taken from ref. 2 (Copyright 2000 by the Society for Neuroscience).

To determine whether D2L-/- mice express a functional D2S receptor, we assessed the electrophysiological effect of dopamine (DA) on the neurons of the zona compacta of the substantia nigra (SNc) since the inhibitory effect of dopamine on SNc dopaminergic neurons is mediated by the D2 receptor. Spontaneous firing of SNc neurons was recorded extracellularly in brain slices.[5] Dopamine (3 - 100 μM), applied by superfusion, elicited a concentration-dependent reduction in the spontaneous firing rate of SNc neurons (Fig. 4). There was no significant difference in the dose-response of DA-induced inhibition in the SNc between WT (n = 8) and D2L-/- (n = 11) mice [$F(1,53) = 1.19$; $p > 0.28$; two-way ANOVA].[2] These studies demonstrated that D2S was functional in D2L-/- mice and suggest that D2S can function well as an impulse-modulating autoreceptor.

D2 receptor antagonists are commonly used to alleviate the psychotic symptoms of schizophrenia; however, some of them (e.g., haloperidol) produce extrapyramidal side effects.[6, 7] On the other hand, a dopamine precursor or D2 agonists effectively relieve the motor symptoms of Parkinson's disease, but they can also produce psychotic side effects in parkinsonian patients.[8, 9] Since regulation of D2 receptor function is required for alleviating either the psychotic symptoms of schizophrenia or motor symptoms of Parkinson's disease, there is a possibility that the D2L and D2S receptors may contribute differentially to the therapeutic ratio of clinically used D2 drugs. In addition, central stimulants such as amphetamine and methylphenidate, which increase dopaminergic neurotransmission, are beneficial to children with attention-deficit hyperactivity disorder (ADHD). Our findings suggest that D2L may have distinct functional roles from D2S in motor behavior, aggression, autoreceptor function, and drug-induced parkinsonism. Thus, understanding of the specific functions of D2L and D2S receptors may provide insights toward developing improved pharmacological agents for the treatment of disorders related to central dopaminergic systems, including schizophrenia, Parkinson's disease, and ADHD. Our studies indicate that the D2L receptor knockout mouse provides a unique model system for dissecting the functions of individual D2 isoforms in intact animals as well as at the cellular and molecular levels in the mammalian central nervous system.

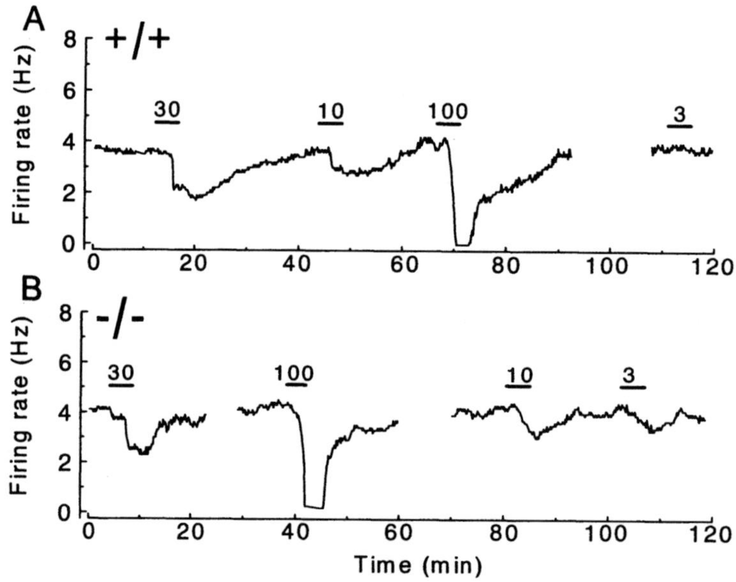

Figure 4. Effect of dopamine on the firing rate of substantia nigra compacta neurons. *(A)* Dopamine-induced inhibition of a representative SNc neuron in +/+ slice. *(B)* Dopamine-induced inhibition of a representative SNc neuron in -/- slice. Firing rate calculated over 10-sec intervals is plotted as a function of time. Horizontal lines above the data indicate duration of application of DA. Numbers above the horizontal lines represent the concentration of DA (μM). Data collection by the computer was sometimes discontinuous to save disk space; this accounts for the gaps in the graph. On the average, there was no significant difference in the response to 3 μM DA between +/+ and -/- mice (see ref. 2).

REFERENCES

1. D. R. Sibley, F. J. Monsma Jr., and Y. Shen, Molecular neurobiology of dopaminergic receptors, *Int. Rev. Neurobiol.* **35**, 391-415 (1993).
2. Y. Wang, R. Xu, T. Sasaoka, S. Tonegawa, M.-P. Kung, and E.-B. Sankoorikal, Dopamine D2 long receptor-deficient mice display alterations in striatum-dependent functions, *J. Neurosci.* **20**, 8305-8314 (2000).
3. K.-L. Vukhac, E.-B. Sankoorikal, and Y. Wang, Dopamine D2L receptor- and age-related reduction in offensive aggression, *NeuroReport* **12**, 1035-1038 (2001).
4. A. Usiello, J.-H. Baik, F. Rouge-Pont, R. Picetti, A. Dierich, M. LeMeur, P. V. Piazza, and E. Borrelli, Distinct functions of the two isoforms of dopamine D2 receptors. *Nature* **408**, 199-203 (2000).
5. Y. Wang and G. K. Aghajanian, Excitation of locus coeruleus neurons by vasoactive intestinal peptide: role of cAMP and protein kinase A, *J. Neurosci.* **10**, 3335-3343 (1990).
6. M. Carlsson and A. Carlsson, Interactions between glutamatergic and monoaminergic systems within the basal ganglia – implications for schizophrenia and Parkinson's disease, *TINS* **13**, 274-276 (1990).
7. P. Seeman and H. H. M. Van Tol, Dopamine receptor pharmacology, *Curr. Opin. Neurol. Neurosurg.* **6**, 602-608 (1993).
8. S. H. Snyder, Catecholamines in the brain as mediators of amphetamine psychosis, *Arch. Gen. Psychiat.* **27**, 169-179 (1972).
9. J. W. Langston, I. Irwin, and G. A. Ricaurte, Neurotoxins, Parkinsonism and Parkinson's Disease, *Pharmac. Ther.* **32**, 19-49 (1987).

NEUROBIOLOGICAL FEATURES OF DOPAMINE D2 RECEPTORS

Distinct functions of D2L and D2S in vivo

Alessandro Usiello, Ciro Iaccarino, and Emiliana Borrelli*

1. INTRODUCTION

Two different isoforms of the dopamine D2 receptor exist *in vivo*, named D2L and D2S. These isoforms are derived by alternative splicing of the sixth exon of the gene (Dal Toso et al., 1989; Giros et al., 1989). D2L is the longer isoform composed of 444 amino acids (aa) in rat (Monsma et al., 1989; Chio et al., 1990) and mouse (Montmayeur et al., 1991) and 443 in human (Dal Toso et al., 1989; Giros et al., 1989; Grandy et al., 1989). D2S is instead composed of 415 aa in rat (Bunzow et al., 1988) and mouse (Montmayeur et al., 1991) and 414 in human (Dal Toso et al., 1989; Giros et al., 1989).

D2L differs from D2S only by a 29 aa insertion in the putative third intracellular loop of the receptor. Interestingly, both isoforms are colocalized in the same cells and in all the regions where D2 receptors are expressed. The two isoforms share also the same pharmacological profile (Dal Toso et al., 1989; Giros et al., 1989; Monsma et al., 1989; Montmayeur et al., 1991). The longer isoform is predominantly expressed, although the ratio of the two isoforms can vary in some locations (Montmayeur et al., 1991).
The 29 aa insert confers to D2L a different G-protein coupling specificity from D2S (Montmayeur and Borrelli, 1991; Montmayeur et al., 1993). Guiramand et al. generated amino acid mutations in the D2L insert which demonstrated a key role of the 29 aa insertion in the selection of the G-protein with which D2 receptor isoforms interact (Guiramand et al., 1995).

These data suggest that the differential G-protein coupling characteristics existing between the two isoforms underlies functional differences and distinct regulations of the signalling pathways activated by D2L and D2S *in vivo* (Picetti et al., 1997).

* Institut de Génétique et Biologie Moléculaire et Cellulaire (IGBMC), Illkirch, France, 67404.

2. D2 RECEPTOR-DEFICIENT MICE

We generated mice lacking the D2 receptor (Baik et al., 1995), and the most striking phenotype of D2 receptor-deficient mice was their motor behaviour impairment.These mice have slower movements and abnormal posture. Moreover, they exhibit an abnormal gait, with sprawled hind legs. When these phenotypes were studied and quantified in locomotor behavioural tests such as the open field, ring test and rotarod test, the homozygous D2 receptor-deficient mice revealed severe impairments. Rearing, one of the stereotyped behaviours is also absent in D2 receptor-deficient mice. Tested on the rotarod apparatus, these mice had difficulty coordinating movements; the ring test revealed a reduction of spontaneous movements.

D2 receptor-deficient mice show a slight reduction of their body weight resulting from a 10-15% reduction in water and food intake. This suggests an involvement of D2 receptors in the control of feeding behaviour.

Interestingly, we showed that no compensation by other members of the D2-like subfamily nor of the D1-like take place in these animals. Thus suggesting a unique role for D2 receptors *in vivo*.

According to previous pharmacological studies, D2 receptor absence leads to a strong elevation of the enkephalin levels in the striatum of D2 receptor-deficient mice. We also observed a slight decrease of substance P expression, whereas the level of expression of dynorphin seemed to be unaltered (Baik et al., 1995).

Strikingly, these findings are reminiscent of the modifications observed in the striatum of 6-hydroxydopamine (6-OHDA) lesioned animals (Gerfen, 2000). Thus, the reduction of dopamine obtained in these lesioned animals seems to preferentially affect D2 receptor-mediated functions.

Reduced fertility was observed in the breeding of homozygous D2 receptor-deficient mice indicating alterations of the hypothalamic-pituitary-gonadal axis. We have assessed that D2 receptor-deficient mice have very high level of circulating prolactin, which may be the cause of fertility problems (Saiardi et al., 1997). In addition, large hyperplasia are found in pituitaries of D2 receptor-deficient mice. These findings suggest a role for D2 receptors not only in the regulation of prolactin synthesis and release but also in the control of pituitary cell proliferation.

Furthermore, D2R-like receptors have been shown to possess autoreceptor functions by pharmacological, neurochemical and electrophysiological means. Indeed, both D2R and D3R are localised in DAergic neurons (Jackson and Westlind-Danielsson, 1994).
In addition, the Kd for most agonists and antagonists are within one order of magnitude between D2 and D3 receptors, they cannot be used to discriminate, at least in a physiologically useful manner, among the members within the D2R-class family (Seeman and Van Tol, 1994). This has raised the question of whether both receptors may have autoreceptor functions. Thus, the precise subtype of the somatodendritic and terminal nerve autoreceptors remains unclear.

Therefore, knockout mice represent a great tool to test whether DA D2 and D3 receptors share this function *in vivo* or if only one of them is responsible for it. Interestingly, D2R knockout mice present a total abrogation of the inhibitory effects of DA on the firing of DAergic neurones in the substantia nigra (Mercuri et al., 1997). In addition, also the DA release assessed in the striatal synaptosomes (L'Hirondel et al., 1998) in D2R deficient mice suggest a pivotal role of D2 receptors in the control of dopamine release. Taken together these results indicate that D2R is the major DA autoreceptor. Importantly, the neurochemical analysis of D3R knockout mice supports this view. Indeed, in these mice autoreceptor functions are maintained in spite of lack of D3 receptors. However, in this study a potential role for D3R in the control of post-synaptic long-loop feedback modulating DA release has been evoked (Koeltzow et al., 1998).

3. D2L RECEPTOR-DEFICIENT MICE

The presence *in vivo* of two almost identical forms of D2R, D2L and D2S, represents an interesting paradigm to understand whether these proteins have redundant or specific functions (Usiello et al., 2000). Targeted distruption of exon 6 does not alter the normal level of expression of the D2 gene, which by default encodes only the D2S mRNA. This also results in an unvariant number of D2 sites, which in D2L-/- are constituted exclusively by D2S sites.

D2L-null mice grow and reproduce normally. In spite of the involvement of D2R in motor functions, D2L -/- mice did not differ from WT littermates in any of the spontaneous motor behaviors tested, suggesting a functional compensation in basal conditions.

Pre-synaptic functions related to the regulation of dopamine release and the locomotor attenuating effects of D2 agonists are preserved in D2L-/- mice. Indeed, pre-synaptic doses of quinpirole induce suppression of locomotion in both D2L-/- and WT. Furthermore, this drug is even more potent in D2L-/- than in WT siblings as established at behavioral and neurochemical levels by *in vivo* microdialysis, probably as a consequence of the higher number of D2S sites. In contrast, lack of D2L impairs the post-synaptic responses mediated by dopaminergic ligands, thereby indicating that D2L is the D2 post-synaptic receptor. This is supported by the lack of a cataleptic response to haloperidol in D2L-/- mice, a behavior very similar to what has been observed in D2R-/- mice under the same challenge. In agreement with loss of D2R post-synaptic functions we also observed the absence of the locomotor-stimulating effect of apomorphine and SKF 82958, two mixed D1/D2 agonists. In conclusion, ablation of the D2L isoform of D2R reveals, for the first time, that D2L and D2S have different and likely antagonistic functions on locomotion *in vivo*. Thus, the observed differential coupling of these receptors to different G-proteins is one mechanism able to explain the diversity of function between D2L and D2S.

We propose that the D2S is principally the D2 autoreceptor and that the DA postsynaptic effects classically attributed to D2R are mediated by D2L. Importantly, D2S receptors expressed at post-synaptic sites negatively modulate D1R-dependent responses as observed using SKF 81297 a D1-like agonist.

Finally, loss of the behavioral effects of haloperidol, a widely used antipsychotic, suggests a role for D2L in neuropsychiatric disorders such as Tourette's syndrome and schizophrenia (Seeman and Kapur, 2000).

REFERENCES

Baik JH, Picetti R, Saiardi A, Thiriet G, Dierich A, Depaulis A, Le Meur M, Borrelli E (1995) Parkinsonian-like locomotor impairment in mice lacking dopamine D2 receptors. Nature **377**:424-428.

Bunzow JR, Van Tol HH, Grandy DK, Albert P, Salon J, Christie M, Machida CA, Neve KA, Civelli O (1988) Cloning and expression of a rat D2 dopamine receptor cDNA. Nature **336**:783-787.

Chio CL, Hess GF, Graham RS, Huff RM (1990) A second molecular form of D2 dopamine receptor in rat and bovine caudate nucleus. Nature **343**:266-269.

Dal Toso R, Sommer B, Ewert M, Herb A, Pritchett DB, Bach A, Shivers BD, Seeburg PH (1989) The dopamine D2 receptor: two molecular forms generated by alternative splicing. Embo J **8**:4025-4034.

Gerfen CR (2000) Molecular effects of dopamine on striatal-projection pathways. Trends Neurosci **23**:S64-70.

Giros B, Sokoloff P, Martres MP, Riou JF, Emorine LJ, Schwartz JC (1989) Alternative splicing directs the expression of two D2 dopamine receptor isoforms. Nature **342**:923-926.

Grandy DK, Marchionni MA, Makam H, Stofko RE, Alfano M, Frothingham L, Fischer JB, Burke-Howie KJ, Bunzow JR, Server AC, et al. (1989) Cloning of the cDNA and gene for a human D2 dopamine receptor. Proc Natl Acad Sci USA **86**:9762-9766.

Guiramand J, Montmayeur JP, Ceraline J, Bhatia M, Borrelli E (1995) Alternative splicing of the dopamine D2 receptor directs specificity of coupling to G-proteins. J Biol Chem **270**:7354-7358.

Jackson DM, Westlind-Danielsson A (1994) Dopamine receptors: molecular biology, biochemistry and behavioural aspects. Pharmacol Ther **64**:291-370.

Koeltzow TE, Xu M, Cooper DC, Hu XT, Tonegawa S, Wolf ME, White FJ (1998) Alterations in dopamine release but not dopamine autoreceptor function in dopamine D3 receptor mutant mice. J Neurosci **18**:2231-2238.

L'Hirondel M, Cheramy A, Godeheu G, Artaud F, Saiardi A, Borrelli E, Glowinski J (1998) Lack of autoreceptor-mediated inhibitory control of dopamine release in striatal synaptosomes of D2 receptor-deficient mice. Brain Res **792**:253-262.

Mercuri NB, Saiardi A, Bonci A, Picetti R, Calabresi P, Bernardi G, Borrelli E (1997) Loss of autoreceptor function in dopaminergic neurons from dopamine D2 receptor deficient mice. Neuroscience **79**:323-327.

Monsma FJ, Jr., McVittie LD, Gerfen CR, Mahan LC, Sibley DR (1989) Multiple D2 dopamine receptors produced by alternative RNA splicing. Nature **342**:926-929.

Montmayeur JP, Borrelli E (1991) Transcription mediated by a cAMP-responsive promoter element is reduced upon activation of dopamine D2 receptors. Proc Natl Acad Sci U S A **88**:3135-3139.

Montmayeur JP, Guiramand J, Borrelli E (1993) Preferential coupling between dopamine D2 receptors and G-proteins. Mol Endocrinol **7**:161-170.

Montmayeur JP, Bausero P, Amlaiky N, Maroteaux L, Hen R, Borrelli E (1991) Differential expression of the mouse D2 dopamine receptor isoforms. FEBS Lett **278**:239-243.

Picetti R, Saiardi A, Abdel Samad T, Bozzi Y, Baik JH, Borrelli E (1997) Dopamine D2 receptors in signal transduction and behavior. Crit Rev Neurobiol **11**:121-142.

Saiardi A, Bozzi Y, Baik JH, Borrelli E (1997) Antiproliferative role of dopamine: loss of D2 receptors causes hormonal dysfunction and pituitary hyperplasia. Neuron **19**:115-126.

Seeman P, Van Tol HH (1994) Dopamine receptor pharmacology. Trends Pharmacol Sci 15:264-270.

Seeman P, Kapur S (2000) Schizophrenia: more dopamine, more D2 receptors. Proc Natl Acad Sci USA **97**:7673-7675.

Usiello A, Baik JH, Rouge-Pont F, Picetti R, Dierich A, LeMeur M, Piazza PV, Borrelli E (2000) Distinct functions of the two isoforms of dopamine D2 receptors. Nature **408**:199-203.

ROLE OF PRESYNAPTIC DOPAMINE D_2-LIKE RECEPTORS IN MIDBRAIN DOPAMINERGIC SYSTEM

Toshihiko Momiyama*

1. INTRODUCTION

Information has been accumulated little by little on the modulatory role of dopamine (DA) in central synaptic transmission. The studies during these several years have reported modulation of central synaptic transmission by D_1-like DA receptors located on the presynaptic terminals of GABAergic or glutamatergic fibres.[1] More recently, presynaptic D_2-like receptor-mediated modulation of central synaptic transmission has been analyzed using electrophysiological techniques. In this chapter, the roles of presynaptic D_2-like DA receptors in midbrain DA systems are reviewed, focused mainly on non-NMDA glutamatergic synaptic transmission onto DA-containing neurons in the ventral tegmental area (VTA), and GABAergic transmission onto cholinergic interneurons in the striatum.

2. D_2-LIKE RECEPTOR-MEDIATED INHIBITION OF THE SYNAPTIC CURRENTS

Whole-cell recordings were made using pipettes containing CsCl-based internal solution from neurons in the brain slices obtained from young rats (postnatal 7-20 days). Synaptic currents were evoked by focal stimulation, or spontaneous miniature postsynaptic currents were recorded in the presence of tetrodotoxin (TTX, 0.3 μM). Drugs were applied in the bath.

2.1 EPSCs

Non-NMDA glutamatergic excitatory postsynaptic currents (EPSCs) were evoked in VTA DA neurons by focal stimulation in the presence of bicuculline (20 μM), strychnine (0.5 μM) and D-AP5 (25 μM) to block GABAergic, glycinergic and NMDA

* Division of Cerebral Structure, National Institute for Physiological Sciences, Myodaiji, Okazaki 444-8585, Japan

Catecholamine Research, Edited by Nagatsu *et al.*
Kluwer Academic/Plenum Publishers, 2002

glutamatergic components, respectively. The EPSCs were reversibly blocked by CNQX (5 μM).

DA suppressed the amplitude of the evoked EPSCs in a concentration-dependent manner between 0.3-1000 μM (IC_{50}, 16.0 μM, Table 1) without affecting the holding current at -60 mV. A D_2-like receptor agonist, quinpirole (30 μM) also suppressed the EPSCs, whereas a D_1-like receptor agonist, SKF 81297 (30 μM) did not affect EPSCs. A D_2-like receptor antagonist, sulpiride (5 μM) blocked the DA-induced inhibition of the EPSCs and shifted the concentration-response curve to the right (an apparent dissociation constant, K_B, 0.96 μM, Table 1), while a D_1-like receptor antagonist, SCH 23390 (10 μM) had no effect.[2]

2.2 IPSCs

$GABA_A$ receptor-mediated inhibitory postsynaptic currents (IPSCs) were evoked in cholinergic interneurons in the striatum in the presence of strychnine (0.5 μM), CNQX (5 μM) and D-AP5 (25 μM) to eliminate glycinergic, non-NMDA and NMDA glutamatergic components, respectively. The IPSCs were reversibly blocked by bicuculline (10 μM).

DA reversibly reduced the amplitude of the evoked IPSCs in a concentration-dependent manner (IC_{50}, 10.0 μM, Table 1). The effect of DA was mimicked by a D_2-like receptor agonist, quinpirole (3-30 μM) but not by a D_1-like receptor agonist, SKF 81297 (30 μM). DA's effect was antagonized by a D_2-like receptor antagonist, sulpiride (5 μM) with a K_B value of 0.36 μM, but was not antagonized by SCH 23390 (10 μM), a D_1-like receptor antagonist (Table 1).[3]

These results suggest that the inhibitory effects of DA on the excitatory transmission onto VTA DA neurons and inhibitory transmission onto striatal cholinergic interneurons are mediated by D_2-like receptors.

3. PRESYNAPTIC SITE OF ACTION

The effects of DA on the miniature postsynaptic currents were examined in the presence of tetrodotoxin (TTX, 0.3 μM) in order to determine the site of DA-induced action. The frequency and amplitude of miniature postsynaptic currents reflects pre- and postsynaptic changes, respectively.

Non-NMDA glutamatergic spontaneous miniature excitatory postsynaptic currents (mEPSCs) were recorded from VTA DA neurons in the presence of TTX in addition to bicuculline, strychnine and D-AP5. The mEPSCs were reversibly blocked by CNQX (5 μM). DA (30 μM) reduced the frequency of mEPSCs without changing their mean amplitude. DA-induced effect on mEPSC frequency was diminished when external Ca^{2+} was replaced by Mg^{2+} (5 mM).[2]

GABAergic spontaneous miniature inhibitory postsynaptic currents (mIPSCs) were recorded from cholinergic interneurons in the striatum in the presence of TTX in addition to strychnine, CNQX and D-AP5. The mIPSCs were reversibly blocked by bicuculline (10 μM). DA (30 μM) also reduced the frequency of mIPSCs, whereas it had no effect on their amplitude distribution.[3] The effect of DA on the frequency of mIPSCs also diminished in nominally Ca^{2+}-free, high Mg^{2+} solution.[3]

These results suggest that DA's action on excitatory transmission onto VTA DA neurons and inhibitory transmission onto striatal cholinergic interneurons are both presynaptically mediated, and that DA affects Ca^{2+} entry into presynaptic terminals (Table 1).

4. COUPLING OF PRESYNAPTIC D_2-LIKE RECEPTORS TO N-TYPE CALCIUM CHANNELS

Multiple types of Ca^{2+} channels regulate fast synaptic transmission in the central nervous system.[4, 5] In the GABAergic synapses onto striatal cholinergic interneurons, the relative contribution of N-, P/Q- and other types of Ca^{2+} channels could be calculated using the data for inhibition of the IPSCs by N-type Ca^{2+} channel selective blocker, the P/Q type Ca^{2+} channel selective blocker, and by simultaneous application of both blockers. Assuming a third or fourth power relationship between the presynaptic Ca^{2+} concentration and the amplitude of synaptic currents, the relative contribution of each of N-, P/Q- or other types of Ca^{2+} channels were calculated to be 30-40%.[3] Therefore, this GABAergic synaptic transmission is also mediated by multiple types of Ca^{2+} channels.

DA (30 μM) no longer suppressed the IPSCs after the blockade of N-type Ca^{2+} channels, whereas DA still had inhibitory effects on the IPSCs after P/Q-type Ca^{2+} channels were blocked. These results suggest that DA, acting on presynaptic D_2-like receptors, inhibits GABA release by selectively blocking N-type Ca^{2+} channels (Table 1). This is the first work that has demonstrated a selective coupling between a certain subtype of Ca^{2+} channels and presynaptic DA receptors regulating transmitter release.[3] Ca^{2+} channel subtypes coupling to presynaptic D_2-like receptors in the glutamatergic terminals onto VTA DA neurons remains to be elucidated. In addition, presynaptic D_1-like receptor-mediated inhibition of EPSCs in magnocellular neurons in the basal forebrain has been reported to be also Ca^{2+}-dependent.[1] It's also a future interest to determine which types of Ca^{2+} channels are involved in presynaptic D_1-like receptor-mediated inhibition of transmitter release.

The findings demonstrating a selective coupling of D_2-like receptor to N-type Ca^{2+} channels in GABAergic terminals in spite of the abundance of P/Q- and other types of Ca^{2+} channels raise a question regarding the spatial localization of these molecules in the terminal in relation to the release sites. For example, are N-type Ca^{2+} channels located closer to D_2-like receptors than P/Q or other types of Ca^{2+} channels (Figure 1)? Morphological studies using ultrastructural techniques are necessary to address such a question.

A recent study has shown that the contribution of N-type Ca^{2+} channels to the regulation of transmitter release disappears within around postnatal 2 weeks in some central synapses of the rat[6]. Therefore, another question arises to be resolved in the future study whether N-type Ca^{2+} channels mediating GABA release onto striatal cholinergic interneurons are also subject to such a developmental elimination, in relation to the developmental change in the DA-induced inhibition. The issue might have pathophysiological implications in terms of the relation between age and basal ganglia-related disorders such as Parkinson's disease.

Table 1. Summary of presynaptic D_2-like dopamine (DA) receptor-mediated effects on the non-NMDA glutamatergic EPSCs in the VTA DA neurons and GABAergic IPSCs in striatal cholinergic interneurons

	EPSCs (VTA DA neurons)	IPSCs (Striatal cholinergic interneurons)
Effect of DA	Inhibition	Inhibition
IC_{50} of DA (μM)	16.0	10.0
Dissociation constant of sulpiride (μM)	0.96	0.36
Presynaptic mechanism	Ca^{2+}	Ca^{2+}
Ca^{2+} channel subtype	?	N

sulpiride, a D_2-like receptor antagonist
Ca^{2+}, to block Ca^{2+} entry into the presynaptic terminals
N, N-type Ca^{2+} channels

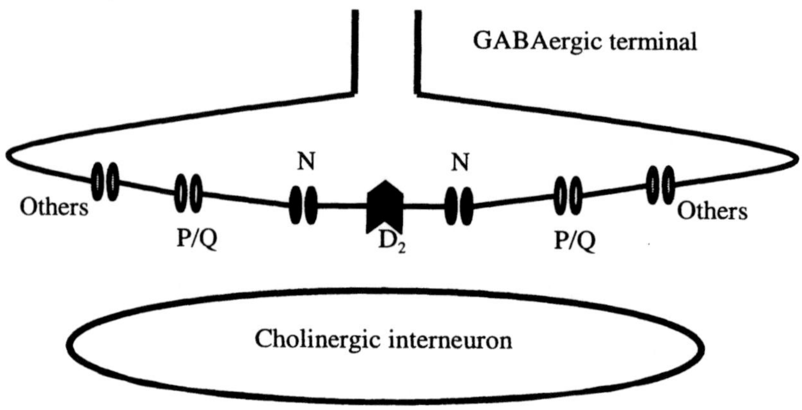

Figure 1. A hypothetical schema showing the localization of D_2-like DA receptors (D_2) and multiple types of Ca^{2+} channels in the GABAergic terminal onto a striatal cholinergic interneuron. N, N-type Ca^{2+} channels; P/Q, P/Q-type Ca^{2+} channels; Others, other types of Ca^{2+} channels.

REFERENCES

1. T. Momiyama, in: *Slow Synaptic Responses and Modulation,* edited by K. Kuba, H. Higashida, D.A. Brown and T. Yoshioka (Springer-Verlag Tokyo 2000), pp. 410-415.
2. E. Koga and T. Momiyama, Presynaptic dopamine D_2-like receptors inhibit excitatory transmission onto rat ventral tegmental dopaminergic neurones, *J. Physiol. (Lond.)* **523** (1), 163-173 (2000).
3. T. Momiyama and E. Koga, Dopamine D_2-like receptors selectively block N-type Ca^{2+} channels to reduce GABA release onto rat striatal cholinergic interneurones, *J. Physiol. (Lond.)* **533** (2), 479-492 (2001).
4. T. Takahashi and A. Momiyama, Different types of calcium channels mediate central synaptic transmission, *Nature* **366**, 156-158 (1993).
5. D.B. Wheeler, A. Randall and R.W. Tsien, Roles of N-type and Q-type Ca^{2+} channels in supporting hippocampal synaptic transmission, *Science* **264**, 107-111 (1994).
6. S. Iwasaki, A. Momiyama, O.D. Uchitel and T. Takahashi, Developmental changes in calcium channel types mediating central synaptic transmission, *J. Neurosci.* **20**, 59-65 (2000).

DOPAMINE-DEPENDENT LONG-TERM POTENTIATION INDUCED BY 3-NITROPROPIONIC ACID IN STRIATAL MEDIUM SPINY NEURONS

Antonio Pisani, Paola Bonsi, Paolo Gubellini, Diego Centonze, Massimo Tolu, Giorgio Bernardi, and Paolo Calabresi*

1. INTRODUCTION

The classical clinical symptoms of Huntington's disease (HD) include abnormal involuntary movements (chorea) and cognitive impairment. This genetically determined disorder selectively involves degeneration of striatal spiny neurons while sparing striatal large cholinergic interneurons.[1] HD is caused by an expansion of CAG repeats near the 5' end of the IT15 gene. IT15 encodes an ubiquitously expressed protein called huntingtin. Moreover, a remarkable decrease in the activity of mitochondrial complex II (succinate dehydrogenase, SD) has been found in brains of HD patients.[2] Indeed, the link between bioenergetic defects and excitotoxic mechanisms, two pathological events which seems to play a major role in HD[3,4] to the mutated huntingtin, remains unknown. The corticostriatal projection represents one of the major glutamatergic pathways in the brain and an abnormal release of glutamate from this pathway seems to play a pathogenic role in HD. The complex II inhibitors 3-nitropropionic acid (3-NP) and methylmalonic acid (MMA) mimic the pathology of HD.[5,6] Thus, enhanced glutamatergic transmission may trigger neurodegeneration in neurons, the energy metabolism of which is compromised due to impaired SD activity. We studied the electrophysiological effects of the pharmacological blockade of SD by either 3-NP or MMA on glutamatergic excitatory postsynaptic potentials (EPSPs), in order to investigate the link between metabolism impairment and glutamatergic transmission both in striatal spiny neurons and cholinergic interneurons. The t-LTP might play a key role in the regional and cell-type specific neuronal death observed in HD.

---

* Antonio Pisani and coauthors, Clinica Neurologica, Dipartimento di Neuroscienze, Università "Tor Vergata", Rome, Italy and Fondazione Santa Lucia, I.R.C.C.S., Via Ardeatina 306, 00179, Rome, Italy.

2. METHODS

Corticostriatal slices (200 μm thick) were prepared from Wistar rats and kept in saline Krebs' solution as previously shown.[7] Intracellular recording electrodes were filled with 2M KCl. Striatal spiny and cholinergic aspiny neurons were distinguished by their electrophysiological and morphological characteristics as previously described.[7,8] For synaptic stimulation, bipolar electrodes were located in the cortex. Data are means ± SEM expressed as percentage of EPSP amplitude. To obtain unilateral nigrostriatal lesions, rats (anaesthetized with pentobarbitone i. p.) were injected with 6-OHDA (8 μg/4 μL of saline containing 0.1% ascorbic acid) via a Hamilton syringe through a cannula inserted just rostral to the *substantia nigra* using stereotaxic coordinates.[9] For combined electrophysiological and optical recordings, sharp microelectrodes were filled with 1 mM bis-fura-2 in 200 mM KCl and backfilled with a 2 M KCl solution. Fluorescence was elicited by a 75 W Xenon lamp bandpass-filtered at 340 and 380 nm. Emission light was filtered (500 nm) and then detected by a CCD camera. Ratio images were obtained from background-subtracted pairs of 340 nm and 380 nm images. Intracellular calcium concentration are expressed as ratio values.

In striatal spiny neurons, NMDA receptor-mediated EPSPs, evoked in the absence of external magnesium and in the presence of 10 μM CNQX, an AMPA receptor antagonist[7] were enhanced by bath application of 100 μM 3-NP or 300 μM MMA while the AMPA-mediated EPSPs were unaffected (Fig.1A). Conversely, striatal large cholinergic interneurons showed no change of NMDA- or AMPA-mediated EPSP neither in 3-NP or in MMA. NMDA receptors mediated the increase in EPSP amplitude, since APV fully suppressed the EPSPs after t-LTP induction in spiny neurons.

3. ROLE OF INTRACELLULAR CALCIUM

Disruption of mitochondrial activity by SD inhibitors results in a disregulation of calcium buffering mechanisms.[10] Treatment with the calcium chelator BAPTA (100 mM). completely prevented t-LTP confirming that calcium is critical for t-LTP. In TTX (1 μM), bath-applied NMDA (30 μM, 1 min) elicited a membrane depolarization coupled to a transient and reversible elevation in intracellular calcium. After pretreatment with 100 μM 3-NP (20 min) NMDA evoked a marked membrane depolarization up to three-fold higher than in controls and intracellular calcium concentrations steadily increased (Fig. 1B).

5. ROLE OF ENDOGENOUS DOPAMINE

Endogenous DA and energy metabolism impairment might cooperatively produce the neuronal damage induced by 3-NP in the striatum by favoring t-LTP induction. We found that t-LTP was absent in slices obtained from 6-hydroxydopamine (6-OHDA)-denervated rats. In DA-denervated slices, t-LTP was restored by 3 μM quinpirole, a D2-like DA receptor agonist, but not by 10 μM SKF38393 a D1-like receptor agonist. To further evaluate the role of DA receptors, slices were incubated in 1 μM L-sulpiride, a D2-like receptor antagonist or 10 μM SCH23390, a D1-like receptor antagonist. L-sulpiride, but not SCH23390, fully prevented t-LTP.

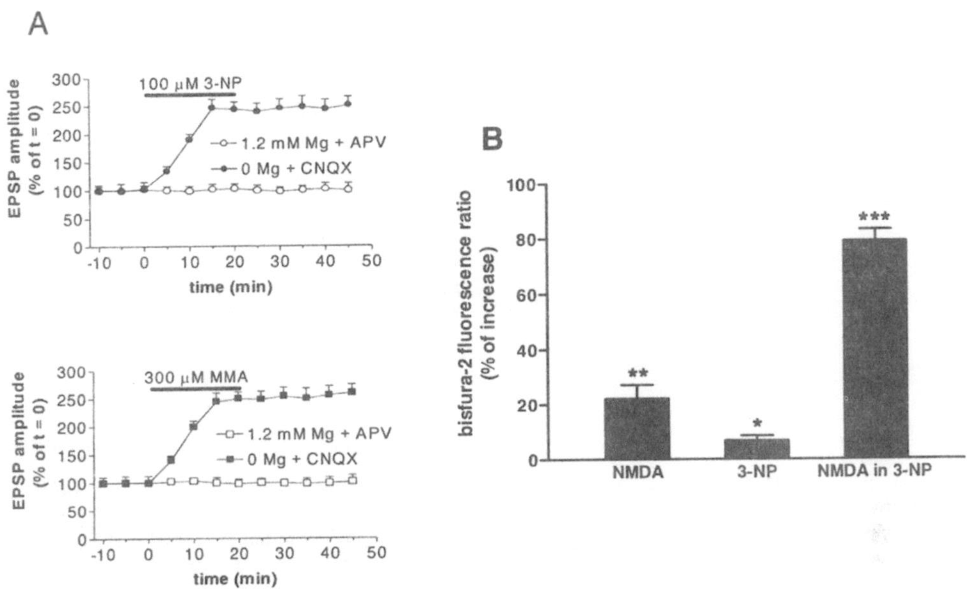

Figure 1. A. Summary plots of the t-LTP induced by 3-NP and MMA in striatal neurons. **B**. The histogram shows that the increase in calcium levels induced by NMDA is dramatically enhanced after pretreatment with 3-NP.

6. DISCUSSION

The discovery of the link between an expanded polyglutamine repeat in the IT15 gene and the synthesis of an altered huntingtin protein represented a key step towards the definition of the pathogenesis of this disorder. Nevertheless it remains unclear how the inhibition of mitochondrial activity leads to the regional and cell-type specific vulnerability in HD. Our experimental model suggests that the striatum is selectively vulnerable in HD because this structure is the main recipient of DA, a neurotransmitter that is critical for t-LTP. Several experimental findings support this hypothesis. First, t-LTP was absent in 6-OHDA-treated rats and was restored by the administration of a D2 receptor agonist. Second, blockade of D2 but not of D1 DA receptors prevented the expression of t-LTP. The first step required for t-LTP expression is an intracellular calcium rise. This is in agreement with the view that mitochondrial activity is of crucial importance for the maintenance of calcium homeostasis.[10] Indeed, 3-NP treatment significantly enhanced NMDA-dependent membrane depolarization and calcium rise. This calcium elevation following inhibition of mitochondrial complex II might initiate the cascade of events leading to neuronal death in the striatum. Clinical studies have shown that in HD patients, binding of [^{11}C]raclopride, a D2 DA receptor ligand, correlates with the loss of striatal spiny neurons preferentially expressing D2 DA receptors.[11] Striatal spiny neurons projecting to the globus pallidus more selectively express D2-like DA receptors, while spiny neurons projecting to the substantia nigra bear D1-like receptors.

Interestingly, spiny neurons projecting to the globus pallidus preferentially degenerate in HD compared to those projecting to the substantia nigra. Thus, neurons preferentially expressing D2-like DA receptors might undergo t-LTP more than those expressing D1 receptors, activating a vicious cycle leading to selective neuronal death in the striatum.

7. ACKNOWLEDGEMENTS

This work was supported by: MURST Cofin 2000 to G.B.; Min. Sanità (Finalizzato Alzheimer) to G.B. and A.P.; Telethon (N° E.0930) to A.P.

REFERENCES

1. R.J. Ferrante, N.J. Kowall, M.F. Beal, E.P. Richardson, and J.B. Martin, Selective sparing of a class of striatal neurons in Huntington's disease. *Science* **320**, 561-563. (1985)
2. M. Gu, M.T. Gash, V.M. Mann, F. Javoy-Agid, J.M. Cooper, and A.H. Schapira, Mitochondrial defect in Huntington's disease caudate nucleus. *Ann. Neurol.* **39**, 385-389 (1996).
3. S.E. Browne, A.C. Bowling, U. MacGarvey, M.J. Baik, S.C. Berger, MMK Muqit, E.D. Bird, and M.F. Beal, Oxidative damage and metabolic dysfunction in Huntington's disease: selective vulnerability of the basal ganglia. *Ann. Neurol.* **41**, 646-653 (1997).
4. J.G. Greene, J.T. Greenamyre, Bioenergetics and glutamate excitotoxicity. *Prog. Neurobiol.* 48, 613-634 (1996).
5. U. Wullner, A.B. Young, J.B. Penney, and M.F. Beal, 3-Nitropropionic acid toxicity in the striatum. *J. Neurochem.* **63**, 1772-1781 (1994).
6. J.G. Greene, R.H. Porter, R.V. Eller, and JT. Greenamyre, Inhibition of succinate dehydrogenase by malonic acid produces an "excitotoxic" lesion in rat striatum. *J. Neurochem.* **61**, 1151-1154 (1993).
7. P. Calabresi, D. Centonze, P. Gubellini, GA. Marfia, A. Pisani, G. Sancesario, G. Bernardi, Synaptic transmission in the striatum: from plasticity to neurodegeneration. *Prog. Neurobiol.* **61**, 231-265 (2000).
8. Y. Kawaguchi, C.J. Wilson, S.J. Augood, P.C. Emson, Striatal interneurons: chemical, physiological and morphological characterization. *Trends Neurosci.* **18**, 527-535, (1995).
9. G. Paxinos, and C. Watson, The rat brain in stereotaxic coordinates. Sydney, Australia: *Academic*.(1986).
10. A.N. Murphy, G. Fiskum, and M.F. Beal, Mitochondria in neurodegeneration: bioenergetic function in cell life and death. *J. Cereb. Blood Flow Metab.* **19**, 231-245. (1999).
11. A. Antonini, K.L. Leenders, and D. Eidelberg, [11C]raclopride-PET studies of the Huntington's disease rate of progression: relevance of the trinucleotide repeat length. *Ann. Neurol.* **43**, 253-255 (1998).

DOPAMINERGIC REGULATION OF SYNAPTIC PLASTICITY OF STRIATAL CHOLINERGIC INTERNEURONS

Masami Miura, Takeo Suzuki, and Toshihiko Aosaki*

1. INTRODUCTION

Striatal cholinergic interneurons have been implicated in associative learning.[1] In behaving monkeys, the tonically active cholinergic interneurons (TANs) acquire dopamine (DA)-dependent conditioned responses during sensorimotor learning.[2, 3] The conditioned stimulus leads to a transient suppression in the ongoing tonic activity of TANs, which is often preceded by spike firing. However, the underlying cellular mechanism has not been elucidated. We recently found that tetanic stimulation of cortico/thalamostriatal fibers induces simultaneous long-term potentiation (LTP) of excitatory postsynaptic potential (EPSP) and long-term decrease in the failure rate of disynaptically mediated GABAergic inhibitory postsynaptic potential (IPSP) in the cholinergic interneurons of the mouse studied *in vitro*.[4] This potentiation of EPSPs required a rise in postsynaptic Ca^{2+}, possibly through voltage-dependent Ca^{2+} channels and Ca^{2+}-permeable α-amino-3-hydroxy-5-methyl-4-isoxazolepropionic acid (AMPA) receptors. Moreover, activation of DA D_1-like, but not D_2-like receptors, was a prerequisite for the LTP of EPSPs. We propose that the dual mechanisms act cooperatively to produce the conditioned pause responses of TANs, which serve to coordinate striatal global activity.

2. TANS MAY CORRESPOND TO CHOLINERGIC INTERNEURONS

Extracellular unit recording technique can discriminate only two cell types in the primate's striatum, tonically active neurons (TANs) and phasically active neurons. The latter are very silent and discharge phasically when the animals move, and are thought to contain medium spiny projection neurons. By contrast, the TANs always fire tonically and respond to conditioned sensory stimuli that elicit behavioral reactions. They are now

* Tokyo Metropolitan Institute of Gerontology, Tokyo 173-0015, Japan.

Catecholamine Research, Edited by Nagatsu *et al.*
Kluwer Academic/Plenum Publishers, 2002

considered to be cholinergic interneurons in the striatum for the following reasons.[5] 1) TANs are preferentially located in the matrix, especially in the border regions between the matrix and striosomal compartments of the striatum, as are the cholinergic interneurons. 2) The density of the cells is very similar to that of the cholinergic interneurons. 3) TANs are considered to be very large, because we can detect the cells in the long distance (up to 500 μm) along an extracellular electrode track. 4) They are considered to be interneurons because they do not respond to antidromic stimulation of the globus pallidus.[6] 5) TANs fire tonically *in vivo* with wide action potentials just as cholinergic interneurons do *in vivo* as well as *in vitro*. All these criteria for the TANs fit very well with the properties of cholinergic interneurons.

3. LTP OF THE CORTICO- AND THALAMOSTRIATAL EPSP

The next question is wherher the pause response of the TANs resulted from a change of synaptic efficacy of their own. To address this question, we placed stimulation electrode in the subcortical white matter in brain slice preparation of the mouse (14 to 21 days old) and studied directly whether these cholinergic neurons develop long-term plastic changes after tetanic stimulation of the cortico and thalamostriatal pathways. This neuronal type is easily identified because it is the largest in the striatum and shows characteristic membrane properties.[7]

Postsynaptic potentials were composed of a small depolarizing EPSP and a subsequent hyperpolarizing component. This latter component was GABAergic, because it was blocked completely by a $GABA_A$ receptor antagonist, bicuculline methiodide (BMI, 30 μM). Also, it was considered disynaptically-mediated, because an AMPA antagonist, CNQX (10 μM), completely abolished not only EPSP but also the GABA component. In other words, stimulation of cortico/thalamostriatal fibers evoked monosynaptic EPSP and disynaptic IPSP in the cholinergic interneurons.

We analyzed first the roles of EPSP by continuous application of BMI and found that tetanic stimuli composed of 100 pulses at 100 Hz evoked LTP of EPSP in the cholinergic interneurons. This LTP induction was blocked entirely by inclusion of the Ca^{2+} chelator, BAPTA (20 mM), in the whole-cell-recording pipette, indicating that LTP induction requires a postsynaptic rise in intracellular Ca^{2+}. One possible source of this Ca^{2+} is the voltage-sensitive Ca^{2+} channels (VSCC). Two types of Ca^{2+} channel blockers, Ni^{2+} (50 μM, for potential T-type) and nimodipine (10 μM, for L-type) significantly blocked the induction of LTP. For the involvement of glutamate receptors, D-APV (for NMDA receptors) and MCPG (for metabotropic glutamate receptors) were not effective in blocking the LTP induction. But, there is another possible source in the glutamate receptor family. It is a Ca^{2+}-perameable AMPA receptor. Indeed, averaged I-V curve taken from 15 cells showed inward rectification. Two kinds of antagonists, NHPP-spermine and JSTX-3 significantly blocked the AMPA EPSPs. Therefore, the cholinergic interneurons possess the Ca^{2+}-perameable AMPA receptors and these channels must play a key role in LTP induction.

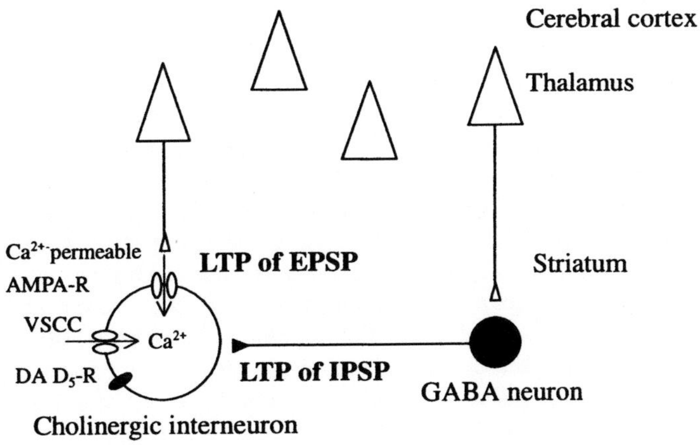

Figure 1. The cellular mechanism of synaptic plasticity of striatal cholinergic interneurons

Then, what about the DA receptors? The cholinergic neurons are known to contain the DA D_2 and D_5 receptors.[8, 9] Because it was reported that tetanic stimulation itself induced the DA release,[10] we applied DA antagonists for 10 min before tetanus to study the role of DA receptors. Sulpiride was first tested (for DA D_2 receptors) without any significant effect. SCH23390 (for the DA D_5 receptors) significantly blocked the LTP. Involvement of the serotonin receptors in its actions was excluded due to the ineffectiveness of three kinds of serotonin receptor blockers, ketanserin, RS39604 and methiothepin. Activation of DA D_5 receptors is therefore necessary for LTP induction.

4. LTP OF DISYNAPTIC IPSP/IPSC

In a second series of experiments, we examined whether disynaptic IPSP (dsIPSP) was involved in the synaptic plasticity observed in the cholinergic interneurons *in vivo*. The whole-cell current clamp recordings in the absence of BMI revealed that dsIPSP was significantly enhanced 30 min after tetanic stimulation, measured at a resting membrane potential range (-58.5 to -62.5 mV). The dsIPSPs were proved to be $GABA_A$-receptor mediated, since the reversal potential was estimated to be around –60 mV, which was close to the equilibrium potential of Cl^-, a charge carrier of $GABA_A$ receptor channels. Voltage clamp recording analysis further revealed that LTP of IPSC was mostly presynaptically-mediated because the amplitude of each unitary IPSC was almost the same as before, whereas the frequency increased after tetanus.

5. CONCLUDING REMARKS

In summary, tetanic stimulation evoked a Ca^{2+}-dependent as well as DA D_5 receptor-dependent LTP of EPSP in the cholinergic interneurons. But, at the same time yet unidentified GABA neurons in the striatum began firing more frequently and send inhibitory inputs to the cholinergic interneurons after the tetanus (Figure). LTP of the EPSP would therefore facilitate spike generation precisely at the time of synaptic inputs, while presynaptic LTP of IPSP might inhibit subsequent firing of the cells. This situation might underlie the conditioned pause response of the TANs in monkey striatum. Interestingly, recent investigation revealed that the inactivation of the centre-médian/parafascicular nucleus (CM-Pf) of the thalamus, which send glutamatergic inputs to the striatal cholinergic interneurons, spared the initial excitation but suppressed significantly the following pause and rebound facilitation of the TAN response in behaving monkeys[11]. This implies that the conditioned TAN response may be composed of at least two distinct mechanisms.

REFERENCES

1. N. M. White, Mnemonic functions of the basal ganglia. *Curr. Opin. Neurobiol.* **7**:164-169 (1997).
2. T. Aosaki, H. Tsubokawa, A. Ishida, K. Watanabe, A. M. Graybiel, and M. Kimura, Responses of tonically active neurons in the primate's striatum undergo systematic changes during behavioral sensorimotor conditioning. *J. Neurosci.* **14**:3969-3984 (1994).
3. T. Aosaki, A. M. Graybiel, and M. Kimura, Effect of the nigrostriatal dopamine system on acquired neural responses in the striatum of behaving monkeys. *Science* **265**:412-415 (1994).
4. T. Suzuki, M. Miura, K. Nishimura, and T. Aosaki, Dopamine-dependent synaptic plasticity in the striatal cholinergic interneurons. *J. Neurosci.*, in press.
5. T. Aosaki, M. Kimura, and A. M. Graybiel, Temporal and spatial characteristics of tonically active neurons of the primate's striatum. *J. Neurophysiol.* **73**:1234-1252 (1995).
6. M. Kimura, M. Kato, and H. Shimazaki, Physiological properties of projection neurons in the monkey striatum to the globus pallidus. *Exp. Brain Res.* **82**:672-676 (1990).
7. Y. Kawaguchi, Large aspiny cells in the matrix of the rat neostriatum in vitro: physiological identification, relation to the compartments and excitatory postsynaptic currents. *J. Neurophysiol.* **67**:1669-1682 (1992).
8. T. Aosaki, K. Kiuchi, and Y. Kawaguchi, Dopamine D1-like receptor activation excites rat striatal large aspiny neurons *in vitro*. *J. Neurosci.* **18**:5180-5190 (1998).
9. Z. Yan, and D. J. Surmeier, D_5 dopamine receptors enhance Zn^{2+}-sensitive $GABA_A$ currents in striatal cholinergic interneurons through a PKA/PP1 cascade. *Neuron* **19**:1115-1126 (1997).
10. P. Calabresi, E. Fedele, A. Pisani, G. Fontana, N. B. Mercuri, G. Bernardi, and M. Raiteri, Transmitter release associated with long-term synaptic depression in rat corticostriatal slices. *Eur. J. Neurosci.* **7**:1889-1894 (1995).
11. N. Matsumoto, T. Minamimoto, A. M. Graybiel, and M. Kimura, Neurons in the thalamic CM-Pf complex supply neurons in the striatum with information about behaviorally significant sensory events. *J. Neurophysiol.* **85**: 960-976 (2001).

ROLE OF Shp2 TYROSINE PHOSPHATASE IN TROPHIC EFFECTS OF BDNF IN CULTURED DOPAMINERGIC NEURONS OF SUBSTANTIA NIGRA

Satomi Takai, Toshiyuki Araki, Hisatsugu Koshimizu, Hiroshi Hatanaka, and Masashi Yamada*

1. INTRODUCTION

Brain-derived neurotrophic factor (BDNF), a member of the neurotrophin family, works as a neurotrophic factor, promoting differentiation and survival in a variety of CNS neurons.[1,5] BDNF binds to a receptor tyrosine kinase, TrkB, a member of the Trk family, to exert its effects. Shp2, a protein tyrosine phosphatase, is activated by various growth factors, and is activated by association with tyrosine-phosphorylated proteins via its two SH2 domains. Studies using phosphatase-inactive and phosphatase domain deficient mutants of Shp2 suggested that Shp2 functions as a positive regulator of MAPK activation and proliferation in various types of mitogenic cells. Shp2 is highly expressed in the brain. Recently, we observed that Shp2 is involved in BDNF signaling in cultured cerebral cortical neurons.[2,7] However, it is not clearly understood how Shp2 functions in response to BDNF in neurons.

2. Shp2 IN CULTURED CEREBRAL CORTICAL NEURONS

To examine the roles of Shp2 in BDNF signaling in cultured cerebral cortical neurons, we generated three adenoviruses expressing wild-type, phosphatase-inactive mutant Shp2 (cysteine 463 was substituted with serine, C/S mutant) or phosphatase domain deficient mutant of Shp2 (delta P mutant). The C/S and delta P mutants of Shp2 act in a dominant negative fashion. In immunoprecipitates with anti-Shp2 antibody, we observed several proteins that displayed enhanced phosphorylation in response to BDNF in cultured cerebral

* Division of Protein Biosynthesis, Institute for Protein Research, Osaka University, 3-2 Yamadaoka, Suita, Osaka 565-0871, Japan.

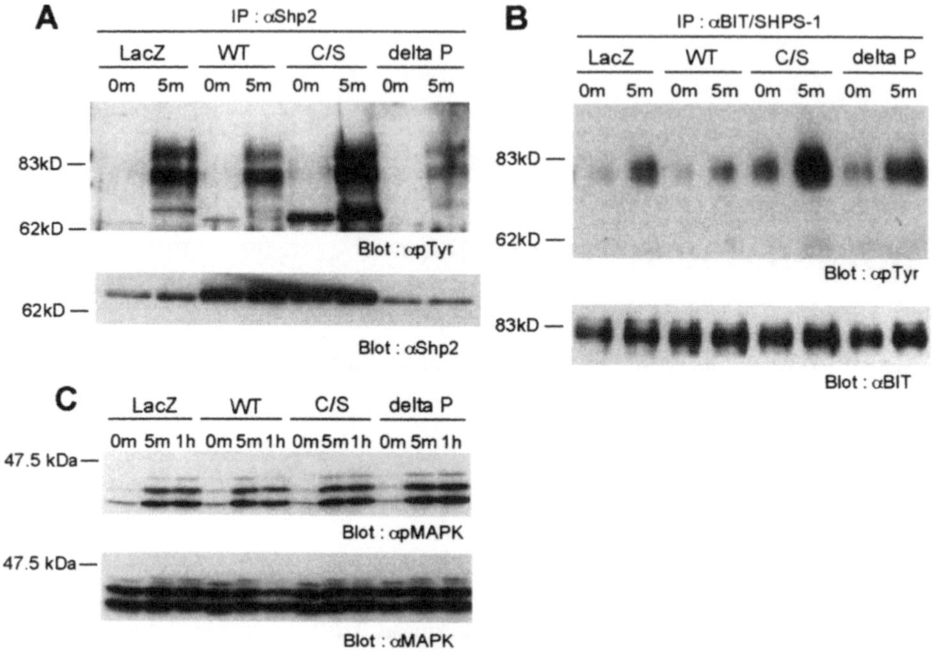

Fig 1. Role of Shp2 in BDNF-induced intracellular signaling in cultured cerebral cortical neurons. The cell lysates were prepared at the indicated times after addition of BDNF (a kind gift of Regeneron) from cultured neurons infected with adenoviruses expressing LacZ or wild-type, C/S mutant or delta P mutant of Shp2. (A and B) The lysates were immunoprecipitated with anti-Shp2 (A) or anti-BIT polyclonal antibodies (B), and then the immunoprecipitates were analyzed by immunoblotting analysis with anti-phosphotyrosine monoclonal antibody (upper panels) and with anti-Shp2 monoclonal antibody in (A) or anti-BIT monoclonal (1D4) antibody in (B) (lower panels). (C) The lysates were subjected to Western blotting analysis using anti-phospho-MAPK polyclonal antibody (upper pannel) or anti-MAPK polyclonal antibody (lower pannel). Molecular masses are shown on the left of the immunoblots.

neurons infected with the C/S mutant adenovirus (Fig. 1A). We found that the Shp2-binding protein, BIT/SHPS-1, showed enhanced tyrosine-phosphorylation in response to BDNF in cultured neurons expressing the C/S or delta P mutant of Shp2 (Fig. 1B). Recently, we have reported that BIT/SHPS-1 is tyrosine-phosphorylated and binds to Shp2 in response to neurotrophins in cultured cerebral cortical neurons and PC12 cells.[2, 6] On the other hand, BDNF-induced activation of MAPK was not affected among neurons expressing wild type, C/S mutant or delta P mutant of Shp2 (Fig. 1C), although many reports indicate that Shp2 is a positive regulator of activation of the Ras-MAPK pathway. These results indicated that in cultured cerebral cortical neurons Shp2 is specifically involved in regulation of several tyrosine-phosphorylated proteins, including BIT/SHPS-1, in the BDNF signaling pathway. In addition, the phosphatase Shp2 may not influence the level of BDNF-induced activation of MAPK in cultured cortical neurons.

To investigate whether Shp2 is involved in a survival-promoting pathway stimulated by BDNF, we analyzed survival of cultured cortical neurons expressing the wild type, C/S mutant, or delta P mutant of Shp2 (Fig. 2). As a result, BDNF-induced survival effects on cultured cerebral cortical neurons were not influenced by expression of the C/S mutant of Shp2 compared with control neurons infected with LacZ-expressing adenovirus. However, cultured cortical neurons expressing the wild-type of Shp2

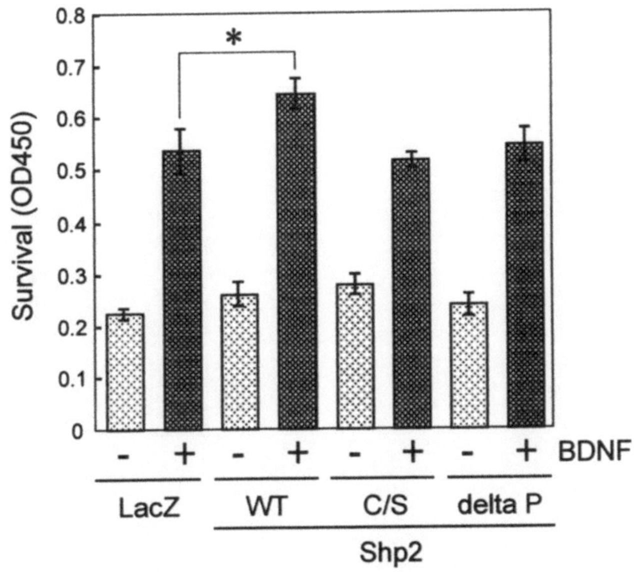

Fig. 2. Role of Shp-2 in BDNF-promoted survival of cultured cerebral cortical neurons. Two days after infection with adenoviruses expressing LacZ or wild-type, C/S mutant or delta P mutant of Shp-2, medium was changed to supplement-free medium not containing (-) or containing 50 ng/ml BDNF (+). After three days, neuronal survival was determined by WST-8 assays. Values are means ± S. D. (n=4). *P<0.01

showed an weak, but statistically significant, enhancement of survival effect of BDNF compared with control. These results suggested that Shp2 may positively, but not dominantly, regulate survival signaling of BDNF in cultured cortical neurons.

3. Shp2 IN CULTURED MESENCEPHALIC DOPAMINERGIC NEURONS

In contrast to the cultured cerebral cortical neurons, we found that Shp2 was strongly involved in neurotrophic effects of BDNF on cultured mesencephalic dopaminergic neurons (Fig. 3). We infected the cultured mesencephalic neurons with LacZ adenovirus or wild-type, C/S mutant or delta P mutant of the Shp2 adenovirus. One day after infection, BDNF was added, and the neurons were cultured for an additional 3 days. Then, the neurons were immunostained with an anti-tyrosine hydroxylase (TH) monoclonal antibody.[4] The BDNF-induced increase in the number of TH-positive neurons was enhanced in the culture infected with the wild-type Shp2 adenovirus. In addition, TH-positive neurons looked healthy and had long neurites, large cell bodies and an increased number of branches and varicosities in the culture infected with the wild-type Shp2 adenovirus. In contrast, these neurotrophic effects of BDNF on the TH-positive neurons were attenuated in the culture infected with the C/S or delta P mutant adenovirus. On the other hand, the three Shp2 adenoviruses did not affect survival and morphology of microtubule-associated protein 2 (MAP2)-positive neurons (data not shown). These results indicate that Shp2 positively regulates BDNF-exerted neurotrophic effects specifically in cultured mesencephalic dopaminergic neurons.

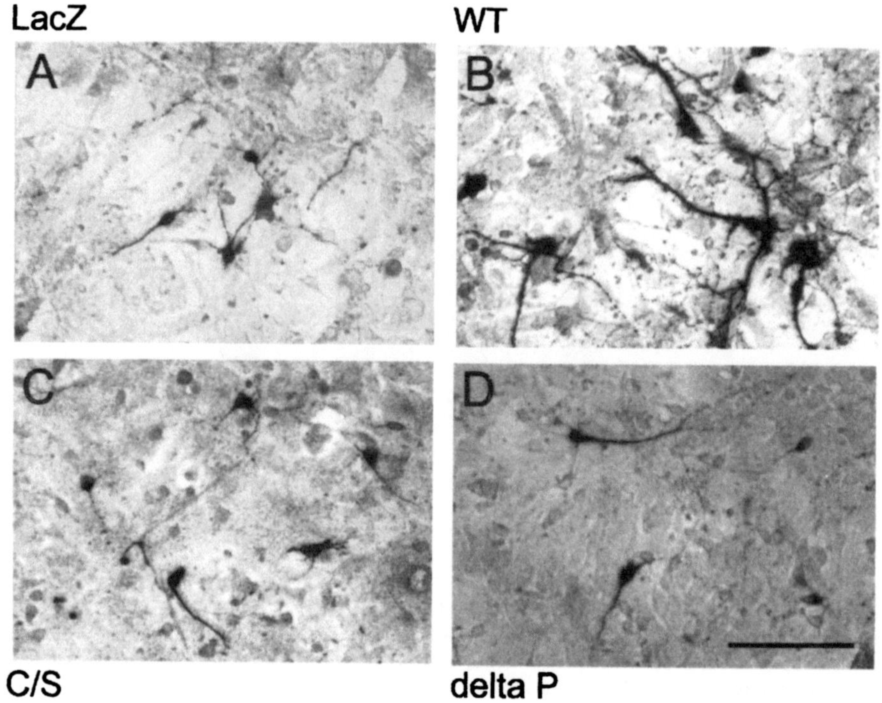

Fig. 3. Role of Shp-2 in BDNF-exerted neurotrophic action in cultured mesencephalic dopaminergic neurons. Mesencephalic neurons infected with adenoviruses expressing LacZ (A) or wild-type (B), C/S mutant (C) or delta P mutant (D) of Shp-2 were cultured in serum-containing medium supplemented with cytosine arabinoside in the presence of 50 ng/ml BDNF. Three days after addition of BDNF, the neurons were stained with anti-TH monoclonal antibody (PCTH-7)(4). Bar represents 100 μm.

REFERENCES

1. Abiru Y., Nishio C., Hatanaka H. (1996) The survival of striatal cholinergic neurons cultured from postnatal 2-week-old rats is promoted by neurotrophins. *Brain Res.Dev.Brain Res.* **91,** 260-267.
2. Araki T., Yamada M., Ohnishi H., Sano S., Uetsuki T., Hatanaka H. (2000) Shp-2 specifically regulates several tyrosine-phosphorylated proteins in brain-derived neurotrophic factor signaling in cultured cerebral cortical neurons. *J. Neurochem.* **74,** 659-668.
3. Araki T., Yamada M., Ohnishi H., Sano S. I.,Hatanaka H. (2000) BIT/SHPS-1 enhances brain-derived neurotrophic factor-promoted neuronal survival in cultured cerebral cortical neurons. *J. Neurochem.* **75,** 1502-1510.
4. Hatanaka H., Arimatsu Y. (1984) Monoclonal antibodies to tyrosine hydroxylase from rat pheochromocytoma PC12h cells with special reference to nerve growth factor-mediated increase of the immunoprecipitable enzymes *Neurosci. Res.* **1,** 253-263.
5. Kubo T., Nonomura T., Enokido Y., Hatanaka H. (1995) Brain-derived neurotrophic factor (BDNF) can prevent apoptosis of rat cerebellar granule neurons in culture. *Brain Res.Dev.Brain Res.* **85,** 249-258.
6. Ohnishi H., Yamada M., Kubota M., Hatanaka H., Sano S. (1999) Tyrosine phosphorylation and association of BIT with SHP-2 induced by neurotrophins. *J. Neurochem.* **72,** 1402-1408
7. Yamada M., Ohnishi H., Sano S., Araki T., Ikeuchi T., Hatanaka H. (1999) Brain-derived neurotrophic factor stimulates interactions of Shp2 with phosphatidylinositol 3-kinase and Grb2 in cultured cerebral cortical neurons. *J. Neurochem.***73,** 41-49.

HETEROMERIZATION OF ADENOSINE AND DOPAMINE RECEPTOR SUBTYPES: RELEVANCE FOR NEURONAL INTEGRATION IN NORMAL AND PATHOLOGICAL STATES

Kjell Fuxe, Sergi Ferré, Maria Torvinen, Jöelle Hillion, Ingrid Strömberg, Ove Franzén, Carlos Ibanéz, Michele Zoli, Carmen Lluis, Luigi F. Agnati and Rafael Franco*

1. INTRODUCTION

Ever since the beginning of the 1980's indications have been obtained for the existence of intramembrane receptors–receptor interactions among different types of G protein coupled receptors (GPCR) in the brain (Agnati et al., 1982, Fuxe et al., 1983) and it was suggested in 1993 that the structural basis for these intramembrane interactions is represented by the formation of GPCR heterodimers (Zoli et al., 1993). It must also be noticed that already in the early 80's crosslinking and radiation inactivation experiments and experiments on gel filtration and analytical sucrose density gradient ultracentrifugation indicated that GPCR could exist as homodimers (see Salahpour et al., 2000). There now exists strong evidence for homo- and heteromerization of GPCR from several groups (see, e.g., Salahpour et al., 2000, Bouvier 2001, Devi and Brady 2000, Tallman 2000, Franco et al., 2000), including evidence for the existence of dimers and oligomers of D1 and D2 receptors and of A1 receptors (Ciruela et al., 1995). In the present review the evidence for adenosine and dopamine receptor subtype-specific heteromerizations will be presented and their role in mediating A1/D1 and A2A/D2 antagonistic receptor interactions and in receptor trafficking will be discussed (Gines et al., 2000, Hillion et al., 2001).

* Kjell Fuxe, Sergi Ferré, Maria Torvinen, Jöelle Hillion, Ingrid Strömberg, Ove Franzén, Carlos Ibañez, Department of Neuroscience, Karolinska Institutet, 171 77 Stockholm, Sweden. Michele Zoli, Luigi F. Agnati, Department of Biomedical Sciences, University of Modena and Reggio Emilia, 41100 Modena, Italy. Carmen Lluis, Rafael Franco, Department of Biochemistry and Molecular Biology, IDIBAPS, University of Barcelona, Barcelona, Spain

2. DISCUSSION

2.1. Heteromerization of Adenosine A1 and Dopamine D1 Receptors

Adenosine acts as a negative feedback regulator of neuronal activity. In fact, adenosine is released in the brain extracellular fluid by active neurons and operates as a volume transmission signal by activating extrasynaptically located high affinity A1 and A2A receptors which put the glial-neuronal networks to rest (Ferré and Fuxe 2000). Extracellular adenosine originates mainly from intracellular breakdown of ATP and then reaches the extracellular space via its transporter. The activation of A1 and A2A receptors leads to motor depression and sleep, and these effects may to a substantial degree involve heteromerization with D1 and D2 receptors, respectively. Indirect evidence for A1/D1 receptor heteromers in the brain comes from morphological and neurochemical experiments. In fact, A1 and D1 receptors colocalize in primary cultures of rat neurons (Gines et al., 2000) and A1 and D1 mRNAs are coexpressed in many strioentopeduncular and strionigral GABA neurons (see Fuxe et al., 1998). Indeed, observations on striatal membrane preparations demonstrate that A1 receptor activation uncouples D1 receptors as seen from the disappearance of high affinity D1 binding, a phenomenon that can be explained by an A1/D1 interaction in the A1/D1 heteromer (see Ferré et al., 1998, Fuxe et al., 1998). In 2000 it became for the first time possible to demonstrate that dopamine D1 and adenosine A1 receptors form functionally interacting heteromeric complexes as studied on co-transfected A1/D1 fibroblast Ltk$^-$ cells (Gines et al., 2000). Already in the basal state coimmunoprecipitation experiments demonstrated that A1 and D1 receptors, but not A1 and D2 receptors, exist as heteromers. The extent of A1/D1 heteromerization was reduced by exposure to D1 agonist treatment, but this disruption was blocked by combined A1 and D1 agonist treatment. Thus, this heteromerization is agonist regulated and combined activation of A1 and D1 agonist binding sites seem to favor its maintenance. It therefore seems possible that A1 induced uncoupling of D1 receptors found in the fibroblast cells in biochemical binding experiments only takes place under conditions of maintained A1/D1 heteromerization. It is likely therefore that the previous demonstration of acute A1/D1 antagonistic interactions in the brain, in terms of D1 signaling at the behavioral level, at the level of neuronal functions using microdialysis, at the adenylate cyclase level and at the protooncogene level could to a substantial degree involve the A1/D1 heteromers expressed in striatal GABA neurons of the direct pathways (Ferré et al., 1997, Fuxe et al., 1998). The study by Gines et al., (2000) also provided evidence that A1/D1 heteromerization has a role in receptor trafficking. Thus, an A1 agonist (R-PIA, 100 nM) after a 3 hour exposure produced a coaggregation of A1 and D1 receptors, while a D1 agonist (SKF38393, 10 μM) after the same exposure time only produced aggregation of D1 receptors but no coaggregation in line with the demonstration that the D1 agonist SKF 38393 caused a disruption of the A1/D1 heteromerization (see above). Either with D1 or A1/D1 aggregation the D1 signaling remained unaffected; namely the D1 agonist (SKF 38393, 10 μM) induced under these conditions an unaltered cAMP accumulation and thus desensitization of D1 receptors was not observed. In contrast, combined A1 (R-PIA) and D1 (SKF 38393) pretreatment for 3 hours did not lead to the formation of coaggregates of A1/D1 receptors, although maintained high levels of colocalization (see above), and was associated with a desensitization of D1 receptor signaling as shown by the reductions of D1 agonist induced cAMP accumulation. Thus, the maintenance of A1/D1 receptor heteromerization, but not induction of aggregates, and a prolonged coactivation of A1 receptors and D1 receptors may be essential for development of D1 receptor desensitization. Under these conditions with no indications that interalization took place since the colocalization remained diffuse, a prolonged allosteric change may develop in the D1 receptor related to its possible phosphorylation and/or association with e.g. receptor activity modulating proteins. Overall the results indicate a role of A1/D1 heteromers also in D1 desensitization mechanisms and in receptor trafficking whereby a combined agonist activation blocks the aggregation of A1 and D1 receptors induced by separate agonist activation.

2.2. Heteromerization of Adenosine A2A and D2 Receptors

Hillion et al., (2001) have recently been able in coimmunoprecipitation experiments to demonstrate heteromerization of human A2A and human D2L receptors in neuroblastoma SH-SY5Y cells stably transfected with D2L R but containing native A2A receptors as well as in fibroblast cells stably transfected with human D2L receptor and transiently transfected with tagged A2A receptors. In these transfected cells as well as in primary cultures of rat striatal neurons, there is a high degree of colocalization between D2 and A2A receptors (Hillion et al., 2001). This pattern of receptor expression mimics the striopallidal GABA neurons (Zoli et al., 1993), where the vast majority of the A2A receptors are found and coexist with the D2 receptors (Rosin et al., 1998; Schiffmann et al., 1991). A2A/D2 receptor interaction is antagonistic in nature, since A2A agonists but not A1 agonists reduce the affinity of D2 receptors in striatal membrane preparations, especially the high affinity state, without changing the density of the D2 receptors. A2A receptors also interfer with the G protein coupling of the D2 receptor (Fuxe et al., 1998; Ferré et al., 1997). The same antagonistic modulation of D2 recognition mechanism by A2A receptor activation has been observed in membrane preparations from human A2A/D2L, dog A2A/human D2L and human A2A/rat D2s containing cell lines (Ltk- fibroblast cell lines, CHO cell lines and neuroblastoma cell lines, respectively) demonstrating that this is a mechanism operating in all types of cells.

Combined activation of the two receptors in the A2A/D2 heteromeric complexes seems to be of importance for cointernalization and codesensitization of D2 receptors (Hillion et al., 2001). These observations may help understand the phenomena of cross-tolerance and cross-sensitization found in behavioural experiments with DA agonists and adenosine drugs (Fenu et al., 2000). It seems possible that D2 signaling is dependent on a balance between at least the D2 and A2A homodimers and A2A/D2 heterodimers.

2.3. The Role of A2A/D2 Heteromeric Complexes in Models of Parkinson's disease. Relevance for Treatment.

It was early demonstrated that the antagonistic modulation of D2 receptor recognition by A2A receptors in membrane preparations was more pronounced in the DA denervated striatum containing supersensitive DA receptors. Thus, it seems likely that the ability of A2A antagonists to preferentially enhance D2 action in the DA denervated striatum, as seen in studies on contralateral turning behaviours and in neurophysiological experiments (Fuxe et al., 2001; Strömberg et al., 2000), is to a major degree caused by removal of A2A antagonism of D2 signaling in the A2A/D2 heteromeric complex. It was therefore suggested that A2A antagonists could represent a novel treatment of Parkinson's disease (see Fuxe et al., 2001). It was also shown in Parkinson models in the monkey that A2A antagonists can exert antiparkinsonian actions without development of dyskinesias. This may be related to the preferential enhancement of D2 signaling in the A2A/D2 heteromers resulting in enhanced inhibition of the indirect GABA pathway and removal of motor inhibition without a major enhancement in the D1 regulated direct GABA pathway mediating motor initiation. After chronic L-dopa treatment alone in Parkinsonian monkeys leading to dyskinesias, A2A receptor expression and signaling may instead be increased, favouring activity in the direct vs the indirect pathway and thus development of dyskinesias (see Chase and Oh, 2000). The use of A2A antagonists in combination with L-dopa in the treatment of Parkinson's disease will also allow the lowering of the L-dopa dose, further reducing the induction of dyskinesias via D1 over-activation in response to the intermittent L-dopa treatment (see Chase and Oh, 2000).

The findings by Hillion et al. (2001) on the A2A and D2 agonist regulation of the A2A/D2 heteromers with coaggregation and cointernalization of the A2A/D2 heteromeric complexes leading to losses of A2A and D2 receptor signaling may give us a new understanding of the on/off phenomenon and of the reduction of the antiparkinsonian actions seen with chronic intermittent L-dopa treatment. Thus, it is known that the extracellular levels of adenosine are increased in MPTP models of Parkinson's disease that may be further increased by chronic intermittent treatment with L-dopa. A combination of L-dopa treatment and endogenous adenosine could markedly enhance the internalization of the A2A/D2 heteromeric complexes in the human striatum resulting in a reduction of D2 receptor function. Furthermore, in the study of the A1/D1 heteromeric complexes in the fibroblast cells, combined and prolonged agonist treatment leads to the desensitization of D1 signaling onto the adenylate cyclase (Gines et al., 2000). It seems possible that the cointernalization and codesensitization of A2A/D2 and A1/D1 receptor complexes is a partial explanation for the wearing off of the therapeutic action of L-dopa treatment in Parkinson's disease. Thus, A1 and A2A antagonists preferentially acting on A1/D1 and A2A/D2 heteromers may have a substantial role in the treatment of Parkinson's disease.

2.4. On the Putative Role of the A2A/D2 Heteromers in Animal Models of Schizophrenia

Adenosine A2A agonists have been shown to represent a novel type of atypical antipsychotic drug based on their ability to counteract amphetamine and PCP-induced motor activity in rats in doses far below those necessary to produce catalepsy (Rimondini et al., 1997; Ferré 1997). Such preferential actions of A2A agonists on certain motor systems may be produced by preferential effects on the ventral vs the dorsal striopallidal GABA pathway since A2A and D2 receptors are colocated also in the ventral component, and probably form heteromers there. The ventral pathway is considered a major target for antipsychotic drugs, namely D2 antagonists (Ferré, 1997).

In line with this view is the evidence from receptor autoradiographic experiments that the A2A agonists reduce the affinity of D2 like receptors in the nucleus accumbens (Diaz-Cabiale et al., unpublished data), the major origin of the ventral striopallidal GABA pathway, more strongly than in the caudate putamen. This finding can be explained by the possible existence also of a strong antagonistic A2A/D3 receptor interaction, the D3 receptors being enriched in the ventral striatum. It is commonly assumed that D2 antagonists exert their antipsychotic effects mainly by blocking the inhibitory effects of D2 receptors on the activity of the ventral striopallidal GABA pathway. Part of the postulated antipsychotic action of A2A agonists would consist in mimicking D2 antagonists since they can counteract this D2 signaling by an antagonistic A2A/D2 receptor interaction in the heteromeric receptor complex. However, A2A receptors may also have an excitatory action on this pathway independent of D2 receptors, via a strong synergistic interaction with mGluR5. Thus, A2A agonists would cause a much stronger increase in the activity of this GABA pathway than can be seen with D2 antagonists alone (see Ferré et al., 1997; Diaz-Cabiale et al., unpublished data), which correlates with counteraction of PCP-induced motor behaviours. It is therefore postulated that A2A agonists can represent novel atypical neuroleptic drugs for treatment of schizophrenia. In conclusion, drugs acting on A2A/D2 heteromerization may have a fundamental role in the therapy of both Parkinson's disease and schizophrenia.

2.5. On the Involvement of Intramembrane Receptor-Receptor Interactions in Learning and Memory Processes through Formation of Heteromeric Complexes

In the early 1980s a theory was introduced that intramembrane receptor-receptor interactions may represent one molecular basis for the formation of memory traces (Agnati et al., 1982; Zoli et al., 1996; Franco et al., 2000). The theory was called the receptor mosaic hypothesis of the engram and the essence of it was that a memory trace of synaptic activity can be created when the receptor cluster formed by multiple receptor-receptor interactions becomes stabilized. These receptor clusters will then result in biased decoding and form memory traces in a neuronal network by producing long term changes in the synaptic weights of the network. Today heteromeric receptor complexes and multiple receptor-receptor interactions have been demonstrated (see Bouvier 2001; Rimondini et al., 1999; Franco et al., 2000) and this theory has therefore been further developed leading to the proposal that long lived heteromeric complexes of higher order formed by intramembrane receptor-receptor interactions represent an important part of the molecular basis of learning and memory (Agnati, Franzén and Fuxe, 2001). Together with cytoplasmatic protein phosphorylation and dephosphorylation pathways (see Greengard et al., 1998) the membrane receptor circuits involving the receptor mosaics form discrete 3 D molecular circuits that handle information at the cellular level and the contribution of each cell to network learning and memory will depend on the types of molecular circuits activated and how they are integrated (Agnati, Franzén, Fuxe, 2001).

REFERENCES

Agnati L. F, Fuxe K, Zoli M, Rondanini C. and Ögren S.-O, 1982, New vistas on synaptic plasticity: the receptor mosaic hypothesis on the engram, *Med Biol.* **60**: 183-190.

Agnati L. F, Benfenati F, Solfrini V, Biagini G, Fuxe K, Guidolin D, Carani C. and Zini I, 1993, Intramembrane receptor-receptor interactions: integration of signal transduction pathways in the nervous system, *Neurochem Int.* **22**: 213-222.

Agnati L. F, Franzén O. and Fuxe K, 2001, On the molecular basis of learning and memory. Possible role of intramembrane receptor-receptor interactions via formation of long-lived heteromeric complexes of higher order, *PNAS (submitted)*.

Bouvier M, 2001, Oligomerization of G-protein coupled transmitter receptors, *Nat. Rev. Neurosci.* **2**: 274-286.

Chase T. N. and Oh J. D, 2000, Striatal dopamine- and glutamate-mediated dysregulation in experimental parkinsonism. *Trends Neurosci.* **23**: S86-S91.

Ciruela F., Casado V., Mallol J., Canela E. I., Lluis C. and Franco R., 1995, Immunological identification of A1 adenosine receptors in brain cortex., *J. Neurosci. Res.* **42**: 818-828.

Devi L. and Brady L, 2000, Dimerization of G-protein coupled receptors, *Neuropsychopharmacology* **23**: S3-S4.

Fenu S, Cauli O. and Morelli M, 2000, Cross-sensitization between the motor activating effects of bromocriptine and cafffeine: role of adenosine A2A receptors. *Behav. Brain Res.* **114**: 97-105.

Ferré S, 1997, Adenosine-dopamine interactions in the ventral striatum. Implications for the treatment of schizophrenia, *Psychopharmacology.* **133**: 107-120.

Ferré S, Fredholm B, Morelli M, Popoli P. and Fuxe K, 1997, Adenosine-dopamine receptor-receptor interactions as an integrative mechanism in the basal ganglia, *Trends Neurosci.* **20**: 482-487.

Ferré S. and Fuxe K, 2000, Adenosine as a volume transmission signal. A feedback detector of neuronal activation. In *Progress in Brain Research,* L. F. Agnati, K. Fuxe, C. Nicholson and E. Syková, eds, Vol.125, Elsevier, Amsterdam, pp. 353-361.

Ferré S, Torvinen M, Antoniou K, Irenius E, Civelli O, Arenas E, Fredholm B. B. and Fuxe K, 1998, Adenosine A1 receptor-mediated modulation of dopamine D1 receptors in stably cotransfected fibroblast cells, *J. Biol. Chem.* **273**: 4718-4724.

Franco R, Ferré S, Agnati L. F, Torvinen M, Gines S, Hillion J, Casado V, Lledo P, Zoli M, Lluis C. and Fuxe K, 2000, Evidence for adenosine/dopamine receptor interactions. Indications for heteromerization, *Neuropsychopharmacology.* **23**: 50-59.

Fuxe K, Agnati L. F, Benfenati F, Celani M. F, Zini I, Zoli M. and Mutt V, 1983, Evidence for the existence of receptor- receptor interactions in the central nervous system. Studies on the regulation of monoamine receptors by neuropeptides, *J. Neural Transm.* **18**: 165-179.

Fuxe K, Ferré S, Zoli M. and Agnati L. F, 1998, Integrated events in central dopamine transmission as analyzed at multiple levels. Evidence for intramembrane adenosine A2A/dopamine D2 and adenosine A1/dopamine D1 receptor interactions in the basal ganglia, *Brain Res Rev.* **26**: 258-273.

Fuxe K, Strömberg I, Popoli P, Rimondini-Giorgini R, Torvinen M, Ögren S. O, Franco R, Agnati L. F. and Ferré S, 2001, Adenosine receptors and Parkinson's Disease. Relevance of antagonistic adenosine and dopamine receptor interactions in the striatum. In *Parkinson's Disease,* D. Calne and S. Calne, eds, Vol.86, Lippincott Williams & Wilkins, Philadelphia, pp. 345-.

Ginés S, Hillion J, Torvinen M, LeCrom S, Casado V, Canela E, Rondin S, Lew J, Watson S, Zoli M, Agnati L, Vernier P, Lluis C, Ferré S, Fuxe K. and Franco R, 2000, Dopamine D1 and adenosine A1 receptors assemble into functionally interacting heteromeric complexes, *Proc. Natl. Acad. Sci.* **97**: 8606-8611.

Greengard P, Nairn A. c, Girault J.-A, Ouimet C. C, Snyder G. L, Fisone G, Allen P. B, Fienberg A. and Nishi A, 1998, The DARPP-32/protein phosphatase-1 cascade: a model for signal integration, *Brain Res. Rev.* **26**: 274-284.

Hillion J, Canals M, Torvinen M, Casadó V, Scott R, Terasmaa A, Hansson A, Watson S, Olah M. E, Mallol J, Canela E. I, Zoli M, Agnati L. F, Ibáñez C. F, Lluis c, Franco R, Ferré S. and Fuxe K, 2001, Coaggregation, cointernalization and codesensitization of adenosine A2A receptors and dopamine D2 receptors, *PNAS (submitted).*

Rimondini R, Ferré S, Ögren S. O. and Fuxe K, 1997, Adenosine A2A agonists: A potential new type of atypical antipsychotic, *Neuropsychopharmacology.* **17**: 82-91.

Rimondini R, Fuxe K. and Ferré S, 1999, Multiple intramembrane receptor-receptor interactions in the regulation of striatal dopamine D2 receptors, *NeuroReport.* **10**: 2051-2054.

Rosin D. L, Robeva A. S, Woodard R. L, Guyenet P. G. and Linden J, 1998, Immunohistochemical localization of adenosine A2A receptors in the rat central nervous system, *J. Comp. Neurol.* **401**: 163-186.

Salahpour A., Angers S. and Bouvier M., 2000, Functional significance of oligomerization of G protein coupled receptors., *TEM.* **11**: 163-168.

Schiffmann S, Jacobs O. and Vanderhaeghen J.-J, 1991, Striatal restricted adenosine A2 receptor (RDC8) is expressed by enkephalin but not by substance P neurons: an in situ hybridization histochemistry study, *J Neurochem.* **57**: 1062-1067.

Strömberg I, Popoli P, Müller C. E, Ferré S. and Fuxe K, 2000, Electrophysiological and behavioural evidence for an antagonistic modulatory role of adenosine A2A receptors in dopamine D2 receptor regulation in the rat dopamine-denervated striatum, *Eur. J. Neurosci.* **12**: 4033-4037.

Tallman J, 2000, Dimerization of G-protein coupled receptors. Implications for drug design and signaling, *Neuropsychopharmacology.* **23**: S1-S2.

Zoli M, Agnati L, Hedlund P, Li X, Ferré S. and Fuxe K, 1993, Receptor-receptor interactions as an integrative mechanism in nerve cells, *Mol Neurobiol.* **7**: 293-334.

Zoli M, Guidolin D, Fuxe K. and Agnati L, 1996, The receptor mosaic hypothesis of the engram: Possible relevance of Boolean network modeling, *Int. J. Neural Systems.* **7**: 363-368.

NEURONAL GENE REGULATION BY THE NEURAL – RESTRICTIVE SILENCER (NRS)

Is the catecholaminergic system excluded from control by NRS?

Nozomu Mori*

1. INTRODUCTION

Neural restrictive silencer (NRS) is a 22-bp-long cell type-specific transcriptional silencer element that is found in a variety of neural-specific genes.[1,2] NRS represses transcription in non-neuronal cells, thereby allowing gene expression only in neurons. The sequence-specific DNA-binding protein that recognizes NRS (NRSF, also named REST) is expressed essentially in undifferentiated neuronal precursors and non-neuronal cells and functions as a transcriptional repressor.[3] NRSF represents the first example of a vertebrate silencer factor that regulates a large battery of neuronal phenotype-specific genes, and therefore may function as a general negative regulator of neuronal terminal differentiation. However, in contrast to what was originally thought, recent evidence suggests that NRS is found in many genes involved in the cholinergic, GABAergic, and glutamatergic neuronal transmission, but may not be evident in the genes that relate to monoaminergic neurotransmission. Here I would like to discuss the possibility that NRS represses neurons to acquire monoaminergic function and that the role of NRS could be somehow limited in monoaminergic or catecholaminergic systems.

2. TRANSCRIPTION FACTOR CASCADES IN NEURAL DEVELOPMENT AND DIFFERENTIATION

Neural development and subsequent neuronal differentiation are controlled by both positive and negative transcriptional regulators. For example, Mash-1 and NeuroD are positive regulators of neural development, whereas HES-1 is a negative one. These HLH-

* Department of Molecular Genetics, National Institute for Longevity Sciences (NILS), and the Program of "Protecting the Brain", CREST, JST, Oobu, Aichi 474-8522, JAPAN.

Catecholamine Research, Edited by Nagatsu *et al.*
Kluwer Academic/Plenum Publishers, 2002

containing transcription factors have been shown to play essential roles in early stages of neural development in mammals. For example, mice lacking the HES-1 gene had severe defects in fusing neural tubes and, therefore were unable to form brain and became embryonic lethal. Mash-1 knockout mice had defects in developing sympathetic ganglia, and did not produce cells positive for neuronal markers such as SCG10, peripherin, and neuron-specific enolase. Thus, HES-1 and Mash-1 functions are essential in early stages of neural development. Similarly, NeuroD plays essential roles in neuronal differentiation at a little later stage in the embryo. When NeuroD is overexpressed in Xenopus embryos, cells positive for various neuronal markers appeared in the abdomen. Thus NeuroD is a positive regulator of neuronal differentiation. However, the NeuroD-driven cells are not fully mature neurons; it seems that there are other factors that are important for terminal neuronal differentiation.

In contrast to the critical roles of these HLH-type transcription factors in early phases of neuronal differentiation, genetic defects in the zinc-finger containing transcription factor regulating NRS, i. e., NRSF/REST, do not interfere with the formation of neural networks, but cause premature neuronal differentiation and ultimately lead to embryonic lethality.[4] Over-expression of dominant-negative forms of NRSF/REST also leads to abnormality in neural development.[5] Thus, NRSF/REST seems to be crucial for terminal neuronal differentiation and the formation of fully mature neural networks.

3. DIVERSITY OF NEURAL-RESTRICTIVE SILENCER (NRS)

A search in the GenBank data base using the 22-bp NRS core sequence as a probe revealed that more than 30 neuron-specific genes contain sequences related to the NRS (data not shown; see Mori, 1997[6]). The genes containing NRS included those for neuronal growth-associated proteins, neurotrophins, neurotransmitter synthetases, receptors, ion channels, cell adhesion proteins, and molecules involved in synaptic functions, all of which relate to mature neuronal function. The collective NRS in these neural-specific genes could be classified into two subclasses: conserved and degenerated NRSs. Conserved NRSs were seen in pan-neuronal genes such as those encoding synaptosomal proteins, neural growth factors, cell adhesion molecules, transcription factors as well as in genes of the cholinergic and glutamatergic systems (Table 1a,b). In contrast, degenerative NRSs were seen in genes of the monoaminergic system (see Table 1c). The database search also revealed NRS in several genes that function outside of the nervous system. For example, the atrial natriuretic peptide (ANP) gene contained a significant NRS homologous sequence in its 3' portion. While the expression of ANP occurs in both neurons and ventricular myocytes, it has recently been shown that the NRS in the ANP gene functions indeed in ventricular myocytes, suggesting that NRS could play roles in a particular non-neural tissue such as the heart tissue.

A part of the mechanisms of transcriptional repression has recently been elucidated.[7-10] NRSF/REST has two independent repression domains, RD-1 and RD-2, in the N-and C-terminal portions, respectively, of the 120-kD protein. The N-terminally located RD-1 binds to corepressor mSin3 and recruits histone deacetylase (HDAC). In contrast, the C-terminally located RD-2 seems to be independent of the HDAC mechanism. Thus NRSF/REST is able to expedite chromatin-dependent and –independent repression.

Table 1. Selected list of the neural-restrictive silencer (NRS) elements found in the genes specific to neurochemical phenotypes

Species	Gene	Location	Orientation[a]	Sequence[b]
				TTTCAGCACCACGGACAGCGCC
(a) Genes of Cholinergic System				
rat	choline acetyltransferase (ChAT)	-1.6 kb upstream	>	G-C---------------TT--
human	nicotinic ACh receptor β2 subunit	5'noncoding region	>	C---------------------
rat/mouse	nicotinic ACh receptor β2 subunit	5'noncoding region	>	C--------------------T
rat	M4 muscarinic receptor	5'noncoding region	<	-----------T-------T--
(fb Genes of Glutamatergic System				
human	NMDA receptor (NR1-1)	5'noncoding region	>	C---------G-----------
rat	NMDA receptor (NR1-1)	5'noncoding region	>	C---------T--------AT-
mouse	AMPA receptor	promoter (-0.2kb)	<	---T------G----------T
(c) Genes of Monoaminergic System				
rat	tyrosine hydroxylase (TH)	-1 kb upstream	>	C--G-.-T---G-----TA-AA
rat	dopamine β-hydroxylase (DBH)	-0.4 kb upstream	>	-C----.G-G----GGT-GCAT
bovine	phenyl N-metyltransferase (PNMT)	-0.75 kb upstream	>	CC--C--TG----CTT---C--
		-0.25 kb upstream	<	-C-GCC-T---A-AC----T--
rat	dopamine transporter	-0.1 kb upstream	<	CC---C-T--C---G-C-----
human	dopamine D4 receptor (DRD4)		<	GC-G------G------A----

[a] > denotes the sense orientation, and < the antisense.

[b] Consensus sequence is given at the top of the right-hand column, and a nucleotide residue is indicated only when it differs from the consensus one.

4. NRS IN THE CATECHOLAMINERGIC SYSTEM

Following the discovery of the significant roles of NRS in cell type-specific regulation of pan-neuronal genes such as SCG10, sodium channel, and synapsin I, several studies have elucidated the potential roles of NRS (or its related) elements found in the genes of the catecholaminergic system. For example, Kim and his colleagues reported a potential involvement of NRS-like elements in the promoter regulation of the rat dopamine β-hydroxylase (DBH) gene.[11] Similarly, there were several reports indicating NRS-involvement in the gene regulation of phenyl ethanolamine N-metyltransferase (PNMT). In addition, the presence of NRS-like sequences was frequently pointed out in discussions about the genes for tyrosine hydroxylase (TH)[12] and the dopamine transporter (DAT), although functional examinations were limited. However, sequence divergence of NRS-like elements in these monoaminergic genes seems to suggest that their direct relevance to NRSF/REST regulation needs to be reevaluated.

5. ROLES OF THE NRS-NRSF SYSTEM IN DEVELOPMENTAL ACQUISITION AND PLASTICITY OF NEUROCHEMICAL PHENOTYPE

The fact that NRS is found in both pan-neuronal and subtype-specific genes, e.g., cholinergic and glutamatergic genes, seems to suggest that the NRS-NRSF system works in several steps during neural development and subsequent neuronal differentiation. The presence of two independent transcriptional repression mechanisms via NRSF/REST may be relevant to this initial (e.g., neuronal or glial vs. pan-neuronal) and subset-specific (e.g.

monoaminergic vs. cholinergic) neuronal differentiation. Otherwise, the two repression mechanisms could be utilized to secure the repression, or to adjust precisely the levels of repression in various spatio-temporal circumstances. Thus the balance between repression and de-repression of the NRSF function could contribute to the acquisition of various neuronal phenotypes. In this respect, recent studies on the promoter regulation of the NRSF/REST gene[13,14] as well as the roles of neuron-specific splicing variants of NRSF/REST, e.g., REST4,[15,16] could be an initial step toward further understanding of the roles of the NRS-NRSF system in the gene regulation of catecholaminergic and other neurons. The results of the studies cited and others indicate that regulation by NRSF is rather more complicated than once expected. Thus, further investigation and insights are needed for a clearer understanding of the involvement or disinvolvement of the NRS-NRSF system in the gene regulation of the catecholaminergic system.

REFERENCES*

1. C. J, Schoenherr, D. J. Anderson Silencing is golden: negative regulation in the control of neuronal gene transcription. Curr. Opin. Neurobiol. 5, 566-571 (1995).
2. K. Miyaguchi, Y. Maeda, T. Kojima, Y. Setoguchi, N. Mori, Neuron-targeted gene transfer by adenovirus carrying neural-restrictive silencer element. Neuroreport 10, 2349-2353 (1999).
3. J. A. Chong, J. Tapia-Ramirez, S. Kim, J. J. Toledo-Aral, Y. Zheng, M. C. Boutros, Y. M. Altshuller, M. A. Frohman, S. D. Kraner, G. Mandel, REST: a mammalian silencer protein that restricts sodium channel gene expression to neurons. Cell 80, 949-957 (1995).
4. Z. F. Chen, A. J. Paquette, D. J. Anderson, NRSF/REST is required in vivo for repression of multiple neuronal target genes during embryogenesis. Nature Genet. 20, 136-142 (1998).
5. A. J. Paquette, S. E. Perez, D. J. Anderson, Constitutive expression of the NRSF/REST in differentiating neurons disrupts neuronal gene expression and causes axon pathfinding errors in vivo. Proc. Natl. Acad. Sci. U S A 97, 12318-12323 (2000).
6. N. Mori, Neural-restrictive silencer (NRS) and its regulatory factor NRSF/REST, Exp. Med. (in Japanese) 15, 446-454 (1997)
7. M. E. Andres, C. Burger, M. J. Peral-Rubio, E. Battaglioli, M. E. Anderson, J. Grimes, J. Dallman, N. Ballas, and G. Mandel, CoREST: a functional corepressor required for regulation of neural-specific gene expression. Proc. Natl. Acad. Sci. U S A 96, 9873-9878 (1999).
8. Y. Naruse, T. Aoki, T. Kojima, N. Mori, NRSF recruits mSin3 and histone deacetylase complex to repress neuron-specific target genes. Proc. Natl. Acad. Sci. U S A 96, 13691-13696 (1999).
9. Y. Huang, S. J. Myers, R. Dingledine, Transcriptional repression by REST: recruitment of Sin3A and histone deacetylase to neuronal genes. Nature Neurosci. 2, 867-872 (1999).
10. A. Roopra, L. Sharling, I. C. Wood, T. Briggs, U. Bachfischer, A. J. Paquette, N. J. Buckley, Transcriptional repression by NRSF is mediated via the Sin3-histone deacetylase complex. Mol. Cell Biol. 20, 2147-2157 (2000).
11. H. Ishiguro, K. S. Kim, T. H. Joh, Identification of a negative regulatory element in the 5'-flanking region of the human dopamine beta-hydroxylase gene. Brain Res. Mol. Brain Res. 34, 251-261 (1995).
12. J. J. Schimmel, L. Crew, S. Roffler-Tarlov, and D. M. Chikaraichi, 4.5 kb of the rat tyrosine hydroxylase 5' flanking sequence directs tissue specific expression during development and contains consensus sites for multiple transcription factors, Mol. Brain Res. 74, 1-14 (1999).
13. C. Koenigsberger, J. J. Chicca, M. C. Amoureux, G. M. Edelman, F. S. Jones, Differential regulation by multiple promoters of the gene encoding the NRSF. Proc. Natl. Acad. Sci. U S A 97, 2291-2296 (2000).
14. T. Kojima, K. Murai, Y. Naruse, N. Takahashi, and N. Mori, Cell-type non-selective transcription of mouse and human genes encoding NRSF, Mol. Brain Res. 90, 174-186 (2001)
15. K. Palm, N. Belluardo, M. Metsis, T. Timmusk, Neuronal expression of zinc finger transcription factor REST/NRSF/XBR gene. J. Neurosci. 18, 1280-1296 (1998).
16. M. Shimojo, A. J. Paquette, D. J. Anderson, L. B. Hersh, Protein kinase A regulates cholinergic gene expression in PC12 Cells: REST4 silences the silencing activity of NRSF/REST. Mol. Cell. Biol. 19, 6788-6795 (1999).

*The author apologizes for the very limited citation of references due to strict space limitation.

PRENATAL LIPOPOLYSACCHARIDE ALTERS POSTNATAL DOPAMINE IN THE LABORATORY RAT

Zaodung Ling, Dave A. Gayle, Jack W. Lipton, and Paul M. Carvey*

1. INTRODUCTION

Evidence links both genetic factors and environmental toxins to the development of PD. Although a genetic basis accounts for some cases of PD, the vast majority of PD is idiopathic and therefore of unknown etiology. Other environmental factors including exposure to viruses (post-encephalitic PD) and consumption of foods thought to contain excitotoxins (ALS-PD-Dementia Complex of Guam) have also been proposed. In the mid-1980s the involvement of environmental DA neurotoxins in PD was brought into focus when drug abusers inadvertently consumed a synthetic heroin eventually found to contain the DA specific neurotoxin MPTP. As a result Calne and Langston[1] hypothesized that an environmental neurotoxin in combination with normal aging was responsible for PD. They proposed that toxin exposure killed DA neurons, but that the DA neuron loss was not, by itself, sufficient to bring the affected patient to the symptom threshold in a normal life span (symptom threshold is widely believed to be loss of ~80% of the striatum's DA and ~50% of the DA neurons in the substantia nigra compacta (SNc)). They thus proposed that normal age-related losses of DA neurons resulted in further DA neuron loss sufficient to produce PD symptoms. In some respects this hypothesis has been borne out since abusers who took large dosages of MPTP developed symptoms within a week while others exposed to lower dosages, developed symptoms years later. This has led most investigators supporting the involvement of environmental toxins as risk factors in PD to study middle-life events. However, if such exposure alone were capable of inducing PD, it is likely that epidemiological studies would have identified it as a risk factor. It is therefore probable that idiopathic disease is a consequence of multiple factors. Several factors have been identified in addition to the traditional environmental neurotoxins including oxidant stress, aging, and more recently, proinflammatory cytokines.[2-5]

Several studies have now shown that the proinflammatory cytokines IL-1∃ and

* Departments of Pharmacology and Neurological Sciences, Rush-Presbyterian-St. Luke's Medical Center 2242 West Harrison St. (Suite 260), Chicago, IL 60612 USA; (312) 455-8327 Fax (312) 455-9556; e-mail pcarvey@rush.edu

Catecholamine Research, Edited by Nagatsu *et al.*
Kluwer Academic/Plenum Publishers, 2002

TNF-∀ are elevated in the SNc of patients with PD[3]. As predicted by their function, the increases in these proinflammatory cytokines were coupled with increases in apoptosis-related proteins and oxidative stress in PD patients. In animals and mesencephalic cultures LPS has also been shown to reduce TH and DA. Similarly, TNF-∀ and IL-1∃ are neurotoxic to THir cells in the E14 rat fetus. Since it is perceived by many that oxidant stress is, at least in part, involved in the pathogenesis of PD and further, that oxidant stress and proinflammatory cytokines interact to enhance tissue damage, it could be assumed that increases in proinflammatory cytokines secondary to oxidant stress may be involved in initiating the neurodegenerative process in PD. We therefore decided to examine the potential role of proinflammatory cytokines in DA cell death. Since we had already shown that TNF∀ was selectively neurotoxic to developing DA neurons in tissue culture,[6] we opted to examine the effects of elevating TNF∀ as well as another important proinflammatory cytokine IL-1∃ in the developing fetus just prior to the birth of DA neurons.

2. METHODS

Sprague-Dawley (Zivic-Miller, Allison Park, PA) female rats were delivered to our facility at embryonic (E) day 8 and injected i.p with 10,000 units/kg LPS (L8274) at E10.5. Control animals received normal saline (1 ml/kg). This dosage is known to produce a robust proinflammatory response. The females (5 LPS and 5 saline) were allowed to give birth and the offspring were sacrificed at postnatal (P) day 10. One LPS treated female failed to give birth. Two animals from each litter were randomly assigned to one of 3 assessment groups destined for striatal biochemistry, proinflammatory cytokine assessment, or assessment of DA cell counts using tyrosine hydroxylase immunoreactive (THir) cell counts lateral to the accessory optic tract as a marker for DA neurons.

3. RESULTS AND DISCUSSION

3.1. DA Biochemistry

The striata were dissected from animals that were perfused with ice-cold saline and assessed using HPLC for DA, its acid metabolite homovanillic acid (HVA), as well as DA Activity ratio ([HVA]/[DA]). Figure 1A and B shows that animals exposed to LPS prenatally had significantly lower levels of DA ($F_{1,20}$ = 8.394; p = 0.009) and significantly higher DA activity ratios ($F_{(1,20)}$ = 9.452; p = 0.006) relative to controls. These data suggest that prenatal exposure to a single dose of LPS led to a reduction of DA that lasted for 22 days (12 prenatal days and 10 postnatal days). The increase in DA activity suggests that the nigro-striatal DA system was attempting to compensate for lost DA by increasing synthesis and release.

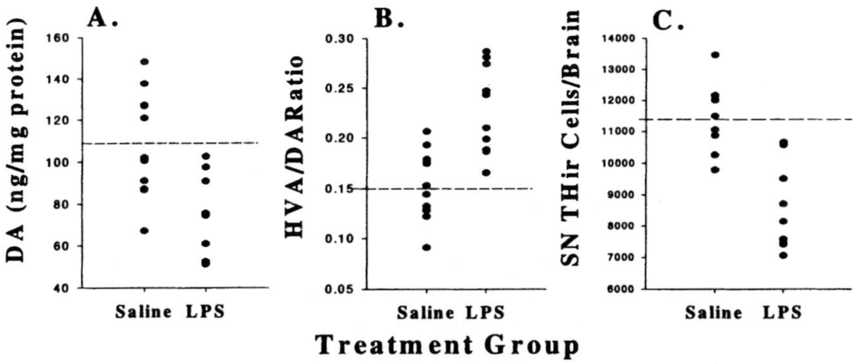

Figure 1. Alterations in DA (A), DA activity (B), and DA cell counts (C) in the Substantia nigra (SN) lateral to the accessory optic track in animals exposed to LPS prenatally compared with saline exposed controls.

3.2. DA Cell Counts

In those pups destined for morphological assessment, the animals were perfusion-fixed at P10 using Zamboni's fix as previously described.[7] The mesencephalons were sectioned and processed for tyrosine hydroxylase immunohistochemistry. In each section, the accessory optic tract was identified and the THir cells lateral to that tract on each side of the brain were counted as described previously[7]. The results are depicted in Figure 1C. Consistent with the DA biochemistry, animals exposed to LPS prenatally had significantly fewer THir cells ($F_{(1,16)} = 20.37$; $p = 0.001$). This reduction in DA cell counts suggests that prenatal LPS led to a long-term loss of DA neurons. However, it is important to point out that P10 is prior to the second phase of DA neuron loss that occurs as part of the normal developmental process (P2 and again at P14). It is therefore possible that the prenatal LPS could have simply altered this DA cell loss and that if the later phase of cell loss at P14 did not occur, cell counts might return to normal. This issue is currently being evaluated and remains an important caveat to this experiment.

3.3. TNF∀ and IL-1∃

The striata, mesencephalons, and cerebellums from the animals were assessed for TNF∀ and IL-1∃ using ELISA kits provided by R & D Systems. The procedures employed were those described in the kit. The levels of TNF∀ were elevated in all three regions (data not shown) with TNF∀ levels elevated by 940% relative to control ($F_{(1,20)} = 52.600$; $p < 0.0001$). A similar finding (all three regions significantly elevated) was seen with IL-1∃ although the magnitude of the effect was not as pronounced. These data suggest that prenatal LPS increases production of proinflammatory cytokines and further, that the elevations of these cytokines may interfere with the development of the DA system.

3.4. Bacterial Vaginosis, LPS and PD

While we were performing these studies, we discovered an extensive literature on a common complication of pregnancy called Bacterial Vaginosis (BV). BV commonly occurs in pregnancy (incidence of 14%) and is associated with the overgrowth of Gram (-) bacteria which flourish in the vaginal environment when its pH exceeds 4.5 as often occurs in pregnancy.[8] It is associated with increased levels of IL-1∀ and ∃, IL-6, and TNF∀ in the chorioamniotic environment and is presumed to be the leading cause of premature delivery and low birth weight. Thus, BV can lead to elevations in prenatal LPS as was true in the animal experiment just described, and because it is a common complication of pregnancy, could compromise DA neuron development in humans as it did in rats. If this were to occur, humans would be born with fewer DA neurons than normal, placing them at greater risk for DA neuron loss following subsequent DA neurotoxin exposure. Based on these data, consideration must therefore be given to the possibility that prenatal infections such as BV may act as risk factors for the subsequent development of PD.

REFERENCES

1. Calne,D.B. and Langston,J.W., Aetiology of Parkinson's disease, Lancet, **2** (1983) 1457-1459.
2. Mogi,M., Harada,M., Riederer,P., Narabayashi,H., Fujita,K., and Nagatsu,T., Tumor necrosis factor-alpha (TNF-alpha) increases both in the brain and in the cerebrospinal fluid from parkinsonian patients, Neurosci. Lett., **165** (1994) 208-210.
3. Nagatsu,T., Mogi,M., Ichinose,H., and Togari,A., Changes in cytokines and neurotrophins in Parkinson's disease, J. Neural Transm. **Suppl**, (2000) 277-290.
4. Mogi,M., Togari,A., Kondo,T., Mizuno,Y., Komure,O., Kuno,S., Ichinose,H., and Nagatsu,T., Caspase activities and tumor necrosis factor receptor R1 (p55) level are elevated in the substantia nigra from parkinsonian brain, J. Neural Transm., **107** (2000) 335-341.
5. Hunot,S., Dugas,N., Faucheux,B., Hartmann,A., Tardieu,M., Debre,P., Agid,Y., Dugas,B., and Hirsch,E.C., FcepsilonRII/CD23 is expressed in Parkinson's disease and induces, in vitro, production of nitric oxide and tumor necrosis factor-alpha in glial cells, J. Neurosci., **19** (1999) 3440-3447.
6. McGuire,S.O. Tumor necrosis factor is toxic to mesencephalic dopamine neurons. Carvey, P. M. and Ling, Z. D. Exp.Neurol. **169**(2):219-230. 2001.
7. Vu,T.Q., Ling,Z.D., Ma,S.Y., Robie,H.C., Tong,C.W., Chen,E.Y., Lipton,J.W., and Carvey,P.M., Pramipexole attenuates the dopaminergic cell loss induced by intraventricular 6-hydroxydopamine, J. Neural Transm., **107** (2000) 159-176.
8. Dammann,O. and Leviton,A., Does prepregnancy bacterial vaginosis increase a mother's risk of having a preterm infant with cerebral palsy?, Dev. Med. Child Neurol., **39** (1997) 836-840.

TRANSCRIPTIONAL CASCADES TO REGULATE NEURONAL DIFFERENTIATION BY MAMMALIAN *Barh1* (*MBH1*)

Tetsuichiro Saito*

1. INTRODUCTION

The vertebrate nervous system comprises an enormous number of cell types, of which differentiation is controlled by a combination of regulators. Members of basic-helix-loop-helix (bHLH) transcription factors play important roles in vertebrate neurogenesis. A mammalian homologue of *achaete-scute*, *Mash1*, is required for the differentiation of autonomic, olfactory sensory, and some ventral forebrain neurons. *neurogenins* (*ngns*), which encode Atonal-related bHLH factors, control the differentiation of some central nervous system (CNS) neurons and sensory neurons in the cranial and dorsal root ganglia. *Mash1* and *ngns* are expressed in distinct subsets or areas of cells giving rise to neurons. The distinct expression patterns of these genes suggest that they play important roles in generating the diversity of neuronal cell types. However, molecular mechanisms of neuronal differentiation at early stages, such as upstream regulators of the bHLH genes, are less well characterized. In order to find genes which function at the early stages, we have applied the differential PCR method to developing CNS tissues and cloned several genes. One of the genes was a Bar-class homeobox (*Barh*) gene and named *MBH1*. *MBH1* is most closely related to the *Drosophila BarH1* and *BarH2* genes. *MBH1* expression was restricted to the nervous system and detected at two distinct stages of neuronal differentiation. Forced expression of *MBH1* in differentiating P19 cells up-regulated *ngn2* and down-regulated *Mash1*, suggesting that *MBH1* is a potential regulator of the bHLH genes. This function of *MBH1* matched expression patterns of the three genes in the developing diencephalon. *MBH1* was also expressed in subsets of post-mitotic neurons in the cerebellum and spinal cord, following *Math1* expression. These suggest that *MBH1* has two functions in neuronal differentiation.

* Department of Development and Differentiation, Institute for Frontier Medical Sciences, Kyoto University, Kyoto 606-8507, Japan

2. IDENTIFICATION OF *MBH1*

We identified a DNA fragment which showed a differential expression pattern in the developing nervous system by the differential PCR method (Saito et al., 1995). By using the DNA fragment as a probe, the *MBH1* gene was cloned from rat and mouse (Saito et al., 1998 and unpublished). Nucleotide sequences of the human *MBH1* gene were obtained from a database (Accession number, NT_004686.3). Comparison of amino acid sequences of the three MBH1 proteins showed their remarkable conservation (Fig. 1A). There are 100% amino acid identities for the homeodomain and more than 90% identities for the whole sequence. The homeodomain of the MBH1 proteins exhibits about 80% amino acid identities to that of *Drosophila* Barh proteins, BarH1 and BarH2. 20 amino acid residues preceded by the homeodomain are also conserved among the Barh proteins, distinguishing the Barh family from another Bar-related Barx one (Fig. 1B). Functional conservation of the *Barh* genes was examined using a *Drosophila* system. Forced expression of *MBH1* generated *Bar*-like eyes in *Drosophila*, as caused by overexpression of the *BarH1* or *BarH2* gene, suggesting that *MBH1* has the same activity as *BarH1* and *BarH2* (Saito et al., 1998).

3. EXPRESSION OF *MBH1*

MBH1 expression was analyzed by Northern blotting. A single mRNA species of ~2.3 kb was detected at early stages (E10.5 for mouse and E11.5 for rat). The amount of the *MBH1* mRNA decreased as embryonic development proceeded and was then elevated at postnatal stages. In adult animals, *MBH1* expression was detected at lower levels only in the brain.

The expression pattern of *MBH1* was examined by in situ hybridization (summarized in Fig. 2). *MBH1* was expressed in various developing nervous tissues. The states of cells expressing *MBH1* were not the same. In the dorsal diencephalon, *MBH1* was expressed in the ventricular zone, which contained mitotic cells and generated neurons. The domain of *MBH1* expression was complementary to that of *Mash1* and overlapping to that of *ngn2* (Saito et al., 1998).

In contrast, *MBH1* was expressed in subsets of post-mitotic neurons in the ventral telencephalon, mesencephalon, spinal cord, and retina. In the postnatal cerebellum, *MBH1* was expressed in the external germinal layer (EGL) and internal granular layer (IGL). During cerebellar development, granule cells are generated in the EGL and migrate inward to form the IGL, where they differentiate into mature neurons. Granule cell precursors in the EGL expressed *Math1*, a mammalian homologue of *atonal*. Following *Math1* expression, *MBH1* was expressed in granule cells which had started migration and settled down in the IGL. In E10.5 mouse spinal cord, *Math1* was expressed in the cells adjacent to the roof plate. The cells migrate ventrally to give rise to commissural neurons (Helms and Johnson, 1998). *MBH1* was expressed in the cells which had started migration. These findings suggest that *MBH1* is expressed in lineages of some migratory cells downstream of *Math1* in the cerebellum and spinal cord.

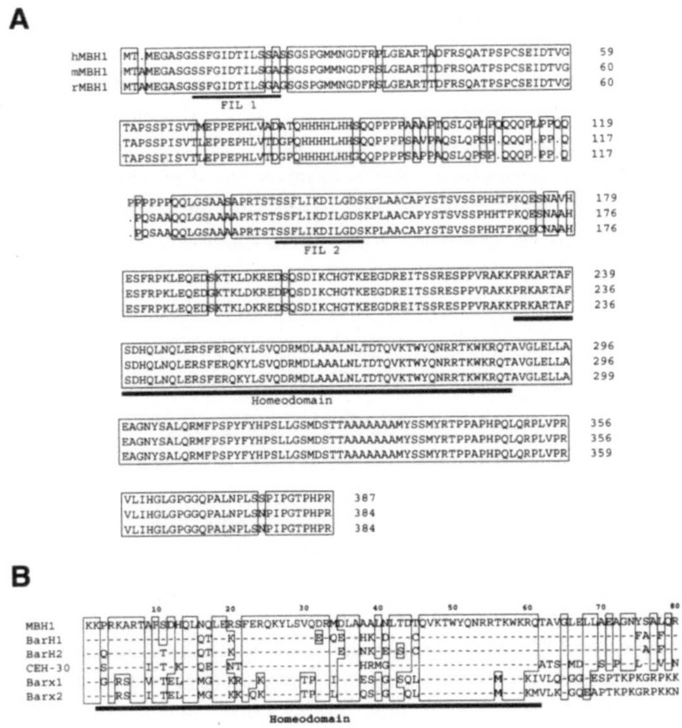

Figure 1. (A) Comparison of MBH1 protein among mammalian species. Homeodomain and FIL-peptide motif, which are conserved among some homeobox genes are shown by bars. (B) Homeodomain sequences of Barh- and Barx-family proteins. CEH-30 is a Barh protein of *C. elegans*. Mouse Barx1 and Barx2 proteins, though Bar-related, form a distinct Barx family.

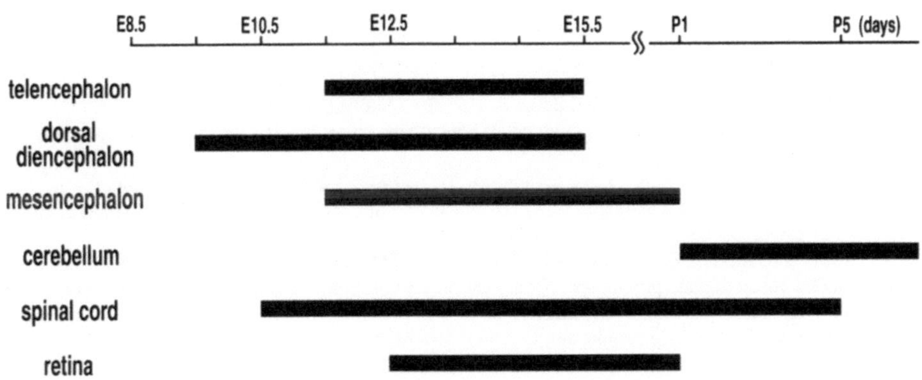

Figure 2. Schematic representation of *MBH1* expression during mouse embryonic (E) and postnatal (P) development. Bars indicate stages at which *MBH1* expression is detected.

4. REGULATION OF bHLH GENES BY *MBH1*

P19 teratocarcinoma cells are differentiated into neurons with retinoic acid treatment. During their differentiation, *Mash1* is transiently expressed (Johnson et al., 1992), but not *MBH1*. We established several *MBH1*-expressing P19 cell lines by transfecting the *MBH1* gene. In the *MBH1*-expressing cells, expression of *Mash1* is down-regulated and that of *ngn2* was up-regulated. This matched expression patterns of the three genes in the developing diencephalon. These suggest that *MBH1* functions as an upstream regulator of the bHLH genes in the diencephalon.

5. DISCUSSION

Remarkable conservation of the *MBH1* genes among mammals and functional conservation of the *Barh* genes among species suggest that *MBH1* may play an important role in the development of the nervous system.

Expression and functional analyses of *MBH1* suggest that *MBH1* has two different functions in neuronal differentiation. Some *Drosophila* homeobox genes, such as *fushi tarazu* and *even-skipped*, are known to function at two distinct stages. They are required for both segmentation of embryos at early stages and specification of some neuronal cell types at later stages. *MBH1* expression in the developing diencephalon at early stages is restricted to a neuromere and may correspond to expression of the *Drosophila* genes for the segmentation. *Drosophila BarH1* and *BarH2* genes are functionally redundant but required for the development of various organs, including the retina (Higashijima et al., 1992a) and peripheral nervous system (Higashijima et al., 1992b). Similarly, *MBH1* may have multiple functions.

REFERENCES

Helms, A. W. and Johnson, J. E. (1998) Progenitors of dorsal commissural interneurons are defined by MATH1 expression. Development **125**, 919-928.

Higashijima, S., Kojima, T., Michiue T., Ishimaru, S., Emori, Y. and Saigo, K. (1992a) Dual *Bar* homeo box genes of *Drosophila* required in two photoreceptor cells, R1 and R6, and primary pigment cells for normal eye development. Genes Dev. **6**, 50-60.

Higashijima, S., Michiue, T., Emori, Y. and Saigo K. (1992b) Subtype determination of *Drosophila* embryonic external sensory organs by redundant homeo box genes *BarH1* and *BarH2*. Genes Dev. **6**, 1005-1018.

Johnson, J. E., Zimmerman, K., Saito, T., and Anderson, D. J. (1992) Induction and repression of mammalian achaete-scute homologue (MASH) gene expression during neuronal differentiation of P19 embryonal carcinoma cells. Development **114**, 75-87.

Saito, T., Greenwood, A., Sun, Q. and Anderson, D. J. (1995) Identification by differential RT-PCR of a novel paired homeodomain protein specifically expressed in sensory neurons and a subset of their CNS targets. Mol. Cell. Neurosci. **6**, 280-292.

Saito, T., Sawamoto, K., Okano, H., Anderson, D. J. and Mikoshiba, K. (1998) Mammalian BarH homologue is a potential regulator of neural bHLH genes. Dev. Biol. **199**, 216-225.

BRAIN-SPECIFIC MIGRATION AND PROTECTIVE ROLES IN BRAIN DAMAGE OF MICROGLIA

A new therapeutic approach for catecholamine neuronal disfunctions

Makoto Sawada, Fumihiro Imai, and Hiromi Suzuki*

1. INTRODUCTION

The ability to manipulate the expression of genes within the mammalian brain provide unique opportunities to study and to potentially treat neurological disorders. The introduction of certain genes specifically in brain has been done with viral vectors or cells carrying DNA. However, these methods require surgery. Recently, we found that primary isolated microglia specifically entered the brain from blood flow when the cells were injected intra-arterially.[1] Microglia, macrophage-like cells in the brain, are multi-functional cells; they play important roles in the development, differentiation and maintenance of neural cells via their phagocytic activity and production of enzymes, cytokines and trophic factors.[2] Since intra-arterially-injected microglia were labeled with fluorescent dye microparticles by their phagocytic activity, this system could apply to a brain-specific delivery for medicines, or other bioactive materials, such as proteins or genes.[3] To apply this brain-specific bio-targeting system for treatment of neurological disorders, vehicle microglia should migrate to the region of insult and not augment the insult. Therefore, we investigated migration of systemically injected microglia into the ischemic brain.

2. GENE DELIVERY TO THE BRAIN BY MICROGLIA

To investigate the possibility that microglia can deliver the gene of interest to brain without any effects to other organs, we transfected β-galactosidase gene expression vector

* Makoto Sawada, PRESTO, Japan Science and Technology Corporation and Institute for Comprehensive Medical Science, Fujita Health University, Toyoake, Aichi 470-1192, Japan. Fumihiro Imai, Department of Neurosurgery, Fujita Health University, Toyoake, Aichi 470-1192, Japan. Hiromi Suzuki, Institute for Comprehensive Medical Science, Fujita Health University, Toyoake, Aichi 470-1192, Japan.

to the purified microglia, then injected them into a vertebral artery of the animals. At 48 h after injection, a frozen brain section were stained with X gal as a substrate for exogenously introduced β-galactosidase, many lacZ-positive cells were observed in the brain. X gal stained cells have multiple processes which are very similar to endogenous ramified microglia. Therefore, intra-arterially injected microglia can express the genes transfected *in vitro* and translate them into biologically active proteins in the brain.

The specificity of microglia migration was determined by measuring β−galactosidase activity in the brain and in other tissues. Using highly sensitive detection methods for β−galactosidase activity with a chemiluminescent substrate we could detect β−galactosidase activity in frozen sections of brain and other tissues. β−Galactosidase activity in tissues derived from microglia at 48h following intra-arterial injection was highest in the brain; over 30-fold higher than that in the liver and spleen and was not detected in the lung sections. Therefore, we concluded that most of the injected microglia migrated to the brain.

Many types of methods and techniques for in *vivo* gene transfer have been developed, and some of them have already been applied in clinical trials. The retroviral system, the most widely accepted gene transfer method to date, can achieve highly efficient integration, providing the potential for permanent gene expression. However, the system has some major disadvantages such as the typically low titer, instability of the viral vector obtained, and the requirement for target cell division for integration and expression. The adenoviral system can provide more efficient gene transfer and stability of virus, however, the difficulties in control of target cells and of re-administrations necessitated by the strong antigenicity of the virus are serious problems. The *in vivo* electroporation has been demonstrated in highly efficient gene transfer into the brain. The systems using adeno-associated viruses and HIV-based vectors as potential vectors are also available for gene therapy. But all of these methods require a major surgical procedure to transfer cells carrying genes, or to insert a stainless steel electrode when they were attempted to apply for brain disorders. We found that an intra-arterially injected microglial cell line migrated specifically into the brain. Using this cell line we established a technique to carry and express a recombinant gene specifically into the brain, and not to other tissues, using only a cannulation in the vertebral artery. With our method, no major surgical procedure is required. Since microglia are normally negative for MHC class I and II antigens *in vivo*, this brain-targeting gene delivery system can facilitate gene therapy of several brain disorders, including brain tumors, metabolic disorders, and degenerative disorders, as well as investigations into the roles of particular genes in brain function and development.

3. INTRA-ARTERIAL INJECTION OF MICROGLIA IN TRANSIENT FOREBRAIN ISCHEMIA OF GERBILS

Twenty-four hours before actual or sham transient forebrain global ischemia, fluorescently labeled microglia were injected as a bolus over 30 s into the subclavian artery of adult male Mongolian gerbils 10-12 weeks old and weighing about 70 g. Global forebrain ischemia was produced transiently by occluding both common carotid arteries for 5 min using aneurysm clips. Rectal temperature was monitored and maintained at 37°C. Seven days following ischemia, gerbils were trans cardically perfused with approximately 200 ml of isotonic saline. Then brains were removed, frozen in liquid nitrogen, and embedded in O.C.T. compound. Sections (8 μm) were cut with a cryostat microtome, transferred to glass slides, and dried. Exogenously migrated microglia were identified by a double labeling technique with a phagocytic fluorescent dye and histofluorescent staining with microglia specific antibody (anti-thymosin) or microglia specific lectin (isolectin B4). Neuronal damage and/or death were identified by Nissl staining or TUNEL staining.

4. SPECIFIC MIGRATION OF EXOGENOUS MICROGLIA FOLLOWING BRAIN INSULT

From 48 h to 7 days after the fluorescent labeled purified microglia were injected intra-arterially, many fluorescent cells were observed in brain sections from normal animals. A large portion of the fluorescent cells was in the brain parenchyma including cerebral cortex, hippocampal region, brainstem and cerebellum, apart from blood capillaries and some of these fluorescent cells had large processes. The migration and clustering of labeled microglia into ischemic hippocampus was observed in all animals 7 days after the insult. The fluorescent microglia were observed in damaged CA1 region; exogenous microglia were associated with TUNEL-positive pyramidal neurons. In contrast, in the sham-operated control gerbils that exhibited hippocampal migration of microglia, the distribution of microglia was uniform between the pyramidal layer and its neighboring regions.

5. PROTECTIVE ROLES OF EXOGENOUSLY MIGRATED MICROGLIA IN ISCHEMIC BRAIN DAMAGE

Exogenous microglia exhibited specific affinity for ischemic brain lesions and did not exacerbate ischemic neuronal injury in the present model.

Microglia respond to varied types of insults to the central nervous system, showing recruitment to the site of injury. Activated microglia have been shown in vitro to release potentially cytotoxic substances, such as free oxygen intermediates, nitric oxide, proteases, arachidonic acid derivates, excitatory amino acids, quinolinic acid and cytokines. However, the cytotoxic properties of activated microglia have not always been confirmed in vivo. On the other hand, microglia can produce growth factors and/or neurotrophic factors. We demonstrated that isolated gerbil microglia expressed BDNF and GDNF in culture by immunohistochemical and western analyses. Furthermore, we found that production of these neurotrophins by microglia preserved in in vivo condition when cultivated microglia were migrated into brain by injection into blood vessel.

In the present study, the migration of exogenous microglia did not augment ischemic hippocampus neuronal injury, rather we found protective effects of exogenous microglia on delayed death of pyramidal neurons. Quantitative analysis showed that activated microglia introduction into CA1 neuronal damaged area protects pyramidal neuron loss significantly.

Neurotrophins including BDNF and GDNF have been shown to protect against neuronal injuries in culture. In nigrostriatal dopaminergic injury, activated microglia quickly accumulate and express increasing amounts of GDNF and BDNF around the wound site. Production of these factors occurs in a similar spatial distribution to sprouting dopaminergic fibers suggesting that the dopaminergic sprouting that accompanies striatal injury results from neurotrophic factor secretion by activated microglia at the wound site. We showed that peripherally injected microglia exhibit specific affinity for brain lesions and do not exacerbate ischemic neuronal injury in the brain damage model, suggesting that microglia may have a potential to protect brain insult. Therefore, we suggest that microglia may have a potential to be used as a vehicle to deliver therapeutic genes and/or drugs for CNS repair in catecholamine disfunctions.

REFERENCES

1. F.,Imai, M. Sawada, H. Suzuki, N. Kiya, M. Hayakawa, T. Nagatsu, T. Marunouchi, and T. Kanno, Migration activity of microglia and macrophages into rat brain, *Neurosci. Lett.*, **237**, 49-52 (1997).
2. M. Sawada, A. Suzumura, and T. Marunouchi, Cytokine network in the central nervous system and its roles in growth and differentiation of glial and neuronal cells, *Int. J. Dev. Neurosci.*, **13**, 253-264 (1995).
3. M. Sawada, F. Imai, H. Suzuki, M. Hayakawa, T. Kanno, and T. Nagatsu, Brain-specific gene expression by immortalized microglial cell-mediated gene transfer in the mammalian brain, *FEBS Lett.*, **433**, 37-40 (1998).

MECHANISMS UNDERLYING NEURONAL DEATH INDUCED BY ACTIVATED MICROGLIA

Hiroshi Nakanishi*

1. INTRODUCTION

Microglia are considered to usually support the survival of neurons. However, in some pathological circumstances, microglia are activated and believed to kill neurons by producing neurotoxic and pro-inflammatory molecules such as nitric oxide, reactive oxygen intermediates and tumor necrosis factor-*a*. In Parkinson's disease, an activation of microglia has an essentially important pathogenic implication, because the substantia nigra has a high density of microglia. In fact, activated microglia have been shown to induce degeneration of dopaminergic neurons both in vitro[1] and in vivo studies[2]. Despite the clinical importance, the precise intracellular mechanism for neuronal death triggered by activated microglia is currently unknown.

There are two distinct types of cell death, necrosis and apoptosis. Necrosis is essentially accidental in its occurrence, being the outcome of severe injurious changes in the environment of the affected cells. By contrast, apoptosis is distinguished from lytic or necrotic cell death by specific biochemical and structural events. The most characteristic features of apoptosis are the activation of caspase cascades, the nuclear and cytoplasmic condensation, the blebbing of cytoplasmic membranes, and the apoptotic body formation. It is generally accepted that apoptosis is a fundamental cell death pathway contributing to the regulation of tissue development and homeostasis and that dysregulated apoptosis may cause a variety of pathologic states such as autoimmune disease and malignancy. Therefore, apoptosis appears to be highly controlled by a complex interplay between regulatory proteins such as Bcl-2, a 26 kDa integral membrane protein, and its family members. Bcl-2 has been shown to inhibit the activation of caspases by blocking cytochrome *c* release from mitochondria. However, the biochemical mechanism involved in the function of these proteins have not been fully elucidated.

In the first part of this chapter, I focus on the mechanism underlying neuronal death induced by activated microglia by using coculture system consisting of primary cultured

* Laboratory of Oral Aging Science, Division of Oral Biological Sciences, Faculty of Dental Sciences, Kyushu University, Fukuoka 812-8582, Japan. E-mail: nakandeg@mbox.nc.kyushu-u.ac.jp

Catecholamine Research, Edited by Nagatsu *et al.*
Kluwer Academic/Plenum Publishers, 2002

rat microglia and control or neuronal Bcl-2-transfected PC12 cells as a model of cathecholaminergic neurons[3]. In the second half of this chapter, I focus on the possible involvement of activated microglia in neuropathological changes in cathepsin D-deficient mice who show a novel type of lysosomal storage disease associated with massive neurodegeneration[4].

2. MICROGLIA-INDUCED NEURONAL DEATH IN CO-CULTURE SYSTEM

2.1 Activation of Caspase-3-like Protease During Activated Microglia-Induced Neuronal Death

The cell viability of neuronal PC12 cells following coculture with rat microglia immunostimulated with interferon-γ(IFN-γ, 100 units/ml)/LPS (1 ng/ml) was examined in the serum-free and NGF-containing medium. When the neuronal cells were cultured for 72 h in the absence of activated microglia, approximately 20% of the total cells died. The neuronal cells, when treated with IFN-γ/LPS in the absence of microglia, exhibited no significant increase of their death. However, when the neuronal PC12 cells were treated with activated microglia, approximately 60 % of the neuronal cells were led to death at 72 h after coculture. Importantly, no significant increase of the neuronal death was induced by the treatment with either unstimulated microglia or activated microglia plated on culture inserts in order to avoid direct contact with the neuronal cells. The activated microglia-induced neuronal death was accompanied by the significant increase of the caspase-3-like protease activity determined with Ac-DEVD-MCA as substrate. When the neuronal cells were pretreated with the caspase-3-like protease inhibitor Ac-DEVD-CHO (100 μM) for 24 h, the caspase-3-like protease activity in the cells was completely inhibited, but the microglia-induced neuronal death was only partially inhibited, as compared with that of the untreated neuronal cells. The morphological characteristics of neuronal death induced by activated microglia were examined by TUNEL staining. At 72 h after treatment with activated microglia, the majority of neuronal cells became TUNEL-positive and contained various sizes of multiple apoptotic bodies, indicating the occurrence of apoptotic cell death. When Ac-DEVD-CHO (100 μM) was added to the culture medium at 24 h before and during treatment with activated microglia, the majority of neuronal cells was not stained with TUNEL. However, damage of neuronal cells was clearly evidenced by morphological alterations such as their retracted neurites and accumulation of vacuoles in their perikarya. These results indicate that activated microglia induce caspase-3-like protease dependent apoptosis in neuronal PC12 cells. The caspase-3-like protease activity is essential for DNA fragmentation but not for cell death itself.

2.2 Effects of Overexpression of Human Bcl-2 on Activated Microglia-induced Neuronal Death

Next we examined whether high levels of Bcl-2 protein in neuronal PC12 cells inhibited the activated microglia-induced neuronal apoptosis because Bcl-2 is known to inhibit activation of caspase-3. The effect of Bcl-2 overexpression on caspase-3-like protease activity was first examined. In the neuronal Bcl-2 N2 cells, the proteolytic activity of caspase-3-like proteases was determined after the treatment with activated microglia by use of the fluorogenic substrate Ac-DEVD-MCA, respectively. Only a slight increase in the activity was observed around 42 h after the coculture. In contrast, the extract of neuronal Vector N2 cells showed a gradual increase in the hydrolytic activity of Ac-DEVD-MCA to attain the peak value at around 42 h after the treatment with activated microglia. At 48 h after treatment with activated microglia, majority of neuronal Vector N2 cells became TUNEL-positive and contained various sizes of multiple apoptotic bodies. Approximately 40 % of the total TUNEL-positive apoptotic bodies was engulfed by microglia. On the other hand, the majority of neuronal Bcl-2 N2 cells were not stained by TUNEL even at 48 hr after treatment with activated microglia, but were nevertheless clearly damaged by the treatment as evidenced by their retracted neurites and accumulated vacuoles in their perikarya. These observations indicated that Bcl-2 overexpression could completely suppress caspase-3-like protease activity. However, Bcl-2 overexpression could not abrogate the activated microglia-induced neuronal death. At the electron microscopic level, degenerating Bcl-2 N2 cells showed slightly condensed chromatins, severely disintegrated perikarya, dilated rough endoplasmic reticulum and swollen mitochondria, suggesting that the degenerating process is similar to the necrotic death pathway rather than the apoptotic death pathway.

These results strongly suggest that activated microglia predominantly induce the caspase-3-like protease-executed apoptosis to neuronal PC12 cells but alternatively trigger non-apoptotic cell death when the caspase-3-like protease cascade is inhibited by the protease inhibitor and Bcl-2 overexpression. Thus, it is more likely that whether activated microglia induce apoptosis or non-apoptosis to the neuronal cells depends on the prominence of caspase-3-like protease activation in the death process. These findings also suggest that adaptation to one type of death pathway may render cells more susceptible to alternative death pathways and that the hypersensitive response of the neuronal cells may be caused by the process of adaptation to the microglia-induced apoptosis. This could be an effective backup mechanism to execute the activated microglia-induced neuronal death.

3. MICROGLIA-INDUCED NEURONAL DEATH IN CATHEPSIN D-DEFICIENT MICE

3.1 Phagocytosis of Storage Neurons by Microglia

The deficiency for the lysosomal protease cathepsin D causes a novel type of lysosomal storage disease associated with massive neural cell death[4]. In the course of experiments to clarify the mechanism underlying this neuronal death, there were marked morphological alterations and marked expression of iNOS in microglia. The morphological alterations of microglia was first observed at P16 in the thalamus of cathepsin D-deficient mice. Processes of F4/80-positive microglia were usually attached to the neuronal cell bodies. At P24, microglia formed an almost continuous rim, suggesting that microglia phagocytozed neuronal cells. The majority of microglia was found to encircle neurons which were occupied with ceroid/lipofuscin. At the electron microscopic level, microglia often enclosed a large area of neuronal cell bodies which were characterized by the presence of a number of autophagosome-like structures. These observations indicate that phagocytosis of storage neurons is responsible for morphological alterations of microglia.

3.2 Involvement of Microglial NO in Neuronal Apoptosis

There was a marked accumulation of nitrite, a major metabolite of NO, in the cytosolic fractions of the brain. The concentration of nitrite in cathepsin D-deficient mice was significantly higher than their wild-type littermates. Furthermore, a single band corresponding to iNOS was observed in the soluble fraction of the brain from cathepsin D-deficient mice. However, no band corresponding to iNOS was detectable in the wild-type littermates. On the other hand, there was no significant difference in the level of nNOS. Immunoreactivity for iNOS was observed in the thalamus of cathepsin D-deficient mice. iNOS-positive cells corresponded well with microglia stained with isolectin, whereas iNOS was not expressed in astrocytes identified by anti-GFAP antibody. Phagocytosis of neurons and iNOS expression in microglia were found to precede the appearance of TUNEL-positive cells. To directly assess the role of microglial NO in neuropathological changes, the competitive NOS inhibitor L-N^G-nitroarginine methylester (L-NAME) or the specific iNOS inhibitor S-methylisothiourea (SMT) was administered to cathepsin D-deficient mice. The chronic treatment of L-NAME or SMT significantly reduced the total number of TUNEL-positive cells[5].

These results strongly suggest that deficiency for cathepsin D causes lysosomal storage in neurons. Microglia are activated to express iNOS by binding and/or phagocytosis of damaged storage neurons. Microglial NO induces apoptosis in adjacent neurons and an inflammatory response in the central nervous system.

REFERENCES

1. M. K. McMillian, P. J. Vanino, and R. K. Tuominen, Role of protein kinase C in microglia-induced neurotoxicity in mesencephalic cultures, *J. Neuropathol. Exp. Neurol.* **56,** 301-307 (1997).
2. W. -G. Kim, R. P. Mohney, B. Wilson, G. -H. Jeohn, B. Liu, and J. -S. Hong, Regional difference in susceptibility to lipopolysaccharide-induced neurotoxicity in the rat brain: role of microglia, *J. Neurosci.* **20,** 6309-6316 (2000).
3. K. Tanabe, H. Nakanishi, H. Maeda, T. Nishioke, K. Hashimoto, S. -Y. Liou, A. Akamine, and K. Yamamoto, A predominant apoptotic death pathway of neuronal PC12 cells induced by activated microglia is displaced by non-apoptotic death pathway following blockage of caspase-3-dependent cascade, *J. Biol. Chem.* **274,** 15725-15731 (1999).
4. M. Koike, H. Nakanishi, P. Saftig, J. Ezaki, K. Isahara, Y. Phsawa, W. Schulz-Schaeffer, T. Watanabe, S. Waguri, S. Kametaka, M. Shibata, K. Yamamoto, E. Kominami, C. Peters, K. von Figura, and Y. Uchiyama, Cathepsin D deficiency induces lysosomal storage with ceroid lipofuscin in mouse CNS neurons, *J. Neurosci.* **20,** 6898-6906 (2000).
5. H. Nakanishi, J. Zhang, M. Koike, T. Nishioku, Y. Okamoto, E. Kominami, K. von Figura, C. Peters, K. Yamamoto, P. Saftig, and Y. Uchiyama, Involvement of nitric oxide released from microglia/macrophages in pathological changes of cathepsin D-deficient mice, *J. Neurosci.* in press

ADRENERGIC CONTROL OF MICROGLIA

Junya Tanaka*, Kohji Mori, Emi Ozaki, Yoshihiro Konishi

1. INTRODUCTION

Microglial cells, which are responsible for immunological surveillance of the brain, may intimately interact with neurons, since they become activated at the earliest phase of neuropathological processes in response to even minor damage of neurons.[1] Some diffusible factors from neurons may be important in the rapid functional changes in microglial cells, because most microglial cells are normally surrounded by astrocytes.[2] Among the diffusible factors, neurotransmitters such as norepinephrine (NE) are possible candidates responsible for regulating the interactions between neurons and microglial cells. As has been suggested on the basis of electron microscopic study, noradrenergic synapses seem designed to release NE into the extracellular fluid so that they cannot be confined to the vicinity of the synaptic cleft. Microglial cells are known to respond to adrenergic agonists.[3] Yet, the expression of adrenergic receptors (AR) has not been identified. Therefore, we first aimed to reveal what types of ARs were expressed in cultured rat microglial cells by RT-PCR in the present study.

2. EXPRESSION OF mRNAs ENCODING ADRENERGIC RECEPTORS

Cultured rat microglial cells were found to express mRNAs encoding α1A, α2A, β1 and β2 ARs as revealed by RT-PCR with specific cDNA primers.[4] Expression of mRNA for β3 AR was not detectable. Microglial cells are known to change their phenotypes in response to many kinds of stimuli or culture conditions.[5] In the present study, α1A AR mRNA was upregulated in the absence of lipopolysaccharide (LPS), a well-known activator for microglial cells, while α2A AR mRNA was upregulated in the presence of LPS. A nonessential amino acid L-Ser, which is necessary for microglial activation,[6] characteristically upregulated mRNA for α1A AR in the absence of LPS. The expression of α2A AR mRNA was markedly increased in the presence of fetal calf serum and L-Ser. Expression of β1 AR mRNA, which was perhaps the weakest among the four ARs, was

* Junya Tanaka et al., Ehime University, School of Medicine, Department of Physiology, Shigenobu, Ehime 791-0295, JAPAN.

detectable only in serum-, L-Ser-free medium containing LPS. mRNA for β2 AR could be detected with the fewest number of PCR cycles, suggesting the abundant expression of β2 AR. Thus, microglial cells express mRNAs for α1A, α2A, β1 and β2 ARs. Marked expressional changes in AR-mRNAs may imply that the influence of NE on microglial cells greatly varies with changes in the conditions in the brain.

3. EFFECTS OF ADRENERGIC AGONISTS ON INTRACELLULAR cAMP LEVELS AND MACROPHAGE-LIKE PROPERTIES OF ACTIVATED MICROGLIAL CELLS

Production of proinflammatory cytokines is one of the important hallmarks of activated microglial cells. A rise in intracellular cAMP was shown to lead to suppression of such macrophage-like properties.[7] We investigated the effects of adrenergic agonists on intracellular cAMP levels by using ELISA. As expected, a β1 agonist dobutamine (Dob) and a β2 agonist terbutaline (Ter) as well as NE elevated cAMP levels, while the effect of Dob was less significant. This was perhaps due to the scarce expression of β1 AR.

Effects of adrenergic agonists on release of tumor necrosis factor α (TNFα) from lipopolysaccharide-treated microglial cells were investigated by ELISA. As a result, not only Dob and Ter but also an α1A agonist phenylephrine (Phe) markedly suppressed the release of TNFα. These three agents similarly inhibited nitric oxide production. The results indicate that not only cAMP-dependent pathway but also other signaling systems are responsible for the suppression of macrophage-like functions of microglial cells.

4. ADRENERGIC AGONISTS ELICIT A VARIETY OF EFFECTS ON EXPRESSION OF mRNAs IN MICROGLIAL CELLS

In a series of the present experiments, we found that ARs mediate a variety of effects on expression of mRNAs encoding a number of substances.

COMT is an important enzyme responsible for degrading catecholamines. Expression of COMT-mRNA was found to upregulate in response to Clo. α2 AR mediates the suppression of NE release from pre-synaptic terminals by decreasing cAMP levels in pre-synaptic neurons. The upregulated expression of COMT mRNA in microglial cells may be in line with the overall suppressive effects of α2 AR on NE systems in the brain. It may be possible that when NE release becomes too great, overflow of NE from synaptic clefts may reach microglial cells, followed by the activation of α2 AR, resulting in the rapid degradation of NE by microglial cells with increased COMT activity.

3PGDH is an enzyme catalyzing the first step of synthesizing pathway for L-Ser and Gly from 3-phosphoglycerate, a glycolysis intermediate. Since neurons and microglial cells express this enzyme at much lower levels than astrocytes,[5] they must depend on the supply of both amino acids from astrocytes.[6, 8] Especially, neurons cannot survive in a culture medium devoid of L-Ser and Gly.[9] The present result indicates the possibility that the demand of microglial cells for both amino acids may become less if they are under the influence of NE from neurons. Then, neurons would be able to utilize more L-Ser and

Gly from astrocytes in the brain at the arousal state due to the activated NE system.

Accompanied by oxidation of catecholamines, free radicals may be generated and harm neurons.[10] The following mRNAs were upregulated when incubated with AR agonists; Mn SOD, Cu/Zn SOD, catalase, hemoxygenase-1, metallothionen-1, γ-glutamyl cystein synthetase, a rate limiting enzyme for glutathione synthesis, glutathione reductase. Furthermore, NE, Phe and Clo upregulated expression of a mRNA encoding anti-apoptotic factor Bcl-xL. Upregulated expression of Bcl-xL would protect microglial cells themselves against free-radical-mediated damage. Taken together, released NE from neurons may strengthen the free-radical scavenging activity of microglial cells.

5. ADRENERGIC AGONISTS POTENTIATE THE NEUROPROTECTIVE ACTIONS OF MICROGLIAL CELLS

If neurons are cultured with microglial cells, neurons can survive even in a culture medium devoid of L-Ser and Gly [6]. Furthermore, microglial cells have been shown to protect neurons against free radical induced damage.[11] Such neuroprotective actions of microglial cells can be clearly observed if microglial cells were seeded at the density being more than one seventh of neuronal density. But we found that the neuroprotection by microglial cells was markedly potentiated if adrenergic agonists were supplemented. This potentiation may be partly related to the upregulated expression of anti-oxidants, anti-oxidant enzymes and Bcl-xL as described above.

Unless catecholamines were supplemented, almost all neurons died in an apoptotic manner when one-twentieth number of microglial cells was cocultured. But in the presence of NE, Phe, Clo, and Ter, most neurons appeared healthy. Thus, catecholamines may have neurotoxic effects as has been indicated, but also potentiate the neuroprotective actions of microglial cells, then the neurotoxicity may be overcome.

6. CONCLUSION

Microglial cells expressed four kinds of adrenergic receptors and actually responded to adrenergic agonists. This suggests that microglia are another target of norepinephrine in the brain. Furthermore, the data also suggest the possibility that microglial cells participate in the regulation of the higher brain functions not only in pathologic but also physiologic conditions by receiving information from noradrenergic neurons.

REFERENCES

1. G.W. Kreutzberg, Microglia: a sensor for pathological events in the CNS, Trends Neurosci. **19** (1996) 312-318.
2. K. Blinzinger, G. Kreutzberg, Displacement of synaptic terminals from regenerating motoneurons by microglial cells, Z. Zellforsch. **85** (1968) 145-157.
3. H. Fujita, J. Tanaka, N. Maeda, et al., Adrenergic agonists suppress the proliferation of microglia through β2-adrenergic receptor, Neurosci. Lett. **242** (1998) 37-40.
4. K. Mori, E. Ozaki, B. Zhang, et al., Effects of norepinephrine on rat cultured microglial cells that express α1, α2, β1, and β2 adrenergic receptors. (update)
5. H. Sugishita, Y. Kuwabara, K. Toku, et al., L-Serine regulates the activities of microglial cells that express

very low level of 3-phosphoglycerate dehydrogenase, an enzyme for L serine biosynthesis, J. Neurosci. Res. **64** (2001) 392-401.
6. J. Tanaka, K. Toku, S. Matsuda, et al., Induction of resting microglia in culture medium devoid of glycine and serine, Glia **24** (1998) 198-215.
7. B, Zhang, L. Yang, Y. Konishi, et al., Suppressive effects of phosphodiesterase type IV inhibitors on rat cultured microglial cells: Comparison with other types of cAMP-elevating agents. (update)
8. L. Yang, J. Tanaka, B. Zhang, et al., Astrocytes modulate nitric oxide production by microglial cells through secretion of serine and glycine, Biochem. Biophys. Res. Comm. **251** (1998) 277-282.
9. L. Yang, B. Zhang, K. Toku, et al., Improvement of the viability of cultured rat neurons by the non- essential amino acids L-serine and glycine that upregulates expression of the anti-apoptotic gene product Bcl-w, Neurosci. Lett. **295** (2000) 97-100.
10. B. Kalyanaraman, C.C. Felix, R.C. Sealy, Electron spin resonance spin stabilization of semiquinones produced during oxidation of epinephrine and its analogues, J. Biol. Chem. 259 (1984) 354-358.
11. K. Toku, J. Tanaka, H. Yano, et al., Microglial cells prevent nitric oxide induced neuronal apoptosis in vitro, J. Neurosci. Res. **53** (1998) 415-425.

DOPA IS A CAUSAL FACTOR FOR GLUTAMATE RELEASE AND DELAYED NEURON DEATH BY TRANSIENT ISCHEMIA IN STRIATUM AND HIPPOCAMPAL CA1 OF CONSCIOUS RATS

Y. Misu, Y. Goshima, N. Furukawa, N. Arai, and T. Miyamae*

1. INTRODUCTION

DOPA is believed to be an inert amino acid that affects Parkinson's disease via conversion to dopamine (DA) by aromatic L-amino acid decarboxylase (AADC). We proposed that DOPA is a transmitter and/or modulator as well as a DA precursor.[1] DOPA seems to be a transmitter of the primary baroreceptor afferents terminating in the nucleus tractus solitarii (NTS).[2] In the NTS, neurons showing immunocytochemically tyrosine hydroxylase-(+), AADC-(-), DOPA-(+) and DA-(-)-reactivity exist.[1, 2] Such neurons may have DOPA as an end product. DOPA is released by aortic depressor nerve (ADN) stimulation.[2] DOPA esters competitively antagonize hypotension and bradycardia by ADN stimulation and by DOPA microinjected.[1-3] DOPA cyclohexyl ester (CHE) is the most potent and relatively stable competitive antagonist among DOPA esters.[3] Responses to DOPA occur under inhibition of central AADC.[1, 2]

No evidence for neurons having DOPA as an end product is proven in striatum and hippocampal CA1.[1] In striata, however, DOPA is released in a transmitter-like manner and may act as a modulator. Sensitivity of immunocytochemical analysis to find DOPAergic neurons seems to be lower, compared to that of biochemical or functional approaches.[1] DOPA potentiates activities of presynaptic β-adrenoceptors to facilitate DA release and of postsynaptic D_2-receptors related to locomotion, and inhibits acetylcholine release in a rat Parkinson's model. These responses and conversion to DA may cooperate in the effectiveness for Parkinson's disease. Meanwhile, micromolar DOPA releases neuronal glutamate from slices.[4] This glutamate release may be related to neuroexcitatory side effects at least such as dyskinesia seen during chronic therapy of Parkinson's disease.

* Y. Misu, Shinobu Hospital, Fukushima, Japan 960-1101. Y. Goshima, N. Furukawa, and T. Miyamae, Yokohama City University School of Medicine, Japan 236-0004. N. Arai, Tokyo Metropolitan Institute of Neuroscience, Tokyo, Japan 183-8526.

Catecholamine Research, Edited by Nagatsu *et al.*
Kluwer Academic/Plenum Publishers, 2002

Competitive antagonism by DOPA esters against DOPA suggests existence of DOPA recognition sites. Binding studies in rat brain membranes show that DOPA sites differ from DA D_1 and D_2 receptors.[1, 3] Furthermore, DOPA sites differ from ionotropic glutamate receptors, because DOPAergic agonist and antagonist do not interact on these receptors. Among binding sites labeled with tritiated ionotropic glutamate ligands, DOPA acts only on AMPA receptors with a low affinity, while DOPA CHE acts only on NMDA ion channel domain with mM IC_{50}.[5] Herein, we survey an important interaction between DOPA and glutamate in neuronal cell death.[6, 7]

2. DOPA-INDUCED NEUROTOXICITY IN RAT STRIATA IN VITRO

Neurons in the dorsolateral striata and hippocampal CA1 pyramidal cell layers are vulnerable against ischemia, precipitating into cell death.[8, 9] Excitotoxicity due to increases in extracellular glutamate is proposed to be a causal factor for neuron death and implicated in a final common pathway for neurologic disorders.[8]

For neuron death in striata, clear relationships are seen between DOPA and glutamate. DOPA releases stereoselectively neuronal glutamate with ED_{50} of 140 μM from slices.[4] This release is seen under inhibition of AADC and is competitively antagonized by a DOPA ester. Meanwhile, DA (300 μM) releases no glutamate. DOPA and DA elicit differential toxicity in primary neuron culture.[10] DA elicits neuron death in 3 and 10 day cultures. It is protected by ascorbic acid, an antioxidant, but not by glutamate antagonists. Hydroxyl radicals play an important role in toxicity induced by DA. DOPA elicits toxicity via two different pathways. One is antioxidant-sensitive toxicity by DOPA in 3 day cultures. D-DOPA also elicits toxicity. DOPA itself or converted DA produces hydroxyl radicals. In addition, DOPA auto-oxidizes to toxic 3,4,6-trihydroxyphenylalanine (TOPA, 6-hydroxy-DOPA) and TOPA quinones.[1] The other is stereoselective toxicity in 10 day cultures. It is protected by CNQX, a non-NMDA antagonist, MK-801, a NMDA ion channel domain antagonist, Mg^{2+} addition and Ca^{2+} removal, but not by ascorbic acid.[10, 11] It is mediated by release of neuronal glutamate.[11] This type of neuron death is explained as follows. DOPA releases neuronal glutamate,[4, 11] which acts on non-NMDA receptors to depolarize neurons.[10] These neurons facilitate the relief of NMDA ion channel function from Mg^{2+}-block.[10] Glutamate in turn acts on NMDA receptors to elicit Ca^{2+}-influx.[8, 10] Ca^{2+}, forming the complex with calmodulin, activates neuronal nitric oxide (NO) synthase (nNOS), which produces NO with a detrimental property to cellular and mitchondrial membranes.[12]

3. ENDOGENOUSLY RELEASED DOPA IS A CAUSAL FACTOR FOR GLUTAMATE RELEASE AND RESULTANT DELAYED NEURON DEATH BY TRANSIENT ISCHEMIA IN STRIATA AND HIPPOCAMPAL CA1 OF RATS

At first, we attempted to clarify whether endogenous DOPA is released by 10 min ischemia due to four-vessel occlusion during striatal microdialysis, and further whether DOPA, when released, functions to cause glutamate release and resultant delayed neuron death by ischemia.[6] If it is the case, AADC inhibition should exaggerate these events, because it increases markedly extracellular DOPA,[1] whereas DOPA CHE, a potent and relatively stable competitive antagonist,[3] ought to protect these events.

Extracellular DOPA, DA and glutamate were stabilized 2-3 h after perfusion with Ringer solution. Ischemia increased extracellular DOPA, DA and glutamate. Peak increases in these substances were seen in the second sample immediately after ischemia. Some parallelism was seen between DOPA and glutamate, because the peak release ratio of DOPA : DA : glutamate was 6 : 220 : 9.

Compared to sham ischemia, dorsolateral striatum and hippocampal CA1 showed tissue damages such as reactive astrocytosis, atrophic acidophilic cytoplasm and pyknotic nuclei 96 h after reperfusion. Ischemic damage was mild to moderate in the striatum and severe in the hippocampal CA1. By a quantitative analysis, 20% and 50% of neurons showed cell damage in striatum and hippocampal CA1, respectively.

DOPA released may function to cause glutamate release and delayed neuron death by ischemia. Inhibition of intrastriatal AADC with perfusion of 30 μM NSD-1015 increased markedly basal DOPA, tripled the peak glutamate release with a tendency of decrease in DA release by ischemia, exaggerated delayed neuron death and increased the density of ischemic neurons. Furthermore, intrastriatal perfusion of 30-100 nM DOPA CHE decreased by 80% the peak glutamate release without modification of DA release by ischemia. The antagonist at 100 nM protected neuron death and decreased the density of ischemic neurons. In the hippocampal CA1, the degree and density of ischemic neurons were not modified by intrastriatal perfusion of either drug. There is concern that DOPA therapy may accelerate neuronal degeneration process especially at progressive stages of Parkinson's disease.[1] DA seems to be not related to these events by ischemia.

Then, we tried to clarify whether DOPA CHE antagonizes glutamate release and delayed neuron death in the hippocampal CA1[7] most vulnerable against brain ischemia.[8, 9] The antagonist is neuroprotective under a mild ischemic condition. DOPA CHE is a mother compound to develop neuroprotectants. Five to 10 min ischemia caused slight to mild glutamate release in 10 min samples during microdialysis and 20% to 50% neuron damages 96 h after reperfusion. Compared to striata,[6] the release is lower (1/4), but the density of ischemic neurons is higher (2.5-fold), showing the highly vulnerable property.[8, 9] DOPA and DA release is under assay sensitivity in this design, but the release is seen by 20 min ischemia. We can not measure extracellular DOPA less than 0.4 nM, as calculated from the assay sensitivity and recovery of probes used.[6]

Intrahippocampal perfusion of 100 nM DOPA CHE did not antagonize some low DOPA released to release the 2-fold higher amount of glutamate by 10 min ischemia, compared to 5 min ischemia, but it abolished glutamate release and protected neurons from cell death by 5 min ischemia. It ought to exert protection via antagonism for DOPA. DOPA released less than 0.4 nM seems to function and to clear over a threshold to trigger glutamate release by ischemia, because extremely low DOPA exerts responses.[13] At 3-10 pM, it potentiates activities of presynaptic β-adrenoceptors to facilitate noradrenaline release, being antagonized by a DOPA ester. DOPA released less than 0.4 nM seems to be a causal factor for glutamate release and delayed neuron death by ischemia in hippocampal CA1, which also suggests a highly vulnerable property. DOPA CHE acts on NMDA ion channel domain with mM IC_{50}. The antagonist, however, is unlikely to elicit protection via this site, because the nM dose was effective. Neuroprotection by systemic MK-801 is partially attributed to hypothermia.[14] DOPA CHE, however, elicits no hypothermia during and after ischemia at least by intrastriatal perfusion.[6]

We showed a delicate property of the hippocampal CA1 highly vulnerable against ischemia. An analogy was seen in the NTS.[2] A DOPA ester antagonizes depressor responses to mild stimulation of the ADN, but not to the strong stimulation. DOPA release is revealed by the stronger intermittent stimulation.

A pathway of DOPA-kynurenate-sensitive ionotropic glutamate receptors-glutamate release-nNOS activation-NO production seems to be involved in depressor responses to DOPA microinjected in the NTS.[15] This pathway seems to be common for delayed neuron death by brain ischemia and baroreflex neurotransmission.

REFERENCES

1. Y. Misu, Y. Goshima, H. Ueda, and H. Okamura, Neurobiology of L-DOPAergic systems, *Prog. Neurobiol.* **49**(5), 415-454 (1996).
2. J.-L. Yue, H. Okamura, Y. Goshima, S. Nakamura, M. Geffard, and Y. Misu, Baroreceptor-aortic nerve-mediated release of endogenous L-3,4-dihydroxyphenylalanine and its tonic function in the nucleus tractus solitarii of rats, *Neuroscience* **62**(1), 145-161 (1994).
3. N. Furukawa, Y. Goshima, T. Miyamae, Y. Sugiyama, M. Shimizu, E. Ohshima, F. Suzuki, N. Arai, K. Fujita, and Y. Misu, L-DOPA cyclohexyl ester is a novel potent and relatively stable competitive antagonist against L-DOPA among several L-DOPA ester compounds in rats, *Jpn. J. Pharmacol.* **82**(1), 40-47 (2000).
4. Y. Goshima, K. Ohno, S. Nakamura, T. Miyamae, Y. Misu, and A. Akaike, L-DOPA induces Ca^{2+}-dependent and tetrodotoxin-sensitive release of endogenous glutamate from rat striatal slices, *Brain Res.* **617**(1), 167-170 (1993).
5. T. Miyamae, Y. Goshima, M. Shimizu, T. Shibata, K. Kawashima, E. Ohshima, F. Suzuki, and Y. Misu, Some interactions of L-DOPA and its related compounds with glutamate receptors, *Life Sci.* **64**(12), 1045-1054 (1999).
6. N. Furukawa, N. Arai, Y. Goshima, T. Miyamae, E. Ohshima, F. Suzuki, K. Fujita, and Y. Misu, Endogenously released DOPA is a causal factor for glutamate release and resultant delayed neuronal cell death by transient ischemia in rat striata, *J. Neurochem.* **76**(3), 815-824 (2001).
7. N. Arai, N. Furukawa, T. Miyamae, Y. Goshima, Y. Sasaki, E. Ohshima, F. Suzuki, K. Fujita, and Y. Misu, DOPA cyclohexyl ester, a competitive antagonist, protects glutamate release and resultant delayed neuron death by transient ischemia in hippocampus CA1 of conscious rats, *Neurosci. Lett.* **299**(3), 213-216 (2001).
8. D. W. Choi, Glutamate neurotoxicity and diseases of the nervous system, *Neuron* **1**(8), 623-634 (1988).
9. T. P. Obrenovitch, and D. A. Richards, Extracellular neurotransmitter changes in cerebral ischaemia, *Cerebrovasc. Brain Metab. Rev.* **7**(1), 1-54 (1995).
10. N.-N. Cheng, T. Maeda, T. Kume, S. Kaneko, H. Kochiyama, A. Akaike, Y. Goshima, and Y. Misu, Differential neurotoxicity induced by L-DOPA and dopamine in cultured striatal neurons, *Brain Res.* **743**(1-2), 278-283 (1996).
11. T. Maeda, N.-N. Cheng, T. Kume, S. Kaneko, H. Kouchiyama, A. Akaike, M. Ueda, M. Satoh, Y. Goshima, and Y. Misu, L-DOPA neurotoxicity is mediated by glutamate release in cultured rat striatal neurons, *Brain Res.* **771**(1), 159-162 (1997).
12. V. L. Dawson, T. M. Dawson, D. A. Bartley, G. R. Uhl, and S. H. Snyder, Mechanisms of nitric oxide-mediated neurotoxicity in primary brain cultures, *J. Neurosci.* **13**(6), 2651-2661 (1993).
13. Y. Goshima, S. Nakamura, K. Ohno, and Y. Misu, Picomolar concentrations of L-DOPA stereoselectively potentiate activities of presynaptic β-adrenoceptors to facilitate the release of endogenous noradrenaline from rat hypothalamic slices, *Neurosci. Lett.* **129**(2), 214-216 (1991).
14. L. Zhang, A. Mitani, H. Yanase, and K. Kataoka, Continuous monitoring and regulating of brain temperature in the conscious and freely moving ischemic gerbil: effect of MK-801 on delayed neuronal death in hippocampal CA1, J. Neurosci. Res. **47**(4), 440-448 (1997).
15. Y. Goshima, K. Yamanashi, T. Miyamae, Y. Sasaki, M. Maeda, H. Hirano, and Y. Misu, Involvement of nitric oxide production via ionotropic glutamate receptor in DOPA-induced depressor responses in the nucleus tractus solitarii of anesthetized rats, Proceedings of the 9th International Catecholamine Symposium, 22 (2001).

STIMULATION OF NEUROTROPHIN SYNTHESIS BY 4-METHYL CATECHOL: AN APPROACH TO THE TREATMENT OF NEURODEGENERATION

Shoei Furukawa, Atsumi Nitta, and Yoshiko Furukawa*

1. INTRODUCTION

The family of neurotrophins, including nerve growth factor (NGF), brain-derived neurotrophic factor (BDNF), neurotrophin (NT)-3,and NT-4/5 plays a crucial role in the differentiation, maintenance, and survival of distinct and overlapping neuronal populations within the central and peripheral nervous systems, respectively. In addition, neurotrophins are related to neuronal plasticity. Neurotrophins are widely distributed in the CNS, and are expressed at the highest level in the hippocampus and cerebral cortex. Expression of both $mRNA^{NGF}$ and $mRNA^{BDNF}$ is known to be regulated by neurotransmission (1), and evoked in association with neural injuries. These observations suggest an involvement of neurotrophins in the process of neuronal degeneration and regeneration. Indeed, intraventricular administration of BDNF prevents neuronal death of the nigral dopaminergic neurons induced by infusion of neurotoxins or axotomy of the nigrostriatal pathway (2). Likewise, administration of NGF or BDNF suppresses neuronal death in the hippocampal pyramidal neurons following transient forebrain ischemia (3). Therefore, BDNF in particular, which has much wider action spectrum than NGF, is expected as a therapeutic agent for neurological disorders, such as Parkinson's disease, amyotrophic lateral screlosis, and Alzheimer's disease. However, there are at least two obstacles against therapeutic application of neurotrophins to brain diseases. First, neurotrophins are macromolecules that cannot pass through the blood-brain barrier (BBB). Second, neurotrophins may be rapidly incorporated into the liver due to their cationic charge, resulting in a short-term circulation in the bloodstream. Finally, an intraventricular infusion of neurotrophins as therapy involves serious technical and ethical problems. Furthermore, transfection of BDNF gene with viral vectors, and transplantation of the cells engineered with BDNF gene may be promising approaches.

---

* Shoei Furukawa and Atsumi Nitta, Laboratory of Molecular Biology, Gifu Pharmaceutical University, Mitahora-higashi,-Gifu 502-8585, Japan. Yoshiko Furukawa, Aichi Bunkyo Women's College, 1-1-41 Nishi-Machi, Inazawa, Aichi 492-8521, Japan. This work was supported in part by a grant from the Smoking Research Foundation.

However, the clinical security of these applications have not yet been fully established. A promising approach is thought to be the stimulation of synthesis of neurotrophic factors. In this study, we describe the effects of 4-methylcatechol (4MC) on the synthesis of BDNF in the CNS, and discuss its potential as a neuroprotective agent.

2. ENHANCEMENT OF BDNF SYNTHESIS *IN VITRO* AND *IN VIVO* BY 4MC

During investigation of regulatory mechanism(s) of NGF secretion in these cells, a potent stimulatory activity of a series of catechol compounds, including 4MC, on NGF synthesis was found (4). Subsequent in vivo experiment demonstrated that 4MC stimulated motor and sensory nerve regeneration after the sciatic nerve transection (5). These observations suggested a possibility that 4MC could stimulate synthesis of neurotrophic factors other than NGF such as BDNF and/or NT-3.

Effects of 4MC on neurotrophin production were examined using neurons and astrocytes cultured from embryonic rat hippocampus. Protein level and mRNA expression of neurotrophins were evaluated by enzyme immunoassays and RT-PCRs, respectively. In astrocytes, 4MC significantly enhanced levels of mRNA and secreted protein of all neurotrophins tested, NGF, BDNF or NT-3. However, in neuronal cultures, 4MC stimulated those of only BDNF (6), suggesting a selective action of 4MC on neuronal gene expression.

We next examined *in vivo* effects of 4MC. 4MC was injected into the left ventricle of adult rat brain, and the animals were processed in 24 hr. We defined areas that are differently apart from the 4MC-injected ventricle. Coronal brain slices including the injection site were cut into four particular portions, and the pieces were sequentially numbered as indicated in Fig. 1A. Content of NGF, BDNF or NT-3 in each piece was measured by the enzyme immunoassays specific for the respective neurotrophins. 4MC markedly elevated NGF and BDNF contents in particular regions, but not NT-3 content. NGF and BDNF levels were highest in the portion E, which includes the 4MC-injected ventricle, and the portion B including contralateral ventricle ranked next to the E (Fig. 1B). The portion F caudally adjacent to the E also elevated BDNF significantly. These observations demonstrated that 4MC facilitated an increase of BDNF content in the brain in association with the diffused concentration of 4MC (6).

In the next experiments, infant rats less than 10 days old, at which time the blood-brain barrier (BBB) is not yet fully established, were used, so that 4MC might better penetrate into the brain. Dose-dependent enhancement of $mRNA^{BDNF}$ expression, estimated by *in situ* hybridization, was observed in neurons of the whole brain, including the cerebral cortex, hippocampus, thalamus, and cerebellar Purkinje cells (7) (Fig. 2). However, the expression of $mRNA^{NGF}$ in the infant rat brain was not detected, irrespective of the period and dosage of 4MC administration. Time-related changes in $mRNA^{BDNF}$ and BDNF-like immunoreactivity (BDNF-LI) were chased in the cerebral cortex following a single intraperitoneal injection of 4MC (7). $mRNA^{BDNF}$ was maximally elevated at 1 h, decreased from 3 h, and recovered to the pretreatment level by 12 h post-4MC injection in layers II/III and V of the cerebral cortex. BDNF-LI is elevated markedly in layer V and slightly in layer II/III 3 h after the injection, and the increased levels were sustained even at 12 h in layers II/III and V. These findings demonstrate that 4MC, penetrating across the BBB, strongly stimulates brain BDNF synthesis in the rat brain.

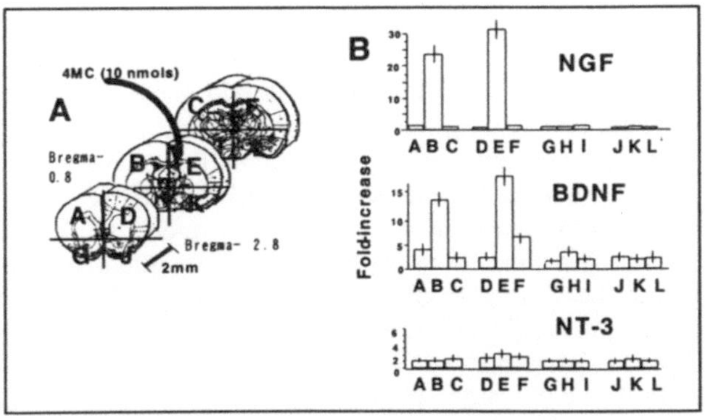

Fig. 1 Effects of ventricular administration of 4MC on neurotrophin content. **A**: Illustration of the 4MC injection site (arrow) and alphabetical order of the tissue sections for the measurement. **B**: Content of NGF, BDNF or NT-3 in each section of tissue.

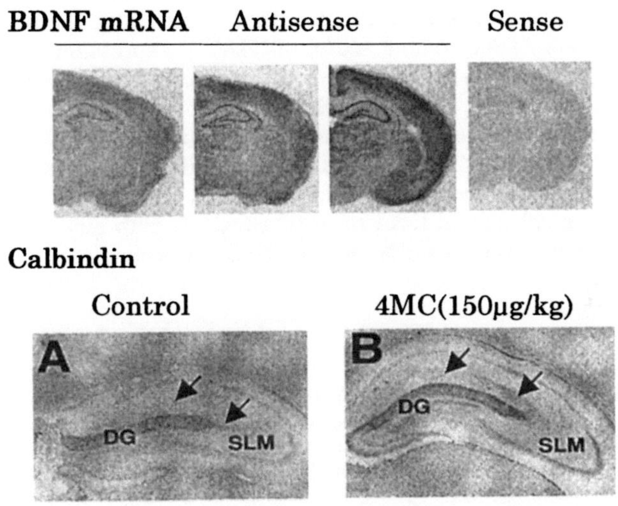

Fig. 2 Effects of intraperitoneal administration of 4MC on BDNF and calbindin expression in infant rat brain. $mRNA^{BDNF}$ expression was evaluated by *in situ* hybridization, and calbindin D-28 (arrows) was visualized using its specific antibody.

3. INFLUENCES ON CALBINDIN D-28 EXPRESSION

Brain BDNF synthesis induced by 4MC may affect certain neuronal functions. This was evaluated by monitoring the expression of calbindin D-28. 4MC that was intraperitoneally administered for 10 days to newborn rats elicited significant increases in calbindin D-28 immunoreactivity in the dentate granule cells, mossy fibers, CA3 stratum lucidum of the hippocampus, and certain neuronal populations in the pyriform cortex (Fig. 2). These findings suggest that subchronic 4MC administration accelerates physiological neuronal differentiation, probably through enhanced BDNF production.

4. POTENTIAL OF 4MC AS A THERAPEUTIC AGENT

4MC is believed to be incorporated into cells *via* a mechanism similar to that for uptake 2 of catecholamines. A long-lasting enhancement of *c-jun* mRNA expression is also caused by 4MC, which generates AP-1 proteins that drive NGF gene expression. However, AP-1 is not required for the activation of the BDNF gene. The stimulatory effect on $mRNA^{BDNF}$ expression is, so far, reported for agents that increase cAMP levels in astrocytes, for lipopolysaccharide in microglia, and for glutamate receptor agonists in neurons (1). Although these suggest an involvement of c-AMP dependent- and/or Ca^{2+}-induced signaling, at present, there are no plausible mechanisms that explain 4MC actions on BDNF gene expression.

Recent investigations have added novel roles of BDNF, such as facilitation of neural transmission, synapse formation or plasticity, regulation of growth of neurites, and also expression of genes during brain development. Furthermore, a stimulating environment exerts positive effects on cerebral health *via* increased BDNF expression. These observations demonstrate the importance of BDNF for brain development, maintenance of functions, and protection of neurons from various insults, and suggest that drug-induced enhancement of brain BDNF synthesis is profitable for prevention and amelioration of degenerative neurological disorders. Specifically, the findings that 4MC can elevate *in vivo* brain BDNF content and/or $mRNA^{BDNF}$ expression should be noticed.

The most serious problem of 4MC for therapeutic use is a difficulty to cross the BBB of mature brains. It is reported that the BBB is partially destroyed in some neurological disorders, such as multiple sclerosis and Alzheimer's disease (8). Otherwise, chemical modifications that can deliver 4MC into the brain would be promising for patients with healthy BBB function. By peripheral administration, Kourounakis *et al* (9) have succeeded to deliver a substantial level of 4MC esterized with dihydropyridine into the brain, and observed significant elevation of brain NGF content.

REFERENCES

1. F. Zafra, E. Castren, H. Thoenen and Lindholm D, Interplay between glutamate and γ-amino butyric acid transmitter systems in the physiological regulation of brain-derived neurotrophic factor and nerve growth factor synthesis in hippocampal neurons, *Proc Natl Acad Sci USA* **88**,10037-10041 (1991)
2. M. B. Spina, S. P. Squinto, J. Miller, R. M. Lindsay and C. Hymann, Brain-derived neurotrophic factor protects dopamine neurons against 6-hydroxydopamine and N-methyl-4-phenylpyridinium ion toxicity: involvement of glutathione system, *J Neurochem* **59**, 99-106 (1992)
3. T. Beck, D. Lindholm, E. Castren and A. Wree, Brain-derived neurotrophic factor protects against ischemic cell damage in rat hippocampus *J Cereb Blood Flow Metab* **14**, 689-692 (1994)
4. Y. Furukawa, S. Furukawa, E. Satoyoshi and K. Hayashi, Catecholamines induce an increase in nerve growth factor content in the medium of mouse L-M cells, *J Biol Chem* **261**, 6039-6047 (1986)
5. K. Kaechi, R. Ikegami, N. Nakamura, M. Nakajima, Y. Furukawa and S. Furukawa, 4-Methylcatechol, an inducer of NGF synthesis, enhances peripheral nerve regeneration across nerve gaps, *J Pharmacol Exp Ther* **272**, 1300-1304 (1995)
6. A. Nitta, M. Itoh, H. Fukumitsu, M. Ohmiya, A. Sometani, H. Nomoto, Y. Furukawa and S. Furukawa, 4-Methylcatechol increases brain-derived neurotrophic factor content and mRNA expression in cultured brain cells or in vivo brain of rats, *J Pharmacol Exp Ther* **291**, 1276-1283 (1999)
7. H. Fukumitsu, A. Sometani, M. Ohmiya, A. Nitta, H. Nomoto, Y. Furukawa and S. Furukawa, Induction of a physiologically active brain-derived neurotrophic factor in the infant rat brain by peripheral administration of 4-methylcatechol, *Neurosci Lett* **274**, 115-118 (1999)
8. I. Elovaara, A. Icen, J. Palo and T. Erkinjuntti, CSF in Alzheimer's disease. Studies on blood-brain barrier function and intrathecal protein synthesis, *J Neurol Sci* **70**, 73-80 (1985)
9. A. Kourounakis, N. Bodor, J. and Simpkins, Synthesis and evaluation of brain-derived targeted chemical delivery systems for the neurotrophomodulator 4-methylcatechol, *J Pharm Pharmacol* **49**, 1-9 (1997)

NEUROTOXIC FACTORS TARGETING THE DOPAMINERGIC SYSTEM

Michael A. Collins*

1. INTRODUCTION

The past several decades have seen an outburst of research on dopaminergic neurotoxins that might underlie Parkinson's disease (PD). These are summarized as endogenous, primarily dopamine (DA)-derived, and exogenous (environmental) agents. In some cases, important metabolic steps for neurotoxin activation, brain accumulation and selective uptake, i.e., oxidation (nitrogen or sulfur), N-methylation (quaternization) and/or conjugation routes, are necessitated.

2. ENDOGENOUS DA-RELATED NEUROTOXIC SPECIES

DA itself nonenzymatically generates transient reactive oxygen species (ROS), particularly in the presence of metals, that mediate oxidative stress in cell cultures. However, oxidized and/or cyclic derivatives of DA have been postulated as endogenous PD toxins because of their neurotoxic actions. Most prominent is 6-hydroxy-DA (6HODA), which is well-known as a catecholamine neurotoxin and potentially arises in brain under oxidative stress conditions (Linert et al., 1996; Pezzella et al., 1997). During its facile autoxidation, 6HODA forms ROS (Cohen and Heikkila, 1978) and quinones that covalently modify neuronal proteins. Dopaminachrome, an oxidized DA precursor of neuromelanin, also kills neuroblastoma cells (Galzigna et al., 2000), but the mechanism is obscure. Excess neuromelanin itself is neurotoxic in the presence of iron, perhaps through ROS generation (Double et al., 2000). Thiocatechol adducts of DA, DOPA and DOPAC have received attention; they are elevated in PD brain (Spencer et al., 1998) and their derived benzothiazines and mercapturates are neurotoxic (Shen et al., 2000; Montine et al., 2000). MAO-mediated deamination of DA produces the aldehyde intermediate, dopaldehyde, which has received attention because of its neurotoxic effects in rats and in neural cultures (Burke

* Dept. CBNA, Loyola University Medical School, Maywood IL 60153 USA mcollin@lumc.edu.

Catecholamine Research, Edited by Nagatsu *et al.*
Kluwer Academic/Plenum Publishers, 2002

et al., 2000; Lamensdorf et al., 2000). Though a DA precursor, tyrosine, when 3-nitrosylated **via** nitric oxide mechanisms, has been found to exert nigrostriatal toxicity (Mihm et al., 2001).

Isoquinolines (IQ's) constitute a structurally interesting group primarily arising from endogenous DA condensations, and they are also plant alkaloids along with β-carbolines (βC's) (below) (Collins, 1983). The pertinent IQ's derived from DA are N-methyl-norsalsolinol, elevated in PD CSF (Moser and Kompf, 1992), and N-methyl-salsolinol (esp. the R-stereoisomer). The latter IQ, formed centrally from salsolinol by brain N-methyltransferase activity, is neurotoxic in vivo and in vitro (Maruyama et al., 2000a). A third trace DA-derived IQ found in human CNS, tetrahydropapaveroline, can induce apoptotic/necrotic neuronal loss, possibly **via** oxidation products (Maruyama et al., 2000b; Surh, 1999). One further IQ derived not from DA but from phenethylamine, 1-benzyl-tetrahydro-IQ, is present in PD brain and induces parkinsonism in animal models (Ohta, 2000).

3. ENVIRONMENTAL NEUROTOXIC FACTORS

Pesticides/herbicides have been associated with idiopathic PD; the most well-known may be paraquat, an N-methylated dipyridinium herbicide in the same chemical group as MPP^+ that has been linked epidemiologically (Jenner, 2001). The bication penetrates the BBB (Brooks et al., 1999), and its dopaminergic neurotoxicity is greatly potentiated in mice when given with maneb, a dithiocarbamate pesticide that alone is not neurotoxic (Thiruchelvam et al., 2000). Rotenone, a natural pesticide with potent complex I inhibitory potency, is neurotoxic in neuroblastoma cells (Sherer et al., 2001), and in low concentrations, induces PD-like CNS deficits along with Lewy bodies in rats (Betarbet et al., 2000; Sherer et al., this volume).

Among persistent chlorinated pesticides, dieldrin is elevated in PD brain (Fleming et al., 1994; Corrigan et al., 1998), and toxic to DA neurons in mesencephalic cultures (Song et al., 2000). A second ubiquitous chlorinated contaminant, ortho-polychlorinated biphenyl (o-PCB), displays selective toxicity toward DA neurons in culture (particularly in combination with methylmercury), and also in utero, causing DA deficits into adulthood (Seegal et al., 1999). Parenthetically, combinations of any of the above agents with the also ubiquitous heptachlor may be important to consider; recent studies indicate that heptachlor alone can exert selective striatal neurotoxic effects in mice (Kirby et al., 2001), and it could potentiate the actions of other agents by increasing DA transporters and inhibiting mitochondrial respiration (Miller et al., 1999).

Certain environmental heterocyclics, nonpesticidal in nature and structurally analogous to MPTP or MPP^+, are possible PD factors. One class of interest, the βC's, and specifically norharman and harman, form pyrolytically during grilling or broiling of protein foods, and are possible co-mutagens in GI and stomach cancer (Wakabayashi et al., 1997). In terms of neurodegeneration, with one exception βC's are viewed as pro-toxins activated (N-methylated) in CNS to form MPP^+-like N-methyl quaternary compounds. Two of these, the 2,9-N,N-dimethylnorharman and 2,9-N,N-dimethylharman cations, are elevated in PD CSF (Matsubara et al., 1995); an N-methylating enzyme activity that forms them is increased in PD brain also (Gearhart et al., 2000). Compared to MPP^+, the above cations are relatively potent, albeit less selective, DA neurotoxins in cultures and in vivo (Collins and Neafsey,

2000; Matsubara et al., 1998). A further βC reported to be produced from tryptamine condensations with chloral, TaClo (1-trichloromethyl- 1,2,3,4- tetrahydro-βC), is found in chloral-exposed individuals and displays DA neurotoxicity without N-methylation (Grote et al., 1995; Bringmann et al., 1999).

Overall, it appears that chronic exposure to diverse chemical agents like pesticides, metals, isoquinolines and β-carbolines, either directly or through bioactivation, can promote a common process of nigrostriatal loss. If this occurs in individuals with particular genetic susceptibility (e.g., reduced mitochondrial function), it might contribute to the substratum of idiopathic PD.

REFERENCES

Betarbet, R., Sherer, T.B., MacKenzie, G., Garcia-Osuna, M., Panov, A.V., and Greenamyre, J.T., 2000, Chronic systemic pesticide exposure reproduces features of Parkinson's disease, *Nature Neurosci.* **3**:1301-1306.

Bringmann, G., God, R., Fahr, S., Feineis, D., Fornadi, K., and Fornadi, F., 1999, Identification of the dopaminergic neurotoxin 1-trichloromethyl-1,2, 3,4-tetrahydro-beta-carboline in human blood after intake of the hypnotic chloral hydrate, *Analyt. Biochem.* **270**:167-175.

Brooks, A.I., Chadwick, C.A., Gelbard, H.A., Cory-Slechta, D.A., and Federoff, H.J., 1999, Paraquat elicited neurobehavioral syndrome caused by dopaminergic neuron loss, *Brain Res.* **823**:1-10.

Burke, W.J., 2000, Catecholamine-derived aldehyde neurotoxins, in: *Neurotoxic Factors in PD and Related Disorders*, A. Storch and M.A. Collins, eds., Kluwer-Plenum Publ., NYC, pp.167-180.

Cohen, G. and Heikkila R.E., 1978, Mechanisms of action of hydroxylated phenylethylamine and indoleamine neurotoxins, *Ann. N.Y. Acad. Sci.* **305**:74-84.

Collins, M.A., 1983, Mammalian alkaloids, in: *Alkaloids XXI*, A. Brossi, ed., Academic Press, NYC, pp. 321-350.

Collins, M.A. and Neafsey, E.J , 2000, β-Carboline analogues of MPP^+ as environmental neurotoxins, in: *Neurotoxic Factors in Parkinson's Disease and Related Disorders*, A. Storch and M.A. Collins, eds., Kluwer-Plenum Publ., NYC, pp. 115-130.

Corrigan, F.M., Murray, L., Wyatt, C.L., and Shore R.F., 1998, Diorthosubstituted polychlorinated biphenyls in caudate nucleus in Parkinson's disease, *Exp. Neurol.* **150**:339-342.

Double, K.L., Zecca, L., Ben-Shachar, D., Youdim, M.B., Riederer, P., Gerlach, M., 2000, Neuromelanin may mediate neurotoxicity via its interaction with iron, in: *Neurotoxic Factors in Parkinson's Disease and Related Disorders*, A. Storch and M.A. Collins, eds., Kluwer-Plenum Publ., NYC, pp.211-218.

Fleming, L., Mann, J.B., Bean, J., Briggle, T., and Sanchez-Ramos, J.R., 1994, Parkinson's disease and brain levels of organochlorine pesticides, *Ann. Neurol.* **36**:100-103.

Galzigna, L., De Iuliis, A., and Zanatta, L., 2000, Enzymatic dopamine peroxidation in substantia nigra of human brain, *Clin. Chim. Acta* **300**:131-138.

Gearhart, D.A., Collins, M.A., Lee, J.M., and Neafsey, E.J., 2000, Increased β-carboline 9N-methyltransferase activity in the frontal cortex in Parkinson's disease, *Neurobiol. Dis.* **7**:201-211.

Grote, C., Clement, H.W., Wesemann, W., Bringmann, G., Feineis, D., Riederer, P., and Sontag, K.H., 1995, Biochemical lesions of the nigrostriatal system by TaClo (1-trichloromethyl-1,2,3,4- tetrahydro-β-carboline) and derivatives, *J. Neural Trans. Suppl.* **46**:275-281.

Jenner, P., 2001, Parkinson's disease, pesticides and mitochondrial dysfunction, Trends Neurosci. 24:245-246.

Kirby, M.L., Barlow, R.L., and Bloomquist, J.R., 2001, Neurotoxicity of the organochlorine insecticide heptachlor to murine striatal dopaminergic pathways, *Tox. Sci.* **61**:100-106.

Lamensdorf, I., Eisenhofer, G., Harvey-White, J., Hayakawa, Y., Kirk, K., and Kopin, I.J., 2000, Metabolic stress in PC12 cells induces the formation of the endogenous dopaminergic neurotoxin, 3,4-dihydroxyphenylacetaldehyde, *J. Neurosci. Res.* **60**:552-558.

Linert, W., Herlinger, E., Jameson, R.F., Kienzl, E., Jellinger, K., Youdim, M.B., 1996, Dopamine, 6-hydroxydopamine, iron, and dioxygen—their mutual interactions and possible implication in the development of Parkinson's disease, *Biochim. Biophys. Acta.* 1316:160-8.

Maruyama, W., Strolin-Benedetti, M., and Naoi, M., 2000a, N-methyl(R)salsolinol and a neutral N-methyltransferase as pathogenic factors in Parkinson's disease, *Neurobiol.* **8**:55-68.

Maruyama, W., Sango, K., Iwasa, K., Minami, C., Dostert, P., Kawai, M., Moriyasu, M., and Naoi, M., (2000b) Dopaminergic neurotoxins, 6,7-dihydroxy-1-(3', 4'-dihydroxybenzyl)-isoquinolines, cause different types of

cell death in SH-SY5Y cells: apoptosis was induced by oxidized papaverolines and necrosis by reduced tetrahydropapaverolines, *Neurosci. Lett.* **291**:89-92.
Matsubara, K., Kobayashi, S., Kobayashi, Y., Yamashita, K., Koide, H., Hatta, M., Iwamoto, K., Tanaka, O., and Kimura, K., 1995, β-Carbolinium cations, endogenous MPP+ analogs, in the lumbar cerebrospinal fluid of patients with Parkinson's disease, *Neurol.* **45**:2240-2245.
Matsubara, K., Gonda, T., Sawada, H., Uezono, T., Kobayashi, Y., Kawamura, T., Ohtaki, K., Kimura, K., and Akaike, A., 1998, Endogenously occurring beta-carboline induces parkinsonism in nonprimate animals: a possible causative protoxin in idiopathic Parkinson's disease, *J. Neurochem.* 70:727-735.
Miller, G.W., Kirby, M.L., Levey, A.I., and Bloomquist, J.R., 1999, Heptachlor alters expression and function of dopamine transporters, *Neurotoxicol.* **20**:631-637.
Mihm, M.J., Schanbacher, B.L., Wallace, B.L., Wallace, L.J., Uretsky, N.J., and Bauer, J.A., 2001, Free 3-nitrotyrosine causes striatal neurodegeneration in vivo, *J. Neurosci.* **21**: RC149 (1-5).
Montine, T.J., Amarnath, V., Picklo, M.J., Sidell, K.R., Zhang, J., and Graham, D.G., 2000, Endogenous brain catechol thioethers in dopaminergic neurodegeneration, in: *Neurotoxic Factors in Parkinson's Disease and Related Disorders*, A. Storch and M.A. Collins, eds., Kluwer-Plenum Publ., NYC, pp. 155-166.
Moser, A. and Kompf, D., 1992, Presence of methyl-6,7-dihydroxy-1,2,3,4- tetrahydroisoquinolines, derivatives of the neurotoxin isoquinoline, in parkinsonian lumbar CSF, *Life Sci.* **50**:1885-1891.
Ohta, S., 2000, Isoquinolines in Parkinson's disease, in: *Neurotoxic Factors in Parkinson's Disease and Related Disorders*, A. Storch and M.A. Collins, eds., Kluwer-Plenum Publ., NYC, pp. 91-100.
Pezzella, A., d'Ischia, M., Napolitano, A., Misuraca, G., and Prota, G., 1997, Iron-mediated generation of the neurotoxin 6-HODA quinone by reaction of fatty acid hydroperoxides with dopamine: a possible contributory mechanism for neuronal degeneration in Parkinson's disease, *J. Med. Chem.* **40**:2211-2216.
Seegal, R.F., 1999, Polychlorinated biphenyls and methylmercury act synergistically to reduce rat brain dopamine content in vitro, Environ. *Health Perspect.* **107**:879-885.
Shen, X.M., Li, H., and Dryhurst, G., 2000, Oxidative metabolites of 5-S-cysteinyldopamine inhibit the α-ketoglutarate dehydrogenase complex: possible relevance to the pathogenesis of Parkinson's disease, *J. Neural Trans.* **107**:959-978.
Sherer, T.B., Trimmer, P.A., Borland, K., Parks, J.K., Bennett, J.P. Jr., and Tuttle, J.B., 2001, Chronic reduction in complex I function alters calcium signaling in SH-SY5Y neuroblastoma cells. *Brain Res.* **891**:94-105.
Song, S., Cardozo-Pelaez, F., and Sanchez-Ramos, J., 2000, Relationship to organochlorine pesticides to PD, in: Neurotoxic Factors in *Parkinson's Disease and Related Disorders*, A. Storch and M.A. Collins, eds., Kluwer-Plenum Publ., NYC, pp. 237-245.
Spencer, J.P., Jenner, P., Daniel, S.E., Lees, A.J., Marsden, D.C., and Halliwell, B. 1998, Conjugates of catecholamines with cysteine and GSH in Parkinson's disease: possible mechanisms of formation involving reactive oxygen species, *J. Neurochem.* **71**:2112-2122.
Surh, Y.J., 1999, Tetrahydropapaveroline, a dopamine-derived isoquinoline alkaloid, undergoes oxidation: implications for DNA damage and neuronal cell death, *Europ. J. Clin. Invest.* **29**:650-651.
Thiruchelvam, M., Richfield, E.K., Baggs, R.B., Tank, A.W., and Cory-Slechta, D.A., 2000, The nigrostriatal dopaminergic system as a preferential target of repeated exposures to combined paraquat and maneb: implications for Parkinson's disease, *J. Neurosci.* **20**:9207-9214.
Wakabayashi, K., Totsuka, Y., Fukutome, K., Oguri, A., Ushiyama, H., and Sugimura, T., 1997, Human exposure to mutagenic/carcinogenic heterocyclic amines and comutagenic β-carbolines, *Mut. Res.* **376**:253-259.

ANTI-APOPTOTIC FUNCTION OF PROPARGYLAMINES

Wakako Maruyama,* Alan A. Boulton, Moussa B.H. Youdim, and Makoto Naoi

1. INTRODUCTION

(-)Deprenyl (selegiline), an inhibitor specific to type B monoamine oxidase (MAO-B), is now clinically applied for Parkinson's disease (PD), as an adjunct of L-DOPA therapy. Recent results have shown that selegiline (Parkinson Study Group, 1989) and structurally related propargylamines, such as (*R*)(+)-*N*-propargyl-1-aminoindan (rasagiline) and (*R*)-*N*-(2-heptyl)-*N*-methylpropargylamine (*R*-2HMP), delay the progress of the disease. However, it remains to be clarified, whether the effects are due to the neuroprotective or neurorescue effects, or only symptomatic. On the other hand, the well-controlled apoptotic cascade has been proposed to be a target of neuroprotection (Thompson, 1995), even though it requires further evidence whether apoptosis is a major type of cell death in PD (Jellinger, 2000). To elucidate this point, the mechanism underlying neuroprotection by propargylamines was examined using a cell model of apoptosis induced by endogenous *N*-methyl(*R*)salsolinol [*N*M(*R*)Sal] and peroxynitrite generating N-morpholino-sydnonimine (SIN-1) in human dopaminergic SH-SY5Y cells.

2. CHEMICAL STRUCTURE REQUESTED FOR ANTI-APOPTOTIC ACTIVITY OF SELEGILINE DERIVATIVES

Cells were pre-treated with propargylamines, then with *N*M(*R*)Sal or SIN-1. The viability and the mode of cell death, apoptosis and necrosis, were assessed by nuclear staining using propidium iodide and Hoechst 33342 (Maruyama et al., 2000a). Apoptotic cells were estimated by typical condensed and fragmented nuclei.

The chemical structures of selegiline, and related compounds are shown in Fig. 1.

* W. Maruyama, National Institute for Longevity Sciences, Obu, Aichi 474-8511. Japan. A. A. Boulton, University of Saskatchewan, Saskatoon, Canada. M. B. H. Youdim, Technion-Israel Institute Technology Faculty of Medicine, Haifa, Israel. M. Naoi, Institute of Applied Biochemistry, Mitake, Gifu, Japan.

Catecholamine Research, Edited by Nagatsu *et al.*
Kluwer Academic/Plenum Publishers, 2002

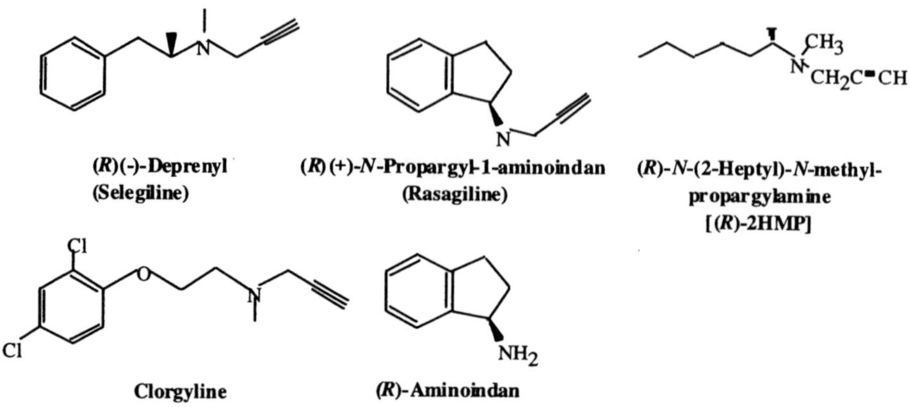

Fig. 1. Chemical structure of selegiline and related compounds with or without propargylamine moiety.

Selegiline, desmethylselegiline, rasagiline and *R*-2HMP protected the cells from apoptosis induced by the neurotoxin and peroxynitrite (Maruyama et al., 2000b, 2001a), but neither aminoindane without a propargyl moiety, nor clorigyline, an MAO-A inhibitor, protected the cells. These results indicate that a propargylamine moiety with adequate hydrophobic structure is essential for anti-apoptotic activity. It should be emphasized that a benzylamine, rasagiline, not metabolized in amphetamine derivatives as in the case of selegiline, was most effective and prevented apoptosis even at 1 nM (Naoi and Maruyama, 2001).

3. MECHANISM OF NEUROPROTECTIVE EFFECT BY PROPARGYLAMINES

Clinical and experimental data suggest the involvement of *N*M(*R*)Sal in the pathogenesis of Parkinson's disease (Naoi et al., 1999). *N*M(*R*)Sal induced apoptosis in SH-SY5Y cells, by step-wise activation of apoptotic cascade (Akao et al., 1999; Maruyama et al., 2001a). Mitochondrial membrane potential, $\Delta\Psi m$, visualized with JC-1 declined in 3 h of the incubation, and nuclear translocation of glyceraldehyde-3-phosphate dehydrogenase (GAPDH) and activation of caspase 3 occurred after 12 h. The mechanism of the anti-apoptotic effect by propargylamines was ascribed to the preservation of $\Delta\Psi m$ and the suppression of following apoptotic events (Maruyama et al., 2000b, 2001a). To clarify the effect of propargylamines on $\Delta\Psi m$, anti-apoptotic Bcl-2, which regulates mitochondrial permeability transition (PT) and following events of apoptosis (Tsujimoto and Shimizu, 2000), was overexpressed by transfection in SH-SY5Y cells. Bcl-2 transfected cells were markedly resistant to apoptosis induced by *N*M(*R*)Sal, and disruption of $\Delta\Psi m$ and GAPDH nuclear translocation was prevented (Fig. 2). These results suggest propargylamines may regulate mitochondrial death signal transduction in a similar way as the overexpression of Bcl-2. Indeed, this hypothesis was proved by our recent findings that rasagiline derivatives increased the amounts of mRNA and protein of Bcl-2 in a dose- and time-dependent way (Akao et al., in preparation). Further studies will clarify how rasagiline induces anti-apoptotic Bcl-2 protein and regulates the survival and death in the cells.

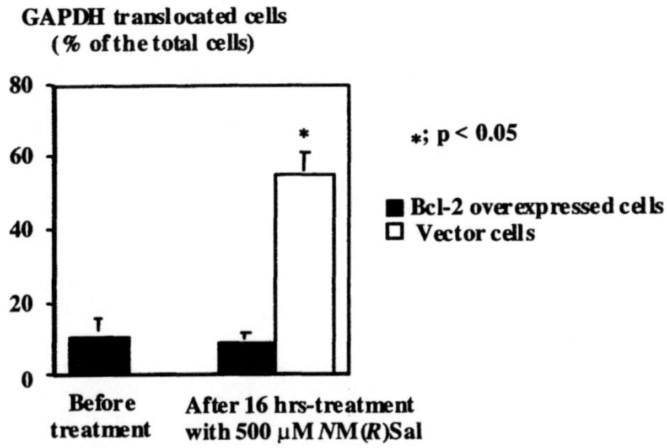

Fig. 2. Inhibition of GAPDH nuclear translocation by *N*M(*R*)Sal in Bcl-2 overexpressed SH-SY5Y cells.

4. CONCLUSION

Our data suggest that propargylamine derivatives have potency to protect or rescue declining neurons in neurodegenerative disorders, such as Parkinson's and Alzheimer's disease. As a mechanism, propargylamines were confirmed to increase the level of anti-apoptotic Bcl-2 protein family, an up-stream regulator of apoptotic cascade. It was suggested that they can protect not only DA but also other neurons from apoptosis induced by various stimuli, DA neurons by MPTP (Tatton and Greenwood, 1991), noradrenergic neurons by DSP-4 (Finnegan et al., 1990) and cholinergic neurons by AF64A (Bronzetti et al., 1992).

In addition, *in vivo* experiments reported that selegiline prolongs the life span of animals (Kitani et al., 1998), and increases activities of anti-oxidant enzymes, superoxide dismutase and catalase (Carrilo et al., 2000; Kitani et al., 2000). These results suggest that propargylamines may activate transcription of specific gene(s), and antagonize cell death induced in neurons by oxidative stress and neurotoxins, and may prolong the life span, and ameliorate the deterioration of psychomotor function in aging.

5. ACKNOWLEDGEMENT

This study was supported by a Grant-in-Aid for Scientific Research on Priority Areas (M. N.) and a Grant-in-Aid for Scientific Research (C) (W. M.) from the Ministry of Education, Science, Sports and Culture, Grant-in-Aid for Scientific Research on Priority Areas (C) from Japan Society for the Promotion of Science (W. M.), Grant for Longevity

Sciences (W. M.), and Grant for Comprehensive Research on Aging and Health, (W. M. and M. N.) and Medical Frontier Strategy Research (W. M., M. N. and Y. A.) from the Ministry of Health, Labour and Welfare, Japan.

REFERENCES

Akao, Y., Nakagawa, Y., Maruyama, W., Takahashi, T., Naoi, M., 1999, Apoptosis induced by endogenous neurotoxin, *N*-methyl(*R*)salsolinol, is mediated by activation of caspase 3. *Neurosci. Lett.*, **267**: 153-156.

Bronzetti, E., Felci, L., Ferrante, F., and Vasecchi, B., 1992, Effect of ethylene mustard aziridinium (AF64A) and of monoamine oxidase-B-inhibitor l-deprenyl on the morphology of the rat hippocampus. *Int. J. Tissue React.*, **14**: 175-181.

Carrilo, C. M., Minami, C. Kitani, K., Maruyama, W., Ohashi, K., Yamamoto, Y., Naoi, M., Kanai, S., and Youdim, M. B. H., 2000, Enhancing effect of rasagiline on superoxide dismutase and catalase activities in the dopaminergic system in rat. *Life Sci.* **67**: 577-585.

Finnegan, K. T., Skratt, J. J., Irwin, I., DeLanney, L. E., and Langston, J. W., 1990, Protection against DSP-4 induced neurotoxicity by deprenyl is not related to its inhibition of MAAO B. *Eur. J. Pharmacol.*, **184**: 119-126.

Jellinger, K. A., 2000, Cell death mechanisms in Parkinson's disease. *J. Neural Transm.* **107**: 1-29.

Kitani, K., Kanai, S., Ivy, G. O., and Carrillo, M. C., 1998, Assessing the effects of deprenyl on longevity and antioxidant defenses in different animal models. *Ann. N.Y. Acad. Sci.* **854**: 290-306.

Kitani, K., Minami, C., Maruyama, W., Kanai, S., Ivy, G. O., and Carrillo, M-C, 2000, Common properties for propargylamines of enhancing superoxide dismutase and catalase activities in the dopaminergic system in the rat: implications for the life prolonging effect of (-)deprenyl. *J. Neural Transm. (Suppl)* **60**, 139-156.

Maruyama, W., Yamamoto, T., Kitani, K., Carrillo, M. C., Youdim, M., and Naoi. M., 2000b, Mechanism underlying anti-apoptotic activity of a (-)deprenyl-related propargylamine, rasagiline. *Mech. Aging Dev.* **116**:181-191.

Maruyama, W., Akao, Y., Youdim, M., and Naoi, M., 2001b, Neurotoxin induced apoptosis in dopamine neurons: protection by *N*-propargylamine-1(*R*)- and (*S*)-aminoindan, rasagiline and TV1022. *J. Neural Transm.* **108**: 11-24..

Maruyama, W., Boulton, A. A., Davis, B. A., Dostert, P., and Naoi. M., 2001b, Enantio-specific induction of apoptosis by an endogenous neurotoxin, *N*-methyl(*R*)salsolinol, in dopaminergic SH-SY5Y cells: Suppression of apoptosis by *N*-(2-heptyl)-*N*-methylpropargylamine *J. Neural Transm.* **108**:11-24 .

Maruyama, W., Akao, Y., Youdim, M. B. H., Davis, B. A., and Naoi, M., 2001c, Transfection-enforced Bcl-2 overexpression and an anti-Parkinson drug, rasagiline, prevent nuclear accumulation of glyceraldehyde-3-phosphate dehydrogenase induced by an endogenous dopaminergic neurotoxin, *N*-methyl(*R*)salsolinol. *J. Neurochem.* In press.

Naoi, M. and Maruyama, W., 1999, *N*-Methyl(*R*)salsolinol, a dopamine neurotoxin in Parkinson's disease. *Adv. Neurol*, **180**: 259-264.

Naoi, M., and Maruyama, W., Models of Parkinson's disease, "Catecholamine Research: From Molecular Insights to Clinical Medicine", KLUWER ACADEMIC/PLENUM PUBLISHERS, New York, In press.

Parkinson Study Group, 1989, DATATOP: A multicenter controlled clinical trial in early Parkinson's Disease. *Arch. Neurol.* **46**: 1052-1060.

Tatton, W. G. and Greenwood, C. E., 1991, Rescue of dying neurons: a new action of deprenyl in MPTP parkinsonism. *J. Neurosci. Res.*, **30**: 666-672.

Thompson C. B., 1995, Apoptosis in the pathogenesis and treatment of diseases. *Science*, **267**: 1456-1462.

Tsujimoto, Y. and Shimizu, S., 2000, Bcl-2 family: Life-or-death switch. *FEBS Lett.* **466**, 6-10.

MELATONIN, NITRIC OXIDE SYNTHASE AND OXIDATIVE STRESS: IMPLICATIONS IN AGING

D.K. Lahiri, Y-.W. Ge, J.I. Nurnberger, Jr., E.Y. Yang, S.C. Bondy[1]

1. INTRODUCTION

Age-related changes in cerebral functions are believed to be partially due to an imbalance between pro-oxidant and antioxidant factors.[1] To compensate for this imbalance, dietary supplementation with various antioxidant agents is being tested.[2] We have been studying the neuroprotective and biochemical role of melatonin.[3] Melatonin, an indoleamine derivative of serotonin, is synthesized within the pineal body, and melatonin levels in the plasma have been noted to decrease with age.[4] The effects of addition of melatonin to the diet over an extended time have been studied with emphasis on age-related changes in two indices of oxidant activity: the rate of formation of reactive oxygen species (ROS) and the activity of nitric oxide synthase (NOS). Ages of mice that reflected middle adulthood and pre-senescent periods rather than stages of early development or extreme age were selected.

2. METHODS

2.1. Animals and Diets

Male B/6C3F1 mice, a hybrid between C57BL/6 and C3H from Harlan Labs, Indianapolis, IN, were maintained on a 12 hour light/ dark cycle in a temperature controlled (20±1°C) room.[2] Diets and water were provided ad lib. The minimal basal diet (#101101, Dyets Inc., PA) consisted of 50% sucrose and 26% casein (w/w) with a minimal salt and vitamin mix that include 110 ppm α-tocopherol acetate. This diet was supplemented with 0.004% melatonin and diets were fed to mice from age 3 to 9 months.

[1] D.K. Lahiri, Y-.W. Ge, J.I. Nurnberger, Jr., Institute of Psychiatric Research, Indiana University School of Medicine, 791 Union Drive Indianapolis, IN 42602; E.Y. Yang, S.C. Bondy, Department of Community and Environmental Medicine, University of California, Irvine, CA-92697. [1] D.K. Lahiri- Corresponding author

Catecholamine Research, Edited by Nagatsu *et al.*
Kluwer Academic/Plenum Publishers, 2002

2.2. Assay of NOS activity in mice brain extracts

NOS activity from the mice cerebellum was assayed using L-[^{3}H]-arginine (Amersham, NJ) and a kit (Stratagene, CA) as described previously.[5] A ratio of radioactive citrulline to unreacted arginine gives the percent conversion of arginine to citrulline and reflects a measure of net NOS activity under the condition used here.

2.3. Measurement of the rate of formation of reactive oxygen species

The cerebrocortical synaptosomal/ mitochondrial fraction was prepared to measure the rate of formation of ROS using 2',7'-dichlorofluorescin diacetate (DCFH-DA).[6] Briefly, DCFH-DA is de-esterified within cells to the ionized free acid, which was oxidized to the fluorescent 2',7'-dichlorofluorescein (DCF) by ROS[6].

3. RESULTS

3.1 Optimization of NOS activity in mouse brain extracts

The assay used was based on monitoring the biochemical conversion of L-arginine to L-citrulline by NOS. This reaction involves a five-electron oxidation of the guanidino-nitrogen of L-arginine to nitric oxide, together with the stoichiometric production of L-citrulline. The reaction requires NADPH, molecular oxygen, calcium, calmodulin and tetrahydrobiopterin. Since this assay is based on the use of radioactive arginine, the sensitivity of the assay is in the picomole level. The standard curve for the NOS assay indicates that NOS activity was linear with 5 to 30 μg of cerebellar extracts in our laboratory. Protein samples within the linear range of NOS activity were used.

3.2. Effect of age and dietary supplementation with melatonin on NOS activity

The activity of NOS in different treatment groups was compared (Table 1). The activity was greatly reduced in older 9-month mice (46.3%) relative to younger 3-month mice. When younger mice were fed a melatonin supplemented diet (40 ppm) for 6 months the original activity was almost completely maintained (87.5%) (Table 1). Results were very similar when NOS activity was adjusted to per mg wet weight of the original tissue. In this case, NOS activity was reduced to 51.4 % in older mice relative to corresponding values for younger mice, and the original activity was maintained to 98.3% when younger mice were fed diets supplemented with melatonin for 6 months.

Table 1. Assay of the activity of NOS in mice cerebellar extracts[a]

Age of mice	Treatment	NOS (% conversion) [b]	S.E.M	% original activity
3 months	None	7.98	0.87	100
9 months	None	3.69*	0.51	46.3
9 months	Melatonin	7.01	1.06	87.5

[a] NOS activity is expressed as the percent conversion of arginine to citrulline per 10 μg of protein samples. Seven animals were used/group.

[b] * Value differs significantly from corresponding value for 3 month-old mice receiving basal diet.

Table 2. Measurement of the formation of ROS[a]

Age of mice	Treatment	ROS [b]	S.E.M	% activity
3 months	None	422.8	2.18	100
9 months	None	294.3*	1.31	69.6
9 months	Melatonin	412.7	3.17	97.6

[a] The rate of formation of ROS was quantitated using a DCF standard curve and results were expressed as nmol DCF formed/ mg protein/ hour. Seven animals were used/group.
[b] * Value differs significantly from corresponding value for 3 month-old mice receiving basal diet.

3.3. Effect of age and dietary supplementation with melatonin on ROS generation

A similar trend to that of NOS activity was observed when the rate of generation of reactive oxygen species within cerebral cortex was measured. However, the age-related decline of ROS was somewhat less than was the case with NOS. Melatonin supplementation in the diets for 6 months was able to maintain the rate of ROS formation to 97.6% of the corresponding 3 month-old value (Table 2).

4. DISCUSSION

The goal of this work was to study the alteration of cerebral redox status during aging, and to use a specific dietary supplement in an attempt to maintain oxidant parameters at levels paralleling those in younger animals. A major age-related decrease in the rate of ROS generation and NOS activity was found but addition of melatonin to the basal diet restored both NOS and ROS generation rates to levels not significantly different from those found in younger animals. These data suggest that melatonin supplementation may result in maintaining a younger metabolic rate profile in chronologically older mice.

Because of its ability to retard age-related changes, melatonin supplementation may have potential value in the treatment of neurological disease closely associated with senescence such as Alzheimer's disease (AD). This is supported by reports that melatonin may play a role in several neuropsychiatric conditions.[7, 8] Abnormalities in melatonin secretion are found in AD[7] and bipolar I affective disorder subjects.[8] It is possible that dietary supplementation of melatonin would retard the progressive imbalance of homeostatic cellular mechanisms characterizing these disorders. The reason to select 3- and 9-month old mice was to avoid age-related extremes but also to determine the changes associated with transition from full maturity to early aging. Major reductions in both NOS levels and ROS production rate occurred with aging despite the relatively short age span covered. The current findings are in contrast to other reports of increased levels of oxidative damage to macromolecules in aging brain.[9] This may be due to the fact that these latter results represented a summation of events over an extended duration. In contrast, our assays reflect a snapshot of actual, ongoing dynamic pro-oxidant status in the aging animal at death, a single instant in time. The present findings are consistent with an earlier report of an age-related depression of short-lived oxidative species.[1]

The reason to select measurement of redox processes, such as ROS generation rate and NOS activity, as biochemical endpoints is because of their involvement in maintaining cellular homeostasis in the body. The major source of the intracellular formation of ROS is believed to be mitochondria. Therefore, reductions of both ROS formation rate and NOS activity in aged mice may reflect a slowing of the functioning of the electron transport chain with age as also shown in another study.[10] Melatonin supplementation was chosen because this agent can readily traverse the blood-brain barrier and some data suggest that it is neuroprotective[4]. Our present results argue strongly for a neuroendocrine role of melatonin during aging. Thus the increase in ROS and NOS found in melatonin-treated older animals may reflect maintenance of levels of mitochondrial activity characterizing that of younger animals, rather than implying deleterious events.

5. CONCLUSION

Levels of cerebellar NOS and rates of generation of cortical ROS were significantly reduced in 9 month-old mice relative to 3 month-old mice. Since most intracellular ROS have a mitochondrial origin, this may reflect a slowing down of the electron transport chain with age. Following six months treatment with dietary melatonin, the NOS activity and ROS formation rate were restored to the corresponding values found in the younger group of mice. Melatonin's ability to maintain the original levels of ROS and NOS may be due to increased endocrine function, rather than any direct antioxidant property. The dietary presence of melatonin may help maintain oxidant parameters at levels characteristic of younger animals by compensating for some of the loss of mitochondrial activity during aging.

ACKNOWLEDGMENTS

This work was supported in part by grants from the National Institutes of Health (ES 7992 and AG 16794) and the Indiana Division of Mental Health.

REFERENCES

1. LeBel CP and Bondy SC. Oxidative damage and cerebral aging. *Prog. Neurobiol.* **38**, 601-602 (1992).
2. Bondy SC, Yang YE, Walsh TJ, Ge Y and Lahiri DK. Dietary modulation of age-related changes in cerebral pro-oxidant status. *Neurochem. Int.*- In Press
3. Lahiri DK and Ghosh C. Interactions between melatonin, reactive oxygen species(ROS) and nitric oxide. *Ann. N.Y. Acad. Sci.* **893**, 325-330 (1999).
4. Reiter RJ. The role of the neurohormone melatonin against macromolecular oxidative damage. *Neurochem. Int.* **6**, 453-460 (1995).
5. Ghosh C and Lahiri DK. Increased vulnerability of neuronal cell lines to sodium nitroprusside-mediated toxicity is caused by the decreased level of nitric oxide metabolites. *J. Mol. Neurosci.* **13**, 77-92 (1999).
6. Bass DA, Parce JW, Dechatelet LR, et al. Flow cytometric studies of oxidative product formation by neutrophils: A graded dose response to membrane stimulation. *J. Immunol.* **130**, 1910 (1983).
7. Witting W, Kwa IH, Eikelenboom P, Mirmiran M and Swaab DF. Alterations in the circadian rest-activity rhythm in aging and Alzheimer's disease. *Biol. Psychiat.* **27**, 563-572 (1990).
8. Nurnberger Jr, JI, Adkins S, Lahiri DK, Mayeda A, et al.. Melatonin suppression by light in euthymic bipolar and unipolar patients. *Arch. Gen. Psychiat.* **57**, 572-579 (2000).
9. Smith CD, Carney JM, Starke-Reed PE, et al.. Excess protein oxidation and enzyme dysfunction in normal aging and in Alzheimer disease. *Proc. Natl. Acad. Sci.* **88**, 10540-10543 (1991).
10 Sharman EH and Bondy SC. Effects of age and dietary antioxidants on cortical electron transport chain activity. *Neurobiol. Aging* **22**, 629-634 (2001).

REGULATION OF CELL DEATH BY THE BCL-2 FAMILY OF PROTEINS

Yoshihide Tsujimoto*

1. INTRODUCTION

One of the promising approaches for the treatment or prevention of neurodegenerative diseases, including Parkinson's disease, is to investigate the molecular basis of neurodegeneration or neuronal death and elucidate possible therapeutic molecular targets. The Bcl-2 family of proteins is a major cell death regulator and is implicated in determining the survival or death of neurons under physiological as well as pathological conditions. I will provide an overview of the functions of this family of proteins and their targets, and will discuss a possible molecular approach for the prevention of neurodegenerative diseases.

2. THE BCL-2 FAMILY OF PROTEINS: THE MAJOR REGULATOR OF APOPTOSIS

Bcl-2 family consists of two functionally distinct groups of proteins: anti-apoptotic proteins such as Bcl-2 or Bcl-x_L and pro-apoptotic proteins, including multi-domain members such as Bax or Bak and single-domain members (so-called BH3-only proteins) such as Bid or Bik[1, 2]. Studies in knockout mice have clearly established an important role of these proteins in regulating the survival or death of neurons under physiological conditions (3 for review). Bcl-2-deficiency leads to significant degeneration of sensory, sympathetic, and motor neurons, as well as degeneration of cerebellar granule neurons. Bcl-x_L is involved in the survival of postmitotic immature neurons, while Bax is required for naturally occurring death of neurons during embryogenesis.

* Yoshihide Tsujimoto, Osaka University Graduate School of Medicine, Lab. Molecular Genetics, 2-2 Yamadaoka, Suita, Osaka 565-0871 Japan

3. BCL-2 FAMILY AND NEURODEGENERATION

Although it is still a major unsolved issue how neurons die in patients with neurodegenerative diseases, there is considerable evidence that death occurs by mechanism(s) that can be modulated by Bcl-2 family members (3 for review). For example, Bcl-2/Bcl-x_L prevents neuronal death induced by ischemia and also by axotomy. Bax is implicated in the axotomy-induced death of motoneurons. It has also been shown that Bcl-2 delays the onset of ALS in a mouse model carrying a mutant SOD gene. The best example of a human neurodegenerative disease involving Bcl-2 is spinal muscular atrophy (SMA), an autosomal disorder that features degeneration of lower motoneurons. Although Smn, the product of the disease-determining gene, plays an important role in spliceosomal regulation, we have shown that Smn also has the ability to bind with Bcl-2 and enhance its anti-apoptotic activity, while Smn mutants in SMA patients do not show this ability[4], suggesting that failure of Bcl-2-Smn interaction underlies the pathogenesis of this disease.

4. BCL-2 and BCL-X_L PREVENT BOTH APOPTOSIS AND NECROSIS

In some neurodegenerative diseases, neuronal death might not proceed in a Bcl-2-sensitive fashion but rather be related to necrosis. It should be noted, however, that Bcl-2 and Bcl-x_L are able to inhibit not only apoptosis but also necrotic death, particularly necrosis induced by mitochondrial damage[5]. Interestingly, mitochondrial damage has been implicated in pathological neurodegeneration. Bax is known to trigger two death pathways, a typical caspase-dependent pathway leading to apoptosis and a caspase-independent pathway that involves loss of the mitochondrial membrane potential (similar to necrotic death). One of the determinants of the mode of cell death, i.e., apoptosis or necrosis with similar mitochondrial dysfunction, is the intracellular ATP level[6, 7]. Thus, Bcl-2 family members and their target molecules seem to be a promising therapeutic target for neuroprotection in various diseases.

5. BCL-2 FAMILY MEMBERS FUNCTION AT THE MITOCHONDRIA

One of the major sites of activity for Bcl-2 family members is the mitochondria, where they modulate the permeability of the mitochondrial membrane. During apoptosis, the permeability of the outer membrane increases and allows some apoptogenic factors (cytochrome c and Smac/Diablo, etc.) to escape into the cytoplasm. Once in the cytoplasm, cytochrome c induces oligomerization of Apaf-1 in the presence of ATP (or dATP), and oligomerized Apaf-1 recruits and induces the activation of apical caspase-9. Smac/Diablo counteracts Xaip, which is a potent caspase inhibitor.

Regarding the process of cytochrome c release, three models have been proposed[2]. One is that cytochrome c release is mediated by physical rupture of the mitochondrial outer membrane, as a result of mitochondrial swelling or membrane instability induced by pro-apoptotic Bcl-2 family members, although the rupture model seems to be more relevant to necrotic death. The other models involve specific cytochrome c-conducting channels such as the VDAC (voltage-dependent anion channel)-Bax channel[8, 9] or a channel constituted by Bax (or another pro-apoptotic multi-domain member) alone.

6. BCL-2 FAMILY MEMBERS INTERACT WITH VDAC, AN ESSENTIAL COMPONENT OF APOPTOSIS

We have shown that some Bcl-2 family members, such as Bcl-2, Bcl-x_L, Bax and Bak can functionally interact with the VDAC[8]: Bcl-x_L closes the VDAC channel on liposomes, whereas Bax/Bak promotes the channel opening so that cytochrome c passes though it (Figure 1).

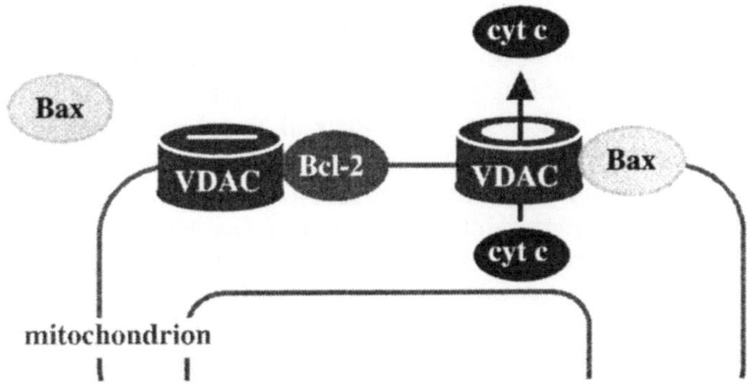

Figure 1. Bcl-2 family of proteins targets VDAC to regulate apoptotic cytochrome c release
Anti-apoptotic Bcl-2 family members such as Bcl-2 and Bcl-x_L close this channel, whereas pro-apoptotic members such as Bax and Bak enhance the channel to induce cytochrome c release

An essential role of the VDAC in Bax/Bak-mediated cytochrome c release has been confirmed by the observation that Bax/Bak induces cytochrome c release from wild-type, but not VDAC-1-deficient, yeast mitochondria[8] and that overexpression of Bax is unable to induce cytochrome c release in VDAC1-deficient yeast cells[10] . More recently, specific antibodies directed against human VDAC1 (which inhibit VDAC activity on liposomes) have been shown to inhibit Bax/Bak-induced cytochrome c release from isolated mitochondria[11]. When microinjected into cells, the same antibodies inhibited apoptosis induced by recombinant Bax and also by other apoptosis inducers, including etoposide, paclitaxel, and staurosporin[11]. Furthermore, these anti-VDAC antibodies inhibited mitochondrial membrane potential loss[11], which is characteristic of necrotic death.

7. INTRINSIC ANTI-CELL DEATH ACTIVITY IS ASSOCIATED WITH THE BH4 DOMAIN OF BCL-2 AND BCL-X_L

The BH4 domain is essential for the anti-cell death activity of Bcl-2 and Bcl-x_L, although the mechanism is still not fully understood. We recently showed that the BH4 domain is not only essential, but also sufficient, for the inhibition of VDAC activity in a liposome system[12], suggesting that the BH4 domain *per se* possesses anti-cell death activity. A BH4 oligopeptide is able to inhibit mitochondrial dysfunction, such as membrane potential loss and cytochrome c release from isolated mitochondria[12]. We have also shown that, when the protein-transduction domain of HIV tat protein was fused to the N terminus of the BH4 oligopeptide, allowing the product to cross the cytoplasmic membrane, the tat-BH4 fusion peptide was able to inhibit apoptotic cell death, although to a lesser extent when compared with full-length Bcl-2 or Bcl-x_L[12].

8. VDAC, A PROMISING THERAPEUTIC TARGET FOR NEUROPROTECTION

The VDAC plays an essential role in apoptotic cell death as well as in some forms of necrotic cell death, and it is modulated by Bcl-2 family members via direct binding. We have shown that this channel is also inhibited by gelsolin[13], an actin regulatory protein that possesses anti-cell death activity. VDAC is a component of the oligo-protein channel, known as the permeability transition pore complex, which is formed at the site of contact between outer and inner mitochondrial membranes and is considered to regulate membrane permeability. The complex is sensitive to changes of Ca^{2+} and redox state, and seems to be involved in a variety of signals modulating the survival or death of cells. Therefore, in addition to the Bcl-2 family of proteins, VDAC might be a promising therapeutic target for neuroprotection in patients with neurodegenerative diseases.

REFERNCES (our works cited)

1. Tsujimoto, Y. Role of Bcl-2 family proteins in apoptosis: apoptosomes or mitochondria. Genes to Cells **3**: 697-707, 1998
2. Tsujimoto, Y. and Shimizu, S.: Bcl-2 family: Life-or-death switch, FEBS lett. **466**: 6-10, 2000
3. Tsujimoto, Y. Prevention of neuronal cell death by Bcl-2. Results Probl. Cell Differ. **24**: 137-55, 1998
4. Iwahashi, H., Eguchi, Y., Yasuhara, N., Hanafusa, T., Matsuzawa, Y. and Tsujimoto, Y.: Synergistic anti-apoptotic activity between Bcl-2 and SMN implicated in spinal muscular atrophy. Nature **390**: 413-417, 1997
5. Shimizu, S., Eguchi, Y., Kamiike, W., Waguri, S., Uchiyama, Y., Matsuda, H. and Tsujimoto, Y.: Retardation of chemical hypoxia-induced necrotic cell death by Bcl-2 and ICE inhibitors: Possible common steps in apoptosis and necrosis. Oncogene **12**: 2045-2050, 1996.
6. Eguchi, Y., Shimizu, S. and Tsujimoto, Y.: Intracellular ATP levels determine cell death fate by apoptosis or necrosis. Cancer Res. **57**: 1835-1840, 1997
7. Eguchi, Y., Srinivasan, A., Tomaselli, K. J., Shimizu, S. and Tsujimoto, Y. ATP-dependent steps in apoptotic signal transduction. Cancer Res. **59**: 2174-2181, 1999
8. Shimizu, S., Narita, M. and Tsujimoto, Y. Bcl-2 family proteins regulate the release of apoptogenic cytochrome c by the mitochondrial channel VDAC. Nature **399**: 483-487, 1999
9. Tsujimoto, Y. and Shimizu, S. VDAC regulation by the Bcl-2 family of proteins. Cell Death Diff. **7**: 1174-1181, 2000
10. Shimizu, S., Shinohara, Y. and Tsujimoto, Y. Bax and Bcl-xL independently regulate apoptotic changes of yeast mitochondria that require VDAC but not adenine nucleotide translocator. Oncogene **19**: 4309-4318, 2000
11. Shimizu, S., Matsuoka, Y., Shinohara, Y., Yoneda, Y. and Tsujimoto, Y. Essential role of voltage-dependent anion channel in various forms of apoptosis in mammalian cells. J. Cell Biol. **152**: 237-250 (2001)
12. Shimizu, S., Konishi, A., Kodama, T. and Tsujimoto, Y.: BH4 domain of anti-apoptotic Bcl-2 family members closes VDAC, and inhibits apoptotic mitochondrial changes and cell death. Proc. Natl. Acad. Sci. USA **97**: 3100-3105, 2000
13. Kusano, H., Shimizu, S., Koya, R.-C., Fujita, H., Kamada, S., Kuzumaki, N. and Tsujimoto, Y. Human gelsolin prevents apoptosis by inhibiting apoptotic mitochondrial changes via closing VDAC. Oncogene **19**: 4807-4814, 2000

G-PROTEIN-MEDIATED REGULATION OF P/Q-TYPE CALCIUM CHANNELS BY α_2 AND β ADRENERGIC RECEPTORS EXOGENOUSLY EXPRESSED IN *XENOPUS* OOCYTES

Shuji Kaneko*, Mariko Kinoshita, Masaki Kanadzu, Akinori Akaike and Masamichi Satoh

1. INTRODUCTION

In the CNS, opening of voltage-dependent Ca^{2+} channels (VDCCs) in response to action potentials triggers synaptic neurotransmission. The presynaptic VDCCs consist of three types of Ca_v2 channel family, formerly known as N-, P/Q- and R-types. Although the modulation of N-type channels by G protein-coupled receptors has been extensively studied so far, little is known about the regulation of P/Q-type channels. To elucidate this point, we have investigated the interaction of α_2 and β adrenergic receptors with P/Q-type channels by functional coexpression in *Xenopus* oocytes.

2. EXPERIMENTAL PROCEDURES

Xenopus oocytes were injected with cRNAs for VDCC α_1 + β_{1b} subunits and G-protein-coupled receptors (GPCRs). Electrophysiological recording of VDCC current was conducted by two-electrode voltage-clamp method using 50 mM Ba^{2+} as charge carrier. Binding of $G\alpha_o$ to the VDCC domain was evaluated using glutathione S-transferase (GST) fusion proteins of VDCC C-termini and maltose-binding protein (MBP) fusion protein of $G\alpha_o$ N-termini.

* Department of Neuropharmacology, Graduate School of Pharmaceutical Sciences, Kyoto University, Sakyo-ku, Kyoto 606-8501, Japan

3. RESULTS

In oocytes co-expressing $G\alpha_o$, α_{2A} adrenergic receptors, δ-opioid receptors and α_{1A} channels, application of α_2-agonist or δ-agonist induced a transient decrease in the current amplitude by 20–25% at maximum (Fig. 1A). The inhibition was only slightly reversed by depolarizing prepulse, which is known to relieve Gβγ-mediated inhibition of N-type channels (1), showing that the voltage-resistant inhibition is predominant in the P/Q-type, α_{1A} channels when $G\alpha_o$ is co-expressed. The inhibition was additive when α_2- and δ- agonists were applied together (Fig.1B), indicating that the inhibition was mediated by a common mechanism. We have done similar experiments in all combinations of three subtypes of α_2 adrenergic receptors and three subtypes of opioid μ, δ- and κ-receptors, however the results were essentially the same, and there was no synergistic effect among them.

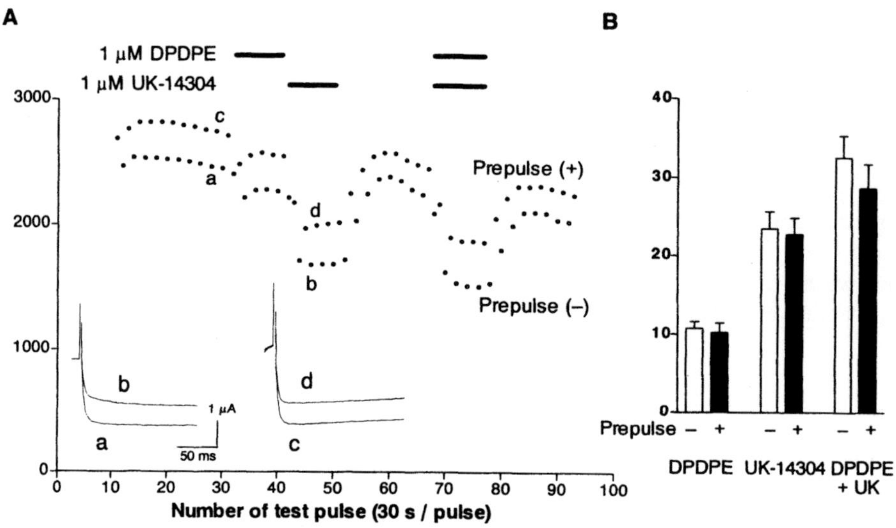

Figure 1. GPCR-mediated inhibition of α_{1A} channel current exogenously expressed in *Xenopus* oocytes with $G\alpha_o$, α_{2A}-adrenergic and δ-opioid receptors. A, Inhibition of VDCC current amplitude by α_2-agonist UK-14304 and δ-agonist DPDPE. VDCC response was evoked by 200-ms step depolarization from -60 mV to +20 mV at a 30-s interval alternately without or with a conditioning 100-ms prepulse to +100 mV which was preceded by 20-ms to the test pulse. Typical current traces at times a-d are shown . B, Inhibition of α_{1A} channel current was insensitive to depolarizing prepulse, and the inhibitions by α_2-agonist and δ-agonist were additive when applied together.

We then investigated the mechanism of voltage-resistant inhibition. In oocytes preinjected with the 20-amino acid peptide PPQT1, which includes a part of predicted α_{1A} C-terminal domain sequence at amino acids 2028-2046, the agonist-evoked inhibition of α_{1A} channel current was significantly diminished either in the presence or absence of prepulses. However, neither the equivalent α_{1B} C-terminal peptide PB3T4 nor the α_{1A}

peptide PPQT1 exerted a significant effect on the inhibition of α_{1B} current. These results suggest that the blocking effect of PPQT1 is preferential to the $G\alpha_o$-mediated, voltage-resistant inhibition of α_{1A} channel current. The agonist-induced inhibition of α_{1A} channel current was also attenuated by an intracellular application of GC/2 (anti-$G\alpha_o$ N-terminal antiserum). The reduction in the agonist-induced inhibition was significant when prepulse was applied, showing that the voltage-resistant component was more selectively suppressed. In constrast, an anti-$G\alpha_o$ C-terminal antiserum EC/2 (anti-$G\alpha_o$(C)) showed much smaller effects on agonist-induced inhibition. On α_{1B} channel current, preinjection of anti-$G\alpha_o$(N) significantly decreased the agonist-induced inhibition either in the presence or absence of prepulse, showing that both voltage-dependent and -resistant components of inhibition were simultaneously attenuated. Preinjection of anti-$G\alpha_o$(C) did not significantly affect the agonist-induced inhibition of α_{1B} channel current. Also, injection of a small amount of GTPγS (5 pmol) evoked an irreversible, long-lasting inhibition of α_{1A} channel currents. When anti-$G\alpha_o$(N) was preloaded, the voltage-resistant component of the GTPγS-induced inhibition was significantly diminished. Since no apparent difference was observed between the effects of anti-$G\alpha_o$(N) on the GTPγS-induced and agonist-induced inhibitions of α_{1A} channel current, these results indicate that anti-$G\alpha_o$(N) uncouples G-proteins from α_{1A} channels by masking the N-terminal region of $G\alpha_o$. Although the N-terminal region of Gα is known to be involved in Gβγ binding (2), these results might suggest a significance of $G\alpha_o$ N-terminal region, rather than C-terminal region, in the interaction of $G\alpha_o$ with VDCC C-terminal domain, which may be responsible for the voltage-resistant regulation of α_{1A} channels.

To evaluate the abilities of anti-$G\alpha_o$ antisera in blocking the interaction between $G\alpha_o$ and the VDCC C-terminal domains in vitro, we have evaluated the effects of antisera in the binding between GST-tagged VDCC C-terminal proteins and $G\alpha_o$ purified from the bovine brain, and found that anti-$G\alpha_o$(N) antiserum, but not anti-$G\alpha_o$(C), inhibits the direct interaction of brain $G\alpha_o$ with the C-terminal domains of α_{1A} and α_{1B} channels. The fact that anti-$G\alpha_o$(N) inhibits the interaction between VDCC C-terminal domain and $G\alpha_o$ N-terminus may imply that synthetic small peptides having the epitopic $G\alpha_o$ N-terminal sequence show similar blocking effects as 'decoy' peptides for the VDCC binding site. To investigate this possibility, we synthesized two peptides, $G\alpha_o$ N-terminal PGON1 and $G\alpha_o$ C-terminal PGOC1 and have tested the effects on voltage-dependent and -resistant inhibitions of α_{1A} and α_{1B} channel currents. In α_{1A} channels, the agonist-induced inhibition of the peak current amplitude was partially but significantly inhibited by 15 h-preinjection of the synthetic $G\alpha_o$(2-17) peptide, PGON1, at a postulated intracellular concentration of 10 μM. The control peptide, PGOC1, having rat $G\alpha_{i3}$ (350-359) sequence, in which 3 amino acids are different from the corresponding region of bovine $G\alpha_o$, did not affect the agonist-induced inhibition. In α_{1B} channels, the agonist-induced inhibition in the absence of prepulses was slightly reversed by preinjection of PGON1. However, being different from α_{1A} channels, the voltage-resistant component of inhibition of α_{1B} currents was not significantly affected. These results suggest that $G\alpha_o$ N-terminus plays a significant role in the interactions with VDCC C-terminal domains in evoking the voltage-resistant inhibition of α_{1A} channels.

Finally, we produced two MBP-tagged $G\alpha_o$ N-terminal proteins, MBP-$G\alpha_o$(1-88) and MBP-$G\alpha_o$(7-88), and examined their binding abilities with GST fusion proteins of α_{1A} and α_{1B} C-terminal domains in vitro. In the tubes containing GST-B1T encoding α_{1A}

C-terminal domain, MBP-Gα_o(1-88) was detected in the glutathione-eluate, whereas MBP-Gα_o(7-88) did not bind, indicating that the N-terminal 6 amino residues of Gα_o are necessary for the interaction with the α_{1A} C-terminal domain. In the presence of PGON1 or PPQT1, at a concentration of 50 μM which is 5-fold of putative cytosolic concentration in electrophysiological assays, the binding of these MBP fusion proteins was completely inhibited. These results demonstrate that small regions of Gα_o N-terminal and α_{1A} C-terminal domains are critical for the interaction between Gα_o and α_{1A} proteins. Similar results were obtained from the experiments using GST-B3T that encodes the α_{1B} C-terminal domain. MBP-Gα_o(1-88), but not MBP-Gα_o(7-88), bound to GST-B3T, and the binding was completely inhibited in the presence of 50 μM PGON1 or PB3T4.

4. DISCUSSION

The present study represents a new mechanism of GPCR-mediated VDCC regulation, in which a direct interaction between N-terminal region of Gα_o and C-terminal domain of α_{1A} or P/Q-type channel causes a voltage-resistant inhibition of channel opening. The binding of Gα_o to α_{1A} channels does not require the presence of G$\beta\gamma$, as shown in the binding study of fusion proteins. In many native neurons, α_{1A} channel currents are regulated by GPCR mainly in a voltage-resistant manner (3). A direct coupling of Gα_o with α_{1A} channel has an important physiological impact, which is additional to the voltage-sensitive regulation of neuronal VDCC function by G$\beta\gamma$. On the other hand, it is most likely that both Gα and G$\beta\gamma$ are required for the voltage-resistant inhibition of α_{1B} channels. Since the α_{1B} subunit has multiple G$\beta\gamma$ binding sites and at least one Gα binding site (4), further studies are necessary to elucidate the molecular mechanism of voltage-resistant inhibition of α_{1B} channels. The voltage-resistant regulation of Ca^{2+} influx will exert a consistent regulation of transmitter release insensitive to frequent firing or sustained depolarization at synapses.

ACKNOWELDGMENTS

We thank Dr. Yasuo Mori (Natl. Inst. Physiol. Sci., Okazaki, Japan) and Dr. Toshihide Nukada (Tokyo Inst. Psychiatry, Tokyo, Japan) for providing us with cDNAs for VDCC and G proteins, respectively.

REFERENCES

1. G. W. Zamponi and T. P. Snutch, Modulation of voltage-dependent calcium channels by G proteins, *Curr. Opin. Neurobiol.* **8**, 351-356 (1998).
2. B. M. Denker, E. J. Neer, and C. J. Schmidt, Mutagenesis of the amino terminus of the α subunit of the G protein G_o, *J. Biol. Chem.* **267**, 6272-6277 (1992).
3. K. P. M. Currie and A. P. Fox, Comparison of N- and P/Q-type voltage-gated calcium channel current inhibition, *J. Neurosci.* **17**, 4570-4579 (1997).
4. T. Furukawa, R. Miura, Y. Mori, M. Strobeck, K. Suzuki, Y. Ogihara, T. Asano, R. Morishita, M. Hashii, H. Higashida, M. Yoshii, and T. Nukada, Differential interactions of the C terminus and the cytoplasmic I-II loop of neuronal Ca^{2+} channels with G-protein α and βγ subunits. II. evidence for direct binding, *J. Biol. Chem.* **273**, 17595-17603 (1998).

α_1-ADRENERGIC RECEPTORS OF MDCK-D_1 CELLS UTILIZE MULTIPLE SIGNALLING COMPONENTS

Paul A. Insel, Rennolds S. Ostrom, Martin C. Michel, and Rainer Büscher*

ABSTRACT

We have examined mechanisms for signal transduction by α_1-adrenergic receptors in a clonal cell line, Madin Darby canine kidney, MDCK-D_1. These cells possess a pure population of α_{1B}-adrenergic receptors (α_{1B}-AR) that activate multiple phospholipases. We show here that these receptors have 2 other actions: release of cellular ATP and increase in cyclic AMP formation. The increase in cyclic AMP formation depends on formation of cyclo-oxygenase-dependent products, in particular, PGE_2. We also show here that P2Y receptor agonists down-regulate α_1-AR in a non-additive manner with AR agonist-promoted down-regulation. Taken together, the data indicate that α_{1B}-AR in MDCK-D_1 cells use multiple mechanisms for activation and deactivation of signalling. Cellular release of and response to ATP may modulate α_{1B}-AR signalling.

1. INTRODUCTION

The three α_1-adrenergic receptors (α_1-AR), α_{1A}-, α_{1B}-, α_{1D}-, are heterotrimeric G-protein-coupled receptors (GPCRs) that preferentially interact with the $G_{q/11}$ proteins to activate phospholipase C (PLC). PLC hydrolyzes phosphoinositides to inositol phosphates and diacylglycerol, which act respectively to increase intracellular Ca^{++} and activate isoforms of protein kinase C (PKC) (as reviewed in 1, 2). Precise, detailed understanding as to how α_1-AR's regulate cellular responses ("activation") is not yet available. Additional signalling pathways may be involved.[1-3] In addition, relatively little is known regarding mechanisms involved in deactivation of signalling by these receptors.[4] In order to assess such issues, we have studied a clonal cell line, Madin-Darby Canine Kidney (MDCK)-D_1, isolated from renal tubular cells, and have examined the

* Paul A. Insel, Rennolds S. Ostrom, and Rainer Büscher, Department of Pharmacology, University of California, San Diego, La Jolla, CA 92093 USA. Martin C. Michel and Rainer Büscher, Departments of Medicine and Pediatrics, University of Essen, D45122, Germany

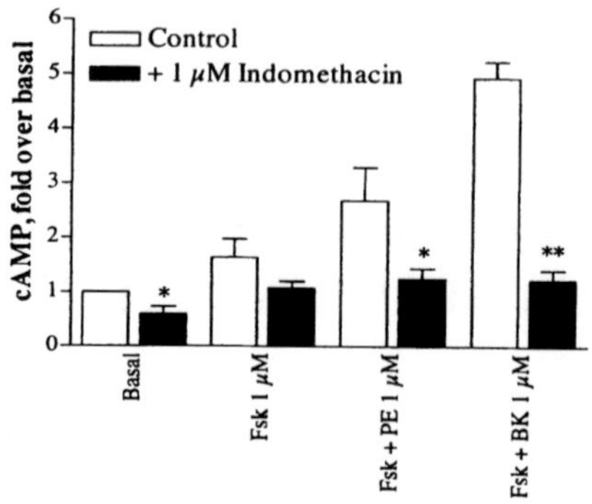

Figure 1. Cyclic AMP accumulation in MDCK-D_1 cells was determined by a 10 minute incubation of cells with 0.2 mM isobutylmethylxanthine alone ("basal"), or together with 1 μM forskolin (Fsk), or 1 μM forskolin (Fsk) + 1 μM phenylephrine (PE), or + 1 μM bradykinin (BK) in the absence (open bars) or presence (solid bars) of 1 μM indomethacin. * = $p < 0.05$; ** = $p < 0.01$ compared to absence of indomethacin.

nature of the α_1-AR on MDCK-D_1 cells and signalling mechanisms utilized by these receptors[3, 5-8]. In the current work, we have examined several additional aspects of α_1-AR action: release of cellular ATP, increase in cyclic AMP (cAMP) levels, and receptor down-regulation.

2. METHODS

Methods for growing cells, assessing receptor expression and signalling, and ATP release, and receptor desensitization were performed as described previously.[3, 5-8]

3. RESULTS AND DISCUSSION

Previous work documented that MDCK-D_1 cells express a pure population of α_{1B}-receptors and that these receptors promote events such as activation of phosphoinositide hydrolysis and $cPLA_2$ with generation of arachidonic acid (AA) and arachidonic acid metabolites, in particular PGE_2.[3, 8] Our recent studies document that α_1-AR-agonists, such as phenylephrine, can increase levels of cAMP, a second messenger not typically associated with signalling by G_q-dependent pathways (Fig. 1). Use of β-AR antagonists or α_1-AR antagonists confirms that the increase in cAMP occurs in response to α_1-AR and not β-AR activation (data not shown). When we tested if the increase in cAMP might be secondary to ability of α_1-AR to generate cyclooxygenase-dependent products (e.g., PGE_2), we found that the COX inhibitor, indomethacin, blocked phenylephrine (or bradykinin) stimulated cAMP formation (tested in the presence of the diterpene, forskolin, in order to amplify response), thus implicating COX-derived products in α_1-AR-mediated cAMP formation (Fig. 1).

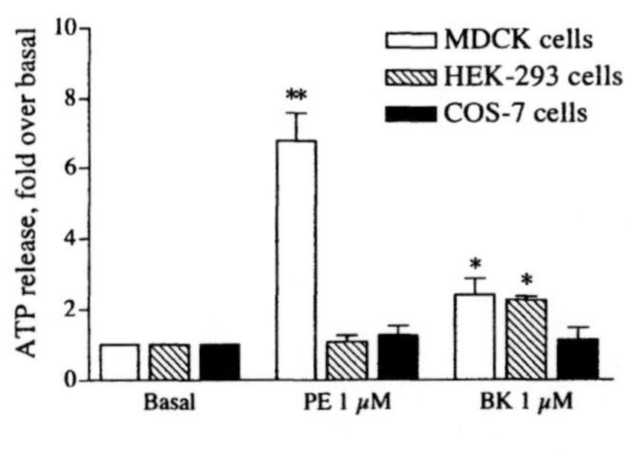

Figure 2. MDCK-D_1, HEK 293, or COS-7 cells were incubated in the absence ("basal") or presence of phenylephrine (1 μM) or bradykinin (1 μM) for 5 minutes and ATP released into extracellular media was assayed by a luciferin/luciferase method. Data are expressed as fold over basal levels (1.6 ± 1.1 nM for MDCK-D_1 cells, 54 ± 6 nM for HEK 293 cells, and 88 ± 22 nM for COS-7 cells. * = $p < 0.05$, ** = $p < 0.01$ compared to basal levels.

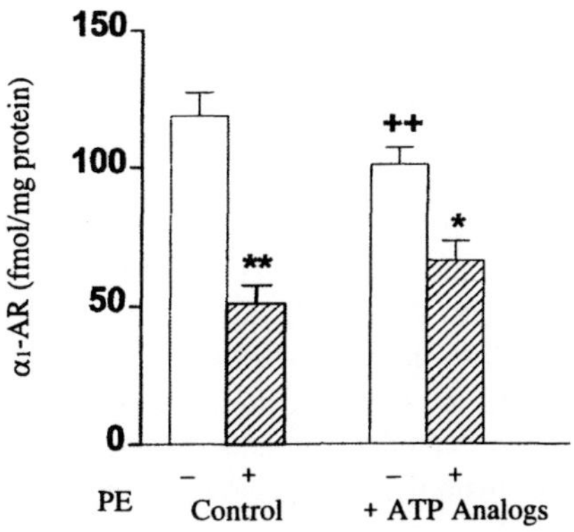

Figure 3. Down-regulation of MDCK-D_1 cells, α_1-AR by 24 hr incubation in the absence (open bars) or presence (hatched bars) of phenylephrine (PE, 100 μM), in the absence (Control) or presence (+ ATP Analogs) of 300 μM each ATPγs and 2-methyl-thio-ATP. * = $p < 0.05$, ++ or ** = $p < 0.01$.

In addition, we have observed another action of α_1-AR in MDCK-D_1 cells: stimulation of release of ATP (Fig 2). Physical perturbation of cells (e.g., changing the media) can increase release of intracellular ATP via poorly understood mechanisms.[9] We have found that treating MDCK-D_1, but not HEK 293 or COS-7, cells with phenylephrine increases release of ATP (Fig. 2); bradykinin also increased ATP release in MDCK-D_1 and HEK 293 cells. Given the ability of ATP to activate $cPLA_2$, increase formation of AA metabolites and cellular levels of cAMP, we wondered whether the released ATP might be responsible for the α_1-AR-stimulated increase in cAMP, but found this is not the case.[7]

Nevertheless, released ATP has another action that can modulate α_1-AR signalling in MDCK-D_1 cells: incubation of these cells with ATP analogs promote a heterologous down-regulation of α_1-AR expression, which was non-additive with phenylephrine-promoted down-regulation of α_1-AR (Fig. 3).

Studies of α_1-AR in cell-culture model systems, such as MDCK-D_1, provide the opportunity to explore activation and deactivation of signalling pathways in more defined settings than exist *in vivo*. By reducing the variables involved in assessment of receptor signalling, one can characterize cellular and biochemical mechanisms in a "cleaner" background. Our work has emphasized studies of endogenously expressed receptors, so as to eliminate the possibly altered location, coupling, signalling, and regulation that may occur with heterologously-expressing systems. GPCRs, such as α_1-AR, show considerable "promiscuity" in coupling to G proteins and downstream pathways, albeit some selectivity is observed for the α_1-AR subtypes.[1,2]

The current studies revealed that even in clonal cell systems, previously unsuspected variables can influence α_1-AR activation and deactivation. We found that MDCK-D_1 cells release ATP in response to phenylephrine, and that ATP analogs can down-regulate

α_1-AR expression. Whether ATP and α_1-agonists share similar mechanisms for receptor down-regulation will require further study. Other data indicate that released ATP interacts with one or more P2Y receptors on MDCK-D_1 cells and helps establish basal levels of signalling in the cells (see[10] for recent review).

Previous work in other cell systems has shown that α_1-AR agonists are able to increase cAMP levels (e.g.,[11, 12]). Theoretically an increase might occur through a variety of mechanisms, such as altered receptor coupling to G proteins, regulation of adenylyl cyclase or of phosphodiesterase isoforms. Our data implicate another key mechanism: generation of a "downstream," COX-derived product for which cells possess receptors that link to stimulation of adenylyl cyclase. Ability of α_1-AR to generate COX products that are capable of autocrine/paracrine regulation provides a potentially important mechanism for regulation of α_1-AR expressing cells.

REFERENCES

1. Graham RM, Perez DM, Hwa J, and Piascik MT, α_1-adrenergic receptor subtypes. Molecular structure, function, and signaling, *Circ Res* **78**, 737-749 (1996).
2. Zhong H, and Minneman KP, Alpha1-adrenoceptor subtypes, *Eur J Pharmacol* **375**, 261-276 (1999).
3. Insel PA, Balboa MA, Mochizuki N, Post SR, Urasawa K, and Xing M, Mechanisms for activation of multiple effectors by alpha 1-adrenergic receptors, *Adv Pharmacol* **42**, 451-453 (1998).
4. García-Sáinz JA, Vazquez-Prado J, and del Carmen Medina L, α_1-adrenoceptors: function and phosphorylation, *Eur J Pharmacol* **389**, 1-12 (2000).
5. Balboa MA, and Insel PA, Stimulation of phospholipase D via α_1-adrenergic receptors in Madin-Darby canine kidney cells is independent of PKCalpha and -epsilon activation, *Mol Pharmacol* **53**, 221-227 (1998).
6. Yang M, Büscher R, Taguchi K, Greubbel B, Insel PA, and Michel MC, Protein kinase C does not mediate phenylephrine-induced down-regulation of Madin-Darby canine kidney cell alpha-1B adrenoceptors, *J Pharmacol Exp Ther* **286**, 36-43 (1998).
7. Ostrom RS, Gregorian C, and Insel PA, Cellular release of and response to ATP as key determinants of the set-point of signal transduction pathways, *J Biol Chem* **275**, 11735-11739 (2000).
8. Meier KE, Sperling DM, and Insel PA, Agonist-mediated regulation of α_1- and β_2-adrenergic receptors in cloned MDCK cells, *Am J Physiol* **249**, C69-77 (1985).
9. Schwiebert EM, ABC transporter-facilitated ATP conductive transport, *Am J Physiol* **276**, C1-8 (1999).
10. Insel PA, Ostrom RS, Zambon AC, Hughes RJ, Balboa MA, Shehnaz D, Gregorian C, Torres B, Firestein BL, Xing M, and Post SR, P2Y receptors of MDCK cells: epithelial cell regulation by extracellular nucleotides, *Clin Exp Pharmacol Physiol* **28**, 351-354 (2001).
11. Lin RZ, Chen J, Hu ZW, and Hoffman BB, Phosphorylation of the cAMP response element-binding protein and activation of transcription by α_1-adrenergic receptors, *J Biol Chem* **273**, 30033-30038 (1998).
12. Ruan Y, Kan H, Parmentier JH, Fatima S, Allen LF, and Malik KU, α_{1A}-adrenergic receptor stimulation with phenylephrine promotes arachidonic acid release by activation of phospholipase D in rat-1 fibroblasts: inhibition by protein kinase A, *J Pharmacol Exp Ther* **284**, 576-585 (1998).

DIFFERENTIAL REGULATION OF MYOCARDIAL CELL APOPTOSIS BY α– AND β–ADRENERGIC PATHWAYS

Koji Hasegawa, Eri Iwai-Kanai, and Shigetake Sasayama*

1. INTRODUCTION

The apoptosis of cardiac myocytes may play a role in the development of heart failure.[1-3] Norepinephrine is one of factors activated in heart failure[4] and can induce myocardial cell apoptosis in culture.[5] However, it is unknown if α- and β-adrenergic pathways coordinately or differentially regulate apoptosis and if this apoptotic pathway utilizes common or cell type-specific apoptotic signals.

2. β-ADRENERGIC STIMULATION ACTIVATES AN APOPTOTIC PROGRAM IN CARDIAC MYOCYTES

For the investigation of the possible contribution of a β-adrenergic pathway to the development of myocardial cell apoptosis, neonatal rat cardiac myocytes were treated with a β-adrenergic agonist (isoproterenol, Iso) or saline as a control in the serum-free condition. Iso markedly increased the number of TUNEL-positive cells (% TUNEL-positive nuclei > 40%) compared with saline stimulation (< 10%). These positive cells may specifically indicate the presence of internucleosomal DNA fragmentation, since no positive cells were found when we omitted the terminal deoxytransferase treatment. The cells stimulated with Iso displayed small condensed nuclei, cell shrinkage, and nuclear fragmentation, consistent with the morphologic features of apoptosis.[6-8] These cells were derived from cardiac myocytes, as evidenced by the positive immunostaining for HHF35, which reacts with the α-actin of cardiac myocytes[9] but not that of fibroblasts. DNA fragmentation was also confirmed by ladder formation in agarose gels. The Iso-induced apoptosis was almost completely blocked by a β-adrenergic receptor antagonist, propranolol (10^{-4}M). We found that 10^{-4}M propranolol did not affect the 8-Br-cAMP-stimulated increase of TUNEL-positive cells. These findings provide evidence that a β-

* Department of Cardiovascular Medicine, Graduate School of Medicine, Kyoto University, Kyoto, Japan

Catecholamine Research, Edited by Nagatsu *et al.*
Kluwer Academic/Plenum Publishers, 2002

adrenergic receptor-dependent pathway activates an apoptotic program in cardiac myocytes.

3. A cAMP/PKA PATHWAY MEDIATES ISO-INDUCED APOPTOSIS

Stimulation of the β-adrenergic receptor activates adenylate cyclase, which increases intracellular cAMP and activates cAMP-dependent protein kinase A (PKA). To determine whether the activation of PKA is required for Iso-induced apoptosis, we examined the effect of KT5720, a highly selective inhibitor of PKA. The Iso-induced apoptosis was almost completely inhibited by the protein kinase A (PKA)-specific inhibitor KT5720. In accord with these results, the administration of a cell-permeable cAMP analogue, 8-Br-cAMP (30mM) also induced apoptosis in cardiac myocytes to an extent similar to that shown by Iso. Taken together, these results demonstrate that cAMP-PKA is necessary and sufficient for Iso-induced apoptosis.

4. α1-ADRENERGIC PATHWAY INHIBITS cAMP-INDUCED CARDIAC APOPTOSICS

To determine the effects of α1-adrenergic stimulation on myocardial cell apoptosis, neonatal rat cardiac myocytes were treated with an α1-adrenergic agonist (phenylephrine, PE). In contrast to Iso, stimulation with PE did not increase the number of TUNEL-positive cells, even at a high concentration (10^{-4}M). PE inhibited the 8-Br-cAMP-induced myocardial cell apoptosis. These results provide evidence that PE has an antiapoptotic effect in cultured cardiac myocytes. An α1-adrenergic receptor antagonist prazocin negated the PE-mediated inhibition of apoptosis, which indicated that the inhibition of myocardial apoptosis by PE is mediated through an α1-adrenergic receptor pathway.

α1-adrenergic stimulation has been shown to activate a MAP kinase cascade in cardiac myocytes. To determine whether the activation of MAP kinase is required for the PE inhibition of apoptosis in cardiac myocytes, we utilized PD098059, a specific MEK inhibitor that selectively inhibits MEK-1 activity. 10^{-5} M of PD098059 negated the inhibitory effects of PE on myocardial cell apoptosis, compared with saline stimulation. This result suggested that PD098059 might block the downstream signaling pathway by which PE prevents apoptosis. Taken together, these results provide evidence that PE has an antiapoptotic effect in cultured cardiac myocytes and that a MAP kinase pathway appears to be involved in this process.

5. cAMP-MEDIATED APOPTOTIC PATHWAY IS NOT FUNCTIONAL IN PC12 CELLS

In cardiac myocytes, serum deprivation alone barely induced internucleosomal cleavage of genomic DNA compared with the condition of 10% fetal bovine serum. Cell-permeable 8-Br-cAMP induced marked apoptosis in the serum-free condition. To determine whether cAMP-induced apoptosis is a conserved phenomenon among different cell types or a phenomenon unique to cardiac myocytes, we examined the effect of 8-Br-cAMP on apoptosis in different cell lines. In PC12 cells, rat pheochromocytoma cells, 12 hours after serum deprivation, marked internucleosomal cleavage of genomic DNA was observed. Notably, the administration of 30 mM of 8-Br-cAMP completely inhibited the serum deprivation-induced apoptosis in PC12 cells. Antiapoptotic effects of 8-Br-cAMP were also observed in other mitotic cell lines such as mouse fibroblast cells (NIH3T3 cells). These findings indicate that a cAMP pathway is involved in promoting cell survival in these cell lines. Thus, 8-Br-cAMP displayed opposite effects on apoptosis in cardiac myocytes and PC12 cells.

6. CONCLUSION AND DISCUSSION

The present study[10] demonstrated that a β–adrenergic agonist induced apoptosis in cardiac myocytes and that this apoptosis was mediated through a cAMP/PKA pathway. In contrast, an α1-adrenergic agonist antagonized cAMP-induced apoptosis. Several lines of evidence suggest that the activation of the sympathetic nervous system exerts a direct deleterious effect on the heart that is independent of the hemodynamic actions of these endogenous mechanisms. Therapeutic interventions by β-adrenergic receptor blockers favorably alter the natural history of heart failure, and such benefits cannot be explained by the effect of these agents on cardiac contractility and ejection fraction. Our findings suggest a mechanism for the beneficial effects of β-adrenergic receptor blockers in patients with heart failure. Interestingly, cAMP had opposite effects on apoptosis in cardiac myocytes and PC12 cells, suggesting that cAMP involves cardiac-specific signal transduction mechanisms. In addition, the elucidation of precise signaling pathways leading to myocardial cell apoptosis may contribute to novel strategies for heart failure therapy in humans.

REFERENCES

1. Y. Liu, E. Cigola, W. Cheng, J. Kajstura, G. Olivetti, TH. Hintze, P. Anversa, Myocyte nuclear mitotic division and programmed myocyte cell death characterize the cardiac myopathy induced by rapid ventricular pacing in dogs, *Lab. Invest.* **73**, 771-787(1995).
2. VG. Sharov, HN. Sabbah, H. Shimoyama, AV. Goussev, M. Lesch, S. Goldstein, Evidence of cardiocyte apoptosis in myocardium of dogs with chronic heart failure, *Am. J. Pathol.* **148**, 141-149 (1996).
3. E. Teiger, VD. Than, L. Richard, C. Wisnewsky, BS. Tea, L. Gadoury, J. Tremblay, K. Schwartz, P. Hamet, Apoptosis in pressure overload-induced heart hypertrophy in the rat, *J. Clin. Invest.* **97**,2891-2897 (1996).
4. GS. Francis, JN. Chon, G. Johnson, TS. Rector, S. Goldman, A. Simon, Plasma norepinephrine, plasma renin activity, and congestive heart failure. Relations to survival and the effects of therapy in V-HeFT II. The V-HeFT VA Cooperative Studies Group, *Circulation.* **87** (6 Suppl),VI40- VI 48 (1993).
5. C. Communal, K. Singh, DR. Pimentel, WS. Colucci, Norepinephrine stimulates apoptosis in adult rat ventricular myocytes by activation of the β-adrenergic pathway. *Circulation*. 98, 1329-1334 (1998).
6. MD. Jacobson, M. Weil, MC. Raff, Programmed cell death in animal development, *Cell.* **88**, 347-354 (1997).
7. S. Nagata, Apoptosis by death factor, *Cell.* **88**, 355-365 (1997).
8. AH. Wyllie, Glucocorticoid-induced thymocyte apoptosis is associated with endogenous endonuclease activation, *Nature.* **284**, 555-556 (1980).
9. K. Hasegawa, H. Fujiwara, H. Itoh, K. Nakao, T. Fujiwara, H. Imura, C. Kawai, Light and electron microscopic localization of brain natriuretic peptide in relation to atrial natriuretic peptide in porcine atrium-Immunohistochemical study using specific monoclonal antibodies, *Circulation.* **84**, 1203-1209 (1991).

10. E. Iwai-Kanai, K. Hasegawa, M. Araki, T. Kakita, T. Morimoto, S. Sasayama, α- and β-adrenergic pathways differentially regulate cell type-specific apoptosis in rat cardiac myocytes, *Circulation.* **100**, 305-311 (1999).

BEHAVIORAL AND HISTOCHEMICAL STUDIES IN REPEATED CEREBRAL ISCHEMIA IN RATS

Katsunori Iwasaki*, Kenichi Mishima, and Michihiro Fujiwara

1. INTRODUCTION

In the behavioral study, we have reported that the 10 min transient cerebral ischemia of the rat caused the deficit of spatial cognition in 8-arm radial maze task 24 hours after re-perfusion.[1] Using this model, we have developed new drugs, so called nootropic drugs. But, unfortunately, many of them were not effective in the clinical use. We considered the reasons why this model did not work as an animal model for cerebrovascular dementia and resulted that the 10 min cerebral ischemia was too weak to produce hippocampal cell death. Recently, Kato et al.[2] suggested that the hippocampal cell death was suppressed when the brief time ischemia was repeated with 6 hours or more intervals and was enhanced, on the contrary, by less than 6 hours intervals. In the present study, we applied the repeated 10 min ischemia with 1 hour interval for producing more severe ischemic insult and evaluated whether this model was more useful as a new animal model for cerebrovascular dementia.

2. MATERIALS AND METHODS

2.1 Animals

Male Wistar rats (250g - 300g, Kyu-do Co. Saga, Japan) were used. The animals were kept on a constant light-dark cycle (light 07:00-19:00) with a restricted diet (CE-2, Clea Japan) in an air-conditioned room.

2.2 8-Arm radial maze

For the behavioral analysis, an image motion analyzer, AXIS-30 (Neuroscience Co., Tokyo, Japan), was used to quantify the task performance of rats in the 8-arm radial maze. The behavioral observation was discontinued after 10 min even if the animal did not finish

* Katsunori Iwasaki, Department of Physiology & Pharmacology, Faculty of Pharmaceutical Sciences, Fukuoka University, 8-19-1 Nanakuma Jonan-ku, Fukuoka 814-0180,Japan

the task. The data were analyzed by a two way (with repeated measures) analysis of variance (2ANOVA) followed by Tukey's test.

2.3 Surgery for cerebral ischemia

A four-vessel occlusion was performed according to the method described by Pulsinelli and Brierley.[3] In brief, a rat was anesthetized and bilateral vertebral arteries were electrocauterized with a bipolar coagulator. Then the bilateral common carotid arteries were exposed and a hydraulic pressure vascular occluder (Technical Supply Co. Osaka, Japan) was fitted to each exposed arteries. On the next day, the common carotid arteries were compressed with the hydraulic pressure occluders for interrupted the cerebral blood flow. Any rats which failed to demonstrate a loss of their righting reflex were excluded from the following experiment.

2.4 Histochemistry

Following observation of the behavior on 7 days after ischemia, the animals were decapitated and the brains were removed. The brains were immersed in the fixative and then embedded in paraffin. The 5 μm of brain slices were cut and stained with MAP-2 antibody. Apoptosis was detected using the TUNEL method (TACSII, Trevigen, USA). The neuronal density and the number of apoptotic cells were analyzed by the Kruskal-Wallis test followed by the Mann-Whitney U-test.

3. RESULTS

3.1 Memory disturbances and apoptosis after repeated ischemia

Three minute cerebral ischemia shows no disruption of spatial cognition in 8-arm radial maze task even if the ischemia was repeated. Twice of 5 min repeated cerebral ischemia with 1 hour interval induced the disruption of spatial cognition 7 days after final re-perfusion, but this disruption recovered to normal at the next day. Single 10 min cerebral ischemia did not show the disruption of spatial, but twice of 10 min cerebral ischemia with 1 hour interval induced significant disruption of spatial cognition and this disruption continued during next 3 days. Three times of 10 min repeated cerebral ischemia also induced the disruption of spatial cognition, but these rats showed severe motor paralysis and couldn't use for behavioral experiments (Fig.1 left).
The number of TUNEL positive cells, indicating apoptotic cells in the hippocampus CA1 areas, did not change by 10min single cerebral ischemia, while significantly increased by twice of 10 min repeated cerebral ischemia (Fig.1, right).

3.2 Effects of several drugs on the disruption of spatial cognition and hippocampal apoptosis after repeated cerebral ischemia

Repeated ischemia (10min x 2) induced disruption of spatial cognition and it was significantly improved by YM-90K , a selective AMPA receptor antagonist, at the

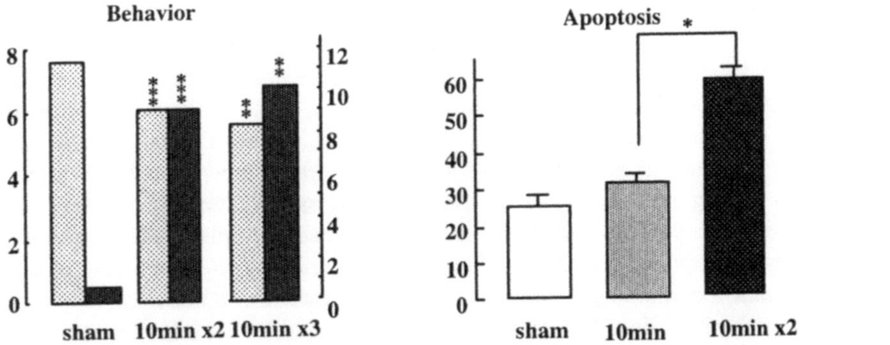

vertical line: left : no. of correct choices
right: no. of errors

vertical line: no. of TUNEL-positive cells/mm^2

*p<0.05, **p<0.01, ***p<0.001 vs sham

Figure 1 The disruption of spatial cognition induced by repeated cerebral ischemia 7days after the final re- perfusion (left) and incidence of apoptosis in the hippocampal CA1 layers (right).

dose of 15 mg/kg,i.p. when it was injected before or 6 hours after ischemia episode. YM-90K also significantly suppressed the apoptosis in the hippocampal CA1 layers. But MK-801, a NMDA receptor antagonist, at the dose of 0.1 mg/kg,i.p. did not improve both the disruption of spatial cognition and hippocampal apoptosis (Table.1).

Apoptosis relating agents, 3-aminobenzamide of PARP inhibitor(10mg/kg,i.v.), z-Asp-Ch_2-DCB of ICE inhibitor (100μg/brain,i.c.v.), Ac-DEVD-CHO of caspase-3 inhibitor (2.1μg/brain,i.c.v.), nilvadipine, a newly developed L-type Ca channel blocker, significantly improved both the disruption of spatial cognition and incidence of hippocampal apoptosis by repeated cerebral ischemia (Table.1).

Tablel. Effects of several drugs on the repeated cerebral ischemia-induced disruption of spatial cognition and hippocampal apoptosis in rats.

Drugs	Spatial Memory	Apoptosis
MK-801 (NMDA-R antagonist)	—	—
YM-90K (AMPA-R antagonist)	+++	+++
3-aminobenzamide (PARP inhibitor)	+	++
z-Asp-CH_2-DCB (ICE inhibitor)	+	++
Ac-DEVD-CHO (Caspase-3 inhibitor)	+	++
Nilvadipine (Ca channel blocker)	+++	++

: no effect; +: slight suppression; ++: moderate suppression; +++: strong suppression

4. DISCUSSION AND COMMENTS

During last 10 years, we used the Pulsinelli's brief cerebral ischemia[3] and evaluated many new drugs for cerebrovascular dementia. Therefore, many of them were assessed to be not effective in the clinical use. In the present study, we considered the reason why this model did not work as an animal model and developed a new model which showed relatively long lasting impairment of spatial cognition and accompanying hippocampal cell death. From the present results, table2 summarizes the relationship between the incidence of the disruption of spatial cognition and the hippocampal cell loss in the rat. This suggests that the spatial cognition is maintained until the number of hippocampal cells falls below the 25% line. YM-90K or calcium channel blocker inhibited the decrease of cell by inhibiting the calcium dependent pathway for apoptosis via AMPA receptor.[4] Caspase-3 inhibitor also inhibited both memory deficit and apoptosis. This was thought to result from the blockade of final signal to apoptosis processing. From these results, comparing with the single cerebral ischemia, repeated cerebral ischemia is a much more useful animal model for cerebrovascular dementia because of its behavioral, pathological and pharmacological similarity to human cerebrovascular dementia.

Table2. Summary of the relationship between spatial cognition and apoptosis

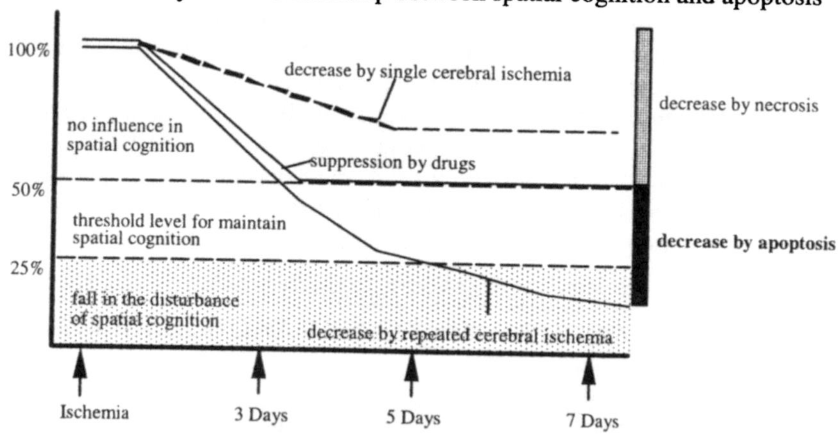

REFERENCES

1. K.Iwasaki, Y.Kitamura, Y.Ohgami, K.Mishima, and M.Fujiwara, The disruption of spatial cognition and changes in brain amino acid,monoamine and acetylcholine in rats with transient cerebral ischemia, *rain Res.* **709**, 163-172 (1996).
2. H.Kato, Y.Liu, T.Araki and K. Kogure, Temporal profile of the effects of effects of pretreatment with brief cerebral ischemia on the neuronal damage following secondary ischemia insult in the gerbil : Cumulative damage and protective effects, *Brain Res*, **553**, 238 – 242 (1991).
3. W.A.Pulsinelli and J.B.Brierley, A new model of bilateral hemispheric ischemia in the unanesthetized rat, *Stroke*, **10** , 267 – 272 (1979).
4. B.Nellgard and T.Wieloch, Postischemic blockade of AMPA but not NMDA receptors mitigates neuronal damage in the rat brain following transient severe cerebral ischemia. *J.Cereb.Blood Flow Metab.*, **12**, 2-11(1992).

STRUCTURE AND FUNCTION OF NEUROMELANIN

Shosuke Ito, Kazumasa Wakamatsu, and Luigi Zecca*

1. INTRODUCTION

Why is the structure of neuromelanin important? It has been known for many years that neuromelanin is present in the dopaminergic neurons of the substantia nigra and may be associated with the pathogenesis of nigral cell death in the Parkinson's disease. Recently, it was shown that the vulnerability of the dopaminergic neurons in Parkinson's disease is related to their high neuromelanin content in the substantia nigra. Dopamine is suggested to be a major precursor of neuromelanin. However, direct evidence to support this assumption is scarce. This lack of knowledge in the structure of neuromelanin hampered studies on its biological function and possible role in the pathogenesis of Parkinson's disease.

Two types of melanin pigments are present in the skin, hair, and eyes.[1] They are black to brown eumelanin and yellow to reddish pheomelanin. Both of these melanin pigments are derived from dopaquinone formed by tyrosinase oxidation of tyrosine. In the absence of cysteine, dopaquinone cyclizes to form dopachrome. Rearrangement of dopachrome gives two dihydroxyindoles which are then oxidized to form eumelanin. On the other hand, in the presence of cysteine, dopaquinone undergoes a rapid addition reaction with cysteine to form cysteinyldopas. Further oxidation of cysteinyldopas leads to the production of pheomelanin through benzothiazine intermediates. Therefore, the relative ratio of eumelanin and pheomelanin in mixed melanin can be determined by the availability of cysteine.

For cutaneous melanin, most is known about the site of synthesis, precursor, oxidant, biological model, synthetic model, and specific degradation products. On the other hand, much less is known about those properties of neuromelanin.[2] For example, we are not sure about the precursors of neuromelanin. This is a serious drawback when one wishes to study the biological significance of neuromelanin in substantia nigra of normal and Parkinson's brains.

Dopamine was originally suggested to be a precursor of neuromelanin. However in

* Shosuke Ito and Kazumasa Wakamatsu, Fujita Health University School of Health Sciences, Toyoake, Aichi 470-1192, Japan. Luigi Zecca, Institute of Advanced Biomedical Technologies – CNR, 20090 Segrate, Italy.

1985, 5-S-cysteinyldopamine was detected in the human substantia nigra by Rosengren et al[3], suggesting its possible involvement in the biosynthesis of neuromelanin. Subsequently in 1991, our study indicated that cysteinyldopamine may not be incorporated into neuromelanin,[4] while a Swedish group suggested that neuromelanin consists of about equal amounts of cysteinyldopamine-derived and dopamine-derived units.[5,6] This discrepancy remained unclarified.

Oxidative stress in substantia nigra results in the production of dopaminequinone. In analogy to cutaneous melanin, dopaminequinone cyclizes to form 5,6-dihydroxyindole (DHI), or it reacts with cysteine to form cysteinyldopamines. Then cysteinyldopamines are oxidized to form benzothiazine and benzothiazole intermediates. These intermediates are oxidatively polymerized to form neuromelanin. Therefore, to know the structure of neuromelanin, it is very important to determine the relative ratio of these precursors in neuromelanin.

Figure 1 shows degradation products of dopamine-melanin and cysteinyldopamine-melanin. Alkaline peroxide oxidation of dopamine-melanin produces pyrrole-2,3-dicarboxylic acid (PDCA) and pyrrole-2,3,5-tricarboxylic acid (PTCA).[7] PDCA is a major degradation product from dopamine-melanin. Thus, PDCA may serve as a specific indicator of dopamine-derived melanin. HI hydrolysis of cysteinyldopamine-melanin produces aminohydroxyphenylethylamine (AHPEA) isomers. The 4-amino isomer is a major degradation product, and thus may serve as a specific indicator of cysteinyldopamine-derived melanin.[4,7]

2. CHEMICAL CHARACTERIZATION OF NEUROMELANIN

The purpose of this study was to chemically characterize neuromelanin isolated from human substantia nigra, in particular, to determine how much cysteine is incorporated into neuromelanin.

Neuromelanin was isolated from pooled samples of human substantia nigra by homogenization, washing with a buffer, followed by a SDS-buffer, and dialysis.[8] Synthetic melanins were prepared from mixtures of dopamine and cysteine in various ratios by tyrosinase oxidation.[4] We performed alkaline peroxide oxidation, HI hydrolysis, and Soluene-350 solubilization.[7]

Figure 1 summarizes what kind of products can be expected from chemical degradation. Alkaline peroxide oxidation of cysteinyldopamine-derived melanin would give, in addition to PDCA and PTCA, thiazole-4,5-dicarboxylic acid (TDCA) and thiazole-2,4,5-tricarboxylic acid (TTCA). HI hydrolysis would give, in addition to AHPEA isomers, benzothiazole-amines.

HPLC analysis of neuromelanin and cysteinyldopamine-melanin after alkaline peroxide oxidation gave TDCA, TTCA, PDCA, and PTCA. Interestingly, PDCA was a major product from neuromelanin, while TTCA was a major product from cysteinyldopamine-melanin. Therefore, we carried out more detailed analyses using melanins prepared from various ratios of dopamine and cysteine. We plotted the yields of PDCA and TTCA and their ratio against the sulfur to nitrogen ratio (Figure 2). The yield of PDCA was fairly constant, while the yield of TTCA was proportional to the sulfur to nitrogen ratio. As a result, the ratio of TTCA to PDCA increased almost linearly as the sulfur to nitrogen ratio increased. The isolated neuromelanin gave a TTCA to PDCA ratio of 0.35, indicating that it has a sulfur to nitrogen ratio of 0.09. This

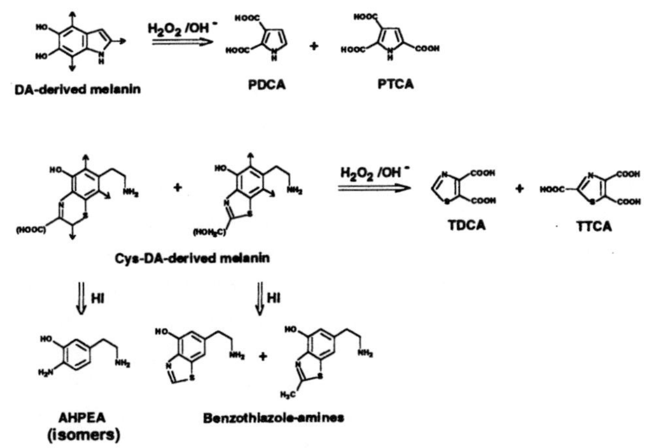

Figure 1. Expected products from chemical degradation of neuromelanin

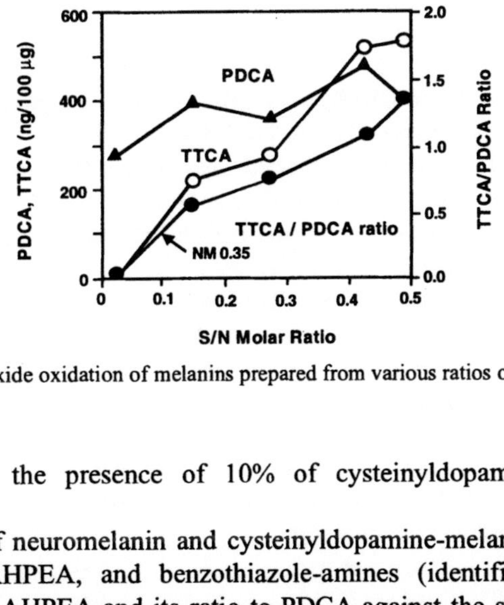

Figure 2. Alkaline peroxide oxidation of melanins prepared from various ratios of dopamine and cysteine.

ratio corresponds to the presence of 10% of cysteinyldopamine-derived units in neuromelanin.

HPLC analysis of neuromelanin and cysteinyldopamine-melanin after HI hydrolysis gave 3-AHPEA, 4-AHPEA, and benzothiazole-amines (identification tentative). We plotted the yield of 4-AHPEA and its ratio to PDCA against the sulfur to nitrogen ratio. The ratio of 4-AHPEA to PDCA increased exponentially as the sulfur to nitrogen ratio increased. Neuromelanin gave a ratio of 3.4, indicating that it has a sulfur to nitrogen ratio of 0.15. This ratio corresponds to the presence of 18% of cysteinyldopamine-derived units.

Absorbance of neuromelanin after Soluene-350 solubilization showed that the isolateneuromelanin contained only 14% of melanin moiety. Our previous study indicates that the same preparation of neuromelanin contains 17% of proteins, 21% of lipids, and 4% of inorganic salts.[8] When combined, these constituents make up about 56% of the total weight. What components constitute the remaining parts of the isolated neuromelanin is not known at present.

3. CONCLUSIONS

This study shows that the TTCA to PDCA ratio and the 4-AHPEA to PDCA ratio are useful in characterizing neuromelanin and that melanin moiety of neuromelanin consists mostly of dopamine-derived units with 10-20% incorporation of cysteinyldopamine-derived units. It is thus proposed that dopamine-melanin, rather than cysteinyldopamine-melanin, should be used as a synthetic model of neuromelanin.

Finally, content of melanin moiety in the isolated neuromelanin was found to be 14% in this study. Content of melanin in substantia nigra is thus calculated to be about 180 µg/g wet weight on the basis of the content of isolated neuromelanin (1.3 mg/g).[8] On the other hand, content of dopamine is only 0.4 µg/g wet weight,[9] which is less than 1% of neuromelanin. Furthermore, as compared to dopamine-melanin, dopamine was found to be only 5-fold more toxic to mice cerebellar granular cells and PC12 cells.[10] We therefore suggest that neuromelanin, rather than dopamine itself, plays a major role in the degeneration of nigral cells.

REFERENCES

1. S. Ito, K. Wakamatsu, and H. Ozeki, Chemical analysis of melanins and its application to the study of the regulation of melanogenesis. *Pigment Cell Res.* **Suppl. 8**, 103-109 (2000).
2. M. D'Ischia and G. Prota, Biosynthesis, strucutre, and function of neuromelanin and its relation to Parkinson's diesease: a critical update. *Pigment Cell Res.* **10**, 370-376 (1997).
3. E. Rosengren, E. Linder-Eliasson, and A. Carlsson, Detection of 5-S-cysteinyldopamine in human brain. *J Neural Transm.* **63**, 247-253, 1985.
4. K. Wakamatsu, S. Ito, and T. Nagatsu, Cysteinyldopamine is not incorporated into neuromelanin. *Neurosci. Lett.* **131**, 57-60 (1991).
5. R, Carstam, C. Brinck, A. Hindemith-Augustsson, H. Rorsman, and E. Rosengren, The neuromelanin of the human substantia nigra. *Biochim. Biophys. Acta* **1097**, 152-160 (1991).
6. G. Odh, R. Carstam, J. Paulson, A. Wittbjer, E. Rosengren, and H. Rorsman, Neuromelanin of the human substantia nigra: a mixed-type melanin. *J. Neurochem.* **62**, 2030-2036 (1994).
7. S. Ito and K. Wakamatsu, Chemical degradation of melanins: application to identification of dopamine-melanin. *Pigment Cell Res.* **11**, 120-126 (1998).
8. L. Zecca, P. Costi, C. Mecacci, S. Ito, M. Terreni, and S. Sonnino, Interaction of human substantia nigra neuromelanin with lipids and peptides. *J. Neurochem.* **74**, 1758-1765 (2000).
9. B. Fornstedt, A. Brun, E. Rosengren, and A. Carlsson, The apparent autoxidation rate of catechols in dopamine-rich regions of human brains increases with the degree of depigmentation of substantia nigra. *J Neural Transm. (P-D Sect.)* **1**, 279-295 (1989).
10. D. Offen, S. Gorodin, E. Melamed, J. Hanania, and Z. Malik, Dopamine-melanin is actively phagocytized by PC12 and cerebellar granular cells: possible implications for the etiology of Parkinson's disease. *Neurosci. Lett.* **260**, 101-104 (1999).

BIOSYNTHESIS OF NEUROMELANIN AND MELANIN: THE POTENTIAL INVOLVEMENT OF MACROPHAGE INHIBITORY FACTOR AND DOPACHROME TAUTOMERASE AS RESCUE ENZYMES

Jun Matsunaga, Patrick A. Riley, Francisco Solano, and Vincent J. Hearing*

1. INTRODUCTION

This chapter summarizes our recent studies on the regulation of melanin synthesis, a chemical pathway which bears a striking resemblance to that proposed to be involved in neuromelanin production from Dopamine (DN).[1] Two key enzymes, DOPAchrome tautomerase (DCT) and macrophage migration inhibitory factor (MIF), have analogous catalytic actions in those biosynthetic pathways which may be linked to their roles as survival enzymes in melanocytes and in catecholaminergic neurons, respectively.

DCT was identified as an enzyme about 20 years ago.[2] Melanin production depends on the catalytic function of tyrosinase (Tyr), the critical and rate-limiting step in the pathway (Scheme 1). Without DCT, the DOPAchrome (DC) generated from DOPA-quinone (DQ) will spontaneously decarboxylate, giving rise to 5,6-dihydroxyindole (DHI), which oxidizes and polymerizes to generate dark, insoluble and high molecular weight melanins. But that default pathway can be diverted by DCT, which tautomerizes DC to DHI-2-carboxylic acid (DHICA), a carboxylated intermediate that is less toxic than DHI.[3] Mice mutant at the locus encoding DCT (termed *slaty*) undergo premature death of follicular melanocytes; although the precise mechanism(s) for that are unknown, it is an interesting speculation that DCT prevents cytotoxicity. DCT is also the melanosomal protein expressed earliest during embryological development,[4] suggesting it may have an important function.

MIF on the other hand, has only recently been shown to function enzymatically.[5] It was identified originally more than 4 decades ago as a lymphokine which inhibits

* Jun Matsunaga and Vincent J Hearing, Laboratory of Cell Biology, National Cancer Institute, National Institutes of Health, Bethesda, MD, 20892 USA; Patrick A. Riley, Gray Laboratory Cancer Research Trust, Mount Vernon Hospital, Northwood, Middlesex HA6 2JR United Kingdom; Francisco Solano, Department of Biochemistry, Molecular Biology B and Immunology, University of Murcia School of Medicine, Campus de Espinardo, E-30100 Murcia Spain

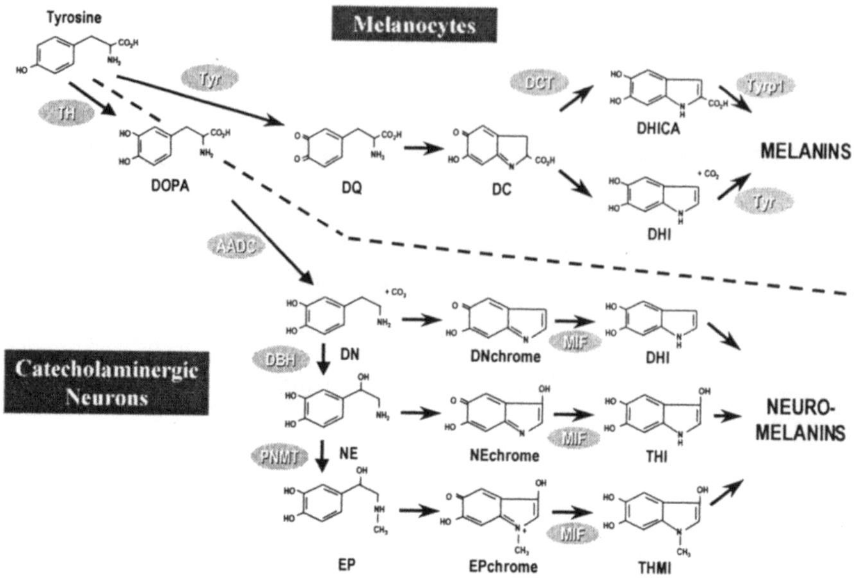

Scheme 1. Similarity of Biosynthetic Pathways in Melanocytes and Catecholaminergic Neurons. Enzymes are abbreviated as noted in the text, plus Tyrp1 = DHICA oxidase; TH = tyrosine hydroxylase; AADC = aromatic amino acid decarboxylase; DBH = dopamine β-hydroxylase; PNMT = phenylethanolamine-N-methyltransferase

monocyte migration (hence the name).[6] The gene encoding MIF was cloned and mapped to chromosome 22q11.2 in humans; it encodes a 115 amino acid, 12.5 kDa protein.[7] Expression of MIF was found to be ubiquitous in newborn mouse tissues and MIF levels generally decrease with age in all tissues, except the eye and brain.[8] The ability of MIF to use only the D-stereoisomer of DC as a substrate was reported a few years ago and was an unexpected finding in several regards, but most intriguingly since (1) it utilized the D-form of DC that is not a physiologically relevant isoform, and (2) DC is found only in melanocytes while the distribution of MIF is virtually ubiquitous. We felt that the continued high expression of MIF in the brain, particularly in the substantia nigra, provided an important clue as to its possible function on a physiological substrate in the catecholamine pathway in that tissue.

2. PHYSIOLOGICAL ROLES FOR MIF AND DCT

There are some obvious differences in the regulation of melanin production in melanocytes and of neuromelanin production in the substantia nigra. Melanin synthesis by melanocytes in the skin is relatively quick. Melanins produced by melanocytes are exported to neighboring keratinocytes and are transported to the surface where they are eventually lost by desquamation or by hair growth. Neuromelanin on the other hand is produced extremely slowly and accumulates in the substantia nigra only over the course of many years. Nevertheless, the pathways leading to melanin production and to neuromelanin production are strikingly similar (Scheme 1). Most relevant to the function of MIF, the conversion of DOPaminechrome (DNchrome) to DHI is very similar to the conversion of L- (or D-) DC to DHI. Even more interestingly, since DNchrome has no carboxylic acid, it is sterically inactive and we hypothesized that it might actually serve as a physiological substrate for MIF in the brain. A series of biochemical assays showed that this was correct.[9, 10] MIF was found to be quite active towards DNchrome, and was also able to use the other 'chrome' derivatives in the catecholamine pathway (epinephrinechrome (EPchrome) and norepinephrinechrome (NEchrome)) as substrates, albeit with diminished vigor. As noted above, DN is thought to represent the major precursor of neuromelanin[1, 11,12].

3. CYTOTOXIC IMPLICATIONS

Many reactions involved in the synthesis of melanins give rise to toxic intermediates; these include quinones, semiquinones, hydroxyl radicals and hydrogen peroxide, among others.[13, 14] Early studies had shown that DHI was toxic to melanocytes[2] and that DHICA was significantly less toxic,[3] which suggested that DCT might play a role in melanocytes to ensure their survival. However, it was of great interest to examine the relative toxicities of other intermediates in both reaction pathways. To this end, we used colony formation assays to assess the effects of various intermediates on B16 mouse melanoma cells and Neuro2A neuroblastoma cells *in vitro.*[15] L- and D-DOPA were only slightly toxic to both types of cells, with an LD_{50} of >1 mM. Dopamine was slightly more toxic (with an LD_{50} of ~1 mM). As previously reported,[3] DHI was significantly more toxic than those precursors, with an LD_{50} of ~0.3 mM, DHICA having only slight toxicity at that concentration. We were quite surprised to find that DC (both the L-form and the D-form) were highly toxic to melanoma cells and to neuroblastoma cells, each with an LD_{50} of 10 μM; DNchrome was only slightly less toxic, with an LD_{50} of ~25 μM. The quinone precursors of the 'chromes' are expected to be the most toxic but are not stable enough to be assessed using this approach. Thus, the importance of DCT and MIF in the respective pathways may not be so much to determine the eventual product of the reaction, but to quickly reduce concentrations of the toxic substrate. In particular this might be relevant for MIF function in neurons, since DN is known to trigger apoptosis of cells when excess DN accumulates.[16] Neuromelanin may simply serve as a relatively inert product that can accumulate in catecholaminergic neurons; the relative merits of neuromelanin as an antioxidant, along with its other favorable biochemical properties, may be an unexpected bonus to those cells.[17-19] Since MIF expression normally remains high with age in the human brain, an interesting consideration is whether the decreased neuromelanin seen in the brains of Parkinson's patients results from an abnormal decrease of MIF function in those patients. Such a decrease would result in increased levels of DN and its metabolites, which may signal those cells to undergo apoptosis.

4. CONCLUSIONS

Many melanogenic intermediates have significant toxicity to melanocytes and to catecholaminergic neurons. DCT promotes the formation of the less toxic DHICA rather than DHI, but more importantly, metabolizes DC. MIF similarly accelerates the conversion of catecholaminechromes (e.g. DNchrome) to their less toxic dihydroxyindole derivatives. MIF may therefore contribute to the biosynthesis of neuromelanin in a manner similar to that of DCT for melanin. MIF and DCT may play important protective roles in the brain and in the skin to maintain the viability of the melanin-producing cells. Future study will need to address whether the expression or function of MIF plays any role in the decrease in neuromelanin synthesis and in the neurodegeneration that is associated with Parkinson's disease.

REFERENCES

1. W.S. Enochs, M.J. Nilges, and H.M. Swartz. Purified human neuromelanin, synthetic dopamine melanin as a potential model pigment, and the normal human substantia nigra: characterization by electron paramagnetic resonance spectroscopy, *J. Neurochem.* **61**, 68-79 (1993).
2. J.M. Pawelek, A.M. Körner, A. Bergstrom, and J. Bolognia. New regulators of melanin biosynthesis and the autodestruction of melanoma cells, *Nature* **286**, 617-619 (1980).
3. K. Urabe, P. Aroca, K. Tsukamoto, D. Mascagna, A. Palumbo, G. Prota, and V.J. Hearing. The inherent cytotoxicity of melanogenic intermediates: a revision, *Biochim. Biophys. Acta* **1221**, 272-278 (1994).
4. K.P. Steel, D.R. Davidson, and I.J. Jackson. TRP2/DT, a new early melanoblast marker, shows that steel growth factor (c-kit ligand) is a survival factor, *Development* **115**, 1111-1119 (1992).
5. E. Rosengren, R. Bucala, P. Aman, L. Jacobsson, G. Odh, C.N. Metz, and H. Rorsman. The immunoregulatory mediator macrophage migration inhibitory factor (MIF) catalyzes a tautomerization reaction, *Mol. Med.* **2**, 143-149 (1996).
6. J.R. David. Delayed hypersensitivity in vitro: its mediation by cell-free substances formed by lymphoic cell-antigen interaction, *Proc. Natl. Acad. Sci. USA* **56**, 72-77 (1966).
7. W.Y. Weiser, P.A. Temple, J.S. Witek-Giannotti, H.G. Remold, S.C. Clark, and J.R. David. Molecular cloning of a cDNA encoding a human macrophage migration inhibitory factor, *Proc. Natl. Acad. Sci. USA* **86**, 7522-7526 (1989).
8. N. Esumi, M. Budarf, L. Ciccarelli, B. Sellinger, C.A. Kozak, and G.J. Wistow. Conserved gene structure and genomic linkage for D-dopachrome tautomerase (DDT) and MIF, *Mamm. Genome* **9**, 753-757 (1998).
9. J. Matsunaga, D. Sinha, L. Pannell, C. Santis, F. Solano, G.J. Wistow, and V.J. Hearing. Enzyme activity of macrophage migration inhibitory factor (MIF) towards oxidized catecholamines, *J. Biol. Chem.* **274**, 3268-3271 (1999).
10. J. Matsunaga, D. Sinha, F. Solano, C. Santis, G.J. Wistow, and V.J. Hearing. Macrophage migration inhibitory factor (MIF) - Its role in catecholamine metabolism, *Cell Mol. Biol.* **45**, 1035-1041 (1999).
11. S. Ito, K. Wakamatsu, and L. Zecca. Structure and function of neuromelanin, in: *Catecholamine Research: From Molecular Insights to Clinical Medicine,* edited by T. Nagatsu, R. Nabeshima, R. McCarthy, and D. Goldstein (Kluwer Academic/Plenum Pulbishers (New York, 2001) pp. (in this volume).
12. K.L. Double, L. Zecca, P. Costi, M. Mauer, C. Griesinger, S. Ito, D. Ben-Shacher, G. Bringmann, R.G. Fariello, P. Riederer, and M. Gerlach. Structural characteristics of human substantia nigra neuromelanin and synthetic dopamine melanins, *J. Neurochem.* **75**, 2583-2589 (2000).
13. D.G. Graham. Oxidative pathways for catecholamines in the genesis of neuromelanin and cytotoxic quinones, *Mol. Pharmacol.* **14**, 633-643 (1978).
14. F. Solano, V.J. Hearing, and J.C. García-Borrón. Neurotoxicity due to o-quinones: neuromelanin formation and possible mechanisms for o-quinone detoxification, *Neurotoxicity Res.* **1**, 153-169 (2000).
15. J. Matsunaga, P.A. Riley, N. Kobayashi, N. Matsunaga, I. Takeuchi, H. Tagami, and V.J. Hearing. Cytotoxicity of intermediates in the melanin and neuromelanin biosynthetic pathways on B16F10 and Neuro2A cells, (submitted), (2001).
16. D. Offen, I. Ziv, A. Barzilai, S. Gorodin, E. Glater, A. Hochman, and E. Melamed. Dopamine-melanin induces apoptosis in PC12 cells; possible implications for the etiology of Parkinson's disease, *Neurochem. Int.* **31**, 207-216 (1997).
17. N.G. Lindquist, B.S. Larsson, and A. Lyden-Sokolowski. Neuromelanin and its possible protective and destructive properties, *Pigment Cell Res.* **1**, 133-136 (1987).
18. G. Prota, and M. d'Ischia. Neuromelanin: a key to Parkinson's disease, *Pigment Cell Res.* **6**, 333-335 (1993).
19. M. d'Ischia, and G. Prota. Biosynthesis, structure and function of neuromelanin and its relation to Parkinson's disease: a critical update, *Pigment Cell Res.* **10**, 370-376 (1997).

HYPOTHALAMO-SYMPATHETIC CONTROL OF CELLULAR IMMUNITY

Toshihiko Katafuchi, Zhichun Shi, Sachiko Take, and Tetsuro Hori*

1. INTRODUCTION

In addition to the presence of β- and α-adrenergic receptors on lymphocytes and other immunocompetent cells (Sanders et al., 2001), anatomical studies have revealed the direct contact between tyrosine hydroxylase positive nerve terminals and lymphocytes in the spleen (Bellinger et al., 2001). However, the effects of noradrenaline (NA) on the immune responses *in vitro* are rather complicated depending on Th predominance, and the dose of NA (for review see Hori et al., 1995; Madden, 2001).

Natural killer (NK) cells serve as effectors of immunosurveillance against viral infection and tumor growth without any prior recognition with a specific antigen. In the present article, to assess the role of the sympathetic NA innervation in the modulation of NK cell activity, we investigated the central mechanisms of the immobilization (IMB) stress-induced suppression of splenic NK cell activity, and found a possible involvement of proinflammatory cytokines that were induced in the brain during IMB.

2. SYMPATHETIC NERVE IN STRESS-INDUCED IMMUNOSUPPRESSION

It has been shown that foot shock-induced suppression of proliferative responses of splenocytes was blocked by surgical denervation of the splenic sympathetic nerve (Wan et al., 1993), suggesting that the nerve may be a significant mediator of the immunosuppression. Shimizu et al. (1996) have demonstrated that IMB for 90 min resulted in a reduction of the splenic NK cell activity in rats, which was attenuated by splenic denervation. Extracellular concentration of NA in the spleen measured by *in vivo* microdialysis markedly increased during IMB, and was decreased by splenic denervation, indicating that the increased NA was derived from splenic sympathetic nerve mediating the central stress message to the spleen.

Irwin et al. (1990) have demonstrated that the foot shock stress-induced reduction of splenic NK cell activity was mediated by brain corticotropin-releasing factor (CRF) independently of the activation of the hypothalamic-pituitary-adrenal (HPA) axis. An intracere

* Toshihiko Katafuchi, Department of Integrative Physiology, Graduate School of Medical Sciences, Kyushu University, Fukuoka, 812-8582, Japan.

broventricular (ICV) injection of CRF has been shown to increase the electrical activity of the splenic sympathetic nerves in rats (Ichijo et al., 1994; Katafuchi et al., 1997). Finally, electrical stimulation of the splenic sympathetic nerve resulted in the suppression of splenic NK cell activity through a β-adrenergic receptor-mediated process (Katafuchi et al., 1993a). These findings, taken together, suggest that the stress-induced suppression of NK cell activity is mediated, at least partly, by the activation of the splenic sympathetic nerve.

3. MECHANISM OF BRAIN CYTOKINE-INDUCED IMMUNOSUPPRESSION

An ICV injection of interleukin-1β (IL-1β) in rats has been shown to suppress various immune responses such NK cell activity, mitogenic response to phytohemagglutinin and IL-2 production of lymphocytes (Sundar et al., 1989; Sundar et al., 1990), specific anti body production (Irwin, 1993), and secretion of IL-1 from splenic macrophages (Brown et al., 1991). The brain IL-1-induced suppression of cellular immunity was mediated by both the activation of HPA axis and the sympathetic nervous system, since chemical (Sundar et al., 1990) and surgical (Brown et al., 1991) sympathectomy and adrenalectomy (Sundar et al., 1989) only partially attenuated the immunosuppression. On the other hand, inter feron-α (IFN-α) administered either ICV or into the medial preoptic area (MPO) in the hypothalamus reduced the splenic NK cell activity, exclusively by activating the sympa thetic nervous system (Take et al., 1993; Take et al., 1995). As expected, an ICV injection of IFN-α (Katafuchi et al., 1993a) and IL-1β (Ichijo et al., 1994), or MPO injection of IFN-α (Katafuchi et al., 1993b) increased the electrical activity of the splenic sympathetic nerve. The IL-1β-induced activation of the nerve was blocked by a cyclooxygenase inhibi tor, sodium salicylate (Ichijo et al., 1994). The splenic sympathetic nerve activity was increased by an ICV injection of prostaglandin E_2 (PGE_2) through an EP1 receptor subtype (Ando et al., 1995). An increase in nerve activity induced by ICV PGE_2 was blocked by an ICV pretreatment with a CRF antagonist, α-helical CRF, whereas the CRF-induced en hancement of nerve activity was not affected by sodium salicylate (Katafuchi et al., 1997). Therefore, it is likely that an action of PGE_2 is followed by activation of CRF system in the brain but not *vice versa*, resulting in the increase in splenic nerve activity.

4. INVOLVEMENT OF BRAIN CYTOKINE IN IMB-INDUCED IMMUNOSUPPRESSION

It is well known that the brain produces cytokines and possesses their receptors. Such cytokines are usually not found, or only are at very low levels, in the normal brain. The production of most cytokines is induced during pathological and inflammatory events in the brain, such as injury, ischemia, infection and degeneration (Hopkins and Rothwell, 1995). In addition, non-inflammatory stress such as IMB has also shown to induce expression of IL-1β and IFN-α in the brain. In agreement with a previous finding by Northern blot analysis (Minami et al., 1991), we have found that the amount of mRNA for IFN-α and IL-1β is increased by IMB stress for 1 hr in the rat hypothalamus as assessed quantitatively by real-time capillary RT-PCR (Fig. 1). As shown in Fig. 1, the amount of IFN-α mRNA increased after IMB in the MPO, the lateral preoptic area (LPO) and ven tromedial nucleus (VMH), while IL-1β mRNA increased in the MPO, the paraventricular nucleus (PVN) and VMH.

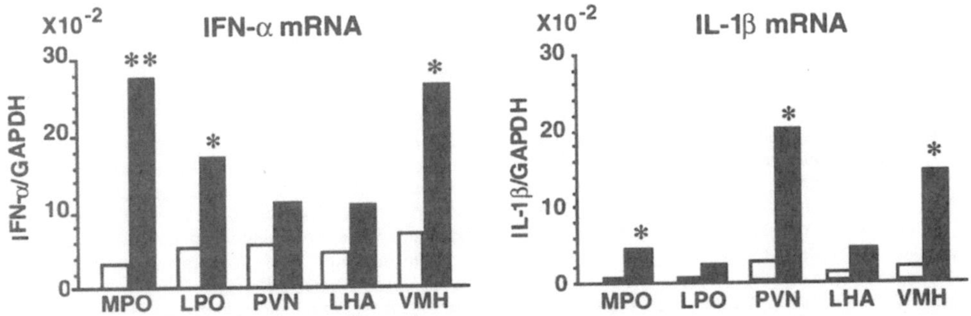

Figure 1. IFN-α and IL-1β mRNA content in the rat hypothalamus before (open column) and 1 hr after IMB (filled column) measured by real-time capillary RT-PCR. The amount of each cytokine mRNA was divided by that of mRNA for a housekeeping gene, glyceraldehyde 3-phosphate dehydrogenase (GAPDH). For abbreviations for each hypothalamic nucleus, see text. Each group, n=4; *, $P<0.05$, **, $P<0.01$

The suppression of the splenic NK cell activity produced by IMB in mice was blocked by pretreatment with anti-IL-1β neutralizing antibody administered ICV 10 min before IMB, while the IMB-induced suppression was not affected by pretreatment with non-specific IgG. In addition, the reduction of lymphocyte proliferation induced by foot-shock stress was also attenuated by ICV pretreatment with anti-IL-1 antibody (Saperstein et al., 1992). These findings suggest that IL-1β produced in the brain during IMB is involved, at least in part, in the IMB-induced suppression of cellular immunity.

5. HYPOTHALAMIC SITES INVOLVED IN IMB-INDUCED IMMUNOSUPPRESSION

After 90 min of IMB, Fos protein encoded by an immediate early gene, c-*fos*, was found most heavily in the parvocellular region of the PVN and moderately in the lateral hypothalamic area (LHA), VMH and MPO. Neurons in this region of the PVN are known to send axons to the brainstem and the spinal cord regulating sympathetic nerve activity. Microinjection of an excitatory amino acid, glutamate, into the PVN and VMH in anesthetized rats enhanced the splenic sympathetic nerve activity, while the one into the MPO and LHA decreased it (Katafuchi et al., 1993b; Shi et al., 1997), indicating an excitatory influence of the PVN and VMH on the nerve activity, and an inhibitory effect of the MPO and LHA.

These findings suggested that the activation of the PVN contributes to the suppression of splenic NK cell activity following IMB. In addition, the MPO was also involved in NK suppression, since microinjection of IFN-α was effective only in the MPO (Take et al., 1995), in which IFN-α mRNA markedly increased after IMB (Fig. 1). IFN-α was demon strated to reduce the firing rates of the MPO neurons (Nakashima et al., 1988). Since the MPO neurons send inhibitory projections to the PVN (Saphier and Feldman, 1986), the suppression of the MPO would disinhibit the PVN, thereby inducing the enhancement of nerve activity. In accordance with this finding, electrical lesion of the MPO activated the splenic sympathetic nerve to induce a suppression of splenic NK cell activity, which was completely blocked by denervation (Katafuchi et al., 1993b). In conclusion, our results suggest that the neuronal network of the hypothalamic-sympathetic nervous system is one of the important communication channels that mediates the central modulation of cellular immunity during IMB.

References

Sanders, V. M., Kasprowicz, D. J., Kohm, A. P., and Swanson, M.A., 2001, Neurotransmitter receptors on lym phocytes and other lymphoid cells, in: *Psychoneuroimmunology*, third edition, R. Ader, D. L. Felten, and N. Cohen, eds., Academic Press, San Diego, pp. 161-196.

Bellinger, D. L., Lorton, D., Lubahn, C., and Felten, D.L., 2001, Innervation of lymphoid organs—Association of nerves with cells of the immune system and their implications in disease, in: *Psychoneuroimmunol ogy*, third edition, R. Ader, D. L. Felten, and N. Cohen, eds., Academic Press, San Diego, pp. 55-111.

Hori, T., Katafuchi, T., Take, S., Shimizu, N., and Niijima, A., 1995, The autonomic nervous system as a communication channel between the brain and the immune system, *Neuroimmunomodulation* **2**: 203-215.

Madden, K. S., 2001, Catecholamines, sympathetic nerves, and immunity, in: *Psychoneuroimmunology*, third edition, R. Ader, D. L. Felten, and N. Cohen, eds., Academic Press, San Diego, pp. 197-216.

Wan, W., Vriend, C. Y., Wetmore, L., Gartner, J. G., Greenberg, A. H., and Nance, D. M., 1993, The effects of stress on splenic immune function are mediated by the splenic nerve, *Brain Res. Bull.* **30**: 101-105.

Shimizu, N., Kaizuka, Y., Hori, T., and Nakane, H., 1996, Immobilization increases norepinephrine release an d reduces NK cytotoxicity in spleen of conscious rat, *Am. J. Physiol.* **271**: R537-R544.

Irwin, M., Vale, W., and Rivier, C., 1990, Central corticotropin-releasing factor mediates the suppressive effect of stress on natural killer cytotoxicity, *Endocrinology* **126**: 2837-2844.

Ichijo, T., Katafuchi, T., and Hori, T., 1994, Central administration of interleukin-1β enhances splenic sympa thetic nerve activity in rats, *Brain Res. Bull.* **34**:547-553.

Katafuchi, T., Ichijo, T., and Hori, T., 1997, Sequential relationship between actions of CRF and PGE_2 in the brain on splenic sympathetic nerve activity in rats, *J. Auton. Nerv. Syst.* **67**: 200-206.

Katafuchi, T., Take, S., and Hori, T., 1993a, Roles of sympathetic nervous system in the suppression of cytotox icity of splenic natural killer cells in the rat, *J. Physiol. (Lond.)* **465**: 343-357.

Sundar, S. K., Becker, K. J., Cierpial, M. A., Carpenter, M. D., Rankin, L. A., Fleener, S. L., Ritochie, J. C., Simson, P. E., and Weiss, J. M., 1989, Intracerebroventricular infusion of interleukin 1 rapidly decreases peripheral cellular immune responses, *Proc. Natl. Acad. Sci. USA* **86**: 6398-6402.

Sundar, S. K., Cierpial, M. A., Kilts, C., Ritchie, J. C., and Weiss, J. M., 1990, Brain IL-1-induced immunosup pression occurs through activation of both pituitary-adrenal axis and sympathetic nervous system by corticotropin-releasing factor, *J. Neurosci.* **10**: 3701-3706.

Irwin, M., 1993, Brain corticotropin-releasing hormone- and interleukin-1β-induced suppression of specific antibody production, *Endocrinology* **133**: 1352-1360.

Brown, R., Li, Z., Vriend, C. Y., Nirula, R., Janz, L., Falk, J., Nance, D. M., Dyck, D. G., and Greenberg, A . H., 1991, Suppression of splenic macrophage interleukin-1 secretion following intracerebroventricular injec tion of interleukin-1β, *Cell. Immunol.* **132**: 84-93.

Take, S., Mori, T., Katafuchi, T., and Hori, T., 1993, Central interferonα inhibits natural killer cytotoxic ity through sympathetic innervation, *Am. J. Physiol.* **265**: R453-R459.

Take, S., Uchimura, D., Kanemitsu, Y., Katafuchi, T., and Hori, T., 1995, Interferon-α acts at the preoptic hy pothalamus to reduce natural killer cytotoxicity in rats, *Am. J. Physiol.* **268**: R1406-R1410.

Katafuchi, T., Ichijo, T., Take, S., and Hori, T., 1993b, Hypothalamic modulation of natural killer cell activity in rats, *J. Physiol. (Lond.)* **471**: 209-221.

Ando, T., Ichijo, T., Katafuchi, T., and Hori, T., 1995, Intracerebroventricular injection of prostaglandin E_2 increases splenic sympathetic nerve activity in rats, *Am. J. Physiol.* **269**: R662-R668.

Hopkins, S. J., and Rothwell, N. J., 1995, Cytokines and the nervous system. I. Expression and recogni tion, *Trends Neurosci.* **18**: 83-88.

Minami, M., Kuraishi, Y., Yamaguchi, T., Nakai, S., Hirai, Y., and Satoh, M., 1991, Immobilization stress in duces interleukin-1β mRNA in the rat hypothalamus, *Neurosci. Lett.* **123**: 254-256.

Saperstein, A., Brand, H., Audhya, T., Nabriski, D., Hutchinson, B., Rosenzweig, S., and Hollander, C. S., 1992, Interleukin 1β mediates stress-induced immunosuppression via corticotropin-releasing factor, *Endocrinology* **130**: 152-158.

Shi, Z., Katafuchi, T., Take, S., and Hori, T., 1997, Responses of splenic sympathetic nerve activity in rats to chemical stimulation of the hypothalamus, *Jpn. J. Physiol.* **47** (suppl. 2): S227, 699.

Nakashima, T., Hori, T., Kuriyama, K., and Matsuda, T., 1988, Effects of interferon-α on the activity of preop tic thermosensitive neurons in tissue slices, *Brain Res.* **454**: 361-367.

Saphier, D., Feldman, S., 1986, Effects of stimulation of the preoptic area on hypothalamic paraventricular nu cleus unit activity and corticosterone secretion in freely moving rats, *Neuroendocrinology* **42**: 167-173.

VASOACTIVE HORMONES AND REGULATION OF THE HYPOTHALAMIC-PITUITARY-ADRENAL AXIS

Greti Aguilera, Cristina Rabadan-Diehl, Alexander Kiss, and Tomazs Ochedalski*

INTRODUCTION

Survival under stress situations requires coordinated behavioral, autonomic and hormonal responses in order to maintain homeostasis. Activation of the autonomic system, the hypothalamic pituitary adrenal (HPA) axis with increase in plasma glucocorticoids is essential for this adaptation (1). The secretion of glucocorticoids is regulated by the pituitary peptide adrenocorticotropic hormone (ACTH), which in turn is regulated by the hypothalamic peptides, corticotropin-releasing hormone (CRH) and vasopressin (VP) (2,3). It is clear that reciprocal interactions exist between the HPA axis and vasoactive hormones. For example, the hypothalamic peptides CRH and VP have recognized vasodilator and vasoconstrictor actions, respectively. Also, catecholamines and a number of vasoactive hormones including angiotensin II (Ang II), atrial natriuretic peptide, adrenomedullin and neuropeptide Y, are involved in the regulation of hypothalamic CRH (3). This discussion will address the interactions between some of these vasoactive hormones and different components of the HPA axis, and their physiological implications during stress adaptation.

GLUCOCORTICOIDS AND VASOACTIVE HORMONES

Glucocorticoids, the end product of activation of the HPA axis, have important metabolic actions and are indispensable for survival under stress conditions (4). The effects of glucocorticoids include gluconeogenesis, protein degradation and inhibition of synthesis, redistribution of lipids, immuno- and neuro-modulation. The steroids also affect the secretion and effects of vasoactive hormones. For example, glucocorticoids stimulate the expression of enzymes of the catecholamine biosynthetic pathway, especially phenylethanolamine N-

* Section on Endocrine Physiology, National Institute of Child Health and Human Development, National Institutes of Heath, Bethesda, MD, 20892

Catecholamine Research, Edited by Nagatsu *et al.*
Kluwer Academic/Plenum Publishers, 2002

methyl transferase, a key enzyme in norepinephrine synthesis (5). The synthesis of angiotensinogen, a component of the renin angiotensin system, and neuropeptide Y, are also stimulated by glucocorticoids. In addition, it has been reported that glucocorticoids enhance the pressor effect of vasoactive hormones including catecholamines, VP and Ang II (6).

VASOPRESSIN AND REGULATION OF THE HPA AXIS

The nonapeptide, VP, is secreted into the peripheral circulation from axonal terminals of magnocellular neurons of the hypothalamic supraoptic (SON) and paraventricular (PVN) nuclei in the neural pituitary lobe. This circulating VP is responsible for its well-known effects on water conservation in the kidney and vasopressor actions (7). VP is also co-secreted with CRH by parvocellular neurons of the PVN into the pituitary portal circulation, where it regulates ACTH secretion by acting upon V1b type receptors in the pituitary corticotroph. Although VP on its own is a weak stimulus of ACTH secretion, it markedly potentiates the stimulatory effect of CRH. In addition, its action is relatively refractory to inhibition by glucocorticoid feedback, a property which may contribute to preservation of ACTH responses during stress in spite of high glucocorticoid levels (2). Moreover, experiments in rats have shown that VP plays a major role in regulating ACTH secretion during chronic stress. The expression of VP mRNA in parvocellular neurons, as well as the peptide content and release in the median eminence, all increase in a number of chronic stress paradigms (8). Secondly, the number of VP receptors in the pituitary increases during stress and there is a good correlation between VP receptor number and ACTH responsiveness. In basal conditions, the expression of V1b receptors in the pituitary is low but rapidly increase during stress. In some stress conditions changes in binding are preceded by parallel changes in V1b-R mRNA. However, in some conditions binding is low in spite of high V1b-R mRNA levels suggesting that receptor synthesis is controlled at the translational level (8).

CATECHOLAMINES AND REGULATION OF THE HPA AXIS

The mechanisms controlling the expression and release of hypothalamic CRH and VP are complex, and include neurotransmitters such as catecholamines, GABA, serotonin, acetylcholine and several neuropeptides (3). Of these factors, catecholamines play a pre-eminent role (9). CRH neurons receive dense adrenergic and noradrenergic innervation from the brain stem, and adrenergic receptors have been demonstrated in the PVN (10). During acute stress, dependence of ACTH responses on catecholaminergic innervation of the PVN is stressor-specific (11). For example, while ACTH responses to insulin-induced hypoglycemia are unaffected by lesions of the ventral noradrenergic bundle, responses to ether exposure are highly dependent on intact catecholaminergic pathways (9). Lesions of the ventral noradrenergic bundle, which carries most noradrenergic inputs to the PVN (9), or central $\alpha 1$-adrenergic blockade with prazosin (12), blunt the increases in plasma ACTH, as well as reduce the number of CRH neurons and CRH release from the median eminence observed after stress induced by i.c.v. colchicine injection. Microdialysis studies in the rat showing that noradrenergic activity in the PVN is increased after repeated immobilization, are consistent with a dominant role of catecholamines in the control of PVN function during the late phase of chronic stress (11).

In contrast to the established role of catecholamines on CRH release during acute and chronic stress, there is evidence that catecholamines contribute to the stress-induced increases in CRH mRNA in the PVN only during chronic stress. Thus, the increase in CRH mRNA 4 h after acute stress by immobilization, i.p. hypertonic saline injection, or i.c.v. colchicine injection, cannot be reproduced by i.c.v injection of the α1-adrenergic agonist, methoxamine, or prevented by central blockade with the antagonist,prazosin (13). Similarly, interruption of the ventral noradrenergic bundle had no effect on the increases in CRH mRNA and proenkephalin mRNA in the PVN induced by acute stress. In contrast, repeated i.c.v. injection of methoxamine every 8 hr for 24 hr was shown to increase CRH mRNA levels in the PVN. Similarly, the increase in CRH mRNA 48 hr after i.c.v colchicine, repeated i.p. hypertonic saline injection or repeated immobilization every 8 hr for 24 h, was significantly reduced by central α1-adrenergic blockade (13).

Catecholamines have also been implicated in the regulation of ACTH secretion directly in the pituitary, though the physiological significance of this effect is controversial (3). The overall evidence shows that catecholamines play an important role in regulating the HPA axis at the central level.

THE RENIN-ANGIOTENSIN SYSTEM AND THE HPA AXIS

The octapeptide, angiotensin II, the active component of the renin-angiotensin system, has important regulatory effects on vascular contactility, mineralocorticoid secretion, autonomic system activity, and fluid and electrolyte balance,all important components of the stress response. The peptide is also involved in the regulation of CRH, and possibly vasopressin, from parvocellular neurons (14,15). Intracerebroventricular injection of low doses of Ang II elicits large increases in plasma ACTH. The stimulatory effect of Ang II on ACTH secretion is mediated at least in part by CRH, as demonstrated by the ability of CRH antiserum to block this action. Ang II can stimulate CRH secretion directly through interaction with receptors in the PVN or median eminence, or indirectly by stimulating afferent neural pathways from the circumventricular organs (15). Icv injection of Ang II increases CRH expression in the PVN, while type 1 Ang II receptor (AT1-R) antagonists attenuate the stimulatory effect of stress on CRH mRNA and sympathetic activity (16). On the other hand, the increases in plasma ACTH following stress are unaffected by Ang II antagonists suggesting that Ang II can regulate CRH mRNA expression but not release of the peptide to the pituitary portal circulation. These effects are at least partially direct in the PVN, since double staining in situ hybridization studies have shown co-localization of AT1-R mRNA in CRH and VP neurons (15,17). The involvement of Ang II in the hypothalamic regulation of the HPA axis and autonomic activity is also supported by reports indicating regulation of the expression of AT1-R in parvocellular neurons by glucocorticoids and stress (15).

SUMMARY AND CONCLUSIONS

Reciprocal interactions exist between the HPA axis and vasoactive hormones including catecholamines, and peptides such as Ang II and VP. Glucocorticoids, the end hormone of the HPA axis modulate the peripheral effects of vasoactive hormones, while the latter regulate the HPA axis. Activation of noradrenergic pathways during stress increases the expression and secretion of the corticotropic releasing peptides, CRH and VP, from parvocellular neurons of the PVN. Parvocellular neurons express type-1 Ang II receptors (AT1-R), the content of which increases during stress. AT1-R blockade blunts the increases in CRH mRNA but not ACTH secretion during stress, suggesting that Ang II plays a role regulating CRH mRNA expression but not on release of the peptide to the pituitary portal circulation. The nonapeptide VP, well known for its vasopressor and antidiuretic actions, is co-secreted with CRH and regulates ACTH secretion by acting upon V1b type receptors in the pituitary corticotroph. VP is critical for adaptation of the HPA axis to repeated stress as indicated by preferential expression of VP over CRH in the PVN, and the association of upregulation of pituitary VP receptors with corticotroph hyperresponsiveness. The data shows that vasoactive hormones are involved in the regulation of the HPA axis at the hypothalamic and pituitary levels, and play a critical role in mediating the hormonal responses to stress.

REFERENCES

1. G.P. Chrousos, and P.W. Gold, The concepts of stress and stress system disorders. JAMA **267**, 1244-1252 (1992)
2. G. Aguilera, Regulation of ACTH secretion during chronic stress. Front.Neuroendocrinol. **15**, 321-350 (1994)
3. M.H. Whitnall, Regulation of the hypothalamic corticotropin-releasing hormone neurosecretory system. *Prog. Neurobiol.* **40**, 573-629 (1993).
4. A. Munck, P.M. Guyre and N.J. Holbrook Physiological functions ofglucocorticoids in stress and their relation to pharmacological actions. Endocrine Rev.**5**, 25-44 (1984).
5. D.L. Wong, B. Siddall and W. Wang, Hormonal control of rat adrenal phenylethanolamine N-methytransferase: enzyme activity, the final critical pathway. Neuropsycopharmacology 13, 223-234 (1995)
6. J.A. Whitworth, Mechanisms of glucocorticoid induced hypertension. Kidney Int. **31**, 1213-1224 (1987)
7. K. Toba, M. Ohta, T. Kimura, K. Nagano, S. Ito and Y. Ouchi. Role of brain vasopressin in regulation of blood pressure. Prog. Brain. Res. **119**, 337-49 (1998).
8. G. Aguilera, C. and Rabadan-Diehl. Vasopressinergic regulation of the hypothalamic-pituitary adrenal axis: implications for stress adaptation. Regulatory Peptides **96**, 23-29 (2000)
9. I. Assenmacher, G. Barbanel, S. Gaillet, L. Givalois, G. Ixart, F. Malaval, M. Mekaouche, P. Siaud, A. Szafarczyk. Central regulation of ACTH release in stress. Ann. N. Y. Acad. Sci. 771, 41-54 (1995).
10. M. Palkovits, R.L. Zaborsky, A. Feminger, E. Mezey, M.L.K. Fekete, J.P. Herman, B. Kanyicska and M. Szabo. Noradrenergic innervation of the rat hypothalamus: experimental biochemical and electron microscopic studies. Brain Res. **191**, 161-171 (1990).
11. K. Pacak, I. Armando, K. Fukuhara, R. Kvetnansky, M. Palkovits, I.J. Kopin and D.S. Goldstein. Noradrenergic activation in the paraventricular nucleus during acute and chronic stress in rats : An in vivo microdialysis study. Brain. Res. **589**, 91-96 (1992).
12. A. Kiss, G. Aguilera, Participation of $\alpha 1$-adrenergic receptors in the secretion of hypothalamiccorticotropin releasing hormone during stress. Neuroendocrinology **56** 153-160 (1992).
13. A. Kiss and G. Aguilera. Role of alpha-1-adrenergic receptors in the regulation of CRH mRNA in the paraventricular nucleus of the hypothalamus during stress. Cell. Mol. Neurobiol. 20, 683-694 (2000).
14. Saavedra JM. Brain and pituitary angiotensin. Endocr. Rev. **13**, 329-80 (1992)
15. G. Aguilera, A. Kiss, X. Luo and B. Sunar-Akbasak. The renin-angiotensin system and the stress response. Ann. N.Y. Acad. Sci. 771, 173-186 (1995)
16. D. Jezova, T. Ochedalski, A. Kiss, G. Aguilera. Modulation of the sympathetic nervous system and hypothalamic-pituitary-adrenal axis during acute immobilization stress by centralangiotensin II receptor antagonism. J. Neuroendocrinol. 9, 689-697 (1997)
17. Lenkei Z, Corvol P, Llorens-Cortes C. Comparative expression ofvasopressin and angiotensin type-1 receptor mRNA in rat hypothalamic nuclei: a double in situ hybridization study. BrainRes Mol Brain Res. **34**, 135-42 (1995).

α_{1D}-ADRENERGIC RECEPTOR KNOCKOUT MICE DISPLAY IMPAIRED VASCULAR CONTRACTILE RESPONSE TO CATECHOLAMINES

Gozoh Tsujimoto and Akito Tanoue

1. INTRODUCTION

The sympathetic nervous system plays an important role in regulating the tone of the peripheral circulation, and hence in the control of blood pressure. Catecholamines cause vascular smooth muscle contraction by activating α_1-adrenergic receptors (α_1-ARs).[1]) Recent extensive efforts has been spent in classifying the three known α_1-AR subtypes (α_{1A}, α_{1B}, and α_{1D}) via molecular cloning[2] and pharmacological analyses[3]; however, still the contribution of each α_1-AR subtype in catecholamine-induced physiological responses has not been well characterized.

In this study, we describe the gene targeting of the mouse α_{1D}-AR and the initial functional characterization of knockout mice lacking this receptor subtype. Since clinical efficacy of α_1-AR antagonists as antihypertensive drugs reflects the important physiological role of α_1-AR in vascular function and in the maintenance of arterial blood pressure, we focused on the functional characterization of the α_{1D}-AR knockout model on the *in vitro* vascular contractility. Our study shows that the α_{1D}-AR is a mediator of the vasoconstrictive responses to catecholamines.

2. METHODS

2.1. Gene Targeting

The murine α_{1D}-AR gene was cloned,[4] and mutant mice lacking the α_{1D}-AR gene were generated, according to a standard method

2.2. Measurement of Aortic Contraction.

The thoracic aorta was excised, cleaned, and cut into 1 mm-long segments. These segments were suspended in isolated tissue baths filled with 10 ml Krebs-Henseleit bicarbonate buffer containing timolol (3 μM). One end of the aortic segment was connected to a tissue holder and the other to an isometric force transducer. Aortic segments were equilibrated for 60 min under resting tension of 0.5 g, and the buffer was replaced every 15 min. In a preliminary experiment, the length of the smooth muscle was increased stepwise during the equilibration period to adjust passive wall tension to 0.5 g; this resting tension was found to be optimal for aorta contraction of 22-26g mice induced by KCl (40 mM). Care was taken to avoid endothelial damage; functional integrity was assessed using acetylcholine (10 μM). Only intact segments were used for further analysis.

3. RESULTS

3.1. Targeted Disruption of the Mouse α_{1D}-AR Gene

Several clones from positive ES clones confirmed by Southern blot analysis with the 5', 3', and Neo probe were independently microinjected into C57Black/6J blastocyst-stage embryos. Chimeric mice that were mated showed germline transmission of the disrupted allele. Analysis of the α_{1D}-AR genotype frequencies after intercrosses of heterozygous mutant mice did not reveal any deviation from Mendelian expectations (results not shown). Monitoring of mice body weight at 4 weeks old did not reveal any significant difference in growth among mice of different α_{1D}-AR genotypes (results not shown). Thus, disruption of the α_{1D}-AR gene does not seem to have any major effect on mouse development, fertility, growth, or feeding behavior under standard breeding conditions.

3.2. Aortic Contraction in α_{1D}+/+ and -/- Mice

To assess whether α_{1D}-AR is involved in the vascular smooth muscle contraction, we measured the effect of norepinephrine on the contraction of isolated aortic segments from male α_{1D}+/+ and -/- mice. As shown in Figure, norepinephrine induced concentration-dependent contractile responses in aortic segments from α_{1D}+/+ and α_{1D}-/- mice. However, the potency of the response was markedly reduced in aortic segments from α_{1D}-/- as compared with α_{1D}+/+ mice. The EC_{50} values of norepinephrine was increased approximately 50-fold in α_{1D}-/- mice compared with α_{1D}+/+ mice. The contractile response induced by serotonin was not decreased. Rather the concentration-response curve of serotonin-induced contraction was slightly shifted to the left in α_{1D}-/- mice compared to α_{1D}+/+ mice.

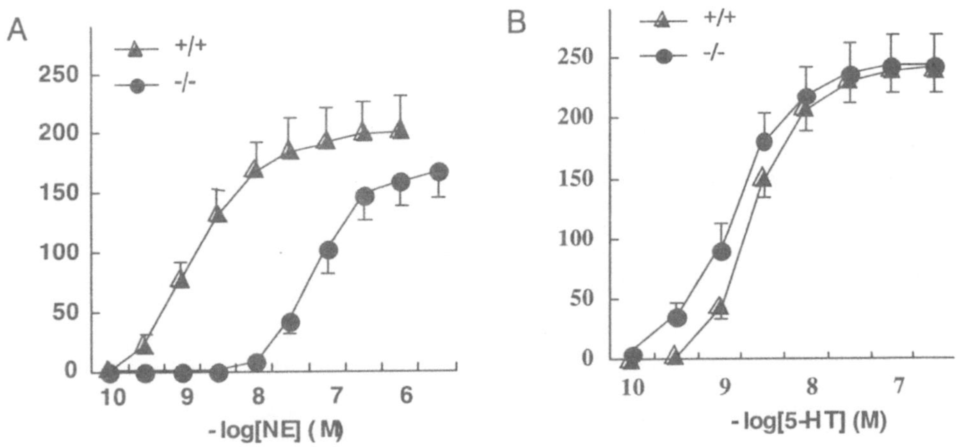

Figure. Aortic contraction in α_{1D}+/+ and -/- mice.
The contractile response to norepinephrine (NE) and serotonin (5-HT).
Concentration-response curves for norepinephrine- (A) and serotonin-induced contractions (B) in aortic segments from -/- or +/+ mice. Aortic segments were pretreated with 40 mM KCL and 0.1 μM norepinephrine prior to exposure to each agent. The results are the mean±S.E.M. of 15-27 preparations for norepinephrine (NE) or serotonin (5-HT).

4. DISCUSSION

Using a gene targeting to create a mouse model lacking the α_{1D}-AR, we investigated the functional role of the α_{1D}-AR subtype in the aorta. Although *in vitro* as well as *in vivo* pharmacological studies[5] have implicated a predominant role of α_{1D}-AR in the vascular contractions caused by α_1-AR agonists, our present study, for the first time, clearly shows that α_{1D}-AR directly mediates the α_1-AR-stimulated vascular smooth muscle contraction. Together with previous observations in α_{1B}-AR knockout mice,[6] our study supports the idea that α_{1B}- as well as α_{1D}-ARs contribute to the α_1-AR-mediated contractile response to catecholamines.

5. ACKNOWLEDGEMENTS

This work was supported in part by research grants from the Scientific Fund of the Ministry of Education, Science, and Culture of Japan, the Japan Health Science Foundation and Ministry of Human Health and Welfare, the Organization for Pharmaceutical Safety and Research (OPSR), and a Grant for Liberal Harmonious Research Promotion System from the Science and Technology Agency.

REFERENCES

1. Hoffman, B. B., and Lefkowitz, R. J. 1996. Catecholamines, sympathomimetic drugs, and adrenergic receptor antagonists in *The pharmacological basis of the therapeutics*. J. G. Hardman, E. LimbirdL, P. B. Molinoff and R. W. Ruddon, editors. McGraw-Hill, New York, USA, 199-248.
2. Hieble, J. P., Bylund. D. B., Clarke, D. E., Eikenburg, D. C., Langer, S. Z., Lefkowitz, R. J., Minneman, K. P. & Ruffolo, R. R. Jr. 1995. International Union of Pharmacology. X. Recommendation for nomenclature of α_1-adrenoceptors: consensus update. *Pharmacol. Rev.* **47**: 267-270.
3. Guarino, R.D., Perez, D.M., and Piascik, M.T. 1996. Recent advances in the molecular pharmacology of the α_1-adrenergic receptors. *Cell Signal.* **8**:323-333.
4. Arai, K., Tanoue, A., Goda, N., Takeda, M., Takahashi, K., and Tsujimoto, G. 1999. Characterization of the mouse α_{1D}-adrenergic receptor gene. *Jpn. J. Pharmacol.* **81**: 271-278.
5. Kenny, B. A., Chalmers, D. H., Philpott, P. C. and Naylor, A. M. 1995. Characterization of an α_{1D}-adrenoceptor mediating the contractile response of rat aorta to noradrenaline. *Br. J. Pharmacol.* **115**: 981-986
6. Cavalli, A., Lattion, A. L., Hummler, E., Nenniger, M., Pedrazzini, T., Aubert, J. F., Michel, M. C., Yang, M., Lembo, G., Vecchione, C., et al. 1997. Decreased blood pressure response in mice deficient of the α_{1b}-adrenergic receptor. *Proc. Natl. Acad. Sci. U S A.* **94**: 11589-11594.

CATHECOLAMINERGIC NEURONS IN THE OLFACTORY BULB

Kazunori Toida*, Katsuko Kosaka, Yusuke Aika, and Toshio Kosaka

1. INTRODUCTION

The olfactory bulb (OB) is an expedient region for analyzing neuronal organization in the central nervous system, as it displays a simple and distinctly laminar cytoarchitecture consisting of a relatively small number of neuron types, but is also so rich in chemical and neuroactive substances located in a variety of neuron types in distinct layers.[1]

Electron microscopy (EM) revealed synaptic connections among inputs, outputs and interneurons of the OB, consulting with classical light microscopic images of them. The olfactory nerves (ON) made asymmetrical synapses with projection neurons, mitral/tufted (M/T) cells. M/T cells made dendro-dendritic reciprocal synapses with periglomerular (PG) cells in the glomerular layer and with granule cells in the external plexiform layer, namely asymmetrical synapses from M/T cells to the interneurons and symmetrical ones conversely to M/T cells. This diagram of synaptic organization has been widely accepted and functional studies used as a morphological basis for interpretation of their results.[1]

Recently, we have re-examined the neuronal organization of the OB, especially for the PG neurons with three-dimensional confocal laser scanning light microscopy (3D-CLSM) and stereology. Then the PG neurons could be classified into at least three types; GABA-like-, calbindin (CB)-, and calretinin (CR)-like-immunoreactive (IR) neuron groups. The PG neurons had been *a priori* believed to be GABAergic[1]; however, so far examined surprisingly, CB- and CR-IR PG neurons were GABA negative in the rat OB.[2]

Moreover, these chemically-defined neurons were further classified into two groups with regard to the spatial relationship of their intraglomerular dendrites with the ON terminals. GABA and tyrosine hydroxylase (TH) neurons (type 1) exhibited close apposition to ON terminals[1], but CR- and CB-neurons (type 2) had rare contacts with them.[3] As the PG neurons had been traditionally believed as GABAergic neurons synapsing from ON terminals, our findings strongly suggested the necessity to re-examine

* Kazunori Toida, Department of Anatomy and Cell Biology, The University of Tokushima School of Medicine, Tokushima, 770-8503, Japan. E-Mail: toida@ basic.med.tokushima-u.ac.jp Fax: 81-88-633-9426, Katsuko Kosaka, Yusuke Aika, Toshio Kosaka, Department of Anatomy and Neurobiology, Graduate School of Medical Sciences, Kyushu University, Fukuoka, 812-8582, Japan.

the synaptic organization traditionally believed.[4] Present study thus aims to reveal TH-IR PG neurons with comparing to other chemically-defined neurons.

2. MATERIAL AND METHODS

Adult male specific pathogen free Wistar rats (5-6 weeks, 110-150 grams) are used in the present study. All experimental procedures generally followed our previous studies,[2-7] except high-voltage electron microscopy (HVEM) correlated with CLSM.

After fixation, Vibratome sections were processed for immuno-labelling with mouse monoclonal (Chemicon MAB318; 1:10,000) or rabbit polyclonal (gift from Prof. Nagatu, I.; 1:10,000[8]) anti-TH antibodies. After immunostaining with the primary antibodies for 1-2 days and subsequently with secondary antibodies (biotinylated anti-mouse or rabbit IgG; 1:200) for 2 hours both at 20°C, sections were incubated with streptavidin conjugated with fluoro-nanogold (NanoProbe; 1:100) for 2 hours at 20°C and examined with CLSM. After examination, sections were processed using ABC method with Cobalt/Nickel-enhanced DAB and treated by silver enhancement (HQ Silver, NanoProbe) for 5 minutes at 20°C to develop nano-gold to 20-30 nm particles. Subsequent procedures for HVEM (Hitachi H-1250M, Okazaki, Japan) generally followed our previous studies.[6]

3. RESULTS

3.1. Chemical and morphological features

TH-IR somata were almost completely localized in the glomerular layer of the main OB (arrowheads in Fig.1), but not encountered in the accessory OB (asterisk in Fig.1).

From CLSM examinations of sections triple-immunostained with GABA, TH and CB, TH-neurons were confirmed to be a major subpopulation of GABAergic neurons but a distinct neuronal group from CB-neurons. In contrast with honeycomb-like features of intraglomerular dendrites of the CB-neurons, those of TH-neurons were more complexly branching and widely distributed, which could not be traced as a single image (Fig.2).

On the other hand, HVEM revealed individual TH-processes labeled with silver-enhanced gold particles (Fig.3). They were so complexly intermingled with one another, ran rather straightly, sometimes branched at relatively thick portions, and therefore, were apparantly different from the CB-neurons in their dendritic features in the glomeruli.[6]

3.2. Synapses

Immuno-EM revealed that TH-processes frequently received asymmetrical synapses directly from ON terminals, and usually made symmetrical synapses to M/T dendrites, forming serial synapses.[7], Furthermore, TH-processes received asymmetrical synapses from M/T dendrites and made symmetrical synapses to other M/T dendrites nearby, thus also forming another type of serial synapse.[7] On the other hand, TH-neurons received symmetrical GABAergic synapses in the periglomerular regions where ON was absent.[7]

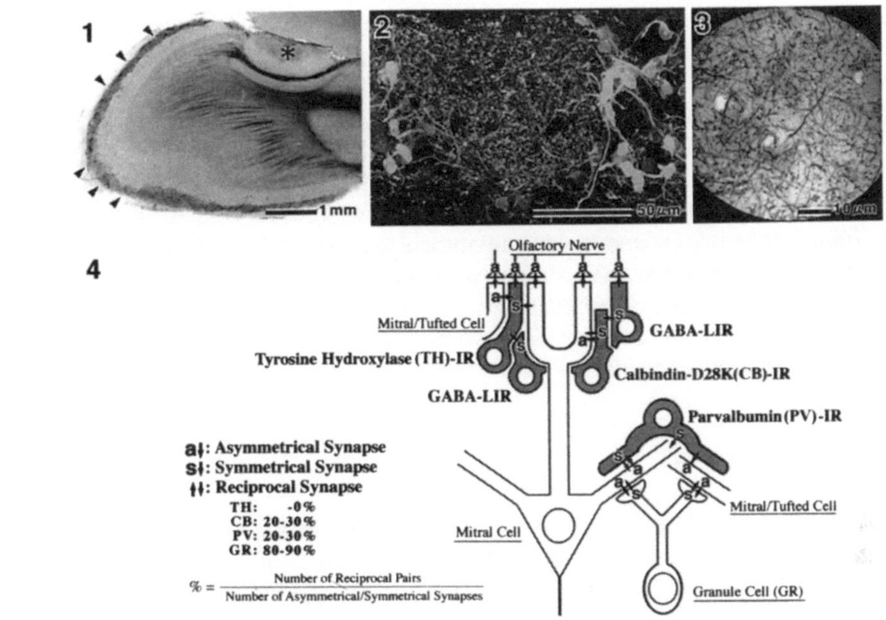

Fig.1; Parasagittal section of the rat OB . TH is localized in the glomerular layer of main OB (arrowheads) but not in accessory OB (asterisk). Fig.2; TH-profiles examined by CLSM. Fig.3; TH-processes examined by HVEM. Fig.4; Heterogeneity in synaptic connections of chemically-defined interneurons in the rat OB, so far clarified.[5-7]

From combined CLSM-EM 3D-analyses of triple fluoro-labeling by immunostaining for TH and CB and by marking particular mitral dendrites with biotinylated dextran amine, TH-neurons were confirmed to be involved in forming serial synapses among different M/T cells and ON terminals. In addition, computer-assisted 3D-reconstructions from serial EM and optical CLSM sections revealed the differences in intraglomerular dendritic features; TH-processes ran across M/T dendrites making serial synapses, whereas CB-processes ran in parallel to them with pleural synapses.[7] So far we encountered many reciprocal synapses on M/T dendrites with CB-neurons but none with TH-neurons, which have been considered to be typical type of synapse between GABAergic PG neurons and M/T cells.

The heterogeneity in synaptic connections of chemically-defined interneurons, so far analyzed by our serial-sectioning EM, is summarized (Fig.4).

4. DISCUSSION

In the glomerular layer, at least two types of chemically-defined neurons, namely the TH- and the CB-IR neurons belonging to type 1 and type 2 PG neurons, respectively, show different chemical/morphological features and synaptic connections.[4]

Characteristically GABAergic TH-neurons make two types of serial synapses, (1) asymmetrical synapses from ON terminals to TH-neurons and symmetrical ones from TH-neurons to M/T cells and (2) asymmetrical synapses from M/T cells to TH-neurons and symmetrical ones to different M/T cells nearby, but they make no reciprocal synapses. It suggests that the TH-neurons are involved in feed-forward inhibition from ON to M/T cells and in lateral inhibition between different M/T cells. In contrast, GABA-negative CB-neurons form reciprocal synapses with M/T cells, but have no synapses from ON terminals[6], thus might be implicated to function as recurent modulation to M/T cells.

So-called PG neurons, thus far considered to play an important role in olfactory processing in the glomeruli, have been still *a priori* believed to be GABAergic, receiving synapses from ON terminals and making reciprocal synapses with M/T cells. However, our combined 3D-CLSM and EM studies revealed that OB interneurons were more heterogenous than traditionally believed not only in their chemical and morphological features but also in synaptic contacts. It is worth re-examining synaptic organization to establish the basis for understanding olfactory information processing in the OB.

ACKNOWLEDGEMENT

The authors thank Prof. Ikuko Nagatsu of Fujita Health University School of Medicine for her gift of the antibody for TH, Prof. Kazunori Ishimura of The University of Tokushima School of Medicine for his encouragement, and Dr. Tsuo Arii of National Insitute for Physiological Sciences, Okazaki, Japan, with his help for HVEM. This work was supported by the Japan Society for the Promotion of Sciences, Grant-in-aids for Scientific Research (11680741 to K. T., 09680746 to K. K., 1170242 to T. K.).

REFERENCES

1. Shepherd G. M. and Greer C. A., Olfactory bulb, in: *Synaptic organization of the brain,* edited by G. M. Shepherd (Oxford, 1998), pp. 159-203.
2. K. Kosaka, Y. Aika, K. Toida, W. Hunziker, D. M. Jacobowitz, I. Nagatsu, P. Streit, T. J. Visser, and T. Kosaka, Chemically defined neuron groups and their subpopulations in the glomerular layer of the rat main olfactory bulb, *Neurosci. Res.* **23**, 73-88 (1995).
3. K. Kosaka, K. Toida, F. L. Margolis, and T. Kosaka, Chemically defined neuron groups and their subpopulations in the glomerular layer of the rat main olfactory bulb: II. prominent differences in the intraglomerular dendritic arborization and their relationship to olfactory nerve terminals, *Neuroscience* **76**, 775-786 (1997).
4. K. Kosaka, K. Toida, Y. Aika, and T. Kosaka, How simple is the organization of the olfactory glomerulus ?: the heterogeneity of so-called periglomerular cells, *Neurosci. Res.* **30**, 101-110 (1998).
5. K. Toida, K. Kosaka, C. W. Heizmann, and T. Kosaka, Electron microscopic serial-sectioning / reconstruction study of parvalbumin-containing neurons in the external plexiform layer of the rat olfactory bulb, *Neuroscience* **72**, 449-466 (1996).
6. K. Toida, K. Kosaka, C. W. Heizmann, T. Kosaka, Chemically defined neuron groups and their subpopulations in the glomerular layer of the rat main olfactory bulb: III. Structural features of calbindin-D28K-immunoreactive neurons, *J. Comp. Neurol.* **392**, 179-198 (1998).
7. K. Toida, K. Kosaka, Y. Aika, T. Kosaka, Chemically defined neuron groups and their subpopulations in the glomerular layer of the rat main olfactory bulb: IV. Intraglomerular synapses of tyrosine hydroxylase-immunoreactive neurons, *Neuroscience* **101**, 11-17 (2000).
8. Nagatsu I. Immunohistochemistry of biogenic amines and immunoenzyme- histocytochemistry of catecholamine-synthesizing enzymes, in: *Methods in biogenic amine research*, edited by S. Parvez, T. Nagatsu, I. Nagatsu, H. Parvez (Elsevier Science Amsterdam), pp. 873-909.

INVOLVEMENT OF IONOTROPIC GLUTAMATE RECEPTORS AND NITRIC OXIDE IN DOPA-INDUCED DEPRESSOR RESPONSES IN THE NUCLEUS TRACTUS SOLITARII

Y. Goshima, K. Yamanashi, T. Miyamae, Y. Sasaki, M. Maeda, H. Hirano, and Y. Misu*

1. INTRODUCTION

L-Dihydroxyphenylalanine (DOPA) is proposed to be a neurotransmitter for the primary baroreceptor afferents terminating in the nucleus tractus solitarii (NTS),[1] and it produces glutamate-like depressor responses when microinjected into the NTS of anesthetized rats. Little is known about the mechanism of such an excitatory action of DOPA. In striata, exogenous DOPA induces glutamate release and endogenous DOPA is a causal factor for the ischemic glutamate release and resultant delayed neuronal death.[2] In the present study, we tried to clarify whether glutamate receptors and/or nitric oxide (NO), an important modulator for central cardiovascular regulation, are involved in the DOPA-induced cardiovascular responses in the NTS of anesthetized rats.

2. MATERIALS AND METHODS

Wistar rats weighting 250-350 g were anesthetized with urethane. Blood pressure (BP) and heart rate (HR) were monitored as described.[3] In brief, after the trachea was cannulated, the animal was paralyzed with D-tubocurarine and artificially ventilated at a rate of 70-75 beats/min. The femoral artery was cannulated for recording BP and HR. A glass micropipette pulled to an outside diameter of approximately 50 μm was inserted into the unilateral NTS. Each compound dissolved was given in a volume of 50 nl in 2s.

* Y. Goshima, K. Yamanashi, T. Miyamae, Y. and Sasaki, Yokohama City University School of Medicine, Yokohama, Japan 236-0004. M. Maeda, Wakayama Medical College, Wakayama 641-0012, Japan. H. Hirano, University of Occupational and Environmental Health, Kitakyusyu 807-8555, Japan. Y. Misu, Shinobu Hospital, Fukushima 960-1101, Japan.

Antisense, sense and scrambled oligo-thionucleotides for the rat neuronal nitric oxide synthase (nNOS) mRNA were synthesized as described previously.[4]

3. RESULTS

Under our experimental conditions, resting mean BP was approximately 80-100 mmHg, and HR was 360-420 beats/min. DOPA at 300 pmol microinjected into the NTS produced depressor and bradycardic responses. Kynurenic acid at 60 nmol microinjected produced no effect on resting BP and HR. Injection of kynurenic acid 2 min prior to DOPA reduced by approximately 80% depressor and bradycardic responses to DOPA (Table 1).

GN-monomethyl-L-arginine (L-NMMA), a competitive inhibitor of NOS, at 100 nmol microinjected into the NTS produced a slight pressor response: the maximal change in BP was 11 ± 6 mmHg. After recovery to the resting level, DOPA (300 pmol) was microinjected into the NTS. Injection of L-NMMA 10 min prior to DOPA reduced reversibly depressor and bradycardic responses to DOPA by approximately 90% (Table 1). Prior injection of D-NMMA produced no effect on the responses to DOPA microinjected.

Antisence, sense or scrambled neuronal NOS (nNOS) oligos each 5 nmol microinjected into the unilateral NTS produced no effect on the resting BP and HR. Forty-five min after microinjection of nNOS oligos, DOPA was microinjected into the NTS. nNOS antisence oligos reduced reversibly the depressor and bradycardic responses to DOPA (Table 2). Sense or scrambled nNOS oligos produced no effect on the responses to DOPA microinjected.

L-Arginine at 30 nmol microinjected into the NTS produced depressor and bradycardic responses. Injection of kynurenic acid 2 min prior to L-arginine failed to reduce the responses to L-arginine (data not shown).

4. DISCUSSION

We demonstrated that exogenously microinjected DOPA into the NTS produced depressor responses via activation of kynurenic acid-sensitive ionotropic glutamate receptors and nNOS. Kynurenic acid, a broad-spectrum antagonist for ionotropic glutamate receptors, has been shown to abolish the depressor response to ADN stimulation.[1] In the present study, kynurenic acid reduced by approximately 80% depressor responses to DOPA. This suggests that a major pathway for DOPA is mediated by kynurenic acid-sensitive ionotropic glutamate receptors. We previously tested possible interactions of DOPA-related compounds with ionotropic glutamate receptors.[5] Among binding sites labeled with available glutamatergic [^{3}H]-ligands in rat brain membrane, DOPA acts only on AMPA receptors with low affinity (IC_{50} =260 μM), whereas competitive DOPA antagonists DOPA esters act only on NMDA ion channel domain with mM IC_{50}s. This indicates that DOPA and DOPA antagonists do not interact with ionotropic glutamate receptors: DOPA recognition sites differ from these receptors. Furthermore, DOPA esters do not displace kynurenic acid-sensitive binding of the ^{3}H-ligands. Thus, DOPA does not produce responses by interacting directly with kynurenic acid-sensitive ionotropic glutamate receptors.

Table 1. Effects of prior injection of kynurenic acid, L-or D-NMMA on depressor and bradycardic responses to DOPA microinjected into depressor sites of the unilateral NTS

		ΔBP (mmHg)		ΔHR (beats/min)	
Pretreatment	n	before	after	before	after
Kynurenic acid	5	-40 ± 5	-11 ± 6*	-70 ± 18	-8 ± 4†
L-NMMA	5	-33 ± 5	-6 ± 3†	-53 ± 19	-8 ± 5*
D-NMMA	4	-38 ± 5	-33 ± 5	-66 ± 11	-40 ± 4

Kynurenic acid (60 nmol), L- or D-NMMA (100 nmol) was microinjected into the NTS prior to DOPA (300 pmol). Data are mean ± SEM. *P < 0.01, †P < 0.05 (Student's *t*-test) vs before.

Table 2. Effects of prior injection of nNOS antisense oligos on depressor and bradycardic responses to DOPA microinjected into depressor sites of the unilateral NTS

		ΔBP (mmHg)		ΔHR (beats/min)	
Pretreatment	n	before	after	before	after
Antisence oligos	6	-48 ± 4	-15 ± 9*	-142 ± 29	-45 ± 31†
Sense	4	-39 ± 6	-39 ± 5	-79 ± 22	-75 ± 22
Scrambled	4	-41 ± 5	-42 ± 6	-97 ± 26	-113 ± 36

After depressor responses to DOPA (300 pmol) microinjected into the unilateral NTS, nNOS oligos (5 nmol) were microinjected into the ipsilateral NTS. The second DOPA into the NTS was challenged 45 min after nNOS oligos. Data are mean ± SEM. *P < 0.01, †P < 0.05 (Student's *t*-test) vs before.

NMMA stereoselectively reduced the cardiovascular responses to DOPA. NNOS antisense oligos reduced the depressor responses to DOPA. These results indicate that nNOS is involved in DOPA-induced depressor responses in the NTS. Microinjection of L-arginine into the NTS produced depressor and bradycardic responses. Kynurenic acid at the dose, which reduced the cardiovascular response to DOPA, failed to affect the responses to L-arginine. This result suggests that activation of kynurenic acid-sensitive glutamate receptors is not involved in the depressor and bradycardic responses to NO converted from L-arginine, and that the NO production is a downstream event of activation of the glutamate receptors involved in depressor responses to DOPA.

Several lines of evidence support the idea that some interaction exists between glutamate receptors and NOS in the NTS.[6] There is an anatomical link between glutamate and NO systems on immunohistochemical studies in the NTS. Our present findings indicate that the cardiovascular responses to DOPA are mediated via ionotropic glutamate receptors and nNOS. Probably, the interaction between glutamate and NO also exists in the cardiovascular responses to DOPA. L-NMMA reduces the depressor responses to glutamate in the NTS. In the rat brain, activation of NMDA glutamate receptors leads to synthesis and release of NO. Our present findings indicate primarily the importance of kynurenic acid-sensitive ionotropic glutamate receptors, because kynurenic acid almost completely blocked depressor responses to DOPA. Our preliminary experiments indicate that D(-)-2-amino-5-phosphonopentanoate, a selective NMDA antagonist, reduces by approximately 50% depressor and bradycardic responses to DOPA microinjected into the NTS (unpublished observation). It is likely that DOPA microinjected into the NTS induces depressor responses via kynurenic acid-sensitive glutamate receptors. DOPA microinjected into the NTS probably releases glutamate via DOPA recognition sites, and glutamate released accelerates synthesis and release of NO. Consistently, DOPA releases endogenous glutamate in striatum in vitro and in vivo.[2,7,8]

We thus propose a new pathway that DOPA is a transmitter of the primary ADN, and glutamate is that of secondary order neurons in neuronal microcircuits of the NTS. DOPA probably releases endogenous glutamate from the secondary neurons via DOPA recognition, glutamate released elicits NO production *via* kynurenic acid-sensitive glutamate receptors, and resultant NO production induces the depressor and bradycardic responses.

REFERENCES

1. Y. Misu, Y. Goshima, H. Ueda, and H. Okamura, Neurobiology of L-DOPAergic systems, *Prog. Neurobiol* .**49,** 415-454 (1996).
2. N. Furukawa, N. Arai, Y. Goshima, T. Miyamae, E. Ohshima, F. Suzuki, K. Fujita, and Y. Misu, Endogenously released DOPA is a causal factor for glutamate release and resultant delayed neuronal cell death by transient ischemia in rat striata, *J. Neurochem.* **76,** 815-824 (2001).
3. J-L. Yue, H. Okamura, Y. Goshima, S. Nakamura, M. Geffard, and Y. Misu, Baroreceptor-aortic nerve-mediated release of endogenous L-3,4-dihydroxyphenylalanine and its tonic depressor function in the nucleus tractus solitarii of rats, *Neuroscience* **62,** 145-161 (1994).
4. M. Maeda., H. Hirano, H. Kudo, Y. Doi, K. Higashi, and S. Fujimoto, Injection of antisense oligos to nNOS into nucleus tractus solitarii increases blood pressure, *NeuroReport* **10,** 1957-1960 (1997).
5. T. Miyamae, Y. Goshima, M. Shimizu, T. Shibata, K. Kawashima, E. Ohshima, F. Suzuki, and Y. Misu, Some interactions of L-DOPA and its related compounds with glutamate receptors, *Life Sci.* **64** , 1045- 1054 (1999).
6. L.-H. Lin, P.C. Emson, and W.T. Talman, Apposition of neuronal elements containing nitric oxide synthase and glutamate in the nucleus tractus solitarii of rat: a confocal microscopic analysis, *Neuroscience* **96,** 341-350 (2000).
7. T. Maeda, N.-N. Cheng, T. Kume, S. Kaneko, H. Kouchiyama, A. Akaike, M. Ueda, M. Satoh, Y. Goshima, and Y. Misu, L-DOPA neurotoxicity is mediated by glutamate release in cultured rat striatal neurons, *Brain Res.* **771,** 159-162 (1997).
8. Y. Goshima, K. Ohno, S. Nakamura, T. Miyamae, Y. Misu, and A. Akaike, L-DOPA induces Ca^{2+}-dependent and tetrodotoxin-sensitive release of endogenous glutamate from rat striatal slices, *Brain Res.* **617,** 167-170 (1993).

EFFICIENT GENE TRANSFER TO NEURAL STEM CELLS BY HIGH TITER RETROVIRAL VECTORS

Hiroko Baba[1,2,5], Hideki Hida[1,2,3], Yuji Kodama[1,2], Cha-Gyun Jung[1,2,3], Chun-Zhen Wu[1,2], Kouji Nanmoku[6], Kazuhiro Ikenaka[3,4], and Hitoo Nishino[1,2,3]

1. INTRODUCTION

Neural stem cells are self-renewing and multipotent cells, and are thought to be ideal sources for neural transplantation.[1] However, controlled conversion of these cells into certain types of neurons is difficult. To establish a system for cell fate control or to employ gene manipulation to assign specific functions, the efficiency of retroviral gene transfer was examined. Retroviral transduction offers the advantage of stable gene expression as well as low cytotoxicity. Yet, the vector is limited by the low efficiency of infection. In this study, high titer retrovirus (up to 10^{12} cfu/ml) was prepared and neural stem cells were transduced efficiently. To test the application of this system in the therapy of Parkinson disease, virus producer cells containing the human tyrosine hydroxylase (HTH) gene were established to prepare high titer TH retrovirus.

2. MATERIALS AND METHODS

Two types of retroviral vectors were used for the present studies (Fig. 1). DNA was transfected into the packaging cell line (φMP34) to establish viral producer cells.[4] The stable transformants were selected using FDG-FACS. High titer retrovirus was prepared by centrifuging the culture media at 6,000 x g for 16 hr at 4°C as described before[2, 5, 6].

Rat fetal brain was dissociated and cultured in the proliferating medium (D/F with N2 and FGF-2). The secondary neurospheres were cultured with virus overnight. The secretion of L-Dopa was assayed using a HPLC with ECD as described previously[4].

[1] Department of Physiology, Nagoya City University Medical School, Nagoya 467-8601, [2]Strategic Promotion System for Brain Research (SPSBS), Ministry of Education, Culture, Sports, Science and Technology, [3]Research for the Future, Japan Society for the Promotion Science, [4]National Institute for Physiological Sciences, Okazaki 444-8585, [5]Tokyo University of Pharmacy and Life Science, Hachioji 192-0392, [6]Hirosaki University, School of Medicine, Hirosaki, Japan

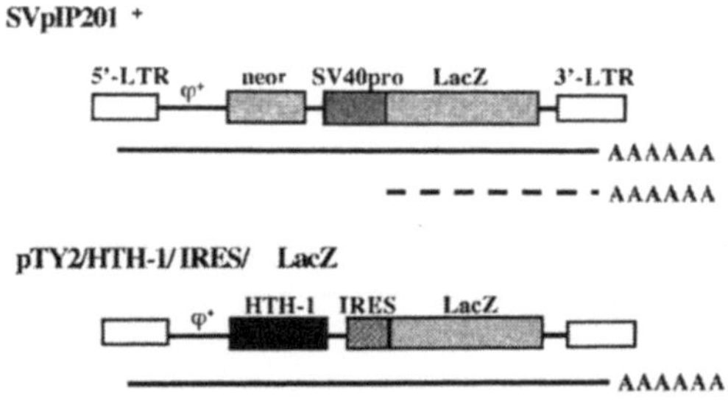

Figure 1. Schematic diagrams of retroviral vectors. SvpIP201⁺ virus contains two transcriptional units under the control of the LTR viral promoter (solid line) and the SV40 promoter (dotted line). The protein rendering neomycin resistance is produced from the longer mRNA, while β-galactosidase is produced from the shorter one. pTY2/HTH-1/IRES/LacZ transcribes only one mRNA, from which human tyrosine hydroxylase-1 (HTH-1) and β-galactosidase are produced independently by the presence of the IRES sequence.

3. RESULTS

Using φMP34 and concentration of the virus, high titer viruses up to 1.0×10^{12} cfu/ml were obtained[2]. First, the transduction efficiencies to neurospheres were examined. The neurospheres from the E12.5 rat ventral mesencephalon were prepared.[7] Figure 2A shows the X-Gal staining of neurospheres transduced by 10 μl of SVpIP201⁺ (1.0×10^{12} cfu/ml). Many cells on the surface as well as the differentiated cells migrating out from the spheres were X-Gal positive. The same efficiency was observed by the virus of 1.0×10^9 cfu/ml (Fig 2B). In contrast, the X-Gal-positive cells decreased to one tenth when the diluted virus (1.0×10^8 cfu/ml) was added (Fig 2C). Thus, while neural stem cells are relatively resistant to retroviral infection, the preparation of high titer retrovirus surmounted this problem.

After transfection of pTY2/TH/IRES/LacZ DNA into φMP34, the cells producing both TH and β-galactosidase were selected by FDG-FACS. Finally, almost all the cells were X-Gal positive after the second selection (φMP34/TH-*lacZ*). TH-*lacZ* virus titer was 1×10^7 cfu/ml. The virus was concentrated successfully up to 1×10^9 cfu/ml. The

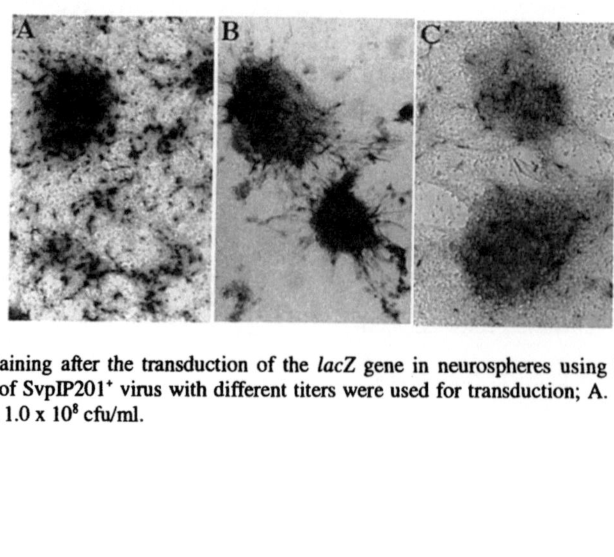

Figure 2. X-Gal staining after the transduction of the *lacZ* gene in neurospheres using high titer retrovirus. Three preparations of SvpIP201⁺ virus with different titers were used for transduction; A. 1.0×10^{12} cfu/ml, B. 1.0×10^9 cfu/ml, C. 1.0×10^8 cfu/ml.

φMP34/TH-*lacZ* cells stably produced the same titer of TH-*lacZ* virus even after the cells were thawed from frozen stock in liquid nitrogen.

The neurospheres were infected by TH-*lacZ* virus. Then the medium was changed into the differentiation medium with 1% FCS. By X-Gal staining, most of blue cells had flat shapes, which apparently were astrocytes (Fig. 3A). In addition, some small neuron-like blue cells colonized on the X-Gal negative astrocytic sheets (Fig. 3B). In these cells, L-Dopa was detected when BH_4 was added to the medium, and the levels of L-Dopa were increased significantly depending on the loading levels of BH_4 (Fig. 3C). The L-Dopa in the medium was 120 nM when 250 μM of BH_4 was present in the medium.

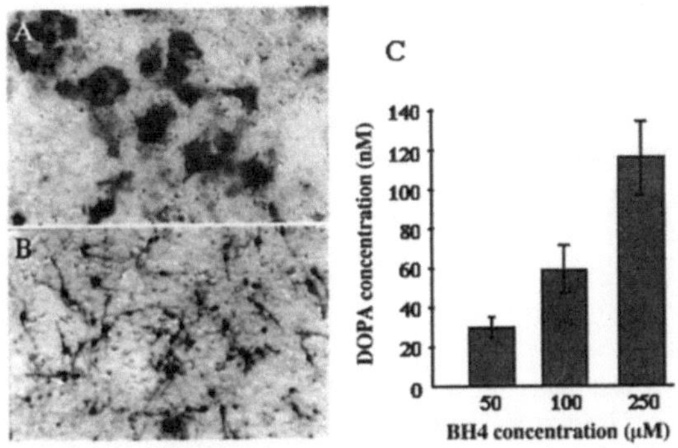

Figure 3. X-Gal staining of the transduced cells after the induction of differentiation (A, B). (C) The transduced cells released L-DOPA into the culture medium when the coenzyme BH_4 was added. The levels of L-DOPA increased in a BH_4- dose dependent manner.

4. DISCUSSION

Previously, we generated Dopa-producing astrocytes by adenoviral transduction of the TH gene. Using these cells, one-third of the transplanted hemiparkinsonian rats exhibited a reduction of methamphetamine-induced rotations.[3] However, the TH-positive astrocytes began to be eliminated after 6 weeks, since cells transduced by adenovirus often induced the late immune response in the host.[3, 8] In contrast, cells treated by retrovirus do not induce such response. Though neurospheres are relatively resistant to retrovirus, the present study clearly demonstrated that using the high-titer retrovirus, the foreign gene was transferred at sufficient quantities to the cells as shown in Fig. 2 and 3.

For clinical applications, simple preparation and a large amount of stable virus supply are important practical factors. For this, retrovirus is the best system. The vector DNA is constructed as easily as standard plasmid, and viral producer cells are simply prepared by DNA transfection into the packaging cell lines. Once the stable viral producer cells are established, retrovirus can be collected easily.[4] Now, φMP34/TH-*lacZ* cells should be a practical viral source for future gene therapy trials.

Many trials to obtain dopaminergic cells are currently ongoing for the cell therapy of Parkinson's diseases. The clinical trial of mesencephalic cells from human embryos demonstrated the effectiveness,[9] however, numbers of embryos required for each patient set the limitation for further clinical usage. Recent study reported that ES cells were able to differentiate to dopaminergic neurons when they were co-cultured with stromal cells.[10] This may be one approach for clinical use, because the cell supply will be unlimited. However, the purification of dopaminergic neurons would be critical for the prevention of tumorigenesis. The genetically manipulated neural progenitor cells circumvents such concerns, and they can be engrafted as integral members of normal structures in the CNS. Thus, the transplantation of neural stem cells that have been genetically manipulated *ex vivo* by retrovirus may provide a strategy for the long-term treatment of CNS manifestations of Parkinson's disease as well as a number of neurodegenerative diseases.

ACKNOWLEDGEMENTS

This study was performed through Special Coordination Funds, No13073-2125-14, of the Ministry of Education, Culture, Sports, Science and Technology of the Japanese Government (HH, HN), and Research for the Future of Japan Society for the Promotion Science, No 00L01507 (HB, IK).

REFERENCES

1. F. H. Gage, Mammalian neural stem cells, *Science*, **287**, 1433-1438 (2000).
2. K. Nanmoku, Y. Miyao, M. Tamura, K. Tamura, T. Yoshimatsu, K. Shimizu, and K. Ikenaka, Gene transfer to the nervous system using high-titer retroviral vector, *Mol. Biol. of Hematopoiesis* **6**, 299-305 (1999).
3. H. Hida, M. Hashimoto, I. Fujimoto, K. Nakajima, Y. Shimano, T. Nagatsu, K. Mikoshiba, and H. Nishino, Dopa-producing astrocytes generated by adenoviral transduction of human tyrosine hydroxylase gene: in vitro study and transplantation to hemiparkinsonian model rats, *Neurosci. Res.* **35**, 101-112 (1999).
4. T. Yoshimatsu, M. Tamura, S. Kuriyama, and K. Ikenaka, Improvement of retroviral packaging cell lines by introducing the polyomavirus early region, *Hum. Gene Ther.* **9**, 161-172 (1998).
5. K. Ikenaka, K. Nakahira, K. Nakajima, I Fujimoto, T. Kagawa, M. Ogawa, and K. Mikoshiba, Detection of brain-specific gene expression in brain cells in primary culture: A novel promoter assay based on the use of a retrovirus vector. *New Biol.* **4**, 53-60 (1992).
6. N.E. Bowles, R. C. Eisensmith, R. Mohuiddin, M. Pyron, and S. L. C. Woo, A simple and efficient method for the concentration and purification of recombinant retrovirus for increased hepatocyte transduction in vivo, *Hum. Gene Ther.* **7**, 1735-1742 (1996).
7. H. Nishino, H. Hida, N. Takei, M. Kumazaki, K. Nakajima, and H. Baba, Mesencephalic neural stem (progenitor) cells develop to dopaminergic neurons more strongly in dopamin-depleted striatum than in intact striatum, *Exp. Neurol.* **164**, 209-214 (2000).
8. M. J. A. Wood, H. M. Charlton, K..J.Wood, K. Kajiwara, and A. P. Byrnes, Immune responses to adenovirus vectors in the nervous system, *Trends Neurosci.* **19**, 497-501 (1996).
9. C. R. Reed, R. E. Breeze, N. L. Rosenberg, S. A. Schneck, T. H. Wells, J. N. Barrett, S. T. Grafton, S. C. Huang, D. Eidelberg, and D. A. Rottenberg, Transplantation of human fetal dopamine cells for Parkinson's disease. Results at 1 year, *Arch Neurol* **47**, 505-12 (1990).
10. H. Kawasaki, K. Mizuseki, S. Nishikawa, S. Kaneko, Y. Kuwana, S. Nakanishi, S. Nishikawa, and Y. Sasai, Induction of midbrain dopaminergic neurons from ES cells by stromal cell-derived inducing activity, *Neuron* **28**, 31-40 (2000).

THE EVOLUTION OF IN VIVO VOLTAMMETRY AND MICRODIALYSIS

Joseph B. Justice, Jr.*

1. INTRODUCTION

In vivo voltammetry and microdialysis are two leading techniques for monitoring neurochemistry in the brain. This paper addresses the state of the art of voltammetry and microdialysis as they have evolved from their respective beginnings. Voltammetry is an old and well established electrochemical method. However, the development and application of in vivo voltammetry and microdialysis for measurement of neurotransmitters began about the same time in the 1970's. In R. N. Adams laboratory at the University of Kansas, carbon paste electrodes were first put in ventricles of the rat brain and then in the brain itself.[1, 2] The developments that followed have been reviewed.[3] About this same time, Urban Ungerstedt, developed the microdialysis probe, based on the idea of exchange of substances across a blood capillary.[4] This was an extension of the earlier push-pull device of Gaddum[5] and the dialysis devices of Bito et al.[6] and Delgado

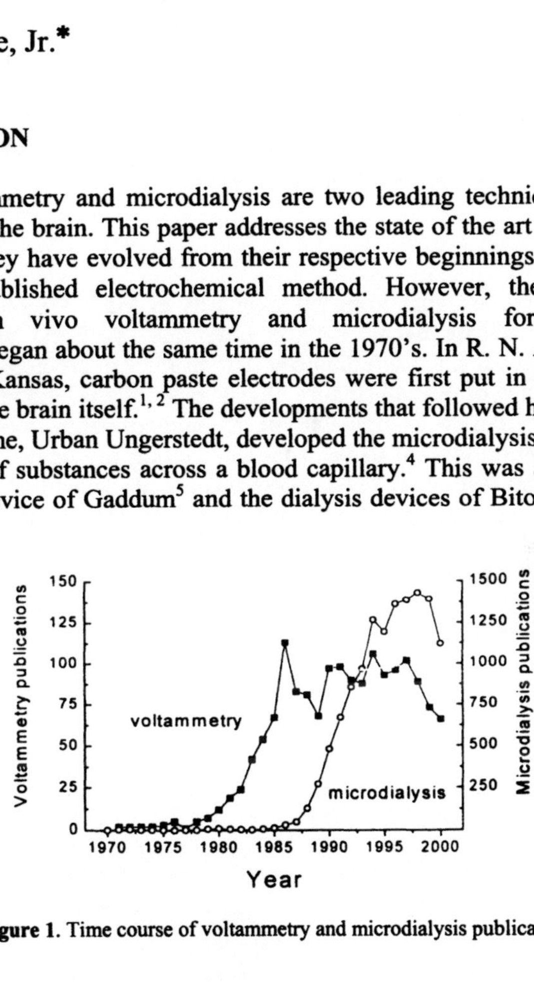

Figure 1. Time course of voltammetry and microdialysis publications.

* Department of Chemistry, Emory University, Atlanta, Georgia, 30322

et al.[7] for sampling the extracellular environment of the brain. More than a thousand papers a year are now published using microdialysis.

This paper focuses on the current state of the art in voltammetry and microdialysis. Four examples are presented illustrating the information that can be obtained.

2. IN VIVO VOLTAMMETRY

In vivo voltammetry has been primarily concerned with monitoring catecholamines. The two catecholaminergic processes receiving the most attention have been release and uptake. Examples illustrating the state of the art voltammetric applications in each area are shown. Recent developments leading to the current state of the art have focused on observation of events at single cells. A particularly successful example is the recording of vesicular release. New catecholamine transport phenomena have also been observed with combined voltammetry and patch-clamp techniques.

2.1 Release

Recent state of the art work from the laboratory of Mark Wightman has shown that individual vesicular release events can be observed.[8] Working with bovine adrenal medullary chromaffin cells, a 5 um diameter carbon fiber microelectrode is brought into contact with the surface of the cell (Figure 2). The electrode is held at a fixed potential of 0.6V so that the electrode is operated in an amperometric mode, allowing rapid fluctuations in oxidation current to be recorded. During the course of the experiment, background-subtracted fast cyclic voltammograms are also obtained for confirmation of identity of chemical species. Individual current spikes are observed corresponding to vesicle fusion and release at the plasma membrane with a half-life of about 5 ms, a time longer than would be predicted for a diffusion limited process. A rate-limiting process appears to be dissociation of the vesicular matrix.[9, 10] The role of additional regulatory mechanisms in exocytosis has also been investigated voltammetrically.[11, 12]

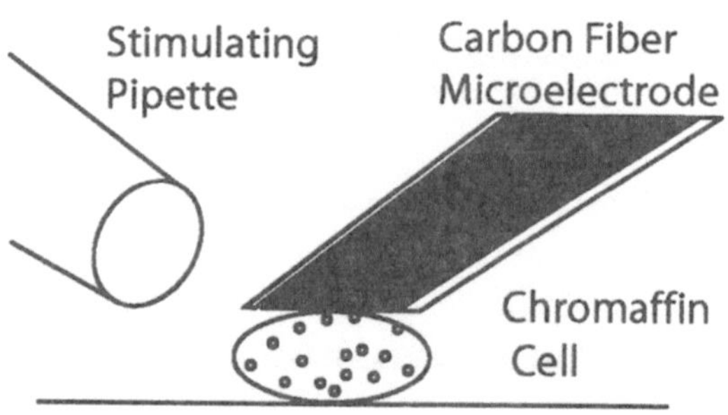

Figure 2. Diagram of electrochemical measurement of exocytosis. The carbon fiber microelectrode is in direct contact with the surface of a single chromaffin cell. A pressure ejection pipette is used to deliver stimulating solutions.

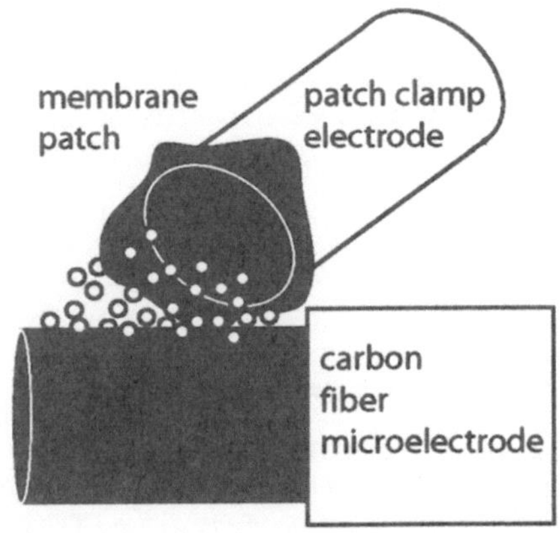

Figure 3. Diagram of patch-clamp and carbon fiber electrode. The membrane contains hNET with uptake occurring in an outward direction. An amperometric oxidation current from NE transported across the inside out membrane is detected upon stepping the membrane voltage to –100 mV.

2.2 Uptake

Voltammetry has recently been extended to study catecholamine transporters on inside-out patches with synchronous recording of flux and current.[13] Inside-out patches of membrane from HEK cells stably expressing the human norepinephrine (NE) transporter (hNET) were used in a voltage-clamped pipette containing NE. The tip of the pipette is placed near a cylindrical carbon fiber microelectrode operated amperometrically, as shown in figure 3. When the voltage across the membrane is clamped from resting 0 mV to –100 mV, an increase in oxidation current is seen that is attributable to transporter mediated flux of NE across the membrane. By monitoring a patch containing fewer transporters, temporally resolved events could be observed in which Galli et al. demonstrated a temporal correlation between NE flux and NE-induced current. The magnitude of the NE flux required hundreds to thousands of NE molecules to cross the membrane rapidly, which suggests the transporter can operate in a channel mode in addition to a mode described by an alternating access model. While a channel cannot concentrate against a concentration gradient, it is possible that the channel mechanism may dominate in the early stages of clearance, given the high concentrations of transmitter that occur in the synapse during release.

3. MICRODIALYSIS

Microdialysis, which also began in the 1970s, in the laboratory of Urban Ungerstedt, differs from voltammetry in that it is sampling technique, and developments have been in increased sampling sensitivity, speed and range of analytes. While catecholamines continue to be analyzed by HPLC with electrochemical detection, capillary electrophoresis and mass spectrometry have found increasing use for other transmitters amino acids, peptides and other substances in the brain. A recent development of particular note is the use of microdialysis in human studies.[14] Analysis of the microdialysis sampling process itself has continued to evolve. The work of Michael and colleagues, reviewed in Peters et al.[15], advances the understanding of the differential influence of clearance processes such as uptake on the microdialysis extraction and recovery.

REFERENCES

1. R. N. Adams, Probing brain chemistry with electroanalytical techniques, *Anal. Chem.* **48**, 1128A-1138A (1976).
2. R. M. Wightman, E. Strope, P. M. Plotsky, R. N. Adams, Monitoring of transmitter metabolites by voltammetry in cerebrospinal fluid following neural pathway stimulation, *Nature* **262**, 145-147 (1976).
3. J. A. Stamford and J. B. Justice, Jr., Probing Brain Chemistry: Voltammetry comes of age, *Anal. Chem.* **68**, 359A-363A (1996).
4. U. Ungerstedt and C. Pycock, Functional correlates of dopaminergic neurotransmission, *Bull. Schweiz. Akad. Med. Wiss.* **1278**, 1-13 (1974).
5. J. H. Gaddum, Push-pull cannulae, *J. Physiol.* **155**, 1-2 (1961).
6. L. Bito, H. Davson, E. M. Levin, M. Murray, N. Snider, The concentration of free amino acids and other electrolytes in cerebrospinal fluid, in vivo dialysate of brain, and blood plasma of the dog, *J. Neurochem.* **13**, 1057-1067 (1966).
7. J. M. R. Delgado, F. V. DeFeudis, R. H. Roth, D. K. Ryugo, B. M. Mitruka, Dialytrode for long term intracerebral perfusion in awake monkeys, *Arch. Int. Pharmacodyn.* **198**, 9-21 (1972).
8. D. J. Leszczyszyn, J. A. Jankowski, O. H. Viveros, E. J. Diliberto, Jr., J. A. Near, and R. M. Wightman, Secretion of catecholamines from individual adrenal medullary chromaffin cells, *J. Neurochem.* **56**, 1855-1863 (1991).
9. T.J. Schroeder, R. Borges, J. M. Finnegan, K. Pihel, C. Amatore, R. M. Wightman, Temporally resolved, independent stages of individual exocytotic secretion events, *Biophys. J.* **70**, 1061-1068 (1996).
10. E. R. Davis and R. M. Wightman, Spatio-temporal resolution of exocytosis from individual cells, *Annu. Rev. Biophys. Biomol. Struct.* **27**, 77-103 (1998).
11. R. H. Chow, J. Klingauf, C. Heineman, R. S. Zucker, E. Neher, Mechanisms determining the time course of secretion in neuroendocrine cells. *Neuron* **16**, 369-376 (1996).
12. R. Rahmimoff, J. M. Fernandez, Pre- and postfusion regulation of transmitter release, Neuron **18**, 17-27 (1997).
13. A. Galli, R. D. Blakely, L. J. DeFelice, Patch-clamp and amperometric recordings from norepinephrine transporters: Channel activity and voltage-dependent uptake, *Proc. Natl. Acad. Sci. USA* **95**, 13260-13265 (1998).
14. O. G. Nilsson, L. Brandt, U. Understedt, H. Saveland, Bedside detection of brain ischemia using intracebral microdialysis: subarachnoid hemorrhage and delayed ischemic deterioration, *Neurosurgery* **45**, 1176-84 (1999).
15. J. L. Peters, H. Yang, and A. C. Michael, Quantitative aspects of brain microdialysis, *Anal. Chim. Acta* **412**, 1-12 (2000).

FAST-SCAN CYCLIC VOLTAMMETRY IN FREELY-MOVING RATS

Paul E.M. Phillips, Evgeny A. Budygin, Donita L. Robinson and R. Mark Wightman*

1. INTRODUCTION

Microdialysis has been the tool of choice for the measurement of bioactive compounds in the brain of behaving animals for a number of years. Quantitative measurements of the extracellular composition can be made with precise chemical specificity over relatively long periods of time in pathological, pharmacological and behaviorally salient situations. Many investigators have utilized this technique for the measurement of forebrain dopamine under a variety of conditions.

Recently, voltammetric methods have also been used to measure extracellular dopamine in freely-moving animals (Garris et al., 1997; Yavich and Tiihonen, 2000). However, rather than sampling the medium, these techniques make direct measurements at carbon microelectrodes implanted in the brain. Measurements are made by applying a potential to the microelectrode that is sufficient to oxidize any dopamine in its immediate vicinity, and recording the current that is produced by the liberation of electrons during this process. The size of the current is directly proportional to the number of molecules oxidized, and hence dopamine can be quantified. Fast-scan cyclic voltammetry has the advantage over other voltammetric techniques in that it has the inherent capacity to chemically identify analytes, while achieving millisecond temporal resolution.

This chapter describes two examples where dopamine recordings using fast-scan cyclic voltammetry in freely-moving rats allow an insight into biological mechanisms that could not be obtained with microdialysis, but compliments data obtained therewith.

* Department of Chemistry and Neuroscience Center, University of North Carolina, Chapel Hill, NC 27599-3290, USA

2. INVESTIGATION OF MECHANISMS

The mean dopamine concentration in the extracellular compartment with respect to time is the net result of its rates of appearance and removal. Appearance of dopamine is due to neuronal release following delivery of an action potential or local depolarizing agents (Grace, 1991), and its removal is predominantly through uptake by the dopamine transporter (Giros et al., 1996). These parameters can be modulated at a number of physiologically relevant levels. Dopamine release can be altered by changing 1) the rate or pattern of afferent action potential traffic 2) the level of tonic releasing factors 3) the number of dopamine vesicles released by an action potential or local depolarization or 4) the average amount of dopamine in a vesicle. The rate of uptake can also be altered.

Data obtained from microdialysis recordings encompass all of the above integrated over time (minutes), and so although a net change in dopamine can be identified, little information is provided on the mechanisms of the change. However, using fast-scan cyclic voltammetry coupled with electrical stimulation of afferent dopamine pathways, more information can be obtained. By repeating stimulations throughout a manipulation (e.g. pharmacological) or across animals of different chronic states, the mechanisms altering dopamine function can be examined. The response is a rise in dopamine concentration during the stimulation and a return to baseline after its termination. The rising phase is reciprocally controlled by dopamine release and uptake, and the falling phase by uptake alone. Therefore, assessment of the rate of uptake can be made by examination of the falling phase of the response. Then, accounting for this, release for a fixed number of action potentials (the delivered stimulus) can be determined from the rising phase. Hence, conditions that alter release mechanisms, uptake or cell firing can be distinguished.

A recent examination into the effects of ethanol on striatal dopamine (Budygin et al., 2001) used this approach and is summarized in figure 1. For ethanol, microdialysis studies reveal a biphasic dose-response (Blanchard et al., 1993), where low doses elevate extracellular dopamine and higher doses decrease it. This is consistent with the biphasic behavioral activation produced by ethanol since dopamine is positively linked to "behavioral alertness" (Schultz, 1994), but it does not give insight into which dopamine regulating parameters are altered. When fast-scan cyclic voltammetry was used to measure electrically evoked dopamine following ethanol administration, no change in uptake was apparent, but there was a dose-dependent decrease in dopamine release (see figure 1). This result can only describe part of the story seen with microdialysis. However, the firing rate of dopamine neurons is increased by ethanol in a dose-dependent manner (Mereu et al., 1984; figure 1). When this is considered together with the voltammetric data the source of the biphasic nature is revealed. The effect of ethanol on dopamine electrophysiology is more potent than that on dopamine release. Hence, for low ethanol doses there is an increase in cell firing with little change in dopamine release producing a net increase in extracellular dopamine, and at higher doses there is still increased firing but much less dopamine released per impulse producing a net decrease in extracellular dopamine. Therefore, we can conclude that extracellular dopamine is altered by ethanol, by an increase in dopamine cell firing and a less potent decrease in dopamine release, producing a net biphasic dose-response.

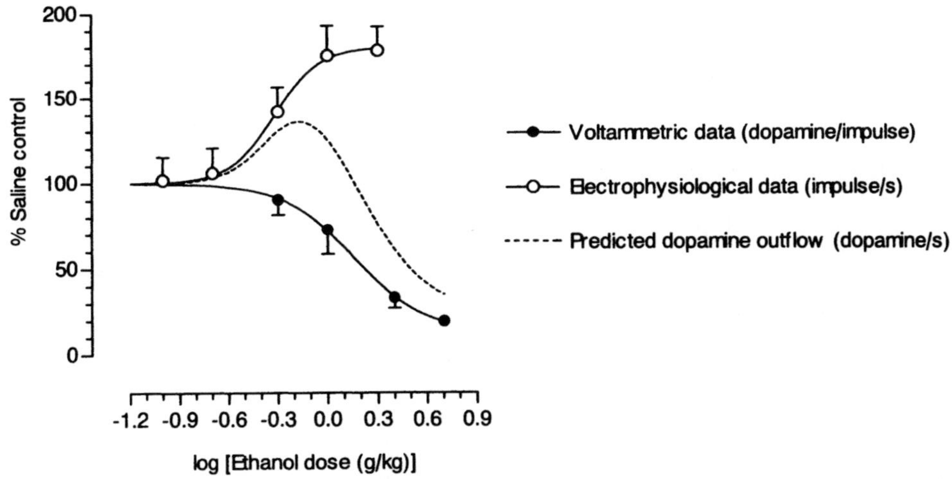

Figure 1. Measures of dopamine neurotransmission as a function of ethanol dose (means ± s.e.m.). The voltammetric data (closed circles) are evoked dopamine concentration after ethanol normalized by the time-matched controls. The electrophysiological data (open circles) are adapted from Mereu et al. (1984). The solid lines are sigmoidal curves obtained by nonlinear regression of the data points. The predicted rate of dopamine outflow (dashed line), and thus extracellular dopamine concentration, is the product of the two data sets.

3. CORRELATION WITH BEHAVIOR

The previous example examined changes in dopamine evoked by electrical stimulation. Although this parameter was demonstrated to be biologically meaningful, its relevance to behavior is not clear. To address this, the effect of GBR 12909, a selective dopamine uptake antagonist, on both stimulated dopamine and behavior was measured (Budygin et al., 2000; figure 2). This drug increases the amplitude of the evoked dopamine signal because uptake is the mechanism that lowers dopamine during the rising phase of the stimulated response. By repeating stimulations at regular intervals, "snapshots" of the status of the extracellular environment were obtained. The increase in evoked dopamine produced by GBR 12909 was temporally linked to its induction of stereotypy. Indeed, when the behavioral effect was ranked over time by stereotypy score, it was tightly correlated to the increase in evoked dopamine (figure 2). Furthermore, the temporal profile of the behavioral change was paralleled much closer by the voltammetric data than by data from microdialysis in the same study (data not shown). This is not altogether surprising since with microdialysis, dopamine has to diffuse through tissue to reach the probe and undergo partition across the dialysis membrane, whereas with the voltammetric approach the small size of the carbon-fiber microelectrode (diameter ~7 μm) allows its placement closer to release sites and the measurement occurs as soon as dopamine encounters the electrode. Although microdialysis has the requisite sensitivity and selectivity to reveal the steady-state changes in basal dopamine levels that occur in response to a pharmacological challenge, it may not be so well suited to measuring the dynamic changes during the onset of drug action. Fast-scan cyclic voltammetry is able to achieve this goal, in circumstances where evoked dopamine is an appropriate parameter to monitor the drug's action. In the example shown here, stimulations were repeated every 10 min, however, for faster acting drugs it is possible to reduce this interval to improve the temporal resolution of the data.

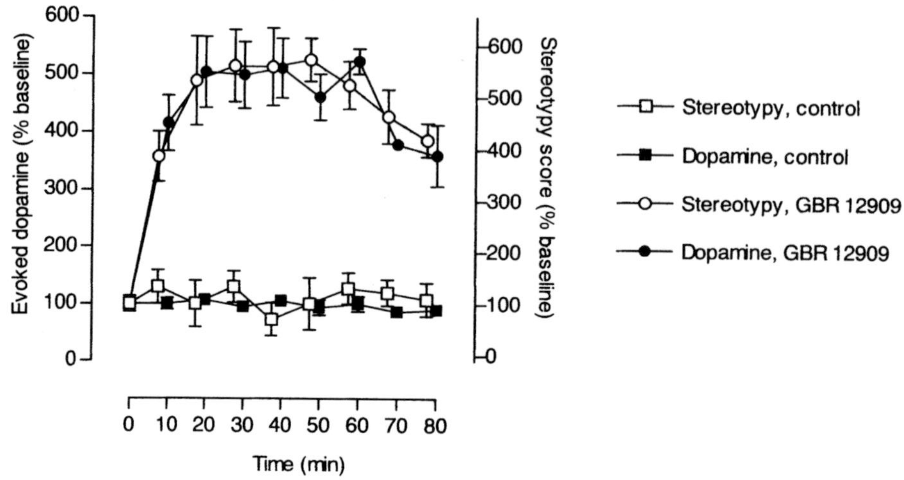

Figure 2. GBR 12909 (20 mg/kg, i.p.) induced increases in evoked dopamine and stereotypy (means ± s.e.m.). Stereotypy was assessed using the scale described by Murray and Waddington (1990).

REFERENCES

Blanchard, B.A., Steindorf, S., Wang, S. and Glick, S.D. (1993) Sex differences in ethanol-induced dopamine release in nucleus accumbens and in ethanol consumption in rats. *Alcohol Clin. Exp. Res.* **17**:968–973.

Budygin, E.A., Kilpatrick, M.R., Gainetdinov, R.R. and Wightman, R.M. (2000) Correlation between behavior and extracellular dopamine levels in rat striatum: Comparison of microdialysis and fast-scan cyclic voltammetry. *Neurosci. Lett.* **281**:9–12.

Budygin, E.A., Phillips, P.E.M., Robinson, D.L., Kennedy, A.P., Gainetdinov, R.R. and Wightman, R.M. (2001) Effect of acute ethanol on striatal dopamine neurotransmission in ambulatory rats. *J Pharmacol Exp Ther.* **297**:27-34.

Garris, P.A., Christensen, J.R.C., Rebec, G.V. and Wightman, R.M. (1997) Real-time measurement of electrically evoked extracellular dopamine in the striatum of freely moving rats. *J. Neurochem.* **68**:152–161.

Giros, B., Jaber, M., Jones, S.R., Wightman, R.M. and Caron, M.G. (1996) Hyperlocomotion and indifference to cocaine and amphetamine in mice lacking the dopamine transporter. *Nature* **379**:606–612.

Grace, A.A. (1991) Phasic versus tonic dopamine release and the modulation of dopamine system responsivity: A hypothesis for the etiology of schizophrenia. *Neuroscience* **41**:1–24.

Mereu, G., Fadda, F. and Gessa, G.L. (1984) Ethanol stimulates the firing rate of nigral dopaminergic neurons in unanesthetized rats. *Brain Res.* **292**:63–69.

Murray, A.M. and Waddington, J.L. (1990) The interaction of clozapine with dopamine D_1 versus dopamine D_2 receptor-mediated function: behavioral indices. *Eur. J. Pharmacol.*, **186**:79-86.

Schultz, W. (1994) Behavior-related activity of primate dopamine neurons. *Rev. Neurol.* **150**:634–639.

Yavich, L. and Tiihonen, J. (2000) Patterns of dopamine overflow in mouse nucleus accumbens during intracranial self-stimulation. *Neurosci. Lett.* **293**:41-4.

DETERMINATION OF SUBNANOMOLAR CONCENTRATIONS OF DOPAMINE AND NOREPINEPHRINE IN NANOVOLUME SAMPLES USING AN AUTOMATED CAPILLARY ZONE ELECTROPHORESIS WITH LASER INDUCED FLUORESCENCE DETECTION

Lionel Bert, Valérie Martin, Laura Lambas-Señas, Marie-Françoise Suaud-Chagny, and Bernard Renaud*

1. INTRODUCTION

Two main types of methods are available for the *'in vivo"* monitoring of catecholamines (CA) in the brain (review in Kehr, 1999): i) *"in vivo"* electrochemistry, improperly often referred to as *"in vivo"* voltammetry, which belongs to the *"in situ"* methods since they rely on a carbon fiber microlectrode measuring directly the CA within the cerebral tissue and ii) *"in vivo"* microdialysis technique, which belongs to the *"ex situ"* methods since it requires the collection of a dialysate which is then analyzed with one or several of the suitable techniques, e.g. HPLC with electrochemical detection for CA.

However these two types of techniques do not exhibit the same features. For example, the biosensor-based techniques, such as *"in vivo"* electrochemistry, are usually characterized by a very good temporal resolution while microdialysis combined with HPLC has a good biochemical specificity and allows the specific determination of several compounds simultaneously. However, both types of techniques have a moderate mass sensitivity.

In order to obtain a better description of the temporal changes in neurotransmitters occurring in the brain extracellar space in the minute range, i.e. 10-30 times faster than with the usual HPLC determination combined with microdialysis, we developed over the past few years the use of capillary electrophoresis with laser-induced fluorescence detection (CE-LIFD) for the determination of CA and other neurotransmitters, such as excitatory amino acids (EAA), in brain microdialysates.

---

* All the authors belong to the Laboratory of Neuropharmacology and Neurochemistry, INSERM U512 and Claude Bernard University, School of Pharmacy, 69008, Lyon, France.

2. ANALYTICAL ASPECTS

2.1. General Principles

Capillary electrophoresis allows the rapid separation of various analytes including CA (Robert *et al.*, 1995) and EAA (Bert *et al.*, 1996; Lada *et al.*, 1996; Rada *et al.*, 1997) from nanoliter volumes of microdialysate. The migration of the molecules is the resultant of two components: the electrophoretic migration itself and the electro-osmotic flow.

However, since none of these compounds is naturally fluorescent at the wavelength of the usual lasers, a derivatization is necessary prior to the separation itself. Naphthalene-2,3-dicarboxaldehyde (NDA) is the fluorescent tag we choose because it reacts rapidly to give stable fluorescent derivatives (cyanobenzoisoindol or CBI derivates) for CA and for EAA.

2.2. Improvements made

The on-line derivatization system ("microreactor") initially developed (Bert *et al.*, 1996) was modified in order to reduce the dead volume of the microdialysate outlet from 2 µl to 500 nL (i.e., 30 seconds microdialysis samples). In these conditions, we have been able to continuously derivatize nanoliter volumes of microdialysate containing CA at a subnanomolar concentration, while preventing evaporation, sample loss and contamination.

Stacking procedures were used for preconcentration and separation of both CA and EAA. While the usual injection volume in capillary electrophoresis must not exceed 1% of the effective column volume, a percentage as high as 10% could be reached using stacking, i.e. 100 nL injected from sample volumes as low as 500 nL.

The stability of the CA CBI derivates at +4°C was increased by the addition of 1.25 mM ascorbate to the derivatization medium.

2.3. Summary of the technique and its performances

Beside the above mentioned improvements, the main characteristics of the analytical procedures we have used have been published previously both for CA and for EAA (Robert *et al.*, 1995 ; Bert et al., 1996 ; Robert *et al.*, 1998 ; Parrot *et al.*, 2001 ; Bert *et al*, submitted).

Different separation procedures were used for CA and EAA analyses: i) CA, 110 mM phosphate buffer pH 7.05 ± 0.02, applied voltage of 25 kV (i ≈ 200 µA), hydrodynamic injection (20 s at 2 psi; i.e., 100 nL injected) of the sample followed by the injection of 200 mM orthophosphoric acid (40 nL); the capillary was flushed with NaOH 0.1 M, H_2O and running buffer between each analysis; ii) EAA, 75 mM borate buffer pH 9.2 ± 0.1, applied voltage of 25 kV (i ≈ 80 µA), hydrodynamic injection (10 s at 0.2 psi) of the sample (i.e., 5 nL injected) followed by the injection of 100 mM orthophosphoric acid (2.5 nL); the capillary was flushed with NaOH 0.1 M, H_2O and running buffer between each analysis.

The capillary electrophoresis apparatus we are presently using is a PACE/MDQ (Beckman-Coulter, USA) combined with a ZETALIF laser-induced fluorescence confocal detector (Picometrics, France) using a Helium-Cadmium laser ($\lambda_{ex.}$ 442 nm, $\lambda_{em.}$ 490 nm) from Liconix (USA).

In our standard conditions, for CA determination, an intra-assay reproducibility of 8-9 % was obtained for 48 injections corresponding to a 6 hour-analysis time.

Concentration detection limits were 6.3 10^{-11} and 3.0 10^{-11} M (i.e. 0.063 nM and 0.030 nM) for dopamine (DA) and norepinephrine, respectively. Due to the small volume injected into the capillary tube, the mass detection limits were in the attomole range both for CA and EAA.

The linearity was excellent in all cases extending into the subnanomolar range (i.e. beyond 10^{-10} M) for both DA and norepinephrine.

3. NEUROPHARMACOLOGICAL ASPECTS

3.1 Expected Results

Recent results illustrate the interest of the combination of microdialysis with CE-LIFD: the effect of various drugs administered through the dialysis probe in the anesthetized rat striatum can be described with a time resolution much better than in the previous studies. For example, the DA uptake inhibitor nomifensine (20 µM for 10 min) was shown to induce an increase in extracellular DA which was significant as early as within the first minute following the beginning of the drug infusion and which was dramatic at its maximum (+3100 %) 4 min. later. In general, although they were expected, this result and the others showed always much higher amplitudes as well as earlier and shorter variations (Bert *et al*, submitted) as those described with conventional microdialysis approaches, likely because the drug effects were not "diluted" by lengthy collection time of the microdialysates (classically 10-30 min. for HPLC instead of 1 min. for CE-LIFD in the present case).

3.2 Unexpected results

The high sampling rate allowed by the combination of CE-LIFD with microdialysis not only improved the description of changes already known, but even led in some cases to the discovery of effects never described before. For example, the striatal administration of NMDA (1 mM for 10 min.), not only induced a rapid increase in extracellular DA, glutamate and aspartate during the drug perfusion, but also evoked a surprising rebound in glutamate 5 min. after the end of the NMDA infusion through the probe (Bert *et al*, submitted). This biphasic change in glutamate could not have been evidenced without a 1 min. sampling allowed by using CE-LIFD. By decreasing this sampling rate from 10 min. to 10 s, we have even been able to show that glutamate increased 30 s earlier than DA after striatal infusion of the EAA uptake blocker PDC. These conditions reveal for the first time the *in vivo* chronology of rapid changes in DA, Glu and Asp efflux, which strongly suggests that, during PDC administration, the increase in DA efflux is a consequence of the increase in EAA concentration.

4. CONCLUSION

Because of its unique features CE/LIFD coupled with intra-cerebral microdialysis has a high potential for the *"in vivo"* simultaneous monitoring of CA but also of several other neurotransmitters which can interact with them.

REFERENCES

Bert, L., Robert, F., Denoroy, L., Stoppini, L., and Renaud, B., 1996, Enhanced temporal resolution for the microdialysis monitoring of catecholamines and excitatory amino acids using capillary electrophoresis with laser-induced fluorescence detection. Analytical developments and in vitro validation, *J. Chromatogr. A.* **755**:99.

Bert, L., Parrot, S., Robert, F., Desvignes, C., Suaud-Chagny, M.F., Denoroy, L., and Renaud, B., Submitted, *In vivo* chronology of rat striatal glutamate, aspartate and dopamine efflux during apomorphine, nomifensine, NMDA and PDC infusion, *Neuropharmacology.*

Kehr, J., 1999, Monitoring Chemistry of brain microenvironment: Biosensors, microdialysis and related techniques, in: *Modern Techniques in Neuroscience Research,* U. Windhorst and H. Johansson eds., Springer-Verlag, Berlin, Heidelberg, pp.1149-1149..

Lada, M.W., and Kennedy, R. T., 1996, Quantitative in vivo monitoring of primary amines in rat caudate nucleus using microdialysis coupled by a flow gated interface to capillary electrophoresis with laser-induced fluorescence detection, *Anal. Chem.* **68**:2790.

Morari, M., O'Connor, W. T., Ungerstedt, U., Bianchi, C., and Fuxe, K., 1996, Functional neuroanatomy of the nigrostriatal and striatonigral pathways as studies with dual probe microdialysis in the awake rat--II. Evidence for striatal N-methyl-D-aspartate receptor regulation of striatonigral GABAergic transmission and moto function, *Neuroscience* **72**:89.

Newton, A.T., and Justice, J. B., Jr., 1994, Temporal response of microdialysis probes to local perfusion of dopamine and cocaine followed with one-minute sampling, *Anal. Chem.* **66**:1468.

Parrot, S., Bert, L., Renaud, B., and Denoroy, L., 2001, Large inter-experiment variations in microdialysate aspartate and glutamate in rat striatum may reflect a circannual rhythm, *Synapse*, **39**:267.

Rada, P., Tucci, S., Murzi, E., and Hernandez, L., 1997, Extracellular glutamate increases in the lateral hypothalamus and decreases in the nucleus accumbens during feeding, *Brain Res.* **768**:338.

Robert, F., Bert, L., Denoroy, L., and Renaud, B., 1995, Capillary zone electrophoresis with laser-induced fluorescence detection for the determination of nanomolar concentrations of noradrenaline and dopamine: Application to brain microdialysate analysis. *Anal. Chem.* **67**:1838.

Robert, F., Bert, L., Lambas-Senas, L., Denoroy, L., and Renaud B., 1996, In vivo monitoring of extracellular noradrenaline and glutamate from rat brain cortex with 2-min microdialysis sampling using capillary electrophoresis with laser-induce fluorescence detection, *J. Neurosci. Methods* **70**:153.

Robert, F., Bert, L., Parrot, S., Denoroy, L., Stoppini, L., and Renaud, B., 1998, Coupling on-line brain microdialysis, precolumm derivatization and capillary electrophoresis for routine minute sampling of Ophosphoethanolamine and excitatory amino acids, *J. Chromatogr. A.* **817**:195.

Segovia, G., and Mora, F., 1997, Endogenous glutamate increases extracellular concentrations of dopamine, GABA, and taurine through NMDA and AMPA/Kainate receptors in striatum of the freely moving rat: a microdialysis study, *J. Neurochem.* **69**:1476.

Westerink, B., 1995, Brain microdialysis and its application for the study of animal behavior, *Behav. Brain Res.* **70**:103.

Zilkha, E., Obrenovitch, T. P., Koshy, A., Kusakabe, H., and Bennetto, H. P., 1995, Extracellular glutamate: on-line monitoring using microdialysis coupled to enzyme-amperometric analysis, *J. Neurosci. Methods.* **60**:1.

THE ROLE OF ANGIOTENSIN II AT_1 RECEPTORS IN THE SYMPATHOADRENAL RESPONSE TO STRESS

Ines Armando,[1, 2] Andrea Carranza,[2] Yasuaki Nishimura,[1] Kwang Lae Hoe,[1] Marta Barontini[2] and Juan M. Saavedra[1]

1. INTRODUCTION

Angiotensin II (Ang II) is a hormone and brain neuropeptide implicated in the response of the hypothalamic-pituitary-adrenal axis to stress (Saavedra, 1992). Stress increases circulating Ang II (Xang et al., 1993) and the expression of Ang II receptors in brain areas such as the hypothalamic paraventricular nucleus (PVN) is crucial for central control of the stress reaction (Castrén and Saavedra, 1988). Ang II stimulates CRH formation in the PVN during stress (Sumitomo et al., 1991; Aguilera et al., 1995) and this results in increased ACTH and adrenal corticoid release. Ang II has direct effects on adrenal function, stimulating aldosterone secretion from the zona glomerulosa and catecholamine release from the medulla (Aguilera, 1993; Livett et al., 1990). Ang II stimulates vasopressin (AVP) formation and release in the PVN (Saavedra, 1992). Released from the median eminence, AVP participates with CRH and enhances ACTH production (Antoni, 1993).

Because Ang II contributes to regulate the hormonal and sympathoadrenal system, and Ang II receptors are highly expressed at all levels of the hypothalamic-pituitary-adrenal axis (Saavedra, 1992), it was reasonable to postulate that blockade of the physiologically active Ang II AT_1 receptors could influence the outcome of the hormonal and sympathetic response to stress. To further clarify the role of Ang II AT_1 receptors in stress, we selected the AT_1 antagonist candesartan, an insurmountable antagonist which, when administered peripherally, readily crosses the blood-brain barrier inhibiting not only peripheral but also central AT_1 receptors (Nishimura et al., 2000). After peripheral administration of candesartan, we studied a clinically relevant form of stress, the stress of isolation. We asked the question of how pretreatment with a peripheral and central Ang II AT_1 receptor antagonist modulated the hypothalamo-pituitary-adrenal response to isolation stress. We found that pretreatment with candesartan almost completely abolished the hormonal and sympathoadrenal response to isolation, suggesting that

[1] Section on Pharmacology, National Institute of Mental Health, NIH, Bethesda, MD USA.
[2] Centro de Investigaciones Endocrinologicas, CONICET, Buenos Aires, Argentina

simultaneous antagonism of peripheral and brain AT_1 receptors is an advantage in the control of the stress reaction.

2. MATERIALS AND METHODS

Wistar male rats, 10 weeks old, were kept in groups of six under standard conditions, and were implanted with osmotic minipumps to deliver candesartan at 1.0 mg/Kg/day or vehicle for 13 days. On day 14, two sets of 6 animals treated with candesartan or vehicle were submitted to isolation for 24 hours in standard, 50 sq. in., plastic metabolic cages. A 24-hour urine production was collected, the animals were sacrificed by decapitation and the brain, pituitary gland and adrenal glands were removed, frozen and stored until used. Pituitary and adrenal hormones were assayed by RIA, adrenal Epinephrine (E) and Norepinephrine (NE) by HPLC and binding to Ang II receptor types by quantitative autoradiography in coronal brain sections as described (Tsutsumi and Saavedra, 1991). *In situ* hybridization of tyrosine hydroxylase (TH) mRNA in the adrenal gland was performed using a specific oligonucleotide probe (Grima et al., 1985). Statistical analysis was performed using the one-way ANOVA followed by post-hoc analysis using the Newman-Keuls multiple comparison test at the $P<0.05$ level of significance.

3. RESULTS

Isolation stress increased AT_1 receptors in the PVN, and pretreatment with the AT_1 antagonist prevented this increase (Table 1). Similarly, pretreatment with candesartan prevented the increase in pituitary ACTH and reversed the decrease in pituitary AVP produced by isolation (Table 1). The stress-induced increase in adrenal corticosterone and aldosterone was also prevented by the AT_1 antagonist.

In the adrenal gland, pretreatment with the AT1 receptor antagonist abolished the increase in TH mRNA (Figure 1, Table 1) and the increase in adrenal E and NE concentrations induced by stress (Table 1).

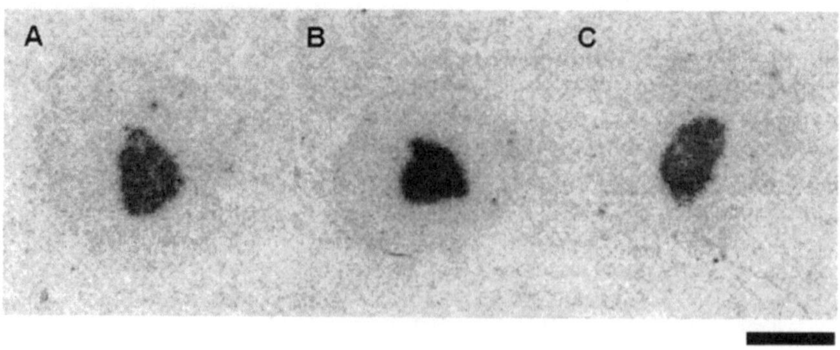

Figure 1. TH mRNA in adrenal medulla. A: Grouped; B: Isolation: C: Isolation + AT_1 blocker. Note the marked increase in TH mRNA during isolation and the decrease during isolation after AT_1 blockade.

Table 1. AT_1 binding, pituitary and adrenal hormones, and adrenal catecholamines after isolation and pretreatment with candesartan

	Grouped	Isolation	Isolation + AT_1 blocker
AT_1 binding in PVN(fmol/mg prot)	13.0±3.0	25.0±3.1*	1.90±1.20
Pituitary ACTH (ng/mg)	3.90±0.20	13.50±3.40*	4.60±1.00
Pituitary AVP (µg/gland)	1.30±0.08	1.00±0.08*	1.33±0.10
Adrenal corticosterone (ng/adrenal)	129±30	394±66*	200±44
Adrenal Adosterone (ng/adrenal)	7.20±1.60	15.30±1.70*	6.90±1.30
Adrenal Epinephrine (µg/adrenal)	13.0±1.2	19.0 ±1.8*	13.0±0.5
Adrenal Norepinephrine (µg/adrenal)	2.8±0.3	4.3±0.4*	3.3±0.3
TH mRNA (nCi/mg)	3.8±0.2	4.5±0.3*	2.7±0.4

* Significantly different from grouped and isolation + AT_1 blocker

4. DISCUSSION

Isolation is a powerful stress for rats, resulting in increased ACTH and adrenal corticoids and increased adrenomedullary catecholamine synthesis. Decreased pituitary AVP can be interpreted as depletion after enhanced hormone release (Gibbs, 1984). Isolation increases AT_1 receptor binding in the PVN, supporting the hypothesis of a role of AT_1 receptors in the regulation of the stress reaction. If Ang II, through AT_1 receptor stimulation in the PVN, plays an important role in CRH formation and sympathoadrenal discharge during stress, AT_1 receptor antagonists pretreatment should attenuate the sympathoadrenal and hormonal response to stress. This was precisely the case in our studies.

We used an inhibitor of AT_1 receptors that is able to block binding to both peripheral and central AT_1 sites (Nishimura et al., 2000). We interpret our results as the consequence of central blockade of AT_1 receptors in the PVN, resulting in diminished CRH formation and release, decreased ACTH release, and decrease mineralo- and glucocorticoid formation and release during stress. Decreased corticoid formation can also be the result of AT_1 receptor inhibition in the adrenal gland (Nishimura et al., 2000). Blockade of AVP alterations during stress could also be predominantly attributed to central AT_1 receptor inhibition. The blockade of increased adrenal catecholamine formation during stress by candesartan pretreatment could be the result of a dual mechanism, blockade of brain AT_1 receptors regulating the sympathoadrenal response to stress (Saavedra, 1992) and blockade of adrenomedullary AT_1 receptors, that are involved in catecholamine release (Wong et al., 1990).

Our results demonstrate that AT_1 receptor antagonism is sufficient to completely suppress the hypothalamic-pituitary-adrenal response to isolation stress. This observation strongly supports the hypothesis that Ang II is an important central and peripheral stress hormone, and that AT_1 receptor antagonism could have a place in the therapy of stress-related disorders.

5. ACKNOWLEDGEMENTS

We thank Dr. Bernd Friedrich, ASTRA GmbH Germany, for his generous supply of candesartan.

REFERENCES

1. Aguilera G 1993 Factors controlling steroid biosynthesis in the zona glomerulosa of the adrenal. J Steroid Biochem Molec Biol **45**:147-151
2. Aguilera G, Young WS, Kiss A, Bathia A 1995 Direct regulation of hypothalamic corticotropin-releasing-hormone neurons by angiotensin II. Neuroendocrinology **61**:437-444.
3. Antoni FA 1993 Vasopressinergic control of pituitary adrenocorticotropin secretion comes of age. Front Neuroendocrinol **14**:76-122.
4. Castrén E, Saavedra JM 1988 Repeated stress increases the density of angiotensin II binding sites in the rat paraventricular nucleus and subfornical organ. Endocrinology **122**:370-372.
5. Grima B, Lamouroux A, Blanot F, Biguet NF, Mallet J 1985 Complete coding sequence of rat tyrosine hydroxylase mRNA. Proc. Natl Acad Sci USA **82**: 617-621.
6. Jöhren O, Inagami T, Saavedra JM 1995 AT_{1A}, AT_{1B}, and AT_2 angiotensin II receptor subtype gene expression in rat brain. Neuro-Report **6**:2549-2552.
7. Livett BG, Marley PD, Wan DC, Zhou XF 1990 Peptide regulation of adrenal medullary function. J Neurol Transm Suppl **29**:77-89.
8. Nishimura Y, Ito T, Hoe K-L, Saavedra JM 2000 Chronic peripheral administration of the angiotensin II AT_1 receptor antagonist Candesartan blocks brain AT_1 receptors. Brain Res **871**:29-38.
9. Saavedra JM 1992 Brain and pituitary angiotensin. Endocr Rev **13**:329-380.
10. Sumitomo T, Suda T, Nakano Y, Tozawa F, Yamada M, Demura H 1991 Angiotensin II increases the corticotropin-releasing factor messenger ribonucleic acid level in the rat hypothalamus. Endocrinology **128**:2248-2252.
11. Tsutsumi K, Saavedra JM 1991 Characterization and development of angiotensin II receptor subtypes (AT_1 and AT_2) in rat brain. Am J Physiol **261**:R209-R216.
12. Wisden W, Morris BJ 1994 In situ hybridization with synthetic oligonucletide probes. In: W.Wisden and BJ Morris (eds) In situ hybridization protocols for the brain. Academic Press Inc. San Diego, CA, USA pp. 9-34.
13. Wong PC, Hart SD, Zaspel AM, Chiu AT, Ardecky J, Smith, RD, Timmermans PBMWM 1990 Functional studies of nonpeptide angiotensin II receptor subtype-specific ligands. Dup 753 (AII-1) and PD 123177 (AII-2). J Pharm Exp Ther **255**:584-592
14. Xang G, Xi Z-X, Wan Y, Wang H, Bi G 1993 Changes in circulating and tissue angiotensin II during acute and chronic stress. Biol Signals **2**:166-172

DIFFERENTIAL GENE EXPRESSION OF TYROSINE HYDROXYLASE IN RATS EXPOSED LONG-TERM TO VARIOUS STRESSORS

Richard Kvetnansky, Bistra B. Nankova, Milan Rusnak, Lucia Micutkova, Lucia Kubovcakova, Sladjana Dronjak, Olga Krizanova and Esther L. Sabban*

1. INTRODUCTION

Activity of the sympathetic-adrenomedullary system (SAS) in rats exposed to heterotypic novel stressors after prior chronic stress exposure is poorly understood. A potentiation of SAS activity evaluated by plasma catecholamine levels in chronically stressed rats exposed to novel stressors has been reported (Dronjak et al., 2001). The exaggerated effect of novel stressors is an interesting phenomenon, but its mechanism at the level of adrenomedullary gene expression of catecholamine biosynthetic enzymes is not known.

Our previous data showed that in rats exposed to long-term cold (4°C) the initially increased activity of adrenal medullary tyrosine hydroxylase (TH) was no longer significantly elevated, whereas the activity remained elevated in rats exposed to repeated immobilization (IMO) (see reviews by Kvetnansky and Sabban, 1993; Sabban and Kvetnansky, 2001).

The aim of the present work was to investigate changes in adrenomedullary TH gene expression in rats exposed to a single and repeated IMO (42-times) and subsequent changes in long-term immobilization adapted rats exposed to heterotypic novel stressors.

2. METHODS

2.1. Animals and Stressors

Male Sprague-Dawley rats (350-400g) underwent immobilization stress for 120 min either once or for 41 consecutive days. On the final day of the experiment, groups of

* Richard Kvetnansky, Milan Rusnak, Lucia Micutkova, Lucia Kubovcakova, Sladjana Dronjak, Olga Krizanova, Institute of Experimental Endocrinology, Slovak Academy of Sciences, 833 06 Bratislava, Slovak Republic. Bistra B.Nankova, Esther L.Sabban, New York Medical College, Valhalla, New York 10595, USA.

repeatedly IMO rats were exposed to the homotypic stressor (42nd IMO) or to heterotypic novel stressors: cold 4°C, administration of insulin (INS, 5IU) or 2-deoxyglucose (2DG, 500 mg/kg). Animals were decapitated 3 h after the 42nd IMO or after 5 h of cold exposure or INS or 2DG administration.

2.2. Molecular Biological Methods

Adrenal medullae were dissected and used for determination of TH mRNA levels by Northern blot, TH and FRA2 immunoreactive protein levels by Western blot (Nankova et al., 1994) and TH activity by an adaptation of the micromethod (Waymire et al., 1971).

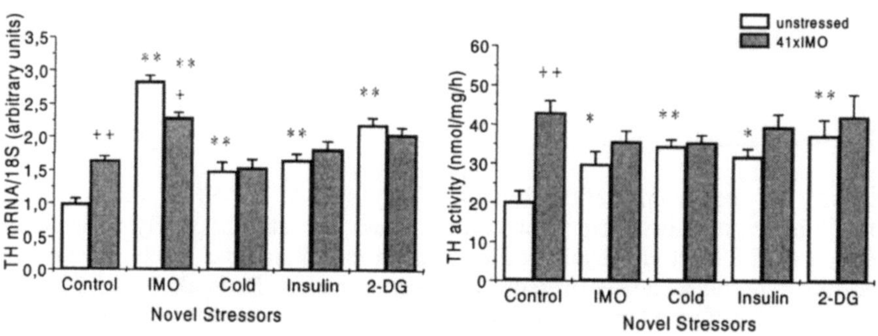

Figure 1. Effects of novel stressors (cold, insulin, 2-deoxy-glucose; 2DG) on adrenomedullary TH mRNA levels and TH activity in rats exposed to repeated IMO (41 times) vs. unstressed animals. Values are mean ± S.E.M. of 7-8 rats. Significance determined by one-factor ANOVA followed by Fisher's post-hoc test: (*) $p<0.05$; (**) $p<0.01$ vs. appropriate control group. (+) $p<0.05$; (++) $p<0.01$ vs. chronically unstressed groups.

3. RESULTS AND DISCUSSION

3.1. Effect of Long-term Immobilization on TH mRNA Levels

A single IMO produced a several-fold elevation of adrenal medullary TH mRNA levels and a small but significant increase in TH activity (Fig. 1; first four bars from the left). Animals repeatedly IMO for 41 times and decapitated the next morning without exposure to any stressor ("adapted control") showed significantly ($p<0.001$) elevated TH mRNA levels and TH activity. An exposure of such animals to 42nd IMO did not show any changes in TH activity and only a small exaggeration of TH mRNA levels (Fig. 1). Huge increases in TH immunoreactive protein, epinephrine and norepinephrine concentrations were also found in the adrenal medulla of "adapted control" (data not shown).

Differences in TH mRNA levels in the adrenal medulla of rats exposed to a short and long-term cold stress suggest an adaptation at the level of TH gene expression. We found a significant increase in adrenal TH mRNA levels after a short-term cold exposure; however, in rats exposed to cold for 28 days, the adrenal TH mRNA levels returned to near control values (Kvetnansky et al., 2001). The reduced TH mRNA levels correlate very well with the reduced TH activity in the adrenal medulla of "cold-adapted rats". Thus, it is evident that animals exposed long-term to different homotypic stressors - immobilization or cold -

develop different mechanisms of regulation of TH gene expression in the adrenal medulla. Possible mechanisms of transcriptional regulation of TH gene are discussed in another chapter of this book (Sabban et al.)

3.2. Effect of Heterotypic Novel Stressors on TH mRNA Levels in Rats Exposed to Long-term Homotypic Immobilization Stress

A single exposure to cold, or administration of INS or 2DG produced significant increases in adrenal TH mRNA levels even if not so marked as IMO stress (Fig.1). TH activity was also increased by all these stressors. Permanently elevated TH mRNA levels in the adrenal medulla of long-term IMO rats is not further changed after exposure to the heterotypic novel stressors employed (Fig. 1). These data show that the adaptation to IMO, which represents a rather severe stressor, depends on molecular mechanism(s) which are not further activated after exposure to less severe stressors such as cold, insulin or 2DG administration. Rats adapted to cold displayed exaggerated responses of adrenal TH mRNA levels after exposure to the heterotypic novel stressors when compared to naive rats exposed to the same stressors (Kvetnansky et al., 2001). Rats exposed to long-term physical activity did not increase adrenal TH gene expression; however, after exposure to a novel heterotypic stressor (shaking) TH mRNA levels were almost three-times elevated compared to non-exercised animals (Levenson and Moore, 1998). Thus, long-term exposure to cold or physical activity may enhance the ability to synthesize catecholamines under exposure of such animals to heterotypic novel stressors. It is possible to speculate that during long-term cold or exercise, adaptation mechanisms are induced that allow rapid and enhanced molecular responses to heterotypic stressors. This could reflect an activation of additional signaling pathways leading to modification of already accumulated transcription factors or an activation of different synergistic pathways.

The adaptation mechanism developed in repeatedly IMO rats must be completely different since such animals have permanently increased TH gene expression and they do not show any exaggerated response to heterotypic novel stressors.

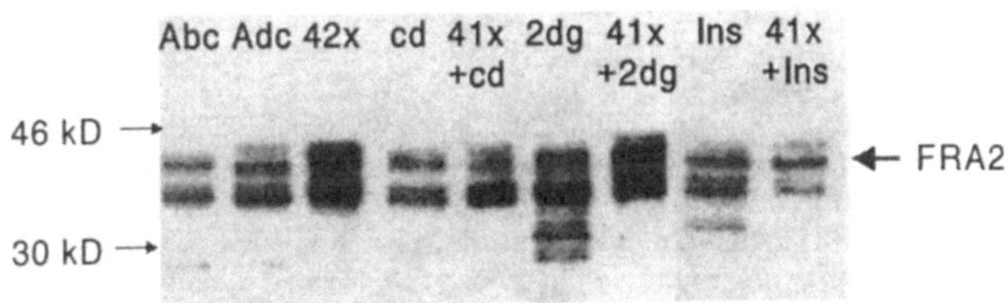

Figure 2. A Western blot of adrenal medullary FRA family proteins in repeatedly IMO rats exposed to novel stressors. Abc= control; Adc=adapted control; 42x=42x IMO; cd=cold; 2dg=2-deoxy-glucose; Ins=insulin.

3.3. Possible Mechanism of Increased TH mRNA Levels in Adrenals of Long-term Immobilized Rats

In repeatedly IMO rats, the transcriptional activation of TH gene expression remained above control levels even one day after the termination of IMO (Nankova et al., 1999). It is important to ascertain which transcription factors mediate this effect. There was little induction of cfos compared to the first IMO; however, some fos-family members were induced (Sabban and Kvetnansky, 2001). One of the fos-family members was identified as Fra-2 and its adrenal immunoreactivity was evident even in "adapted control" group rats decapitated one day after the 41st IMO (Fig. 2). FRA2 binds to the AP1 site of the TH promoter and is able to elevate transcription of the TH gene (Nankova et al., 2000). The induction of this long-lasting fos-family member, which was even potentiated by subsequent IMO (42nd), might be one of the mechanisms involved in permanently elevated TH mRNA levels in the adrenal medulla of repeatedly IMO rats.

3.4. Summary

The observed differences in TH mRNA levels in the adrenal medulla suggest an adaptation of TH gene expression in rats exposed to long-term cold but not in rats exposed to long-term repeated immobilization. The exposure of such rats to heterotypic novel stressors induces exaggerated responses of TH mRNA levels in cold-adapted but not in repeatedly immobilized rats. The data suggest that different stressors regulate adrenal TH gene expression by specific mechanisms especially in chronically stressed rats. Our data suggest an involvement of Fra-2, a fos family member, in TH gene regulation in the adrenal medulla of repeatedly immobilized rats.

ACKNOWLEDGMENT

Supported by Slovak Grant Agency for Sciences, Grant (2-6109); FIRCA Grant (TW00984), and European Commission ICA1-CT-2000-70008.

REFERENCES

Dronjak, S., Ondriska, M., Svetlovska, D., Jezova, D. and Kvetnansky, R., 2001, Effect of novel stressors on plasma catecholamine levels in rats exposed to long-term cold, in: *Stress: Neural, Endocrine and Molecular Studies,* R. McCarty, et al., eds. Gordon and Breach, New York (in press).

Kvetnansky, R., Jelokova, J., Rusnak, M., Dronjak, S., Serova, L., Nankova, B.B. and Sabban, E.L., 2001, Novel stressors exaggerate the tyrosine hydroxylase gene expression in the adrenal medulla of rats exposed to long-term cold stress, in: *Stress: Neural, Endocrine and Molecular Studies.* R. McCarty, et al., eds. Gordon and Breach, New York (in press).

Kvetnansky, R. and Sabban, E.L., 1993, Stress-induced changes in tyrosine hydroxylase and other catecholamine biosynthetic enzymes, in: *Tyrosine Hydroxylase: From Discovery To Cloning.* M. Naoi, and S.H. Parvez, eds., Utrecht, VSP Press, pp. 253-281.

Levenson, C.W. and Moore, J.B., 1998, Response of rat adrenal neuropeptide Y and tyrosine hydroxylase mRNA to acute stress is enhanced by long-term voluntary exercise, *Neurosci. Lett.* **242**, 177-179.

Nankova, B.B., Kvetnansky, R., McMahon, A., Viskupic, E., Hiremagalur, B., Frankle, W.G., Fukuhara, K., Kopin, I.J. and Sabban, E.L., 1994, Induction of tyrosine hydroxylase gene expression by a non-neuronal non-pituitary mediated mechanism in immobilization stress, *Proc. Natl. Acad. Sci, USA,* **91**, 5937-5941.

Nankova, B.B., Rivkin, M., Kelz, M., Nestler, E.J. and Sabban, E.L., 2000, Fos-related antigen 2: potential mediator of the transcriptional activation in rat adrenal medulla evoked by repeated immobilization stress, *J. Neurosci.* **20**, 5647-5653.

Nankova, B.B., Tank, A.W. and Sabban, E.L., 1999, Transient or sustained transcriptional activation of the genes encoding rat adrenomedullary catecholamine biosynthetic enzymes by different durations of immobilization stress, *Neuroscience* **94**, 803-808.

Sabban, E.L. and Kvetnansky, R., 2001, Stress-triggered activation of gene expression in catecholaminergic systems: dynamics of transcriptional event, *Trends in Neurosciences* **24**, 91-98.

Waymire, J.C., Bjur, R. and Weiner, N.,1971, Assay of tyrosine hydroxylase by coupled decarboxylation of DOPA formed from L-(1-^{14}C) tyrosine, *Analyt. Biochem.* **43**, 588-600.

DYNAMICS OF TRANSCRIPTIONAL EVENTS IN STRESS-TRIGGERED REGULATION OF ADRENAL CATECHOLAMINE BIOSYNTHETIC ENZYMES

Esther L. Sabban, Bistra B. Nankova, Lidia I. Serova, Richard Kvetnansky, Nikolaos A. Papanikolaou, and Akira Nakashima*

1. INTRODUCTION

One of the most important aspects of stress research is to understand why the same type of stress stimulus does not always elicit an identical response. Many different factors, including prior experience have been proposed to affect the response to stress (McEwen, 1998).

This is especially evident regarding the stress-triggered changes in gene expression of adrenal catecholamine biosynthetic enzymes with various amounts of immobilization stress (IMO) (Nankova et al, 1994; Sabban and Kvetnansky, 2001). IMO elicited a robust elevation of tyrosine hydroxylase (TH) mRNA in naive animals or those with prior exposure to IMO. However, the rise in naïve animals is not sustained upon termination of the stressor and is not reflected in changes in TH protein or TH activity. Thus, there would be no long lasting effect of exposure to this stressor, were it not followed by subsequent episodes of stress. When animals that had prior experience with IMO were subjected to the same stressor, TH mRNA levels remained high much longer. Several days are required after cessation of the stressor for a return of TH mRNA to basal levels. In addition, TH immunoreactive protein and activity levels werenow significantly higher than in controls.

In this study, we investigated the molecular mechanisms that are likely responsible for these effects. We will concentrate on the adrenal medulla, where the stress-triggered elevation in gene expression of catecholamine biosynthetic enzymes was shown to be transcriptionally mediated (Nankova et al, 1994, 1999; Osterhout et al, 1997).

* Esther L. Sabban, Bistra B. Nankova, Lidia I. Serova, Nikolaos A. Papanikolaou, New York Medical College, Valhalla, New York 10595. Richard Kvetnansky, Slovak Academy of Sciences, Bratislava, Slovak Republic. Akira Nakashima, New York Medical College, Valhalla, New York, USA and Fujita Health University, Toyoake, Aichi, Japan 470-1192.

2. METHODS

2.1. Exposure of Rats to Different Durations or Repetitions of Stress

Sprague-Dawley male rats (250-300g) underwent immobilization stress (IMO) for different durations (up to 2 hr) once or on several consecutive days. Adrenal medullae were dissected and used for Northern blots, run-on assays of transcription, mobility shift assays or Western blots as previously described (Nankova et al, 1999; 2000).

2.2.TH and DBH Promoter-Driven Reporter Activity

The rat TH (-272/+27) promoter (intact or with mutant AP1 or Egr1 sites) (from Dr. D. Chikaraishi, Duke University) and DBH (-248/+21) were subcloned into pGL3-Basic vector (Promega). They were co-transfected into PC12 cells (Papanikolaou and Sabban, 2000) with Fra-2 or Egr1 expression vectors (from Drs. S. Yuspa, NIH and D.Wong, Harvard Medical School).

3. RESULTS AND DISCUSSION

3.1. Different Durations of Stress

The effects of brief or intermediate durations of a single IMO, or repeated stress IMO on several transcription factors, and on transcription of catecholamine biosynthetic enzymes are summarized in Figure 1.

3.1.1.Brief exposure to stress – post- translational modification of pre-existing factors

Surprisingly, even a brief (5-min) single IMO event was found to elicit more than a 3-fold elevation in rate of transcription of the TH and dopamine β-hydroxylase (DBH) genes. This induction was transient and did not result in any significant accumulation of the corresponding mRNAs (Nankova et al, 1999). The rise in transcription was too rapid for de novo synthesis and most probably was mediated by post-translational modifications of pre-existing factors. Likely candidates include CREB, c-Jun, Elk-1 and ATF-2.

CREB: CREB is an important convergence point for a number of signaling pathways. It can be phosphorylated by several kinases including protein kinase A, CaM kinase II and IV, p38 mitogen acitvated kinase, protein kinase C and protein kinase B (De Cesare et al, 1999). We found increased CREB phosphorylation with just 5 min of IMO.

JNK: Targets of phosphorylation by JNK (c-Jun N-terminal kinase) are also likely candidates to mediate the response to a brief stressor. Activation of JNK is associated with phosphorylation of several transcription factors including c-Jun, JunB, JunD, ATF-2 and Elk1 (Ip and Davis, 1998). Since c-jun is an especially important component of the complex formed on the AP-1 element of TH and DBH promoters under basal conditions, we suspected that activation of JNK might be involved in their regulation *in vivo*. As little as 5 min of IMO was found to increased JNK activity in adrenal medulla by about 5-fold (Nankova et al, 1998).

Single IMO		**Repeated IMO**
Brief (5 min)	**Intermediate (15-30 min)**	**2 hr Daily**
↑ P-CREB	↑ cFos	↑ Fra-2
↑ JNK	↑ Fras (including Fra2)	↑ Egr-1
↑ TH, DBH Transcription (Transient)	↑ Egr1	↑ JNK
	↑ P-CREB	↑ TH, DBH Transcription (stabilized)
	↑ JNK	
	↑ TH, DBH Transcription (Transient DBH)	↑ TH, DBH protein and enzyme activity
		CREB not examined

Figure 1. Summary of responses of adrenal medulla to different durations or repetitions of immobilization stress.

3.1.2 Intermediate duration of a single stressor – de novo synthesis of short-lasting transcription factors

With longer durations of a single stressor (15-30 min), there was more persistent activation of TH transcription (still elevated by the end of the 2-hr session) that may be associated with *de novo* synthesis of transcription factors. Candidates for mediating this more sustained elevation of TH transcription are cfos and Egr1.

cFos: cfos is an immediate early gene that forms heterodimers with members of the Jun family to bind to AP-1 sites and regulate transcription (Angel and Karin, 1991). cfos immunoreactive protein levels, undetectable in the adrenal medullae of control rats, increased markedly within 15 min to 2 hr of a single exposure to IMO stress. This increase correlated with its elevated binding to the AP-1 site on the TH promoter and elevation of TH mRNA levels (Nankova et al, 1994). The *de novo* synthesis of cfos can be modulated by transcription factors, which can be activated by the phosphorylation elicited with brief stress (especially CREB and Elk-1)

Egr1: We have recently found that the TH promoter contains a functional Egr1/Sp1 site (Papanikolaou and Sabban, 2000). Egr1 binding to the TH promoter was elevated with IMO stress (Papanikolaou and Sabban, 1999).

3.1.3 IMO of previously stressed animals

With repeated IMO, the extent of transcriptional activation of TH and DBH genes is the same as with the first exposure to stress (about 3-fold). However, the elevations of transcription are more sustained, remaining significantly above control values even one day after the termination of the IMO (Nankova et al, 1999). In contrast to the first IMO, there was little induction of cfos. However, other fos-family members were induced. One of them was identified as Fra-2. Increased Fra-2 immunoreactivity was evident even 1 day after cessation of stress. The induction of long-lasting fos-family members was correlated with increased interaction with the TH-AP1 promoter element. In addition, binding of Egr1 was also elevated by stress (Papanikolaou and Sabban, 1999).

3.2. Reporter Activation

To validate the involvement of Fra-2 and Egr1 in stress-elicited activation of TH and DBH transcription, PC12 cells were transfected with plasmids in which TH and DBH promoters control luciferase (Luc) reporter activity. Co-transfection with Fra2 or Egr1-expression vectors elicited a several-fold increase in reporter activity. The induction of luciferase activity by Fra-2 was abolished by mutation of the AP-1 like motif in the TH promoter. Intact Egr1/Sp1 and AP1 elements of the TH promoter were both required for an Egr1-triggered rise in reporter activity (Papanikolaou and Sabban, 2000).

3.3. Summary

Different signaling pathways and/or transcription factors are activated depending on the duration of the stress signal and prior experience with stress. The persistence of the rise in transcription seems to depend on the consecutive switching of transcription factors activated either by phosphorylation or *de novo* synthesis. These include not only cfos, but also other fos family members, such as Fra-2, as well as members of CRE/ATF and Egr1 families. Mutations in any of these systems can affect the progression from a transient response to a sustained, and potentially maladaptive, response.

ACKNOWLEDGMENTS

We are grateful for support of NIH grant NS28869 and FIRCA award TW00984.

REFERENCES

1. Angel, P., and Karin, M. 1991, The role of Jun, Fos and the AP-1 complex in cell-proliferation and transformation. Biochim. Biophys. Acta. **1072**, 129-157.
2. Ip, Y.P. and Davis, R.J., 1998, Signal transduction by the c-Jun N-terminal kinase (JNK)--from inflammation to development. Curr. Opin. Cell Biol. **10**, 205-219.
3. DeCesarem D., Fimia, G.M. and Sassone-Corse, P., 1999, Signaling routes to CREM and CREB: plasticity in transcriptional activation. Trends Biochem Sci., **24**,281-285.
4. McEwen, B. S., 1998, Protective and damaging effects of stress mediators. New Eng. J. Med **338**, 171-180.
5. Nankova, B. B., Fuchs, S. Y., Serova, L. I., Ronai, Z., Wild; D. and Sabban, E. L., 1998, Selective in vivo stimulation of stress-activated protein kinase in different rat tissues by immobilization stress. Stress **28**, 1-9.
6. Nankova, B.; Kvetnansky, R. ; McMahon, A. ; Viskupic, E.; Hiremagalur, B.; Frankle, G.; Fukuhara, K.; Kopin, I.J. and Sabban, E. L.. 1994, Induction of tyrosine hydroxylase gene expression by a nonneuronal nonpituitary-mediated mechanism in immobilization stress. Proc. Natl. Acad. Sci USA **91**, 5937-5941.
7. Nankova, B. B., Tank, A. W.; and Sabban, E. L., 1999, Transient or sustained transcriptional activation of the genes encoding rat adrenomedullary catecholamine biosynthetic enzymes by varying times of immobilization stress. Neuroscience **94**, 803-808.
8. Osterhout, C. A., Chikaraishi, D. M. and Tank, A. W, 1997, Induction of tyrosine hydroxylase protein and a transgene containing tyrosine hydroxylase 5' flanking sequences by stress in mouse adrenal gland. J. Neurochem. **68**, 1071-1077.
9. Papanikolaou, N. A.; and Sabban, E. L., 1999, Sp1/Egr1 motif: a new candidate in the regulation of rat tyrosine hydroxylase gene transcription by immobilization stress. J. Neurochem. **73**, 433-436.
10. Papanikolaou, N. A., and Sabban, E. L., 2000, Ability of egr1 to activate tyrosine hydroxylase transcription in PC12 Cells: cross-talk with AP-1 Factors. J. Biol. Chem. **275**, 26683-26689.
11. Sabban, E.L., and Kvetnansky, R., 2001, Stress-triggered activation of gene expression in catecholaminergic systems: dynamics of Transcriptional Events, Trends in Neuroscience, **24**, 91-97.

THE ROLE OF THE CENTRAL CATECHOLAMINE SYSTEM IN THE ORGANIZATION OF NEURONAL AND NEUROHUMORAL RESPONSES TO VARIOUS STRESSFUL STIMULI

Miklós Palkovits*

1. INTRODUCTION

Neuroanatomically, the routes of neurogenic stressors are constituted by neuronal pathways for stressful signals to brain centers (ascending pathways), and neuronal pathways carrying stress response signals to the periphery (descending pathways). They may consist of short loop reflexes (spinal and supraspinal) or long loop circuits (ascending neuronal pathways - higher centers [hypothalamus, limbic system. neocortex] - descending neuronal pathways with outputs to the periphery). Both are involved in stress responses with or without the participation of central catecholaminergic (CA) neurons. Different stressors may act on different central CA neurons, and *vice versa*, different CA neurons are diversely involved in stress responses. Their anatomical organization, however, is similar: CA neurons participate both in the ascending and descending stress pathways.

2. METHODS

A great deal of neurochemical (measuring of the levels of CAs and CA enzymes in brain nuclei), immunohistochemical (*in situ* hybridization histochemistry for mRNA of CAs, CA enzymes and CA transporters), neuroanatomical (chemical and viral tract-tracings) and neurosurgical (lesioning of CA neurons, transections of CA pathways) techniques were applied to demonstrate the stressor-specific participation of central CA neurons in the organization of responses to various stressors.

Different acute stressful stimuli, such as pain, restraint (immobilization), thermal stress, hemorrhage, toxic agents, audiogenic and emotional stress were employed in this study.

---

* Department of Anatomy, Semmelweis University, Budapest, Hungary H-1094

3. NEURONAL PATHWAYS OF STRESS SIGNALS

Stressful stimuli from internal (body and viscera) and external environments to the central nervous system are carried by viscero- and somatosensory fibers, respectively. The signals are conducted *via* spinal and cranial sensory nerves into the dorsal horn of the spinal cord or medullary sensory nuclei (spinal trigenimal nucleus and nucleus of the solitary tract [NTS],[3] respectively). From sensory neurons of these areas, the stress signals may reach higher centers (hypothalamus, limbic system, neocortex) through 3 avenues (Fig.1): (1) direct pathways through the spinothalamic, spinoreticulothalamic and spinohypothalamic[1] tracts; (2) signals are relayed by peptidergic neurons in the NTS and brainstem reticular formation, or (3) through the activation of brainstem catecholaminergic neurons.

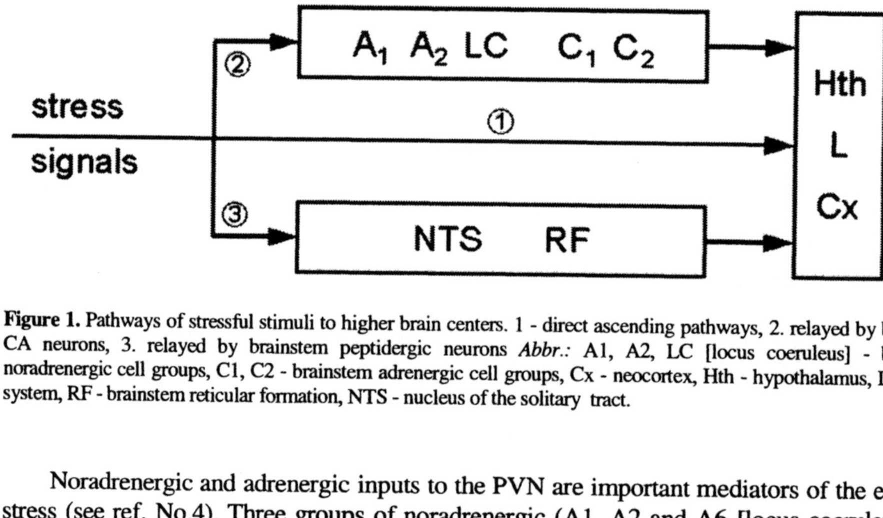

Figure 1. Pathways of stressful stimuli to higher brain centers. 1 - direct ascending pathways, 2. relayed by brainstem CA neurons, 3. relayed by brainstem peptidergic neurons *Abbr.:* A1, A2, LC [locus coeruleus] - brainstem noradrenergic cell groups, C1, C2 - brainstem adrenergic cell groups, Cx - neocortex, Hth - hypothalamus, L - limbic system, RF - brainstem reticular formation, NTS - nucleus of the solitary tract.

Noradrenergic and adrenergic inputs to the PVN are important mediators of the effect of stress (see ref. No.4). Three groups of noradrenergic (A1, A2 and A6 [locus coeruleus] CA cell groups) and three groups of adrenergic (C1, C2 and C3) cells with hypothalamic and telencephalic projections are present in the pons-medulla. Viscerosensory fibers entering to the medulla may terminate on CA neurons in the dorsomedial medulla (A2, C2 and C3 cell groups) which, in turn, project to the hypothalamus. Somatosensory signals relayed by spinal cord neurons may innervate the caudal ventrolateral medulla[2] including A1 noradrenergic and C1 adrenergic neurons[7] (see ref. No.6 for further references). The axons of these neurons or their collaterals may cross over at the level of the spinal cord or the lower part of the medulla, from where they ascend bilaterally and innervate both sides of the hypothalamus and the telencephalon. Pain induced by 4% formalin injected unilaterally into the hindlimb elicited *c-fos* activation on both sides of brainstem CA neurons and the hypothalamus,[7] and the subsequent release of norepinephrine from the paraventricular nucleus (PVN) was also bilateral.[4, 8] Direct synaptic connections between CA nerve terminals and corticotropin releasing hormone (CRH) synthesizing neurons in the PVN were demonstrated at an electron microscopic level. Light and electron microscopic investigations have demonstrated that ascending CA fibers from the brainstem terminate in the median eminence,[9] where they may exert their actions on hypothalamic nerve terminals, influencing the release of hypothalamic releasing or inhibiting hormones into the portal circulation.

4. NEURONAL PATHWAYS OF STRESS RESPONSE

In the organization of responses to stress, the hypothalamus, the central amygdaloid nucleus and some cortical areas (insular and infralimbic cortex) are prominent. None of them project directly to the periphery, they activate either neurohumoral (hypothalamo-pituitary-adrenal axis) or neuronal pathways descending to preganglionic neurons in the dorsal vagal motor nucleus (DVN) or in the intermediolateral cell column (IML) in the thoracic spinal cord (Fig.2).

The descending hypothalamic fibers arise mainly in the PVN, but neurons in the arcuate, perifornical, medial preoptic nucleus and in the lateral hypothalamic area also project to the brain stem and the spinal cord. The PVN projections can be divided into three groups (Fig.2): (1) projections to autonomic preganglionic neurons, (2) projections to those brainstem CA neurons which innervate preganglionic neurons (A5, A6 [ventral part of the locus coeruleus and subcoeruleus neurons], A7, and rostral part of the C1 cell group), and (3) projections to sensory neurons in the medulla (NTS and spinal trigeminal nucleus) and the spinal cord (dorsal horn) carrying feedback signals from the PVN to these brain regions.[6] The descending PVN projections are bilateral with ipsilateral dominance.[10]

Descending fibers from the central nucleus of the amygdala are partly relayed in the bed nucleus of the stria terminalis. From here, fibers run to the hypothalamus and innervate the PVN (minor portion), or descend to the brainstem. A significant percentage of these descending fibers are relayed in the lateral parabrachial nucleus (Fig.2). The connection of descending central amygdaloid axons with brainstem CA neurons is minor, if any.

Axons arising in the insular or infralimbic cortex may reach the brainstem directly, but the majority of the fibers, as it has been demonstrated by tract-tracing techniques, terminate either in the amygdala or in the PVN (Fig.2). No direct cortico-brainstem CA interconnection has been reported yet.

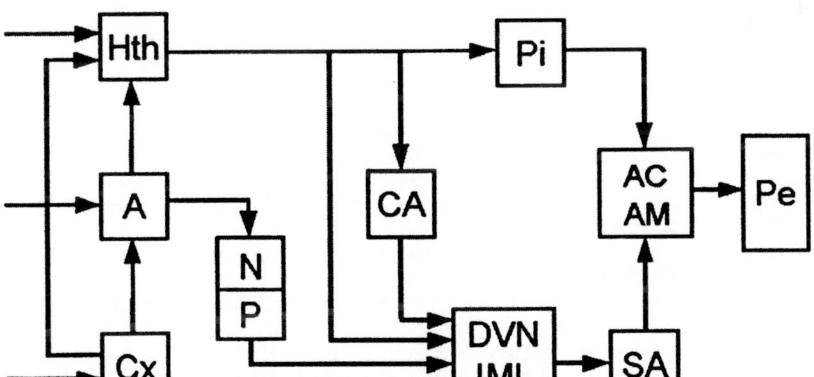

Figure 2. Possible neuronal routes in stress response from the "higher centers" to the periphery. *Abbr.:* A - central amygdaloid nucleus, AC - adrenal cortex, AM - adrenal medulla, CA - brainstem CA neurons with descending projections (A5,A6,A7,C1 cell groups), Cx - insular and infralimbic cortex, DVN - dorsal motor vagal nucleus, Hth - hypothalamus (mainly PVN), IML - intermediolateral cell column in the thoracic spinal cord, N - bed nucleus of the stria terminalis, P - parabrachial nucleus, Pe - periphery, Pi -anterior pituitary, SA - sympatho-adrenomedullary system (noradrenergic sympathetic ganglionic cells).

5. HETEROGENEITY IN STRESS RESPONSE

Cold stress rapidly elevated norepinephrine levels in the plasma, most probably by the activation of the hypothalamo-sympathoadrenal path with minor or no actions through the amygdala or cortical areas. It is more likely, that central CAs have a limited role in the organization of responses to cold stress.

The high epinephrine levels in the plasma in insulin hypoglycemia indicate the activity of the adrenal medulla. Signals may arise from hypothalamic nuclei most probably by neuronal ways.

Responses to acute *hemorrhage* are organized in the hypothalamus. The high plasma ACTH level indicates the activation of the hypothalamo-pituitary-adrenal axis.

As it is indicated by rapid *c-fos* expression, formalin-induced *pain* activates neurons in the PVN and cortical areas. The descending signals may run from the hypothalamus down in all three pathways shown on Fig.2.

Restraint (immobilization) is a very strong, complex stressful stimulus that activates the three "high" centers (shown by both *c-fos* and microdialysis studies[4, 5]). Signals run in all pathways down (Fig.2) to the adrenocortical and sympatho-adrenal systems. Brainstem CAs are fully involved in the organization of this stress response.

In conclusion, our results confirmed a marked heterogeneity of responses to acute stressors.[5] Each stressor has its own distinct central mechanism and pathway. Depending on the nature, the intensity and the duration of the stressful stimuli, central CA neurons show certain specificity to influence different neuronal and neurohumoral mechanisms.

REFERENCES

1. R. Burstein, K.D. Cliffer, and G.J.J. Giesler, Direct somatosensory projections from the spinal cord to the hypothalamus and telencephalon, *J. Neurosci.* **7**, 4159-4164 (1987).
2. D. Lima, and A. Coimbra, Neurons in the substantia gelatinosa rolandi (lamina II.) project to the caudal ventrolateral reticular formation of the medulla oblongata in the rat, *Neurosci. Lett.* **132**, 16-18 (1991).
3. D. Menétrey, and A.I. Basbaum, Spinal and trigeminal projections to the nucleus of the solitary tract: A possible substrate for somatovisceral and viscerovisceral reflex activation, *J. Comp. Neurol.* **255**, 439-450 (1987).
4. K. Pacak, M. Palkovits, I.J. Kopin, and D.S. Goldstein, Stress-induced norepinephrine release in the hypothalamic paraventricular nucleus and pituitary-adrenocortical and sympathoadrenal activity. In vivo microdialysis studies, *Front. Neuroendocrinol.* **16**, 89-150 (1995).
5. K. Pacak, M. Palkovits, G. Yadid, R. Kvetnansky, I.J. Kopin, and D.S. Goldstein, Heterogeneous neurochemical responses to different stressors: A test of Selye's doctrine of nonspecificity, *Amer. J. Physiol.* **275**, R1247-R1255 (1998).
6. M. Palkovits, Interconnections between the neuroendocrine hypothalamus and the central autonomic system, *Front. Neuroendocrinol.* **20**, 270-295 (1999).
7. M. Palkovits, J.S. Baffi, and S. Dvori, Neuronal organization of stress response: Pain-induced *c-fos* expression in brain stem catecholaminergic cell groups, *Ann. N.Y. Acad. Sci.* **771**, 313-326 (1995).
8. M. Palkovits, J.S. Baffi, and K. Pacak, The role of ascending neuronal pathways in stress-induced release of norepinephrine in the hypothalamic paraventricular nucleus of rats, *J. Neuroendocrinol.* **11**, 529-539 (1999).
9. M. Palkovits, G.B. Makara, Cs. Léránth, and H. Van Cuc, Intrahypothalamic terminals of stress conducting fibers, *Brain Res.* **190**, 399-407 (1980).
10. Zs. Tóth, K. Gallatz, M. Fodor, M. Palkovits, Decussations of the descending paraventricular pathways to the brainstem and spinal cord autonomic centers, *J. Comp. Neurol.* **414**, 255-266 (1999).

EMOTIONAL STRESS- AND CATECHOLAMINE-INDUCED MOLECULAR AND ELECTROCARDIOGRAPHIC RESPONSES IN THE HEART

Takashi Ueyama[1], Ken-ichi Yoshida[3], and Emiko Senba[2]

1. INTRODUCTION

Emotional stress, and associated activation of the sympatho-adrenergic system underlie cardiac accidents. Immobilization (IMO) of rats is a useful model of emotional stress, since it activates the sympatho-adrenergic system and hypothalamo-pituitary-adrenocortical axis.[1] When cells are confronted with the change of their environment, a set of genes and their proteins are rapidly induced. Immediate early genes (IEGs)/proto-oncogenes such as c-*fos* and c-*jun*[2] and heat shock proteins (HSP)[3] are the typical stress-induced substances. Fos family and Jun family proteins form hetero-(Fos-Jun) or homo-(Jun-Jun) dimers (AP-1), which may elicit expression of target genes and contribute to the cellular response to the primary stimulus. HSPs are also important since they act as molecular chaperones to prevent the misfolding of cellular proteins. *In situ* hybridization histochemistry (ISH) is a very effective method to detect discrete cells activated by stress. Monitoring mRNAs for IEGs or HSPs has demonstrated the spatial and temporal spread of excitation from the whole body to the intracellular molecules[2] (Fig. 1). By use of ISH and IMO models, we demonstrated the molecular and electrocardiographic changes, which are mediated by activation of α- and β- adrenoceptors in the heart.

2. ELECTROCARDIOGRAPHIC RESPONSES

Elevation of the ST segment at II, III, aVF and precordial leads was observed soon after the onset of IMO.[4] ECG returned to the pre-IMO basal level soon after the end of IMO, suggesting that IMO induces a transient change just like vasospastic angina.

[1] Department of Anatomy and Cell Biology, [2] Department of Anatomy and Neurobiology, Wakayama Medical University, Kimiidera, Wakayama, 641-8509, Japan. [3]Department of Legal Medicine, Graduate School of Medicine, University of Tokyo, Tokyo 113-0033, Japan

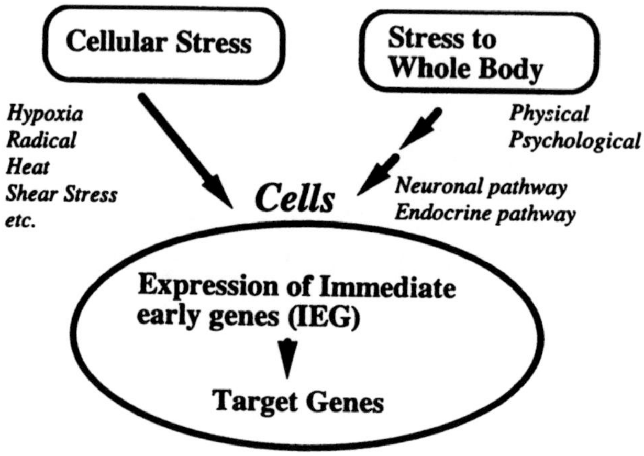

Figure 1. Concept of immediate early genes.

Pretreatment with a combination of α1-blocker, prazosin, and β1-blocker, metoprolol, or with αβ-blocker, amosulalol, completely abolished the stress-induced tachycardia and the elevation of ST segment. Vasodilators such as nitroglycerine or diltiazem were ineffective.

3. MOLECULAR RESPONSES

Signals for c-*fos* or other IEG mRNAs were observed from 15 min to 45 min and they returned to the control level at 60 min under IMO.[5] They were strongly expressed in the myocardium in the area surrounding the left ventricular cavities and in the smooth muscle cells of coronary arteries. Pretreatment with prazosin and metoprolol, or with amosulalol, abolished the stress-induced expression of IEG mRNAs. Treatment of perfused heart with the α-agonist, phenylephrine, β-agonist, isoproterenol, or their combination, elicited a strong expression of IEG mRNAs.[6] IMO stress activated p44/p42 Mitogen-Activated Protein kinase (p44/p42 MAP kinase, Erk1/Erk2). Pretreatment with metoprolol did not inhibit the activation of stress-induced MAP kinase, while blockage of the α1-adrenoceptor by pretreatment with prazosin or amosulalol attenuated the activation. Application of phenylephrine, but not isoproterenol, to the perfused rat heart elicited activation of MAP kinase. HSP70 mRNA was observed from 30 min to 6h, and the maximum expression was at 3h. Strong signals were detected in the smooth muscle cells of the aorta and coronary arteries. They were scattered in the myocardium of left ventricle. Signals for atrial natriuretic peptide (ANP) and brain natriuretic peptide (BNP) mRNAs were observed in the control heart, while they were upregulated in response to IMO and reached a maximum at 3h following the onset of IMO.

4. DISCUSSION

IMO of rats induced a transient change in ECG just like a vasospastic angina and the upregulation of IEG mRNAs in the discrete types of cardiac cells. Induction of IEG mRNAs indicates the cellular activation, suggesting that this stress activates individual types of cardiac cells. Failure of the potent vasodilator to inhibit these ECG changes and expression of IEG mRNAs suggests that ischemia is not responsible for them. Complete inhibition of the stress-induced changes by combined blockade of α- and β-adrenoceptors indicates that blocking both α1- and β1-adrenoceptors is essential. Perfusion studies indicate that activation of either α- or β-adrenoceptors is sufficient to mimic the response evoked by IMO in the rat. IMO stress activates the α1-adrenoceptor-mediated MAP kinase pathway, whereas the pathway of the response mediated by the β-adrenoceptor remains unknown. IMO induced the upregulation of ANP and BNP mRNAs in the left ventricle, where induction of IEGs was also observed. In addition, colocalization of ANP mRNA and c-Fos protein was observed, suggesting that c-Fos is involved in the upregulation of these natriuretic peptides. Activation of α- or β-adrenoceptors is the primary trigger of stress-induced molecular and electrophysiological changes in the heart. Sequential upregulation of genes can be considered as an adaptive response (Fig. 2).

5. CLINICAL IMPLICATIONS

These cardiac changes induced by IMO stress of rats are sometimes observed in patients suffering from emotional stress or an episode of pheochromocytoma. Dote *et al.* (Hiroshima Asa City Hospital, Hiroshima, Japan) proposed a special type of non-ischemic acute cardiomyopathy.[7,8] Its clinical features are: 1) sudden onset of chest pain, 2) elevation of ST segment in ECG with normal coronary angiogram, 3) a very remarkable sign in left ventriculogram, that is to say, akinesis around the left ventricle (LV) apex and hyperkinesis around the LV outflow. Dr. Dote named this as "tako-tsubo (a bottle for trapping an octopus traditionally used by Japanese Fishermen) cardiomyopathy" according to this unique shape. These changes usually normalize within 1 week. Another interesting character is the background of patients. In most of the cases, emotional stress precedes their attack. More interestingly, most of the patients are post-menopausal women. Though the pathogenesis of this "tako-tsubo cardiomyopathy" is not clear, activation of the cardiac adrenoceptors may underlie these changes.

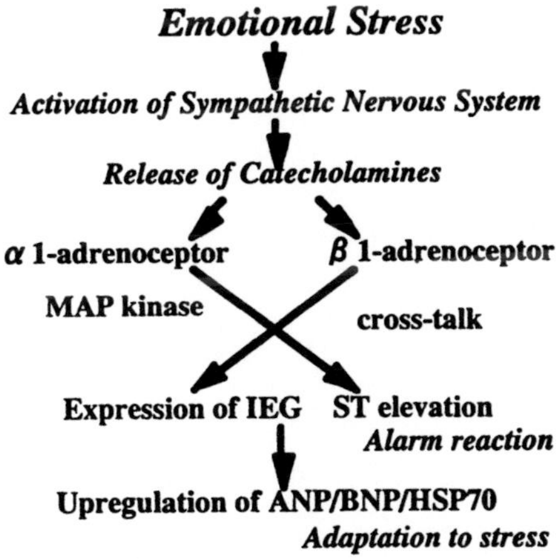

Figure 2 A simplified cascade of events that take place in the heart of emotionally stressed rats

REFERENCES

1. R. Kvetnansky, R. McCarty, N.B. Thoa, C.R. Lake, and I.J. Kopin. Sympatho-adrenal responses of spontaneously hypertensive rats to immobilization stress. *Am. J. Physiol.* **236,** H457-462 (1979).
2. E. Senba, and T. Ueyama. Stress-induced expression of immediate early genes in the brain and peripheral organs of the rat. *Neuroscience Res.* **29,** 183-207 (1997)
3. H. Itoh, and Y. Tashima. The stress (heat shock) proteins. *Int. J. Biochem.* **23,** 1185-1191 (1991).
4. T. Ueyama, K. Yoshida, and E. Senba. ST segment elevation in ECG induced by emotional stress is improved by combined blockade of alpha- and beta-adrenoceptors. *Clin. Exp. Pharamacol. Physiol.* **27,** 384-386 (2000).
5. T. Ueyama, S. Umemoto, and E. Senba. Immobilization stress induced c-*fos* and c-*jun* immediate early genes expression in the heart. *Life Sci.* **59,** 339-347 (1996).
6. T. Ueyama, K. Yoshida, and E. Senba. Emotional stress induces expression of immediate early genes in rat heart via activation of alpha- and beta-adrenoceptors. *Am. J. Physiol.* **277,** H1553-H1561 (1999).
7. K. Dote, H. Mituda, M. Ninomiya, and T. Okuhara. Acute reversible catecholamine cardiomyopathy. (in Japanese) *Ryoikibetsu Shokogun Shirizu* **14,** 166-169 (1996).
8. K. Tsuchihashi, K. Ueshima, T. Uchida, N. Oh-mura, K. Kimura, M. Owa, M. Yoshiyama, S. Miyazaki, K. Haze, H. Ogawa, T. Honda, M. Hase, R. Kai, and I. Morii. Transient left ventricular apical ballooning without coronary artery stenosis: a novel heart syndrome mimicking acute myocardial infarction. Angina Pectoris-Myocardial Infarction Investigations in Japan. *J. Am. Coll. Cardiol.* **38,** 11-18 (2001).

STRESSOR-SPECIFIC ACTIVATION OF CATECHOLAMINE SYSTEMS IN HUMANS

David S. Goldstein, Courtney S. Holmes, and Steven M. Frank*

1. INTRODUCTION

Hans Selye introduced and popularized stress as a medical scientific idea. While a medical student he came to the view that all sick patients looked about the same. Eventually he developed a theory of stress that enjoyed wide acceptance.

According to Selye's "doctrine of nonspecificity," all stressors elicit the same "stress response," characterized by adrenal enlargement, immunosuppression, and gastrointestinal ulceration or bleeding.[1, 2]

Recent studies of laboratory animals have shown that without certain assumptions, the doctrine of nonspecificity is impossible to disprove, and that given those assumptions experimental findings fail to support nonspecificity of stress responses.[3] These findings mean that the doctrine of nonspecificity has limited scientific value.

Acute glucoprivation, such as produced by administration of insulin or 2-deoxyglucose, evokes profound adrenomedullary activation, with relatively little activation of sympathetic neuronal outflows. In contrast, exposure to cold evokes prominent sympathoneural activation, with relatively little of adrenomedullary hormonal activation.[4] These findings support an alternative "homeostat" theory of stress, which includes the derivative concept of "primitive specificity" of neuroendocrine responses to different stressors.[5]

According to the homeostat theory, in stress the organism senses a disruption or a threat of disruption of homeostasis. This sensation requires a comparative process, where the brain compares available information with setpoints for responding. "Homeostats" are homeostatic comparators. Each homeostat compares information with a setpoint for responding, determined by a regulator (Figure 1). The homeostat uses one or more effectors to change values for the controlled variable. The negative feedback loop is closed by monitoring changes in the levels of the controlled variable, via one or more

* David S. Goldstein and Courtney S. Holmes, Clinical Neurocardiology Section, National Institute of Neurological Disorders and Stroke, National Institutes of Health, Bethesda, MD 20892-1620 USA; Steven M. Frank, Department of Anesthesiology and Critical Care Medicine, The Johns Hopkins Medical Institutions, Baltimore, MD 21205 USA.

monitored variables. Inclusion of multiple effectors in homeostatic systems extends the range of control, enables compensatory activation of alternative effectors if one is inactivated, and promotes development of stressor-specific patterns. Thus, activities of stress effector systems are coordinated in relatively specific patterns, including neuroendocrine patterns. These patterns, produced by the actions of different homeostats, serve different homeostatic needs.

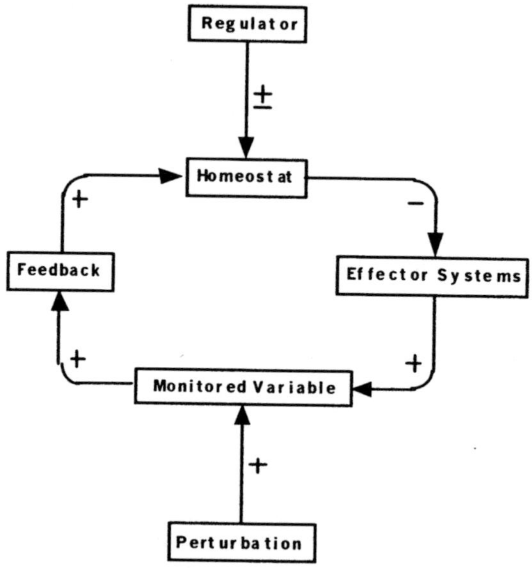

Figure 1. A homeostatic system.

In this study we tested in humans whether acute glucoprivation induced by 2-deoxyglucose, and mild core hypothermia induced by cold saline infusion, differentially affect sympathoneural and adrenomedullary outflows, as indicated by changes in plasma levels of norepinephrine, the sympathetic neurotransmitter, and of epinephrine, the main adrenomedullary hormone.

We also asked in patients with chronic orthostatic intolerance whether neurally mediated hypotension evoked by head-up tilt table testing or occurring spontaneously is associated with differential changes in plasma catecholamine concentrations. Tilt-evoked syncope is known to be associated with prominent increases in plasma epinephrine levels[6] and withdrawal of sympathetic neuronal outflows;[7] however, whether differential adrenomedullary and sympathoneural alterations precede tilt-evoked or spontaneous syncope has been unknown.

2. METHODS

Normal volunteers underwent i.v. administration of 2-deoxyglucose, with measurements of plasma levels of catecholamines before and after drug administration.

Some subjects were re-studied after sedation with alprazolam. Data from these experiments have been published previously.[8] Other normal volunteers underwent central i.v. infusion of warm and then cold saline, with measurements of total body spillovers of norepinephrine and epinephrine, by a previously published tracer-kinetic method.[9] Finally, 10 patients referred for orthostatic intolerance or frequent presyncope had plasma levels of catecholamines measured before (0-5 minutes) syncope evoked by tilt. Four others had catecholamine measurements before spontaneous syncope while monitored supine, and 4 did not have tilt-evoked syncope.

3. RESULTS

Glucoprivation evoked profound adrenomedullary activation, as indicated by the plasma epinephrine response, with little or no sympathoneural activation, as indicated by the plasma norepinephrine response. In contrast, mild core hypothermia evoked mainly sympathoneural activation, as indicated by the estimated rate of entry of norepinephrine into arterial plasma (total body norepinephrine spillover), with less adrenomedullary activation, as indicated by total body epinephrine spillover (Figure 1).

All patients with syncope evoked by tilt had marked increases in plasma epinephrine levels (mean 13 times baseline) before the loss of consciousness ($p<0.001$). Simultaneously obtained plasma norepinephrine levels were unchanged or decreased. Before syncope, forearm vascular resistance consistently decreased (mean decrease 28% from baseline, $p<0.001$), and the change in forearm vascular resistance correlated significantly negatively with changes in plasma epinephrine ($r=-0.70$, $p<0.001$) and the epinephrine:norepinephrine ratio ($r=-0.73$, $p<0.001$). Patients without syncope did not have this neuroendocrine and hemodynamic pattern.

All 4 patients with spontaneous syncope while supine also had increases in the plasma epinephrine:norepinephrine ratio before or at the time of syncope (mean 5 times baseline), including one patient where arterial plasma epinephrine levels clearly increased and norepinephrine levels decreased progressively for several minutes before syncope.

4. CONCLUSIONS

In humans, glucoprivation evokes compensatory adrenomedullary activation, which increases epinephrine-induced release of glucose by the liver, whereas mild core hypothermia evokes compensatory sympathoneural activation, which redistributes blood volume away from the skin. Thus, stress responses in humans have a degree of "primitive specificity," with patterned responses of multiple effectors tending to maintain levels of monitored variables at pre-set apparent steady-states ("allostasis").

Since adrenomedullary stimulation and relative sympathoinhibition precede both tilt-evoked and spontaneously occurring neurocardiogenic syncop, sympathoadrenal imbalance may explain skeletal vasodilation that attends the syncope.

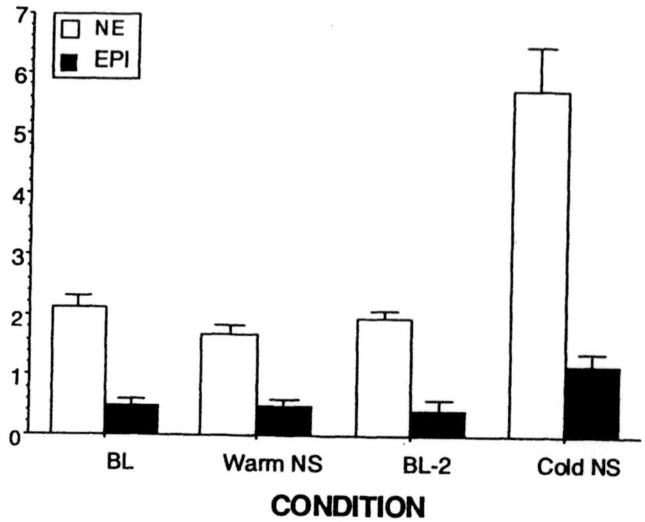

Figure 2. Total body spillovers of norepinephrine (NE) and epinephrine (EPI) at baseline (BL), during central i.v. infusion of warm normal saline (NS), during a second baseline (BL-2), and during central i.v. infusion of cold NS in normal volunteers.

REFERENCES

1. D. S. Goldstein, *Stress, Catecholamines, and Cardiovascular Disease* (Oxford University Press, New York, 1995).
2. D. S. Goldstein, *The Autonomic Nervous System in Health and Disease* (Marcel Dekker, Inc., New York, 2001).
3. K. Pacak, M. Palkovits, G. Yadid, R. Kvetnansky, I. J. Kopin, and D. S. Goldstein, Heterogeneous neuroendocrine responses to various stressors: A test of Selye's doctrine of non-specificity., *American Journal of Physiology* **(275)**, R1247-R1255 (1998).
4. K. Fukuhara, R. Kvetnansky, G. Cizza, K. Pacak, H. Ohara, D. S. Goldstein, and I. J. Kopin, Interrelations between sympathoadrenal system and hypothalamo-pituitary-adrenocortical/thyroid systems in rats exposed to cold stress, *J Neuroendocrinol* **(8)**, 533-541 (1996).
5. D. S. Goldstein, Stress as a scientific idea: A homeostatic theory of stress and distress., *Homeostasis* **(4)**, 177-215 (1995).
6. D. A. Robertson, G. A. Johnson, R. M. Robertson, A. S. Nies, D. G. Shand, and J. A. Oates, Comparative assessment of stimuli that release neuronal and adrenomedullary catecholamines in man, *Circulation* **(59)**, 637-643 (1979).
7. M. G. Ziegler, C. Echon, K. D. Wilner, P. Specho, C. R. Lake, and J. A. McCutchen, Sympathetic nervous withdrawal in the vasodepressor (vasovagal) reaction, *J.Auton.Nerv.Sys.* **(17)**, 273-278 (1986).
8. A. Breier, O. Davis, R. Buchanan, S. Listwak, C. Holmes, D. Pickard, and D. Goldstein, Effects of alprazolam on pituitary-adrenal and catecholaminergic responses to metabolic stress in humans., *Biol Psychiatry* **(32)**, 880-890 (1992).
9. M. Esler, G. Jennings, G. Lambert, I. Meredith, M. Horne, and G. Eisenhofer, Overflow of catecholamine neurotransmitters to the circulation: Source, fate, and functions, *Physiol.Rev.* **(70)**, 963-985 (1990).

BRAIN NOREPINEPHRINE SYSTEMS AND ANXIETY

Masatoshi Tanaka*, Shigeto Yamada**, Masami Yoshida*, Hiroyuki Emoto*, Hideo Ishii**

1. INTRODUCTION

For investigations of the neurochemical bases of the provocation of mechanisms of anxietyand/or fear and the reduction mechanism of anxiety by anxiolytic drugs, stress paradigms are a very useful tool, since a variety of stressful stimuli have been known to cause marked changes in negative emotions including anxiety and/or fear.

In this paper, we suggest that the activity of the brain norepinephrine sysytem, in particular, in brain regions such as the hypothalamus, amygdala and locus coeruleus, is in part, involved in the provocation and attenuation of anxiety and/or fear.

2. STRATEGIES FOR NEUROCHEMICAL STUDIES OF ANXIETY IN ANIMALS

There are various types of stressors involving both physical and psychological factors, as shown by immobilization stress and electric shock stress, and involving mainly psychological factors, as shown by psychological stress and conditioned fear stress. Thus, one of the neurochemical strategies is to determine the neurochemical changes when the rats are exposed to stress situations and show anxiety and/or fear, by measuring the levels of neurotransmitters and their metabolites in discrete brain regions or in perfusates obtained by in vivo microdialysis. Another approach is to measure neurochemical changes in animals treated with anxiogenic drugs which have been reported to cause anxiety in humans, such as yohimbine and b-carbolines. When a certain neurochemical change is found to occur during provocation of anxiety, it should be confirmed whether such neurochemical changes can be reduced or abolished by the in the animals.

---

* Department of Pharmacology , Kurume University School of Meddine, Kurume 830-0011, Japan,
** Department of Psychiatry, Saga Medical School, Nabeshima, Saga 849-8501, Japan

Catecholamine Research, Edited by Nagatsu *et al.*
Kluwer Academic/Plenum Publishers, 2002

3. STRATEGIES FOR ESTABLISHING THE NOREPINEPHRINE HYPOTHESIS OF ANXIETY

While physical factors are involved in immobilization stress, negative emotions, such as anxiety and/or fear, may be provoked because animals stressed by immobilization show emotional responses such as defecation, vocalization, and struggling. We have reported that immobilization stress causes significant and marked increases in norepinephrine release in extended brain regions (Tanaka M. et al., 1982). These findings have been confirmed by intracerebral microdialysis in the hypothalamus (Yokoo H. et al., 1990) and amygdala. Among the brain regions examined, the increase in norepinephrine release was shown to be very marked in the hypothalamus, amygdala and locus coeruleus, one of the biggest cell body sites of brain norepinephrine systems. Theses findings raise the possibility that increased release of norepinephrine in specific brain regions (hypothalamus, amygdala and locus coeruleus) is the basic neurochemical mechanism for the provocation of anxiety and/or fear in rats during exposure to stress. According to this strategy, we examined the following five questions.

4. INCREASE IN BRAIN NOREPINEPHRINE RELEASE AND ANXIETY

In all experiments, male Wistar rats weighing around 200 g were used. Pretreatment with diazepam at 2 and 5 mg/kg i.p. significantly attenuated increases in 3-methoxy-4-hydroxyphenylethyleneglycol sulfate (MHPG-SO4) levels caused by 1 h immobilization stress in the hypothalamus, amygdala, locus coeruleus, hippocampus and cerebral cortex in a dose-dependent manner as compared to the effect of vehicle in a flumazenil, an antagonist of benzodiazepine receptor, reversible manner. Diazepam at 5 mg/kg significantly decreased both the frequency of vocalization and the number of feces deposited during the stress exposure in a flumazenil reversible manner (Ida et al., 1985). Further, it is confirmed in the study using intracerebral microdialysis technique that the increase in norepinephrine release in the anterior hypot halamus caused by immobilization stress was significantly attenuated by pretreatment with diazepam at 5 mg/kg i.p. before stress exposure in a flumazenil reversible manner. Pretreatment with morphine (3 or 6 mg/kg) significantly attenuated not only the stress-induced increases in MHPG-SO4 levels in the hypothalamus, amygdala, hippocampus, thalamus and midbrain but also emotional responses including vocalization, struggling, defecation and weight loss of the rats exposed to stress in a naloxone reversible manner. We also reported that Met-enkephalin at 100 mg/rat or b-endorphin at 10 mg/rat injected i.c.v. before stress exposure significantly attenuated both stress-induced increases in MHPG-SO4 levels in the hypothalamus, amygdala, locus coeruleus, thalamus, hippocampus, and midbrain and the emotional changes induced by stress in a naloxone reversible manner.

Psychological stress and conditioned fear stress, wherein the psychological factor is predominantly involved, caused significant increases in MHPG-SO4 levels preferentially in the hypothalamus, amygdala and locus coeruleus. A microdialysis study further confirmed that conditioned fear increased norepinephrine release in the anterior hypothalamus (Yokoo H. et al., 1990). Furthermore, the increases in norepinephrine release caused by these stressors were significantly attenuated by pretreatment with diazepam or alprazolam in a flumazenil reversible manner.

Yohimbine, an a2-adrenoceptor antagonist, was reported to increase norepinephrine release in several brain regions including the hypothalamus, amygdala and locus coeruleus. Anxiolytic drugs such as diazepam selectively suppressed conditioned defensive burying behavior in a dose-dependent fashion, although yohimbine at 0.5 and 2 mg/kg significantly increased both duration and the number of burying episodes similarly as the anxiogenic drug, ethyl-b-carboline-3-carboxylate (b-CCE). Yohimbine at 0.2 and 2 mg/kg and the anxiogenic drug b-CCE at 1 and 2 mg/kg significantly increased straw-climbing, although diazepam at 0.5 and 2 mg/kg, doses which were unlikely to cause muscle relaxation and sedation, significantly decreased the number of attempts at straw climbing in the modified forced-swim test. The findings suggest that yohimbine, which increases norepinephrine release in brain regions such as the hypothalamus, amygdala and locus coeruleus, enhanced anxiety in anxiety-related behavioral tests in rats (Tanaka M. et al., 2000).

Administration of FG 7142 (metyl-b-carboline-3-carboxamide), which has been reported to cause anxiety in human,s and b-CCE significantly increased MHPG-SO4 levels in the hypothalamus, amygdala and locus coeruleus. These findings indicate that anxiogenic drugs increase norepinephrine release in specific brain regions (Tanaka M. et al., 2000).

5. SALIVARY LEVELS OF FREE MHPG IN PSYCHIATRIC OUTPATIENTS WITH ANXIETY

We measured salivary levels of free MHPG using gas chromatography-mass spectrometry in 21 first visit outpatients with anxiety (Yamada S. et al., 1998). The pretreatment levels of salivary MHPG in the patient group were significantly higher than thevalues in the disease control group and control group. One week treatment with alprazolam, an anxiolytic of triazolobenzodiazepines, caused a significant reduction of salivary MHPG levels as well as scores on the Hamilton Anxiety Scale. The finding indicates that salivary MHPG might be a useful marker for evaluationg states of anxiety.

6. THE NEUROCHEMICAL MECHANISM OF THE PROVOCATION AND REDUCTION OF ANXIETY

Based upon these findings, we proposed a scheme to explain the neurochemical mechanism underlying the provocation of anxiety and the reduction of anxiety by some anxiolytics (Tanaka M., 2000). A variety of stressful stimuli, yohimbine acting via a2-adrenoceptors, and b-carbolines acting via benzodiazepine receptors as inverse agonists of benzodiazepines, increase norepinephrine release in the hypothalamus, amygdala, and locus coeruleus, which results in the provocation of anxiety and/or fear. Benzodiazepine anxiolytics acting via benzodiazepine receptors, morphine and opioid peptides such as b-endorphin and Met-enkephalin, acting via opioid receptors, attenuate the increases in norepinephrine release in these brain regions, leading to the relief of anxiety and/or fear.

REFERENCES

Ida Y., Tanaka M., Tsuda A., Tsujimaru S., and Nagasaki N., 1985, Attenuating effect of diazepam on stress-induced increases in noradrenaline turnover in specific brain regions of rats: antagonism by Ro 15-1788. Life Sci, **37**: 2491-2498.

Tanaka M., Kohno Y., Nakagawa R., Ida Y., Iimori K., Hoaki Y., Tsuda A., and Nagasaki N., 1982, Time-related differences in noradrenaline turnover in rat brain regions by stress. Pharmacol Biochem Behav, **16**; 315-319.

Tanaka M., Yoshida M., Emoto H., and Ishii H., 2000, Noradrenaline systems in the hypothalamus, amygdala and locus coeruleus are involved in the provocation of anxiety: basic studies. Europ J Pharmacol, **405**: 397-406.

Yamada S., Yajima, J., Harano M., Miki K., Nakamura J., Tsuda A., Shoji H., Maeda H. and Tanaka M., 1998, Saliva level of free 3-methoxy-4-hydroxyphenylglycol in psychiatric outpatients with anxiety. Int Clin Psychopharmacol, **13**: 213-217.

Yokoo H., Tanaka M., Tanaka T., and Tsuda A., 1990, Stress-induced increases in noradrenaline release in the rat hypothalamus assessed by intracranial microdialysis. Experientia, **46**: 290-292.

MONOAMINES AND FEAR-RELATED BEHAVIORS

Mitsuhiro Yoshioka, Machiko Matsumoto, Kiyoshi Mori and Hiroko Togashi*

1. INTRODUCTION

The mesocortical dopamine (DA) neuronal system has been reported to show regional differences in sensitivity to various pharmacological and behavioral manipulations. For instance, cortical DA neuronal activity was facilitated following application of atypical antipsychotic drugs (Broderick and Piercey, 1998) or anxiogenic agent (Bradberry et al., 1991), but nigrostriatal and mesolimbic DA systems were not influenced by these drugs. Physiological stimuli such as tail-shock stress (Abercrombie et al., 1989) or prolonged handling (Feenstra et al., 1998) produced greater enhancement of DA release in the prefrontal cortex (PFC) than in the striatum and/or the nucleus accumbens. Recently, we have demonstrated that DA release in the PFC was enhanced by conditioned fear stress (CFS), an animal model of anxiety which is regarded as psychological stress without physical stimuli (Fanselow, 1980), whereas any changes of striatal DA levels were not found by exposure to CFS. These regional differences of characteristics of DA neurons may be pivotal for understanding the mechanism underlying the pathophysiology of DA neuronal function. The PFC receives not only dopaminergic inputs from the ventral tegmental area but also reciprocal projections of other neuronal systems. Among these neurotransmitters, serotonin (5-HT) has been focused on the regulation of DA neuronal activity in both cell bodies and terminal regions based on clinical and neurochemical studies (Broderick and Piercey, 1998; Saito et al, 1996). The present study was focused on the possible involvement of 5-HT neuronal system in the characteristics of mesocortical DA neurons with a view towards elucidating the mediation of emotional function.

2. METHODS

Male Wistar-strain rats (10-14 week old) were anesthetized with ketamine (100 mg/kg, i.p.) and 3-mm concentric guide cannulae were stereotaxically implanted into the PFC (rostral 3.2, lateral 0.7, ventral 1.0 mm). Dialysis probes were perfused with

* Department of Pharmacology, Hokkaido University School of Medicine, Sapporo 060-8638, JAPAN

Catecholamine Research, Edited by Nagatsu *et al.*
Kluwer Academic/Plenum Publishers, 2002

Ringer's solution and successive 20 μl samples were collected at 20 min intervals. Extracellular levels of DA and 5-HT were measured using HPLC-ECD as previously described (Yoshioka et al., 1996).

3. RESULTS

3.1. Effects of anxiogenic agent, idazoxan on extracellular levels of DA

The dosage of idazoxan was chosen according to previous reports that it showed an anxiogenic profile in the rat maze exploration (Handley and Mithani, 1984; Wright et al., 1992). Idazoxan (0.25 mg/kg, i.p.) produced significant increases in DA levels and the maximum response was observed 80 min after drug administration (Fig. 1A). Idazoxan-induced increases in DA levels were prevented by a typical benzodiazepine anxiolytic drug, diazepam (0.5 mg/kg, i.p). Pretreatment with the $5\text{-HT}_3/_4$ receptor antagonist tropisetrone (0.1 mg/kg, i.p.), a putative anxiolytic agent, also tended to abolish the enhancement of DA release (Fig. 1B).

Fig. 1

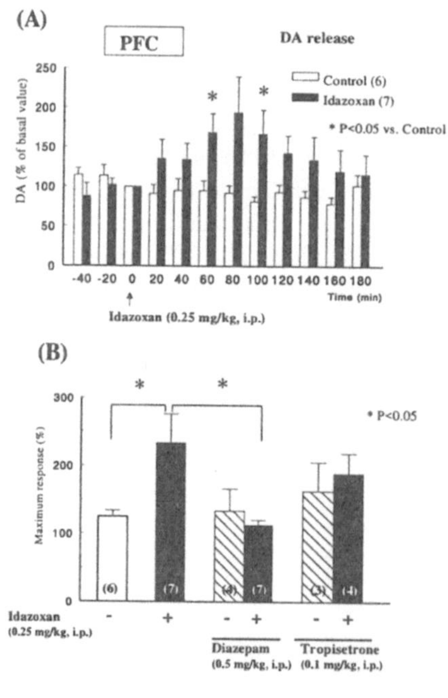

Fig. 2

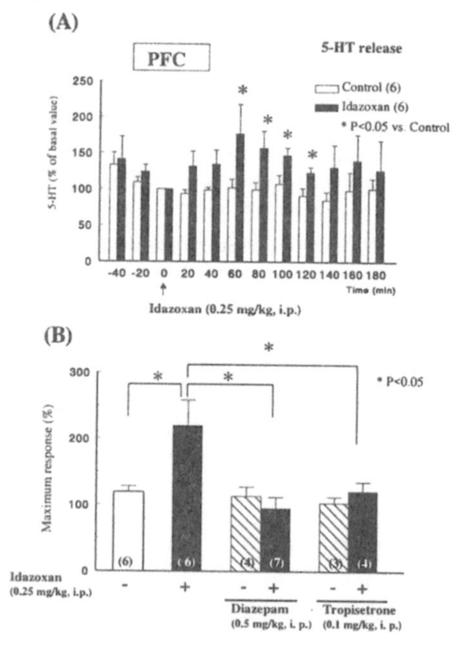

Fig 1. Time course response of idazoxan-induced (0.25 mg/kg, i.p.) changes in DA release from the rat PFC (A). Maximum responses of idazoxan-induced increases in DA release in the absence or in the presence of anxiolytic drugs, diazepam (0.5 mg/kg, i.p.) and tropisetrone (0.1 mg/kg, i.p).

Fig 2. Time course response of idazoxan-induced changes in 5-HT release from the rat PFC (A). Maximum responses of idazoxan-induced increases in 5-HT release in the absence or in the presence of diazepam and tropisetrone. These anxiolytic drugs were injected 10 min before idazoxan administration. All results are given as mean ± S.E.M.

3.2. Effects of idazoxan on extracellular levels of 5-HT

Extracellular 5-HT levels were significantly increased by administration of idazoxan (0.25 mg/kg, i.p.)(Fig. 2A). Idazoxan-induced increases in 5-HT release were completely abolished by pretreatment with either diazepam or tropisetrone (Fig. 2B). Neither diazepam nor tropisetrone by itself significantly altered the spontaneous 5-HT or DA levels.

3.3. Effects of pretreatment with 5,7-dihydroxytryptamine (5,7-DHT) on idazoxan-induced increases in DA levels

As shown in Fig. 3, idazoxan (0.25 mg/kg, i.p.) failed to increase DA release in 5,7-DHT (200 μg/rat, i.c.v.) pretreated rats. The mean value of maximum response of DA release in 5,7-DHT pretreated rats was significantly reduced compared with those in intact rats.

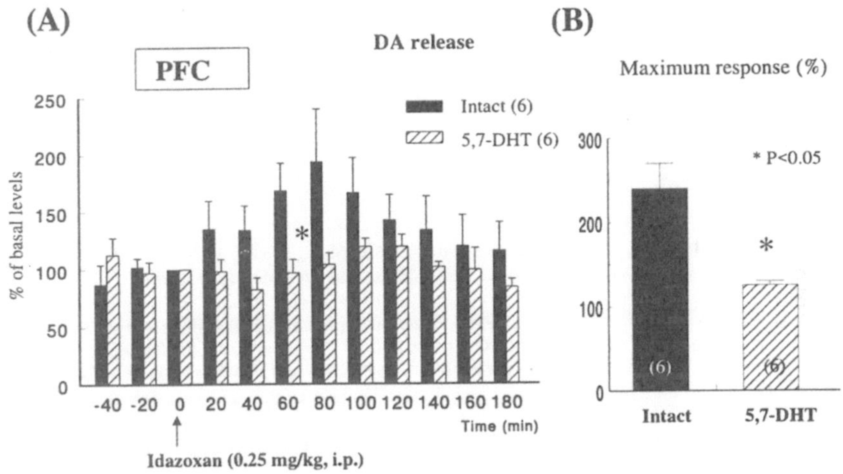

Figure 3. (A) Time course response of idazoxan-induced (0.25 mg/kg, i.p.) changes in DA release from the PFC in rats pretreated with 5,7-DHT (200 μg/kg, i.c.v.). (B) Maximum responses of idazoxan-induced DA release in intact rats and 5,7-DHT treated rats. All results are given as mean ± S.E.M.

4. DISCUSSION

In the present study, idazoxan possessing an anxiogenic properties produced increases in DA levels in the PFC. These facilitatory effects were prevented by pretreatment with anxiolytic drugs, diazepam and tropisetrone, suggesting that excessive DA neuronal activity is associated with the etiology of anxiety disorders. This finding was supported by the previous reports showing that CFS caused increases in DA release from the PFC (Yoshioka et al., 1996) in addition to facilitation of 5-HT neuronal activity (Yoshioka et al., 1995). Interestingly, extracellular levels of DA in the striatum did not show any changes following exposure to CFS. Not only physical stress but also psychological stress without physical stimuli preferentially activated cortical DA release, indicating the regional specificity of mesocortical DA neurons.

We observed furthermore that idazoxan failed to increase DA release in 5-HT depleted rats. Thus, intact serotonergic neurons appear to be required for the facilitation of DA release. These results indicate the possibility that endogenous 5-HT contributes to enhance DA release, i.e., excessive 5-HT neuronal activity may have an essential influence on the mesocortical DA neuronal activity at least to the psychological stimuli. In other words, 5-HT neuronal system may play an important role in the mediation of emotional function. Indeed, numerous studies provided evidence that 5-HT related agents, for instance, tandospirone (5-HT_{1A} receptor agonists), tropisetrone ($5\text{-HT}_{3/4}$ receptor antagonists) and fluvoxamine (selective 5-HT reuptake inhibitor) exert anxiolytic effects. We have recently shown that not only diazepam but also tandospirone and fluvoxamine suppressed long-term potentiation (LTP) in the hippocampal CA1 fields, an electrophysiological basis of the memory processes, which has been focused on the possible relationship in the anxiety or fear. However, no influence of LTP formation was observed following DA reuptake inhibitor, GBR 12909.

In summary, our research suggests that DA release in the PFC was enhanced by psychological stress in the absence of physiological stimuli, and that the intact serotonergic neuronal system was required for the excessive DA neuronal activity. The functional interaction between dopaminergic and serotonergic neuronal systems therefore might be associated with an impairment of emotional functions.

REFERENCES

Abercrombie E. D., Keefe K. A., DiFrischia D. S., and Zigmond M. J., 1989, Differential effect of stress on in vivo dopamine release in striatum, nucleus accumbens, and medial frontal cortex. *J. Neurochem.* **52:** 1655-1658.

Bradberry C. W., Lory J. D., and Roth R. J. , 1991, The anxiogenic–carboline FG 7142 selectively increases dopamine release in the rat prefrontal cortex as measured by microdialysis. *J. Neurochem.* **56:** 748-752.

Broderick P. A., and Piercey M. F ., 1998, Clozapine, haloperidol, and the D_4 antagonist PNU-101387G: in vivo effects on mesocortical, mesolimbic, and nigrostriatal dopamine and serotonin release. *J. Neural Transm.* **105:** 749-767.

Fanselow M. S., 1980, Conditional and unconditional components of postshock freezing. *Pavlov J. Biol. Sci.* **15:** 177-180.

Feenstra M. G. P., Botterblom M. H. A., and van Uum J. F. M., 1998, Local activation of metabotropic glutamate receptors inhibits the handling-induced increased release of dopamine in the nucleus accumbens but not that of dopamine or noradrenaline in the prefrontal cortex: Comparison with inhibition of ionotropic receptors. *J. Neurochem.* **70:** 1104-1113.

Handley, S.L., Mithani, S., 1984, Effects of alpha-adrenoceptor agonists and antagonists in a maze-exploration model of 'fear'-motivated behaviour. *Naunyn-Schmiedeberg's Arch Pharmacol.* 327: 1-5.

Saito, H., Matsumoto, M., Togashi, H., Yoshioka, M., 1996, Functional interaction between serotonin and other neuronal system focus on in vivo microdialysis studies. *Jpn. J. Pharmacol.* **70:** 203-225

Yoshioka, M., Matsumoto, M., Togashi, H., Saito, H., 1995, Effects of conditioned fear stress 5-HT release in the rat prefrontal cortex. *Pharmacol. Biochem. Behav.* **51:** 515-519.

Yoshioka, M., Matsumoto, M., Togashi, H., Saito H., 1996, Effect of conditioned fear stress on dopamine release in the rat prefrontal cortex. *Neuroscience Lett.* **209:** 201-203.

Wright, I.K., Heaton, M., Upton, N., Marsden, C.A., 1992, Comparison of acute and chronic treatment of various serotonergic agents with those of diazepam and idazoxan in the rat elevated X-maze. *Psychopharmacol.* **107**: 405-414.

DOPAMINERGIC EFFECTS ON ACQUISITION AND CONSOLIDATION: A MICRODIALYSIS STUDY

Masahiko Nomura[1], Kiyoyuki Maruki[2] and Koji Hori[3]

1. INTRODUCTION

While experimental animals are learning on a schedule-controlled discrimination learning task in an operant type Skinner box, one may examine the relationship between learning ability and neurochemical changes by measuring the content of various neurotransmitters using a microdialysis method. In particular, some of the catecholamines have much closer correlations with learning ability compared to other neurotransmitters. Among the catecholamines, dopamine and its metabolites in amygdala, hippocampus, thalamus, and cortex are increased during learning performance. Dopamine appears to play an important role in the acquisition and consolidation stages of learning.

2. MATERIALS AND METHODS

2.1 Subjects

Thirty male Wistar rats, aged 12 weeks at the start of the experiment, were housed in individual cages on a 12 hour light-dark schedule. Rats were maintained at 85% of their free-feeding body weights throughout the experiment.

2.2 Behavioral protocol

The experimental apparatus and protocol for the learning paradigm were the same as those described in detail by Hori et al.[1] Learning experiments were conducted in a one lever operant chamber containing a cue lamp and food dispenser as a reward. One group of fifteen rats received discrimination training on a multiple variable-interval 15 seconds

[1] Department of Physiology, [2] Department of Psychiatry,Saitama Medical School, Moro, Iruma, Saitama 350-0495, [3] Department of Psychology, Rikkyo University, Tokyo 171-8501, Japan

extinction schedule (mult VI EXT: MULT group). Another group of fifteen rats received non-discrimination training on a mixed variable-interval extinction schedule (MIX group). Learning schedules are summarized in Figure 1. Daily sessions consisted of 20 VI and 20 EXT trials and each trial lasted for 20 seconds. Procedures for the MIX group were the same as those for the MULT group, except that the cue lamp was always illuminated irrespective of the trial type, so that there was no programmed cue to discriminate the trial type.[2,3]

2.3 Microdialysis experiments

Following 30 daily sessions of learning, microdialysis probes were implanted stereotaxically in amygdala, hippocampus and anterior thalamus. Dopamine and its metabolites, DOPAC (3,4-dihydroxyphenylacetic acid) and HVA (homovanillic acid) were measured in freely behaving animals in a Skinner box using microdialysis methods[4]. When a rat pressed the lever to get a food pellet reward, catecholamine content was measured.[5,6] Three kinds of samples during three test stages, pre-learning, during learning, and post-learning, were collected every 20 minutes, and catecholamine content in the samples was measured using high performance liquid chromatography (HPLC-ECD).[7,8]

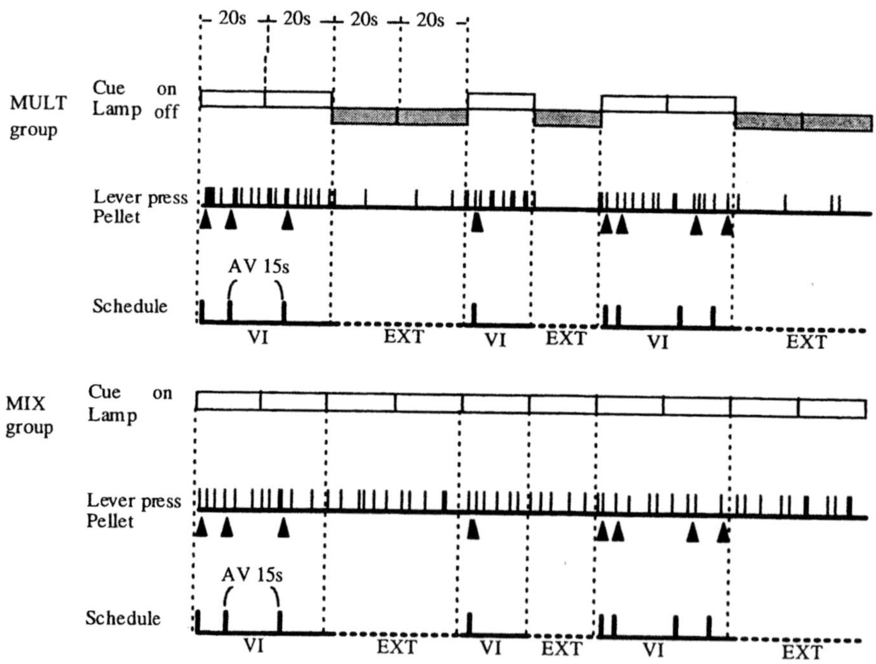

Figure 1. A multiple variable-interval 15-second extinction schedule (multi VI EXT: MULT) and a mixed variable-interval extinction schedule (MIX) as used in the present study.

3. RESULT

Figure 2 shows the discrimination learning ratio for the MULT and MIX learning schedules. The MULT group showed an increase of learning ratio, but the MIX group showed the same level ratio for 30 days of test sessions. This finding indicates that no learning occurred in the MIX group.

In order to clarify the regional changes in brain catecholamines, we measured DA, DOPAC and HVA content before and during learning performance period. In amygdala, hippocampus and anterior thalamus, dopamine and its metabolites showed significant increases (Figure 3). Figure 4 summarizes the time course of changes in DOPAC in amygdala in the MULT and MIX groups. Pre-learning values were set at 100%, and every time point during the learning phase showed in increase in the MULT group but not in the MIX group. The MULT group had a significantly higher discrimination ratio than the MIX group but there were no differences between the two groups in the total number of responses and in the frequency of reinforcement.

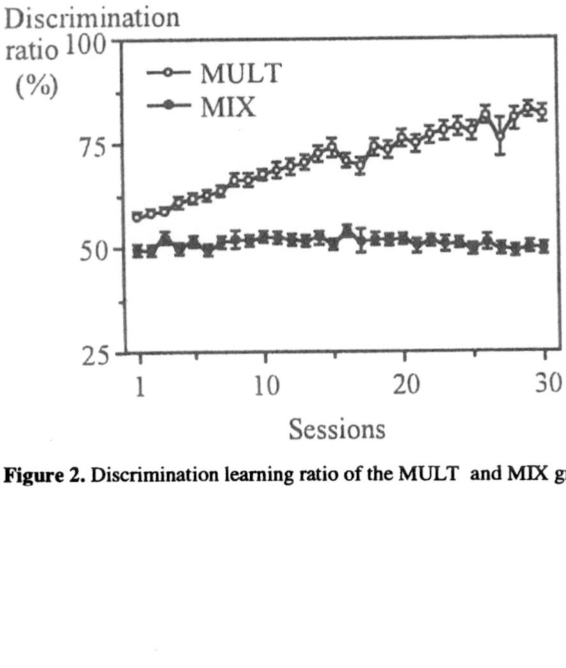

Figure 2. Discrimination learning ratio of the MULT and MIX groups.

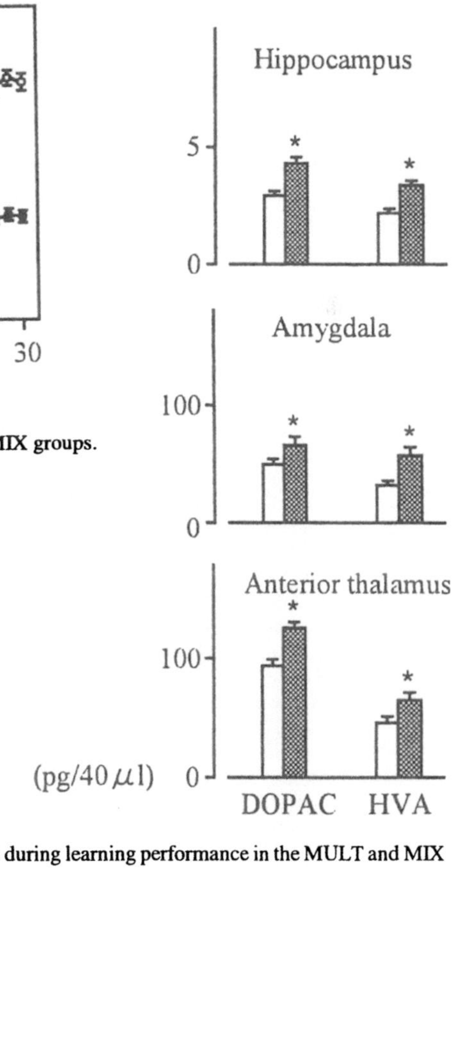

Figure 3. Regional changes in DOPAC and HVA before and during learning performance in the MULT and MIX groups.

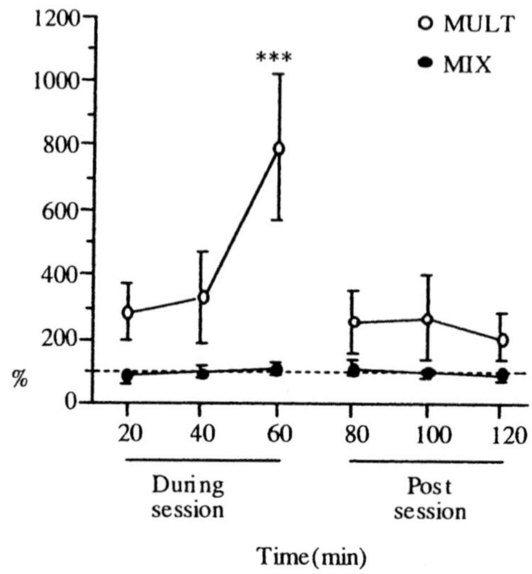

Figure 4. Time course of changes in DOPAC in amygdala of MULT and MIX groups.

4. CONCLUSIONS

Dopaminergic neurons appear to be involved in acquisition of a discrimination learning task in an operant type Skinner box.[1, 2, 3, 6, 7] Our learning task involved no fear, no spatial cues and it was not especially stressful to the animals.[8, 9, 10] By comparing brain catecholamine responses of the MULT and MIX groups, we find that dopaminergic neurons may play an important role in the acquisition and consolidation phases of this task.

REFERENCES

1. K. Hori, J. Tanaka, and M. Nomura, Effects of discrimination learning on the rat amygdala dopamine release : a microdialysis study, *Brain Res.*, **621**, 296-300 (1993).
2. Y. Yamamuro, K. Hori, H. Iwano, and M. Nomura, The relationship between learning performance and dopamine in the prefrontal cortex of the rat, *Neurosci Lett.*, **177**, 83-86 (1994).
3. Y. Yamamuro, K. Hori, J. Tanaka, H. Iwano, and M. Nomura, Septohippocampal cholinergic system under the discrimination learning task in the rat: a microdialysis study with the dual-probe approach, *Brain Res.*, **684**, 1-7 (1995).
4. H. Benveniste, Brain microdialysis, *J Neurochem.*, **52**, 1667-1679 (1989).
5. G. Schoenbaum, A.A.Chiba, and M. Gallagher, Orbitofrontal cortex and basolateral amygdala encode outcomes during learning, *Nature Neurosci.*, **1**, 155-159 (1998).
6. Y. Izaki, K. Hori, and M. Nomura, Dopamine and acetylcholine elevation on lever press acquisition in rat prefrontal cortex, *Neurosci Lett.*, **258**, 33-36 (1998).
7. Y. Izaki, K. Hori, and M. Nomura, Distribution of rat lever-press learning by hippocampo-prefrontal disconnection, *Brain Res.*, **860**, 199-202 (2000).
8. K. Nader, and J.E. LeDoux, Inhibition of the mesoamygdala dopaminergic pathway impairs the retrieval of conditioned fear associations, Behav Neurosci., **113**, 891-901 (1999).
9. M. A. Perinado-Monzano, The role of the amygdala and the hippocampus in working memory for spatial and non-spatial information, Behav Neurosci., **38**, 117-134 (1990).
10. W. Schultz, Multiple reward signals in the brain. Nature Rev Neurosci., **1**, 199-207 (2000)

THE LOCUS COERULEUS/NORADRENERGIC SYSTEM: ITS ROLE IN REGULATING WORKING MEMORY /ATTENTION AND INTERACTION WITH THE CHOLINERGIC SYSTEM

Tsuneyuki Yamamoto* and Shigenori Watanabe

1. INTRODUCTION

There is much evidence that the central cholinergic system plays an important role in the processes of learning and memory. Correlations have been found between the severity of cholinergic deficits and cognitive decline in patients with Alzheimer's disease (AD) in clinical trials for cholinomimetic therapy against deficits associated with this disease. It has been reported that multiple neurotransmitter deficits occur in the brains of AD patients. It is, therefore, interesting that numerous populations of neurons and neurotransmitter systems are damaged simultaneously by cholinergic deficits in the AD brain, contributing to the severity of the memory dysfunctions in this disease. In particular, the AD brain has a severe noradrenergic deficit as evidenced by cell losses in the locus coeruleus (LC) and by reduced levels of noradrenaline, its metabolites, and synthetic enzymic activity.[1] Bondrareff et al[2] suggested that a subgroup of patients with AD characterized by a reduced number of LC neurons showed a more severe cognitive decline than AD patients with LC neuronal counts comparable to those of age-matched control subjects. Therefore, it seems to be important to clarify the roles of interactions between the NA and Ch systems in regulating cognitive function. In the present study, we clarified[1] how the NA system participates in working memory and attention and determined[2] whether NA/Ch interactions play an important role in mediating processes involved in working memory and attention.

* Tsuneyuki Yamamoto, Department of Pharmacology, Graduate School of Pharmaceutical Sciences, Kyushu University, Fukuoka 812-8582, Japan.

2. METHODS

2. 1. Animals

Eight- to ten-week-old male Wistar-strain rats were obtained from Japan SLC. The initial free-feeding weights were 230-250 g, and the rats were placed on a deprivation schedule to maintain their weights at approximately 80% of the free-feeding level.

2. 2. Apparatus and procedures

Working memory was assessed with a 3-panel runway apparatus and a 3-lever operant apparatus, as described in our previous reports.[3, 4] Attention was assessed in a choice reaction time task using a 3-lever operant apparatus, as described in our previous report.[5]

2. 3. Drugs

The following drugs were used in this study: the muscarinic receptor antagonist (-)-scopolamine hydrobromide, the noradrenergic neurotoxin DSP-4 [N-(2-chloroethyl)-N-ethyl-2-bromobenzylamine hydrochloride], the α-adrenoceptor antagonist phentolamine hydrochloride and the β-adrenoceptor antagonist DL-propranolol hydrocholoride.

2. 4. Surgery

Lesions of the locus coeruleus (LC), and the basal forebrain (BF) were produced by a lesion generator (70℃, 15sec), according to the brain atlas of Paxinos and Watson.[6]

3. RESULTS

3. 1. Interactions between the adrenergic and cholinergic systems in regulating working memory performance in a 3-panel runway task and a 3-lever operant task

In a working memory task with a 3-panel runway apparatus, scopolamine at 0.32 mg/kg (i.p.) significantly increased the number of working memory errors, whereas errors were not affected at 0.1 mg/kg. However, a noradrenergic neurotoxin DSP-4 at 50 mg/kg i.p. given 14 days before the test had no effect on the number of errors. In rats with noradrenergic DSP-4 lesions, 0.1 mg/kg scopolamine significantly increased the number of working memory errors. To determine whether an α- or β-adrenergic mechanism participates in this cholinergic interaction, we examine the effects of α- and β-adrenoceptor antagonists on the working memory. Combined administration of propranolol at 10 mg/kg (i.p.) and scopolamine at 0.1 mg/kg significantly increased the number of working memory errors. However, the α-adrenergic antagonist phentolamine (3.2-10 mg/kg, i.p.) had no significant effect on working memory errors, whether it was administered independently or in combination with scopolamine at 0.1 mg/kg.

3. 2. Effects of lesions in the LC and BF on working memory and attentional function in a 3-lever operant task

The LC lesions, in the 1st session, produced an impairment of working memory; they reduced the number of reinforcements in the test trial without affecting the number in the sample trial. There was a return to normal levels by the 5th session. The BF lesions also reduced markedly the number of reinforcements and the impairment lasted until the 9th session. In the 1st session of the sample trial, BF-lesioned rats showed a significant decrease in the number of reinforcements but the impairment was over by the 5th day. LC and BF (LC+BF)-lesioned rats showed a more severe impairment than the single-lesioned groups. In the 1st session of the sample trial, LC+BF-lesioned rats were severely impaired compared to sham controls ($p<0.01$), and this lasted until the 6th session. In the test trial, the decrease continued ($p<0.01$) until 19 days after the surgery ($p<0.05$).

Lesions in only the LC or BF produced a significant delay in the choice reaction time in the 1st session after the surgery ($p<0.01$). At this time, the accuracy of choice was not decreased by single LC or BF lesions and the total number of lever presses was not decreased either. On the other hand, lesions in both LC and BF produced a more significant delay in the choice reaction time in comparison with single lesions. Although the choice accuracy was not reduced by lesions in both LC and BF, suppression of lever-pressing was induced by this surgery (5 of the 8 rats).

4. DISCUSSION

This study demonstrated that noradrenergic DSP-4 impaired working memory, when combined with 0.1 mg/kg scopolamine, which itself did not affect working memory. Decker and Gallagher[7] also reported that dorsal noradrenergic bundle lesions enhance the disruptive effects of scopolamine on spatial working memory in a radial arm maze task. In addition, dorsal noradrenergic bundle lesions decrease the efficacy of physostigmine in reversing passive avoidance deficits in rats with BF lesions.[8] Another major finding of the present study is that scopolamine and propranolol profoundly impaired working memory when administered in combination. Taken together, these results suggest that the working memory processes involved functional interactions between the cholinergic and noradrenergic systems, possibly through mediation of muscarinic/β-adrenergic receptors. Likewise, a behavioral pharmacological study with concurrent administration of scopolamine and propranolol revealed muscarinic/β-adrenergic interactions involved in spatial learning in a Morris water maze task.[9]

In addition, in this study, LC-lesions and BF-lesions produced by thermal coagulation impaired transiently working memory as well as choice reaction time performance in a 3-lever operant task. However, in the 3-panel runway task, working memory performance was not affected by DSP-4. In addition, in the radial maze or T-maze task, working memory and attention were insensitve to noradrenergic depletion by administration of DSP-4 or 6-hydroxydopamine.[10, 11] The differences may be related to the type of working memory task, the level of task difficulty or the method of producing NA dysfunction. At any rate, our data confirm the influence of LC lesions on the processes of working memory and attention, at least in the DMTS task and the CRT task, respectively. In the early stage of impairment of working memory following the formation of lesions in the BF and in both the LC and BF, the number of reinforcements in the sample trial was also significantly reduced. Therefore, nonspecific changes such as motor dysfunction may underlie the impairments of working memory in the early stages in both groups. Lesions in the LC, BF and both areas also resulted in severe impairment of attentional function (delaying the choice reaction time). Judging from these results, the impairments of working memory in LC-, BF- and LC+BF- lesioned rats may be attributed to an attention deficit.

5. CONCLUSION

These results suggested that β-adrenergic/muscarinic interactions play an important role in mediating the processes involved in working memory performance. Furthermore, it is also indicated that LC and BF are responsible for working memory/attention and severe and prolonged impairment is induced by lesions in both areas.

REFERENCES

1. L. Hertz, Is Alzheimer's disease an anterograde degeneration, originating in the brainstem, and disrupting metabolic and functional interactions between neurons and glial cells?, *Brain Res. Rev.* **14**, 335-353 (1989)
2. W. Bondareff, C. Q. Mountjoy, M. Roth, M. N. Rossor, L. L. Iversen, and G. P. Reynolds, Age and histopathological heterogeneity in Alzheimer's disease. Evidence for subtypes, *Arch. Gen. Psychiatry* **44**, 412-417 (1987)
3. Y. Furuya, T. Yamamoto, S. Yatsugi, and S. Ueki, A new method for studying working memory by using the three- panel runway apparatus in rats, *Jpn. J. Pharmacol.* **46**, 183-188 (1988)
4. T. Yamamoto and M. Ohno, Pharmacological application of a three- panel runway task to elucidate learning and memory in rats, *Meth. Find. Exp. Clin. Pharmacol.* **18** (Suppl A), 5-15 (1996)
5. J. Jingji, T. Yamamoto, and S. Watanabe, The involvement of sigma receptors in the choice reaction performance deficits induced by phencyclidine, *Eur. J. Pharmacol.* **319**, 147-152 (1999)
6. G. Paxinos and C. Watson, *The rat brain in stereotaxic coordinates* (Academic Press, Australia, 1989).
7. M. W. Decker, and M. Gallagher, Scopolamine-disruption of radial arm maze performance: Modification by noradrenergic depletion, *Brain Res.* **417**, 59-69 (1987)
8. V. Haroutunian, P. D. Kanof, G. Tsuboyama, and K. L. Davis, Restoration of cholinomimetic activity by clonidine in cholinergic plus noradrenergic lesioned rats, *Brain Res.* **507**, 261-266 (1990)
9. M. W. Decker, T. M. Gill, and J. L. McGaugh, Concurrent muscarinic and beta-adrenergic blockade in rats impairs place-learning in a water maze and retention of inhibitory avoidance, *Brain Res.* **513**, 81-85 (1990)
10. J. J. Chrobak, D. L. Dehaven, and T. J. Walsh, Depletion of brain norepinephrine with DSP-4 does not alter acquisition or performance of a radial-arm maze task, *Behav. Neural. Biol.* **44**, 144-150 (1985)
11. M. Pisa and H. C. Fibiger, Evidence against a role of the rat's dorsal noradrenergic bundle in selective attention and place memory, *Brain Res.* **272**, 319-329 (1983)

NORADRENERGIC CONTROL OF LONG-TERM MEMORY CONSOLIDATION

Kazuto Kobayashi and Yasunobu Yasoshima*

1. INTRODUCTION

Memory formation involves dynamic interactions among many brain structures and their linking pathways. Long-term memory requires de novo protein synthesis and gene expression, and is mediated by some intracellular signaling pathways. The noradrenaline (NA) system in the CNS is known to mediate a wide variety of neurological and psychological functions, such as cognition, attention, and memory formation. Clinical investigations have implicated alterations in the NA system in the pathological states of neuropsychiatric diseases. Tyrosine hydroxylase (TH) is the first and rate-limiting enzyme of catecholamine biosynthesis. The regulatory mechanism of the TH reaction is considered to play a key role in controlling the catecholaminergic actions. In this report, we describe a behavioral analysis of mice heterozygous for the mutation in the TH gene. We provide genetic evidence for an essential role of the NA system in the consolidation process of long-term memory of certain learning paradigms.[1]

2. NA METABOLISM IN TH MUTANT MICE

A knockout mutation in the mouse TH gene results in profound depletion of catecholamines and lethality of the homozygous mutants at the perinatal stage.[2, 3] Mice heterozygous for the TH mutation are apparently normal in their development and gross behavior, but they display a decline in TH activity in the tissues (approximately 40% of the wild type activity).[2] These mice provide a genetic model to investigate the behavioral and physiological role of catecholamines by analyzing phenotypic effects derived from altered catecholamine synthesis. We studied here the behavioral and neuropsychological functions involving the NA system in mice heterozygous for TH mutation.

In the mutant mice, NA accumulation in brain regions was moderately decreased to 73%-80% of the wild type value. Although the spontaneous NA release level was

* Department of Molecular Genetics, Institute of Biomedical Sciences, Fukushima Medical University School of Medicine, 1 Hikarigaoka, Fukushima 960-1295, Japan

Catecholamine Research, Edited by Nagatsu *et al.*
Kluwer Academic/Plenum Publishers, 2002

indistinguishable between the wild type and mutant, the NA release evoked by high K+ condition during repetitive stimulation was decreased in the mutant, suggesting that the NA stores of the mutant were more depleted. These data indicated a moderate reduction in NA metabolism in the mutant as a consequence of the decreased TH activity.

3. ROLE OF NORADRENALINE IN MEMORY AND LEARNING

The TH heterozygous mutant mice were included in several memory and learning tasks. Initially, we tested the performance of latent learning with the water-finding test. In this test, a mouse is placed in an environment containing a water tube during a training trial, and on the next day the animal is placed back into the same environment for the test of the learning performance. Normally, the trained animals can find the water tube faster than the nontrained animals. However, the TH mutant mice did not show a reduction in the finding time in spite of the training, indicating an impairment of latent learning. This impairment was recovered by injecting a NA reuptake inhibitor (desipramine), which stimulates the NA neuron activity, indicating the impaired latent learning was because of the reduced NA metabolism in the mutant.

Next, the performance of conditioned learning was tested with a tone-dependent active avoidance paradigm. This paradigm requires an association between a footshock unconditioned stimulus (US) and a tone conditioned stimulus (CS). During daily training, the wild type mice showed a normal acquisition of the avoidance responses, which reached a plateau on the fifth trial day with 92% success of avoidance. In contrast, the mutant mice learned more slowly, and the rate of the successful response was significantly lower than that of the wild type mice. The success of avoidance in the fifth trial was decreased to 75% in the mutants. The impairment of active avoidance was restored by desipramine treatment, indicating that the reduced NA metabolism in the mutant causes a mild impairment of the active avoidance learning.

4. ROLE OF NORADRENALINE IN MEMORY CONSOLIDATION

To characterize the performance of conditioned learning in the TH mutant, tone-cued fear conditioning was performed. In this paradigm, a foot shock US was associated with a tone CS during the training session, and the response of freezing behavior to the subsequent tone CS was measured (Fig. 1A). To evaluate time-dependent processes of memory consolidation, the freezing behavior was monitored with different time intervals after the conditioning (Fig. 1B). At 1 hr after the conditioning, both wild type and mutant mice exhibited a similar extent of the freezing response. In contrast, at 24 hr after the conditioning the wild type mice sustained the performance, whereas the freezing response in the mutant mice was reduced. In the mutants, short-term memory of conditioned fear was likely normal but long-term memory was mildly impaired. Desipramine was injected after the conditioning to test the dependence of NA in memory consolidation (Fig. 1B). This treatment recovered the learning deficit in the mutant mice, indicating the consolidation of long-term memory is impaired because of reduced NA metabolism.

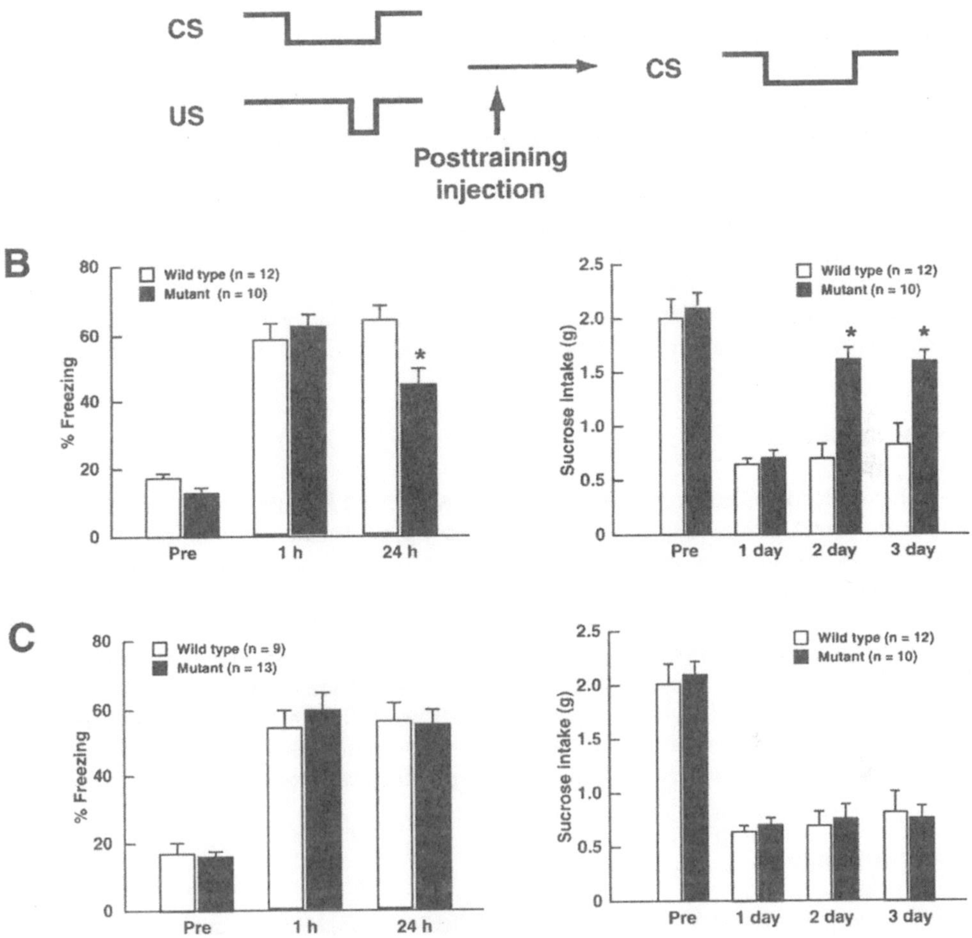

Fig.ure 1. Genetic evidence for an important role of NA in memory consolidation. (A) Procedures for evaluating a role of NA in conditioned learning. During the conditioning phase, mice are exposed to a CS paired with an US. The mice are treated with saline or desipramine after the conditioning. (B) Cued fear conditioning. The freezing response was evaluated as a suppression of locomoter activity by the tone stimulus. (C) Conditioned taste aversion. Sucrose intake during the test period (10 min) was measured. Pre, Preconditioning phase. Values indicate mean + SEM of the data. *$p < 0.01$ according to Newman-Keuls test. (Data are modified from ref. 1.)

To further evaluate the role of NA in long-term memory, we examined the conditioned taste aversion that requires an association between a taste CS and visceral malaise-inducing US. In the training session, the animals receive a novel taste (0.5 M sucrose) from a drinking bottle as a CS, followed by an i.p. injection of LiCl as an US. The US induces a sickness soon after the injection, and thus conditioned animals show aversive behavior to the taste CS. Acquisition and retention of aversive responses were tested after the conditioning (Fig. 1C). On day 1, both wild type and mutant mice exhibited the taste aversion to a similar extent, and the sucrose intake was significantly lower than that in the preconditioning phase. On day 2, the wild type mice continued to exhibit aversive responses, whereas the mutant mice took sucrose to almost the same extent as during the preconditioning. Posttraining desipramine treatment also restored the learning performance in the mutant mice (Fig. 1C). These observations are consistent with the results obtained from the cued fear conditioning, and indicate that the NA system plays an important role in the consolidation of associative long-term memory.

5. CONCLUSION

The results obtained from behavioral analyses of the TH mutant mice indicated a key role of the central NA system in long-term memory consolidation of conditioned learning. This process was highly susceptible to a reduction in NA metabolism. The formation of conditioned learning is considered to be dependent on the amygdala and its linking pathways[4, 5]. In addition, brain lesion studies suggest an important role of the cerebral cortex in long-term memory of conditioned learning. Anatomically, the brain regions involved in conditioned learning receive NA fibers from some nuclei in the brain stem. For a clearer understanding of neuronal mechanisms underlying NA-dependent memory consolidation, it is necessary to determine dynamic changes in NA metabolism and neuronal activity correlated to the behaviors in specific brain regions.

REFERENCES

1. K. Kobayashi, Y. Noda, N. Matsushita, K. Nishii, H. Sawada, et al., Modest neuropsychological deficits caused by reduced noradrenaline metabolism in mice heterozygous for a mutated tyrosine hydroxylase gene. *J. Neurosci.*, **20**, 2418-2426 (2000)
2. K. Kobayashi, S. Morita, H. Sawada, T. Mizuguchi, K. Yamada et al., Targeted disruption of the tyrosine hydroxylase locus results in severe catecholamine depletion and perinatal lethality in mice. *J. Biol. Chem.*, **270**, 27235-27243 (1995).
3. Q.-Y. Zhou, C. J. Quaife, R. D. Palmiter, Targeted disruption of the tyrosine hydroxylase gene reveals that catecholamines are required for mouse fetal development. *Nature*, **374**, 640-643 (1995).
4. J. E. LeDoux, Emotion: clues from the brain. *Annu. Rev. Psychol.*, **46**, 209-235 (1995).
5. J. L. McGaugh, L. Cahill, B. Roozendaal, Involvement of the amygdala in memory storage: interaction with other brain systems. *Proc. Natl. Acad. Sci. USA*, **93**, 13508-13514 (1996).

ROLE OF THE LOCUS COERULEUS-NOREPINEPHRINE SYSTEM IN ATTENTION AND BEHAVIORAL FLEXIBILITY

Gary Aston-Jones, Janusz Rajkowski and Jonathan Cohen*

The ability to respond selectively to certain aspects of the environment, and filter-out others that are irrelevant to the current behavioral plan, is critical for goal-directed behavior. At the same time, behavior must be flexible and adaptive, so as to quickly adjust to new imperative, or unexpected events. Thus, successful behavior in both animals and humans requires the capacity for both selective responding in a stable environment, and rapid adaptive responding in a changing environment. Our recent work on the locus coeruleus-norepinephrine (LC-NE) system has led us to hypothesize that it plays a central role in regulating this balance between focused vs. flexible responding.

1. NEUROPHYSIOLOGY EXPERIMENTS

We recorded LC impulse activity in monkeys during performance of a visual discrimination task. Conditioned stimuli (CSs) were either a vertical or horizontal bar; one orientation was used as the target (CS+) stimulus and the other as the distractor (nontarget, or CS-) stimulus. The animal's task was to release the lever immediately following a target stimulus to receive a juice reward. Incorrect releases to distractor stimuli resulted in a 3-sec time-out. Target stimuli occurred randomly on 20% of trials and intertrial intervals averaged 1.6 sec.

1.1 Phasic activation of LC neurons by meaningful stimuli.

Responses of LC neurons during this task were surprisingly selective. Nearly all of the >300 LC neurons recorded to date were phasically activated preferentially by target stimuli but only weakly or not at all by nontarget stimuli; lever release outside of the task elicited no response.[2] In addition, juice delivery following correct behavioral responses evoked no LC response. Some LC neurons were weakly inhibited by distractor stimuli.

* Gary Aston-Jones, Janusz Rajkowski and Edwin Clayton, University of Pennsylvania, Philadelphia, PA. Jonathan Cohen, Princeton University, Princeton, NJ.

Catecholamine Research, Edited by Nagatsu *et al.*
Kluwer Academic/Plenum Publishers, 2002

The latencies of LC responses to targets were surprisingly short (~100 msec onset), and preceded behavioral responses by 100 msec or more. Moreover, the latencies of response for LC neurons and lever releases were significantly correlated over trials, so that shorter LC responses were associated with shorter behavioral responses to the same cues.[2] These findings indicate that LC target responses could facilitate behavioral responses to target stimuli.

Recordings during reversal training confirmed that LC responses to targets were independent of sensory attributes. Reversal of cue meaning caused LC cells to reverse responsiveness, and become selectively activated by the new target stimulus and lose responsiveness to the old target.[1] Thus, phasic LC activation in this task is specifically related to the meaning of the stimuli, not to their physical attributes.

1.2 Fluctuations in tonic LC activity during task performance

Tonic LC activity often varied during alert task performance, between intermediate (1-2 spikes/sec) and elevated rates (2-4 spikes/sec). These different levels of tonic LC discharge were closely associated with differences in performance on the task. Periods of elevated tonic LC activity were consistently accompanied by more frequent false alarm errors but no changes in misses or hits.[7] Signal detection analyses revealed that during elevated LC activity the animal was less able to discriminate target from non-target stimuli (lower d'), and displayed increased distractibility and a greater tendency to respond to non-target stimuli (lower response beta).

Relationship of tonic to phasic LC activity. Periods of elevated tonic LC activity were also consistently associated with decreased phasic responsiveness of LC neurons to target stimuli. Thus, the LC responses typically seen for target stimuli occurred selectively during epochs of intermediate tonic LC discharge and excellent behavioral performance. Conversely, elevated tonic LC discharge corresponded to both decreased attentional performance and decreased phasic activation of LC neurons by target stimuli.

1.3. Interpretation of results – LC neurophysiology experiments

Overall, these results indicate that there are two modes of LC activity that correspond to different levels of performance on this task that requires focused attention. (i) In the *phasic mode*, LC cells exhibit phasic activation selectively for target stimuli but only a moderate level of tonic discharge. This mode of LC activity is consistently associated with excellent performance on the task with few errors. (ii) In the *tonic mode*, LC cells fail to phasically respond to any task stimuli, but exhibit higher levels of ongoing tonic activity. This mode corresponds to poor performance with many false alarm errors.

At very low levels of LC activity the animal failed to engage sufficiently to perform the task, apparently because of low arousal and drowsiness. At an intermediate level of tonic LC activity with phasic responses to targets, arousal and performance increased. However, at the highest tonic LC activities, performance *decreased*, inconsistent with earlier predictions. These new results for an inverted U relationship between the level of tonic LC discharge and task performance resemble the classic Yerkes-Dodson relationship that has frequently been observed between arousal and performance.

2. COMPUTATIONAL MODELING STUDIES

The data described above pose several questions about the mechanisms underlying LC activity and its relationship to behavioral performance. Two in particular stand out: 1) What mechanisms underlie the different modes of LC firing? 2) What mechanisms explain the relationship between each of these modes and the pattern of behavioral performance with which it was associated? To address these and related questions, we developed a simulation model of LC function and its influence on performance in the vigilance task. The model is described in detail in ref. 7 and outlined briefly below. The model is a hybrid with two components: a simple stimulus discrimination network that simulated performance in the behavioral task, and a more detailed model of LC neuronal activity.

Stimulus discrimination network. The task model consisted of a small number of units, each of which represented cell assemblies supporting stimulus or response representations necessary for performing the task. Thus, there were two input units (for the target and distractor stimuli), two decision units, and one response unit. Connections between units in different processing layers were excitatory (information flow) while those within a layer were inhibitory (competition), and activity of units was subject to small random variations (noise). Finally, the physiological effect of NE was simulated as an increase in the gain parameter of the activation function of processing units. This captures a large body of physiological data indicating that NE increases responsivity or signal-to-noise ratios of target neurons[8] (reviewed in ref. 5).

LC model. This consisted of a population of 250 spiking neurons, each of which was a leaky integrate-and-fire cell that exhibited temporal dynamics similar to those obtained in detailed compartmental models. Each LC cell received input from the target decision unit, as well as noise which was responsible for a weak spontaneous firing rate of about 1 Hz (as observed *in vivo*). LC cells in this model interacted with each other in two ways. First, lateral inhibition simulated the effect of local NE release. Second, we included electrotonic coupling among LC cells, which simulated such coupling found empirically.[6,4] As described below, modulation of electrotonic coupling among LC neurons alone in our model was able to capture the full set of neurophysiological and behavioral findings observed in the monkey experiments above.

Simulation results. The model explained the two modes of LC firing in terms of differences in the degree of electrotonic coupling within LC. High coupling caused stronger, more synchronized (phasic) activation of the LC in response to the target input due to the distribution of the voltage in spiking cells to the remainder of the population across electrotonic links. At the same time, coupling reduced spontaneous firing by averaging uncorrelated noise among the population, resulting in an overall reduction in tonic activity. Thus, our model reveals that changes in phasic and tonic firing properties may be inversely related to one another, and governed by the same mechanism: electrotonic coupling among LC units. As noted above, we simulated the physiological effect of the LC on targets by increasing the gain of the activation function of processing units in the task model. Thus, reduced coupling that produces high tonic LC activity causes higher responsivity of units in the task model, and thus a greater number of false alarms (triggered by intrinsic noise). Conversely, increased coupling leads to a reduction of tonic activity and an increase in phasic activity. The reduction in tonic activity produces a reduction of the overall responsivity of units in the task model, and therefore a concomitant reduction in false alarms. However, on target stimulus trials during high coupling this reduced responsivity is compensated for by an increased phasic LC response to the target, which in turn produces a temporary increase in responsivity and therefore a quick response. Thus, an increase in electrotonic coupling is able to produce an increase in the accuracy of performance (reduced false alarms), without a cost in the response time to targets. These results are described in detail in our recent publication.[7]

3. A NEW THEORY OF LC FUNCTION.

These empirical and modeling studies have led us to a new hypothesis of LC function. We propose that the LC governs the responsivity of its target neuronal assemblies to their afferent inputs. This can manifest overtly as the likelihood that a stimulus will elicit a behavioral response, or have internal consequences, such as on attentional selection. Thus, the influence that LC-evoked changes in responsivity have on performance will depend upon the current task demands (e.g., whether they engage behavioral and/or attentional responses). It will also depend on the magnitude and pattern of LC discharge. In the absence of LC activity, systems modulated by LC input will be relatively unresponsive, and this level of activity is important for facilitating sleep and unresponsiveness to external stimuli. In the phasic LC mode, such systems will become responsive to specific task-relevant stimuli. As the levels of tonic activity increase (with a concomitant reduction in phasic activity), increases in responsivity will occur to a broader class of stimuli, ultimately to the point that behavior (and attention) become relatively indiscriminate and labile. These different modes of LC activity each have adaptive advantages under different circumstances. The phasic mode may support cortically driven behaviors optimized to specific stable environments (e.g., tasks requiring focused attention). While the behavioral lability associated with the higher tonic activity would impair performance under such circumstances, it may be adaptive in changing or unpredictable environments. Thus, the variability associated with high tonic LC activity may facilitate behavioral flexibility and responsiveness to unexpected events.

4. ACKNOWLEDGEMENTS

This work was supported by USPHS grants MH 55309 and MH 59978.

REFERNCES

1. Aston-Jones, G., Rajkowski, J. and Kubiak, P., 1997, Conditioned responses of monkey locus coeruleus neurons anticipate acquisition of discriminative behavior in a vigilance task, *Neuroscience* **80:** 697-715.
2. Aston-Jones, G., Rajkowski, J., Kubiak, P. and Alexinsky, T., 1994, Locus coeruleus neurons in the monkey are selectively activated by attended stimuli in a vigilance task, *J. Neurosci.* **14:** 4467-4480.
3. Carter, C. S., Braver, T. S., Barch, D. M., Botvinick, M. M., Noll, D. and Cohen, J. D., 1998, Anterior cingulate cortex, error detection, and the online monitoring of performance, *Science* **280:** 747-749.
4. Christie, M. J. and Jelinek, H. F., 1993, *Neuroscience* **56:** 129-137.
5. Foote, S. L., Bloom, F. E. and Aston-Jones, G., 1983, Nucleus locus ceruelus: new evidence of anatomical and physiological specificity, *Physiol. Rev.* **63:** 844-914.
6. Ishimatsu, M. and Williams, J. T., 1996, Synchronous activity in locus coeruleus results from dendritic interactions in pericoerulear regions, *J. Neurosci.* **16:** 5196-5204.
7. Usher, M., Cohen, J. D., Rajkowski, J., Kubiak, P. and Aston-Jones, G., 1999, The role of locus coeruleus in the regulation of cognitive performance, *Science* **283:** 549-554.
8. Waterhouse, B. D. and Woodward, D. J., 1980, Interaction of norepinephrine with cerebrocortical activity evoked by stimulation of somatosensory afferent pathways in the rat, *Exp Neurol* **67:** 11-34.

ADJUSTING BEHAVIOR TO CONTEXT: STRIATAL NETWORK FUNCTIONS IN LEARNING AND MEMORY

Ann M. Graybiel and Pablo M. Blazquez

The basal ganglia have been implicated in the neural basis of procedural learning and habit formation both on the basis of clinical and experimental studies.[1-6] It is likely that the basal ganglia participate in such learning and memory mechanisms by influencing activity in cortico-basal ganglia circuits that link the neocortex and underlying structures of the basal ganglia. The striatum sits at a key point in these cortico-basal ganglia loops. The striatum is the largest input structure of the basal ganglia, and it is neurons of the striatum that give rise to the largest output pathways of the basal ganglia. Most of the neurons in the striatum are both input and output neurons. It is thought that patterns of convergent and divergent corticostriatal projection may be critical to the processing of information in the striatum. For example, functionally related zones in the cortex can send inputs convergently to corresponding zones in the striatum, and this projection pattern can occur in multiple such zones within the striatum. This widespread and yet compartmentalized system of input connectivity of the striatum is matched in the patterns of projection of striatal output neurons to the output nuclei of the striatum, the pallidum and substantia nigra. The input-output compartments of the striatum (known as striosomes and matrisomes) may confer on the striatum the capacity to reorganize cortical (and other) inputs to this structure. The resorting and recombining of information that occurs may depend critically on patterns of coordinate activation of particular sets of inputs.[2,7]

In addition to this fundamental compartmental input-output architecture of the striatum, there are systems of striatal interneurons that modulate throughput in the striatum. Five types of interneuron have been identified in the striatum, and each set is broadly but sparsely distributed throughout the entire structure.[8] The best known of these neurons are the so-called tonically active neurons (TANs). These neurons respond mainly to sensory inputs and lack responses clearly time-locked to movement. In this respect, they differ from many of the projection neurons of the striatum, for which movement-related responses have been described. What is particularly interesting about the physiology of TANs is that they can acquire and lose responses to particular sensory stimuli during the course of reward-based sensorimotor conditioning. TANs have thus become a central focus for work on neuroplasticity in the striatum.[9-12]

Catecholamine Research, Edited by Nagatsu *et al.*
Kluwer Academic/Plenum Publishers, 2002

We and our colleagues first studied the activity of TANs in macaque monkeys undergoing training in a simple conditioning task in which the monkeys learned to associate an auditory or visual stimulus with reward delivery.[9] Before training, a relatively small percentage of the TANs responded to the conditioning stimuli. As training proceeded, however, the percentage of responses of TANs increased markedly, to levels as high as 60-70% of the neurons recorded. Much evidence now suggests that these acquired responses of TANs are related to the reward-predictive character of the conditioning stimulus.[11] TANs can also respond to primary rewards themselves.[13] In all of these respects, the firing patterns of the TANs of the striatum resemble the reward-related firing patterns of the dopamine-containing neurons of the substantia nigra. To test whether the striatal TAN responses were dependent on input from the nigral dopaminergic neurons, we and our colleagues depleted striatal dopamine by intrastriatal infusion of the neurotoxin MPTP (1-methyl-4-phenyl-1,2,3,6-tetrahydropyridine) in one cerebral hemisphere of monkeys that had received training in a sensorimotor conditioning task.[10] We found that dopamine depletion largely abolished the acquired responses of the TANs. These experiments suggest that the dopamine-containing neurons of the substantia nigra and the TANs of the striatum form part of the basic neural circuitry for reward-based learning.

Striatal TANs, however, not only respond to reward-related stimuli, but also can respond to neutral and even to aversive stimuli.[14, 15] To address the possibility that TANs might have a broader function than one strictly related to reward, we recently have carried out experiments in which we trained macaque monkeys in a classical delay eyeblink conditioning task[15] in which the monkeys learn to blink in response to a conditioning stimulus that indicated the future (250 or 400 ms later) occurrence of an airpuff directed toward the eye. After training, we carried out extinction testing in which no airpuff was delivered. The delay eyeblink conditioning task has been used extensively to study plasticity in the cerebellum and other brainstem structures, and more recently has been found to activate the striatum as well. We also tested for the effects of habituation training on the responses of the striatal TANs in experiments in which the monkeys were exposed to tones of various intensity for extended periods of time.

We have found striking changes in TAN responses during all of these tasks.[16] As a result of habituation training, the responses of the TANs to the same stimuli gradually fall. During conditioning, the responses of the TANs to the conditioning stimuli increase, and during extinction, the responses to these stimuli decrease. The TANs thus respond to stimuli under conditions extending from those in which the stimuli are associated with reward to those in which the stimuli are related to an upcoming aversive event, and to those in which neutral stimuli are presented. These striatal interneurons thus appear to register both the acquired and the unconditioned salience of stimuli in particular contexts. To test how these responses relate to the behavior of the monkeys in the different tasks, we examined the relationship between the occurrence of TAN responses and the occurrence of behavioral responses. We found a very close relationship between the neural and behavioral responses.[16] This suggests that the TANs carry a signal related to the relevance of particular sensory stimuli for behavior.

TANs are identified in alert behaving monkeys by their slow irregular but tonic firing patterns, which originally gave them their name. It so far has not been possible to identify the neural phenotype of these electrophysiologically identified neurons. Experiments in slice preparations of rodent brain, however, strongly suggests that the tonically firing neurons of the striatum correspond to the cholinergic interneurons of this structure.[17] If so, the TANs could play a crucial role in dopaminergic-cholinergic interactions in the striatum. Acetylcholine in the striatum largely comes from the cholinergic interneurons and can influence the release of dopamine from nigrostriatal terminals and other input activity in the striatum as well as both projection neurons and interneurons.[18] Thus the TANs, if they do indeed correspond to cholinergic interneurons, could have crucial network functions in the striatum.

As a first step toward testing this possibility, we analyzed in macaques the timing characteristics of TAN responses to stimuli acquired through reward-based sensorimotor conditioning.[19] We found that TANs distributed across large parts of the caudate nucleus and putamen exhibited responses that occurred during overlapping time windows. The responses of TANs to conditioning stimuli typically occur ca. 70-130 ms after the appearance of the stimulus and last for ca. 200 ms. This basic "pause" response is typically preceded and followed by excitations. The temporal coordination of these responses implies that the TANs could act as network neurons with a reset or latch-on function, tuning network function in the striatum.[7] In studies with simultaneous recordings of TANs, near synchrony of responsivity has indeed been found,[12] supporting the idea that coordinate action of the TANs may be central to their function.

What could be the function of such coordinate activity? Our initial idea was that TANs might serve a binding function in the motor system.[7] During their coordinate pauses and associated excitations, the network configuration of the striatum might be modified coordinately, and such modifications would be continuously updated as TAN responses increased or decreased as a result of learning. Because TANs are broadly distributed within the striatum, a comparable binding function might also apply to striatal regions that are part of "cognitive" cortico-basal ganglia loops.[2] The TANs thus would have a general function in modifying the throughput of the striatum, binding together different aspects of cognitive and sensorimotor behavior according to ongoing experience. Some support for the generalized binding function is coming from current experiments suggesting that one of the major targets of acetylcholine in the striatum are the parvalbumin-containing interneurons.[20] These form an electronically coupled network of inhibitory interneurons that, in turn, can coordinately influence striatal output neurons. Thus the TANs (if corresponding to cholinergic interneurons) could affect the synchronously acting parvalbumin interneuronal network to change the properties of this network.

One particular characteristic of the anatomical distribution of TANs (and of striatal cholinergic interneurons as well) may directly bear on their function. The cholinergic interneurons and cholinergic neuropil are preferentially distributed in the matrix compartment of the striatum. In experiments aimed at determining the distribution of TANs in the striatum, we and our colleagues marked the sites at which TANs were recorded in acute recording experiments in squirrel monkeys and then mapped the locations of the marks in relation to the anatomically defined boundaries of striosomes and matrix.[19] We found a striking preponderance of TANs in the matrix compartment, and a statistically significant proportion of the TANs lay at the borders between striosomes and matrix. Striosomes are the limbic-affiliated compartment of the striatum. In the monkey, striosomes in the caudate nucleus have been shown to receive dense input from the posterior orbitofrontal cortex and from the caudal medial prefrontal/rostral anterior cingulate cortex.[21] These cortical regions are tightly interconnected with the amygdala and hippocampus and form core regions of limbic circuitry. In contrast to striosomes, the matrix compartment receives strong input from sensorimotor and associative regions of the neocortex, including large parts of the frontal cortex. One possibility raised by these findings is that the TANs modulate information transfer between these two main compartments of the striatum. More generally, TANs, by virtue of the plasticity of their responses to sensory stimuli, may influence the response properties of a range of striatal neurons by signaling experience-based estimates of the relevance of sensory stimuli for behavior. This function could be critical in shaping activity in cortico-basal ganglia circuits.

REFERENCES

1. D. P. Salmon & N. Butters, Neurobiology of skill and habit learning, *Curr. Opin. Neurobiol.* **5**, 184-190 (1995).
2. A. M. Graybiel, The basal ganglia and chunking of action repertoires, *Neurobiol. Learn. Mem.* **70**, 119-136 (1998).
3. A. M. Graybiel, Building action repertoires: Memory and learning functions of the basal ganglia, *Curr. Opin. Neurobiol.* **5**, 733-741 (1995).
4. C. D. Frith, C. A. Bloxham & K. N. Carpenter, Impairments in the learning and performance of a new manual skill in patients with Parkinson's disease, *J. Neurol. Neurosurg. Psychiatry* **49**, 661-668 (1986).
5. J. D. Gabrieli, G. T. Stebbins, J. Singh, D. B. Willingham & C. G. Goetz, Intact mirror-tracing and impaired rotary-pursuit skill learning in patients with Huntington's disease: evidence for dissociable memory systems in skill learning, *Neuropsychology* **11**, 272-81 (1997).
6. P. Soliveri, R. G. Brown, M. Jahanshahi, T. Caracani & C. D. Marsden, Learning manual pursuit tracking skills in patients with Parkinson's disease, *Brain* **120**, 963-976 (1997).
7. A. M. Graybiel, T. Aosaki, A. W. Flaherty & M. Kimura, The basal ganglia and adaptive motor control, *Science* **265**, 1826-1831 (1994).
8. Y. Kawaguchi, C. J. Wilson, S. J. Augood & P. C. Emson, Striatal interneurons: chemical, physiological and morphological characterization, *Trends Neurosci.* **18**, 527-535 (1995).
9. T. Aosaki et al., Responses of tonically active neurons in the primate's striatum undergo systematic changes during behavioral sensorimotor conditioning, *J. Neurosci.* **14**, 3969-3984 (1994).
10. T. Aosaki, A. M. Graybiel & M. Kimura, Effects of the nigrostriatal dopamine system on acquired neural responses in the striatum of behaving monkeys, *Science* **265**, 412-415 (1994).
11. P. Sardo, S. Ravel, E. Legallet & P. Apicella, Influence of the predicted time of stimuli eliciting movements on responses of tonically active neurons in the monkey striatum, *Eur. J. Neurosci.* **12**, 1801-1816 (2000).
12. A. Raz, A. Feingold, V. Zelanskaya, E. Vaadia & H. Bergman, Neuronal synchronization of tonically active neurons in the striatum of normal and parkinsonian primates, *J. Neurophysiol.* **76**, 2083-2088 (1996).
13. P. Apicella, E. Legallet & E. Trouche, Responses of tonically discharging neurons in the monkey striatum to primary rewards delivered during different behavioral states, *Exp. Brain Res.* **116**, 456-466 (1997).
14. S. Ravel, E. Legallet & P. Apicella, Tonically active neurons in the monkey striatum do not preferentially respond to appetitive stimuli, *Exp. Brain Res.* **128**, 531-534 (1999).
15. P. Blazquez, J. Kojima, N. Fujii, R. Courtemanche & A. Graybiel, Aversive conditioning and habituation as well as reward conditioning modify the activity of tonically active neurons in the macaque striatum, *Soc. Neurosci. Abstr.* **26**, 683 (2000).
16. P. Blazquez, N. Fujii, W. DeCoteau & A. Graybiel, Tonically active neurons in the primate striatum have responses that correlate with the probability of behavioral response in reward and aversive Pavlovian conditioning and habituation., *Soc. Neurosci. Abstr., in press.* (2001).
17. B. D. Bennett & C. Wilson, Spontaneous activity of neostriatal cholinergic interneurons *in vitro, J Neurosci* **19**, 5586-5596 (1999).
18. P. Calabresi, D. Centonze, P. Gubellini, A. Pisani & G. Berenardi, Acetylcholine-mediated modulation of striatal function, *Trends Neurosci.* **23**, 120-126 (2000).
19. T. Aosaki, M. Kimura & A. M. Graybiel, Temporal and spatial characteristics of tonically active neurons of the primate's striatum, *J. Neurophysiol.* **73**, 1234-1252 (1995).
20. T. Koos & J. M. Tepper, Inhibitory control of neostriatal projection neurons by GABAergic interneurons, *Nat. Neurosci.* **2**, 467-72. (1999).
21. F. Eblen & A. M. Graybiel, Highly restricted origin of prefrontal cortical inputs to striosomes in the macaque monkey, *J. Neurosci.* **15**, 5999-6013 (1995).

BDNF-DEPENDENT BEHAVIORAL SENSITIZATION IN HEMIPARKINSONIAN RATS

Olivier Guillin, Jorge Diaz, Nathalie Griffon, Jean-Charles Schwartz, and Pierre Sokoloff[1]

1. INTRODUCTION

A diverse role for BDNF (Brain-derived neurotrophic factor) has been inferred from observations that it is anterogradely transported, released upon neuron depolarization and triggers rapid intracellular signals.[1,2] Converging evidence implicates the D_3 receptor (D_3R) in the physiopathology and treatment of various neuropsychiatric disorders including Parkinson's disease,[3,4] schizophrenia,[5, 6] and drug addiction.[7, 8] In a rat model of Parkinson's disease, obtained by unilateral destruction of dopamine neurons by 6-hydroxydopamine (6-OHDA), D_3R expression is decreased in the shell part of nucleus accumbens (AccSh) of the denervated side[3], by deprivation of BDNF[9] synthesized by dopamine neurons. Repeated administration of levodopa triggers D_3R overexpression in the denervated striatum[3], which is responsible for the development of behavioral sensitization to levodopa. Progressive enhancement of levodopa-induced rotations are blocked by a preferential D_3R antagonist[3] and induced by a selective partial D_3R agonist.[8] Here we show that BDNF is also responsible for induction of D_3R overexpression in 6-OHDA-lesioned rats, and thereby for behavioral sensitization.

2. RESULTS

During levodopa treatment of 6-OHDA-lesioned rats, we infused into the denervated striatum IgG-TrkB, a selective BDNF antagonist. This treatment impaired induction of both D_3R mRNA and protein expressions (Fig. 1a, b) and behavioral sensitization (Fig. 1c), indicating that BDNF is necessary for both processes.

[1] Olivier Guillin, Nathalie Griffon, Pierre Sokoloff. Unité de Neurobiologie et Pharmacologie Moléculaire, INSERM U 109, 75014 Paris, France. Jorge Diaz, Université René Descartes, 75006 Paris, France.

Catecholamine Research, Edited by Nagatsu *et al.*
Kluwer Academic/Plenum Publishers, 2002

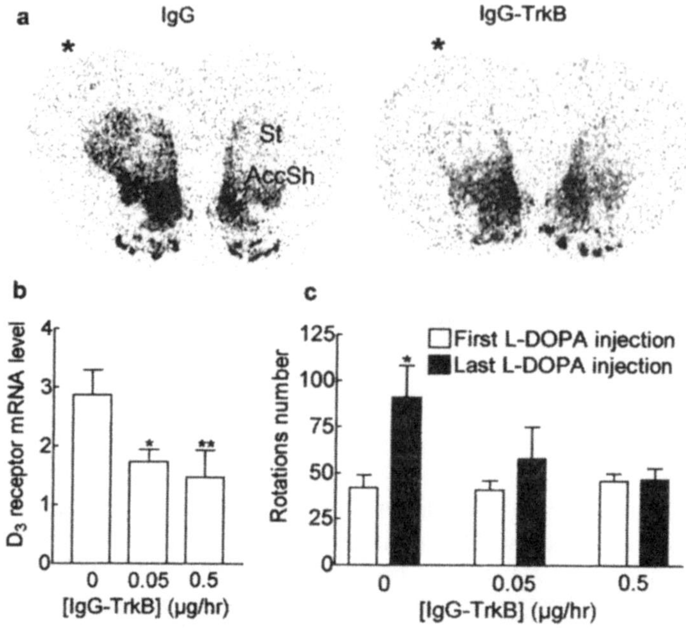

Figure 1. Levodopa-induced D_3 receptor expression and behavioral sensitization in 6-OHDA-lesioned rats require BDNF. The asterisk denotes the lesioned side. **a**, *In situ* hybridisation signals of D_3R mRNA in animals receiving continuous infusions into the striatum of IgG (control animal, left) or IgG-TrkB (right) for 7 days and levodopa for 5 days. **b**, Means $\pm$ S.E.M. of values from 5-7 animals. * $P < 0.05$ and ** $P < 0.01$ versus IgG-treated animals. **c**, Rotations were recorded after the first and last levodopa (L-DOPA) administrations in animals shown in **b**. * $P < 0.05$ versus first L-DOPA injection.

Because striatal BDNF originates mainly from cortical neurons,[10] we next examined the effects of levodopa on cortical BDNF gene expression. A single administration of levodopa progressively induced BDNF mRNA in the frontal, but not cingulate cortex in the 6-OHDA-lesioned side (Fig. 2a). BDNF mRNA was increased by levodopa mainly in cortical deeper layer V containing pyramidal cell bodies and layer VI (Fig. 2a); BDNF mRNA-positive cells in cortical layer V displayed the typical triangular shape of pyramidal cells (Fig. 2b). These cells project to striatal and accumbal areas, suggesting that levodopa may increase striatal BDNF release by cortico-striatal neurons. Levodopa also induced BDNF mRNA in the control side, but this effect was much more limited in amplitude (Fig. 2c). This effect of levodopa on BDNF, like that on D_3R overexpression and behavioral sensitization,[3] critically depends upon activation of a D_1R or D_5R, since it was also produced by SKF 81297, a D_1R/D_5R agonist, but not by quinpirole, a $D_2R/D_3R/D_4R$ agonist (Fig. 2d). These results are consistent with the presence of D_1R on cortical pyramidal cells and with the observation that stimulation of a D_1R or D_5R phosphorylates CREB, a factor activating BDNF gene transcription. We conclude that induction of D_3 receptor expression in striatum is triggered by the combination of a D_1/D_5 receptor stimulation-dependent elevation of BDNF in cortico-striatal neurons together with a denervation-dependent upregulation of striatal TrkB expression. Both processes are prominent in the 6-OHDA-lesioned side as compared to the control side, which accounts for the induction of D_3 receptor expression restricted to the lesioned side. Moreover, BDNF-induced D_3 receptor expression causes a more pronounced disequilibrium in responsiveness to dopamine between the two sides, which leads to enhanced rotational behavior.

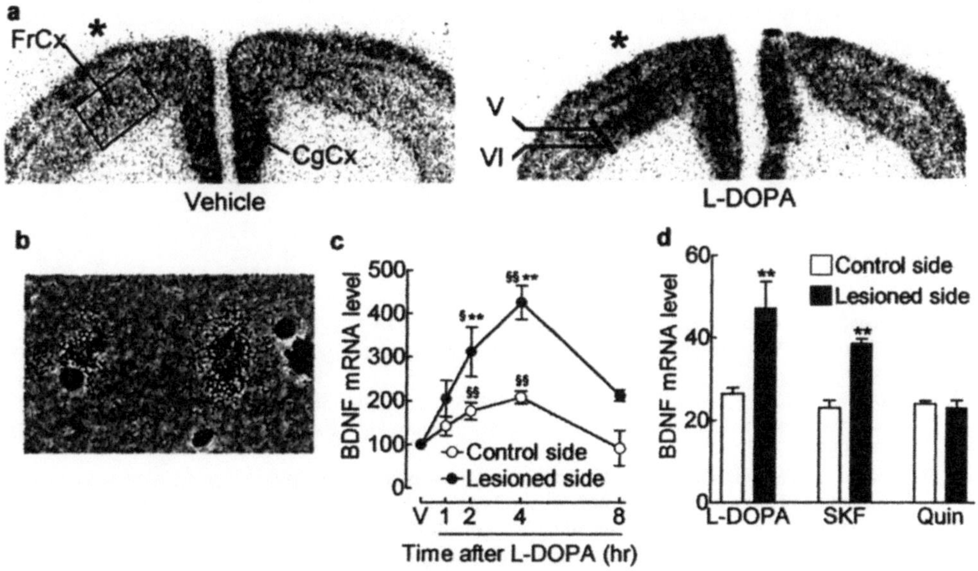

Figure 2. Levodopa upregulates BDNF expression in frontal cortex of 6-OHDA-lesioned rats. **a**, Increase in BDNF mRNA in the lesioned-side (asterisk) of frontal cortex (FrCx, boxed area 4 h after levodopa (L-DOPA). CgCx, cingulate cortex. Roman numerals indicate cortical layers **b**, Two BDNF mRNA-positive cells of deep layer 5 of FrCx of lesioned side of levodopa-treated animals. **c**, Time-course of levodopa effects on BDNF mRNA levels in FrCx. Mean ± S.E.M. of values obtained in 3-4 animals, expressed as percent values of vehicle (V)-treated animals. * $P < 0.05$ and ** $P < 0.01$ vs. control, § $P < 0.05$ and §§ $P < 0.01$ vs. vehicle-treated animals. **d**, Compared effects of a single administration of levodopa, SKF 81297 (SKF) or quinpirole (Quin), on BDNF mRNA levels in FrCx. ** $P < 0.01$ vs. control side).

3. CONCLUSION

These observations, together with those showing that BDNF modulates synaptic transmission, strengthen the view that BDNF elicits adaptive changes in cerebral neurotransmission. Modulation of dopamine responsiveness by BDNF could be an important determinant in the etiopathology and/or treatment of several conditions implicating dopamine. Thus, progressive treatment-dependent elevation of corticostriatal BDNF, which may occur in Parkinson's disease as it does in 6-OHDA-lesioned rats, is a previously unraveled mechanism possibly accounting for the enhanced response to levodopa, that underlies its initial beneficial effect but leads to the abnormal movements often present in long-term treated patients.

The AccSh also receives innervations from areas highly-enriched in BDNF,[11] such as hippocampus, lesions of which also downregulate D_3R expression[12] presumably by a mechanism similar to that described here, and basolateral amygdala, from which BDNF is also anterogradely transported.[11] These brain areas, and notably the amygdala[13] process conditioned aspects of the environment, such as contextual drug taking-associated cues, the reactivity to which is controlled by the D_3R.[8]

We suggest that, via the above pathways and mechanisms, progressive changes in BDNF expression occurring during repeated drug-taking might alter dopamine responsiveness and induce drug-conditioned responses, a key process in drug addiction. In agreement, intra-accumbal infusions of BDNF induce enhancement of behavioral sensitization to cocaine and responding for cocaine-related stimuli in rats[14] and D_3 receptor expression is elevated in accumbens of cocaine addicts[7]. More generally, via the feedback loops AccSh forms with the prefrontal cortex, BDNF could affect control of emotions and sensorimotor gating, that is disturbed in some psychiatric disorders.[15]

REFERENCES

1. H. Thoenen, Neurotrophins and neuronal plasticity, *Science* **270**, 593-598 (1995).
2. C.A. Altar and P.S. DiStefano, Neurotrophin trafficking by anterograde transport, *Trends Neurosci.* **21**, 433-437 (1998).
3. R. Bordet, S. Ridray, S. Carboni, J. Diaz, P. Sokoloff and J.-C. Schwartz, Induction of dopamine D_3 receptor as a mechanism of behavioral sensitization to levodopa, *Proc. Natl. Acad. Sci. USA* **94**, 3363-3367 (1997).
4. H.L. Ryoo, D. Pierrotti and J.N. Joyce, Dopamine D_3 receptor is decreased and D_2 receptor is elevated in the striatum of Parkinson's disease, *Mov. Disord.* **13**, 788-797 (1998).
5. E.V. Gurevich, Y. Bordelon, R.M. Shapiro, S.E. Arnold, R.E. Gur and J.N. Joyce, Mesolimbic dopamine D_3 receptors and use of antipsychotics in patients with schizophrenia, *Arch. Gen. Psychiatry* **54**, 225-232 (1997).
6. J.-C. Schwartz, J. Diaz, C. Pilon and P. Sokoloff, Possible implications of the D_3 receptor in schizophrenia and antipsychotic drug actions, *Brain. Res. Rev.* **31**, 277-287 (2000).
7. J.K. Staley and D.C. Mash, Adaptive increase in D_3 dopamine receptors in the brain reward circuits of human cocaine fatalities, *J. Neurosci.* **16**, 6100-6106 (1996).
8. M. Pilla, S. Perachon, F. Sautel, F. Garrido, A. Mann, C.G. Wermuth, J.-C. Schwartz, B.J. Everitt and P. Sokoloff, Selective inhibition of cocaine-seeking behaviour by a partial dopamine D_3 receptor agonist, *Nature* **400**, 371-375 (1999).
9. O. Guillin, J. Diaz, P. Carroll, N. Griffon, J.-C. Schwartz and P. Sokoloff, BDNF controls dopamine D_3 receptor expression and triggers behavioral sensitization, *Nature* **411**, 86-89 (2001).
10. C.A. Altar, N. Cai, T. Bliven, M. Juhasz, J.M. Conner, A.L. Acheson, R.M. Lindsay and S.J. Wiegand, Anterograde transport of brain-derived neurotrophic factor and its role in the brain, *Nature* **389**, 856-860 (1997).
11. J. Conner and J.C. Lauter, Distribution of Brain-Derived Neurotrophic Factor (BDNF) protein and mRNA in the normal adult rat CNS: evidence for anterograde axonal transport, *J. Neurosci.* **17**, 2295-2313 (1997).
12. G. Flores, D. Barbeau, R. Quirion and L.K. Srivastava, Decreased binding of dopamine D_3 receptors in limbic subregions after neonatal bilateral lesion of rat hippocampus, *J. Neurosci.* **16**, 2020-2026 (1996).
13. R.B. Whitelaw, A. Markou, T.W. Robbins and B.J. Everitt, Excitotoxic lesions of the basolateral amygdala impair the acquisition of cocaine-seeking behaviour under a second-order schedule of reinforcement, *Psychopharmacology* **127**, 213-224 (1996).
14. B.A. Horger, C.A. Iyasere, M.T. Berhow, C.J. Messer, E.J. Nestler and J.R. Taylor, Enhancement of locomotor activity and conditioned reward to cocaine by brain-derived neurotrophic factor, *J. Neurosci.* **19**, 4110-4122 (1999).
15. A. Carlsson, The current status of the dopamine hypothesis of schizophrenia, *Neuropsychopharmacol.* **1**, 179-186 (1988).

REINFORCEMENT NEURONAL SIGNALS AND DIMENSIONALITY REDUCTION IN THE CORTICO-BASAL GANGLIA SYSTEM

Genela Morris, David Arkadir and Hagai Bergman*

1. INTRODUCTION

Although anatomical studies of the basal ganglia show extensive convergence and lateral inhibitory connections (Bolam et al., 2000), physiological studies failed to show correlated pallidal activity (Nini et al., 1995;Raz et al., 2000). These seemingly contradictory results can be explained within a model, which postulates that the basal ganglia reduce the dimensionality of cortical information using optimal extraction methods (Bar-Gad et al., 2000). Such compression can be controlled by the reinforcement neuronal systems of the basal ganglia, i.e., the midbrain dopaminergic neurons (Schultz, 1998) and the cholinergic interneurons of the striatum (Graybiel et al., 1994). Previous studies have shown that midbrain dopaminergic neurons respond to mismatch between the animal's predictions and reality. Thus, dopaminergic neurons respond to primary rewards only when the reward occurs unpredictably (Hollerman and Schultz, 1998). In a trained animal, the dopaminergic neurons increase their discharge with the earliest reward-predicting stimulus, and reduce their discharge when this prediction fails to occur (Mirenowicz and Schultz, 1994). The tonically active neurons (TANs) of the striatum (presumably the cholinergic interneurons of this structure (Wilson et al., 1990;Aosaki et al., 1995)) show responses of a similar nature (Graybiel et al., 1994;Ravel et al., 1999), although with reverse polarity. In this work we report the preliminary results of a study which tested the hypotheses that 1) the reinforcement-related neurons (cholinergic, dopaminergic) and output neurons of the basal ganglia respond to cues predicting reward similarly, in a manner reflecting the amount of information for future reward; and 2) that these neurons emit a single reinforcement signal, in response to the first cue indicating the probability of future reward.

---

* Department of Physiology, the Center for Neural Computation and the Eric Roland Center for Neurodegenerative diseases, The Hebrew University, Jerusalem, ISRAEL

2. METHODS

A macaque (Macaca fasicularis) monkey was trained on a probabilistic visual-motor task in which the probability of receiving reinforcement for correct performance depended on the specific cue. The monkey initiated a trial by touching a central key. After a delay of 2-4 s, a short (0.45 s) cue appeared on either side of a screen situated in front of it. The monkey was over-trained with a set of four different cues associated with different probabilities (0.25, 0.5, 0.75 and 1.0) for reward following correct performance. The monkey had to continue pressing the central key for two additional seconds, to release it when the color of the screen changed from red to yellow (a go signal) and to touch the right or left key (according to the side of the visual cue) within 0.9s. If the monkey touched the correct key within the time limits of the task, it received a reward consisting of a drop of water according to the probability associated with the cue (e.g., one for every two correct trials with the cue associated with a probability of 0.5) one second after the correct touch. The trial was aborted for any violation of the task limits (e.g., release of the central key before the GO signal, late response etc.). An inter-trial interval of 3-6 seconds followed all correct and error trials.

We recorded the simultaneous activity of several neurons in the striatum, pallidum and midbrain neurons while the monkey performs the task. Since we don't have yet a histological confirmation of the midbrain recording, this report will be restricted to the recordings of the TANs and pallidal cells. The monkeys' care and surgical procedures were in accordance with the NIH Guide for the Care and Use of Laboratory Animals (1996), and with the Hebrew University guidelines for the use and care of laboratory animals in research, supervised by the institutional animal care and use committee.

3. RESULTS

Our preliminary results show that both TANs and pallidal neurons respond to the visual cues (Figures 1 and 2). In most cases, a neuron responded to more than one component of the task. Most TANs exhibited a stereotypical response to the cue (a pause in firing, often flanked by brief elevation of the firing rate) and a similar response to the reward (Figure 1). The pallidal responses were of longer duration and were composed of elevation,

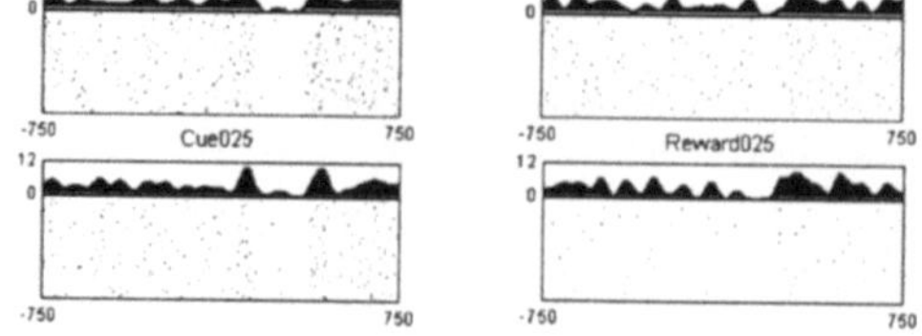

Figure 1: TAN's responses to cues and reward. The figure shows the raster and the PSTH of a TAN in response to cue with different prediction for future rewards (left) and to reward. The probability associated with the cue is given on top margin.

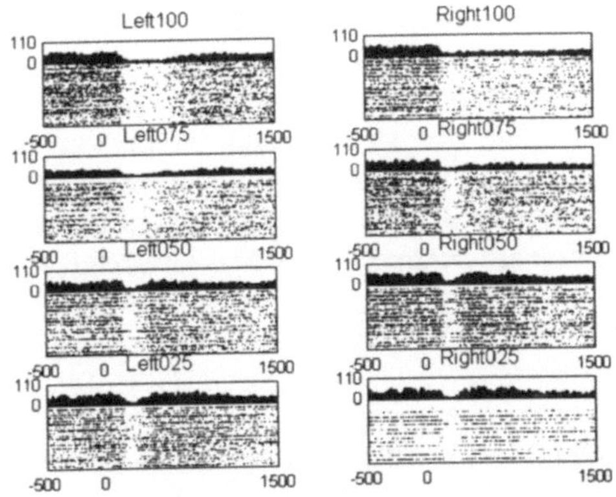

Figure 2: Pallidal responses to left and right cues. The figure shows the raster and the PSTH of a pallidal cell in response to cue with different prediction for future rewards. The probability associated with the cue is given on top margin.

suppression or bipolar changes in firing rate (Figure 2). Pallidal cells showed fewer responses to the reward itself, but showed responses to other components of the task. A small fraction of both TANs and pallidal cells had weak responses to the disappointment, e.g., a change in firing rate at the time of the expected reward when reward was not given. Most task-related pallidal units have a differential response to right or left cues, and differential responses to the different cues associated with the different probabilities (Figure 2). In contrast, the TAN's response was not significantly different between the different cues (right, left, different probabilities).

4. CONCLUSIONS

Previous cross-correlations studies of the simultaneous activity of pallidal neurons confirm the predictions of the dimensionality reduction model of flat pallidal and striato-pallidal correlations in the normal monkeys (Raz et al., 2000;Raz et al., 2001). However, a biological plausible dimensionality reduction system should be controlled by a reinforcement signal.

The two main candidates for providing a reinforcement signal to the dimensionality reduction network of the basal ganglia are the dopaminergic and cholinergic neuromodulator systems. The main target of the dopaminergic neurons is the striatum. The unique triple synaptic arrangement of the cortico-dopaminergic-striatal synapses make the dopaminergic system to be ideally located to influence basal ganglia processing. While cholinergic interneurons constitute only 1-5% of the total population of striatal neurons (Kawaguchi et al., 1995), they give rise to an extensive and dense network of local axonal arbors, permeating the striatum with cholinergic markers (Mesulam et al., 1992;Yelnik et al., 1993). The striatal dopaminergic and cholinergic systems interact with each other (Kitai and Surmeier, 1992;Aosaki et al., 1998;Yan et al., 1997;Pisani et al., 2000;Aosaki et al., 1994;Watanabe and Kimura, 1998)

Our preliminary results indicate that the dopaminergic and the cholinergic systems of the basal ganglia provide a complex signal that can be used by the basal ganglia networks for optimal adjustment the actor loop. The signal is not a simple reinforcement signal since it is related to more than one component of the behavioral task, and moreover different properties are displayed by the different components of the system, e.g., cholinergic, dopaminergic (data not shown) and pallidal neurons. This multifaceted neuronal signal probably enables the basal ganglia is to reduce the dimensionality of the complex cortical information in a manner which is related to the different aspects of the animal's predictions, attentional resources and actual rewards.

REFERENCE LIST

1. Aosaki T, Graybiel AM, Kimura M (1994) Effect of the nigrostriatal dopamine system on acquired neural responses in the striatum of behaving monkeys. Science **265**: 412-415.
2. Aosaki T, Kimura M, Graybiel AM (1995) Temporal and spatial characteristics of tonically active neurons of the primate's striatum. J Neurophysiol **73**: 1234-1252.
3. Aosaki T, Kiuchi K, Kawaguchi Y (1998) Dopamine D1-like receptor activation excites rat striatal large aspiny neurons in vitro. J Neurosci **18**: 5180-5190.
4. Bar-Gad I, Havazelet Heimer G, Goldberg JA, Ruppin E, Bergman H (2000) Reinforcement driven dimensionality reduction - a model for information processing in the basal ganglia. Journal of Basic & Clinical Physiology & Pharmacology **11**: 305-320.
5. Bolam JP, Hanley JJ, Booth PA, Bevan MD (2000) Synaptic organisation of the basal ganglia. J Anat **196**: 527-542.
6. Graybiel AM, Aosaki T, Flaherty AW, Kimura M (1994) The basal ganglia and adaptive motor control. Science **265**: 1826-1831.
7. Hollerman JR, Schultz W (1998) Dopamine neurons report an error in the temporal prediction of reward during learning. Nat Neurosci **1**: 304-309.
8. Kawaguchi Y, Wilson CJ, Augood SJ, Emson PC (1995) Striatal interneurons: chemical, physiological and morphological characterization. TINS **18**: 527-535.
9. Kitai ST, Surmeier DJ (1992) Cholinergic and dopaminergic modulation of potassium conductances in neostriatal neurons. Adv Neurol **60**: 40-52.
10. Mesulam MM, Mash D, Hersh L, Bothwell M, Geula C (1992) Cholinergic innervation of the human striatum, globus pallidus, subthalamic nucleus, substantia nigra, and red nucleus. J Comp Neurol **323**: 252-268.
11. Mirenowicz J, Schultz W (1994) Importance of unpredictability for reward responses in primate dopamine neurons. J Neurophysiol **72**: 1024-1027.
12. Nini A, Feingold A, Slovin H, Bergman H (1995) Neurons in the globus pallidus do not show correlated activity in the normal monkey, but phase-locked oscillations appear in the MPTP model of parkinsonism. J Neurophysiol **74**: 1800-1805.
13. Pisani A, Bonsi P, Centonze D, Calabresi P, Bernardi G (2000) Activation of D2-like dopamine receptors reduces synaptic inputs to striatal cholinergic interneurons. J Neurosci Online 20: RC69.
14. Ravel S, Legallet E, Apicella P (1999) Tonically active neurons in the monkey striatum do not preferentially respond to appetitive stimuli. Exp Brain Res **128**: 531-534.
15. Raz A, Frechter-Mazar V, Feingold A, Abeles M, Vaadia E, Bergman H (2001) Activity of pallidal and striatal tonically active neurons is correlated in MPTP-treated monkeys but not in normal monkeys. The Journal of Neuroscience 21:RC128: 1-5.
16. Raz A, Vaadia E, Bergman H (2000) Firing pattern and correlations of spontaneous discharge of pallidal neurons in the normal and the tremulous MPTP vervet model of parkinsonism. J Neurosci **20**: 8559-8571.
17. Schultz W (1998) Predictive reward signal of dopamine neurons. J Neurophysiol **80**: 1-27.
18. Watanabe K, Kimura M (1998) Dopamine receptor-mediated mechanisms involved in the expression of learned activity of primate striatal neurons. J Neurophysiol **79**: 2568-2580.
19. Wilson CJ, Chang HT, Kitai ST (1990) Firing patterns and synaptic potentials of identified giant aspiny interneurons in the rat neostriatum. J Neurosci **10**: 508-519.
20. Yan Z, Song WJ, Surmeier J (1997) D2 dopamine receptors reduce N-type Ca2+ currents in rat neostriatal cholinergic interneurons through a membrane-delimited, protein-kinase-C-insensitive pathway. J Neurophysiol **77**: 1003-1015.
21. Yelnik J, Percheron G, Francois C, Garnier A (1993) Cholinergic neurons of the rat and primate striatum are morphologically different. Prog Brain Res **99**: 25-34.

EFFECTS OF HISTAMINERGIC DRUGS ON DOPAMINERGIC AND CHOLINERGIC BRAIN ACTIVITY IN THE RAT

Relevance for mechanisms of memory and reward

Rainer K.W. Schwarting*

1. INTRODUCTION

Neuronal histamine in the brain is provided exclusively by a group of neurons located in the tuberomammillary nucleus (TM) of the posterior hypothalamus. The histaminergic fibers originating from this area are highly divergent and project to almost all brain regions, where they exert their effects through H1, H2, and H3 receptors.[1, 2] Brain histamine seems to be involved in a number of functions, including arousal, sleep/wake cycle, feeding, drinking, aggression, learning, memory, reinforcement, and functional recovery after brain damage.[3, 4] Here, neurochemical studies will be reviewed where the effects of H1 antagonists or histamine where studied on extracellular dopamine (DA) and acetylcholine (ACh) in the rat brain. The rationale for these experiments was obtained from behavioral work showing pronounced effects of histaminergic manipulations on learning, memory, and reinforcement. These studies will shortly also be addressed.

2. H1 ANTAGONISTS, LEARNING, MEMORY, AND BRAIN ACETYLCHOLINE

There is substantial evidence that histaminergic manipulations can affect learning and memory.[3, 4] With respect to the involvement of specific receptor subtypes, H1 receptors have turned out to be interesting candidates. Thus, performance in a number of learning tasks is enhanced by TM lesions that produce a loss of histaminergic markers in the TM region. Similar performance-enhancing effects[5] have been observed with administration of the H1 antagonist chlorpheniramine (CHL). The fact that a reduction in

* Experimental and Physiological Psychology, Philipps-University of Marburg, 35032 Marburg, Germany. Email: schwarti@mailer.uni-marburg.de

Catecholamine Research, Edited by Nagatsu *et al.*
Kluwer Academic/Plenum Publishers, 2002

histaminergic transmission (by TM lesions or antagonist treatment) can improve performance in learning and memory tests has led to the suggestion that histamine normally exerts an inhibitory control over learning and memory processes.[3] However, the underlying neural mechanisms involved in these behavioral effects of histaminergic manipulations remain to be investigated. Since learning and memory appears to be closely related to ACh function in the brain[6], we asked whether an H1 antagonist would affect, i.e. enhance, ACh activity in the brain.

Using in-vivo microdialysis, we found that intraperitoneal injections of the H1 antagonist CHL (5-20mg/kg) led to dose-dependent increases of extracellular ACh levels in the frontal cortex and ventral hippocampus of urethane-anesthetized rats.[7] Similar effects were obtained in the awake animal. These cholinergic enhancements were probably not due to the antimuscarinic effects of CHL, since pyrilamine (PYR), which has a low affinity for muscarinic receptors, produced comparable cholinergic increases like CHL (Fig. 1). In avoidance learning,[5] behaviorally effective doses of CHL were in the same range as those which led to cholinergic increases in the frontal cortex and hippocampus. These close relationships between the behavioral and neurochemical data suggest that pro-cholinergic effects of H1 antagonists may contribute to their pro-mnestic effects.

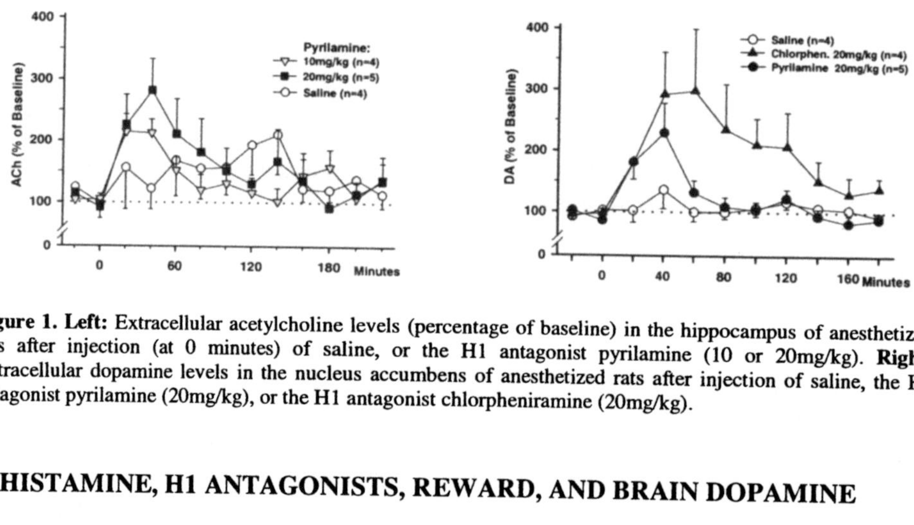

Figure 1. Left: Extracellular acetylcholine levels (percentage of baseline) in the hippocampus of anesthetized rats after injection (at 0 minutes) of saline, or the H1 antagonist pyrilamine (10 or 20mg/kg). **Right:** Extracellular dopamine levels in the nucleus accumbens of anesthetized rats after injection of saline, the H1 antagonist pyrilamine (20mg/kg), or the H1 antagonist chlorpheniramine (20mg/kg).

3. HISTAMINE, H1 ANTAGONISTS, REWARD, AND BRAIN DOPAMINE

It has been shown in various paradigms that elevated histamine activity in the brain can have an inhibitory effect on reward-related behavior, whereas damage of the histaminergic projection system or administration of H1 receptor antagonists can act in a pro-rewarding way.[3, 8] Such data point to a role of histamine in the control of reinforcement, and suggest that it may be of an inhibitory nature.[3] Regarding the neural substrates critical for reward, most evidence has been obtained for a role of the transmitter DA in the mesostriatal and mesolimbic system.[9] Thus, we asked whether H1 antagonists or histamine itself would affect DA activity in these systems.

Using in-vivo microdialyis in urethane-anesthetized rats,[10] we found that the H1 antagonists CHL (5, 20mg/kg) and PYR led to moderate (i.e. <200% of baseline) but sustained increases of extracellular DA levels in the neostriatum. In the NAcc (see Fig. 1), CHL (20mg/kg) produced a transient DA increase to about 300% of baseline. With 20mg/kg of PYR, DA increased to about 230%, and returned to baseline more rapidly than in the case of CHL. Both drugs also led to decreases in extracellular levels of the DA metabolites homovanillic acid and dihydroxyphenylacetic acid. These neurochemical patterns are similar to those commonly seen with drugs of abuse, especially psychostimulants. Similar to the psychostimulants, the DAergic effects of the H1 antagonist may be due to their effectiveness to inhibit re-uptake of DA. Thus, the interaction of the H1 antagonist with striatal DAergic transmission may explain the reinforcing effects and abuse potential associated with such drugs.

Recently,[8] we tested whether local histaminergic intervention is also effective to modulate DAergic activity in the NAcc. Using the technique of reverse dialysis, we found that histamine administration (0.1-10μmol/L) led to a dose-dependent enhancement of extracellular DA levels there. With an even higher concentration of histamine (1mmol/L), a maximal increase to about 500% of baseline was obtained. The H1 antagonist PYR led to decreased extracellular DA levels when locally administered into the NAcc.

The data with histamine extend previous work showing that histamine can enhance DA activity in the brain[11]; however, the mechanism by which histamine exerted these effects is not clear. Histaminergic receptors are located on various neuronal elements within the NAcc. An action through receptors located presynaptically on DA terminals is unlikely, since these H_3 receptors seem to regulate DA activity in an inhibitory way.[12] Alternatively, histamine may have acted on output neurons, which via downstream projections may have indirectly stimulated DAergic cells in the VTA, and thereby, DA release in the NAcc. Furthermore, histamine can interfere with the uptake of DA,[13] which would also lead to increased levels of extracellular DA. Finally, it may be possible that histamine affected DA via an action on glutamatergic inputs to the NAcc. From the hippocampus it is known that histamine can enhance glutamate signaling via NMDA receptors.[2] Most of the inputs to the NAcc are glutamatergic, and can also activate striatal DA release. Thus, it is possible that histamine infusion into the NAcc enhanced extracellular DA levels indirectly, that is, via activation of its glutamatergic afferents.

The DA-decreasing effects of the locally adminstered H1 antagonist PYR are in contrast to its activating effect when given systemically. PYR, however, is known to have local anesthetic actions, which might explain why it decreased DA, when applied into the NAcc.

The analysis of interactions between histaminergic drugs and DAergic mechanisms are especially relevant with respect to reward-modulating effects of these drugs.[3] Most drugs which exert positive reinforcing effects also promote DA activity, and this action is considered to be a critical requirement for their reinforcing effects.[9] Based on this relationship, the present data suggest that intra-NAcc administration of histamine should be reinforcing. This hypothesis was recently supported, since it was found that histamine injections into the caudal shell of the NAcc were effective in producing place preference.[14] PYR, on the other hand, should not be reinforcing when injected into the NAcc. Finally, CHL has been shown to lead to conditioned place preference when injected into the Nacc.[15] Following the DA hypothesis, one should expect increases of DA activity with local administration of this H1 antagonist.

4. CONCLUSIONS

The work outlined here shows that histamine and H_1 antagonists can have profound effects on brain neurotransmission; that is, on extracellular levels of ACh (in frontal cortex and hippocampus) and DA (in neostriatum and NAcc). It is suggested that these neurochemical changes are related to the known effects of such histaminergic manipulations on learning, memory and reinforcement. In general, histamine seems to be a potent modulator of several basic biopsychological functions. Studying the mechanisms by which histamine and its receptors exert these effects should provide further substantial insights regarding the underlying functional mechanisms in the brain.

REFERENCES

1. K. Onodera, A. Yamatodani, T. Watanabe, and H. Wada, Neuropharmacology of the histaminergic neuron system in the brain and its relationship with behavioral disorders. *Prog. Neurobiol.* **42**, 685-702 (1994).
2. R.E. Brown, D.R. Stevens, and H.L. Haas, The physiology of brain histamine. *Prog. Neurobiol.* **63**, 637-672 (2001).
3. J.P. Huston, U. Wagner, and R.U. Hasenöhrl, The tuberomammillary nucleus projections in the control of learning, memory and reinforcement processes: evidence for an inhibitory role. *Behav. Brain Res.* **83**, 97-105 (1997).
4. M.B. Passani, L. Bacciottini, P.F. Mannaioni, and P. Blandina, Central histaminergic system and cognition. *Neurosci. Biobehav. Rev.* **24**, 107-113 (2000).
5. C. Frisch, R.U. Hasenöhrl, and J.P. Huston, The histamine H_1-antagonist chlorpheniramine facilitates learning in aged rats. *Neurosci. Lett.* **229**, 89-92 (1997).
6. A. Blokland, Acetylcholine: a neurotransmitter for learning and memory? *Brain Res. Rev.* **21**, 285-300 (1996).
7. H.C. Dringenberg, M.A. de Souza-Silva, J. Roßmüller, J.P. Huston, and R.K.W. Schwarting, Histamine H_1 receptor antagonists produce increases in extracellular acetylcholine in rat frontal cortex and hippocampus. *J. Neurochem.* **70**, 1750-1758 (1998).
8. R. Galosi, L. Lenard, A. Knoche, J.P. Huston, and R.K.W. Schwarting, Dopaminergic effects of histamine administration in the nucleus accumbens and the impact of H1 receptor blockade. *Neuropharmacology* **40**, 624-633 (2001).
9. G. Di Chiara, E. Acquas, and E. Carboni, in: *The Mesolimbic Dopamine System: From Motivation to Action*, edited by P. Willner and J. Scheel-Krüger (John Wiley, Chichester 1991), pp. 367-384.
10. H.C. Dringenberg, M.A. de Souza-Silva, R.K.W. Schwarting, and J.P. Huston, Increased levels of extracellular dopamine in neostriatum and nucleus accumbens after histamine H1 receptor blockade. *Naunyn-Schmiedeberg´s Arch. Pharmacol.* **358**, 423-429 (1998).
11. N. Subramanian, and A.H. Mulder, Modulation by histamine of the efflux of radiolabeled catecholamines from rat brain slices. *Eur. J. Pharmacol.* **43**, 143-152 (1976).
12. H. Prast, M.H. Tran, C. Lamberti, H., Fischer, M. Kraus, K. Grass, and A. Philippu, Histaminergic neurons modulate acetylcholine release in the ventral striatum: role of H_1 and H_2 histamine receptors. *Naunyn-Schmiedeberg´s Arch. Pharmacol.* **360**, 552-557 (1999).
13. J. Tuomisto, and L. Tuomisto, Effects of histamine and histamine antagonists on the uptake and release of catecholamines and 5-HT in brain synaptosomes. *Med. Biol.* **58**, 33-37 (1980).
14. R.U. Hasenöhrl, A. Kuhlen, P. Zimmermann, and J.P. Huston, Reinforcing and memory-promoting effects of intraaccumbens histamine injection. *Soc. Neurosci. Abstr.*, Vol. 26, Part II, 1742 (2000).
15. P. Zimmermann, C. Privou, and J.P. Huston, Differential sensitivity of the caudal and rostral nucleus accumbens to the rewarding effects of a H1-histaminergic receptor blocker as measured with place-preference and self-stimulation behavior. *Neuroscience* **94**, 93-103 (1999).

INVOLVEMENT OF THE BASAL GANGLIA AND DOPAMINE SYSTEM IN LEARNING AND EXECUTION OF GOAL-DIRECTED BEHAVIOR

Minoru Kimura, Naoyuki Matsumoto, Yasumasa Ueda, Takemasa Satoh, Takafumi Minamimoto and Hiroshi Yamada*

1. INTRODUCTION

To perform any task of our own volition, we execute multiple movements in a specific order based on the likelihood of obtaining a successful outcome. Neurons in the supplementary motor area (SMA), pre-SMA and those in the basal ganglia encode the temporal order, or the sequence of movements used in the tasks (Mushiake and Strick, 1995; Kermadi and Joseph, 1995; Nakamura et al., 1998; Shima and Tanji, 1998, 2000). These neurons must play a crucial role in the mechanisms of planning and execution of temporally organized multiple movements, action.

The neurons in the striatum process not only topographically organized specific inputs from wide cortical areas but also modulatory inputs. One of the modulatory signals is a limbic signal that is related to motivation and reward originating from the amygdala (Russchen et al. 1985; Smith and Parent 1986), orbital frontal cortex (Thorpe et al., 1983) and the nigrostriatal dopaminergic system (Schultz, 1998; Schultz et al., 1997). These inputs to the striatum have been suggested to play a significant role in learning (Aosaki et al., 1994; Kimura, 1995; Schultz et al., 1997; Schultz, 1998; Matsumoto et al., 1999), cognitive processing (Kawagoe et al., 1998; Shidara et al., 1998) and movement (Steinfels et al., 1983; DeLong, 1990).

We propose a point of view that the basal ganglia play a crucial role in learning and execution of goal-directed behavior in a dopamine dependent manner.

* Minoru Kimura, Department of Physiology, Kyoto Prefectural University of Medicine, Kawaramachi-Hirokoji, Kamigyo-ku, Kyoto 602-8566

2. BUILDING UP GOAL-DIRECTED MOTOR SEQUENCE IN DOPAMINE DEPENDENT MANNER

To study the consequences of nigrostriatal dopamine depletion for learning and memory of motor sequences, we used a neurotoxin, 1-methyl-4-phenyl-1,2,3,6-tetrahydropyridine (MPTP), to deplete dopamine unilaterally in the striatum of macaque monkeys either before or after training them on sequential push-button motor tasks (Matsumoto et al., 1999). We compared the monkeys' performance with the arms ipsilateral and contralateral to dopamine depletion. During training and retraining on the tasks, we measured initial and serial movement times and reaction times for the push button movements, electromyographic patterns of arm and orofacial muscle activity during button pushing and reward licking, and saccadic eye movements during the button push sequences.

With the arm ipsilateral to the side of dopamine depletion, each monkey showed progressive shortening of movement times and initial and serial reaction times. When the monkey was asked to learn sequential 3 button push movements for a drop of reward water after the monkey had learned the two button push task, the monkey stopped at the second button and tried to lick the spoon on which reward had been delivered in previously learned two push task. The monkey gradually stopped licking at the second push to complete the third push to get reward in 2 to 3 days of training. In parallel with the change of task strategy, serial reaction time of the button push remarkably decreased. With the arm contralateral to the dopamine depletion, the monkey showed significant qualitative deficits. The monkey continued to stop and lick the spoon at the second push even after two weeks of training. This made the serial reaction times much longer than those with the ipsilateral hand. Anticipatory saccadic eye movements were not well time-locked to individual button pushes made with the contralateral hand; and push and licking movements were not smoothly coordinated. We conclude that the striatum and nigrostriatal dopamine system function in the learning underlying performance of sequences of movements as goal-directed single motor programs.

3. GOAL-DIRECTED ASSEMBLY OF LEARNED MOTOR ELEMENTS INTO AN ACTION

Neural activity was recorded in the striatum by penetrating two to five microelectrodes simultaneously by a manipulator (EPS, Alpha-Omega Corp). Raw cross-correlograms (RCCG) between pairs of neuronal spike trains were computed for individual trials and then averaged across all trials in the same motor sequences; this constituted an analysis block. The cross-correlogram expected by chance was constructed by computing individual cross-correlograms between the spike train of one neuron of the pair in each trial and the spike train of the other neuron in every non-corresponding trial in the same analysis block and then averaging those correlograms. This average chance correlogram (the shift predictor) was subtracted from the average raw correlogram to obtain an estimate of the spike correlation above or below that expected by chance. The final averaged shift predictor- corrected cross-correlogram (SCCG) for each block of trials was then computed by averaging over all corrected correlograms.

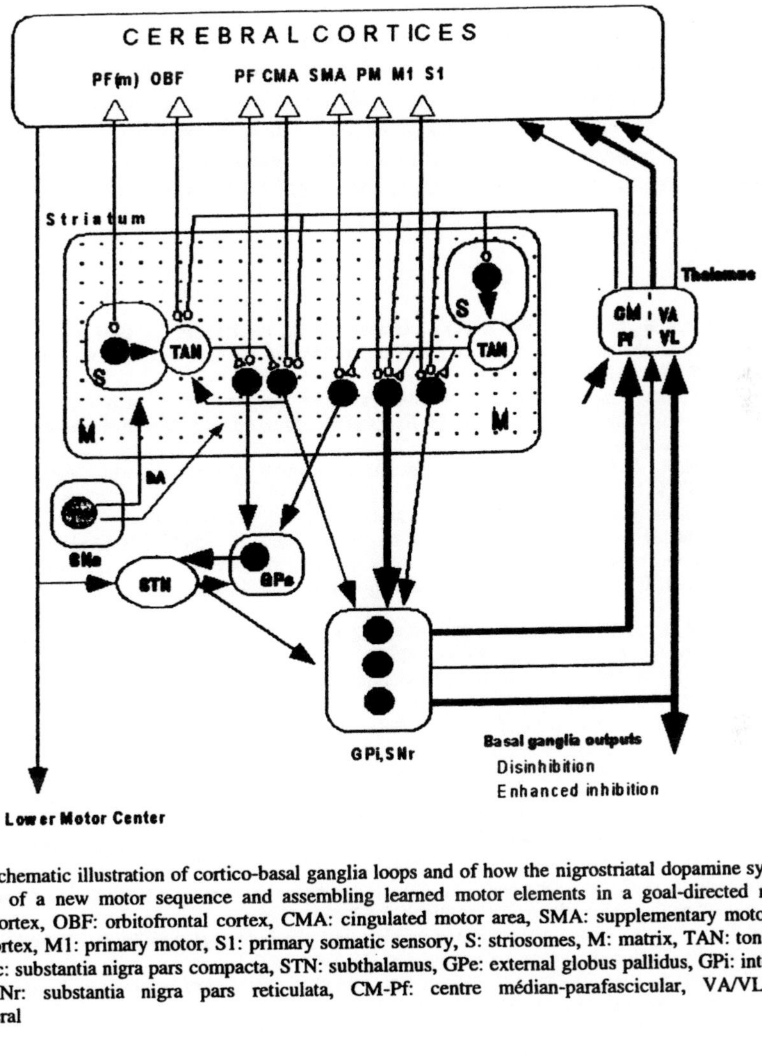

Figure 1. Schematic illustration of cortico-basal ganglia loops and of how the nigrostriatal dopamine system guides building up of a new motor sequence and assembling learned motor elements in a goal-directed manner. PF: prefrontal cortex, OBF: orbitofrontal cortex, CMA: cingulated motor area, SMA: supplementary motor area, PM: premotor cortex, M1: primary motor, S1: primary somatic sensory, S: striosomes, M: matrix, TAN: tonically active neuron, SNc: substantia nigra pars compacta, STN: subthalamus, GPe: external globus pallidus, GPi: internal globus pallidus, SNr: substantia nigra pars reticulata, CM-Pf: centre médian-parafascicular, VA/VL: ventralis-anterior/lateral

We have found that simultaneously recorded projection neurons in the primate striatum whose activity is related to components of a motor sequence showed correlated spike discharges, indicating serial facilitation from one class of neuron to another. The serial facilitation observed between activity of different classes of striate projection neurons does not seem to originate through the local circuit in the striatum but through the cortico-basal ganglia loop, because the striatal projection neurons are inhibitory (Yoshida et al., 1972; Obata and Yoshida 1973) and the facilitation occurred with a long duration up to more than 200 msec. We propose that the facilitatory correlation of spike discharges between the striate projection neurons encoding components of a motor sequence may be a basic mechanism of how the cortically represented specific information about motor sequences is assembled to plan temporally organized multiple movements in a goal-directed manner.

REFERENCES

Aosaki, T., Tsubokawa, H., Ishida, A., Watanabe, K., Graybiel, AM., and Kimura, M. 1994, Responses of tonically active neurons in the primate's striatum undergo systematic changes during behavioral sensory-motor conditioning. *J. Neurosci.* **14**: 3969-3984.

DeLong, M.R. 1990, Primate models of movement disorders of basal ganglia origin. *Trends Neurosci.* **13**: 281-285.

Kawagoe, R., Takikawa, Y., and Hikosaka, O. 1998, Expectation of reward modulates cognitive signals in the basal ganglia. *Nature. Neurosci.* **1**: 411-416.

Kermadi, I. and Joseph, J. P. 1995, Activity in the caudate nucleus of monkey during spatial sequencing. *J. Neurophysiol.* **74**: 911-933.

Kimura, M. 1995, Role of basal ganglia in behavioral learning. *Neurosci. Res.***22**: 353-358.

Matsumoto, N., Hanakawa, T., Maki, S., Graybiel, A.M., and Kimura, M. 1999, Role of nigrostriatal dopamine system in learning to perform sequential motor tasks in a predictive manner. *J. Neurophysiol.* **82**: 978-998.

Mushiake, H. and Strick, P. L. 1995, Pallidal neuron activity during sequential arm movements. *J. Neurophysiol.* **74:** 2754-2758.

Nakamura, K., Sakai, K. and Hikosaka, O. 1998, Neuronal activity in medial frontal cortex during learning of sequential procedures. *J. Neurophysiol.* **80**: 2671-2687.

Obata, K. and Yoshida, M. 1973, Caudate-evoked inhibition and actions of GABA and other substances on cat pallidal neurons. *Brain Res.* **21:** 455-459.

Russchen, F.T., Bakst, I., Amaral. D.G., Price, J.L. 1985, The amygdalostriatal projections in the monkey. An anterograde tracing study. *Brain Res.* **329**: 241-257.

Schultz, W. 1998, Predictive reward signal of dopamine neurons. *J Neurophysiol* **80**: 1-27.

Schultz, W., Dayan, P., and Montague, R.R.A. 1997, Neural substrate of prediction and reward. *Science* **275**: 1593-1599.

Shidara, M., Aigner, T.G/, and Richmond, B.J. 1998, Neural signals in the monkey ventral striatum related to progress through a predictable series of trials. *J. Neurosci.* **18**: 2613-2625.

Shima, K. and Tanji, J. 1998, Both supplementary and presupplementary motor areas are crucial for the temporal organization of multiple movements. *J. Neurophysiol.* **80**: 32473260.

Shima, K. and Tanji, J. 2000, Neural activity in the supplementary and presupplementary motor areas for temporal organization of multiple movements. *J. Neurophysiol.* **84**: 2148-2160.

Smith, Y., and Parent, A. 1986, Differential connections of caudate nucleus and putamen in the squirrel monkey (Saimiri sciureus). *Neuroscience* **18**: 347-371.

Steinfels, G.F., Heym, J., Strecker, R.E., and Jacobs, B.L. 1983, Behavioral correlates of dopaminergic unit activity in freely moving cats. *Brain Res..* **258**: 217-228.

Thorpe, S.J., Rolls, E.T., and Maddison, S. 1983, The orbitofrontal cortex: neuronal activity in the behaving monkey. *Exp. Brain Res.* **49**: 93-115.

Yoshida, M., Rabin, A. and Anderson, M. 1972, Monosynaptic inhibition of pallidal neurons by axon collaterals of caudato-nigral fibers. *Exp. Brain Res.* **15**: 333-347.

MOLECULAR MECHANISM OF THE REGULATION OF THE BIOLOGICAL CLOCK IN MAMMALS

Hitoshi Okamura, Kazuhiro Yagita, and Shun Yamaguchi*

1. INTRODUCTION

Most organisms on earth produce a variety of behavioral or physiological rhythms with periods that are close to 24 hours. Such "circadian" rhythms are genetically determined at least in *cyanobacteria*, *Arabidopsis*, *Neurospora*, *Drosophila* and rodents, and recent progress of the molecular dissection opens up the clock genes and their time-keeping mechanisms. A recent genetic screening of human disease of Utah's group verified the clock generation mechanism is common even in human (Toh et al., 2001). Here we describe our data related to the molecular mechanism of the clock gene in mammals.

2. MOUSE PERIOD GENES

We cloned mammalian *Period* genes which are homologous to *Drosophila period* (*per*) gene (Takumi et al., 1998a; Takumi et al., 1998b; Tei et al., 1997). These genes were also isolated in other laboratory (Albrecht et al., 1997; Shearman et al., 1997; Sun et al., 1997; Zylka et al., 1998). Sequence alignments for mPER1, mPER2 and mPER3 show that the PAS (Period, Arnt and Sim) including PAS-A and PAS-B repeat, which is regarded as a potential PER dimerization domain is highly conserved in *Drosophila Per* genes. The cytoplasmic localization domain (CLD) and elements of the immediately downstream protein sequence were also well conserved. mPER3 contains a classical NLS, and mPER2 contains a bipartite basic type NLS. Although there seems to be no discernible NLS in mPER1, Vielhaber et al. (Vielhaber et al., 2000) recently reported a functional NLS site in the C-terminal region of mPER1. No typical basic helix-loop-helix (bHLH) motif was found in these three mPERs, but there is a possible HLH region without an adjoining basic region in mPER2 (Takumi et al., 1998a). In the carboxyl-terminal region of the predicted protein, several clusters of homologous sequence are also found, yet their functions remain unknown.

* Hitoshi Okamura, Division of Molecular Brain Science, Department of Brain Sciences, Kobe University Graduate School of Medicine, Kobe 650-0017, Japan

Catecholamine Research, Edited by Nagatsu *et al.*
Kluwer Academic/Plenum Publishers, 2002

3. POSSIBLE OSCILLATORY MACHINARY IN THE SUPRACHIASMATIC NUCLEUS

Recent molecular dissection has revealed that a transcription-translation based autoregulatory negative feedback loop works as the basic mechanism of circadian oscillation from cyanobacteria to mice (Dunlap, 1999). Especially, most components of circadian clock in mammals were in common with those in *Drosophila*. In the following paragraph, we will describe the common schema in oscillatory machinery in mammalian oscillating center, the suprachiasmatic nucleus (SCN) (Fig.1).

The mammalian oscillator gene is composed of two genes *mPer1* and *mPer2* (Zheng et al., 2001). The circadian cycle begins the activation of *mPer1/ mPer2* transcription by the heterodimer of positive transcription factors (CLOCK and BMAL1) (Gekkakis et al., 1998). The nuclear entry of negative factors, mPER1 and mPER2 proteins, is a critical step to close the feedback loops. In cytoplasm mPER1 and mPER2 accumulate to critical level and heterodimerize with mPER3 (Yagita et al., 2000).

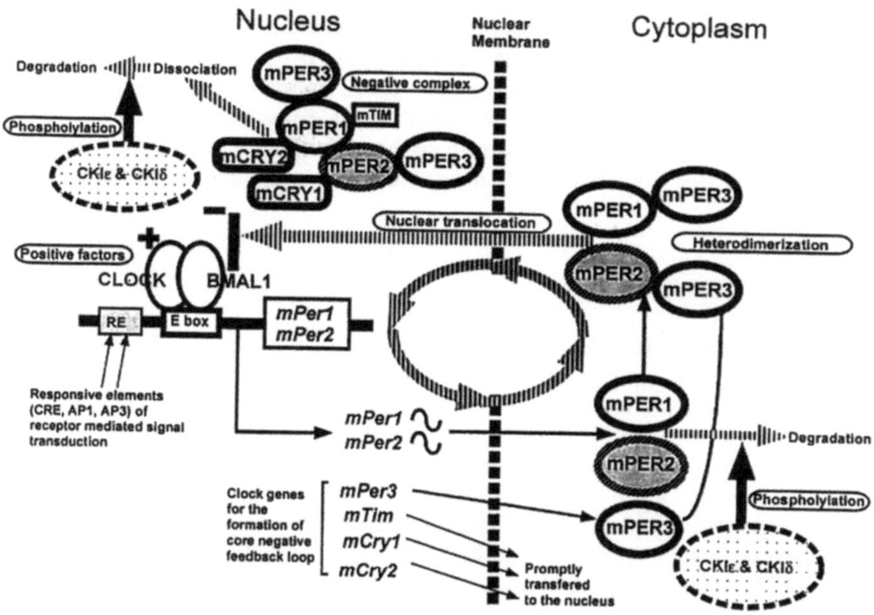

Fig. 1 Mammalian feedback loop of clock genes

Then PER/PER heterodimers enter into the nucleus, and forms negative complex with mCRY1 and mCRY2. This complex inhibits transcription of *mPer1* and *mPer2* through interaction with CLOCK/BMAL1 (Kume et al., 1999). As a result of this inhibition, mPER1 and mPER2 decrease gradually. Again the cycle starts with the activation of CLOCK/BMAL1. In SCN clock oscillating cells, it is well known that accumulation of PER1/PER2 delayed 6-8 hours from accumulation of *mPer1/mPer2* mRNAs, and thought to be essential to generate the self-sustained oscillation. Recently a factor which controls the long time periodicity is revealed. Caseine kinase 1ε, originally found in *Drosophila* as *doubletime* (Price et al., 1998), phosphorylates PER monomers. Phosphorylated PER monomers eventually targeted for degradation. Only unphosphorylated PER monomers are able to dimerize and enter into the nucleus.

4. CIRCADIAN EXPRESSION PROFILES OF CLOCK GENES IN THE SUPRACHIASMATIC NUCLEUS

The mammalian clock genes (positive factors; *clock*, *BMAL1*: negative factors; *Periods*, Crys) were demonstrated to localize highly in the SCN. We examined the daily mRNA expression profiles of these genes by quantitative *in situ* hybridization. *Clock* was constantly expressed in the mouse SCN in any circadian time, although mammalian *mPer1*, *mPer2* and *mPer3* and *BMAL1* were expressed in time-dependent manner: peaks of *Periods* were in the daytime but those of BMAL1 were in the nighttime (Taguchi et al., 2001; Takumi et al., 1998b). Digoxigenin-*in situ* hybridization demonstrated all these clock genes were expressed in most SCN cells. These findings suggest clock genes were expressed in the SCN with the different expression profiles.

5. OSCILLATORY MECHANISM OF PERIPHERAL CLOCKS

In mammals, the central circadian pacemaker resides in the hypothalamic SCN, but circadian clock gene expression was also observed in peripheral tissues. By using wildtype and *mCry1/2*-deficient cell lines, we showed the peripheral oscillator in cultured fibroblasts is identical to the oscillator in the SCN in (i) temporal expression profiles of all known clock genes, (ii) the phase of the various mRNA rhythms (i.e. antiphase oscillation of Bmal1 and mPer genes), (iii) the delay between maximum mRNA levels and appearance of nuclear mPER1 and mPER2 protein, (iv) the inability to produce oscillations in the absence of functional *mCry* genes, and (v) the control of period length by mCRY proteins (Yagita et al., 2001). These data indicate that the molecular make-up of the peripheral circadian oscillator in cultured fibroblasts is similar to that of the master oscillator in the SCN. The same set of circadian genes is assembled into positive and negative transcription translation feedback loops. mRNA expression profiles for these circadian genes display an SCN-like temporal expression profile as well as phase relationship and, at least for mPER1, mPER2 and DBP, the delay between onset of transcription and nuclear appearance of the corresponding gene product is comparable to that in the SCN. Moreover, homozygous inactivation of one or both *mCry* genes, known to accelerate, retard or even abolish the biological clock in the SCN affect the peripheral oscillator to a similar extent. Thus, the peripheral oscillator in immortalised cultured fibroblasts constitutes a bona fide in vitro model for the molecular oscillator in the SCN, and could potentially allow the use of skin fibroblasts from patients with clock gene defects to identify the defect in circadian disorders.

6. A PROPOSED CLOCK MODEL IN MAMMALS

Since most of cells composed of human body have the clock oscillating system, it is now possible that human time keeping system is composed in hierarchical order. SCN, a permanently oscillating central clock, secretes humoral and/or neuronal factors and tunes the time of other brain clocks. Thereafter, brain clock information flows to clocks at lower levels, and transmitted to the clocks in peripheral nervous system. Informations from the nervous system flows by humoral factors to reset the clocks in visceral organs. In the peripheral organs, clocks in some organs (eg, pineal body) receive time information directly from the neuronal transmission, but in the majority of peripheral clocks, time is transmitted by humoral factors. Clocks was adjusted to the environmental time by various time cues; in the SCN, photic information resets strongly to the environmental light-dark cycles. At each level of the hieralchical organization, environamental time cues entrain the phase of the clock, and information of the phase of the peripheral clock is feedbcked to the SCN and change its phase as dark pulse. Other stimuli such as restricted feeding or maternal care entered to the dependent slave clocks in the non-SCN cells in brain and visceral organs such as digestive organs. The feedback information to the central SCN cells came from the non-SCN clocks in visceral organs, pineal body, and certain brain nuclei (raphe nuclei, lateral geniculate nuclei) as non-photic entrainment.

7. ACKNOWLEDGMENTS

This work was supported in part by grants from the Special Coordination Funds of the Science and Technology Agency of Japan, the Grant-in-Aid for the Scientific Research on Priority Areas of the Ministry of Education, Science, Sports and Culture of Japan, Ministry of Welfare of Japan, and SRF.

REFERENCES

1. Albrecht, U., Sun, Z.S., Eichele, G. and Lee, C.C., 1997, A differential response of two putative mammalian circadian regulators, mper1 and mper2, to light. *Cell*, **91**, 1055-1064.
2. Dunlap, J.C., 1999, Molecular bases for circadian clocks. *Cell*, **96**, 271-290.
3. Gekakis, N., Staknis, D., Nguyen, H.., Davis, F., Wilsbacher, L., King, D.P., Takahashi, J.S. and Weitz, C.J., 1998, Role of the CLOCK protein in the mammalian circadian mechanism. Science, **280**, 1564-1569.
4. Kume, K., Zylka, M.J., Sriram, S., Shearman, L.P., Weaver, D.H., Jin, X., Maywood, E., Hastings, M.H. and Reppert, S.M., 1999, mCRY1 and mCRY2 are essential components of the negative limb of the circadian clock feedback loop. *Cell*, **98**, 193-205.
5. Price, J.L., Blau, J., Rothenfluh, A., Abodeely, M., Kloss, B. and Young, M.W., 1998, Double-time is a new Drosophila clock gene that regulates PERIOD protein accumulation. *Cell*, **94**, 83-95.
6. Shearman, L.P., Zylka, M.J., Weaver, D., Kolakowski, L. and Reppert, S.M., 1997, Two period homologs: circadian expression and photic regulation in the suprachiasmatic nuclei. *Neuron*, **19**, 1261-1269.
7. Sun, Z.S., Albrecht, U., Zhuchenko, O., Bailey, J., Eichele, G. and Lee, C.C., 1997, RIGUI, a putative mammalian ortholog of the Drosophila period gene. *Cell*, **90**, 1003-1011.
8. Taguchi, K., Okura, R., Yagita, K., Ishida, Y., Onishi, H., Shigeyoshi, Y., Yamaguchi, S., Nishimura, M., Masubuchi, S. and Okamura, H. 2001, Different circadian expression profiles of clock genes in the mouse suprachiasmatic nucleus. *Biomed. Res.*, in press.
9. Takumi, T., Matsubara, C., Shigeyoshi, Y., Taguchi, K., Yagita, K., Maebayashi, Y., Sakakida, Y., Okumura, K., Takashima, N. and Okamura, H., 1998a, A new mammalian period gene predominantly expressed in the suprachiasmatic nucleus. *Genes Cells*, **3**, 167-176.
10. Takumi, T., Taguchi, K., Miyake, S., Sakakida, Y., Takashima, N., Matsubara, C., Maebayashi, Y., Okumura, K., Takėkida, S., Yamamoto, S., Yagita, K., Yan, L., Young, M.L. and Okamura, H., 1998b, A light independent oscillatory gene *mPer3* in mouse SCN and OVLT. *EMBO J.*, **17**, 4753-4759.
11. Tei, H., Okamura, H., Shigeyoshi, Y., Fukuhara, C., Ozawa, R., Hirose, M. and Sakaki, Y., 1997, Circadian oscillation of a mammalian homologue of the *Drosophila period* gene. *Nature*, **389**, 512-516.
12. Toh, K.L., Jones, C., He, Y., Eide, E., Hinz, W., Virshup, D., Ptacek, L.J. and Fu, Y.-H., 2001, An hPer2 phosphorylationn site mutation in familial advanced sleep phase syndrome. *Science* **291**, 1040-1043.
13. Vielhaber, E., Eide, E., Rivers, A., Gao, Z.-H. and Virshup, D.M., 2000, Nuclear entry of the circadian regulator mPER1 is controlled by mammalian casein kinase 1e. *Mol. Cell. Biol.*, **20**, 4888-4899.
14. Yagita, K., Tamanini, F., van der Horst, G. and Okamura, H., 2001, Molecular mechanisms of the biological clock in cultured fibroblasts. *Science*, **292**, 278-281.
15. Yagita, K., Yamaguchi, S., Tamanini, F., van der Horst, G.T.J., Hoeijmakers, J.H.J., Yasui, A., Loros, J.J., Dunlap, J.C. and Okamura, H., 2000, Dimerization and nuclear entry of mPER proteins in mammalian cells. *Genes Develop.*, **14**, 1353-1363.
16. Zheng, B., Albrecht, U., Kaasik, K., Sage, M., Lu, W., Vaishnav, S., Li, Q., Sun, Z.S., Eichele, G., Bradley, A. and Lee, C.C., 2001, Nonredundant roles of the mPer1 and mPer2 genes in the mammalian circadian clock. *Cell*, **105**, 683-694.
17. Zylka, M.J., Shearman, L.P., Weaver, D.R. and Reppert, S.M., 1998, Three period homolog in mammals: differential light responses in the suprachiasmatic circadian clock and oscillating transcripts outside of brain. *Neuron*, **20**, 1103-1110.

ACTIVATION OF HYPOCRETIN NEURONS AND SLEEP

Véronique Fabre, Patrice Bourgin, J. Gregor Sutcliffe, and Luis de Lecea*

1. THE HYPOCRETINERGIC SYSTEM

The hypocretins (hcrt-1 and hcrt-2), also called orexins (OX-A and OX-B), are neuropeptides first described because of their restricted expression in a subset of hypothalamic neurons (de Lecea et al., 1998). The hypocretin precursor gives raise to two peptides named hypocretin-1 and hypocretin-2, after their sequence similarity to secretin and the prominent expression of preprohypocretin mRNA in the hypothalamus. Independently, Sakurai et al. (1998) identified identical peptides from brain extracts through their ability to bind and activate orphan G-protein coupled receptors. These authors also showed that when injected into the brain ventricles, the peptides stimulate food intake, thus they named them orexins, from the Greek word orexis (appetite).

1.1. Anatomical Distribution Of Hypocretin-containing Neurons

Anatomical studies using *in situ* hybridization and immunohistochemistry indicated that hcrt cell bodies are localized in the lateral hypothalamus and the perifornical area, while hcrt terminals project throughout the brain (Peyron et al., 1998). A dense arborization has been found in several nuclei of the hypothalamus, including the paraventricular nucleus, the arcuate nucleus or the ventromedial nucleus, all known as feeding centers. Outside the hypothalamus, the brainstem shows dense projections, in particular monoaminergic areas such as the locus coreuleus or the raphe nucleus exhibit numerous hcrt fibers. Limbic areas, like the nucleus accumbens or the amygdala, also show dense hcrt innervation. The widespread distribution of hcrt terminals suggests diverse functions for hcrts (Peyron et al., 1998).

* Department of Molecular Biology, MB10. The Scripps Research Institute. 10550 N. Torrey Pines Rd. La Jolla CA 92037

1.2. Hypocretins Bind Two G-Protein Coupled Receptors

Sakurai et al. (1998) identified the hcrt receptors, hcrtr-1 (OX-1) and hcrtr-2 (OX-2), both members of the super family of G-protein coupled receptors. While hcrtr-1 exhibits a higher affinity for hcrt-1, hcrtr-2 binds both hcrt peptides with equal affinity. *In situ* hybridization experiments showed a distinct expression pattern for the two receptors. The hypothalamus reflects such differences since hcrtr-1 is expressed in the ventromedial nucleus, whereas hcrtr-2 mRNA is present in the periventricular and arcuate nuclei. Increases in free intracellular calcium in response to hcrts in neurons or transfected non neural cells suggest an interaction of hcrtr-1 with Gq proteins whereas hcrtr-2 may couple to Gq and/or inhibitory Gi/o proteins (Sakurai et al., 1998).

1.3. Hypocretins Are Neuroexcitatory Peptides

The biological activity of the hcrts was initially described in cultured hypothalamic neurons in which hcrt-2 increases the frequency of post-synaptic events (de Lecea et al., 1998). In the hypothalamus, the hcrts stimulate the firing rate of glutamatergic and GABAergic neurons (van den Pol et al., 1998). The excitatory properties of the hcrts were also reported in the locus coreuleus (LC), a structure that receives a high density of hcrt terminals (Bourgin et al., 2000). Recently, the stimulatory effects of hcrt have been reported in several brain areas such as the raphe nucleus or the laterodorsal tegmentum and dopaminergic neurons in the ventral tegmental area (Hungs and Mignot, 2001).

2. HYPOCRETINS AND SLEEP PHYSIOLOGY

2.1. Narcolepsy Is A Neurodegenerative Disease Of The Hypocretinergic System

While hcrt function was first related to food intake and energy homeostasis and subsequently has been implicated in the regulation of endocrine system and blood pressure (reviewed in Sutcliffe and de Lecea, 2000), the key role of hcrtergic neurotransmission in narcolepsy has been demonstrated by several studies. Thus, a mutation of the gene encoding hcrtr2 is responsible for canine narcolepsy (Lin et al., 1999). In mice, the absence of hcrt expression is associated with cataplectic-like attacks (Chemelli et al., 1999). More recently, a deficit in hcrt expression, detected by *in situ* hybridization (Peyron et al., 2000) and immunohistochemistry (Thannickal et al., 2000), has been reported in brain of human narcoleptics. Despite all these recent advances on the etiology of narcolepsy, the mechanisms by which reduction of hcrt activity causes sleep dysfunction remain to be determined.

2.2. Hypocretins And Arousal

The role of hcrts in the regulation of sleep/wakefulness cycles was first assessed after intracerebroventricular administration of the hcrt peptides. In rats, injection of Hcrt-1 increases the time spent awake and strongly inhibits REM sleep (Hagan et al., 1999). Several brain structures are known to regulate REM sleep and wakefulness, in particular REM-off neurons possess the characteristics to be highly active during wakefulness, while their activity becomes low or quiescent during REM sleep. To understand the arousal properties of hcrts, it is important to identify the structure(s) mediating hcrt action and REM-off structures are attractive candidates.

2.3. The Locus Coeruleus Is A Key Target Of The Hypocretinergic System

Noradrenergic neurons of the LC belong to the anatomical circuitry proposed by Mc Carley and Hobson to be responsible for REM sleep regulation. This model is based on the hypothesis of reciprocal inhibitory interactions between REM-off cells, noradrenegic, serotonergic and histaminergic neurons, and REM-on cells, including cholinergic neurons of the pons. During sleep, a decrease in monoaminergic tone raises the inhibitory control of cholinergic REM-on neurons generating REM sleep.

Several arguments suggest that noradrenergic neurons of the LC may be a key target for hcrts. The LC receives a dense innervation from hcrt neurons of the hypothalamus. The activity of noradrenergic neurons is increased in the presence of hcrts. Moreover, during cataplectic episodes, noradrenergic cells of the LC cease to fire. We demonstrated the ability of local administration in the LC of hcrt-1, but not hcrt-2, to decrease REM sleep amount and to increase wakefulness in a dose-dependent manner. This effect is site-specific, since hcrt-1 administration does not reduce REM sleep in areas near the LC that are important for sleep regulation. Hcrt-1 mediates REM sleep inhibition through the activation of noradrenergic neurons, as indicated by electrophysiological recording and increased *c-fos* expression after hcrt-1 infusion in the LC. Double labeling with antibodies directed against the tyrosine hydroxylase and hcrtr-1-like immunoreactivity indicated that noradrenergic neurons of the LC express hcrtr-1, leading to the conclusion that hcrtr-1 is the main target of hcrt innervation in the LC (Bourgin et al., 2000).

These data strongly support a key role of noradrenergic neurons in mediating the effect of hcrt on the sleep/wakefulness cycle. The neuroexcitatory properties of hcrts in several REM-on/REM-off structures place the hcrtergic system as a main player of arousal and REM sleep regulation.

3. NEW APPROACHES TO ASSESS THE ROLE OF HYPOCRETINERGIC NEURONS

3.1. Transgenic Mice Expressing Cholera Toxin in Hcrt Neurons

To determine the role of hypocretin activation *in vivo*, we have generated transgenic mice with constitutively active hcrt neurons. Cholera toxin is known to activate stimulatory G-protein, increase cAMP levels and neural activity. The sequence encoding the intracellular enzymatic subunit of the cholera toxin (CT-A1) has been used to stimulate neuronal activity in transgenic mice (Burton et al., 1991). To drive the expression of CT-A1 in hcrt cells, we first isolated a 7 kb fragment corresponding to the hcrt promoter from a mouse genomic library. The transgene construct encompasses a 7 kb hcrt mouse promoter fragment followed by the coding sequence of the CT-A1 and a polyadenylation cassette. The hcrt promoter confers spatially restricted expression of the toxin in hcrt neurons. CT-A1 mRNA positive cells colocalize with hcrt neurons, as identified by combined in situ hybridization to CT and immmunohistochemistry to hcrt peptides. In contrast, MCH, which is expressed by a distinct neuronal population in the lateral hypothalamus, does not co-localize with CT-A1 mRNA. Expression of CT-A1 in hcrt cells will constitutively activate hcrt neurons, which may induce release of hcrts and the neurotransmitters co-localized with these peptides.

3.2. Transgenic Mice Expressing Diphtheria Toxin In Hcrt Neurons

Hypocretin-containing neurons are located in a hypothalamic region involved in the regulation of energy balance. Hypocretin-containing neurons are sensitive to glucose, express leptin receptors and are innervated by NPY-containing terminals from the arcuate nucleus, suggesting that they have a role as homoeostatic "sensors". To determine the role of hcrt cells in homoeostatic control, we have used the same promoter to drive the expression of attenuated diphtheria toxin, and ablate hcrt-expressing cells. Analysis of these mice will give important information about the role of the lateral hypothalamus as an integrator of physiological information.

REFERENCES

Bourgin, P., Huitron-Resendiz, S., Spier, A. D., Fabre, V., Morte, B., Criado, J. R., Sutcliffe, J. G., Henriksen, S. J., and de Lecea, L., 2000, Hypocretin-1 modulates rapid eye movement sleep through activation of locus coeruleus neurons, *J Neurosci.* **20**:7760-7765.

Burton, F. H., Hasel, K. W., Bloom, F. E. and Sutcliffe, J. G., 1991, Pituitary hyperplasia and gigantism in mice caused by a cholera toxin transgene, *Nature.* **350** (6313): 74-7.

Chemelli, R. M., Willie, J. T., Sinton, C. M., Elmquist, J. K., Scammell, T., Lee, C., Richardson, J. A., Williams, S. C., Xiong, Y., Kisanuki, Y., Fitch, T.E., Nakazato, M., Hammer, R.E., Saper, C.B., and Yanagisawa, M., 1999, Narcolepsy in orexin knockout mice: molecular genetics of sleep regulation, *Cell.* **98**:437-451.

de Lecea, L., Kilduff, T. S., Peyron, C., Gao, X., Foye, P. E., Danielson, P. E., Fukuhara, C., Battenberg, E. L., Gautvik, V. T., Bartlett, F. S., Frankel, W. N., van den Pol, A. N., Bloom, F. E., Gautvik, V. T. and K. M., Sutcliffe, J. G., 1998, The hypocretins: hypothalamus-specific peptides with neuroexcitatory activity, Proc *Natl Acad Sci U S A.* **95**:322-327.

Hagan, J. J., Leslie, R. A., Patel, S., Evans, M. L., Wattam, T. A., Holmes, S., Benham, C. D., Taylor, S. G., Routledge, C., Hemmati, P., Munton, R. P., Ashmeade, T. E., Shah, A. S., Hatcher, J. P., Hatcher, P. D., Jones, D. N., Smith, M. I., Piper, D. C., Hunter, A. J., Porter, R. A. and Upton, N.,1999, Orexin A activates locus coeruleus cell firing and increases arousal in the rat. *Proc Natl Acad Sci U S A.* **96** (19): 10911-6.

Hungs, M., and Mignot, E., 2001, Hypocretin/orexin, sleep and narcolepsy, *Bioessays.* **23**:397-408.

Lin, L., Faraco, J., Li, R., Kadotani, H., Rogers, W., Lin, X., Qiu, X., de Jong, P. J., Nishino, S., and Mignot, E., 1999, The sleep disorder canine narcolepsy is caused by a mutation in the hypocretin (orexin) receptor 2 gene, *Cell.* **98**:365-376.

Peyron, C., Tighe, D. K., van den Pol, A. N., de Lecea, L., Heller, H. C., Sutcliffe, J. G. and Kilduff, T. S., 1998, Neurons containing hypocretin (orexin) project to multiple neuronal systems, *J Neurosci.* 18: 9996-10015.

Peyron, C., Faraco, J., Rogers, W., Ripley, B., Overeem, S., Charnay, Y., Nevsimalova, S., Aldrich, M., Reynolds, D., Albin, R., Li, R., Hungs, M., Pedrazzoli, M., Padigaru, M., Kucherlapati, M., Fan, J., Maki, R., Lammers, G. J., Bouras, C., Kucherlapati, R., Nishino, S., and Mignot, E., 2000, A mutation in a case of early onset narcolepsy and a generalized absence of hypocretin peptides in human narcoleptic brains, *Nature Med.* **6**:991-997.

Sakurai, T., Amemiya, A., Ishii, M., Matsuzaki, I., Chemelli, R. M., Tanaka, H., Williams, S. C., Richardson, J. A., Kozlowski, G. P., Wilson, S., Arch, J. R., Buckingham, R. E., Haynes, A. C., Carr, S. A., Annan, R. S., McNulty, D. E., Liu, W. S., Terrett, J. A., Elshourbagy, N. A., Bergsma, D. J., and Yanagisawa, M., 1998, Orexins and orexin receptors: a family of hypothalamic neuropeptides and G protein-coupled receptors that regulate feeding behavior, *Cell.* **92**:573-585

Sutcliffe, J.G. and de Lecea, L. 2000. The hypocretins: excitatory neuromodulatory peptides for multiple homeostatic systems, including sleep and feeding. *J. Neurosci. Res.* **62**:161-168

Thannickal, T. C., Moore, R. Y., Nienhuis, R., Ramanathan, L., Gulyani, S., Aldrich, M., Cornford, M., and Siegel, J. M., 2000, Reduced number of hypocretin neurons in human narcolepsy, *Neuron.* **27**:469-474.

van den Pol, A. N., Gao, X. B., Obrietan, K., Kilduff, T. S., and Belousov, A. B., 1998, Presynaptic and postsynaptic actions and modulation of neuroendocrine neurons by a new hypothalamic peptide, hypocretin/orexin, *J Neurosci.* **18**:7962-7971

N-METHYL-D-ASPARTATE RECEPTORS IN OPIOID DEPENDENCE AND WITHDRAWAL

Rob Rockhold[1], Hong Zhu[1], Niansen Liu[2] and Ing K. Ho[1]

1. INTRODUCTION

The etiology of neurochemical, electrophysiological and behavioral expressions that accompany, indeed define, withdrawal from dependence on opioids remains elusive. However, recent evidence has emphasized three key features of opioid withdrawal. First, increased firing rates (hyperactivity) of noradrenergic neurons in the locus ceruleus correlate strongly with behavioral signs and symptoms of withdrawal (Rasmussen *et al.*, 1990). Second, the majority of this hyperactivity (as much as 80%) is elicited as a result of increases in afferent excitatory glutamatergic input to these neurons (Akaoka and Aston-Jones, 1991; Aston-Jones *et al.*, 1997). Third and finally, the predominant (but not necessarily the sole) source for this excitatory input is the nucleus paragigantocellularis (PGi) in the ventrolateral medulla (Akaoka and Aston-Jones, 1991). The receptor or receptors that mediate the glutamatergic excitatory input have yet to be resolved, although the N-methyl-D-aspartate (NMDA) ionotropic glutamate subtype has been implicated. The results presented here summarize our evidence for an involvement of NMDA receptors in opioid dependence and withdrawal phenomena (Zhu and Ho, 1998; Zhu *et al.*, 1999). Furthermore, as our laboratories have recently identified a pattern of opioid withdrawal-like behaviors that can be elicited following discrete electrical stimulation of the PGi (Liu *et al.*, 1999a, b), this communication will present the term, opioid withdrawal-like behavior, review the evidence demonstrating mediation of these responses by glutamatergic neurotransmission within the locus ceruleus, and identify questions concerning putative excitatory opioid receptor involvement in the behavioral effects of PGi stimulation.

---

[1] Department of Pharmacology and Toxicology, University of Mississippi Medical Center (UMC), Jackson, MS 39216-4505. [2] Department of Psychiatry and Human Behavior, UMC.

2. METHODS

Adult male Sprague-Dawley (250-300g) rats were used in all studies. Intracerebroventricular (i.c.v.) drug administration was accomplished following surgical implantation of a stainless steel cannula (21 or 26 gauge, depending on study) into a right lateral cerebral ventricle (Zhu and Ho, 1998; Zhu *et al.*, 1999; Liu *et al.*, 1999b). Opioid dependence was produced by i.c.v. infusion of morphine sulfate (26 nmoles/μL/h, for three days) using an osmotic minipump (Alzet 2001) implanted s.c. at the nape of the neck. Behavioral assessments of withdrawal from opioid dependence were determined over 30 min. periods in which a battery of standard behavioral signs and symptoms (such as jumping, rearing, wet dog shakes, teeth chattering, scratching, stretching, penis licking, vocalization, diarrhea) were recorded to provide withdrawal scores. Withdrawal was precipitated by interruption of morphine infusion, followed two h later with an i.c.v. injection of 48 nmoles/5 μL over a two minute period of an opioid antagonist (naloxone, naltrindole, nor-binaltorphimine, β-funaltrexamine). Phosphothiorated antisense (5'-CAGCAGGTGCATGGTGCT-3') and sense oligonucleotides were obtained from Midland Certified Reagent Co. These agents were injected i.c.v., twice daily in doses of 15 nmoles/5 μL for 6 days, beginning 3 days prior to initiation of dependence on morphine by i.c.v. morphine infusions. Evaluation of withdrawal responses and determination of alterations in the channel binding site of the NMDA receptor were determined by terminating the i.c.v. morphine infusion, waiting two hours and then injecting the opioid antagonist, naloxone, i.c.v. Binding of [^{3}H]MK-801 ([3-^{3}H](+)-5-methyl-10,11-dihydro-5H-dibenzo(*a,d*)cyclohepten-5,10-imine; DuPont NEN) was determined in membrane fractions obtained from frontal cortex of rats decapitated immediately following behavioral tests (conducted 30 min. after i.c.v. injection of naloxone) in morphine dependent rats. Non-specific binding was determined in the presence of 100 μM unlabeled MK-801. Hybridization of [α-^{35}S] dATP 3' end-labeled oligodeoxynucleotide probes to the NMDA-NR1, -NR2A and -NR2B receptor subunits (DuPont-NEN) was accomplished using 14 μm frozen brain sections taken from rats following 3 days of i.c.v. morphine infusion (Zhu *et al.*, 1999). Thus, animals were in a state of dependence and not in withdrawal. The signals were measured using Hyperfilm-βmax (Amersham) film and a digital scanning densitometer (Personal Densitometer, Molecular Dynamics). Separate groups of rats were prepared, each with a chronic unilateral microdialysis guide cannula into the left locus ceruleus and a concentric bipolar stimulating electrode (SNEX100, Rhodes) in the ipsilateral PGi. Some animals were prepared with only a stimulating electrode and an i.c.v. injection cannula for administration of opioid antagonists. Electrical stimulation was provided by a Medical Systems Corp. stimulator (S-90) at frequencies of 0.5-50 Hz, pulse duration of 1 ms, and 0-1 volts. Stimulation trains were delivered for 30 min. periods. Glutamate was measured by HLPC-EC (BAS 200B, Bioanalytical Systems, Inc.) using the method of Ellison *et al.* (1987).

3. RESULTS

Antisense oligonucleotide, but not saline or sense oligonucleotide, significantly attenuated six (jumping, rearing, stretching, teeth chattering, vocalization, penis licking) of nine observed withdrawal signs. Treatment with antisense oligonucleotide, but not sense oligonucleotide or saline, significantly reduced the B_{max} of [^{3}H]MK-801 binding by approximately 13% ($p<0.05$) with no significant changes in K_d. Significant increases (21-28%; $p<0.05$) in the level of the NR1 subunit mRNA were found in the locus ceruleus and the hypothalamic paraventricular nucleus following 3 days of i.c.v. morphine infusion through osmotic minipumps. No changes were detected in expression of the NR1 mRNA in the frontal cortex, caudate-putamen, nucleus accumbens, amygdala, CA1, CA2, or in either the dentate gyrus of the hippocampus, or the central grey after morphine treatment. The expression of NR2A and NR2B subunit mRNAs did not change after morphine treatment in any brain region. A characteristic series of opioid withdrawal-like behaviors (rearing, teeth-chattering, wet-dog shakes, etc.) and increases in extracellular glutamate concentrations within the locus ceruleus were evoked, in a frequency-dependent (0.5-50 Hz) manner, during 30 min. periods of PGi stimulation. The profile of these behaviors resembled qualitatively, but was not quantitatively identical, with that seen during precipitated withdrawal from opioid dependence. This behavioral syndrome has been termed, opioid withdrawal-like behavior. Tolerance to repeated stimulation of the PGi did not develop following a series of 30 min. runs of stimulation over 3.5 h. Reverse dialysis perfusion of the locus ceruleus with the nonspecific glutamate receptor antagonist, kynurenic acid (0.1, 1 mM), reduced the intensity of stimulation-induced behaviors by roughly 50%, but had no effect on the corresponding increases in glutamate concentrations. Similar perfusion with the glutamate transporter inhibitor, l-*trans*-pyrrolidine dicarboxylic acid, at 1, but not at 0.1, mM significantly increased glutamate levels in dialysates. Neither concentration of the transporter inhibitor altered the behavioral score. Intracerebroventricular injections of the nonselective opioid antagonist, naloxone, significantly decreased (by 40-50%) the intensity of stimulation-induced behavioral responses, as did injections of either the μ-selective (β-funaltrexamine) or the δ-selective (naltrindole) opioid antagonists. In contrast, similar i.c.v. injections of the κ-selective antagonist, nor-binaltorphimine, did not block behavioral responses to PGi stimulation.

4. DISCUSSION

A role for glutamate receptors, and in particular the NMDA subtype, has been supported by several lines of evidence (see Trujillo and Akil, 1995), including blockade of opioid withdrawal in dependent animals with the use of selective glutamate receptor antagonists, a similar inhibition of withdrawal following elimination of excitatory amino acid afferents to the locus ceruleus, and demonstration of a correlation between neuronal hyperactivity in the locus ceruleus, concentrations of extracellular glutamate, and behavioral withdrawal responses. Our data, using antisense probes against the obligatory NR1 subunit of the NMDA receptor, further support this involvement and have the additional advantage of blocking NMDA receptor function with a pharmacological tool that does not exert the non-specific behavioral side effects commonly associated with several classes of NMDA receptor antagonists. In addition, increases in expression of an essential NMDA receptor subunit were detected in rats made dependent upon morphine. The fact that these increases were noted only in the locus ceruleus and the hypothalamic paraventricular nucleus suggests that the influence of the NMDA receptor is exceptionally region specific, a finding that should implicate the existence of highly localized neural circuits involving this receptor. Data obtained following electrical stimulation of the PGi, from which the principal excitatory glutamatergic input to the locus ceruleus is derived (Ennis and Aston-Jones, 1988), increasingly substantiate glutamatergic mediation of behavioral withdrawal. Inhibition of withdrawal-like behaviors following reverse dialysis application of the non-selective glutamate receptor blocker, kynurenic acid, and correlation of the intensity of behavioral response with locus ceruleus efflux of glutamate strongly support a fundamental glutamatergic link. More curious, and very intriguing, is the finding that antagonists selective for μ- and δ-, but not κ-opioid, receptors can reduce PGi stimulation evoked opioid withdrawal-like behaviors. We are unaware of excitatory opioidergic influences at the level of the locus ceruleus, or elsewhere in the brain, that might explain this behavior. Since the antagonists were applied i.c.v., it is possible that stimulation of the PGi activates afferents to other brain regions and that trans-synaptic effects trigger endogenous opioid peptide-mediated dysinhibitory events, resulting in the otherwise paradoxical findings. The results are consistent with a fundamental involvement of the glutamate/NMDA receptor in mediation of opioid dependence and in associated withdrawal phenomena. They also demonstrate unexpected complexities to the behavioral withdrawal responses that open new vistas to the neuropharmacology of opioid dependence.

REFERENCES

Aston-Jones, G., Hirata, H., and Akaoka, H., 1997. Local opiate withdrawal in locus coeruleus in vivo, Brain Res. **765**: 331-336.

Akaoka, H., and Aston-Jones, G., 1991. Opiate withdrawal-induced hyperactivity of locus coeruleus neurons is substantially mediated by augmented excitatory amino acid input. J. Neurosci. **11**: 3830-3839.

Ellison, D.W., Beal, M.F., and Martin, J.B., 1987. Amino acid neurotransmitters in postmortem human brain analyzed by high performance liquid chromatography with electrochemical detection. J. Neurosci. Methods **19**: 305-315.

Ennis, M., and Aston-Jones, G., 1988. Activation of locus coeruleus from nucleus paragigantocellularis: A new excitatory amino acid pathway in brain. J. Neurosci. **8**: 3644-3657.

Liu, N., Rockhold, R. W., and Ho, I. K., 1999a. Electrical stimulation of nucleus paragigantocellularis induces opioid withdrawal-like behaviors in the rat. Pharmacol. Biochem. Beh. **62**:263-271.

Liu, N., Ho, I.K., and Rockhold, R.W. , 1999b. Contribution of glutamatergic systems in locus coeruleus to nucleus paragigantocellularis stimulation-evoked behavior. Pharmacol. Biochem. Beh. **63**: 555-567.

Rasmussen, K., Beitner-Johnson, D.B., Krystal, J.H., Aghajanian, G.K., Nestler, E.J., 1990. Opiate withdrawal and the rat locus coeruleus: behavioral, electrophysiological, and biochemical correlates. J. Neurosci., **10**: 2308-2317.

Trujillo, K.A., and Akil, H., 1995. Excitatory amino acids and drug abuse: a role for N-methyl-D-aspartate receptors in drug dependence, sensitization and physical dependence. Drug Alcohol Depend. **38**: 139-154.

Zhu, H., and Ho, I.K., 1998. NMDA-R1 antisense attenuates morphine withdrawal behaviors. Eur. J. Pharmacol. **352**: 151-156.

Zhu, H., Jang, C.-G., Ma, T., Oh, S., Rockhold, R. W. and Ho, I. K., 1999. Region specific expression of NMDA receptor NR1 subunit mRNA in hypothalamus and pons following chronic morphine treatment. Eur. J. Pharmacol., **365**:47-54.

THE MECHANISMS OF MORPHINE DEPENDENCE AND IT'S WITHDRAWAL SYNDROME: STUDY IN MUTANT MICE

Catecholamines and addiction

Yukihiro Noda and Toshitaka Nabeshima*

1. INTRODUCTION

The catecholaminergic systems appear to mediate some of physical dependence of opiates.[1] For example, clonidine, a drug that decreases noradrenergic activity, prevents behavior induced by morphine withdrawal.[2] An increase in the noradrenergic neuron firing rate in the locus coeruleus (LC) [3] and an increase in the turnover of noradrenaline (NA) have been reported during naloxone-precipitated morphine withdrawal.[4]

Recently, the increase in knowledge of intracellular messenger pathways has provided an experimental framework for studies of the molecular mechanisms underlying drug addictions. Investigations have demonstrated that changes in the activity of G-protein and cyclic AMP (cAMP) second messenger and the protein phosphorylation pathway mediate important aspects of opiate addiction in a number of drug-responsive brain regions.[5] Adaptations in the cAMP signal transduction pathway underlie the mechanisms of opioid tolerance and dependence, and up-regulation of these components plays an important role in the onset of the withdrawal syndrome.[6-8] In particular, the activation of the transcription factor cAMP response element (CRE) binding protein (CREB) is implicated in withdrawal syndrome.[9]

It is well known that repeated exposure to opiates such as morphine results in dependence and tolerance. The development of dependence and tolerance has been postulated to be an attempt by the body to compensate for the chronic existence of opiates, however, the exact mechanisms underlying these physiological changes have not been elucidated *in vivo*.

As stated above, a key role for the catecholaminergic and cAMP signal pathways in the development of morphine dependence has been suggested. Here, we report the involvement of both signal pathways in morphine dependence by using two mutant mice; tyrosine hydroxylase (TH) heterozygous mice,[10] in which the catecholamine synthetic

---

* Toshitaka Nabeshima, Department of Neuropsychopharmacology and Hospital Pharmacy, Nagoya University Graduate School of Medicine. 65, Tsuruma-cho, Showa-ku, Nagoya 466-8560, Japan.

pathway is impaired, and CREB binding protein (CBP) heterozygous mice,[11] in which gene transduction through the cAMP pathway is impaired. First, we investigated whether morphine dependence develops in the mutant mice. Second, we examined whether cAMP signal pathways contribute to morphine dependence.

2. CATECHOLAMINERGIC PATHWAYS AND MORPHINE DEPENDENCE

Physical dependence on opiate drugs is characterized by a withdrawal syndrome following either the abrupt termination of morphine intake or precipitated by the administration of a narcotic antagonist. When the wild-type mice pretreated with morphine (10 mg/kg s.c.) twice a day for 5 days were challenged with naloxone (5 mg/kg i.p.), they showed increases of jumping, rearing and forepaw tremor counts as signs of withdrawal (Table 1). In the TH heterozygous mice pretreated with morphine, however, the frequency of jumping and forepaw tremor was significantly lower than in wild-type mice after naloxone challenge (Table 1). These results suggest that the catecholaminergic signal pathway is involved in the naloxone-precipitated morphine withdrawal syndrome.

The LC represents the largest cluster of noradrenergic neurons in the brain.[12] This nucleus possesses a high density of opioid receptors, particularly of the mu and kappa subtypes.[13] Experimental evidence suggests functional interactions in the LC between opioid and noradrenergic systems during opiate withdrawal. A noradrenergic hyperactivity in the LC has been hypothesized to mediate the expression of some components of the morphine withdrawal syndrome. Supporting this hypothesis, an increase in the noradrenergic neuron firing rate in the LC[3] and an increase in the turnover of NA have been reported during naloxone-precipitated morphine withdrawal.[4] Further, clonidine, a drug that decreases noradrenergic activity, prevents behaviors induced by morphine withdrawal.[2] On the other hand, although the dopaminergic systems are a neuronal substrate of opiate-induced reward,[14] it is debatable whether withdrawal shares common neuronal mechanisms with opiate reward. Bozarth and Wise[15] proposed that opiate dependence and withdrawal are mediated by the periaquedutal gray, and Koob et al.[16, 17] have suggested that the nucleus accumbens is a substrate for the aversiveness of opiate withdrawal. In mice lacking dopamine-D2 receptors, further, the behavioral expression of morphine withdrawal was unchanged.[18] There is very little information to date on the specific effects of opiate withdrawal on dopaminergic systems. Thus, noradrenergic, rather than dopaminergic, systems may play an important role in the behavioral expression of morphine withdrawal.

Table 1. Morphine dependence in TH and CBP heterozygous mice

Mice	Withdrawal Syndrome	cAMP levels	CREB protein levels	c-fos mRNA levels
Wild	+	↑	↑	↑
TH heterozygous	±	±	nd	±
CBP heterozygous	±	↑	nd	nd

The numbers of escape jumping, rearing and forepaw tremor were recorded as withdrawal syndrome. TH: tyrosine hydroxylase, CREB: cyclic AMP (cAMP) response element binding protein, CBP: CREB binding protein. +: expression, ±: no development/unchange, ↑: increase, nd: not determined.

3. CYCLIC AMP SIGNAL PATHWAYS AND MORPHINE DEPENDENCE

Several lines of evidence indicate that an up-regulation of intracellular cAMP systems is responsible for the development of morphine dependence. If a compensatory up-regulation of cAMP systems following chronic morphine exposure is essential for the development of dependence, morphine dependence may be ensured by inhibiting alterations of the cAMP systems induced by chronic morphine exposure. In the present biochemical study, the wild-type mice experiencing the naloxone-precipitated morphine withdrawal syndrome showed an elevation of cAMP and c-foc mRNA levels in the thalamus (Table 1). However, the TH heterozygous mice, which did not undergo the naloxone-precipitated withdrawal showed no such change of cAMP and c-foc mRNA levels in the thalamus (Table 1). Therefore, our results suggest that naloxone-precipitated morphine withdrawal syndrome is mediated by the up-regulation of intracellular cAMP, and there is no increase in the cAMP level in the TH heterozygous mice because they lack catecholaminergic-cAMP signal pathways.

Recently, Nestler[19] suggested that the regulation of gene expression is involved in the up-regulation of cAMP systems in the development of morphine dependence. The CRE is a transcription factor, which is activated in response to a second messenger stimulus and mediates many effects of cAMP on gene expression.[20] It has been demonstrated that acute morphine treatment decreases CREB phosphorylation and withdrawal of morphine after chronic exposure increases it.[9] The development of morphine dependence is attenuated in CREB deficient mice.[21] More recently, direct evidence of a role for CREB in mediating up-regulation of cAMP systems in the morphine dependent state was obtained by Lane-Ladd et al.[22] Although the CBP heterozygous mice pretreated with morphine had elevated levels of cAMP in the thalamus following naloxone challenge (Table 1), the frequency of the symptoms of naloxone-precipitated morphine withdrawal was significantly lower compared to wild-type mice (Table 1). Thus, these findings suggest that the expression of transcription factors in the Fos/Jun family mediated by CREB phosphorylation is not high enough in the CBP heterozygous mouse. Further, it can be considered that CBP heterozygous mice developed fewer symptoms of morphine withdrawal due to inhibition of the alteration of CREB phosphorylation resulting from a continuous reduction in the expression level of related genes. The cAMP signal pathway including the expression of family of genes encoding transcription factors may play an important role in the expression of morphine withdrawal. The mechanisms for transcriptional activation/repression are not fully understood, but likely involve the direct or indirect interaction of transcription factors with the RNA polymerase II transcription complex.[23] Further studies are needed to elucidate the role of CBP in the morphine withdrawal.

In conclusion, our findings suggest that the alteration of catecholaminergic biosynthetic and cAMP signal pathways plays a key role in the development of morphine dependence and further that the expression of genes mediated by phosphorylation CREB is involved in the development of morphine withdrawal.

4. ACKNOWLEDGEMENT

These studies were performed in collaboration with Drs. T. Mamiya, X. Ren, and Mrs. T. Nagai in our laboratory. The TH and CBP heterozygous mice were established by Drs. K. Kobayashi (Fukushima Medical University), T. Nagatsu and their collaborators (Fujita

Health University) and by Drs. Y. Oike, K. Yamamura and their collaborators (Kumamoto University), respectively. These works were supported partly by a Grant-in-Aid for Scientific Research (10044260, 12922097) from the Ministry of Education, Science, Sports and Culture of Japan, by Special Coordination Funds for Promoting Science and Technology, the Target-oriented Brain Science Research Program, from the Ministry of Science and Technology of Japan and by Health Scientific Research Grants for Research on Pharmaceutical and Medical Safety from the Ministry of Health and Welfare of Japan.

REFERENCES

1. G.F. Koob and F.E. Bloom, Cellular and molecular mechanisms of drug dependence, *Science*. **242**, 715-723 (1988).
2. L.F. Tseng, H.H. Loh, and E.T. Wei, Effects of clonidine on morphine withdrawal signs in the rat, *Eur. J. Pharmacol.* **30**, 93-99 (1975).
3. G.K. Aghajanian, Tolerance of locus coeruleus neurons to morphine and suppression of withdrawal response by clonidine, *Nature* **267**, 186-188 (1978).
4. R. Laverty and R.H. Roth, Clonidine reverses the increased norepinephrine turnover during morphine withdrawal in rats, *Brain Res.* **182**, 482-485 (1980).
5. E.J. Nestler, Molecular mechanisms of drug addiction, *J. Neurosci.* **12**, 2439-2450 (1992).
6. I. Matsuoka, R. Maldonado, N. Defer, F. Noel, J. Hanoune, and B.P. Roques, Chronic morphine administration causes region-specific increase of brain type VIII adenylyl cyclase mRNA, Eur. J. Pharmacol. **268**, 215-221 (1994).
7. E.J. Nestler, B.T. Hope, and K.L. Widnell, Drug addiction: a model for the molecular basis of neural plasticity, Neuron. **11**, 995-1006 (1993).
8. R.Z. Terwilliger, D. Beitner-Johnson, K.A. Sevarino, S.M. Crain, and E.J. Nestler, A general role for adaptations in G-proteins and the cyclic AMP system in mediating the chronic actions of morphine and cocaine on neuronal function, *Brain Res.* **548**, 100-110 (1991).
9. X. Guitart, M.A. Thompson, C.K. Mirante, M.E. Greenberg, and E.J. Nestler, Regulation of cyclic AMP response element-binding protein (CREB) phosphorylation by acute and chronic morphine in the rat locus coeruleus, *J. Neurochem.* **58**, 1168-1171 (1992).
10. K. Kobayashi, S. Morita, H. Sawada, T. Mizuguchi, K. Yamada, I. Nagatsu, T. Hata, Y. Watanabe, K. Keisuke, and T. Nagatsu, Targeted disruption of the tyrosine hydroxylase locus results in severe catecholamine depletion and perinatal lethality in mice, *J. Biol. Chem.* **270**, 27235-27243 (1995).
11. Y. Oike, A. Hata, T. Mamiya, T. Kaname, Y. Noda, M. Suzuki, H. Yasue, T. Nabeshima, K. Araki, and K. Yamamura, Truncated CBP protein leads to classical Rubinstein-Taybi syndrome phenotypes in mice: Implication of a dominant negative mechanism, *Human Mol. Genet.* **8**, 387-396 (1999).
12. S.L. Foote, F.E. Bloom, and G. Aston-Jones, Nucleus locus ceruleus: new evidence of anatomical and physiological specificity, *Physiol. Rev.* **63**, 844-914 (1983).
13. A. Tempel and R.S. Zukin, Neuroanatomical patterns of the mu, delta, and kappa opioid receptors of rat brain as determined by quantitative in vitro autoradiography, *Proc. Natl. Acad. Sci. USA.* **84**, 4308-4312 (1987).
14. R.A. Wise, Opiate reward: sites and substrates, *Neurosci. Biobehav. Rev.* **13**, 129-133 (1989).
15. M.A. Bozarth and R.A. Wise, in: *Problem of Drug Dependence,* edited by L.S. Harris (NIDA Research Monograph 43, Washington DC, 1982), pp. 171-177.
16. G.F. Koob, T.L. Wall, and F.E. Bloom, Nucleus accumbens as a substrate for the aversive stimulus effects of opiate withdrawal, *Psychopharmacology (Berl).* **98**, 530-534 (1989).
17. L. Stinus, M. Le Moal, and G.F. Koob, Nucleus accumbens and amygdala are possible substrates for the aversive stimulus effects of opiate withdrawal, *Neuroscience.* **37**, 767-773 (1990).
18. R. Maldonado, A. Saiardi, O. Valverde, T.A. Samad, B.P. Roques, and E. Borrelli, Absence of opiate rewarding effects in mice lacking dopamine D2 receptors, Nature. **388**, 586-589 (1997).
19. E.J. Nestler, The brain on opiates, *Neuron.* **16**, 897-900 (1996).
20. M.R. Montminy, G.A. Gonzalez, and K.K. Yamamoto, Regulation of cAMP-inducible genes by CREB, *Trends Neurosci.* **13**, 184-188 (1990).
21. R. Maldonado, J.A. Blendy, E. Tzavara, P. Gass, B.P. Roques, and J. Hanoune, Reduction of morphine abstinence in mice with a mutant in the gene encoding CREB, *Science.* **273**, 657-659 (1996).
22. S.B. Lane-Ladd, J. Pineda, V.A. Boundy, T. Pfeuffer, J. Krupinski, G.K. Aghajanian, and E.J. Nestler, CREB (cAMP response element-binding protein) in the locus coeruleus: biochemical, physiological, and behavioral evidence for a role in opiate dependence, *J. Neurosci.* **17**, 7890-7901 (1997).
23. M. Ptashne and A.A. Gann, Activators and targets, *Nature.* **346**, 329-331 (1990).

INVOLVEMENT OF ENHANCED GENE EXPRESSION OF ANTI-OPIOID SYSTEMS IN MORPHINE TOLERANCE AND DEPENDENCE

Akira Yoshida, Makoto Inoue, and Hiroshi Ueda*

1. INTRODUCTION

Chronic morphine application induces such side effects as tolerance and physical dependence, which have been speculated to be developed through neuronal plasticity of the central nervous system. Based on the assumption that the reduction of morphine analgesia during chronic treatment is mediated through anti-opioid systems in brain, several attempts have been done to characterize morphine analgesic tolerance by use of blocking agents for anti-opioid systems.[1] Glutamaterigic system using NMDA receptor (NMDAR) was the first to be demonstrated as an anti-opioid system to develop morphine tolerance and dependence.[2] On the other hand, nociceptin/orphanin FQ (N/OFQ) was discovered to be an endogenous ligand for opioid receptor-like orphan receptor, ORL1, which is now called NOPR.[3,4] From the initial pharmacological studies N/OFQ has been reported to be an anti-opioid peptide.[5,6] Using this idea, we have firstly clarified that N/OFQ system through NOPR is also involved in the development of morphine tolerance and dependence, by use of $NOPR^{-/-}$ mice, which lack the gene encoding NOPR.[7,8] However, as it is often observed that knock-out mice develop some compensatory changes during development and growth, we have studied the involvement of NOPR in such mechanisms by use of specific antagonist (J-113397).[9] Here, we report the characterization and their possible molecular mechanisms of both NMDAR and NOPR systems involved in the development of morphine tolerance and dependence.

2. MATERIALS AND METHODS

Male ddY mice or mutant mice weighing 20 - 22 g were used. Procedures were approved by Nagasaki University Animal Care Committee and complied with the recommendations of the International Association for the Study of Pain.[10] For the

*Department of Molecular Pharmacology and Neuroscience, Nagasaki University School of Pharmaceutical Sciences, Nagasaki 852-8521, Japan.

evaluation of nociceptive and analgesic activities, Haffner tail-pinch assay and tail flick assay were used. Details including the schedules of morphine and naloxone injections for tolerance and dependence experiments were as previously reported.[11] The gene expression of NOPR was measured by RT-PCR after the isolation of total RNA from various brain regions and spinal cord of mice.[8] The expression of ε_1-subunit in cortical or hippocampal cells was analyzed by flow cytometric analysis after dissociation of cells by papain,[12] fixation and staining with anti-ε_1-subunit antibody, as reported previously.[13]

3. RESULTS

We investigated the action of J-113397 on the morphine tolerance. The intrathecal (i.t.) injection of J-113397 (1 nmol) 60 min prior to subcutaneous (s.c.) morphine injection did not affect the morphine analgesia assessed by tail pinch test in naive ddY mice. Repeated treatments with morphine showed tolerance on the 6th day, and the i.t. treatment with J-113397 significantly attenuated the tolerance. Similar results were observed in tail-flick test. In this test, however, the i.t. injection of J-113397 completely abolished the tolerance, but the intracerebroventricular injection of this compound had no effect. Collectively, these findings indicate that spinal N/OFQ or NOPR system develops anti-opioid plasticity observed on morphine tolerance and dependence.

Naloxone-precipitated withdrawal symptoms, such as jumping, paw tremor, backward locomotion, sniffing and defecation were observed, but the degree varied from one symptom to another. The s.c. injection of 10 mg/kg of J-113397 60 min before the naloxone injection, abolished or partially, but significantly attenuated these symptoms.

The NOPR gene expression throughout brain regions and spinal cord following chronic treatments with morphine (10 mg/kg, s.c., 6 days) was measured (Figure 1A). There was a significant increase only in the spinal cord of morphine-treated mice. The gene expression in the spinal cord increased as the time, and reached a plateau on the 5th day, and declined thereafter. A similar increase in NOPR gene expression in the spinal cord was also observed when increasing morphine doses from 20 to 100 mg/kg s.c. had been used as a paradigm for physical dependence. However, there were no significant

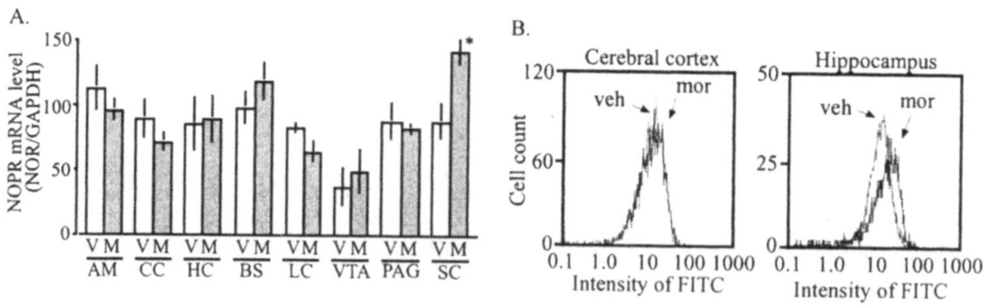

Figure 1. Increase in the expression levels of NOPR and NMDAR ε_1 subunit after chronic morphine application. A. RT-PCR analysis of NOPR mRNA levels in various CNS regions from vehicle (V) or morphine (M)-treated mice. AM; amygdala, CC; cerebral cortex, HC; hippocampus, BS; brain stem, LC; locus coeruleus, VTA; ventral tegmental area, PAG; Periaqueductal gray matter, SC; spinal cord. B. Flow cytometric analysis of NMDAR ε_1 subunit protein levels in cortical or hippocampal cells from control (veh) or dependent (mor) mice

changes in NOPR gene expression in any brain regions tested including locus coeruleus, ventral tegmental area, amygdala and periaqueductal grey matter in both paradigms for morphine tolerance and physical dependence.

The NMDAR ε_1-subunit protein expression was evaluated by the flow cytometric analysis. An increase of ε_1-subunit level was observed in hippocampal cells from morphine-dependent mice, while no change in the level of ε_1-subunit was detected in cortical cells from the dependent mice (Figure 1B).

4. DISCUSSION

The present study demonstrated that J-113397 attenuated the morphine tolerance. This finding is in a good accordance with the data with NOPR$^{-/-}$ mice, and they strongly suggest that NOPR plays a role in the development of such mechanisms. The present results also demonstrated that the spinal NOPR system is likely more important in the morphine tolerance than the supraspinal one. The selective enhancement of NOPR expression in this study strongly supports this view. Similar blockade of naloxone-precipitated withdrawal behaviors by J-113397 and enhancement of NOPR gene expression in morphine-dependent mice were observed. Thus, we propose the schematic model for the development of morphine tolerance and dependence (Figure 2).

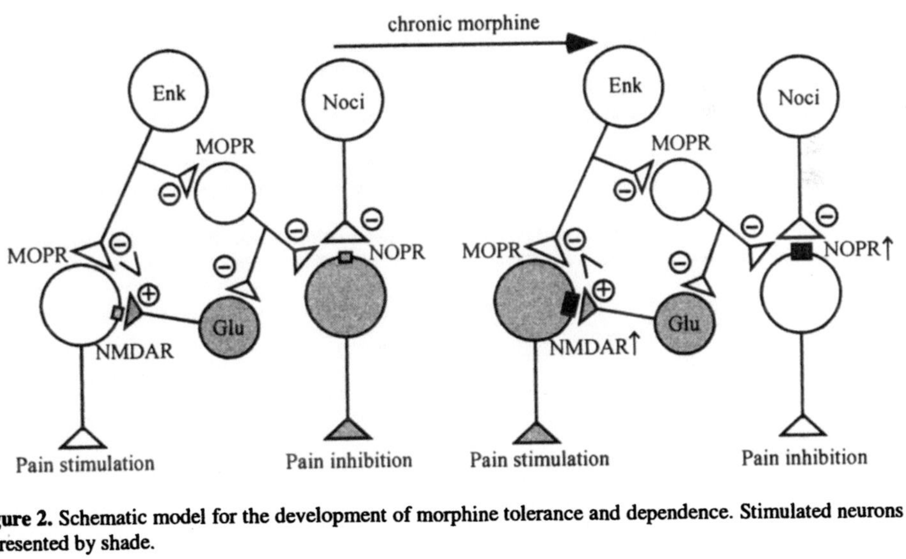

Figure 2. Schematic model for the development of morphine tolerance and dependence. Stimulated neurons are represented by shade.

There are many reports of the involvement of NMDA receptor in the morphine tolerance and dependence. The present study showing enhanced protein expression of ε_1-subunit during chronic morphine treatments supports this view. However, in our preliminary experiments, there was a prolonged duration of morphine in non-treated ε_1-subunit deficient mice. Therefore the contribution of NMDA receptor system to the changes in the plasticity is likely different from that of NOPR (Figure 2).

In summary, the present results suggest that the NOPR and NMDAR systems are important components in some changes in plasticity observed with morphine tolerance and dependence.

REFERENCES

1. J. M. Mitchell, A. I. Basbaum and H. L. Fields, A locus and mechanism of action for associative morphine tolerance, *Nat Neurosci* **3**(1), 47-53. (2000).
2. K. A. Trujillo and H. Akil, Inhibition of morphine tolerance and dependence by the NMDA receptor antagonist MK-801, *Science* **251**(4989), 85-87. (1991).
3. J. C. Meunier, C. Mollereau, L. Toll, C. Suaudeau, C. Moisand, P. Alvinerie, J. L. Butour, J. C. Guillemot, P. Ferrara, B. Monsarrat and et al., Isolation and structure of the endogenous agonist of opioid receptor- like ORL1 receptor, *Nature* **377**(6549), 532-535. (1995).
4. R. K. Reinscheid, H. P. Nothacker, A. Bourson, A. Ardati, R. A. Henningsen, J. R. Bunzow, D. K. Grandy, H. Langen, F. J. Monsma, Jr. and O. Civelli, Orphanin FQ: a neuropeptide that activates an opioidlike G protein- coupled receptor, *Science* **270**(5237), 792-794. (1995).
5. J. H. Tian, W. Xu, Y. Fang, J. S. Mogil, J. E. Grisel, D. K. Grandy and J. S. Han, Bidirectional modulatory effect of orphanin FQ on morphine-induced analgesia: antagonism in brain and potentiation in spinal cord of the rat, *Br J Pharmacol* **120**(4), 676-680. (1997).
6. M. M. Heinricher, S. McGaraughty and D. K. Grandy, Circuitry underlying antiopioid actions of orphanin FQ in the rostral ventromedial medulla, *J Neurophysiol* **78**(6), 3351-3358. (1997).
7. H. Ueda, T. Yamaguchi, S. Tokuyama, M. Inoue, M. Nishi and H. Takeshima, Partial loss of tolerance liability to morphine analgesia in mice lacking the nociceptin receptor gene, *Neurosci Lett* **237**(2-3), 136-138. (1997).
8. H. Ueda, M. Inoue, H. Takeshima and Y. Iwasawa, Enhanced spinal nociceptin receptor expression develops morphine tolerance and dependence, *J Neurosci* **20**(20), 7640-7647. (2000).
9. H. Kawamoto, S. Ozaki, Y. Itoh, M. Miyaji, S. Arai, H. Nakashima, T. Kato, H. Ohta and Y. Iwasawa, Discovery of the first potent and selective small molecule opioid receptor-like (ORL1) antagonist: 1-[(3R,4R)-1-cyclooctylmethyl-3- hydroxymethyl-4-piperidyl]-3-ethyl-1, 3-dihydro-2H-benzimidazol-2-one (J-113397), *J Med Chem* **42**(25), 5061-5063. (1999).
10. M. Zimmermann, Ethical guidelines for investigations of experimental pain in conscious animals, *Pain* **16**(2), 109-110. (1983).
11. H. Ueda, M. Inoue and T. Matsumoto, Protein kinase C-mediated inhibition of mu-opioid receptor internalization and its involvement in the development of acute tolerance to peripheral mu-agonist analgesia, *J Neurosci* **21**(9), 2967-2973. (2001).
12. Y. Kushima, C. Nishio, T. Nonomura and H. Hatanaka, Effects of nerve growth factor and basic fibroblast growth factor on survival of cultured septal cholinergic neurons from adult rats, *Brain Res* **598**(1-2), 264-270. (1992).
13. G. Tezel and M. B. Wax, The mechanisms of hsp27 antibody-mediated apoptosis in retinal neuronal cells, *J Neurosci* **20**(10), 3552-3562. (2000).

NEOCORTICAL GENE EXPRESSION ASSOCIATED WITH BEHAVIORAL SENSITIZATION TO PSYCHOSTIMULANTS

Yasushi Kajii[*‡], Takanori Hashimoto[*], Asami Umino[*] and Toru Nishikawa[*†]

INTRODUCTION

Experience with amphetamine-like psychostimulants, such as amphetamine, methamphetamine (MAP) or cocaine, results in enhanced neuronal and behavioral responses to subsequent drug exposure. This behavioral sensitization, which is believed to be a part of the mechanisms sustaining drug addiction and drug-induced psychosis in human beings[1], is a long-lasting adaptation to drugs of abuse based on persistent cellular and neurochemical changes in some specific brain circuits including the ventral tegmental area (VTA), nucleus accumbens (NAc) and cerebral cortex[2]. This type of brain plasticity seems to require the gene expression that drives the cascade leading to the establishment and maintenance of the sensitized behavioral responsiveness to stimulants or stress because the application of protein synthesis inhibitors blocks the induction of the sensitization[3]. Therefore, identification of the stimulant-responsive gene expression that is specifically observed under sensitization-inducing conditions affords helpful clues to understanding the molecular and neuronal mechanism of this unique behavioral plasticity.

2. NEUROANATOMICAL SUBSTRATES FOR BEHAVIORAL SENSITIZATION TO PSYCHOSTIMULANTS

Some brain regions have been identified as pivotal components of the neuronal circuits that are responsible for the psychostimulant-induced behavior. The VTA where the cell bodies of dopamine neurons projecting to the NAc are located plays a central role in the induction of the sensitization, whereas the NAc has been demonstrated to be

[*] Department of Mental Disorder Research, National Institute of Neuroscience, National Center of Neurology and Psychiatry, Kodaira, Tokyo 187-8502, Japan, and [†]Section of Psychiatry and [†]Behavioral Science, Tokyo Medical and Dental University Graduate School, Bunkyo-ku, Tokyo 113-8519, Japan. [‡]Present address: Drug Discovery Laboratories, Pharmaceutical Research Division, Welfide Co., Iruma, Saitama 358-0026, Japan.

Catecholamine Research, Edited by Nagatsu *et al.*
Kluwer Academic/Plenum Publishers, 2002

critical for expressing the sensitized behavior to the subsequent drug or stress[2]. Dopaminergic neurotransmission through the D1-like receptors is required for inducing the neurochemical adaptations in this neuronal system underlying the behavioral sensitization[2].

In addition to the mesoaccumbens dopamine system, accumulated evidence suggests that other brain regions including the cerebral cortex play an important regulatory role in the induction of the neurochemical and behavioral sensitization to stimulants[2,4]. Anatomically, the medial prefrontal cortex (mPFC) and the amygdala are the major sources of glutamatergic afferent inputs into the mesoaccumbens system[5]. Blocking the glutamate receptors within the VTA and lesions in the mPFC prevent cocaine or amphetamine from inducing the behavioral sensitization[2]. Dopamine in the mPFC has modulatory effects on the extracellular dopamine levels in the NAc and lesioning the mesocortical dopamine projection enhances the development of the amphetamine sensitization[2]. Therefore, a neuronal circuit including the cerebral cortex seems to be implicated in the establishment of the stimulant sensitization.

3. DEVELOPMENT-DEPENDENT CORTICAL GENE EXPRESSION THAT IS RESPONSIVE TO PSYCHOSTIMULANTS

Regulation of the stimulus-responsive gene expression is completed by the combination of various transcription factors in addition to the basic transcription machinery. Psychostimulants activate protein kinase A (PKA) signalling through the G_s-coupled receptors followed by phosphorylation of the cyclic AMP-responsive element binding protein (CREB)[6]. The phosphorylated CREB is an active transcription factor and evokes a transient burst of gene expression including Fos family immediate early genes (IEGs)[6]. However, the consequence of the induction of IEGs remains to be elucidated.

Although several genes have been found to be involved in the adaptive process to drug exposure in the NAc and its anatomically and functionally related brain region, the striatum, the regulatory role of the cerebral cortex in establishing the behavioral sensitization to psychostimulants as described in the previous section suggests that stimulant-induced changes in cortical gene expression also contribute to long-lasting behavioral adaptations. Indeed, we reported the specific induction of the tissue plasminogen activator (tPA) mRNA in corticostriatal neurons by psychotomimetics including MAP, cocaine and phencyclidine[7]. The tPA is the most abundant serine proteinase in the rodent central nervous system and is considered to play an important role in synaptic plasticity by catalyzing extracellular proteolysis that seems to result in synaptic micro-remodelling because the tPA gene is induced as an IEG in hippocampal neurons during seizure, kindling and LTP[8]. In fact, genetic modification of the neuronal tPA expression levels results in alternation of the LTP, LTD and memory, indicating that tPA really plays a critical role in the formation of synaptic plasticity[9,10]. Although the consequence of the drug-induction of the tPA remains unclear, these observations suggest that drugs of abuse induce synaptic plasticity in some corticofugal pathways. This is also supported by the fact that exposure to amphetamine or cocaine results in persistent morphological changes in the dendrites of the accumbal and cortical neurons[11].

The induction of *c-fos* gene expression is a useful tool for examining activated neuronal circuits in response to a wide variety of different stimuli because of its low base-line levels of expression in the brain[12]. The acute administration of drugs of abuse induces a common neuroanatomical pattern of *c-fos* expression, and subchronic administration leads to an increased expression in the somatosensory cortex[13]. In addition, our data demonstrated that the pattern of MAP-induced c-fos expression in developing rats significantly changed from postnatal days (PD) 8 to 50, and shifted to the adult pattern around postnatal week three[14]. The alteration in the gene expression pattern is most prominent in the neocortex, followed by the striatum[14], suggesting that MAP-responsive neuronal circuits including the neocortex alter their function at that time. The levels of neocortical *c-fos* mRNA measured by a semi-quantitative RT-PCR method also indicate that MAP-responsive neuronal activity in the neocortex reaches adult levels around postnatal week three (Fig. 1). It is notable that the induction of stereotypy sensitization to stimulants depends on postnatal development; in the case of rats, the critical period is around postnatal week three and repeated injections of stimulants before this period have no enhancing effect on the stereotypy behavior induced by subsequent drug exposure[15]. Therefore, together with the findings described earlier, the MAP-induced brain activity mapping suggests that the development-dependent corticofugal paths play an important regulatory role in the establishment of stimulant sensitization. In fact, the psychotomimetics-induction of tPA mRNA in corticostriatal neurons is dependent on the postnatal development in rats (unpublished data) and we have identified other genes whose neocortical expression induced by acute MAP administration is specifically observed after the critical period (unpublished data). Such types of genes may contribute to stimulant-induced adaptations in the possible regulatory circuit including the cerebral cortex.

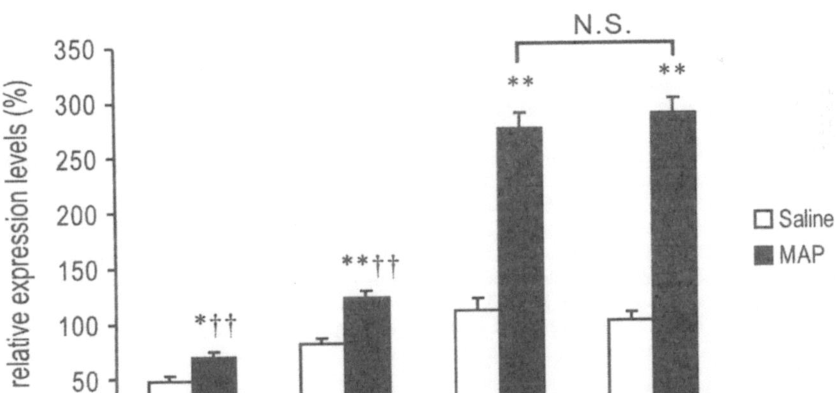

Figure 1. Methamphetamine-induced neocortical *c-fos* expression in developing rats measured by a relative RT-PCR method. Data are expressed as percentages relative to saline-treated rats of PD50 and represent mean values ± s.e.m. (n=6). Statistical differences among groups were estimated by a one-way analysis of variance (ANOVA) followed by the Fisher PLSD post hoc test. *$P<0.05$, **$P<0.0001$ compared to the respective saline-treated controls; ††$P<0.0001$ compared to the MAP treated rats of PD50. N.S., no significant difference.

4. CONCLUSION

The cellular and behavioral changes are ultimately driven by the sequential changes in gene expression. Therefore, although there still exist gaps among the molecular, cellular and behavioral levels, identification of the genes whose expression is specifically modified by the sensitization regimens is not only quite important for elucidating the mechanism of the behavioral sensitization but also helpful for understanding the molecular basis of drug addiction and drug-induced psychosis. However, no change in the accumbal or striatal gene expression associated with any sensitization regimen has so far been demonstrated to account for such a long-lasting functional alteration in the central nervous system. Our approach based on the developmental aspect of the stereotypy sensitization and MAP-evoked changes in neocortical gene expression is so unique that we will be able to find novel clues for clarifying the molecular cascade leading to the behavioral plasticity.

REFERENCES

1. M. Sato, A lasting vulnerability to psychosis in patients with previous methamphetamine psychosis, *Ann N Y Acad Sci* **654,** 160—170 (1992).
2. L. J. M. J. Vanderscuren and P. W. Kalivas, Alternations in dopaminergic and glutamatergic transmission in the induction and expression of behavioral sensitization; a critical review of preclinical studies. *Psychopharmacology* **151,** 99—120 (2000).
3. R. Karler, K. T. Finnegan, and L. D. Calder, Blockade of behavioral sensitization to cocaine and amphetamine by inhibitors of protein synthesis, *Brain Res.* **603,** 19—24 (1993).
4. T. M. Tzschentke, Pharmacology and behavioral pharmacology of the mesocortical dopamine system, *Prog. Neurobiol.* **63,**: 241—320 (2001).
5. S. R. Sesack and V. M. Pickel, Prefrontal cortical efferents in the rat synapse on unlabeled neuronal targets of catecholamine terminals in the nucleus accumbens septi and on dopamine neurons in the ventral tegmental area, *J. Comp. Neurol.* **320,** 145—160 (1992).
6. E. J. Nestler, Molecular basis of long-term plasticity underlying addiction, *Nat. Rev. Neurosci.* **2,** 119—128 (2001).
7. T. Hashimoto, Y. Kajii, and T. Nishikawa, Psychotomimetics-induction of tissue plasminogen activator mRNA in corticostriatal neurons in rat brain, *Eur. J. Neurosci.* **10,** 3387—3399 (1998).
8. Z. Qian, M. E. Gilbert, M. A. Colicos, E. R. Kandel, and D. Kuhl, Tissue-plasminogen activator is induced as an immediate-early gene during seizure, kindling and long-term potentiation, *Nature* **361,** 453—457 (1993).
9. Y.-Y. Huang, M. E. Bach, H.-P. Lipp *et al*, Mice lacking the gene encoding tissue-type plasminogen activator show a selective interference with late-phase long-term potentiation in both Schaffer collateral and mossy fiber pathways, *Proc. Natl. Acad. Sci. USA* **93,** 8699—9704 (1996).
10. R. Madani, S. Hulo, N. Toni *et al*, Enhanced hippocampal long-term potentiation and learning by increased neuronal expression of tissue-type plasminogen activator in transgenic mice, *EMBO J.* **18,** 3007—3012 (1999).
11. T. E. Robinson and B. Kolb, Alternations in the morphology of dendrites and dendritic spines in the nucleus accumbens and prefrontal cortex following repeated treatment with amphetamine or cocaine, *Eur. J. Neurosci.* **11,** 1598—1604 (1999).
12. R. E. Harlan and M. M. Garcia, Drugs of abuse and immediate-early genes in the forebrain. *Mol. Neurobiol.* **16,** 221—267 (1998).
13. E. J. Curran, H. Akil, and S. J. Watson, Psychomotor stimulant- and opiate-induced *c-fos* mRNA expression patterns in the rat forebrain: comparisons between acute drug treatment and a drug challenge in sensitized animals, *Neurochem. Res.* **21,** 1425—1435 (1996).
14. T. Nishikawa, A. Umino, A. Kashiwa, A. Ooshima, N. Nomura, and K. Takahashi, Stimulant-induced behavioral sensitization and cerebral neurotransmission, in: *Neurotransmitters in neuronal plasticity and psychiatric disorders*, edited by M. Toru (Excerpta Medica, Tokyo, 1993), pp53—62.
15. Y. Fujiwara, M. Kazahaya, M. Nakashima, and S. Ootsuki, Behavioral sensitization to methamphetamine in the rat: an ontogenic study, *Psychopharmacology* **91,** 316—319, 1987.

NEUROIMAGING ON METHAMPHETAMINE USERS WITH PSYCHIATRIC SYMPTOMS

Masaomi Iyo, Yoshimoto Sekine and Norio Mori *

SUMMARY

A long-term abuse of methamphetamine causes psychiatric symptoms, such as psychosis and anxiety disorder-like symptoms. Some of the symptoms are persistent for long time of period even after cessation of the methamphetamine use. We employed positron emission tomography (PET) studies to clarify the changes in the brain dopaminergic systems of methamphetamine abusers and the relations of the changes to their psychiatric symptoms. It was indicated that persistent reduction of dopamine transporter density in the caudate/putamen, nucleus accumbens and prefrontal cortex in the abusers, whereas the density of dopamine D2 receptors were not different from the control subjects. The results also suggested the reduction of dopamine transporter density may be closely associated with duration of methamphetamine use and severity of persistent psychiatric symptoms in methamphetamine users.

1. INTRODUCTION

Methamphetamine has been abused over the past six decades worldwide.[1] It is well known that a large dose of methamphetamine induces psychotic state, consisting of mainly hallucination and delusion. A long-term use of methamphetamine also induces susceptibility to psychosis, i.e., a low to moderate dose of methamphetamine or psychological stress induces psychotic episode in the long-term abusers.[2] Some abusers, furthermore, exhibit psychosis prolonging for a long time of period even after stopping the use of methamphetamine. In Japan, the patients with this type of long-lasting psychosis are classified as methamphetamine psychosis, whereas most of those may be diagnosed schizophrenia with methamphetamine abuse.[3-6] Therefore, we have studied to identify the changes in brain dopaminergic systems, closely related to methamphetamine use and contributing genesis of psychotic state, by PET stuies in methamphetamine users.

* Masaomi Iyo, GraduateSchool of Medicine, Chiba University, Chiba, 260-8670, Japan, Yoshimoto Sekine and Norio Mori, Hamamatsu University School of Medicine, Hamamatsu, Japan

Catecholamine Research, Edited by Nagatsu *et al.*
Kluwer Academic/Plenum Publishers, 2002

2. MEASUREMENT OF BRAN DOPAMINE SYSTEMS BY PET

2.1. Striatal Dopamine D2 Receptors

Methamphetamine increases synaptic dopamine levels and blockers of dopamine D2 receptors such as haloperidol ameliorate methamphetamine psychosis. Long-term of high levels of released dopamine in the synaptic clefts may compensate to decrease post-synaptic dopamine receptors. Another idea may be that long-term use of methamphetamine induces long-lasting changes in the receptors, which may be related to psychotic state. We evaluated striatal dopamine D2 receptor density in the methamphetamine uses, who had stopped the use more than 2 months before the PET study using ^{11}C-N-methylspiperone as a tracer, and had past-experiences of methamphetamine psychosis.[7] We selected the subjects following the above criterion because our interest was focused on detection of long-lasting biological changes induced by methamphetamine use and the test should be performed at the time where compensatory changes may be recovered.

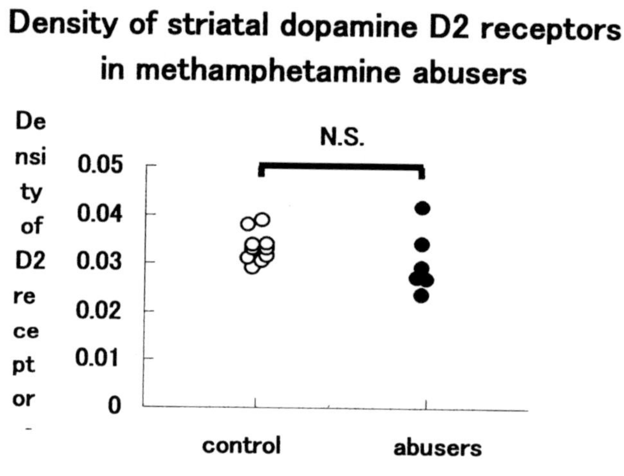

Figure 1. The density of striatal dopamine D2 receptors in methamphetamine abusers. There was no difference between the healthy subjects and the abusers, who had stopped methamphetamine use at least 2 months before the PET study.

The results showed no significant difference in the density of striatal D2 receptors between the controls and the subjects with past-history of methamphetamine abuse and psychosis (Fig.1). It was reported that the decreased density of striatal dopamine receptors in cocaine abusers was recovered about 1 month following cessation of cocaine use, revealed by PET and ^{11}C-N-methylspiperone. It is known that cocaine blocks dopamine transporters (reuptake sites) and then increases dopamine levels in synaptic clefts. Cocaine may transiently reduce dopamine receptors as a compensatory effect for increased dopamine levels and the reduction is recovered following the decrease of dopamine levels after cessation of cocaine.[8] Methamphetamine also increases dopamine levels in synaptic cleft so that no change in dopamine D2 receptor level in previous methamphetamine users in our study may be due to the recover after compensatory reduction for an increase of dopamine levels.

2.2. Density of Dopamine Transporter In The Brain

A recent postmortem study indicated the reduction of dopamine transporter density in the caudate/putamen and nucleus accumbens in chronic methamphetamine abusers,[9] and a positron emission tomography (PET) study also suggested reduction of the dopamine transporter density of the caudate/putamen in methamphetamine (MAP) users.[10] However, the relation of the reduction of psychiatric symptoms in the abusers and patterns of the drug use. The subjects used in these studies were polydrug users, i.e., they used cocaine and marijuana in addition to methamphetamine. Therefore, we measured the density in the nucleus accumbens (NAc) and prefrontal cortex (PFC), in addition to the CP, in MAP users without use of other abused drugs, and assessed the relation of the changes to their clinical characteristics.[11]

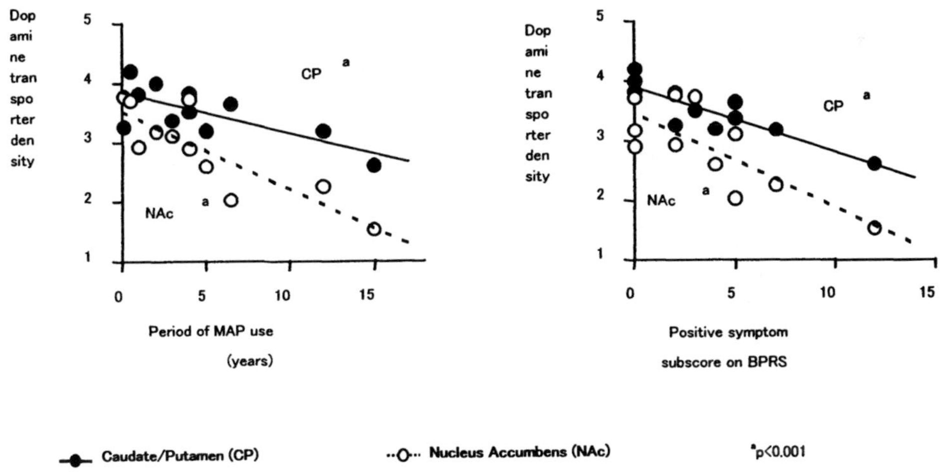

Figure 2. The relation of the dopamine transporter density to the period of methamphetamine use and positive symptoms subscore on BPRS.

PET and [^{11}C]WIN-35,428, a DAT ligand, were used in the study. Psychiatric symptoms in MAP users were evaluated by the Brief Psychiatric Rating. The DAT density in all of the three regions observed was significantly reduced in the MAP users. The severity of psychiatric symptoms was significantly correlated with the duration of MAP use. The DAT reduction in the CP and NAc was significantly associated with the duration of MAP use, and closely related to the severity of persistent psychiatric symptoms (Fig. 2). Our study suggests that longer use of MAP may cause more severe psychiatric symptoms and greater reduction of DAT density in the brain, and the DAT reduction may be long-lasting, even if MAP use ceases. Further, persistent psychiatric symptoms, including psychotic symptoms, in MAP users may be attributable to the reduction of DAT density.

It has been argued whether prolonged psychosis in chronic drug users is a consequence of chronic drug use or merely a co-morbidity of drug use and schizophrenia.[21-23] Our results show close relationship between reduction of the DAT density in the brain, which is associated with the duration of MAP use, and the symptoms of a chronic psychotic state. Our findings suggest that chronic MAP use significantly contributes to the production of prolonged psychosis. This contention is further supported by postmortem and PET studies of schizophrenia, in which no apparent changes in the DAT density of the brains were found.[12, 13]

3. CONCLUSIONS

Methamphetamine abuse reduces dopamine transporter density, which may be closely associated with duration of methamphetamine use and severity of persistent psychiatric symptoms in methamphetamine users. On the other hand, the density of dopamine D2 receptor density in the abusers, who stops the methamphetamine use longer than 2 months, does not differ from the controls.

REFERENCES

1. K. Konuma, S. Hirai, and M. Kasahara, in: *Amphetamine and Its Analogs*, edited by A. K. Cho and D. S. Segal (San Diego, Academic Press, 1994), pp. 459-478.
2. M. Sato, Y. Numach, and T. Hamamura, Relapse of paranoid psychotic state in methamphetamine model of schizophrenia. *Schizophr. Bull.* **18**:115-122 (1992).
3. B. Angrist, in: *Amphetamine and Its Analogs*, edited by A. K. Cho and D. S. Segal (San Diego, Academic Press, 1994), pp. 387-414.
4. P.A. LeDuc, G. Mittleman, Schizophrenia and psychostimulant abuse: a review and re-analysis of clinical evidence, *Psychopharmacology.* **121**:407-427 (1995).
5. J. M. Cleghorn, R. D. Kaplan, B. Szechtman, H. Szechtman, G. M. Brown, and S. Franco, Substance abuse and schizophrenia: effect on symptoms but not on neurocognitive function, *J. Clin. Psychiatry.* **52**:26-30 (1991).
6. B. Kirkpatrick, X. F. Amador, M. Flaum, S. A. Yale, J. M. Gorman, W. T. Jr. Carpenter, M. Tohen, and T. McGlashan, The deficit syndrome in the DSM-IV field trial: I. Alcohol and other drug abuse, *Schizophr. Res.* **20**:69-77 (1996).
7. M. Iyo, M. Nishio, T. Itoh,, H. Fukuda, K. Suzuki, T. Yamasaki, S. Fukui, and Y. Tateno, Dopamine D2 and serotonin S2 receptors in susceptibility to methamphetamine psychosis detected by PET, *Psychiatry. Res.* **50**: 217-231 (1993).
8. N. D. Volkow, J. S. Fowler, A. P. Wolf, D. Schlyer, C. Y. Shiue, R. Alpert, S. L. Dewey, J. Logan, B. Bendriem, D. Christman, R. Hitzemann, and F. Henn, Effects of chronic cocaine abuse on postsynaptic dopamine receptors. *Am. J. Psychiatry.* **147**:719-724 (1990).
9. J. M. Wilson, K. S. Kalasinsky, A. I. Levey, C. Bergeron, G. Reiber, R. M. Anthony, G. A. Schmunk, K. Shannak, J. W. Haycock, and S. J. Kish, Striatal dopamine nerve terminal markers in human, chronic methamphetamine users, *Nat. Med.* **2**:699-703 (1996).
10. U. D. McCann, D. F. Wong, F. Yokoi, V. Villemagne, R. F. Dannals, and G. A. Ricaurte, Reduced striatal dopamine transporter density in abstinent methamphetamine and methcathinone users: Evidence from positron emission tomography studies with [^{11}C]WIN-35,428, *J. Neurosci.* **18**:8417-8422 (1998).
11. Y. Sekine, M. Iyo, Y. Ouchi, T. Matsunaga, H. Tsukada, H. Okada, E. Yoshikawa, M. Futatsubashi, N. Takei, and N. Mori, Methamphetamine-Related Psychiatric Symptoms and Reduced Brain Dopamine Transporters Studied With PET, *Am. J. Psychiatry.* (August, 2001 in press).
12. A. Laakso, H. Vilkman, B. Alakare, M. Haaparanta, J. Bergman, O. Solin, J. Peurasaari, V. Rakkolainen, E. Syvalahti, and J. Hietala, Striatal dopamine transporter binding in neuroleptic-naive patients with schizophrenia studied with positron emission tomography, *Am. J. Psychiatry.* **157**:269-271 (2000).
13. P. Seeman, and H. B. Niznik, Dopamine receptors and transporters in Parkinson's disease and schizophrenia, *F. A. S. E. B. J.* **4**:2737-2744 (1990).

DOPAMINE PSYCHOSIS IN SCHIZOPHRENIA? MOLECULAR GENETIC STUDY ON DELUSIONAL DISORDER

Kiyoshi Morimoto, Ryosuke Miyatake and Mitsuo Nakamura*

1. INTRODUCTION

Delusional disorder is characterized by mono-symptomatic paranoid symptoms without other schizophrenic symptoms. Thus, delusional disorder can be a good clinical model for investigating the molecular genetic mechanisms for paranoid symptoms. In this study, we hypothesized that a hyperdopaminergic state in the brain is responsible for producing paranoid symptoms. To test this hypothesis, we (1) compared neuroleptic responses between patients with delusional disorder and schizophrenia, (2) measured plasma levels of homovanillic acid (pHVA), an indicator of brain dopamine activity, in patients with delusional disorder, and (3) analysed the polymorphism of dopamine receptors (DRs) and its synthesizing enzyme gene.

2. NEUROLEPTIC RESPONSE OF DELUSIONAL DISORDER

Eleven patients with delusional disorder and 15 patients with schizophrenia classified according to ICD-10 (1992), who were admitted to our psychiatric ward, were studied. All patients presented the first delusional episode without previous neuroleptic medication. We compared the neuroleptic responses between those patients.

The onset of delusional disorder appeared to be in middle-age in all patients and was much later than for schizophrenia. A clear gender difference was also seen in delusional disorder (female>male). A relatively smaller dose of haloperidol (4.7 mg/day) was effective in diminishing completely delusional symptoms within a shorter period of administration (65 days) in delusional disorder as compared with schizophrenia (12.7 mg/day, 104 days), suggesting that the neuroleptic response is extremely good in this disorder. At discharge, the Global Assessment of Functioning (GAF) score was significantly higher in delusional disorder, which also suggests a better prognosis.

---

* Department of Neuropsychiatry, Faculty of Medicine, Kagawa Medical University, Kagawa, 761-0793 Japan

Catecholamine Research, Edited by Nagatsu *et al.*
Kluwer Academic/Plenum Publishers, 2002

3. pHVA IN DELUSIONAL DISORDER

To investigate brain dopamine function, pHVA levels were measured in patients with delusional disorder and compared with those of age- and sex-matched controls. Thirteen neuroleptic-naive patients with delusional disorder (age at onset was 48.4 years; male/female ratio was 4/9) were subjects in our in-patient and out-patient clinics. The Brief Psychiatric Rating Scale (BPRS) was used to evaluate the level of delusional symptoms. In addition to the pretreatment examination, in 8 patients, pHVA and BPRS scores were measured 8 weeks after neuroleptic treatment with a small dose of haloperidol (average 2.7 mg/day).

The pretreatment level of pHVA tended to be more elevated in delusional disorder than in normal control individuals (64.5 vs 48.9 pmol/ml, $p<0.1$) and it was correlated significantly to the delusion and hallucination BPRS score ($r=0.44$, $p<0.02$). When the delusional patients were classified into the different subtypes, the persecution-type showed a significantly higher level of pHVA (68.5 pmol/ml, $p<0.05$) than the controls, while the jealousy-type did not (55.5 pmol/ml).

Eight weeks after neuroleptic treatment with a small dose of haloperidol, the delusion and hallucination BPRS scores improved markedly, and the level of pHVA also significantly decreased 8 weeks after the successful neuroleptic treatment.

4. THE DRs AND TH GENE IN DELUSIONAL DISORDER

To investigate the possible genetic basis underlying the hyperfunction of dopamine systems in delusional disorder, we analysed the polymorphism of genes for DRD2 and DRD3 and their synthesizing enzyme tyrosine hydroxylase (TH). Twenty-three patients with delusional disorder, 48 patients with schizophrenia, and 48 normal controls were studied. We analysed polymorphisms of Ser311Cys in the DRD2 gene, Ser9Gly in the DRD3 gene, and variable number of tandem repeat (VNTR) in the first intron of TH gene.

The genotype frequency of DRD2 Ser311Cys was 17.4% in delusional disorder, 6.3% in schizophrenia, and 6.3% in control subjects. There was a significantly higher incidence of Ser311Cys heterozygote among the persecution-type delusional patients (21.1%, $p<0.05$), compared with the controls.

The genotype frequency of the DRD3 Ser9Ser gene was 47.8% in delusional disorder (52.6% in persecution-type and 25.0% in jealousy-type), 47.9% in schizophrenia, and 39.6% in control subjects. Patients who were homozygous for the Ser9Ser had a significantly higher pretreatment level of pHVA than those who were heterozygous (85.6 vs 51.3 pmol/ml, $p<0.05$).

There was a significant positive correlation between the polymorphic $(TCAT)_n$ repeat in the first intron of the TH gene and pretreatment level of pHVA in delusional disorder ($r=0.62$, $p<0.05$). Patients with the $(TCAT)_8$ or $(TCAT)_9$ allele had significantly higher levels of pHVA than those with $(TCAT)_6$ or $(TCAT)_7$ (87.1 vs 47.9 pmol/ml, $p<0.02$).

5. HYPERDOPAMINERGIC STATE RESPONSIBLE FOR PARANOID STATE

While a number of previous pHVA studies have failed to obtain consistent results with regard to schizophrenia, three main lines of evidence have been reported: (1) pHVA is correlated positively with the severity of psychotic symptoms, (2) in good responders pHVA decreases along with neuroleptic treatment, and (3) pHVA is lower in deficit-type than in non-deficit-type. In contrast to these complicated results for pHVA, more consistent results have been observed in psychoses with a good prognosis, which is unrelated to their conventional diagnosis. For example, Garver et al (1997) have demonstrated recently that baseline pHVA is bimodally distributed in acute psychotic patients, and that high-pHVA psychotic patients have a more rapid and complete neuroleptic response than lower-pHVA psychotics. Our findings on pHVA in delusional disorder are in good agreement with Garver's results. In our study, pretreatment pHVA levels were elevated in persecution-type patients, positively correlated with the severity of paranoid symptoms, and normalized after successful neuroleptic treatment. It is concluded from our pHVA study that hyperfunction of the dopamine systems could be at least partly responsible for the brain mechanisms underlying its paranoid symptoms.

6. MOLECULAR GENETIC BASIS OF DELUSIONAL DISORDER

Arinami et al (1994) were the first to report that the allele frequency of the 311Cys of the DRD2 gene was about three times higher in Japanese schizophrenics than in control subjects. These authors have since duplicated their own results in a larger population study (1996), in which it was shown that the clinical characteristics of schizophrenic patients with the 311Cys gene were lacking negative symptoms, going into remission and had a family history of the disorder. Many studies, however, have failed to duplicate these results in various races. Very recently, Serretti et al (2000) demonstrated a significant association between the Ser311Cys variant and delusional features in major psychoses. This indicates that this mutation may be connected with delusional symptomatology independent of diagnoses. Since the Ser311Cys mutation has been shown to cause functional alteration of DRD2 *in vitro*, dopamine signal transmission may be increased in delusional patients with this mutation.

After the first demonstration that homozygous Ser9Ser of the DRD3 gene was significantly higher in patients with schizophrenia by Crocq et al (1992), several studies, that focused especially on the schizophrenics who were good neuroleptic responders, have been able to duplicate it, but many other studies have failed. In our results it appeared that the pretreatment level of pHVA was markedly higher in patients with Ser9Ser than those with Ser9Gly. In delusional disorder, the DRD3 gene homozygous for Ser9Ser may be one of the etiologic genes responsible for producing the high pHVA levels.

It has been reported previously that polymorphism of TH gene VNTR is related to transcription *in vitro* and to catecholamine turnover *in vivo*. Several studies on schizophrenia have indicated a significant association with the polymorphic VNTR region of the TH gene. Consistent with these results, our findings in delusional patients demonstrated that the pretreatment level of pHVA is significantly correlated to the VNTR $(TCNT)_n$ repeat, suggesting that TH VNTR is another etiologic candidate gene for the high pHVA level.

In our study, it is notable that 3 patients possessing a combination of DRD3 Ser9Ser and DRD2 Ser311Cys genes showed abnormally high pretreatment levels of pHVA. In these cases, it is speculated that polygenes of both the pre- and post-synaptic mechanisms of dopamine systems may be involved in the genetic etiology (Figure 1). The former may accelerate dopamine metabolic turnover, whereas the latter may intensify dopamine signal transduction, together resulting in a hyperdopaminergic state. Polymorphism of TH VNTR may also be involved in the pre-synaptic mechanism.

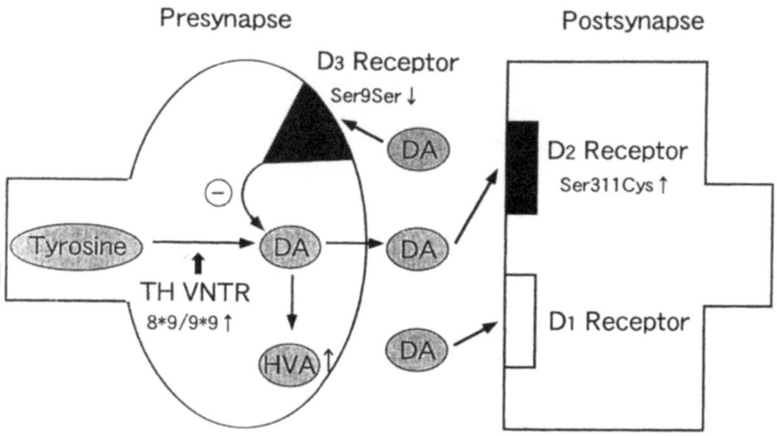

Figure 1. Possible polygenetic involvement in dopamine systems in delusional disorder

7. CONCLUSION

Although further studies are required to investigate the molecular genetic etiology in delusional patients with low pretreatment levels of pHVA, it is concluded that delusional disorder, especially of the persecution-type, includes dopamine psychosis, and that polymorphism of the DRD2, DRD3 and/or TH gene is part of the genetic basis underlying the hyperdopaminergic state that produces paranoid symptoms.

REFERENCES

Garver DL, Steinberg JL, McDermott BE, et al: Etiologic heterogeneity of the psychoses: is there a dopamine psychosis? *Neuropsychopharmacol* **14**:191-201,1997.

Arinami T, Itokawa M, Enguchi H, et al: Association of dopamine D2 receptor molecular variant with schizophrenia. *Lancet* **343**:703-704,1994.

Arinami T, Itokawa M, Aoki J, et al: Further association study on dopamine D2 receptor variant S311C in schizophrenia and affective disorder. *Am J Med Genet* **67**:133-138,1996.

Serretti A, Latuada E, Lorenzi R, et al: Dopamine receptor D2 Ser/Cys311 is associated with delusion and disorganization symptomatology in major psychoses. *Mol Psychiatry* **5**:270-274,2000.

Crocq MA, Mant R, Asherson P, et al: Association between schizophrenia and homozygosity at the dopamine D3 receptor gene. *J Med Genet* **29**:858-860,1992.

DOPAMINE RECEPTORS IN SCHIZOPHRENIA

Tetsuya Suhara* Fumihiko Yasuno, Yoshiro Okubo, Yasuhiko Sudo, Makoto Inoue, Tetsuya Ichimiya and Akihiro Takano

1. INTRODUCTION

Schizophrenia is believed to involve altered dopaminergic transmission. This is supported by the fact that chronic use of dopamine agonists such as amphetamine can cause psychotic symptoms, and dopamine D2 receptor antagonists are the most widely used drugs for the treatment of schizophrenia. However, most of the in vivo imaging studies with drug-free or drug-naïve patients could not detect a significant change in striatal D2 receptors [1]. On the other hand, neuropathological [2] and neuroimaging studies have demonstrated abnormalities in the extrastriatal regions of schizophrenia. Dopamine receptors are classified into five subtypes, and selective ligands have been developed for D1 and D2 subtypes. In the cortical regions, the density of D1 receptor is higher than that of D2 receptor [3,4]. D1 receptor has been implicated in the control of working memory. We have recently reported the abnormality of dopamine D1 receptor binding in the prefrontal cortex of patients with schizophrenia, and its relation to negative symptoms and poor performance in the Wisconsin Card Sorting Test [5]. On the other hand, an increasing body of evidence favors a crucial role of extrastriatal regions in the pathophysiology of positive symptoms, and the extrastriatal D2 receptor is expected to be the common site of action of antipsychotics [6]. We found that extrastriatal dopamine D2 receptor was the site of antipsychotic drug action [7]. However, unexpectedly, reduced D2 receptor binding was observed in the anterior cingulate cortex of patients with schizophrenia, and a significant negative correlation was observed between D2 receptor binding and the positive symptom score on Brief Psychiatric Rating Scale (BPRS) [8]. Alterations in D2 receptor function in the extrastriatal region may underlie the positive symptoms of schizophrenia, and abnormality of D1 receptor function may be at the bottom of the negative symptoms and cognitive impairment of schizophrenia.

2. DOPAMINE D2 RECEPTORS IN SCHIZOPHRENIA

Sixteen male patients who met the DSM-IV criteria for schizophrenia or

* Tetsuya Suhara, Brain Imaging Project, National Institute of Radiological Sciences, 9-1, 4-chome, Anagawa, Inage-ku, Chiba, Japan 263-8555

schizophreniform disorder (11 drug-naïve and 5 drug-free) were examined by positron emission tomography using [^{11}C]FLB 457. Both drug-naïve and drug-free patients (drug-free for at least one month) were included in this analysis. The duration of illness ranged from 1 month to 20 years (average 4.5 years). Symptoms were assessed using the brief psychiatric rating scale (BPRS). Positive and negative symptom scores were calculated as the sum of the following items. Positive symptom subscales: conceptual disorganization, mannerisms and posturing, hostility, grandiosity, suspiciousness, hallucinatory behavior, unusual thought content, and excitement. Negative symptom subscales: emotional withdrawal, motor retardation, and blunted affect. Comparison was made with 18 healthy male controls. Regions of interest were drawn on the anterior cingulate cortex, prefrontal cortex, temporal cortex, thalamus, parietal cortex, occipital cortex, hippocampal cortex and cerebellar cortex. Binding potentials (BP) from the reference tissue method were used for the index of dopamine D2 receptor binding [9].

Parametric images of the BP of [^{11}C]FLB 457 were analyzed using SPM 99. The ligand specific template image for the BP of [^{11}C]FLB 457 was constructed from 12 normal controls and used to define the stereotactic transformation parameters for the BP and R1 images of [^{11}C]FLB 457. Normalized BP images were smoothed with a Gaussian filter to 12 mm full-width half-maximum. Contrasts of the condition effects at each voxel of the BP of [^{11}C]FLB 457 were assessed using the t-value, giving a statistical image for each contrast. P for trend was thresholded at less than 0.01, uncorrected.

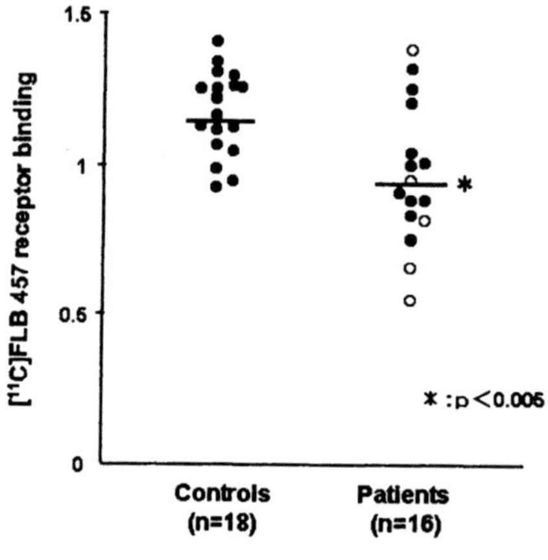

Figure 1. [^{11}C]FLB 457 binding potentials (BP=Bmax/Kd) in the anterior cingulate cortex of 16 patients with schizophrenia [11 drug-naïve (filled circles) and 5 drug-free (circles)] and 18 healthy controls. Values were adjusted for age; adjusted BP = BP + 0.016 x (age − 29.7 yr). Univariate ANOVA analysis revealed that the patient group (mean=0.97, SD=0.23) had significantly reduced D2 binding compared with the control group (mean=1.18, SD=0.14) (P<0.005). Horizontal lines represent group means.

The BP value was significantly lower in the anterior cingulate cortex in patients with schizophrenia than in normal controls (ANCOVA with age as covariate; $F_{1,31}$=10.7, P=0.003) (Fig.1). A significant negative correlation was observed between BP in the anterior cingulate cortex and the positive symptom score on BPRS (r=-0.68, P=0.004) (Fig.2). There was no significant relationship between BP in any of the regions and negative symptoms. The lower BP values could not be explained by the direct competition with increased endogenous dopamine, since our monkey experiments indicated that the extrastriatal [^{11}C]FLB 457 binding was not sensitive to endogenous dopamine [10], and thus fewer D2 receptors could be expected in the anterior cingulate cortex in patients with schizophrenia. A reduced number of receptors might be a result of an increased concentration of endogenous dopamine. On the other hand, cellular localization of dopamine D2 receptors has been demonstrated on both pyramidal neurons and nonpyramidal interneurons, which use GABA as an inhibitory transmitter. Based on the finding of the deficit of inhibitory interneurons in the anterior cingulate cortex, less GABAergic inhibitory transmission has been proposed in the cortical local circuit of schizophrenic brain, and reduced D2 receptor binding could be interpreted as a marker of the disinhibition of dopaminergic transmission [8].

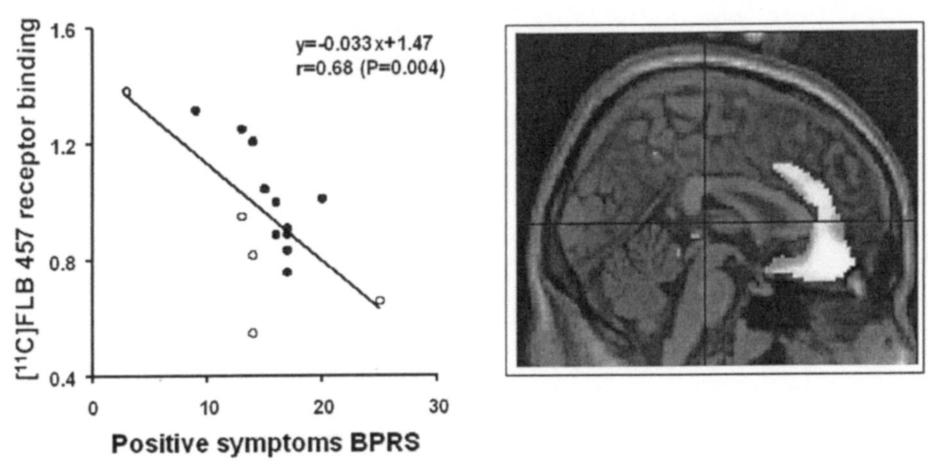

Figure 2. Relationship between [^{11}C]FLB 457 binding potentials in the anterior cingulate cortex and the positive symptom score on BPRS in 16 patients [11 drug-naïve (filled circles) and 5 drug-free (circles)] with schizophrenia. A significant negative correlation was observed between BP values and clinical rating for positive symptom score on BPRS in the anterior cingulate cortex ($r = -0.68$, $P = 0.004$). Right figure shows three-dimensional SPM projections superimposed on representative sagittal (X=2mm) magnetic resonance image brain slices (threshold for display, $P<0.01$).

3. DOPAMINE D1/D2 RECEPTORS IN THE ANTERIOR CINGULATE IN SCHIZOPHRENIA

We have recently reported the abnormality of dopamine D1 receptor binding in the prefrontal cortex of patients with schizophrenia, and its relation to negative symptoms and poor performance in the Wisconsin Card Sorting Test [5]. However, the relation between dopamine D1 receptor and D2 receptor in the same subject is still not clear. Nevertheless, receptor–receptor interaction and functional difference between D1 receptor and D2 receptor have been reported in animal experiments and a postmortem study in schizophrenia.

Thirteen subjects with schizophrenia (8 drug-naïve and 5 drug-free) who were involved in the extrastriatal dopamine D2 receptor study using [^{11}C]FLB 457 were also examined for dopamine D1 receptor using [^{11}C]SCH 23390. The D1 receptor binding was analyzed using the reference tissue model in the anterior cingulate cortex where the D2 receptor binding was significantly lower than that in the control. A significant negative correlation was observed between D1 receptor binding and the negative symptom score ($r=-0.69$, $p=0.009$), whereas there was no relationship with the positive symptom score ($r=0.23$, $p=0.46$) (Fig.3). There was a significant negative correlation between D2 receptor binding and the positive symptom score (r=-0.62, p=0.02), whereas there was no relationship between D2 receptor binding and the negative symptom score (r=0.16, p=0.60). Alterations in D1 and D2 receptor functions might differentially affect the positive and negative symptoms in schizophrenia.

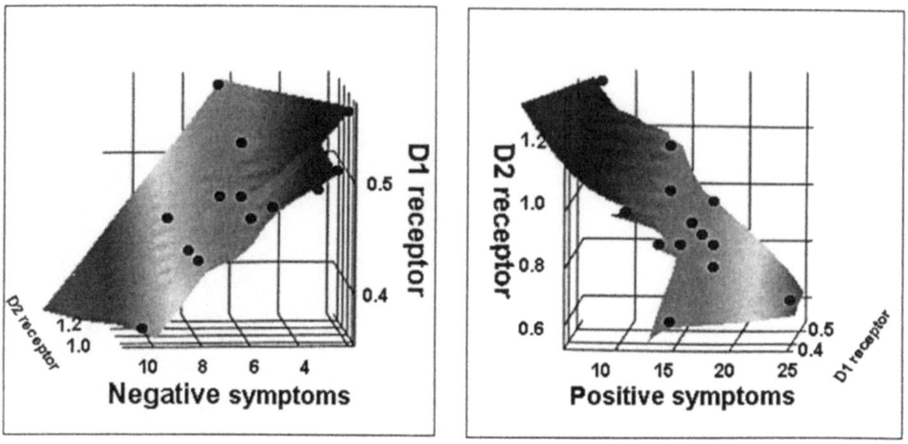

Figure 3. 3D-scattered and plane plots showing the relation among Dland D2 receptor binding in the anterior cingulate cortex and negative (left) and positive (right) symptoms in 13 schizophrenic patients (8 drug-naïve and 5 drug-free).

REFERENCES

1. M. Laruelle, Imaging dopamine transmission in schizophrenia. Q. J. Nucl. Med.**42**, 211-221 (1998).
2. F. M. Benes, J. McSparren, E. D. Bird, J. P. SanGiovanni, S. L. Vincent, Deficits in small interneurons in prefrontal and cingulate cortices of schizophrenic and schizoaffective patients. Arch. Gen. Psych. **48**, 996-1001(1991)
3. T. Suhara, H. Fukuda, O. Inoue, T. Itoh, K.Suzuki, T. Yamasaki, Y. Tateno, Age-related changes in human D1 dopamine receptors measured by positron emission tomography. Psychopharmacology, **103,** 41-45 (1991).
4. T. Suhara, Y. Sudo, T. Okauchi, J. Maeda, K. Kawabe, K. Suzuki, Y. Okubo, Y. Nakashima, H. Ito, S. Tanada, C. Halldin, L. Farde, Extrastriatal dopamine D2 receptor density and affinity in the human brain measured by 3D PET. Int. J. Neuropsychopharmacol. **2**, 73-82 (1999).
5. Y. Okubo, T. Suhara, K. Suzuki, K. Kobayashi, O. Inoue, O. Terasaki, Y. Someya, T. Sassa, Y. Sudo, E. Matsushima, M. Iyo, Y. Tateno, M. Toru. Decreased prefrontal dopamine D1 receptors in schizophrenia revealed by PET. Nature **385**, 634-636 (1997).
6. L. Farde, T. Suhara, S. Nyberg, P. Karlsson, Y. Nakashima, Y. Hietala, C. A. Halldin, A PET-study of [^{11}C]FLB 457 binding to extrastriatal D2-dopamine receptors in healthy subjects and antipsychotic drug-treated patients. Psychopharmacology **133**, 396-404 (1997).
7. F. Yasuno, T. Suhara, Y. Okubo, Y. Sudo, M. Inoue, T. Ichimiya, S. Tanada, Dose relation of limbic-cortical D_2-dopamine receptor occupancy with risperidone. Psychopharmacology **154**,112-114 (2001)
8. T. Suhara, Y. Okubo, F. Yasuno, Y. Sudo, M. Inoue, T. Ichimiya, Y. Nakashima, K. Nakayama, S. Tanada, K. Suzuki, C. Halldin, L. Farde, Decreased dopamine D2 receptor binding in the anterior cingulate cortex in schizophrenia. Arch. Gen. Psych. in press
9. H. Ito, Y. Sudo, T. Suhara, Y. Okubo, C. Halldin, L. Farde, Error analysis for quantification of [^{11}C]FLB 457 binding to extrastriatal dopamine D2 receptor in the human brain. NeuroImage **13**, 531-539 (2001)
10. T. Okauchi, T. Suhara, J. Maeda, K. Kawabe, S. Obayashi, K. Suzuki, Effect of endogenous dopamine on extrastriatal [^{11}C]FLB 457 binding measured by PET. Synapse **41**, 87-95 (2001)

THE DOPAMINE HYPOTHESIS OF SCHIZOPHRENIA: NEW ASPECTS

Arvid Carlsson*

1. ABSTRACT

In spite of its proven heuristic value, the dopamine hypothesis of schizophrenia is now yielding to a multifactorial view, where the other monoamines as well as glutamate and GABA are included, with a focus on neurotransmitter interactions in complex neurocircuits. The primary lesion(s) in schizophrenia do not necessarily involve any of these neurotransmitters directly but could deal with a more general defect, such as a faulty connectivity of developmental origin. Nevertheless, a precise identification of neurotransmitter aberrations in schizophrenia will probably provide clues for a better understanding of the disease and for the development of new treatment and prevention strategies.

2. INTRODUCTION

The dopamine hypothesis of schizophrenia has guided schizophrenia research for several decades and has clearly proven its heuristic value. However, despite emerging direct evidence supporting a dopaminergic dysfunction in schizophrenia, this hypothesis is now yielding to a multifactorial view, where the other monoamines as well as glutamate and GABA are joining up. Research in this area tends to focus on neurotransmitter interactions in complex neurocircuits. The primary lesion(s) in schizophrenia may not even involve any of these neurotransmitters directly but could deal, for example, with a more general defect in connectivity, perhaps of developmental origin.[1] Nevertheless, the identification of neurotransmitter aberrations in schizophrenia, now underway, is likely to bring clues for understanding the fundamental nature of the disease as well as for the development of new treatment and prevention strategies.

* Department of Physiology and Pharmacology, University of Göteborg, Göteborg, Sweden, Box 431, SE 40530 Göteborg, Sweden

Catecholamine Research, Edited by Nagatsu *et al.*
Kluwer Academic/Plenum Publishers, 2002

3. DOPAMINE: STILL A CORNERSTONE

In recent years important progress has been made in basic schizophrenia research. The dopamine hypothesis of schizophrenia, which postulates a dopaminergic dysfunction in this disorder, has for a long time been supported only by indirect pharmacologic evidence,[2] but has now received more direct support by two different lines of research, both using imaging techniques. First, it has been shown that the synthesis of labeled dopamine or fluorodopamine in the brain, measured by means of PET following administration of radiolabeled DOPA or fluoro-DOPA, is increased in drug-naive schizophrenic patients, compared to age-matched controls.[3, 4, 5] Second, SPECT and PET studies, using a sophisticated technique to measure the release of dopamine in the basal ganglia *in vivo*, have shown that following an amphetamine challenge this release is elevated in drug-naive schizophrenic patients compared to age-matched controls, and that this elevation correlates to the induction of positive psychotic symptoms.[6, 7, 8] In an additional series of experiments Laruelle,[9] using the SPECT technique with α-methyltyrosine as a tool, have obtained evidence that the unchallenged release of dopamine is elevated in schizophrenic patients compared to controls.

While these novel data are impressive, a number of caveats should be remembered. First, although statistically significant aberrations of dopamine synthesis and release have been demonstrated in schizophrenic patients, the data show a considerable scatter with some of the values observed within a normal range. Thus a dysfunction of dopamine may be limited to a subpopulation of patients suffering from this probably heterogeneous disorder. (A few observations suggest that the aberration of dopamine synthesis may actually go in the opposite direction in catatonia, compared to other cases of schizophrenia.) Interestingly, the elevated dopamine release correlated to a good response to antipsychotic drugs.

Secondly, it must be remembered that the abnormal values were obtained in patients who were exposed to the stress inevitably caused by the imaging procedures. Whether a dopamine dysfunction occurs also under minimal stress thus remains an open question.

Thirdly, the patients examined were in acute episodes, and the situation may be different in chronic schizophrenic patients between episodes. In fact, recent observations by Laruelle *et al*[9] indicate that the amphetamine-induced release of dopamine in schizophrenic patients in remission is within the normal range. This tallies with the clinical experience that patients in remission complain about the side effects of antidopaminergic drugs more than during an exacerbation. If the level of dopaminergic function is normal in patients in remission, the practical implications are obvious. All agents used today to prevent relapse in schizophrenia are antidopaminergic and should thus induce hypodopaminergia in such patients. This is a most unpleasant and incapacitating condition leading to extrapyramidal side effects and, perhaps more importantly, to a failure of the reward system, resulting in dysphoria and anhedonia, and to a cognitive deficit. To develop drugs capable of preventing relapse without these side effects should be an urgent task. In fact, as will be discussed below, such agents may already be underway.

Finally, this elevation of dopamine release in schizophrenia is not necessarily a primary phenomenon but could be secondary to, for example, hypoglutamatergia. In support of this it has been found in rats as well as in humans that drugs blocking NMDA receptors are capable of enhancing the spontaneous and especially the amphetamine-induced release of dopamine. The latter effect could be due to blockade of a negative feedback mechanism.

On the whole, these recent data would tend to support treatment strategies involving dopamine. However, most of the efforts made in recent years using such strategies have been discouraging. These efforts have largely focused on the D4 and partly also on the D3 and D1 subtypes. A general feeling seems to prevail that the D2 subtype is no longer an attractive target for an improved antipsychotic therapy.

4. DOPAMINERGIC STABILIZERS: A PROMISING NOVEL THERAPEUTIC APPROACH

Is the therapeutic potential of dopaminergic agents exhaused? Several reasons support the view that this is far from the case. First of all, the role of the various subtypes of dopamine receptors need to be further explored. But perhaps even more important will be some ongoing attempts to reach a deeper understanding of the function of these receptors. This may open up entirely new ways to improve this function and to optimize the receptors' abilities to cope with aberrations in neural circuits. In support of this prediction some recent observations in our research group will be briefly mentioned.

We have developed a series of compounds capable of stabilizing the dopaminergic system without inducing the hypodopaminergia so ominous for the currently used antipsychotic drugs. Some of these new drugs are partial dopamine receptor agonists, acting on the D2 family of receptors. A number of partial dopamine receptor agonists, developed by us and by others, are now in clinical trials and seem to offer promise (for recent clinical data on (-)-3PPP, see ref. 10; several of the partial dopamine receptor agonists studied so far are less suitable as probes because of their poor selectvity, e.g the ergolines terguride and SDZ 208-912). Others are pure antagonists, again acting on the D2 family of receptors, and can thus readjust elevated dopamine functions, but in contrast to the currently used antipsychotic agents they do not cause hypodopaminergia. In fact, they rather antagonize subnormal dopamine function. The reason for this aberrant pharmacological profile seems to be that their action on different subpopulations of dopamine receptors differs from that of the currently used drugs. Thus, whereas they exert a strong action on dopaminergic autoreceptors, they have a weaker effect postsynaptically and seem unable to reach a subpopulation of postsynaptic dopamine receptors.[11, 12, 13]

In subhuman primates, in which Parkinsonism had been induced by MPTP, one member of this class, named (-)-OSU6162, given in single doses, could prevent L-DOPA-induced dyskinesias without interfering with the therapeutic movement response, and in subsequent trials on Parkinsonian patients the same kind of response was observed (14, and J. Tedroff *et al*, unpublished data). Subsequent trials on patients with Huntington's disease[15] showed a marked reduction of choreatic movements, considerably outlasting the presence of the drug in the blood. These observations support the view that drugs of this class are capable of stabilizing dopaminergic function, that is, they are able to alleviate signs of hyperdopaminergia without inducing any signs of reduced dopaminergic function. If these findings can be extrapolated from neurology to psychiatry, these agents should possess antipsychotic activity without any concomitant signs of hypodopaminergia. Forthcoming trials with such agents in schizophrenia will answer this question. In fact, a preliminary study on a few schizophrenic patients, using a double blind crossover design, has demonstrated an antipsychotic action of (-)-OSU6162.[16]

5. BEYOND DOPAMINE

In view of the close interaction between neurotransmitters in the brain, it is unlikely that dopamine is the only neurotransmitter showing dysfunction in schizophrenia. As already indicated, the change in dopaminergic function may even be secondary to aberrations elsewhere. In any event there are good reasons to study the function of several other neurotransmitters in schizophrenia, such as noradrenaline, serotonin, acetylcholine, glutamate and GABA. These neurotransmitters are more difficult to study in the living intact brain than dopamine. Least difficult would perhaps be serotonin, because it seems possible to study it using the same kind of approach as for dopamine, that is to administer radiolabelled precursor (5-hydroxytryptophan) and measure the turnover of serotonin. Such a study has been carried out in depressed patients, and an aberration was actually demonstrated.[17]

For several years considerable interest has focussed on the possible role of glutamate in schizophrenia.[18, 19] One reason for this is the discovery that phencyclidine (PCP, "angel dust"), which can induce a psychotic condition mimicking schizophrenia, perhaps even more faithfully than the amphetamines, is a powerful antagonist on one of the glutamate receptor subtypes, namely the NMDA receptor.[20] This receptor is equipped with an ion channel regulating the penetration of calcium and other cations into the neuron. PCP binds to a specific site in this channel, thereby blocking the function of the receptor. A number of other NMDA antagonists are available, binding to different sites of the receptor molecule, such as to the "PCP site", e g MK-801 and ketamine (an "uncompetitive" binding), or (competitively) to the same site as glutamate, e g AP5, D-CPPene and CGS 19755, or to still another site on the NMDA receptor, where glycine functions as an additional agonist. An example of antagonists acting at the glycine site is D-cycloserine (a mixed agonist-antagonist). All these different NMDA antagonists appear to be psychostimulants, at least in rodents, and psychotogenic in humans.

6. CONCLUDING REMARKS - OUTLOOK

As mentioned in the Introduction, the dopamine hypothesis of schizophrenia has received fairly strong both indirect pharmacological and more recently, direct pathophysiological support. Thus a dysfunction of the dopaminergic system is probably somehow involved in the disease. However, the role of dopamine in the pathophysiology or etiology of schizophrenia should not be overemphasized. In our original paper,[21] which is often quoted as the origin of the hypothesis, we only proposed that major antipsychotic agents might act by blocking receptors for catecholamines. Our data pointed to a particularly strong action on dopamine, but we did not emphasize that. Rather, we proposed that serotonin receptors could also be involved in the antipsychotic action. During a subsequent phase several authors presented data more strongly supporting the role of dopamine, which then came to dominate the scene for a couple of decades. It was not until the 1980s that glutamate started to attract attention in this context. This was largely due to the discovery that PCP, a strongly schizophrenia-mimicking agent, is an antagonist on the glutamatergic NMDA receptors. In parallel with this development there was an impressive growth of knowledge regarding the neuroanatomy of pathways of potential psychotogenetic interest. Thus, during the past decade a new direction of research has emerged, focusing upon the interaction of a variety of neurotransmitters in complex neurocircuits engaged in the control of major mental functions, such as affect, emotions, cognition and their involvement in reward-directed and other complex behavior. Rather than trying to identify a single culprit in psychotogenesis the emphasis is now moving towards trying to understand these interactions and how imbalances in such circuitry can arise, perhaps mainly on a multifactorial basis. In such a perspective it seems reasonable to look upon schizophrenia as a syndrome of heterogeneous etiology and pathophysiology. Evidence quoted in this review argues for a role of not only dopamine and glutamate but also serotonin, noradrenaline and GABA.There is now overwhelmingly convincing evidence that a primary disturbance in one neurotransmitter system will inevitably influence several other systems. To try to establish, in a given case, the order in which a complex chain of events has arisen and thus to identify the origin, would in general seem unrealistic. A more modest aim would be to try to describe aberrations in the patterns of physiological and biochemical events and to try to correlate such aberrations to behavioral changes.

Just as reductionism has emphasized the search for an individual culprit among the neurotransmitters, it has led to attempts to identify a particular brain region, such as the prefrontal cortex or the temporal lobe, as the site of the schizophrenic lesion. Rather compelling evidence for a role of both these and a number of other regions has been advanced. Again, to account for all the evidence it will probably be useful to look for a more complex disease model in which many different brain regions in the upper and lower brainstem, the basal ganglia and the limbic cortex as well as the neocortex are engaged and interact in complex circuitries. In order to define the contribution of different brain regions to psychotogenesis it may again prove useful to apply a pattern recognition approach.

For pattern recognition, multivariate analysis is an essential tool. In our own research we use this tool for describing aberrations in neurotransmitter neurochemistry as well as in behavior in animal disease models. We have also used it for describing the biochemical aberrations occurring in schizophrenic brains *post mortem*. Recently we have started to apply multivariate analysis on biochemical measurements related to neurotransmitter function, using PET methodology, in humans belonging to different diagnostic groups, either untreated or following therapeutic interventions. One goal is to correlate the patterns observed in animal models to those observed in different diagnostic categories of patients, in either case with or without therapeutic interventions. We feel confident that this will bridge the gap between animal and human research and speed up the development of novel therapeutic procedures.

REFERENCES

1. Bunney BG, Potkin SP, Bunney WE Jr. 1995. New morphological and neuropathological findings in schizophrenia: a neurodevelopmental perspective. *Clin. Neurosci.* **3**:81-88
2. Carlsson A. 1995. The dopamine theory revisited. In Schizophrenia, eds. SR Hirsch, DR Weinberger, pp. 379-400. Oxford: Blackwell Science
3. Hietala J, Syvälahti E, Vuorio K, Någren K, Lehikoinen P, Ruotsalalainen U *et al.* 1994. Striatal D2-dopamine receptor characteristics in neuroleptic-naive schizophrenic patients studied with positron emission tomography. *Arch. Gen. Psychiatry* **51**:116-123
4. Dao-Costellana M-H, Paillère-Martinot M-L, Hantraye P, Attar-Lévy D, Rémy P, Crouzel C *et al.*1997. Presynaptic dopaminergic function in the striatum of schizophrenic patients. *Schizophrenia Research* **23**:167-174
5. Lindström LH, Gefvert O, Hagberg G, Hagström P, Lundberg T, Bergström P *et al.* 1997. Increased synthesis of dopamine in prefrontal cortex and striatum in drug-naive schizophrenic patients studied by use of C^{11}-labelled l-DOPA and positron emission tomography (PET). Abstract, 36th Annual ACNP Meelting, December 8-12, 1997, p. 290
6. Laruelle M, Abi -Dargham A, van Dyck CH, Gil R, D'Souza CD, Erdos J, McCance E *et al.*1996. Single photon emission computerized tomography imaging of amphetamine-induced dopamine release in drug-free schizophrenic subjects. *Proc. Natl. Acad. Sci.* U S A. 20, **93**(17): 9235-40
7. Breier A, Su T-P, Saunders R, Carson RE, Kolachana BS, de Bartolomeis A *et al.* 1997. Schizophrenia is associated with elevated amphetamine-induced synaptic dopamine concentrations: Evidence from a novel positron emission tomography method. *Proc. Natl. Acad. Sci.* USA **94**:2569-2574
8. Abi-Dargham A, Gil R, Krystal J, Baldwin RM, Seibyl JP, Laruelle M *et al.* 1998. Increased striatal dopamine transmission in schizophrenia: confirmation in a second cohort. *Am. J. Psychiatry* **155**: 761-767
9. Laruelle M. 2000. Imaging dopamine dysregulation in schizophrenia: Implication for treatment. Presented at Workshop Schizophrenia: Pathological Bases and Mechanisms of Antipsychotic action, Chicago
10. Lahti AC, Weiler MA, Corey PK, Lahti RA, Carlsson A, Tamminga CA. 1998. Antipsychotic properties of the partial dopamine agonist (-)-3-(3-hydroxyphenyl)-N-n-propylpiperidine (preclamol) in schizophrenia. *Biol. Psychiat.* **43**:2-11
11. Svensson K, Hjorth S, Clark D, Carlsson A, Wikström H, Andersson B, *et al.*1986. (+)-UH 232 and (+)-UH 242: Novel stereoselective DA receptor antagonists with preferential action on autoreceptors. *J.Neural Transm.* **65**:1-27
12. Sonesson C, Lin C-H, Hansson L, Waters N, Svensson K, Carlsson A *et al.*1994. Substituted (S)-phenylpiperidines and rigid congeners as preferential dopamine autoreceptor antagonists: Synthesis and structure-activity relationships. *J. Med. Chem.* **37**:2735-2753
13. Hansson LO, Waters N, Holm S, Sonesson C.1995. On the quantitative structure-activity relationships of *meta*-substituted (S)-phenylpiperidines, a class of preferential dopamine D2 autoreceptor ligands. Modeling of dopamine synthesis and release in vivo by means of partial least squares regression. *J.Med. Chem.* **38**:3121-3131
14. Ekesbo A, Andrén PE, Gunne LM, Tedroff J. 1997. (-)-OSU6162 inhibits levodopa-induced dyskinesias in a monkey model of Parkinson's disease. *Neuroreport* **8**:2567-2570
15. Tedroff J, Ekesbo A, Sonesson C,Waters N, Carlsson A. 1999. Long-lasting improvement following (-)-OSU6162 in a patient with Huntington's disease. *Neurology* **53**:1605-1606
16. Gefvert O, Lindström LH, Dahlbäck O, Sonesson C, Waters N *et al.* 2000. (-)-OSU6162 induces a rapid onset of antipsychotic effect after a single dose. A double-blind study. Presented at Scand. Soc. Psychopharmacol., 41st meeting. *Nordic J. Psychiat.* **54**/2:93-94
17. Ågren H, Reibring L, Hartvig P, Tedroff J, Bjurling P, Hornfeldt K *et al.* 1991. Low brain uptake of L-[11C]5-hydroxytryptophan in major depression: a positron emission tomography study on patients and healthy volunteers. *Acta Psychiatr. Scand.* **83**:449-455
18. Kim JS, Kornhuber HH, Schmid-Burgk W, Holzmüller. 1980. Low cerebrospinal fluid glutamate in schizophrenic patients and a new hypothesis on schizophrenia. *Neurosci. Lett.* **20**:379-382
19. Garland Bunney B, Bunney WE, Carlsson A. 1995. Schizophrenia and glutamate. In *Psychopharmacology: The Fourth Generation of Progress*, eds. FE Bloom, DJ Kupfer, pp. 1203-1214, New York: Raven Press
20. Lodge D.1989. Modulation of N-methylaspartate receptor channel complexes. *Drugs of Today* **25**:395-411
21. Carlsson A, Lindqvist M.1963. Effect of chlorpromazine or haloperidol on the formation of 3-methoxytyramine and normetanephrine in mouse brain. *Acta Pharmacol. (Kbh.)* **20**:140-144

MOLECULAR GENETIC ANALYSIS OF PLASMA DOPAMINE β-HYDROXYLASE IN DEPRESSION

New implications for an old biomarker

Joseph F. Cubells, Cyrus P. Zabetian, George M. Anderson, Lawrence H. Price, Barnett S. Meyers, Robert T. Malison, J. Craig Nelson, Toshiharu Nagatsu, Hiroshi Ichinose, Joel Gelernter*

1. INTRODUCTION

Dopamine β-hydroxylase (DβH) catalyzes the conversion of dopamine to norepinephrine. It is localized within neurosecretory vesicles of noradrenergic and adrenergic neurons and neurosecretory cells, where it is present in both soluble and membrane bound fractions. DβH enzyme activity can be measured in plasma or serum. Enzyme activity and levels of DβH protein in these body fluids are strongly correlated and essentially reflect the same biochemical trait. DβH levels are strongly influenced by genetic inheritance (Weinshilboum, 1978) and the structural locus encoding DβH protein, *DBH*, is the major locus controlling the trait (Zabetian et al., 2001).

Plasma DβH levels in patients with unipolar major depression with psychotic features are lower than in patients with uncomplicated unipolar depression (Meltzer et al., 1976; Mód et al., 1986; Sapru et al., 1989; Meyers et al., 1999). Given the strong influence of *DBH* genotype on plasma DβH levels, we conducted a *DBH*-genotype-controlled analysis of plasma DβH levels in unipolar major depression.

* Joseph F. Cubells, Cyrus P. Zabetian, George M. Anderson, Robert T. Malison, J. Craig Nelson and Joel Gelernter, Yale University School of Medicine and VA Connecticut, West Haven, CT 06516. Lawrence H. Price, Brown University School of Medicine, Providence RI, 02906; Barnett S. Meyers, Cornell University Medical College, White Plains, NY 10465; Toshiharu Nagatsu and Hiroshi Ichinose, Fujita Health University, Toyoake, Japan.

2. METHODS

2.1. Identification of a single nucleotide polymorphism (SNP) at DBH associated with plasma DβH activity

We employed sequencing-based mutational analysis of DNA samples from individuals with extreme plasma DβH phenotypes to identify a single nucleotide polymorphisms at *DBH*, and to assess their association to plasma DβH levels in African American (AA), European-Americans (EA), and Japanese (Jp) populations, as previously described (Zabetian, et al., 2001).

2.2. Genotype-controlled analysis of plasma DβH activity in unipolar depression

The methods of this study will be described in detail in a forthcoming publication (Cubells et al., manuscript in preparation). Briefly, EA patients satisfying DSM-IIIR criteria for unipolar major depressive disorder, either with or without psychotic features, gave informed consent to participate in molecular genetic studies. Heparinized blood was drawn, aliquots of plasma were assayed for DβH activity (Cubells et al., 1998), and DNA for genotype analysis was extracted from the cellular fractions. Plasma DβH levels in the two patient groups were compared, with each group further subdivided by *DBH* genotype.

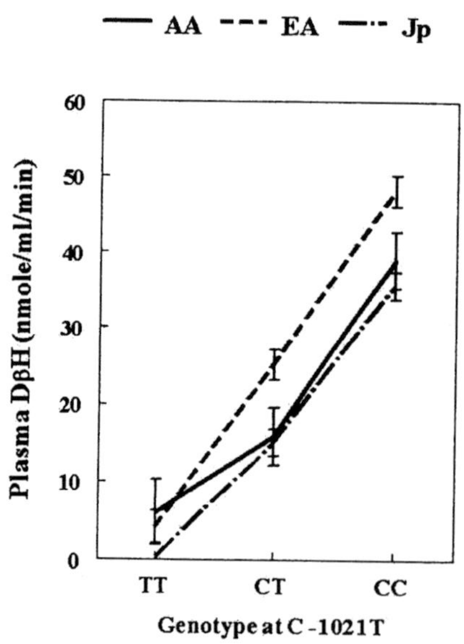

Figure 1. Dependence of mean (± S.E.M.) plasma DβH activity on genotype at C −1021T in subjects from three populations.

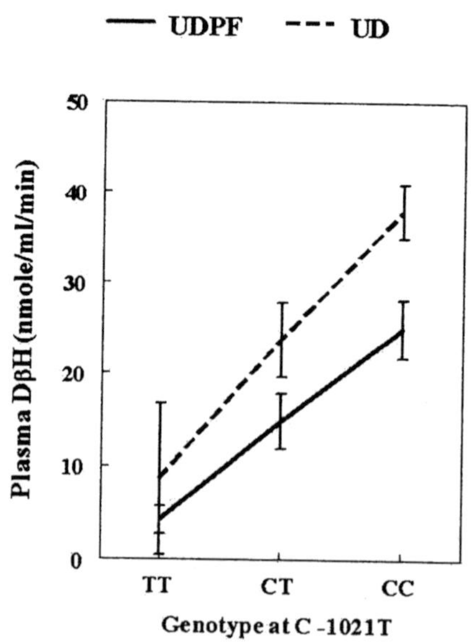

Figure 2. Mean (± S.E.M.) plasma DβH activity in depressed patients with (UDPF) or without (UD) psychotic features, genotyped at C −1021T.

3. RESULTS

3.1. C-1021T associates with plasma DβH activity in three human populations of different geographic origin

Figure 1 summarizes the results of our sequence-based search for polymorphisms in *DBH* that associate with plasma DβH activity in EA, AA, and Jp individuals. For all three groups, genotype at C –1021T strongly predicted plasma DβH activity, accounting for 30-50% of the total variance in the enzyme measure.

3.2 Plasma DβH activity differs in depressed patients with and without psychotic features, independent of genotype at C –1201T

Figure 2 shows plasma DβH activity in depressed patients, with and without psychotic features, grouped according to genotype at C –1021T. Two-way analysis of variance showed significant effects of genotype and diagnosis, but no significant interaction between these factors (see figure legend for statistical analysis). Allele and genotype frequecies at C –1021T did not differ in depressed patients with or without psychotic features (data not shown).

4. DISCUSSION

This report presents two major findings. First, C –1021T strongly associates with plasma DβH levels across three populations of distinct geographic origin (Zabetian et al., 2001). The location of C –1021T in the proximal 5' region of *DBH* suggests the variant could differentially influence gene transcription by a cis-acting mechanism. Several considerations are consistent with the hypothesis that C –1021T influences *DBH* promoter function. First, C –1021T is the only polymorphism at *DBH* identified to date that associates with variation in plasma DβH levels across three populations (EA, AA, Jp; Zabetian et al., 2001). Given the substantial population-based variation in *DBH* allele frequencies (Cubells et al., 1997; Zabetian et al., 2001), such cross-population consistency in genotype-phenotype relationship would be expected only for a functional polymorphism, or one in very tight linkage disequilibrium with a functional polymorphism. Second, plasma DβH activity varies predominantly because of differences in levels of circulating protein, rather than differences in homospecific enzyme activity (Weinshilboum 1978). Thus, the location of C-1021T in the proximal 5' region of *DBH* is consistent with a mechanism in which altered expression of gene product, rather than a change in protein primary structure, accounts for the phenotypic difference. Third, C-1021T resides in a DNase-sensitive area of the DBH gene (Ishiguro et al., 1993). Fourth, transgenic expression of lac-Z reporter constructs driven by DBH 5' sequence in mice suggests that the region between -600 and -1100 bp contains elements critical for DBH gene expression. While none of the foregoing observations prove a cis-acting functional mechanism for the association of C –1021T, the convergence of so many lines of circumstantial evidence certainly warrants further investigation.

The second major finding reported here is that plasma DβH levels differ in unipolar depression, with and without psychotic features, independent of variation at C –1021T.

This result was unexpected, because C −1021T accounts for 30-50% of the variance in plasma DβH (Zabetian et al., 2001). Several alternative mechanisms could account for the observed difference in mean plasma DβH activity between depressed patients with, and without, psychotic features. First, genetic variation elsewhere in the *DBH* gene, or at other loci, could account for some or all of the differences in plasma DβH between the diagnostic groups. Thus, further analysis of genetic mechanisms in unipolar depression seems warranted. Second, altered neurobiological or neuroendocrine function in unipolar depression with psychotic features, resulting in diminished expression of DβH protein, could lower circulating levels of DβH protein. Finally, other nonheritable factors influencing plasma levels of DβH could differ consistently between depressed patients with and without psychotic features. The foregoing hypotheses are not mutually exclusive; i.e., the group differences observed here could result from some combination of two or more of the above mechanisms. Investigating these possibilities should shed new light on our understanding of psychotic depression.

REFERENCES

Cubells, J.F., Kobayashi, K., Nagatsu, T., Kidd, K.K., Kidd, J.R., Calafell, F., Kranzler, H. R., Ichinose, H., and Gelernter, J. (1997) Population genetics of a functional variant of the dopamine β-hydroxylase gene (DBH), *Amer. J. Medical Genet. (Neuropsychiatric Genet.)* **74**: 374-379.

Cubells, J.F., van Kammen, D.P., Kelley, M.E., Anderson, G.M., O'Connor, D.T., Price, L.H., Malison, R., Rao, P.R., Kobayashi, K., Nagatsu, T., and Gelernter, J. (1998) Dopamine β-hydroxylase: two polymorphisms in linkage disequilibrium at the structural gene DBH associate with biochemical phenotypic variation, *Hum. Genet.* **102**: 533-540.

Ishiguro H., Kim K.T., Joh T.H., Kim K.S. (1993) Neuron-specific expression of the human dopamine β-hydroxylase gene requires both the cAMP-response element and a silencer region, *J. Biol. Chem.* **268**: 17987-17994.

Meltzer H.Y., Hyong H.W., Carroll B.J., and Russo P. (1976) Serum dopamine-β-hydroxylase in the affective psychoses and schizophrenia: Decreased activity in unipolar psychotically depressed patients. *Arch. Gen. Psychiatry* **33**: 585-591.

Meyers B.S., Alexopoulos G.S., Kakuma T., Tirumalasetti F., Gabriele M., Alpert S., Bowden C., Meltzer H.Y. (1999) Decreased dopamine beta-hydroxylase activity in unipolar geriatric delusional depression. *Biol. Psychiatry* **45**: 448-452.

Mód, L., Rihmer, Z., Magyar, I., Arató, M., Alföldi, A., and Bagdy, G. (1986) Serum DBH activity in psychotic vs. nonpsychotic unipolar and bipolar depression. *Psychiatry Res. 19:* 331-333.

Sapru M.K., Rao B.S.S.R., Channabasavanna S.M. (1989) Serum dopamine-β-hydroxylase activity in clinical subtypes of depression, *Acta Psychiatr. Scand.* **80**: 474-478.

Weinshilboum R.M. (1978) Serum dopamine-β-hydroxylase. *Pharmacol. Rev.* **30**: 133-166.

Zabetian C.P., Anderson G.M., Buxbaum S.G., Elston R.C., Ichinose H., Nagatsu T., Kim K.S., Kim C.-H., Malison R.T., Gelernter J., and Cubells J.F. (2001) A quantitative trait locus analysis of human dopamine β-hydroxylase activity: evidence for a major functional polymorphism at the *DBH* locus. *Am. J. Hum. Genet.* **68**:515-522.

DOPAMINERGIC HYPERACTIVITY IN STRIATUM IN SCHIZOPHRENIA: A FAILURE OF CORTICAL GLUTAMATERGIC CONTROL?

Lawrence S. Kegeles and Marc Laruelle*

1. INTRODUCTION

The classical dopamine hypothesis of schizophrenia proposes that positive symptoms of the disorder are a result of hyperactivity of dopamine transmission.[1] This hypothesis has received new support from brain imaging studies that have shown an exaggerated response to acute amphetamine challenge in untreated patients with schizophrenia compared with healthy subjects using dopamine D_2 receptor SPECT (single photon emission computed tomography) or PET (positron emission tomography).[2-4]

The mechanism underlying dopaminergic hyperfunction of the striatal terminal projections in schizophrenia remains to be clarified, in view of the lack of evidence for pathology of these neurons. The glutamate hypothesis of schizophrenia[5] holds that the illness is associated with a deficiency of glutamate transmission, which might in turn give rise to dysregulation of dopamine transmission.[6, 7]

Here we present a series of D_2 receptor imaging studies that support an underlying glutamatergic regulatory dysfunction. The scans were combined with pharmacological disruption of cortical-subcortical glutamatergic regulation in healthy volunteers and in nonhuman primates. Two agents were used, the noncompetitive NMDA (*N*-methyl-d-aspartate) glutamate receptor antagonist ketamine, and the group II metabotropic glutamate receptor agonist LY354740.[8]

2. KETAMINE MODULATION OF RESTING DOPAMINE RELEASE

A PET study using the radioligand [^{11}C]raclopride examined the effects of ketamine on resting dopamine activity[9] in healthy humans. This study (n = 20 scans) included 10 two-hour "ketamine condition" scans with a bolus plus constant infusion of [^{11}C]raclopride, and ketamine administration from 50 min to the end of the study (0.12

* Departments of Psychiatry and Radiology, Columbia University, New York, NY 10032

Catecholamine Research, Edited by Nagatsu *et al.*
Kluwer Academic/Plenum Publishers, 2002

mg/kg i.v. bolus plus 0.65 mg/kg/hr i.v. infusion). An additional 10 "control condition" scans were obtained without ketamine. The striatal specific to nonspecific binding ratio (V_3") was quantified prior to and during ketamine infusion. We found significant dissociative, positive, and negative symptom induction, but no significant changes in V_3" induced by ketamine administration. The absence of effect contrasts with other studies in healthy humans,[10-12] but is consistent with microdialysis studies that showed a small effect in rodents[13] or nondetectable effect in nonhuman primates.[14] This study together with the microdialysis data suggests an absence of effect of acute NMDA receptor blockade on resting striatal dopamine transmission.

3. KETAMINE MODULATION OF STIMULATED DOPAMINE RELEASE

In marked contrast, we found a significant modulatory effect of ketamine on amphetamine-stimulated dopamine release.[15] Eight healthy volunteers underwent SPECT imaging (n = 32 scans) with the D_2 receptor radioligand [^{123}I]IBZM on two different days ("control condition" and "ketamine condition"). Each participant underwent two one-hour scans under control conditions (one scan before and one after d–amphetamine 0.25 mg/kg i.v.) and two scans under ketamine conditions, again pre and post-amphetamine. Ketamine was administered i.v. over 4 hours, beginning one hour before the first scan (0.2 mg/kg bolus plus 0.4 mg/kg/hr infusion). Plasma ketamine levels were unchanged by amphetamine injection. The amphetamine-induced decrease in [^{123}I]IBZM V_3" changed from 5.5 ± 3.5% under control conditions to 12.8 ± 8.8% under ketamine conditions, $p = 0.02$ by repeated measures ANOVA. This increase was comparable to the exaggerated response to challenge by amphetamine alone reported in patients with schizophrenia. Clinical ratings showed an absence of induction of either dissociative or psychotic symptoms by amphetamine alone. Ketamine induced significant dissociative and negative symptoms that were diminished acutely by amphetamine bolus. The effect of ketamine alone on [^{123}I]IBZM binding was evaluated by comparing the pre-amphetamine scans under control and under ketamine conditions. The effect of ketamine alone was nondetectable, consistent with the [^{11}C]raclopride PET study described above. This study suggests that ketamine results in excess striatal dopamine transmission under stimulated but not resting conditions.

Sensitization or tolerance to repeated administration of amphetamine was eliminated as a possible explanation of the enhanced response to amphetamine in the second (ketamine) experiment by the high test/retest reliability found in a study of repeated amphetamine injections in healthy humans at a 2-week interval.[16]

4. METABOTROPIC GLUTAMATE RECEPTOR AGONIST MODULATION OF STIMULATED DOPAMINE RELEASE

To clarify the mechanism of the dopamine-enhancing effects of ketamine, we conducted a [^{11}C]raclopride PET study[17] of the modulatory effects of the group II metabotropic glutamate receptor agonist LY354740 on amphetamine-stimulated dopamine release, an agent which acts presynaptically to inhibit stimulated release of glutamate.[8] In this study, each of 3 baboons underwent 2 pairs of PET scans (n = 12 scans). The first pair consisted of a baseline scan followed by amphetamine challenge

(0.5 mg/kg) followed by the second scan. On a separate day two to four weeks later, a second pair of PET scans consisted of a baseline scan followed by injection of LY354740 (20 mg/kg) followed by amphetamine (0.5 mg/kg) followed by the second scan. LY354740 challenge enhanced striatal amphetamine-induced dopamine release significantly (from 28 ± 9% decrease in V_3" to 37 ± 10% decrease, or a 32 ± 8% enhancement, $p = 0.02$ by repeated measures ANOVA).

5. SUMMARY

D_2 receptor imaging was used to show that acute i.v. ketamine administration enhances amphetamine-stimulated (but not resting baseline) dopamine transmission, and does so to a degree similar to the exaggerated response to amphetamine alone previously reported in patients with schizophrenia. These findings leave open the question whether ketamine causes this enhancement by stimulating glutamate release[18] driving the "accelerator" circuitry in the Carlsson model,[19] or instead by impairing the "brake" circuitry. A study in baboons using LY354740 similarly enhanced amphetamine-stimulated dopamine release, suggesting that acute impairment of glutamate transmission, whether by LY354740 or by ketamine, causes impairment of the "brake" circuitry under conditions of stimulated dopamine release. In conclusion, acute disruption of glutamatergic transmission appears to provide a pharmacological model of the striatal dopaminergic hyperactivity found in schizophrenia.

REFERENCES

1. A. Carlsson and M. Lindqvist, Effect of chlorpromazine or haloperidol on formation of 3-methoxytyramine and normetanephrine in mouse brain, *Acta Pharmacol. Toxicol.* **20**, 140-144 (1963).
2. M. Laruelle, A. Abi-Dargham, C. H. van Dyck, R. Gil, C. D. De Souza, J. Erdos, E. McCance, W. Rosenblatt, C. Fingado, S. S. Zoghbi, R. M. Baldwin, J. P. Seibyl, J. H. Krystal, D. S. Charney, and R. B. Innis, Single photon emission computerized tomography imaging of amphetamine-induced dopamine release in drug free schizophrenic subjects, *Proc. Natl. Acad. Sci.* **93**, 9235-9240 (1996).
3. A. Breier, T. P. Su, R. Saunders, R. E. Carson, B. S. Kolachana, A. deBartolomeis, D. R. Weinberger, N. Weisenfeld, A. K. Malhotra, W. C. Eckelman, and D. Pickar, Schizophrenia is associated with elevated amphetamine-induced synaptic dopamine concentrations: Evidence from a novel positron emission tomography method, *Proc. Natl. Acad. Sci.* **94**, 2569-2574 (1997).
4. A. Abi-Dargham, R. Gil, J. Krystal, R. Baldwin, J. Seibyl, M. Bowers, C. van Dyck, D. Charney, R. Innis, and M. Laruelle, Increased striatal dopamine transmission in schizophrenia: confirmation in a second cohort, *Am. J. Psychiatry* **155**, 761-767 (1998).
5. D. C. Javitt and S. R. Zukin, Recent advances in the phencyclidine model of schizophrenia, *American Journal of Psychiatry* **148**, 1301-1308 (1991).
6. A. A. Grace, Phasic versus tonic dopamine release and the modulation of dopamine system responsivity: a hypothesis for the etiology of schizophrenia., *Neuroscience* **41**, 1-24 (1991).
7. J. W. Olney and N. B. Farber, Glutamate receptor dysfunction and schizophrenia, *Arch. Gen. Psychiatry* **52**, 998-1007 (1995).
8. G. Battaglia, J. A. Monn, and D. D. Schoepp, In vivo inhibition of veratridine-evoked release of striatal excitatory amino acids by the group II metabotropic glutamate receptor agonist LY354740 in rats, *Neurosci Lett* **229**, 161-164. (1997).
9. L. Kegeles, D. Martinez, L. Kochan, N. Simpson, D. R. Hwang, O. Mawlawi, T. Cooper, R. Van Heertum, and M. Laruelle, Ketamine-induced dopamine release measured with [11C]raclopride PET in humans, *Soc Neurosc Abst* **26**, 1839 (2000).
10. A. Breier, C. M. Adler, N. Weisenfeld, T. P. Su, I. Elman, L. Picken, A. K. Malhotra, and D. Pickar, Effects of NMDA antagonism on striatal dopamine release in healthy subjects: application of a novel PET approach, *Synapse* **29**, 142-147 (1998).

11. G. S. Smith, R. Schloesser, J. D. Brodie, S. L. Dewey, J. Logan, S. A. Vitkun, P. Simkowitz, A. Hurley, T. Cooper, N. D. Volkow, and R. Cancro, Glutamate modulation of dopamine measured in vivo with positron emission tomography (PET) and 11C-raclopride in normal human subjects, *Neuropsychopharmacology* **18**, 18-25 (1998).
12. F. X. Vollenweider, P. Vontobel, I. Oye, D. Hell, and K. L. Leenders, Effects of (S)-ketamine on striatal dopamine: a [11C]raclopride PET study of a model psychosis in humans, *J Psychiatr Res* **34**, 35-43 (2000).
13. D. W. Miller and E. D. Abercrombie, Effects of MK-801 on spontaneous and amphetamine-stimulated dopamine release in striatum measured with in vivo microdialysis in awake rats, *Brain Research Bulletin* **40**, 57-62 (1996).
14. B. W. Adams, S. R. Rubino, C. W. Bradberry, and B. Moghaddam, Effects of phencyclidine, ketamine,and amphetamine on striatal extracellular dopamine levels of awake primate, *Soc Neurosci Abst* **26**, 728 (2000).
15. L. S. Kegeles, A. Abi-Dargham, Y. Zea-Ponce, J. Rodenhiser-Hill, J. J. Mann, R. L. Van Heertum, T. B. Cooper, A. Carlsson, and M. Laruelle, Modulation of amphetamine-induced striatal dopamine release by ketamine in humans: implications for schizophrenia, *Biol Psychiatry* **48**, 627-640 (2000).
16. L. S. Kegeles, Y. Zea-Ponce, A. Abi-Dargham, J. Rodenhiser, T. Wang, R. Weiss, R. L. Van Heertum, J. J. Mann, and M. Laruelle, Stability of [^{123}I]IBZM SPECT measurement of amphetamine-induced striatal dopamine release in humans, *Synapse* **31**, 302-308 (1999).
17. B. N. Van Berckel, R. N. Waterhouse, D. Hwang, O. Mawlawi, Y. Huang, L. S. Kegeles, R. L. Van Heertum, and M. Laruelle, The group II metabotropic glutamate receptor agonist LY354740 increases the amphetamine induced striatal [11C]raclopride displacement in baboons, *J Nucl Med* **42**, 142P (2001).
18. B. Moghaddam, B. Adams, A. Verma, and D. Daly, Activation of glutamatergic neurotransmission by ketamine: A novel step in the pathway from NMDA receptor blockade to dopaminergic and cognitive disruptions associated with the prefrontal cortex, *J Neurosci* **17**, 2921-2927 (1997).
19. A. Carlsson, N. Waters, and M. L. Carlsson, Neurotransmitter interactions in schizophrenia--therapeutic implications, *Biol Psychiatry* **46**, 1388-1395 (1999).

TETRAHYDROBIOPTERIN DIFFERENTIALLY REGULATES THE BIOSYNTHESIS OF CATECHOLAMINES AND SEROTONIN

Implication in neuropsychiatric disorders

Chiho Sumi-Ichinose, Fumi Urano, Masayo Kojima, Risa Kuroda, Hiroaki Shiraishi, Takahide Nomura, Hiroshi Ichinose*

1. INTRODUCTION

Catecholamines, e.g., dopamine (DA), norepinephrine, and epinephrine, and serotonin (5HT) are important neurotransmitters in the brain. They are well known to be involved in the regulation of various aspects of brain function, including emotion, motivation, addiction, aggressiveness, and involuntary movement. Alteration in the metabolism of these monoamines leads to abnormalities in behavior and neuropsychiatric states.

Biosynthesis of catecholamines and serotonin is mainly regulated by the activity of the hydroxylation reaction catalyzed by tyrosine hydroxylase (TH) in catecholamine synthesis and by tryptophan hydroxylase (TPH) in 5HT-synthesis. 6*R*-L-*erythro*-5,6,7,8-tetrahydrobiopterin (BH4) is an essential cofactor for both TH and TPH. Although many cofactors are required to be dietary supplemented, mammals have a *de novo* biosynthetic pathway for BH4. In this pathway, guanosine triphosphate (GTP) is converted to 7,8-dihydroneopterin triphosphate by GTP cyclohydrolase 1 (GCH), and 7,8-dihydroneopterin triphosphate is then converted to 6-pyruvoyltetrahydropterin by pyruvoyltetrahydropterin synthase (PTS). This tetrahydropterin is subsequently converted to BH4 by sepiapterin reductase. The intracellular concentration of BH4 is mainly regulated by the activity of GCH.

In the present study, we established BH4-deficient mice by disrupting the *Pts* gene to investigate the effects of BH4 depletion on the animals and the involvement of BH4 in regulating neural systems (Sumi-Ichinose et al., 2001).

* Chiho Sumi-Ichinose, Risa Kuroda, Hiroaki Shiraichi, Takahide Nomura, Department of Pharmacology, School of Medicine, Fujita Health University; Fumi Urano, Masayo Kojima, Hiroshi Ichinose, Institute for Comprehensive Medical Science, Fujita Health University, Toyoake, Aichi 470-1192, Japan.

Catecholamine Research, Edited by Nagatsu *et al.*
Kluwer Academic/Plenum Publishers, 2002

2. RESULTS

Homozygous $Pts^{-/-}$ mice were born at the almost expected Mendelian ratio without gross anomaly in major organs (data not shown), however, all homozygotes died within 48 hours after birth. Biochemical analysis revealed a dramatic decrease in the level of biopterin (6.3 % of wild type), the oxidized form of BH4, and the accumulation of neopterin (17.6 times higher than wild type), a metabolite of 7, 8-dihydroneopterin triphosphate. The levels of DA, NE, and 5HT were all decreased in homozygotes (14.6 %, 6.3 % and 7.8 % of wild type, respectively), and there was no significant difference between heterozygotes and wild-type mice. These results show that *in vivo* synthesis of catecholamines and 5HT was diminished by the lack of BH4.

We assayed the activity of TH, TPH, nitric oxide synthase (NOS), GCH, and aromatic L-amino acid decarboxylase (AADC) in the brain of newborn mice. AADC decarboxylates L-DOPA and 5-HTP to form DA and 5HT, respectively, and co-exists with TH and TPH in catecholaminergic and serotonergic neurons. Although both TH and TPH are BH4-dependent enzymes, only TH activity was severely impaired in the homozygotic mutant mice (7.0 % of wild type; Figure 1). TPH, GCH and AADC activities were not altered significantly. Activity of NOS was decreased to 72 % of the wild type, but far less than that of TH (Figure 1). We could not see a big difference in the mRNA level of *Th* by Northern blot analysis (data not shown).

Next, we examined the localization of TH protein by using an immunohistochemical technique. Dopaminergic neural cell bodies are located in the substantia nigra, and ventral tegmental area. Those neural cell bodies in homozygotic mutant mice as well as in wild-type ones were stained with anti-TH antibody (data not shown). Although there was no marked difference in TH staining on cell bodies, we found a striking difference in the terminal region. In the brain of wild-type mice, nerve fibers were strongly stained with anti-TH antibody, and many positive varicosities were easily seen at higher magnification in the striatum, to which dopaminergic neurons in the substantia nigra project (Figure 2A). In the same region of the homozygotic mutant mice, nerve fibers were not stained, and instead, a few unusual TH-immunopositive cells were found in the

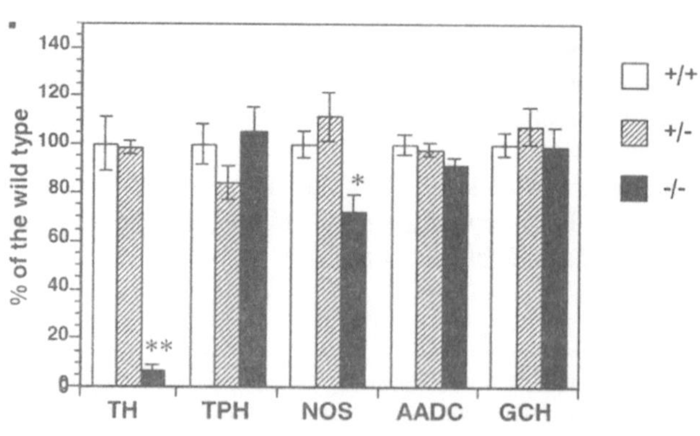

Figure 1. Enzyme activities in the brains of biopterin-deficient mice. Values are the mean ± SEM of a percentage of the corresponding wild-type mice; wild-type (+/+) mice (open bar, n = 5), heterozygous (+/-) mice (hatched bar, n = 6), and homozygous (-/-) mice (solid bar, n = 4; Asterisk indicates $*P < 0.05$ and $**P < 0.01$ for wild-type versus $Pts^{-/-}$ mice, Student's t test).

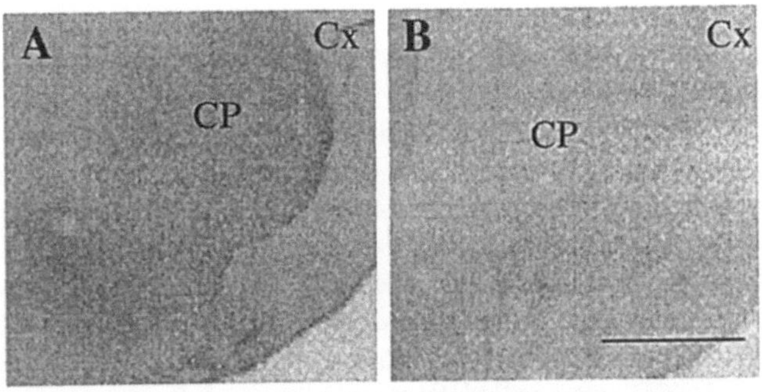

Figure 2. Immunostaining of TH in the brain of wild-type and homozygous mutant mice. (A) is a wild-type mouse (+/+), and (B) is a homozygous mutant (-/-). In the striatum of the wild-type mouse (A), many neural fibers with varicosities were reactive with anti-TH antibody. Neural fibers were not stained with anti-TH antibody in the same region of the homozygous mutant mouse (B), and unusual TH-immunopositive cells could be seen in the mutant striatum. Abbreviations: CP, caudate-putamen; Cx, cortex. The scale bar is 500 μm.

striatum (Figure 2B). In order to examine the presence of dopaminergic fibers, we stained the same sections with anti-AADC antibody. This time nerve fibers in the striatum were positively stained in homozygotes similarly as in the wild type (data not shown). These results indicate that the amount of TH protein was markedly decreased mainly in nerve terminals, but not in cell bodies.

In order to examine whether BH4 itself is responsible for the reduction of the TH protein from the striatum, we repeatedly administered BH4 by the intraperitoneal route at 50 mg/kg once a day 7 times. The knockout mice could survive up to postnatal day 7 (PD 7), although growth retardation was obvious. We stained with anti-TH antibody brain sections of mice rescued by 7 days' administration of 50 mg/kg BH4. Strikingly, the caudate-putamen of the rescued homozygous mice was quite reactive with the anti-TH antibody as was that of the heterozygous control mice (data not shown).

We explored the acute effect of administration of BH4 on DA and 5HT levels in the brain under the BH4-deficient state. We injected newborn mice with BH4 at a dose of 50 mg/kg body weight by the intraperitoneal route on day 0 (postnatal day 0), and sacrificed them 60 min later. Monoamine contents and the monoamine-synthesizing enzyme activities in their brain were analyzed and compared with those of vehicle-injected mice (0.25 % ascorbic acid used to protect BH4 from oxidization). The biopterin content in the brains of both BH4-injected wild type and homozygous mutants was 6 to 7 times higher than that of the ascorbic acid-injected wild-type mice. In spite of the great elevation in the amount of BH4, the DA content in the brain was not altered significantly in the wild-type mice, whereas the 5HT content was slightly increased (1.4 times). In the *Pts* knockout mice, the 5HT level in the brain of BH4-injected animals was markedly elevated to more than 10 times the level for the ascorbic acid-injected ones, reflecting a recovering up to 73 % of the level for the wild type injected with ascorbic acid. To the contrary, the increase in DA levels in the *Pts*-null mice was only 1.5 fold the original level (data not shown).

3. DISCUSSION

Our results showed that BH4-depletion caused a reduction in the content of TH protein, and the reduction in the TH protein was reversible by supplementation with BH4.

This suggests that BH4 regulates the catecholamine-synthesis through altering the amount of TH molecules especially at nerve terminals. This is a novel regulatory mechanism for TH protein by BH4.

One possible explanation for the reduction only at nerve terminals would be that TH forms a stable complex with BH4 and/or catecholamines, and that only the stable complex is carried to nerve terminals through axonal flow. TH has been reported to make a stable and inactive complex with catechol-compounds (Okuno and Fujisawa, 1985). Although more study is required to clarify the molecular mechanism, these data suggest that the turnover of TH protein is rapid in the absence of the biopterin cofactor.

We found that autosomal dominant mutations in the *GCH* gene cause dopa-responsive dystonia (DRD; also called Segawa's disease) (Ichinose et al., 1994). The DRD patients have both mutated- and wild-type alleles of the *GCH* gene, and the resultant partial depletion of BH4 causes a shortage in DA in the brain of DRD patients. In these patients, there are no symptoms other than dystonia/parkinsonism, and dysfunction of serotonergic neurons is not obvious. A reduction in the level of TH protein mainly in the striatum was shown also in patients with DRD (Rajput et al., 1994; Furukawa et al., 1999). Our present results would explain the reason why only dysfunction of nigrostriatal dopaminergic neurons is apparent. At the nerve terminals (striatum), although a small amount of BH4 is present there, the ability to synthesize DA is not sufficient due to the decreased number of the TH molecules. In the case of serotonergic neurons, the TPH activity is not affected by depletion of BH4. Thus, the number of TPH molecules at the nerve terminals is the same as the normal one, and a sufficient amount of 5HT can be synthesized; although the rate of *de novo* synthesis of BH4 is partly impaired.

In summary, we established mice unable to synthesize BH4 by gene targeting. From the analysis of homozygous mutant mice, we showed that the amount of TH protein at nerve terminals was dependent on the intracellular concentration of BH4, and that transient elevation in the BH4 content increased the biosynthesis of 5HT preferentially to that of catecholamines. These findings suggest the involvement of BH4 metabolism in the etiology of monoamine-based neurological and neuropsychiatric disorders.

REFERENCES

Furukawa, Y., Nygaard, T. G., Gütlich, M., Rajput, A. H., Pifl, C., DiStefano, L., Chang, L. J., Price, K., Shimadzu, M., Hornykiewicz, O., et al. 1999, Striatal biopterin and tyrosine hydroxylase protein reduction in dopa- responsive dystonia, *Neurology,* **53**: 1032-1041.

Ichinose, H., Ohye, T., Takahashi, E., Seki, N., Hori, T., Segawa, M., Nomura, Y., Endo, K., Tanaka, H., Tsuji, S., Fujita, K., and Nagatsu, T. 1994, Hereditary progressive dystonia with marked diurnal fluctuation caused by mutations in the GTP cyclohydrolase I gene, *Nature Genetics,* **8**: 236-242.

Okuno, S., and Fujisawa, H. 1985, A new mechanism for regulation of tyrosine 3-monooxygenase by end product and cyclic AMP-dependent protein kinase, *J. Biol. Chem.*, **260**: 2633-2635.

Rajput, A. H., Gibb, W. R., Zhong, X. H., Shannak, K. S., Kish, S., Chang, L. G., and Hornykiewicz, O., 1994, Dopa-responsive dystonia: pathological and biochemical observations in a case, *Ann. Neurol.*, **35**: 396-402.

Sumi-Ichinose, C., Urano, F., Kuroda, R., Ohye, T., Kojima M., Tazawa, M., Shiraishi, H., Hagino, Y., Nagatsu, T., Nomura, T., and Ichinose, H., 2001, Catecholamines and serotonin are differently regulated by tetrahydrobiopterin: a study from 6-pyruvoyltetrahydropterin synthase knockout mice, (submitted for publication).

LITHIUM AND SIGNAL TRANSDUCTION

Shigeru Morinobu, Shigeto Yamawaki, Takuji Fukumoto, Seiichi Tsuji, Takami Suenaga, Jun Takahashi, Kazuhide Tanaka, Koichiro Fujimaki*

Numerous clinical studies have demonstrated that lithium is widely prescribed as a mood stabilizer in the treatment of bipolar mood disorder. An accumulation of evidences, resulted from the pharmacological studies of lithium, postulates several putative actions of this drug.[10] These proposed actions include the inhibition of inositol monophosphatase, the modulation of GTP-binding proteins, and the inhibition of glycogen synthase kinase 3ß. These proposed actions are induced in response to an acute administration of lithium. However, since a long-term administration of lithium is required to obtain optimal therapeutic effects, changes in neuronal gene expression that are induced by an activation of transcription factors may be involved in the clinical efficacy of this drug. Numerous molecular pharmacological studies have demonstrated that several signal transduction pathways, such as the β-adrenoceptor-cAMP-protein kinase A or 5-HT_{2A}-calcium (Ca^{2+})-Ca^{2+}/calmodulin dependent protein kinase II, IV cascade regulate neuronal gene expression through the phosphorylation (activation) of a cAMP response element binding protein (CREB), a major transcription factor in the brain. In this context, the regulation of the expression and phosphorylation of CREB is thought to be an important molecular target of lithium action.

Furthermore, since it is noteworthy that the signal transduction pathways mediated by β-adrenoceptors as well as 5-HT_{2A} receptors are also modulated by an administration of antidepressant drugs, the functional changes in these pathways may also play an important role in the pathogenesis of mood disorders. Previous our studies[5, 13] demonstrated that chronic, but not acute, administration of antidepressants significantly increased the levels of brain-derived neurotrophic factor (BDNF) mRNA in rat brain. It has been well known that CREB is a key element in the transcription of BDNF mRNA in the rat brain.[15, 17] In addition, recent our study[6] demonstrates that chronic, but not acute, treatment with lithium significantly upregulates the levels of BDNF in rat brain. Taken together, examining the

* Shigeru Morinobu, Takuji Fukumoto, Seiichi Tsuji, Shigeto Yamawaki, Hiroshima University, Hiroshima, Japan 734-8551, Jun Takahahi, Kazuhide Tanaka, Koichiro Fujimaki, Shiga University Medical Science, Otsu, Japan 520-2192

Catecholamine Research, Edited by Nagatsu *et al.*
Kluwer Academic/Plenum Publishers, 2002

influence of lithium on the levels of CREB and phospho-CREB in rat brain may promote our understanding of the molecular action of lithium.

Western blot analysis indicates that neither 1 nor 14 day-treatment with lithium has no change in the levels of phospho-CREB in rat frontal cortex or hippocampus. In addition, administration of lithium for either 1 day or 14 days does not affect the levels of CREB in these brain regions. The result from our laboratory is not consistent with the previous study by Ozaki and Chen[14] in which the levels of phospho-CREB was significantly increased in response to chronic lithium treatment. On the other hand, Chen and his colleagues[2] demonstrated that the chronic administration of lithium significantly reduced the levels of phospho-CREB in the rat hippocampus. Although further studies are needed to confirm the effect of chronic lithium treatment on the phosphorylation of CREB, the different transcription factors, such as ATF-2, may be critical in the upregulation of BDNF by lithium.

In addition to the protein phosphorylation, the protein dephosphorylation induced by protein phosphatases (PPs) also plays an important role in alterations of neuronal gene expression. The upregulation of the serine/threonine phosphatase activity of PP2B (calcineurin) directly or indirectly promotes the dephosphorylation of various phospho-proteins. For example, the dephosphorylation of inhibitor 1 by calcineurin (CaN) activates protein phosphatase 1 and subsequently upregulates the dephosphorylation of phospho-CREB.[1] It has been revealed that the change in serine/threonine phosphatase activity of CaN is critical in the regulation of synaptic strength through the alteration in neuronal gene expression, such as in the development of long-term potentiation or long-term depression.[1, 9, 12] In addition, it has been well known that protein phosphatase 2A (PP2A) also plays an important role in the regulation of neuronal gene expression. For example, the activation of serine/threonine phosphatase of PP2A is shown to upregulate the dephosphorylation of Ca^{2+}/calmodulin dependent protein kinase (CaMK) II and IV.[8, 17] Because the phosphorylation of CaMK II and IV leads to the phosphorylation of CREB[7, 11] and subsequently induces the transcription of several genes through the CRE-site,[7] the change in PP2A activity is also associated with the regulation of gene expression. In this context, it is conceivable that examining the influence of lithium on the expression of PPs or serine/threonine phosphatase activity of PPs may also contribute to our understanding of the therapeutic action of lithium.

Northern blot analysis indicates that neither 1 nor 14 day-treatment with lithium change the levels of CaN mRNA in rat frontal cortex or hippocampus. The results of serine/threonine phosphatase assay demonstrate that administration of lithium for either 1 day or 14 days does not affect the activity of CaN in these brain regions. Based on these findings, it is unlikely that the increase in the dephosphorylation of phospho-proteins mediated by CaN is involved in the molecular action of lithium.

On the other hand, the analysis of PP2A activity indicates that 1 as well as 14 day-treatment with lithium significantly upregulates the activity of PP2A in rat frontal cortex. Not only 1 but also 14 day-treatment trends to upregulate the activity of PP2A in hippocampus. The results obtained by the study of PP2A activity demonstrate that acute as well as repeated treatment with lithium upregulates the activity of PP2A in these brain region, suggesting that the inhibitory action of lithium on the phosphorylation of CaMK II and IV through the activation of PP2A may be, at least in part, involved in the therapeutic action of this drug.

The results of the recent our studies demonstrate that chronic administration of therapeutically relevant concentrations of lithium has no change in either the phosphorylation of CREB or serine/threonine phosphatase activity of CaN, suggesting that the regulation of the CREB-mediated gene transcription pathway may be less important in the molecular action of lithium. However, acute as well as repeated treatment with lithium upregulates the activity of PP2A. As these findings indicate that the downregulation of the phosphorylation of CREB mediated by the inhibition of CaMK II and IV phosphorylation in response to lithium, it is likely that the regulation of the CREB phosphorylation may be, at least in part, involved in the molecular action of lithium. To determine whether the upregulation of PP2A activity by lithium is involved in the therapeutic action, the influence of a different type of a mood stabilizer, such as valproate, on the activity of PP2A should be investigated. Since chronic treatment with lithium regulates several neuronal gene expressions, such as BDNF[6] or Bcl-2,[3, 4] it is plausible that the regulation of the expression or activity of the different transcription factors, such as ATF-2, may be involved in the therapeutic action of lithium. Further studies examining the influence of lithium on the function of transcription factors are required to elucidate the precise molecular action of this drug.

REFERENCES

1. 1. H. Bito, K. Desseroth, and R. W. Tsien, CREB phosphorylation and dephosphorylation: a Ca^{2+}-, and stimulus duration-dependent switch for hippocampal gene expression, Cell **87**, 1203-1214 (1996).
2. B. Chen, J. F. Wang, B. C. Hill, and L. T. Young, Lithium and valproate differentially regulate brain regional expression of phosphorylated CREB and c-Fos, *Brain Res. Mol. Brain Res.* **70**, 45-53 (1999).
3. G. Chen, W-Z. Zeng, P-X. Yuan, L-D. Huang, Y-M. Jiang, Z-H. Zhao, and H. K. Manji, The mood-stabilizing agents lithium and valproate robustly increase the levels of the neuroprotective protein bcl-2 in the CNS, J. Neurochem. **72**, 879-882 (1999).
4. R-W. Chen, and D-M. Chuang, Long term lithium treatment suppresses p53 and Bax expression but increases Bcl-2 expression, J. Biol. Chem. **274**, 6039-6042, 1999.
5. K. Fujimaki, S. Morinobu, and R. S. Duman, Administration of a cAMP phosphodiesterase 4 inhibitor enhances antidepressant-induction of BDNF mRNA in rat hippocampus, *Neuropsychopharmacol.* **22**, 42-51 (2000).
6. T. Fukumoto, S. Morinobu, Y. Okamoto, A. Kagaya, and S. Yamawaki, Chronic lithium treatment increases the expression of brain-derived neurotrophic factor (BDNF) in the rat brain, *Psychopharmacol.* (in press).
7. G. E. Hardingham, F. J. Arnold, and H. Bading, Nuclear calcium signaling controls CREB-mediated gene expression triggered by synaptic activity, Nat. Neurosci. **4**, 261-267 (2001).
8. A. Ishida, I. Kameshita, and H. Fujisaki, A novel protein phosphatase that dephosphorylates and regulates Ca^{2+}/calmodulin-dependent protein kinase II, J. Biol. Chem. **273**, 1904-1910 (1998).
9. R. C. Malenka, Synaptic plastisity in the hippocampus, Cell 78, 535-538 (1994).
10. H. K. Manji, G. J. Moore, G. Rajikowska, and G. Chen, Neuroplasticity and cellular resilience in mood disorders, *Mol. Psychiat.* **5**, 578-593 (2000).
11. R. P. Matthews, C. R. Guthrie, L. M. Wailes, X. Zhao, A. R. Means, and G. T. McKnight, Calcium/calmodulin-dependent protein kinase types II and IV differentially regulate CREB-dependent gene expression, Mol. Cell Biol. **14**, 6107-6116 (1994).
12. R. M. Mukey, S. Endo, A. Shenolikar, and R. C. Malenka, Involvement of a calcineurin/inhibitor-1 phosphatase cascade in hippocampal long-term depression, Nature **369**, 486-488 (1994).
13. M. Nibuya, S. Morinobu, and R. S. Duman, Regulation of BDNF and trkB mRNA in rat brain by chronic electroconvulsive seizure and antidepressant drug treatments, *J. Neurosci.* **15**, 7539-7547 (1995).
14. N. Ozaki, and D. M. Chuang, Lithium increases transcription factor binding to AP-1 and cyclic AMP-responsive element in cultured neurons and rat brain, *J. Neurochem.* **69**, 2336-2344 (1997).
15. P. B. Shieh, S. C. Hu, K. Bobb, T. Timmusl , and A. Ghosh, Identification of a signaling pathway involved in calcium regulation of BDNF expression, *Neuron* **20**, 727-740 (1998).
16. X. Tao, S. Finkbeiner, D. B. Arnold, A. J. Shaywitz, and M. E. Greenberg, Ca^{2+} influx regulates BDNF transcription by a CREB family transcription factor-dependent mechanism, *Neuron* **20**, 709-726 (1998).
17. D. M. Virshup, Protein phosphatase 2A: a panoply of enzymes, Curr. Opin. Cell Biol. **12**, 180-185 (2000).

PLATELET MONOAMINE OXIDASE, TRANSCRIPTION FACTOR AP-2 AND PERSONALITY – A FUNCTIONAL TRIANGLE?

Lars Oreland, Mattias Damberg, Håkan Garpenstrand, Jaanus Harro, Jarmila Hallman*

1. INTRODUCTION

Almost thirty years ago it was discovered that low activity of the enzyme monoamine oxidase (MAO; E.C. 1.4.3.4) in blood platelets, which is to about 75% genetically regulated,[1] is strongly linked to personality traits such as impulsiveness and sensation seeking.[2; see 3] Mountaineers[4] and pathological gamblers[5] have lower activity of the enzyme than controls, an observation also made in type 2 alcoholics (males with a heavy genetic disposition) and individuals prone to commit violent crimes (see 3). It has also been shown that new-born babies with low platelet MAO are crying more and are more difficult to console than those with medium activity.[6] Furthermore, platelet MAO activity in free-living monkeys has been shown to be linked to behaviour as well as to voluntary alcohol intake.[7, 8]

Many theories have been presented in order to explain the linkage between platelet MAO activity and personality/behaviour and some of the theoretical possibilities have previously been discussed by us.[3, 9] Our main hypothesis, has, however, already since the start of our work, been that platelet MAO activity is a genetic marker for the "size" or "capacity" of the monoamine systems, preferentially the serotonin system.[10]

2. AP-2 TRANSCRIPTION FACTORS AND THE ASSOCIATION WITH PERSONALITY

Our theory recently got additional support by the finding of a relation between the expression of MAO in platelets and the occurrence of a polymorphism of the AP-2ß gene.[11] AP-2 is a family of closely related transcription factors, which all are of

* Dept. of Neuroscience, Uppsala University, POB 593, S-75124 Uppsala, Sweden and Dept. of Psychology, Tartu University, EE-40510 Tartu, Estonia. "Lars.Oreland@neuro.uu.se"

Catecholamine Research, Edited by Nagatsu *et al.*
Kluwer Academic/Plenum Publishers, 2002

importance for the regulation of a variety of genes of which several are expressing proteins within the central monoamine transmitter systems, e.g. the serotonin transporter, tryptophan and tyrosine hydroxylases. The AP-2 transcription factors seem to be of importance both during embryonic development of the central nervous system, and in the adult brain where AP-2 might have a role as to contribute to maintain a suitable tone of the transmitter systems involved (see 3).

Earlier research has shown that the gene encoding AP-2ß occurs in two allelic forms in humans. In intron 2 there is a variable sequence consisting of a four basepair repeat, i.e. [CAAA], repeated four or five times. In a study on 137 healthy volunteers a specific polymorphism of the AP-2ß gene was analysed in relation to personality traits as estimated by Karolinska Scales of Personality (KSP). Homozygotes for the long variant $[CAAA]_5$ (39% of the individuals) displayed lower scores in anxiety-related traits.[12] Moreover, the same specific AP-2ß genotype was found to correlate to low platelet MAO activity.

In another study on females with binge-eating disorder, a strong overrepresentation of individuals being homozygous for the long AP-2ß allele was found.[3] Thus, 63 % of the binge eating females were homozygous for this allele, while this figure was only 36 % in control females (chi square = 7.47; $p = 0.0063$). Also the personality differed between the groups, the binge eating women having e.g. higher degree of impulsiveness as indicated by the KSP inventory. Furthermore, as expected, those females with binge-eating disorder who were homozygous with regard to the long allele of the AP-2ß gene displayed significantly lower platelet MAO than those who were heterozygous or homozygous for the short allele. We speculate that also other groups of patients characterised by low platelet MAO activity might have a high frequency of the long allele of the AP-2ß gene. Examples of such groups are type 2 alcoholics and individuals with violent criminal behaviour.

3. AP-2 AND MONOAMINE SYSTEMS

Transcription factors are proteins that regulate the expression of one or several genes. The AP-2 family of transcription factors have been shown to be of importance for the development of certain brain regions, e.g. the midbrain and as a consequence they might be important for determining the "size" or "capacity" of central monoamine transmitter systems (see 3). Though several monoaminergic genes have binding sites for AP-2 in regulatory regions, it seems likely that AP-2 regulate the expression of a variety of proteins within the monoamine transmitter systems in the adult brain. The associations between the AP-2ß polymorphism with personality factors and psychopathology mentioned above might suggest a mediation by the 5-HT-ergic system in particular. Indeed, we have recently shown that several antidepressive drugs, including a SSRI, greatly affect AP-2 levels and DNA-binding activity of AP-2 in the rat brain.[13] It can be speculated that those changes are involved in the shifts in monoamine metabolism which are likely to occur for an antidepressive effect. But what these data clearly demonstrate is that drugs with their primary action on monoaminergic neurons affect AP-2 function. In order to exclude that the changes in AP-2 levels are not merely an epiphenomenon of drug treatment, we studied the association of AP-2 levels in the brainstem and monoamine turnover in the forebrain in rats naive to drugs. We found the levels of AP-2 transcription factors to correlate positively with a number of monoaminergic measures in the frontal cortex, hippocampus, and septum of the adult rat brain. Furthermore, a number of behaviours, which are usually linked to central serotonergic activity, such as active social behaviour in the social interaction test, correlated positively both to levels of AP-2 in the brain stem and to serotonin turnover in the frontal cortex.[14]

4. CONCLUSIVE REMARKS

The role of transcription factors for personality and the development of psychiatric disorders has, with a few exceptions, generally been overlooked. A few single gene polymorphisms have until now been linked to e.g. novelty seeking and anxiety traits. In other cases, such as in the case of ADHD/DAMP, a pattern of occurrence of alleles in several genes has been associated with the disorder (see 15). It seems reasonable to assume that polymorphisms in genes for transcription factors might give rise to rather obvious changes in the phenotype since several genes with recognition sites for the same transcription factor might be affected. As we have recently pointed out,[3] another reason for the importance of transcription factors might be their possible importance for the development of the brain, perhaps with regard to the "size" or "capacity" as a result of the number of neurones besides receptors, enzymes and transporter proteins. Detailed knowledge of the role of transcription factors for brain function would also open new possible targets for new classes of drugs.

REFERENCES

1. Pedersen NL, Oreland L, Reynolds C, McClearn GE. Importance of genetic influence on thrombocyte MAO activity in twins reared apart and twins reared together. Psychiat. Res. **46**, 239-251, 1993.
2. Buchsbaum MS, Coursey RD, Murphy DL. The biochemical high-risk paradigm: behavioural and familial correlates of low platelet monoamine oxidase activity. Science **194**, 339-341, 1976.
3. Damberg M, Garpenstrand H, Hallman J, Oreland L. Genetic mechanism of behaviour – don't forget about the transcription factors. Millennium article. Mol. Psychiatry. In press 2001.
4. Fowler CJ, v. Knorring L. and Oreland L. Platelet monoamine oxidase activity in sensation seekers. Psychiat. Res. **3**, 273-279. 1980.
5. Carrasco JL, Saiz-Ruiz J, Hollander E, Cesar J, Lopez-Ibor JJ Jr. Low platelet monoamine oxidase activity in pathological gambling. Acta Psychiatr. Scand. **90**, 427-31, 1994.
6. Sostek AJ, Sostek AM, Murphy DL, Martin EB, Born WS. Cord blood amine oxidase activities relate to arousal and motor functioning in human newborns. Life Sci. **28**, 2561-2568, 1981.
7. Redmond DE, Murphy DL, Baulu J. Platelet monoamine oxidase activity correlates with social affiliative and agonistic behaviors in normal rhesus monkeys. Psychosom. Med. **41**, 87-100, 1979.
8. Åsberg M, Oreland L, Garpenstrand H, Damberg M, Ekblom J, Forslund K, Rylander G, Fahlke C, Higley D. Thrombocyte-MAO and Transcription Factor AP2 are Related to Personality – is There a Connection? Abstract. 38th ACNP Acapulco, Mexico Dec. 1999.
9. Oreland L, Hallman J. The correlation between platelet MAO activity and personality - a short review of findings and a discussion on possible mechanisms. Progr. Brain Res. 106, 77-84, 1995.
10. Oreland L, Shaskan, EG. Monoamine oxidase activity as a biological marker. Trends Pharmacol. Sci., **4**, 339-341, 1983.
11. Damberg M, Garpenstrand H, Berggård C, Åsberg M, Hallman J, Oreland L. The genotype of human transcription factor AP2ß is associated with platelet monoamine oxidase B activity. Neurosci. Lett. **291**, 2000, 204-206.
12. Damberg M, Garpenstrand H, Alfredsson J, Ekblom J, Forslund K, Rylander G, Oreland L. A polymorphic region in the human transcription factor AP-2ß is associated with specific personality traits. Mol. Psychiatry **5**, 220-224, 2000.

13. Damberg M, Ekblom J, Oreland L. Chronic pharmacological treatment with certain antidepressants alters the expression and DNA-binding activity of transcription factor AP-2. Life Sci. **68**, 669-678, 2000.
14. Harro J, Damberg M, Eller M, Tonissaar M, Oreland L. Social behaviour in rats is associated with frontal 5-HT neurotransmission and AP-2 transcription factors. World J Biol. Psychiat. 2, suppl. **1**, 34-35S, 2001.
15. Comings DE. Polygenic inheritance and micro/minisatellites. Mol. Psychiatry. **3**, 21-31, 1998.

CARDIAC SYMPATHETIC DENERVATION IN PARKINSON'S DISEASE

David S. Goldstein*

Chronic autonomic failure has been classified clinically into pure autonomic failure, multiple system atrophy, and Parkinson's disease with autonomic failure, as indicated in Figure 1.[1]

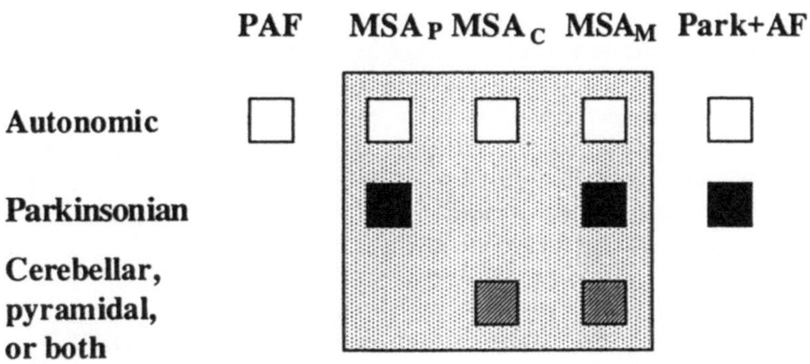

Figure 1. Clinical classification of chronic autonomic failure. Abbreviations: PAF = pure autonomic failure; MSA_P = multiple system atrophy with parkinsonian dominance; MSA_C = MSA with cerebellar predominance; MSA_M = MSA with mixed cerebellar and parkinsonian features; Park+AF = Parkinson's disease with autonomic failure.

Research about chronic autonomic failure has focused on neuropharmacologic or neurochemical comparisons between pure autonomic failure and multiple system atrophy and relatively little on Parkinson's disease.[2]

* David S. Goldstein, Clinical Neurocardiology Section, National Institute of Neurological Disorders and Stroke, National Institutes of Health, Bethesda, MD 20892-1620 USA.

Investigators have attributed orthostatic hypotension in Parkinson's disease to levodopa treatment, rather than to a neuropathologic process,[3] or else patients have been thought to have a different disease, such as striatonigral degeneration or the parkinsonian form of multiple system atrophy. Thus, autonomic failure has been thought to be a rare cause of orthostatic hypotension in Parkinson's disease and probably to result from central dysregulation of sympathetic outflows.

Neurologists often have difficulty distinguishing multiple system atrophy from Parkinson's disease with autonomic failure. Some patients with multiple system atrophy respond to levodopa treatment, and some with Parkinson's disease have severe orthostatic hypotension. A clinicians might not even initiate levodopa treatment in a parkinsonian patient who already has orthostatic hypotension, because of the belief that levodopa produces or worsens orthostatic hypotension in Parkinson's disease.

WHERE IS THE LESION?

Since loss of dopamine cells of the nigrostriatal system produces the movement disorder in Parkinson's disease, one might presume that autonomic failure in Parkinson's disease would reflect an analogous central neurodegenerative process to that in multiple system atrophy—i.e., a preganglionic lesion. After noting that two patients with Parkinson's disease and orthostatic hypotension had evidence of cardiac sympathetic denervation, whereas patients with multiple system atrophy did not,[4] we examined whether multiple system atrophy might differ from Parkinson's disease, in that the former would involve a preganglionic and the latter a postganglionic, presynaptic autonomic lesion. Discoveries made in studying mechanisms of orthostatic hypotension in Parkinson's disease now appear to have implications not only for the etiology of the autonomic failure but also for that of the disease process generally.

ORTHOSTATIC HYPOTENSION FROM SYMPATHETIC FAILURE

We found that orthostatic hypotension in patients with Parkinson's disease occurrred independently of levodopa treatment[5] and invariably was associated with sympathetic neurocirculatory failure, detected by abnormalities of beat-to-beat blood pressure in response to the Valsalva maneuver or after spontaneous premature ventricular contractions.[6, 7] Briefly, in late Phase II of the Valsalva maneuver, the blood pressure normally increases, due to reflexive sympathetically-mediated cardiovascular stimulation in response to decreased cardiac stroke volume; and in Phase IV, the blood pressure increases to above baseline, due to cardiac ejection of a normal stroke volume into the reflexively constricted vasculature. In sympathetic neurocirculatory failure, the blood pressure falls progressively in Phase II, and there is no overshoot of blood pressure in Phase IV.

The finding of abnormal blood pressure responses to the Valsalva maneuver, while identifying sympathetic neurocirculatory failure in all patients with Parkinson's disease and orthostatic hypotension, did not have specificity in distinguishing Parkinson's disease from pure autonomic failure or multiple system atrophy.

NEUROCHEMISTRY AND NEUROIMAGING

Neurochemical and neuroimaging evidence has consistently indicated cardiac sympathetic denervation in patients with Parkinson's disease and sympathetic neurocirculatory failure.

The rate of entry of endogenous norepinephrine into the venous drainage of the heart (cardiac norepinephrine spillover) was markedly decreased in all such patients.[5] The patients also had virtually no cardiac arteriovenous increments in plasma levels of dihydroxyphenylglycol or of endogenous levodopa. Since under resting conditions production of dihydroxyphenylglycol mainly reflects turnover of norepinephrine in sympathetic nerves,[8] and production of endogenous levodopa mainly reflects tyrosine hydroxylation, the rate-limiting enzymatic step in norepinephrine synthesis in sympathetic nerves,[9] the combination of low cardiac norepinephrine spillover and low cardiac production of dihydroxyphenylglycol and endogenous levodopa indicates low rates of norepinephrine release, turnover, and synthesis in the heart. This pattern probably can only result from loss of functional sympathetic nerve terminals.

The same patients had markedly decreased myocardial concentrations of 6-[^{18}F]fluorodopamine-derived radioactivity, detected by positron emission tomographic scanning, after i.v. injection of this sympathoneural imaging agent.[5] The neurochemical and neuroimaging findings therefore indicate a postganglionic, presynaptic sympathetic lesion as the basis for orthostatic hypotension in Parkinson's disease. In contrast, all patients with multiple system atrophy, with or without sympathetic neurocirculatory failure, have had neurochemical and neuroimaging evidence of intact cardiac sympathetic innervation, consistent with a preganglionic lesion.

CARDIOSELECTIVITY OF SYMPATHETIC DENERVATION

From low ratios of cardiac to total body spillover of norepinephrine in patients with Parkinson's disease,[5] we hypothesized that the sympathetic denervation might be cardioselective. To test this hypothesis, we carried out positron emission tomographic scanning of the head, which contains several sympathetically innervated structures. (Because of the blood-brain barrier for catecholamines, systemically administered 6-[^{18}F]fluorodopamine normally does not enter the central nervous system.) In normal volunteers, treatment with desipramine, which attenuates neuronal uptake of catecholamines, decreased tissue concentrations of 6-[^{18}F]fluorodopamine-derived radioactivity, detected by positron emission tomographic scanning, in the nasopharyngeal mucosa, parotid and submandibular salivary glands, and the thyroid gland. These findings supported the view that 6-[^{18}F]fluorodopamine-derived radioactivity has the potential to detect altered density of sympathetic innervation in patients with Parkinson's disease.

Scanning of the head, chest, and abdomen revealed approximately normal innervation in most organs, except for the heart and the thyroid gland. Decreased 6-[^{18}F]fluorodopamine-derived radioactivity in the thyroid seemed to occur independently of orthostatic hypotension.

SYMPATHETIC DENERVATION IN PARKINSON'S DISEASE WITHOUT ORTHOSTATIC HYPOTENSION

We then asked whether cardiac sympathetic denervation might characterize Parkinson's disease in general, regardless of the occurrence of sympathetic neurocirculatory failure. Surprisingly, about half of patients with Parkinson's disease who did not have sympathetic neurocirculatory failure nevertheless had severely decreased 6-[^{18}F]fluorodopamine-derived radioactivity throughout the left ventricular myocardium.[5] Moreover, most of the remaining patients without diffusely decreased radioactivity had localized decreases in the left ventricular apex or lateral or inferior wall, with relative sparing of the interventricular septum.

Moreover, as noted above, patients with Parkinson's disease have had neuroimaging evidence of sympathetic denervation of the thyroid gland, with or without sympathetic neurocirculatory failure. These findings indicate that sympathetic denervation occurs in most patients with Parkinson's disease.

The cardiac sympathetic denervation in patients with Parkinson's disease occurred independently of levodopa treatment or the duration or severity of the parkinsonian signs.

PROGRESSION OF SYMPATHETIC DENERVATION

To examine whether the loss of cardiac sympathetic innervation progresses over time, we have re-studied patients with Parkinson's disease who had loss of 6-[^{18}F]fluorodopamine-derived radioactivity that was localized within the left ventricular myocardium. Loss of cardiac sympathetic innervation can progress over the course of months, even in the absence of levodopa treatment (unpublished observations).

SYMPATHETIC DENERVATION IN FAMILIAL PARKINSON'S DISEASE

Studying patients with familial Parkinson's disease, in whom the underlying mutation has been identified, might provide insights about potential mechanisms of the more common sporadic form. A patient with familial Parkinson's disease from mutation of the gene encoding α-synuclein had orthostatic hypotension, sympathetic neurocirculatory failure, and virtually absent myocardial 6-[^{18}F]fluorodopamine-derived radioactivity. This combination introduces the concept that abnormal α-synuclein can lead to both the motor and autonomic deficits in Parkinson's disease.[10]

PERSPECTIVE

The findings described above provide the basis for a clinical pathophysiologic approach for distinguishing chronic autonomic failure in Parkinson's disease from that in multiple system atrophy, the former reflecting a postganglionic, presynaptic and the latter a preganglionic lesion (Figure 2). More generally, the findings provide new insights about what the etiologic basis of Parkinson's disease might be. Because of the remarkably high frequency of sympathetic denervation, Parkinson's appears to be both a motor and autonomic disease, involving loss of nigrostriatal dopaminergic terminals in the brain and sympathetic noradrenergic terminals in the periphery.

Identifying processes that produce both parkinsonism and sympathetic denervation could lead to discovery of the primary causes of the disease and to effective treatment and prevention strategies. For instance, oxidative injury in catecholamine-producing cells might constitute a common factor producing susceptibility of nigrostriatal dopamine and sympathoneural norepinephrine terminals in Parkinson's disease. Development of appropriate neurochemical assay methodologies might enable determination of whether animal and cellular models and patients with sympathetic denervation have high local rates of production of catecholaldehydes, which are neurotoxic metabolites of catecholamines. Findings supporting this hypothesis would mean that neurotoxins produced within the body itself contribute to Parkinson's disease.

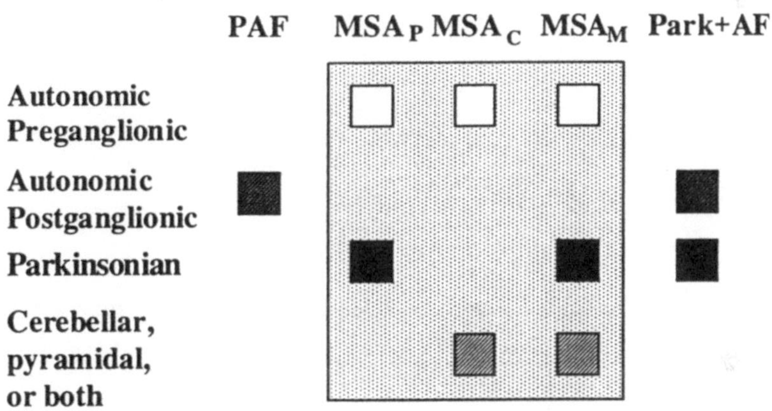

Figure 2. Pathophysiologic classification of chronic autonomic failure. According to this schema, the parkinsonian form of multiple system atrophy (MSAP) differs from Parkinson's disease with autonomic failure (Park+AF), in that the latter but not the former includes an autonomic postanglionic lesion. Other abbreviations: PAF = pure autonomic failure; MSAC = MSA with cerebellar predominance; MSAM = MSA with mixed cerebellar and parkinsonian features.

REFERENCES

1. S. Gilman, P. Low, N. Quinn, A. Albanese, Y. Ben-Shlomo, C. Fowler, H. Kaufmann, T. Kockgether, A. Lang, P. Lantos, I. Litvan, C. Mathias, E. Oliver, D. Robertson, I. Schatz, and G. Wenning, Consensus statement on the diagnosis of multiple system atrophy, *Clin. Auton. Res.* **8**, 359-362 (1998).
2. R. J. Polinsky, I. J. Kopin, M. H. Ebert, and V. Weise, Pharmacologic distinction of different orthostatic hypotension syndromes, *Neurology* **31**, 1-7 (1981).
3. M. M. Hoehn, Levodopa-induced postural hypotension. Treatment with fludrocortisone, *Arch. Neurol.* **32**, 50-51 (1975).
4. D. S. Goldstein, C. Holmes, R. O. Cannon, III, G. Eisenhofer, and I. J. Kopin, Sympathetic cardioneuropathy in dysautonomias, *N. Engl. J. Med.* **336**, 696-702 (1997).

5. D. S. Goldstein, C. Holmes, S. T. Li, S. Bruce, L. V. Metman, and R. O. Cannon, Cardiac sympathetic denervation in Parkinson disease, *Ann. Intern. Med.* **133**, 338-347 (2000).
6. D. S. Goldstein, and C. Tack, Non-invasive detection of sympathetic neurocirculatory failure, *Clin. Auton. Res.* **10**, 285-291 (2000).
7. D. S. Goldstein, A new sign of sympathetic neurocirculatory failure: Premature ventricular contraction as a "one-beat Valsalva maneuver." *Clin. Auton. Res.* **10**, 63-67 (2000).
8. D. S. Goldstein, G. Eisenhofer, R. Stull, C. J. Folio, H. R. Keiser, and I. J. Kopin, Plasma dihydroxyphenylglycol and the intraneuronal disposition of norepinephrine in humans, *J. Clin. Invest.* **81**, 213-220 (1988).
9. D. S. Goldstein, R. Udelsman, G. Eisenhofer, R. Stull, H. R. Keiser, and I. J. Kopin, Neuronal source of plasma dihydroxyphenylalanine, *J. Clin. Endocrinol. Metab.* **64**, 856-861 (1987).
10. D. S. Goldstein, S.-T. Li, and I. J. Kopin, Sympathetic neurocirculatory failure in Parkinson disease: Evidence for an etiologic role of a-synuclein, *Ann. Intern. Med.* (in press).

DOPAMINERGIC IMPAIRMENT IS RELATED TO CORTICAL DYSFUNCTION IN PARKINSON'S DISEASE: A PET study

Kengo Ito, Yutaka Arahata, Atsuko Nagano-Saito, Yuji Abe, Takashi Kato, Akinori Nakamura, Yasuhiro Kawasumi, Kentaro Hatano, Takako Yamada, and Teruhiko Kachi*

1. INTRODUCTION

There may be involvement of extrastriatal dopamine terminals to varying extents in different subgroups of idiopathic Parkinson's disease (PD). Accurate detection of changes in extrastriatal areas is unreliable using conventional ROI analysis. As a consequence the ROI approach is not really suited for use as an exploratory technique with 6-[^{18}F]fluoro-L dopa (FDOPA) PET. For that reason we developed the spatial normalization for FDOPA influx (Ki) images to assess the extrastriatal uptake of FDOPA in the normal and PD.[1-3]

Then we investigated relative differences in striatal and extrastriatal dopaminergic function between Parkinson's disease without (PD) and with dementia (PDD) with FDOPA PET. Between group significant differences in mean regional Ki values were localised with statistical parametric mapping (SPM) on a voxel-by-voxel basis. Compared with the normal group, the PDD group showed reduced FDOPA Ki bilaterally in the striatum, midbrain, and anterior cingulate area. A relative difference in FDOPA uptake between PD and PDD was the decline bilaterally in the anterior cingulate area and in the right caudate nucleus in the latter. We concluded that dementia in PD is associated with impaired cingulate and caudate dopaminergic function.

These findings were consistent with the hypothesis that impairment of dopaminergic

* Kengo Ito, Yutaka Arahata, Atsuko Nagano-Saito, Yuji Abe Y, Takashi Kato, Akinori Nakamura, Yasuhiro Kawasumi, and Kntaro Hatano, National Institute for Longevity Sciences, Obu, Japan, 474-8522. Takako Yamada and Teruhiko Kachi, Chubu National Hospital, Obu, Japan 474-8511.

projections from the medial substantia nigra and ventral tegmental area (VTA) played an important role in the dementia of PD.[4, 5] To reinforce this hypothesis, we investigated the relation between dopaminergic impairment and cortical dysfunction.

2. PATIENTS AND METHODS

2.1. Patients

We studied forty idiopathic PD patients (age: 49-88 years, 66±8 years [mean±SD]) who satisfied the UK Brain Bank criteria for diagnosis of the disorder. The severity of physical symptoms were graded from I to V, according to the Hoehn & Yahr's disability scale (I, 3; II, 6; III, 17; IV, 13; V, 1). A mini-mental state examination (MMSE) was carried out in all except one patient; the scores ranged from 17 to 30 (25.1±4.5). FDOPA and FDG PET examinations were performed in all patients. This study was approved by the Ethical Committee of the Chubu National Hospital and National Institute for Longevity Sciences. All subjects gave written informed consent prior to PET scanning. All medications were withheld for at least 24 and 48 hours before the FDG and FDOPA PET, respectively. Subjects fasted for 6 hours before the PET scan.

2.2. PET study

For FDOPA PET, the protocol included 25 time frames over 94 minutes. The FDOPA influx constant (Ki) image and integrated image were created using the "kronos" software, developed by D. Bailey (MRC Cyclotron Unit, Hammersmith Hospital, London, UK).

The Ki image was co-registered to each individual MR image. Circular ROIs of 10.8 mm in diameter were defined for the caudate, putamen (two circles for each side), and anterior cingulate cortex in each of three planes on the Ki image. The mean values for these three areas of the corresponding bilateral regions were used as the regional value in statistical analysis.

For FDG PET, a qualitative image obtained from 36-60 minutes after the tracer injection was used for analysis. Each FDG image was transformed into standard stereotactic space. After the smoothing procedure, the count of each image was normalised by its mean value of the upper 40% of pixels, which correspond to pixels of almost all gray matter in normal brain, from our preliminary examination.

2.3. Statistical parametric mapping

We used bilateral mean Ki values in the caudate, putamen, and anterior cingulate cortex as covariates. Significant correlations between the Ki values and normalised glucose metabolic images were localised with SPM96, on a voxel-by-voxel basis. Each voxel value was transformed to normal distribution Z scores. Maps of Z scores surviving a height threshold $p < 0.01$ ($Z>2.33$) and extent corrected $p < 0.05$ were applied for significant areas.

3. RESULTS

The Ki value in the caudate nucleus significantly correlated with the normalised glucose metabolic image (GMnorm) in broad areas of two clusters, including the lateral posterior cingulate and the bilateral temporo-parietal association areas, with dominance in the right hemisphere ($p < 0.01$, corrected). In addition, the Ki value in caudate nucleus was inversely correlated with GMnorm in the cerebellar vermis ($p < 0.01$, corrected).

On the other hand, the FDOPA-Ki value in the anterior cingulate gyrus correlated with GMnorm in the bilateral anterior cingulate gyri and a part of the right frontal association area ($p < 0.01$, corrected). The correlated areas are shown in Figure 1.

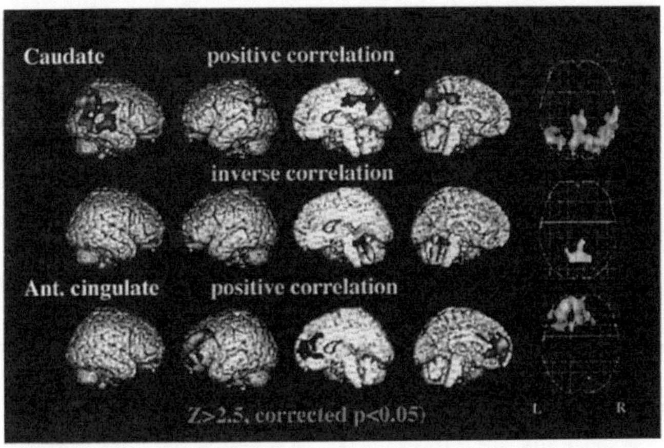

Figure 1 Regions of significantly correlated area ($p < 0.01$, corrected) between influx constant values (Ki) in ^{18}F-DOPA and normalised glucose metabolic value (GMnorm) in ^{18}F-FDG PET. The positive correlation between Ki in the caudate and GMnorm in the bilateral posterior cingulate and temporo-parietal association areas, and between Ki in the anterior cingulate and GMnorm in the bilateral anterior cingulate and right frontal association area are shown. The Ki value in the caudate showed a negative correlation with GMnorm in the cerebellar vermis.

4. DISCUSSION

Pathologically, dementia in PD seems to be related to the coexistence of cortical Lewy body disease (LBD) and Alzheimer's disease, and defects in the cholinergic systems (loss of neurons in the nucleus basalis of Meynert). However, PD with apparent dementia without these pathological changes has also been reported. In some reports, loss of the neurons in the ventral tegmental area (VTA) or medial substantia nigra has been emphasised as the cause of dementia.[4, 5] In addition, PET studies have shown that reductions in caudate ^{18}F-DOPA uptake rate correlate with memory performance in advanced PD patients.[6] Our preliminary PET study showed that in PD with dementia, the Ki value in the mesolimbic dopaminergic system was lower than in PD without dementia.

The caudate nucleus is thought to be involved in an association loop and in cognitive function[7]. The caudate nucleus receives dopaminergic fibers from the dorsal tier of the substantia nigra pars compacta (SNc), and indirectly projects through the thalamus to the associate cortex in the frontal, parietal, temporal, and occipital lobes, and also receives feedback projections from these association areas. Our results show that the decrement of Ki for FDOPA in the caudate nucleus might be related to dysfunction in the parieto-temporal association cortex. Glucose metabolism in the posterior cingulate was also highly correlated with Ki for FDOPA in the caudate nucleus. The posterior cingulate gyrus is known to be the primary area of depressed glucose metabolism in the early stages of Alzheimer's disease; it has connections with association cortices and other limbic structures. These findings suggest to us that the Ki depression in the caudate nucleus may be related to depression of higher cortical functions.

On the other hand, Ki in the anterior cingulate correlated with glucose metabolism in the anterior cingulate and frontal association cortex. Dopaminergic neurons in the VTA directly project to the limbic areas, the anterior cingulate, amygdala, hippocampus, and nucleus accumbens (mesolimbic dopaminergic system).[7] Neurons from the nucleus accumbens (NAc) also indirectly project through the thalamus to the anterior cingulate and frontal association cortex. It is known that there are feedback projections from the anterior cingulate and frontal association cortex to the Nac. The decrease in FDOPA Ki in the anterior cingulate seems to represent a loss of dopaminergic neurons in the VTA, which is the origin of the mesolimbic dopaminergic system. Our results suggest that dopaminergic dysfunction in the mesolimbic dopaminergic system may relate to dysfunction in the anterior cingulate and frontal association cortex. The anterior cingulate participates in the control of social behaviour, cognition, attention, and autonomic functions. These symptoms are also often seen in patients with DLB and PD with dementia.

Our results may simply be the result of a statistical correlation between FDOPA and FDG PET. This is certainly not a direct proof of a neural network in the basal ganglia and cerebral cortex. Nevertheless, it is important for clarifying the pathophysiology and the pattern of progression in PD to investigate the relation of dopaminergic changes to cortical dysfunction in living patients. The present study indicates that the extent of cell loss in the dopaminergic nucleus may change the areas involved in the cerebral cortex in Parkinson's disease.

5. CONCLUSION

These results suggest that impairment of the dopaminergic system - both the nigrostriatal and the mesolimbic projections - is correlated with cortical dysfunction in different areas of the brain in PD.

ACKNOWLEDGEMENTS

This study was supported by the fund for Comprehensive Research of Aging and the fund for Brain Science from the Ministry of Health and Welfare of Japan.

REFERENCES

1. Ito K, Morrish PK, Rakshi JS, et al. Statistical Parametric Mapping with F-18-dopa PET demonstrates bilaterally reduced striatal and nigral dopaminergic function in early Parkinson's disease. J Neurol Neurosurg Psych 1999;**66**:754-758.
2. Rakshi JS, Uema T, Ito K, et al. Frontal, midbrain and striatal dopaminergic function in early and advanced Parkinson's disease A 3D [(18)F]dopa-PET study. Brain 1999;**122**:1637-50.
3. Nagano AS, Ito K, Kato T, et al. Extrastriatal Mean Regional Uptake of Fluorine-18-FDOPA in the Normal Aged Brain - An Approach Using MRI-Aided Spatial Normalization. NeuroImage 2000;**11**:760-766.
4. Rinne JO, Rummukainen J, Paljarvi L, et al. Dementia in Parkinson's disease is related to neuronal loss in the medial substantia nigra. Ann Neurol 1989;**26**:47-50.
5. Torack RM, Morris JC. The association of ventral tegmental area histopathology with adult dementia. Arch Neurol 1988;**45**:497-501.
6. Holthoff-Detto VA, Kessler J, Herholz K, et al. Functional effects of striatal dysfunction in Parkinson disease. Arch Neurol 1997;**54**:145-150.
7. Alexander GE, DeLong MR, Strick PL. Parallel organization of functionally segregated circuits linking basal ganglia and cortex. Ann Rev Neurosci 1986;**9**:357-381.

ALIPHATIC PROPARGYLAMINES AS TREATMENTS FOR NEURODEGENERATIVE DISEASES

Catecholamine Research: From Molecular Insights to Clinical Medicine

Mark D. Berry and Alan A. Boulton*

1. INTRODUCTION

During the past decade, apoptosis has been implicated increasingly as playing a role in a number of human neurodegenerative conditions. The acceptance of a role for apoptosis in human disease states has led to a marked increase in attempts to develop anti-apoptotic agents as potential therapeutics. We have developed a number of novel aliphatic compounds, several of which have been tested in a number of *in vitro* and *in vivo* model systems (Berry, 1999). Here we focus on data obtained from *in vivo* models of two neurodegenerative conditions, stroke and Parkinson's disease.

2. STROKE

Using a modified Levine preparation of unilateral global cerebral ischaemia in rats, (Paterson *et al.*, 1997) we have tested a number of our compounds for their ability to prevent neuronal death seen in the CA1, CA3 and CA4 regions of the hippocampus. Our prototype compounds, R-2HMP and R-2HPA have been compared to the well documented anti-apoptotic agent R-deprenyl. As can be seen in figure 1, following either oral or sub-cutaneous administration, both compounds show a considerably increased potency in comparison to R-deprenyl. Indeed, at the doses tested, R-deprenyl was unable to show a significant reduction in cell death following oral administration. Important to note, is that in all cases, administration of compounds was delayed until 1 hour after lesion, with pre-treatment failing to show any rescue effects (data not shown). A further 11 aliphatic propargylamine related compounds have been identified to be active in this system, following sub-cutaneous administration starting 1 hour after lesion, at a dose of 0.1μmol/kg.

* AL*viva* Biopharmaceuticals Inc., #218 – 111 Research Drive, Saskatoon, Saskatchewan, Canada S7N 3R2

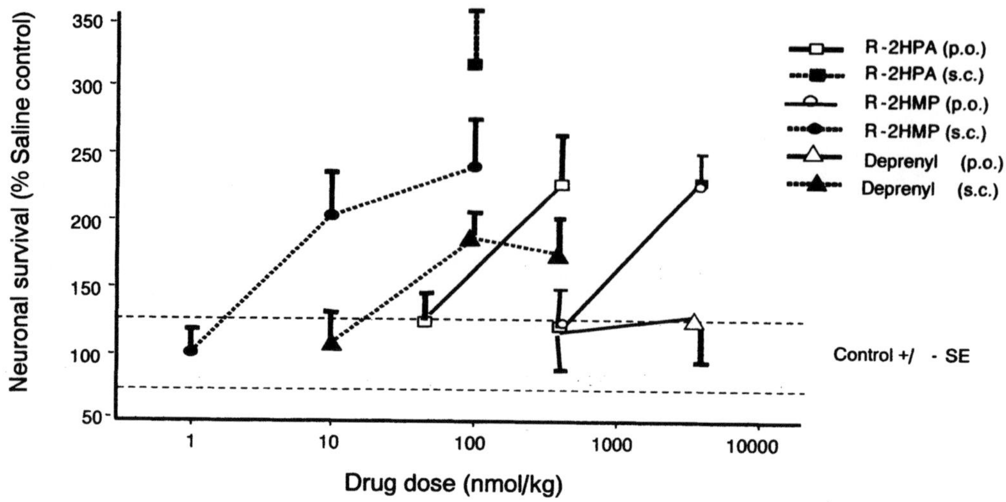

Figure 1. Comparison of the Potency of R-2HMP, R-2HPA and R-deprenyl in a Unilateral Hypoxia/Ischaemia Model of Stroke. Compounds were administered as indicated, with the first dose administered 1 hour following lesion. The number of surviving hippocampal CA1 neurons was assessed, and the percent rescue determined.

Three compounds have been chosen for further study in a permanent focal ischaemia model of stroke. This model involves rupture of all terminal blood vessels present on the brain surface in a 5mm diameter area. Using this model, an approximately conical lesion can be demonstrated 48 hours after surgery. This lesion contains distinct core and penumbral regions, the volumes of which can be readily estimated using the formula for the volume of a cone, where the radius is taken as half the maximal width of the lesion, and the height as the maximal depth of lesion. Using this model we have shown that R-2HMP, R-2HPA and R-2HPcAc all demonstrate a time and dose dependent reduction of lesion volume. Thus, while R-2HMP shows a significant reduction in lesion volume when administered 1 hour after lesion, little effect is seen if administration is delayed until 3 hours following lesion (data not shown). In contrast, R-2HPA and R-2HPcAc show a marked reduction in lesion volume when administration is delayed until 3 hours following lesion (Table 1). Preliminary data indicate that R-2HPA administration can be delayed 5 hours or more, with a significant reduction in lesion volume still observed.

3. PARKINSON'S DISEASE

We have selected R-2HMP for testing in Parkinson's disease model systems on the basis of joint, independent, MAO-B inhibitory (Yu *et al.*, 1992) and neural rescue (Berry 1999) activities. It was reasoned that such a combined activity would offer the possibility of both symptomatic relief of existing symptoms due to MAO-B inhibition and at least a slowing of disease progression due to an anti-apoptotic effect. MPTP neurotoxicity is still regarded as the gold standard animal model of Parkinson's disease.

Table 1. Effect of R-2HPA and R-2HPcAc Administered 3 Hours Following Lesion

	% Rescue (Total Lesion)	
Dose	R-2HPA	R-2HPcAc
Intra-peritoneal 0.05mg/kg	35.4	43.3
0.1mg/kg	81.8	
1mg/kg	80.4	16.7
Oral 0.05 mg/kg	32.7	12
0.3mg/kg	36.6	
1mg/kg	57.1	52.4
3mg/kg	57.9	
10mg/kg		69.2
30mg/kg	6.6	78.9
100mg/kg	3.9	37.2

In a mouse MPTP model, we have investigated the effects of R-2HMP treatment beginning two days following the last dose of MPTP and continuing twice daily for a further 18 days. Drug effects were determined following staining of brain sections containing the substantia nigra for tyrosine hydroxylase immunoreactivity. Using this protocol, R-2HMP administered orally, at a dose of 0.1mg/kg, resulted in a highly significant, 50% reduction in the loss of tyrosine hydroxylase positive neurons caused by MPTP (Berry and Boulton, 2001). It is important to note that by delaying the first administration of R-2HMP until two days following the last dose of MPTP, the confounding effects of MAO-B inhibition preventing MPTP toxification are avoided. Thus, this effect of R-2HMP represents a true neural rescue phenomenon.

The above study has recently been extended to examine the effects of R-2HMP in a non-human primate MPTP model. For this study, we adopted the intra-carotid MPTP administration model of Bankiewicz *et al.* (1986), which results in a predominantly unilateral lesion. Again, as above, R-2HMP administration was delayed until a time (2 hours) at which MPTP has been completely cleared from the brain, and conversion to MPP+ has already occurred (Bankiewicz, pers. comm.). R-2HMP was administered twice daily, intra-muscularly, at a dose of 0.1mg/kg. As in the above mouse study, drug effects were determined by counting tyrosine hydroxylase positive neurons remaining. As can be seen in figure 2, R-2HMP resulted in a marked increase in the number of tyrosine hydroxylase positive neurons surviving MPTP administration. Although preliminary in nature, this pilot study contained 2 animals in each group, thus not allowing statistical analysis, this binding suggests that R-2HMP is a promising candidate compound for development in Parkinson's disease.

4. CONCLUSIONS

The studies outlined above demonstrate that small molecule, aliphatic propargylamine related compounds are highly promising drug candidates for development in the area of human neurodegenerative conditions. Importantly, in all the studies outlined above, drugs were administered after the insult had occurred, and thus showed a true neural rescue effect. As such, these compounds would appear to offer the potential of use in pre-existing neurodegenerative conditions. This is of particular importance in conditions such as Parkinson's disease and stroke, where pre-lesion identification of patients is not currently possible.

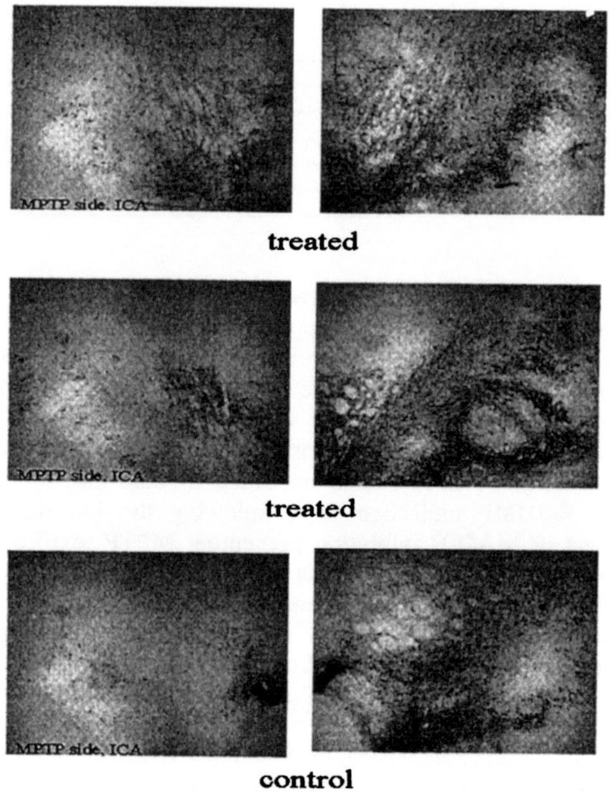

Figure 2. Effect of R-2HMP on Substantia nigra Neuron Survival Following MPTP Administration. MPTP was administered intra-carotid at a dose of 2mg. Two hours following MPTP administration treatment with R-2HMP (0.1mg/kg, i.m., twice per day) began.

REFERENCES

Bankiewicz, K. S., Oldfield, E. H., Chieuh, C. C., Doppman, J. L., Jacobwitz, D. M and Kopin, I. J., 1986, Hemiparkinsonism in monkeys after unilateral internal carotid artery infusion of 1-methyl-4-phenyl-1,2,3,6-tetrahydropyridine (MPTP) *Life Sci.* **39**:7. - 16

Berry, M. D. and Boulton, A. A., 2001, Aliphatic propargylamines as symptomatic and neuroprotective treatments for neurodegenerative diseases, *Neurotoxicol. Teratol.* (in press).

Berry, M. D., 1999, *R*-2HMP: an orally active agent combining independent antiapoptotic and MAO-B-inhibitory activities, *CNS Drug Rev.* **5**:105. - 124

Paterson, I. A., Barber, A. J., Gelowitz, D. L. and Voll, C., 1997, (-)Deprenyl reduces delayed neuronal death of hippocampal pyramidal cells, *Neurosci. Biobehav. Rev.* **21**:181. - 186

Yu, P. H., Davis, B. A. and Boulton, A. A., 1992, Aliphatic propargylamines: potent, selective, irreversible monoamine oxidase B inhibitors, *J. Med. Chem.* **35**:3705. - 3713

ß-CARBOLINES AND N-METHYLATION UNDERLYING PARKINSON'S DISEASE

Kazuo Matsubara and Koji Aoyama*

1. INTRODUCTION

The discovery of MPTP leads to the hypothesis that Parkinson's disease (PD) is may be initiated or precipitated by endogenous toxins by the mechanism similar to that of MPTP in genetically-predisposed individuals. The higher cerebrospinal fluid (CSF) levels of N-methylated azaheterocyclic amines, such as ß-carbolines (ßCs) and tetrahydroisoquinoline, have been found in PD patients. Thus, there is a hypothesis to influence the pathogenesis of PD, that is, the excess enzyme activity to activate neurotoxins, such as N-methyltransferase, might be higher in PD patients (Matsubara et al., 1998a).

2. ß-CARBOLINES

Indoleamine-related ßCs have been proposed as promising endogenous MPTP-like protoxins (Collins et al., 1992). We have previously reported that enzymes in mammalian brain particulate fractions methylate ßCs, sequentially forming 2-mono-[N]-methylated and neurotoxic 2,9-di-[N,N']-methylated ß-carbolinium cations (2,9-$Me_2ßC^+$s) (Matsubara et al., 1992). These $ßC^+$s are structural analogs of MPP^+ with a nitrogen bridge. Among ßC species, norharaman (NH) could be an underlying factor in idiopathic PD that is sequentially methylated to form 2,9-Me_2NH^+ in vitro and in vivo, since a higher CSF level of 2,9-Me_2NH^+ has been found in the CSF of patients with PD (Matsubara et al., 1995a). Systematic administration of simple ß-carboline, NH, induces bradykinesia with the decreased dopamine contents in the striatum and midbrain of C57/BL mouse (Matsubara et al., 1998). In this animal model, tyrosine hydroxylase-positive cells in the substantia nigra pars compacta have been diminished. The formation of a toxic metabolite in the central nervous system, 2,9-Me_2NH^+ has been suggested to be responsible for the NH toxicity. A higher concentration of 2,9-Me_2NH^+

* K. Matsubara, Dept. of Hospital Pharm. & Pharmacol., Asahikawa Med. College, Asahikawa 078-8510, Japan; K. Aoyama, Dept. of Internal Med. III, Shimane Med. Univ., Izumo 693-8501, Japan.

Catecholamine Research, Edited by Nagatsu *et al.*
Kluwer Academic/Plenum Publishers, 2002

localized in the nigra than in the cortex, and observed the S-adenosyl-L-methionine (SAM)-dependent methylation of 2[ß]- and 9[indole]-nitrogens of ßCs in non-PD human brains (Matsubara et al., 1993). N-Methylating enzyme activity for ßCs has been increased in PD brain (Gearhart et al., 2000). $ßC^+$s not only inhibit DA reuptake (Drucker et al., 1990; Matsubara et al., 1996; Matsubara et al., 1998c) and tyrosine hydroxylase (Matsubara et al., 1995b), but also function as NADH-linked respiratory inhibitors in isolated mitochondria (Albores et al., 1990; Collins et al., 1992). The nigrostriatal toxicity of 2,9-Me_2NH^+ is attenuated by the treatment with L-deprenyl (Matsubara et al., 2001).

3. N-METHYLATION ABILITY FOR AZAHETEROCYCLIC AMINES IN PARKINSON'S PATIENTS

There are two metabolic hypotheses to influence the pathogenesis of PD; i) the excess enzyme activity to activate neurotoxins, such as N-methyltransferase, might be higher in PDs. Indeed, in patients with PD, the activity of neutral N-methyltransferase, which can methylate catechol isoquinolines, is higher in comparison with that in controls (Naoi et al., 1998). ii) The activity of cytochrome P450 mono-oxygenase, that could detoxificate endogenous neurotoxins, might be lower in the patients (Smith et al., 1992). Thus, the determination of aberrant metabolic state in PD patients is an intriguing issue.

Nicotinamide has a pyridine ring in its structure and is metabolized through the pathways similar to those for the endogenous neurotoxins. To estimate the N-methylation ability for azaheterocyclic amines in PD patients, nicotinamide was dosed with 100 mg to 26 PDs and 20 controls consisted of 16 other neurogenic disease patients and 4 healthy volunteers (Aoyama et al., 2000). The urine was collected for 4 h, and then analyzed urinary its metabolites by an improved HPLC method. The urinary excretions of nicotinamide metabolites were significantly affected by aging. The excretion of N^1-methylnicotinamide decreased along with aging both in PD patients and controls. In younger (65 years old or younger) PD patients, the excretion amount of N^1-methylnicotinamide was significantly higher than that in younger controls. The decline rate of N^1-methylnicotinamide excretion in PDs was significantly greater than that in controls; the rate is more than 2-fold higher in PD patients. The age-associated decrease in 1-methyl-2-pyridone-5-carboxyamide excretion was observed only in PD patients, but not in controls. The total excreted amount of N-methylated metabolites (N^1-methylnicotinamide plus 1-methyl-2-pyridone-5-carboxyamide) was also observed the age-related decline in both groups. Aging did not influence the urinary excretions of nicotinamide and nicotinamide-N-oxide.

4. NICOTINAMIDE N-METHYLTRANSFEARSE IN THE CSF OF PD PATIENTS

Among several N-methyltransferase that could be involved in N-methylation of azaheterocyclic amines to form neurotoxic products in mammals, nicotinamide N-methyltransferase (NNMT) is the sole enzyme that has been well characterized. NNMT is an SAM-dependent cytosolic enzyme that catalyzes the N-methylation of nicotinamide to form N-methylnicotinamide and exists in many tissues. NNMT catalyzes N-methylation of nicotinamide and other pyridines to form pyridinium ions. The protein amount of NNMT was measured in the lumber CSF of PD patients by immunoblot analysis using anti-human NNMT antibody (Aoyama et al., 2001). In younger (65 years old or younger) PD patients, the relative level of NNMT protein was significantly higher than that in younger controls. The NNMT protein was significantly affected by aging: the amount decreased along with aging in PD patients. These results suggested that the excess NNMT in the central nervous system might be implicated in the PD pathogenesis.

5 CONCLUSION

These findings would indicate that the excess N-methylation ability for azaheterocyclic amines, such as ßCs, before the onset had been implicated in PD pathogenesis. On the other hand, the contribution of aberrant cytochrome P450 or aldehyde oxidase activity acting on the pyridine ring, that could act as detoxification routes of endogenous neurotoxins, would be small in the etiology of PD.

REFERENCES

Albores, R., Neafsey, E. J., Drucker G., Fields J. Z., and Collins M. A., 1990, Mitochondrial respiratory inhibition by N-methylated-ß-carboline derivatives structurally resembling N-methyl-4-phenylpyridine, *Proc. Natl. Acad. Sci. USA* **87**:9368.

Aoyama, K., Matsubara, K., Okada, K., Fukushima, S., Shimizu, K., Yamaguchi, S., Uezono, T., Satomi, M., Hayase, N., Ohta, S., Shiono, H. and Kobayashi, S., 2000, N-Methylation ability for azaheterocyclic amines is higher in Parkinson's disease: nicotinamide loading test, *J. Neural. Transm.* **107**:985.

Aoyama, K., Matsubara, K., Kondo, M., Murakawa, Y., Suno, M., Yamashita, K., Yamaguchi, S. and Kobayashi, S., 2001, Nicotinamide-N-methyltransferase is higher in the lumbar cerebrospinal fluid of patients with Parkinson's disease, *Neurosci. Lett.*,**298**:78.

Collins, M. A., Neafsey, E. J., Matsubara, K., Cobuzzi, R. J. Jr and Rollema, H., 1992, Indole-N-methylated ß-carbolinium ions as potential brain-bioactivated neurotoxins, *Brain Res.* **570**:154.

Drucker, G., Raikoff, K., Neafsey, E. J. and Collins, M. A., 1990, Dopamine uptake inhibitory capacities of ß-carboline and 3,4-dihydro-ß-carboline analogs of N-methyl-4-phenyl-1,2,3,6-tetrahydropyridine (MPTP) oxidation products, *Brain Res.* **509**:125.

Gearhart, D. A., Collins, M. A., Lee, J. M. and Neafsey, E. J., 2000, Increased ß-carboline 9N-methyltransferase activity in the frontal cortex in Parkinson's disease, *Neurobiol. Dis.* **7**:201.

Green, S., Buttrum, S., Molloy, H., Steventon, G., Sturman, S., Waring, R., Pall, H. and Williams, A., 1991, N-methylation of pyridines in Parkinson's disease. *Lancet* **338**:120.

Matsubara, K., Neafsey, E. J. and Collins, M. A., 1992, Novel S-adenosylmethionine-dependent indole-N-methylation of ß-carbolines in brain particulate fractions, *J Neurochem.* **59**:511.

Matsubara, K., Collins, M. A., Akan,e, A., Ikebuchi, J., Neafsey, E. J., Kagawa, M., and Shiono, H., 1993, Potential bioactivated neurotoxicants, N-methylated ß-carbolinium ions, are present in human brain, *Brain Res.* **610**: 90.

Matsubara, K., Kobayashi, S., Kobayashi, Y., Yamashita, K., Koide, H., Hatta, M., Iwamoto, K., Tanaka, O. and Kimura, K., 1995a, ß-Carbolinium cations, endogenous MPP^+ analogs, in the lumber cerebrospinal fluid of parkinsonian patients, Neurology **45**:2240.

Matsubara, K., Idzu, T., Kobayashi, Y., Nakahara, D., Maruyama, W., Kobayashi, S., Kimura, K. and Naoi M., 1995b, N-Methyl-4-phenylpyridinium and an endogenously formed analog, N-methylated ß-carbolinium, inhibit striatal tyrosine hydroxylation in freely moving rats, *Neurosci. Lett.* **199**:19.

Matsubara K., Idzu T., Kobayashi Y., Gonda T., Okunishi H., and Kiumura K., 1996, Differences in dopamine efflux induced by MPP^+ and ß-carbolinium in the striatum of concious rats, *Eur. J. Pharmacol.* 315:145.

Matsubara, K., 1998a, Bioactivation of Azaheterocyclic Amines via S-Adenosyl-L-Methionine-Dependent N-Methyltransferase, in, Pharmacology of Endogenous Neurotoxins, A. Mozel, ed., Birkhäuser, Boston, pp. 191-208.

Matsubara, K., Gonda, T., Sawada, H., Uezono, T., Kobayashi, Y., Kawamura, T., Ohtaki, K., Kimura, K. and Akaike, A., 1998b, Endogenously occurring beta-carboline induces parkinsonism in nonprimate animals: a possible causative protoxin in idiopathic Parkinson's disease, *J. Neurochem.* **70**:72.

Matsubara K., Senda T., Uezono T., Fukushima S., Ohta S., Igarashi K., Naoi M., Yamashita Y., Ohtaki K., Hayase N., Akutsu, S. and Kimura, K., 1998c, Structural significance of azaheterocyclic amines related to Parkinson's disease for dopamine transporter. *Eur. J. Pharmacol.* **348**: 77.

Matsubara, K., Senda, T., Uezono, T., Awaya, T., Ogawa, S., Chiba, K., Shimizu, K., Hayase, N. and Kimura, K., 2001, L-Deprenyl prevents the cell hypoxia induced by dopaminergic neurotoxins, MPP^+ and ß-carbolinium: a microdialysis study in rats, Neurosci. Lett. 302:65.

Naoi, M., Maruyama, W., Nakao, N., Ibi, T, Sahashi, K., Benedetti, M. S., 1998, (R)Salsolinol N-methyltransferase activity increases in parkinsonian lymphocytes, *Ann Neurol.* **43**:212.

NEUROPROTECTIVE ACTIONS OF GREEN TEA POLYPHENOL, (-)-EPIGALLOCATECHIN-3-GALLATE IN MODELS OF PARKINSON'S DISEASE: GENE TARGETS

Silvia Mandel, Yona Levites, Orly Weinreb and Moussa B.H. Youdim*

1. INTRODUCTION

Biochemical evidence points to central roles for oxidative stress (OS) and inflammation in neuronal death in both idiopathic and animal models of Parkinson's disease (PD). Reports from our laboratory, as supported by others, point to the presence of ongoing OS and inflammatory processes occurring selectively in the substantia nigra pars compacta (SNPC) of parkinsonian brains and in animal models of PD.[1-4] Therefore, drugs that exhibit free radical-scavenging and iron chelating properties, may serve as potential candidates for the treatment of PD.

Tea and tea polyphenols have attracted increasing interest because of their well reported biological effects, including radical scavenging,[5, 6] iron chelating,[7] anti-inflammatory,[8, 9] actions. We have recently reported neuroprotective properties of both green tea (GT) extract and GT polyphenol (-)-epigallocatechin-3-gallate (EGCG) in the mouse MPTP model of PD.[10] We extended this study to neuronal cell culture model of PD and determined gene expression as a consequence of EGCG treatment.

2. METHODS

2.1 Animal treatment

Mice were orally administered EGCG alone (2 mg/kg/day) for ten days and the following four days the animals received a combination of EGCG and MPTP (24 mg/kg/day). Catecholamine analysis and striatal homogenate preparation were performed as described.[10]

*Eve Topf and USA National Parkinson Foundation Centers of Excellence for Neurodegenerative Diseases Research, Department of Pharmacology, Technion- Faculty of Medicine, P.O.B. 9697, 31096 Haifa, Israel.

Catecholamine Research, Edited by Nagatsu *et al.*
Kluwer Academic/Plenum Publishers, 2002

2.2 Cell culture and cDNA microarray hybridization

Human neuroblastoma (NB) SH-SY5Y cells were placed in microtiter plates (96 wells) and grown at 37 °C, in a humid 5% CO_2, 95% air environment, in DMEM supplemented with streptomycin/penicillin (100 U/ml), and fetal calf serum (10%) for 24 hours. Drugs were added 15 min before insult with 6-OHDA, for a subsequent 24 hours.

For hybridization studies total RNA was extracted from NB SH-SY5Y cells exposed to EGCG (1 and 50 μM) for 6.5 hours, followed by synthesis of first strand cDNA biotinylated probe using gene-specific primer mix. The probes were hybridized to a small cDNA expression array membrane consisting of 25 genes related to apoptosis and cell cycle pathways.

3. RESULTS AND DISCUSSION

3.1 EGCG prevents striatal DA and metabolites reduction induced by MPTP

Chronic daily administration of EGCG (2 mg/kg) given orally for 2 weeks prevented the decrease in striatal dopamine (DA), DOPAC and HVA, induced by MPTP (Fig.1a). In addition, both EGCG and GT extract prevented the decrease in TH protein and activity caused by MPTP.[10] Similarly, R-APO, an iron chelator and radical scavenger, showed similar effects[11] (Fig. 1b). EGCG (2 mg/kg, orally) seems to be more potent than R-APO (10 mg/kg, s.c.), since a lower dose was required to prevent the damage and it was given by gavage, so the concentrations reached in blood might be even lower.

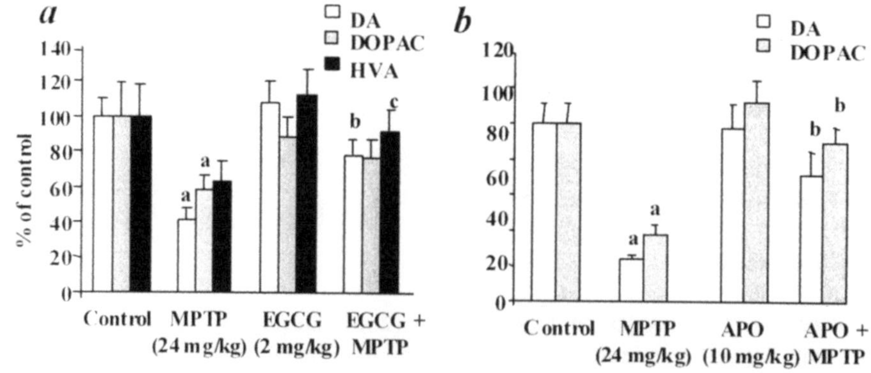

Figure 1. Preventive effect of EGCG* and R-APO** on MPTP-induced decrease in striatal catecholamine content. C57BL mice were injected with either EGCG (2 mg/kg/day, orally) for 10 days alone, followed by a combination with MPTP (24 mg/kg per day) for 4 consecutive days (a), or with R-APO (10 mg/kg, s.c.per day), followed by a dose of MPTP (24 mg/kg per day) for 5 days (b). The results represent the mean±SEM (n= 6-8 mice). ANOVA; ^{a}p<0.01, vs. control. ^{b}p<0.01 vs. MPTP.

*Adapted from Levites et al., 2001; **adapted from Grünblatt et al., 1999

3.2 Neuroprotective effects of tea extracts and EGCG in neuronal cell cultures

NB SH-SY5Y cell survival was dramatically reduced by 24 h treatment with 6-OHDA as assessed by MTT test, while pretreatment with GT or BT extracts (3 μM total polyphenols) doubled and tripled the viability of the cells, respectively, when compared to 6-OHDA alone (Fig. 2*a*). Similarly, EGCG (0.1-10 μM) conferred significant protection (72±2.5%) against 6-OHDA (Fig. 2b), whereas higher concentrations of both tea extracts and either EGCG promoted damage (Fig. 2a and 2*b*). This bell-shaped pattern is typical of antioxidative drugs, such as vitamin C[12] and R-apomorphine,[13] being neuroprotective at few micromolar concentrations, while having pro-oxidant activity at higher doses. These high, toxic concentrations are suppose to be responsible for the widely reported anticancer or apoptotic effects attributed to GT polyphenols.

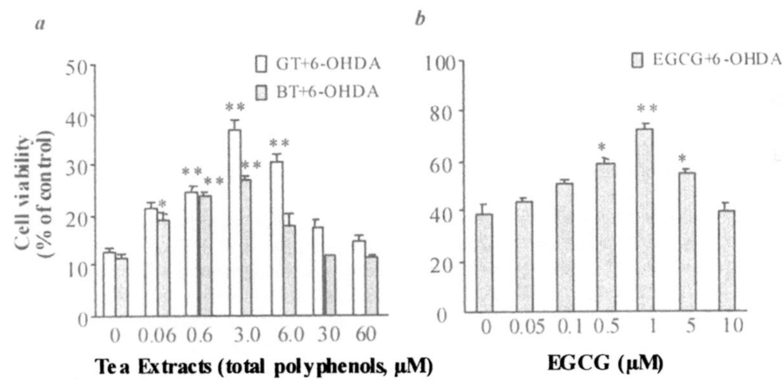

Figure 2. Protection by tea extracts and EGCG against 6-OHDA-induced neurotoxicity in NB SH-SY5Y cells. NB SH-SY5Y cells were pretreated with increasing concentrations of either GT extract, BT tea extract (a), or EGCG (b) 15 min before the addition of 6-OHDA (50 μM) for 24 hours. The results are the mean+SEM, n=8. one-way ANOVA; $^{*}p<0.05$, $^{**}p<0.001$, vs 6-OHDA).
*Adapted from Levites et al., 2001a.

3.3 Regulation of gene expression by EGCG

The neuroprotective and/or the lethal action of EGCG may represent two separate effects. To address this question, a customized array containing a limited number of genes related to apoptotic and cell cycle pathways was applied. NB SH-SY5Y cells treated with the high, 50 μM toxic concentration of EGCG, exhibited a decrease in the expression of Bcl-x mRNA, whereas caspase-6, Bax, and p21 (cell-cycle inhibitor) mRNAs were increased (Table 1). On the contrary, the protective 1μM concentration did not affect the expression of these genes or even down regulated p21 and Bax mRNAs, raising the question whether low EGCG concentrations may induce cell proliferation, in contrast to the high doses. Furthermore, the decreased Bax mRNA levels may contribute to an increase in the ratio of Bcl-2/Bax. This ratio constitutes a very important index of susceptibility to apoptosis.[14] It is more than possible that the in vivo neuroprotective effects observed in this study may result from a combination of the antioxidant properties and interaction of the different polyphenols with OS-downstream signaling events.

It is now apparent that the future approach to neuroprotection must consider either the use of neuroprotective drug combinations[15] or individual drugs having multipharmacological activities. As a consequence, the neuroprotective activities of individual and the combination of the purified major GT polyphenols may provide new insights into their major neuroprotective-related pathways.

Table 1. Effect of EGCG concentration on apoptosis-related gene expression in NB SH-SY5Y cells

EGCG	Bcl-x	Bax	caspase-6	p21
1 μM	Not changed	Down[b]	Not changed	Down[b]
50 μM	Down	Up[a]	Up[a]	Up[a]

[a] up regulated; [b]down regulated vs control (without EGCG).

REFERENCES

1. Youdim M.B.H., Ben-Shachar D. and Riederer P., 1993, The possible role of iron in theetiopathology of Parkinson's disease. *Mov. Disord.* **8**:1-12.
2. Gerlach M., Ben-Shachar D., Riederer P. and Youdim M.B.H., 1994, Altered brain metabolism of iron as a cause of neurodegenerative diseases? *J. Neurochem.* **63**:793-806.
3. Jenner P. and Olanow C.W., 1996, Pathological evidence for oxidative stress in Parkinson's disease and related degenerative disorders. in: *Neurodegeneration and neuroprotection*in *Parkinson's disease*, C.W. Olanow, P. Jenner, M.B.H. Youdim, eds., Academic Press,London, pp. 24-45.
4. Grünblatt E., Mandel S., Maor G. and Youdim M.B.H., 2001, Gene expression analysis in MPTP model of Parkinson's disease using cDNA microarray: effect of R-apomorphine. *J. Neurochem.* **78**:1-12.
5. Nanjo F., Goto K., Seto R., Suzuki M., Sakai M., Hara Y., 1996, Scavenging effects of teacatechins and their derivatives on 1,1-diphenyl-2-picrylhydrazyl radical. *Free Radic. Biol. Med.* **21**:895-902.
6. Salah N., Miller N.J., Paganga G., Tijburg L., Bolwell G.P., Rice-Evans C., 1995, Polyphenolic flavanols as scavengers of aqueous phase radicals and as chain-breaking antioxidants. *Arch. Biochem. Biophy.* **322**:339-346.
7. Guo Q., Zhao B., Li M., Shen S., Xin W., 1996, Studies on protective mechanisms of four components of green tea polyphenols against lipid peroxidation in synaptosomes. *Biochim. Biophys. Acta* **1304**:210-222.
8. Haqqi T.M., Anthony D.D., Gupta S., Ahmad N., Lee M.S., Kumar G.K.,Mukhtar H., 1999, Prevention of collagen-induced arthritis in mice by apolyphenolic fraction from green tea. *Proc. Nat .Acad. Sci. U S A* **96**:4524-4529.
9. Pan M.H., Lin-Shiau S.Y., Ho C.T., Lin J.H., Lin J.K., 2000, Suppression oflypopolysaccharide-induced nuclear factor kappaB activity by theaflavin-3,3'-digallate from tea and otherpolyphenols through down-regulation of IkappaB kinase activity in macrophages. *Biochem. Pharmacol.* **59**:357-367.
10. Levites Y., Weinreb O., Maor G., Youdim M.B.H. and Mandel, S., 2001, Green tea polyphenol (-)-epigallocatechin-3-gallate prevents N-methyl-4-phenyl-1,2,3,6-tetrahydropyridine-induced dopaminergic neurodegeneration *J. Neurochem.* **78**(5): 1073 – 1082
11. Levites Y., Youdim M.B.H., Maor G. and Mandel, S., 2001a, Attenuation of 6-hydroxydopamine (6-OHDA)-induced nuclear factor-kappaB (NF-κB) activation and cell death by tea extracts in neuronal cultures. *Biochem Pharmacol.* In press.
12. Grünblatt E., Mandel S., Berkuzki T. and Youdim M.B.H., 1999, Apomorphine protects against MPTP-induced neurotoxicity in mice. *Mov. Disord.* **14**:612-618.
13. Halliwell B., 1996, Vitamin C: antioxidant or pro-oxidant in vivo? *Free Radic. Res.* **25**:439-454.
14. Gassen M., Gross A. and Youdim M.B.H., 1998, Apomorphineenantiomers protect pheochromocytoma (PC12) cells from oxidative stress induced by hydrogen peroxide and 6-hydroxydopamine*Mov. Disord.* **13**:242-248.
15. Yang E. and Korsmeyer S.J., 1996, Molecular thanatopsis: a discourse on the BCL2 family and cell death. *Blood*, **88**:386-401.
16. Slikker W., Youdim M.B., Palmer G.C., Hall E., Williams C., Trembly B., 1999, The future of neuroprotection. *Ann. NY Acad. Sci.* **890**:529-533.

DOPAMINE TRANSPORTER AND PARKINSON'S DISEASE

Shigenobu Nakamura, Hideshi Kawakami, Hiroyuki Morino, Takatoshi Kawarai, Hiroshi Yamashita, Tetsuya Takahashi, Takako Mitsuoka, and Yumiko Kaseda*

1. INTRODUCTION

Dopaminergic transmission is terminated by the rapid reuptake of dopamine into presynaptic nerve terminals via a dopamine transporter (DAT) localized in presynaptic cytoplasmic membranes. Since Parkinson's disease (PD) is closely related with dopamine metabolism, we investigated the relationship between DAT and PD to develop a new therpeutic approach to the disease.

2. STRUCTURE AND ORGANIZATION OF GENE ENCODING HUMAN DOPAMINE TRANSPORTER (DAT)

DAT is a member of the monoamine transporter subfamily which possesses a common topological structure and significant amino acid sequence homology. We isolated and characterized the human DAT gene including 1 kb of 5'-flanking region[1]. The human DAT gene spans over 64kb, consisting of 15 exons separated by 14 introns. The intron-exon structure of the human DAT gene was similar to that of the human noradrenaline transporter gene. Promoter sequence analysis demonstrated a 'TATA'-less, 'CAT'-less and G+C-rich structure. Two E box and several Sp-1-binding sites exist in the promoter region. Structural features are similar to those of the human D_{1A} dopamine receptor gene and the human monoamine oxidase A gene. The transcription start site was determined by both 5'-RACE and RNase protection assay and we elucidated the 5' end of the mRNA by identifying the 5'-terminal cap-G residue.[1]

* Shigenobu Nakamura, Hiroshima University, Hiroshima 734-8551, Japan.

3. NUCLEOTIDE POLYMORPHISM OF DOPAMINE TRANSPORTER (DAT) GENE ASSOCIATED WITH PARKINSON'S DISEASE (PD)

The site of lesion in PD is similar to the distribution of the DAT and DAT levels are related to neuronal vulnerability. Toxins such as 6-hydroxydopamine and 1-methyl-4-phenylpyridinium (MPP^+) producing PD-like model are incorporated via DAT. Dopamine or MPP^+ uptake is altered in DAT mutants. These results prompted us to analyze the DAT polymorphism in PD.

To identify DAT polymorphism, direct sequence was performed in 24 PD patients and SSCP in 24 controls. We identified 2 polymorphisms out of all coding regions of the DAT gene. One existed in exon 9 (1215A/G), and another in exon 15 (1898T/C).[2]

We studied the distribution of these polymorphisms in all subjects by using dot blot hybridization analysis with sequence-specific oligonucleotide probes. The 1215G in exon 9 was more frequent in the normal control group than in the PD group. χ^2 analysis on the distribution of genotype and allele frequency revealed a significant difference between the PD and normal control groups (genotype: $\chi^2=8.82$, $p=0.0122$, allele: $\chi^2=4.14$, $p=0.0418$, OR=2.07 (95% CI: 1.15-3.78)). In addition, 1898C in exon 15 was also more frequent in normal controls than in the patient group, but no significant difference existed between the two groups (genotype: $\chi^2=1.83$, $p=0.401$, allele: $\chi^2=0.751$, $p=0.386$, OR=1.44 (95% CI: 0.788-2.65)). Otherwise, a significant positive correlation was observed between 1215G and 1898C in both groups.[2]

We screened the polymorphisms out of the entire coding region of the DAT gene, and compared the differences in the distribution between patients with PD and normal controls. Our results indicate that there are 2 polymorphisms in all the exons of the DAT gene, and one (1215G) is significantly more frequent in the controls than in PD. The results suggest that 1215G is associated with decreased susceptibility to PD.

Previous studies failed to prove the association between the polymorphism of the DAT gene and PD. However, only a variable number of tandem repeat (VNTR) polymorphism was examined in the 3'-untranslated region, and did not cover exonal polymorphisms. Exonal polymorphisms will be more related to the function of the gene and more conservative than VNTR in untranslated regions. Thus, examinations on exonal polymorphisms in the gene might lead to determining their contribution to the disease. Polymorphisms of the cytochrome P-450 CYP2D6 gene and glutathione transferase gene have been investigated in PD. Polymorphisms of the cytochrome P-450 CYP2D6 gene were reported in association with PD, which turned out not to be reproducible. Although many studies suggest that toxic factors contribute to the development of the disease, a genetic predisposition cannot fully explain the mechanism for selective neurodegeneration.

At first, we assumed a nucleotide variation with amino acid substitutions causing a change in protein functions of DAT. However, we found only 2 polymorphisms without amino acid substitutions in all coding regions. DAT is much conserved at both the nucleotide and protein levels. Since 1215G is associated with development of PD, there would be a mechanism other than one associated with protein functions of DAT. A couple of possibilities could be raised: 1) 1215G would link to polymorphisms of other genes near the locus. 2) Polymorphisms in other regions of the DAT gene would link to 1215G and might decrease the expression level of DAT mRNA, the latter of which seems more plausible. The lowered expression of DAT mRNA might decrease the uptake of toxic factors via DAT, and prevent the degeneration of dopaminergic neurons. In fact, the angiotensinogen gene has 2 mutations which are linked to each other. The mutation in the 5'-flanking region affects the basal transcription rate of the angiotensinogen gene, and contributes to essential hypertension.

However, our study requires some caution. 1) We enrolled subjects from an ethnic population. 2) Our patients were diagnosed as PD only clinically. All patients in PD group might not have suffered from idiopathic PD in spite of careful clinical selections. Therefore, further and larger studies recruiting subjects from other ethnic populations or pathologically diagnosed patients would reveal a more obvious relationship between DAT and PD, and elucidate the mechanism that would explain how the DAT gene contributes to the pathogenesis of PD.

We found 2 polymorphisms in exons of the DAT gene, and one was associated with decreased susceptibility to PD. The study on the transporter based mechanism might explain selective cell loss of dopaminergic neurons in PD and provide a new strategy for disease prevention.

4. EFFECT OF NICOTINE ON DOPAMINE UPTAKE IN COS CELLS POSSESSING RAT DOPAMINE TRANSPORTER AND IN PC12 CELLS

The effect of nicotine on the uptake of dopamine is not completely understood. We studied the effects of nicotine on PC12 cells and on COS cells transfected with the rat DAT gene[3]. Dopamine uptake by PC12 cells was inhibited by nicotine in a concentration-related fashion. Treatment of PC12 cells with nerve growth factor (NGF) increased dopamine uptake inhibition. The effect was abolished by inhibitors of nicotinic acetylcholine receptors such as hexamethonium and mecamylamine, indicating an inhibition via nicotinic receptors in PC12 cells. This view is also supported by the evidence that acetylcholine reduced the uptake of dopamine in a hexamethonium-, but not atropine-sensitive fashion. In contrast, nicotine failed to inhibit dopamine uptake by COS cells possessing DAT.[3] These results suggest that the inhibitory effect of nicotine on dopamine uptake may play a role in regulating extracellular concentrations of dopamine, when coupled with a nicotinic receptor, leading to a decreased sodium gradient across the membrane transporter, brought about by nicotinic stimulation.

5. EFFECT OF NICOTINE ON THE SUVIVAL OF PC12 CELLS DEPRIVED OF SERUM AND NGF

We examined the protective effect of nicotine against neuronal cell death in the model system of serum- and NGF-free cultures of PC12 cells.[4] Serum deprivation induced a massive death of undifferentiated PC12 cells, which was inhibited by the addition of NGF, but not by nicotine (0.1 mM). Even after PC12 cells had been differentiated by treatment with NGF, serum and NGF deprivation induced a rapid and massive death of the cells. The addition of nicotine (1-100 μM) to the deprivation system prevented such cell death almost completely. The protective effect of nicotine was abolished by hexamethonium or mecamylamine, suggesting the protective effect of nicotine against the neuronal cell death was through nicotinic acetylcholine receptors.[4]

We examined the effect of NGF and nicotine on the expression of nicotinic receptor subunits in PC12 cells using Northern blot analysis[5]. NGF increased the transcription of α5, 7 and β4, but changes in α3 and β2 levels were smaller than α5, 7 or β4. Nicotine also increased the level of α5 and β2 mRNA. The subunit mRNA change was different between NGF and nicotine treatment. Probably, NGF changes the expression of nicotinic receptors in a subtype-specific manner over the course of differentiation, and the disproportionate subunit expressions might be related to neuro-protective effects exerted by nicotine[4] or inhibited dopamine uptake through DAT.[3]

6. THERAPEUTIC EFFECT OF NICOTINE ON PARKINSON'S DISEASE (PD)

Nicotine has been reported to show beneficial effects on PD. We examined the therapeutic effect of nicotine chewing gum in patients with early-onset PD. Subjects were 8 patients with early-onset PD. Four out of 8 patients had a history of smoking (smokers). We examined symptoms of PD, and auditory event-related potentials (ERPs) before and after nicotine administration. In smokers, PD scores improved by more than 10% and P300 latency of auditory ERPs was shortened by more than 30 msec. In contrast, nicotine had no remarkable effects on parkinsonian symptoms or auditory ERPs in non-smokers. We suggest that nicotine may be a possible choice for the treatment of patients with early-onset PD, especially when they are smokers.

As mentioned before, nicotine inhibits DAT indirectly[3] in addition to the release of dopamine. Smokers have been reported to have an increased number of nicotinic receptors in the brain, and might have responded better to the administration of nicotine gum than non-smokers through the inhibition of DAT.[3]

7. CONCLUSION

The decreased density of the DAT in the presynaptic membranes of the striatum in PD patients is well documented. A novel aspect of DAT related with PD was presented both in the vulnerability of the disease and in the inhibition of dopamine uptake by nicotine, which suggests a new clue for the prevention and treatment of PD.

REFERENCES

1. T. Kawarai, H. Kawakami, Y. Yamamura, and S. Nakamura, Structure and organization of the gene encoding human dopamine transporter, *Gene* **195**, 11-18 (1997).
2. H. Morino, T. Kawarai, Y. Izumi, T. Kazuta, M. Oda, O. Komure, F. Udaka, M. Kameyama, S. Nakamura, and H. Kawakami, A single nucleotide polymorphism of dopamine transporter gene is associated with Parkinson's disease, *Ann. Neurol.* **47**, 528-531 (2000).
3. H. Yamashita, S. Kitayama, Y.-X. Zhang, T. Takahashi, T. Dohi, and S. Nakamura, Effect of nicotine on dopamine uptake in COS cells possessing the rat dopamine transporter and PC12, *Biochem. Pharmacol.* **49**, 742-745 (1995).
4. H. Yamashita, and S. Nakamura, Nicotine rescues PC12 cells from death induced by nerve growth factor deprivation, *Neurosci. Lett.* **213**, 145-147 (1996).
5. T. Takahashi, H. Yamashita, S. Nakamura, H. Ishiguro, T. Nagatsu, and H. Kawakami, Effects of nerve growth factor and nicotine on the expression of nicotinic acetylcholine receptor subunits in PC12 cells, *Neurosci. Lett.* **35**, 175-181 (1999).

L-DOPA RESPONSIVE DYSTONIA/INFANTILE PARKINSONISM ASSOCIATED WITH MUTATIONS AT THE HUMAN TYROSINE HYDROXYLASE LOCUS

Torgeir Flatmark*, Bjørg Almås, and Anne P. Døskeland

1. CLINICAL AND METABOLIC PHENOTYPES – DIAGNOSTIC CRITERIA

The L-DOPA responsive dystonias (DRD) represent a group of hereditary neurometabolic disorders with some common clinical features, but with a locus heterogeneity and different genotypes. The first described and best characterized is the hereditary progressive dystonia (HPD/DRD)(for review, see Segawa, 2000), an autosomal dominantly inherited motoric dysfunction caused by mutations at the GTP cyclohydrolase I (GCH 1; EC 3.5.4.16) locus on chromosome 14 (gen map locus 14q22.1-q22.2)(Ichinose et al., 1994). An autosomal recessive form of DRD was later described and found to be associated with a point mutation (Q381K) at the human tyrosine hydroxylase (hTH; EC 1.14.16.2) locus (gene map locus 11p15.5)(Lüdecke et al., 1995; Knappskog et al., 1995). Subsequently, a recessively inherited L-DOPA responsive infantile parkinsonism (DRIP), also caused by a point mutation (L205P) at the hTH locus, was reported (Lüdecke et al., 1996). Both hTH mutations were considered sufficient to explain the deficient function of the hydroxylase, the attendant neurometabolic pattern and the resultant clinical signs and symptoms. A total of twelve different mutations at the hTH locus are now known to be associated with DRD/DRIP (Table 1). The patients with mutations at the hTH locus show larger variations in and often more severe clinical phenotypes as well as an earlier onset than in the classical Segawa's syndrome (HPD/DRD). The disease is manifested only in homozygotes and compound heterozygotes, whereas the autosomal dominant form of HPD/DRD is observed also in heterozygotes. The mild forms of DRD associated with mutations at the hTH locus are characterized by an onset in the first decade with postural dystonia with lower extremity predominance and marked diurnal variations in the symptoms (Lüdecke et al., 1995). Thus, the clinical characteristics are very similar to those described for the classical HPD/DRD. Recently, a case of DRD simulating spastic paraplegia has been reported (Furukawa et al., 2001) with a compound heterozygote mutation at the hTH locus (Table

* Torgeir Flatmark, Department of Biochemistry and Molecular Biology, University of Bergen, 5009 Bergen, Norway.

1). The more severe forms of hTH deficiency, i.e. L-DOPA responsive infantile (early-onset) parkinsonism (DRIP), are characterized by an onset within the first months of life with generalized hypokinetic rigidity, and often reduced facial mimicry and ptosis, but without the diurnal variability in the symptoms observed in DRD (Lüdecke et al., 1996; van den Heuvel et al., 1998; Wevers et al., 1999; Bräutigam et al., 1999; Swaans et al., 2000). Positron emission tomography (PET) scan analyses have so far not been performed systematically in this group of patients, but brain computed tomography (CT) and magnetic resonance imaging (MRI) analyses have been reported to be normal. In an autopsied case of DRD a reduced level of immunoreactive TH and of TH catalytic activity in the striatum has been reported, but with no evidence of a degenerative process in the brain (Rajput et al., 1994). At the metabolite level a reliable diagnosis of hTH deficiency can be made only by analysis of the cerebrospinal fluid (CSF). Decreased CSF levels of homovanillic acid (HVA) and 3-methoxy-4-hydroxyphenylethyleneglycol (MHPG), together with normal levels of pterin, tyrosine and 5-hydroxyindolacetic acid (5-HIAA) are the diagnostic hallmarks of hTH deficiency (Lüdecke et al., 1996; van den Heuvel et al., 1998; Wevers et al., 1999; Bräutigam et al., 1999; Swaans et al, 2000).

2. RESPONSIVENESS TO L-DOPA SUBSTITUTION THERAPY

For all forms of DRD/DRIP (Table 1) a marked and sustained response to substitution therapy with L-DOPA has been observed without any unfavourable long term side effects. In two families the patients (in the fourth decade of life) have been able to live a normal life with a low-dose of L-DOPA for more than 30 years (Swaans et al., 2000). Thus, the time-course of these recessively inherited forms of parkinsonism has so far

Table 1. L-DOPA-responsive dystonia (DRD)/infantile parkinsonism (DRIP) associated with mutations at the human tyrosine hydroxylase (hTH) locus.

Exon	Mutation[a]		Clinical phenotype	Reference
3	1-bp del(T)/	TH4	DRD[b]	Furukawa et al., 2001
	D498G	TH4		
5	L205P[c]	TH1	DRIP	Lüdecke et al., 1996
6	R233H	TH4	DRIP	van den Heuvel et al., 1998
3,6	R233H/	TH4	DRIP	Wevers et al., 1999
	291delC			
8	926T>C	TH4	DRIP	de Lonlay et al., 2000
8,9	T276P/		DRIP	Swaans et al., 2000
	T314M			
9,14	R337H/	TH4	DRD/DRIP	Swaans et al., 2000
	T494M			
10	C359F	TH4	DRIP	Bräutigam et al., 1999
11	Q381K[c]	TH1	DRD	Lüdecke et al., 1995;
				Knappskog et al., 1995

a The residue numbers refer to the amino acid sequence of either isoform hTH1 (497 residues) or hTH 4 (528 residues). b Simulating spastic paraplegia. c L236P and Q412K in hTH4.

given no evidence for neurodegeneration in the basal ganglia. In cases where the metabolite levels in CSF have been measured, the catecholamine (including dopamine) levels were normalized after treatment with L-DOPA (Bräutigam et al., 1999).

3. MOLECULAR GENETICS AND BIOLOGY

Twelve different mutations at the hTH locus associated with extrapyramidal motoric dysfunction have been reported so far (see Table 1). Most of them are missense mutations in the exons coding for the catalytic and tetramerization domains of the hTH structure (Flatmark and Stevens, 1999). Two of the mutations have been subjected to in vitro expression analysis. The Q381K mutation represents a kinetic variant form of the enzyme with an approx. 15 % residual activity (Knappskog et al., 1995) for a structurally rather stable recombinant enzyme (Lüdecke et al., 1996). By contrast, the L205P mutation revealed both a low residual activity (1.5 - 16 %) and a reduced stability when expressed in A293 cells (Lüdecke et al., 1996). For both mutations the reduced function of the hydroxylase (enzymatic phenotype) was sufficient to explain the observed neurometabolic pattern in the CSF (metabolic phenotype) and the resultant clinical characteristics (clinical phenotype). However, as shown for mutations at the hPAH locus (Scriver and Waters, 1999) the clinical and metabolic phenotypes of the mutations in the hTH locus are expected to be modulated in vivo by a number of factors, which are beyond the control of the hTH locus by itself.

4. WHY IS THE OBSERVED MUTATION FREQUENCY AT THE hTH LOCUS SO LOW AS COMPARED TO THAT AT THE hPAH LOCUS?

Elimination of TH in mice (TH-/-) is lethal and indicates that the catecholamines are essential for fetal survival (Kobayashi et al., 1995; Zhou et al., 1995). Therefore, mutations at the hTH locus, which result in an inactive/unstable enzyme, is expected to lead to perinatal lethality also in humans, and thus explain the low number of mutations detected so far at the hTH locus as compared to the structurally and functionally closely related enzyme human phenylalanine hydroxylase (hPAH). Thus, > 400 mutations have been reported at the hPAH locus (frequency about 1 in 10 000 in the Caucasian population) in patients with the autosomal recessive hyperphenylalaninemia/ phenylketonuria (Scriver and Waters, 1999). A large number of these mutations have been shown to represent almost knock-out mutations which, however, can be controled by a low-phenylalanine diet, and by the fact that L-phenylalanine is a substrate for TH in neuroendocrine cells (Fukami et al., 1990). Thus, viable human mutations at the hTH locus must result in a measurable residual catalytic activity and/or a sufficient cellular stability, which so far has been shown to be the case by cloning and expression of the L205P and Q381K mutations (Knappskog et al., 1995; Lüdecke et al., 1996).

5. hTH DEFICIENCY AND TARGET NEURONS

TH is the rate-limiting enzyme in the biosynthesis of the catecholamines (for recent review, see Flatmark, 2000). In the severe forms of DRD/DRIP symptoms have been observed from both the dopaminergic and noradrenergic systems in CNS/PNS (Lüdecke et al., 1996; van den Heuvel et al., 1998; Wevers et al., 1999; Bräutigam et al., 1999; Swaans et al, 2000). The dominating extrapyramidal motoric dysfunctions are related to a DA deficiency in the nigrostratum DA neurons. In a severe form of DRIP cognitive functions have been observed to be affected (Lüdecke et al., 1996).

REFERENCES

Bräutigam, C., Steenbergen-Spanjers, G.C.H., Hoffmann, G.F., Dionisi-Vici, C., van den Heuvel, L.P.W.J., Smeitink, J.A.M., and Wevers, R.A., 1999, Biochemical and molecular genetic characteristics of the severe form of tyrosine hydroxylase deficiency, Clin. Chem. 45:2073.

de Lonlay, P., Nassogne, M.C., van Gennip, A.H., van Cruchten, A.C., Billatte de Villemeur, T., Cretz, M., Stoll, C., Launay, J.M., Steenberger-Spante, G.C.V., van den Heuvel, L.P.W., Wevers, R.A., Saudubray, J.M., and Abeling, N.G.G.M., 2000, Tyrosine hydroxylase deficiency unresponsive to L-dopa treatment with unusual clinical and biochemical presentation, J. Inherit. Metab. Dis. 23:819.

Flatmark, T., and Stevens, R.C., 1999, Structural insight into the aromatic amino acid hydroxylases and their disease-related mutant forms, Chem. Rev. 99:2137.

Flatmark, T., 2000, Catecholamine biosynthesis and physiological regulation in neuroendocrine cells, Acta Physiol. Scand. 168:1.

Fukami. M.H., Haavik, J., and Flatmark, T., 1990, Phenylalanine as substrate for tyrosine hydroxylase in bovine adrenal chromaffin cells, Biochem. J. 268:525.

Furukawa, Y., Graf, W.D., Wong, H., Shimadzu, M., and Kish, S.J., 2001, Dopa-responsive dystonia simulating spastic paraplegia due to tyrosine hydroxylase (TH) gene mutations, Neurology 56:260.

Ichinose, H., Ohye, T., Takahashi, E., Seki, N., Hori, T., Segawa, M., Nomura, Y., Endo, K., Tanaka, H., Tsuji, S., Fujita, K., and Nagatsu, T., 1994, Hereditary progressive dystonia with marked diurnal fluctuation caused by mutations in the GTP cyclohydrolase I gene, Nature Genet. 8:236.

Knappskog, P.M., Flatmark, T., Mallet, J., Lüdecke, B., and Bartholomé, K., 1995, Recessively inherited L-DOPA-responsive dystonia caused by a point mutation (Q381K) in the tyrosine hydroxylase gene, Hum. Mol. Genet. 4:1209.

Kobayashi, K., Morita, S., Sawada, H., Mizuguchi, T., Yamada, K., Nagatsu, I., Hata, T., Watanabe, Y., Fujita, K., and Nagatsu, T., 1995, Targeted disruption of the tyrosine hydroxylase locus results in severe catecholamine depletion and perinatal lethality in mice, J. Biol. Chem. 270:27235.

Lüdecke, B., Dworniczak, B., and Bartholomé, K., 1995, A point mutation in the tyrosine hydroxylase gene associated with Segawa's syndrome, Hum. Genet. 95:123.

Lüdecke, B., Knappskog, P.M., Clayton, P.T., Surtees, R.A.H., Clelland, J.D., Heales, S.J.R., Brand, M.P., Bartholomé, K., and Flatmark, T.,1996, Recessively inherited L-DOPA-responsive parkinsonism in infancy caused by a point mutation (L205P) in the tyrosine hydroxylase gene, Hum. Mol. Genet. 5:1023.

Rajput, A.H., Gibb, W.R., Zhong, X.H., Shannak, K:S., Kish, S., Chang, L.G., and Hornykiewicz, O., 1994, Dopa-responsive dystonia; pathological and biochemical observations in a case, Ann. Neurol. 35:396.

Segawa, M., 2000, Hereditary progressive dystonia with marked diurnal fluctuation, Brain Dev. 22:S65.

Scriver, C.R., and Waters, P.J., 1999, Monogenic traits are not simple: lessons from phenylketonuria, Trends Genet. 15:267.

Swaans, R.J., Rondot, P., Renier, W.O., van den Heuvel, L.P., Steenbergen-Spanjers, G.C., and Wevers, R.A., 2000, Four novel mutations in the tyrosine hydroxylase gene in patients with infantile parkinsonism, Ann. Hum. Genet. 64:25.

van den Heuvel, L.P.W.J., Luiten, B., Smeitink, J.A.M., deRijk-van Andel, J.F., Hyland, K., Steenbergen-Spanjers, G.C.H., Janssen, R.J.T., and Wevers, R.A., 1998, A common point mutation in the tyrosine hydroxylase gene in autosomal recessive L-DOPA-responsive dystonia in the Dutch population, Hum. Genet. 102:644.

Wevers, R.A., deRijk-van Andel, J.F., Bräutigam, C., Geurtz, B., van den Heuvel, L.P., Steenbergen-Spanjers, G.C., Smeitink, J.A., Hoffmann, G.F., and Gabreels, F.J., 1999, A review of biochemical and molecular genetic aspects of tyrosine hydroxylase deficiency including a novel mutation (291delC), J. Inherit. Metab. Dis. 22:364.

Zhou, Q.Y., and Palmiter, R.D., 1995, Dopamine-deficient mice are severely hypoactive, adipsic, and aphagic, Cell 83:1197.

DOPAMINE IS A PATHOGENETIC FACTOR IN HIV-INDUCED NEURO-AIDS

Eleni Koutsilieri, Carsten Scheller, Stefanie Czub, Sieghart Sopper, Volker ter Meulen, and Peter Riederer*

1. INTRODUCTION

During the first years of the AIDS pandemic caused by the human immunodeficiency virus (HIV), a progressive, dementing illness, later called simply Neuro-AIDS, AIDS dementia complex or HIV dementia was defined as one of the consequences of the viral brain infection (Navia et al., 1986a; Navia et al., 1986b). According to WHO report, December 2000, it is estimated that around 36 million people world wide are infected with HIV. One-third of them is expected to develop HIV-dementia. HIV is the leading cause of dementia in people less than 60 years of age (Janssen et al., 1992; McArthur et al., 1993).

HIV-induced dementia is characterized by psychiatric and neurologic symptoms including cognitive dysfunction, behavioural abnormalities, and motor disorders which are attributed to HIV infection *per se* rather than being associated with subsequent opportunistic infections or malignancies (Price, 1996). Although, to date, the pathogenesis of HIV-dementia is incompletely understood and certainly multifactorial, many of the symptoms may be attributed to abnormalities in the dopamine rich-basal ganglia. Neuro-AIDS patients have clinical signs in common with Parkinsonian features (Berger and Arendt, 2000). Many questions concerning the pathogenesis of Neuro-AIDS and the role of the dopaminergic system in the development of HIV-dementia remain unanswered. These issues are difficult to approach in the human host because of the difficulty in identifying acutely infected individuals and the inaccessibility of the human brain for sampling during infection. Use of animal models is required for the understanding of the early asymptomatic stage of infection and the progression towards subacute and finally manifested disease. Therefore, we used the established animal model with simian

* Eleni Koutsilieri, Carsten Scheller, Sieghart Sopper, Volker ter Meulen, Institute for Virology and Immunobiology,, 97078 Würzburg. Stefanie Czub, Institute for Pathology, 97078 Würzburg. Peter Riederer, Institute for Clinical Neurochemistry, Dept. Psychiatry, 97080 Würzburg, Germany.

immunodeficiency virus (SIV)-infected macaques in order to investigate changes in the dopaminergic system during early asymptomatic infection.

2. VULNERABILITY OF THE DOPAMINERGIC SYSTEM IN SIV-INFECTED RHESUS MONKEYS

Animals were infected under ketamine anesthesia (10 mg/kg) with SIVmac 251 MPBMC, a neurotropic virus strain. The monkeys were sacrificed between three and six months after infection.We found that dopamine is reduced already two months following infection with SIV with an accompanying increase in dopamine turnover as indicated by the ratio of HVA/DA and DOPAC/DA (Fig. 1 A, B). These data provided the first direct evidence for a decrease in dopamine levels in the early asymptomatic phase (Czub et al., 2001). Dopamine levels were reduced in putamen, hippocampus and frontal cortex of SIV-infected animals. The substantia nigra, a region rich in dopaminergic cell bodies, showed no dopamine loss, indicating that the virus or viral products affect initially dopaminergic terminals in the postsynaptic dopaminergic areas (data not shown). In contrast, brains of HIV-infected subjects exhibited marked neuronal degeneration in the substantia nigra in the last stage of Neuro-AIDS (Itoh et al., 2000). Whether a further retrograde degeneration of the dopaminergic projections will accompany late stages of SIV infection remains to be elucidated.

3. DOPAMINERGIC DRUGS ACCELERATE THE PROGRESSION OF SIV INFECTION IN CNS

Antiretroviral therapy appears to fail in improving the neurochemical deficits and clinical symptoms of Parkinsonism in patients with HIV infection (Mirsattari et al., 1998). L-DOPA has been shown to reverse some of the Parkinsonian symptoms only in a subset of HIV-infected patients (Kieburtz et al., 1991). Moreover, Selegiline, a monoamine oxidase inhibitor, which retards the catabolism of dopamine in the CNS, improved cognitive functions in patients with Neuro-AIDS (Dana Consortium, 1998). In order to restore the observed dopamine deficiency of SIV-infected monkeys, we administered the dopaminergic and neuroprotective substance Selegiline in SIV-infected rhesus monkeys. We chose a concentration of Selegiline (Tatton and Greenwood, 1991) which can inhibit both monoamine oxidase (MAO)-B and –A (Fig. 1 C). Although the treatment with Selegiline restored dopamine deficits and this seemed to us initially beneficial (Fig.1 A), histological examination revealed extensive pathological changes in the CNS in the form of a diffuse vacuolar degeneration and SIV encephalitis without indications of AIDS (Czub et al., 2001).

Selegiline treatment resulted in such effects because of increased dopamine availability since further experiments with additional monkeys demonstrated that L-DOPA, the precursor of dopamine had similar effects on the induction of pathological lesions. Moreover, Selegiline treatment resulted in a significant increase in SIVmRNA-expressing cells compared with the number in SIV-infected/untreated animals (Fig. 1 D).

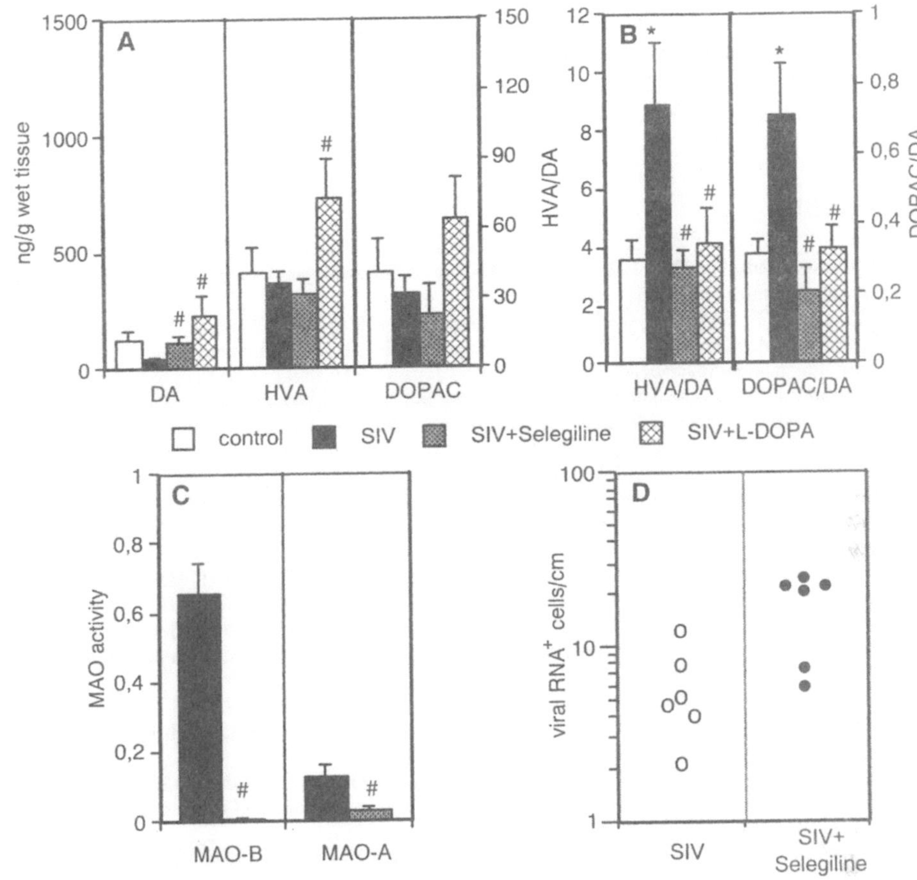

Figure 1. Dopamine metabolism in frontal cortex of SIV-infected and uninfected animals (A); Dopamine turnover (B); MAO activity in frontal cortex following treatment with Selegiline (C); SIV RNA was quantified in frontal cortex, hippocampus, basal ganglia, substantia nigra, brain stem and cerebellum of SIV-infected monkeys (D).

4. DOPAMINE ACTIVATES HIV THROUGH OXIDATIVE STRESS IN VITRO

From our studies in SIV-infected animals, it seems that dopamine accelerates the onset of Neuro-AIDS. Therefore, we conducted in vitro experiments to further clarify the mechanism of action of dopamine on immunodeficiency virus infection. We checked specifically whether dopamine could activate latent HIV in chronically infected CD4 positive T lymphoblasts. We found that dopamine markedly activates HIV in chronically-infected T lymphoblasts in a dose-dependent manner and that glutathione and N-acetylcysteine inhibited dopamine-induced HIV activation (Scheller et al., 2000). It is known that dopamine or its precursor L-DOPA can regulate glutathione (Gabby et al., 1996; Koutsilieri et al., 1997).

Taken together, treatment with dopaminergic drugs may contribute to the pathogenesis and likely exhibit adverse affects on the course of HIV dementia possibly through oxidative stress. The safety of such drugs in the clinical management of HIV-infected patients has to be reconsidered.

REFERENCES

Berger, J. R.and Arendt, G., 2000, HIV dementia: the role of the basal ganglia and dopaminergic systems, *J. Psychopharmacol.* **14**:214-221.

Czub, S., Koutsilieri, E., Sopper, S., Czub, M., Stahl-Hennig, C., Müller, J. G., Pedersen, V., Gsell, W., Heeney, J. L., Gerlach, M., Gosztonyi, G., Riederer, P., and ter Meulen, V., 2001, Enhancement of CNS pathology in early simian immunodeficiency virus infection by dopaminergic drugs, *Acta Neuropathol.* **101**:85-91.

Dana Consortium, 1998, A randomized, double-blind, placebo-controlled trial of deprenyl and thioctic acid in human immunodeficiency virus-associated cognitive impairment. Dana Consortium on the Therapy of HIV Dementia and Related Cognitive Disorders, *Neurology* **50**:645-651.

Gabby, M., Tauber, M., Porat, S., and Simantov, R., 1996, Selective role of glutathione in protecting human neuronal cells from dopamine-induced apoptosis, *Neuropharmacology* **35**:571-578.

Itoh, K., Mehraein, P., and Weis, S., 2000, Neuronal damage of the substantia nigra in HIV-1 infected brains, *Acta Neuropathol.* **99**:376-384.

Janssen, R. S., Nwanyanwu, O. C., Selik, R. M., and Stehr-Green, J. K., 1992, Epidemiology of human immunodeficiency virus encephalopathy in the United States, *Neurology* **42**:1472-1476.

Kieburtz, K. D., Epstein, L. G., Gelbard, H. A., and Greenamyre, J. T., 1991, Excitotoxicity and dopaminergic dysfunction in the acquired immunodeficiency syndrome dementia complex. Therapeutic implications, *Arch. Neurol.* **48**:1281-1284.

Koutsilieri, E., Gotz, M. E., Sopper, S., Sauer, U., Demuth, M., ter Meulen, V., and Riederer, P., 1997, Regulation of glutathione and cell toxicity following exposure to neurotropic substances and human immunodeficiency virus-1 in vitro, *J. Neurovirol.* **3**:342-349.

McArthur, J. C., Hoover, D. R., Bacellar, H., Miller, E. N., Cohen, B. A., Becker, J. T., Graham, N. M., McArthur, J. H., Selnes, O. A., Jacobson, L. P. et al, 1993, Dementia in AIDS patients: incidence and risk factors. Multicenter AIDS Cohort Study, *Neurology* **43**:2245-2252.

Mirsattari, S. M., Power, C., and Nath, A., 1998, Parkinsonism with HIV infection, *Mov. Disord.***13**:684-689.

Navia, B. A., Cho, E. S., Petito, C. K., and Price, R. W., 1986a, The AIDS dementia complex: II. Neuropathology, *Ann. Neurol.* **19**:525-535.

Navia, B. A., Jordan, B. D., and Price, R. W., 1986b, The AIDS dementia complex: I. Clinical features, *Ann. Neurol.* **19**:517-524.

Price, R. W., 1996, Neurological complications of HIV infection, *Lancet* **348**:445-452.

Rohr, O., Sawaya, B. E., Lecestre, D., Aunis, D., and Schaeffer, E., 1999, Dopamine stimulates expression of the human immunodeficiency virus type 1 via NF-kappaB in cells of the immune system, *Nucleic Acids Res.* **27**:3291-3299.

Scheller, C., Sopper, S., Jassoy, C., ter Meulen, V., Riederer, P., and Koutsilieri, E., 2000, Dopamine activates HIV in chronically infected T lymphoblasts, *J. Neural Transm.* **107**:1483-1489.

Tatton, W. G.and Greenwood, C. E., 1991, Rescue of dying neurons: a new action for deprenyl in MPTP parkinsonism, *J. Neurosci. Res.* **30**:666-672.

MODELS OF PARKINSON'S DISEASE

Makoto Naoi* and Wakako Maruyama

1. INTRODUCTION

Parkinson's disease (PD) at cellular level is characterized by selective death of dopamine neurons in nigral pars compacta. The type of cell death remains an enigma, even though recent results suggest that apoptosis may be a common type of cell death. Since the discovery of a neurotoxin selective to dopamine neurons, 1-methyl-4-phenyl-1,2,3,6-tetrahydro-pyridine (MPTP), various types of animal and cell models have been proposed to clarify the pathogenesis of PD. In this paper, an endogenous dopaminergic neurotoxin, *N*-methyl-(*R*)salsolinol [*N*M(*R*)Sal] was used to prepare the animal and cell models, and the mechanism underlying the selective degeneration of dopamine neurons was studied. By use of the cell model, a series of (-)deprenyl (selegiline) related propargylamines showed neuroprotective functions against apoptosis induced by the neurotoxin. These results are discussed in relation to the validity of animal and cell models in the study of PD and related neurodegenerative disorders.

2. ANIMAL MODELS OF PD

Various types of animal models of PD have been devised using extrinsic and intrinsic neurotoxins, MPTP, 6-hydrodopamine (6-OHDA) and endogenous *N*M(*R*)Sal. Of endogenous isoquinolines, only *N*M(*R*)Sal could prepare an appropriate PD model judged from behavioral, biochemical and histopathological changes (Naoi et al., 1996a). The fundamental criteria of neurotoxins in the pathogenesis of PD include the occurrence in human brains, specific uptake from dopaminergic terminal in the striatum and transport to nigral cell bodies, selective accumulation in mitochondria and finally induction of apoptosis. The results from *N*M(*R*)Sal-induced parkinsonism were relevant with those in parkinsonian brains. *N*M(*R*)Sal is synthesized *in situ* from dopamine by 2 step-wise reactions with (*R*)salsolinol synthase (Naoi et al., 1996b), and neutral (*R*)salsolinol *N*-methyltransferase (Naoi et al., 1997). *N*-Methylation of (*R*)salsolinol gives arise to

* Makoto Naoi, Department of Brain Sciences, Institute of Applied Biochemistry, Mitake, Kani-gun, Gifu 505-0116, Japan. Wakako Maruyama, Department of Basic Gerontology, National Institute for Longevity Sciences, Obu, Aichi 474-8511, Japan.

selective localization in the nigro-striatum. The involvement of *N*M(*R*)Sal and neutral (*R*)salsolinol *N*-methyltransferase in the pathogenesis of PD was strongly suggested by their increase in parkinsonian cerebrospinal fluid (Maruyama et al., 1996) and lymphocytes (Naoi et a., 1998), respectively (Figure 1). These results suggest that *N*M(*R*)Sal and its synthesizing enzyme may be endogenous factors involved in the cell death in PD (Naoi and Maruyama, 1999; Naoi et al., 2000).

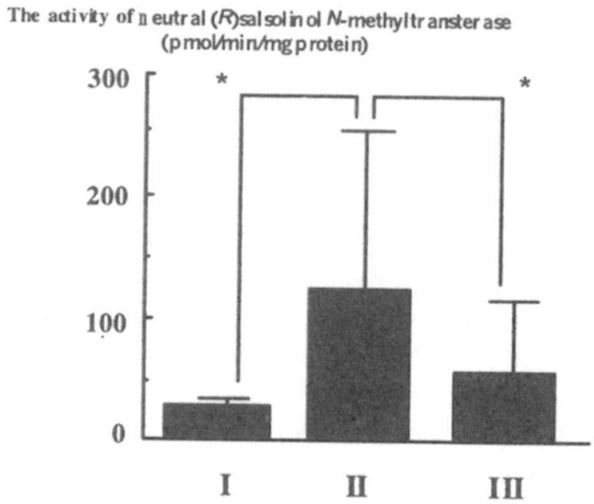

Figure 1. The activity of neutral (*R*)salsolinol *N*-methyltransferase in lymphocytes. I; Control. II; Parkinsonian patients, III; non-parkinsonian patients treated with L-DOPA. * Difference from II is statistically significant, $p < 0.05$.

3. CELLULAR MODELS OF PD

The intracellular mechanism underlying cell death in PD has been studied using various kinds of primarily cultured and clonal cells. *N*M(*R*)Sal was proved to induce apoptosis in human dopaminergic SH-SY5Y cells, and the step-wise activation of the cell death process was clarified; decline in mitochondrial membrane potential, $\Delta\Psi m$, (Maruyama et al., 2000, 2001a), activation of caspase 3 (Akao et al., 1999), translocation of glyceraldehydes-3-phosphate dehydrogenase (GAPDH) into nuclei (Maruyama et al., 2001b) and finally fragmentation of nucleosomal DNA. These findings are quite relevant to the detection of apoptosis-related proteins in parkinsonian nigro-striatum (Hartman et al., 2000).

In addition, a series of genes of cytokines and their receptors were induced in apoptotic cells. In parkinsonian brains, especially substantia nigra, increase in cytokines, such as TNF-alpha, was reported (Nagatsu et al., 2000). Our data from the cell model indicate a new view that increased cytokines may originate from neurons themselves, not from glia as an immune reaction. Further studies using this cell model will clarify the role of cytokines in cell death in PD, as neurotoxic or neuroprotective signals.

(-)Deprenyl (selegiline) and related propargylamine derivatives were proposed to protect or rescue declining neurons in PD (Parkinson's Study Group, 1993), even though the detailed mechanism remains to be clarified. In some types of apoptosis, anti-apoptotic

Bcl-2 protein family regulates apoptotic death (Tsujimoto and Shimizu, 2000), as shown by the suppression of apoptosis by Bcl-2 overexpression in the cells (Maruyama et al., 2001b). The apoptotic cell model using SH-SY5Y cells was successfully applied to study the mechanism of anti-apoptotic function of propargylamines, which suppress apoptosis induced by *N*M(*R*)Sal and peroxynitrite (Maruyama et al., 2000, 2001a and c), as summarized in Table 1. Propargylamines suppress the decline in $\Delta\Psi m$, activation of caspase 3 and nuclear translocation of GAPDH, which is quite similar to the anti-apoptotic function of Bcl-2. Actually, propargylamines can induce the mRNA and protein of anti-apoptotic Bcl-2 (Akao et al., in preparation), suggesting that the anti-apoptotic function of propargylamines is mediated by anti-apoptotic Bcl-2 family. The results provide us an intracellular mechanism underlying neuroprotective effects of (-)deprenyl, as shown by clinical trials.

Table 1. Anti-apoptotic activity of propargylamines

Propargylamine	Concentration required to suppress apoptosis
(-)Deprenyl (selegiline)	1 μM–100 nM
(+)Deprenyl	10 μM
(-)Desmethyldeprenyl	10–1 nM
(R)(+)-N-Propargyl-1-aminoindan (Rasagiline)*	10–1 nM
(S)(-)-N-Propargyl-1-aminoindan (TV 1022)*	1 μM–100 nM
Aminoindan*	Not active
(R)-N-(2-Heptyl)-N-methylpropargylamine	1 μM–100 nM
(S)-N-(2-Heptyl)-N-methylpropargylamine	10 μM
(R)-N-(2-Heptyl)-propargylamine	1 μM–100 nM
(R)-3-(2-Heptylamine)-N-methylpropionic acid	Not active

Propargylaminoindans were kindly supplied by TEVA Pharmaceutical (Tetanya, Israel), and aliphatic propargylamines by Alviva Biopharmaceuticals (Saskatoon, Canada).
SH-SY5Y cells were incubated with propargylamines for 20 min, then incubated with100 μM peroxynitrite-generating SIN-1 for 3 h. The concentration required for the prevention of apoptosis was determined by "COMET" assay.
*The difference shows that a propargylamine moiety is essential for anti-apoptotic function.

4. CONCLUSIONS

Our results clearly show that the animal and cellular models of PD are useful to study the mechanism of cell death of dopamine neurons in substantia nigra. By comparison with the results obtained by analyses of clinical samples, the *in vivo* and *in vitro* experiments suggest that endogenous dopamine-derived *N*M(*R*)Sal may induce apoptosis in dopamine neurons, and its synthesizing enzyme, neutral (*R*)salsolinol *N*-methyltransferase may be an endogenous factor in the pathogenesis of PD. Future studies will bring us clues to understand the death mechanism and also to develop new strategy to find neuroprotective drugs for PD and other neurodegenerative disorders.

5. ACKNOWLEDGMENTS

This study was supported by a Grant-in-Aid for Scientific Research on Priority Areas (M. N.) and a Grant-in-Aid for Scientific Research (C) (W. M.) from the Ministry of Education, Science, Sports and Culture, Grant-in-Aid for Scientific Research on Priority Areas (C) from Japan Society for the Promotion of Science (W. M.), Grant for Longevity Sciences (W. M.), and Grant for Comprehensive Research on Aging and Health, (W. M. and M. N.) and Medical Frontier Strategy Research (W. M., M. N. and Y. A.) from the Ministry of Health, Labour and Welfare, Japan.

REFERENCES

Akao, Y., Nakagawa, Y., Maruyama, W., Takahashi, T., and Naoi, M., 1999, Apoptosis induced by an endogenous neurotoxin, *N*-methyl(*R*)salsolinol, is mediated by activation of caspase 3. *Neurosci. Lett.*, **267**: 153-156.

Hartmann, A., Hunot, S., Michel., P. P., Muriel, M.-P., Vyas, S., Faucheux, B. A., Mouatt-Prigent, A., Turmel, H., Srinivasan, A., Ruberg, M., Evan, G. I., Agis, V., and Hirsch, E. C., 2000, Caspase 3: A vulnerability factor and final effector in apoptotic death of dopaminergic neurons in Parkinson's disease. *Proc. Natl. Acad. Sci. USA*, **97**: 2875-2880.

Maruyama, W., Abe, T., Tohgi, H., Dostert, P., and Naoi, M., 1996, A dopaminergic neurotoxin, (*R*)-*N*-methylsalsolinol, increases in parkinsonian cerebrospinal fluid. *Ann. Neurol.* **40**:119-122.

Maruyama, W., Akao, Y., Youdim, M. B. H., and Naoi, M., 2000, Neurotoxins induce apoptosis in dopamine neurons: protection by *N*-propargylamine-1(*R*)- and (*S*)-aminoindan, rasagiline and TV1022. *J. Neural Transm.* [Suppl] **60**: 171-186.

Maruyama, W., Boulton, A. A., Davis B. A., Dostert, P., and Naoi, M., 2001a, Enantio-specific induction of apoptosis by an endogenous neurotoxin, *N*-methyl(*R*)salsolinol, in dopaminergic SH-SY5Y cells: suppression of apoptosis by *N*-(2-heptyl)-*N*-methylpropargylamine. *J. Neural Transm.* **108**: 11-24.

Maruyama, W., Akao, Y., Youdim, M. B. H., Davis, B. A., and Naoi, M. 2001b, Transfection-enforced Bcl-2 overexpression and an anti-Parkinson drug, rasagiline, prevent nuclear accumulation of glyceraldehydes-3-phosphate dehydrogenase induced by an endogenous dopaminergic neurotoxin, *N*-methyl(*R*)salsolinol. *J. Neurochem.* In press.

Maruyama, W., Youdim, M. B. H., Naoi, M., 2001c, Anti-apoptotic properties of rasagiline, *N*-propargylamine-1(*R*)-aminoindan, and its optical (*S*)-isomer, TV1022. *Ann. N. Y. Acad. Sci.* **939**: 320-329.

Nagatsu, T., Mogi, M., Ichinose, H., and Togari, A., 2000, Changes in cytokines and neurotrophins in Parkinson's disease. *J. Neural Transm.* [Suppl] **60**: 277-290.

Naoi, M., Maruyama, W., Dostert, P., Hashizume, Y., Nakahara, D., Takahashi, T., and Ota, M., 1996a, Dopamine-derived endogenous 1(*R*),2(*N*)-dimethyl-6,7-dihydroxy-1,2,3,4-tetrahydroisoquinoline, *N*-methyl(*R*)salsolinol, induced parkinsonism in rat: biochemical, pathological and behavioral studies. *Brain Res.* **709**: 285-295.

Naoi, M., Maruyama, W., Dostert, P., Kodak, K., and Kaiya, T., 1996b, A novel enzyme enantio-selectively synthesizes (*R*)salsolinol, a precursor of a dopaminergic neurotoxin, *N*-methyl(*R*)salsolinol. *Neurosci. Lett.* **212**: 183-186.

Naoi, M., Maruyama, W., Matsubara, K., and Hashizume Y., 1997, A neutral *N*-methyltransferase activity in the striatum determines the level of an endogenous MPP$^+$-like neurotoxin, 1,2-dimethyl-6,7-dihydroxyisoquinolinium ion, in the substantia nigra of human brains. *Neurosci. Let.t* **235**: 81-84

Naoi, M., Maruyama, W., Nakao, N., Ibi, T., Sahashi, K., and Strolin Benedetti, M., 1998, (*R*)Salsolinol *N*-methyltransferase activity increases in parkinsonian lymphocytes. *Ann. Neurol.***43**: 212-216.

Naoi, M. and Maruyama, W., 1999, *N*-Methyl(*R*)salsolinol, a dopamine neurotoxin, in Parkinson's disease. *Adv. Neurol.* **80**: 259-264.

Naoi, M., Maruyama, W, Akao, Y., Zhang, J., and Parvez, H., 2000, Apoptosis induced by an endogenous neurotoxin, *N*-methyl(*R*)salsolinol, in dopamine neurons. *Toxicology* **153:** 123-141.

Parkinson's Study Group, 1993, Effect of deprenyl on the progression of disability in early Parkinson's disease. *N. Eng. J. Med.* **42**: 176-183.

Tsujimoto, Y. and Shimizu, S., 2000, Bcl-2 family: Life-or-death switch. *FEBS Lett.* **466**: 6-10

THE EFFECT OF ACUTE LOADING WITH SELEGILINE AND SELEGILINE WITH LEVODOPA ON BLOOD PRESSURE AND PLASMA NOREPINEPHRINE LEVELS IN CHRONIC PARKINSONIAN PATIENTS

Jose M. Rabey and Rafael O. Stryjer*

1. INTRODUCTION

Selegiline, an irreversible inhibitor of monoamine oxidase B (MAO B), was introduced as a symptomatic treatment for Parkinson's disease (PD) in the mid-1970s, with the intention of enhancing the effect of levodopa (LD).[1-3]

However, a report by Lees and colleagues[4] refuted this data. In the LD with selegiline group, the mortality was 28% within 5 years, while in the LD treated patients the mortality was 17.7% in the same observation period[4] On the other hand, a recent meta-analysis review published by Olanow et al.[5] has refuted the concept that patients treated with selegiline have a higher mortality rate.

In order to find out if selegiline exerts a deleterious effect on the homeostasis of blood pressure, we conducted a study of acute loading with LD or selegiline and LD, in a group of PD pts chronically treated with these drugs, and a group of healthy subjects. During the test we measured blood pressure, pulse and norepinephrine (NE) plasma levels in response to changing position from supine to standing. (Study approved by local Helsinki committee.)

2. PATIENTS AND METHODS

2.1. Patient's Characteristic and Methods

The study included 28 participants who suffered from idiopathic Parkinson's disease (PD). 10 PD pts chronically treated with LD (mean age 63.9; mean disease duration 7.05;

* Department of Neurology, Assaf Harofeh Medical Center, Zerifin, Israel.

Table 1. Patients' characteristics

Group	Control n=8	Group D n=10	Group S n=10
Mean age and standard deviation	52.6 ±6.68	63.9±10.5	62.2±10.73
Mean dis. Dur.(years) and standard deviation		7.05 ± 3.9	8.5± 4.48
Postural hypotension	3	4	8 (5 AADL) (1 symptomatic)
Staging (Yahr)		Stage II=2 Stage III=7 Stage IV=1	Stage II=3 Stage III=3 Stage IV=4

AADL – after acute drug loading

group D); 10 PD pts chronically treated with selegiline together with LD (mean age 62.2; mean disease duration 8.5 years; group S); 8 healthy volunteers (mean age 52.6; group CTRL) (see table 1 for patients' characteristics). In the first step the orthostatic test was performed in PD pts after 12 hours off medication. In the second step the test was repeated in each PD patient 90 minutes after acute loading with 125mg LD, in group D, and 125mg LD and 5mg of selegiline in group S. The CTRL group underwent the test only once. Orthostatic testing was performed according to the American Academy of Neurology criteria.[6]

The examination included the measurement of blood pressure and pulse and collection of venous blood samples for NE determination (HPLC procedure), obtained after 5 minutes lying (supine) and at 3 minutes after standing.

3. STATISTICS

ANOVA for repeated measures was used to evaluate correlation between changes in blood pressure, pulse, NE levels (dependent variables) and orthostatism, with age as covariance.

4. RESULTS

During the test, in absolute numbers, NE levels (pg\ml) (after 12 hours off medication), increased in group D from 298±182 while supine to 445 ±203 while standing, in comparison to group S, in which the difference was greater, from 346 ± 240 while supine to 1044±321 while standing. In the control group, NE levels (pg/ml) while supine were 322± 223, and increased to 435 ± 251 while standing.

After acute loading with medication, the difference in NE levels (pg/ml) from supine to standing position was 830±488 to 932±1330 in group D (who acutely received LD),

Table 2. Mean blood pressure and pulse before and after treatment

Mean blood pressure (mmHg) ±SD	**Control n=8***	**Group D n=10**		**Group S n=8**	
		Before loading	After loading	Before loading	After loading
Supine: Systolic	123±15	151±26	138±34	130±24	123±15
Diastolic	79±12	88±8	84±16	85±14	79±10
Pulse (per min).	77±8	76±8	77±10	73±12	77±15
Standing: Systolic	118±26	144±36	129±30	124±17	113±21
Diastolic	85±20	93±12	89±20	87±10	78±12
Pulse (per min).	87±11	84±9	88±10	83±13	88±22

* (n= numbers of patients)
Group D – patients treated chronically with LD
Group S – patients treated chronically with LD and selegiline
Loading: Group D: 125 mg Levodopa (+ 12.5g carbidopa)
Group S: 125 mg Levodopa + 5 mg Selegiline (+ 12.5g carbidopa)

while in group S (who acutely received selegiline +LD), the difference was bigger, from 495±451 while supine to 1186± 1357 while standing. Although not statistically significant (due to the SD), a greater difference in NE plasma levels between supine and standing (before and after acute loading) in group S, in comparison to group D, was found.

Analysing all the subjects, POHY was found to be associated with age (F=10.2, P=0.006). In addition no significant difference was detected in radial pulse between the two groups (Group D and S) (Table 2).

5. DISCUSSION

Selegiline is metabolized in the liver to metamphetamine and amphetamine.[7] Both metabolites are known to increase the release of catecholamine stores in sympathetic synaptic terminals.[8, 9] As a consequence of that effect, amphetamine synthesized "de novo" may induce supine hypertension. However, it has been described that amphetamine may also induce orthostatic hypotension by impairing sympathetic function.[10] This effect may be explained by the metabolization of amphetamine to the false neurotransmitter hydroxyephedrine, which depletes nerve terminals of NE.[9] Selegiline may lose some of its selectivity for monoamine oxidase type B at high doses (10mg/d) or after chronic treatment.[11] The inhibition of monoamine oxidase type A may then be expected to increase plasma concentration of the false transmitter tyramine.[12] It is known that when a MAO inhibitor is administered chronically, tyramine may be absorbed through the gastrointestinal system from where it is transported into the adrenergic nerve terminal, where its catabolism is prevented because of the inhibition of MAO A at this site. As a consequence tyramine is beta- hydroxilated to octopamine. This may explain the "hypertensive crises" described after acute ingestion of cheese, beer or red wine (due to displacement of norepinephrine - from the adrenergic nerve terminals). In addition to the information summarized above, selegiline by itself has been shown to inhibit NE reuptake in the central and peripheral nervous system.[13] This may explain the higher NE plasma levels measured at rest (supine position) in group S in comparison to group D.

In summary, our study suggests that patients submitted to chronic selegiline, may seem to suffer from more events of orthostatic hypotension (asymptomatic and symptomatic). The clinical significance of elevated circulating plasma NE levels during chronic treatment with selegiline, leaves an open question concerning its possible deleterious effect on the cardiovascular system.

BIBLIOGRAPHY

1. Birkmayer W, Riederer P, Ambrozi L, et al. Implications of combined treatment with "Madopar" and L-deprenyl in Parkinson's disease. *Lancet* 1977;**1**:439-43.
2. Lees AJ, Shaw KM, Kohout LJ, et al. Deprenyl in Parkinson's disease. *Lancet* 1997;**2**:791-5.
3. Streifler M, Vardi J, Borenstein N, et al. S. B-type Monoamine oxidase (M.A.0.) inhibitors in long-term levodopa treated Parkinsonism. A combined clinical trial with L-deprenyl. *Curr Ther Res Clin Exp* 1980;**27**:643-8.
4. Lees AJ, on behalf of the Parkinsonian's Disease Research Group of the United Kingdom. Comparison of therapeutic effects and mortality data of levodopa and levodopa combined with selegiline in patients with early, mild Parkinson's disease. *BMJ* 1995;**311**:1602-7.
5. Olanow CW, Myllyla VV, Sotaniemi KA, et al. Effects of selegiline on mortality in patients with Parkinson's disease: a meta-analysis. *Neurology* 1998;**51**:825-30.
6. Consensus statement on the definition of orthostatic hypotension, pure autonomic failure, and multiple system atrophy. The Consensus Committee of the American Autonomic Society and the American Academy of Neurology. *Neurology* 1996;**46**:1470.
7. Churchyard A, Mathias CJ, Boonkongchuen P, et al. Autonomic effects of selegiline: possible cardiovascular toxicity in Parkinson's disease. *J Neurol Neurosurg Psychiatry* 1997;**63**:22228-234.
8. Reynold GP, Elsworth JD, Blau K, et al. Deprenyl is metabolized to metamphetamine and amphetamine in man. *Br J Clin Pharmacol* 1978;**6**: 542-4.
9. Gill JR, Mason DT, Bartter FC. Effects of hydroxyamphetamine (paredrine) on the function of the sympathetic nervous system in normotensive subjects. *J Pharmacol Exp Ther* 1967;**155**:288-95.
10. Cavanaugh JH, Griffith JD, Oates JA. Effects of amphetamine on the pressor response to tyramine: formation of p-hydroxynorephedrine from amphetamine in man. *Clin Pharmacol Ther* 1970;**11**:656-64.
11. Waldmeier PC, Felner AE. Deprenyl: loss of selectivity for inhibition of B-type MAO after repeated treatment. *Biochem Pharmacol* 1978;**27**:801-2.
12. Gelowitz DL, Richardson JS, Wishart TB, et al. Chronic L-.deprenyl or L- amphetamine: equal cognitive enhancement, unequal MAO inhibition. *Pharmacol Biochem Behav* 1994;**47**:41-5.
13. Goodman Gilman A, Goodman LS, Rall TW, et al, eds. *Goodman and Gilman's The Pharmacological Basis of Therapeutics, 7th ed.* New York: Macmillan Publishing Company, 1985:1786.

CARDIAC SYMPATHETIC IMAGING WITH RADIOLABELLED METAIODOBENZYLGUANIDINE (MIBG) IN PATIENTS WITH PARKINSON'S DISEASE

Hisato Takatsu*, Kazuo Satomi, Hiroshi Nishida, Hitoshi Matsuo, Sachiro Watanabe, Kenshi Nagashima, Hisayoshi Fujiwara.

1. INTRODUCTION

A radiolabelled norepinephrine analogue, ^{123I} metaiodobenzylguanidine (MIBG) has been widely used to detect cardiac sympathetic denervation or hyperactivity in cardiac diseases. Recently, cardiac ^{123I} MIBG accumulation was reported to be significantly reduced in patients with Parkinson's disease (PD). On the other hand, PD is known to be associated with various sympathetic disorders. Therefore, we sought to examine the relationship between severity of clinical symptoms or of autonomic dysfunction of PD and reduction of cardiac ^{123I} MIBG accumulation.

2. SUBJECTS AND METHODS

Thirty-two patients with PD, 12 with multiple system atrophy (MSA) and 10 control subjects were subjected to cardiac scintigraphy with ^{123I} MIBG. From their clinical activity using Hoehn-Yahr's scale, Parkinsonian patients were classified into five stages. There were two patients with Hoehn-Yahr's stage I, 9 patients with stage II, 14 patients with stage III, 6 patients with stage IV, and one patient with stage V. They had been treated with amantagin, L-dopa, and/or trihexyphenidyl hydrochloride. Their medication for PD was continued during this study for obvious ethical reasons. In addition, five patients with multiple system atrophy (MSA), including three patients with olivopontocerebellar atrophy (OPCA) and two patients with striatonigral degeneration (SND), were included to compare with PD patients. The average age of MSA patients was 69±4 years. Patients were excluded from this study if they had been taking any medication (such as an anti-depressant) which had previously been reported to influence MIBG uptake. None of the patients had a history or signs of cardiac disease revealed by an interview by a cardiologist

* First Department of Internal Medicine, Gifu Municipal Hospital, 7-1 Kashima-cho, Gifu, 500-8513, Japan.

and routine examinations of chest X-ray, electrocardiogram and echocardiography. Cardiac accumulation was determined as heart-to-mediastinum (H/M) ratio on the plannar images.

Orthostatic blood pressure changes were measured in all patients studied. A drop of more than 20 mmHg in systolic blood pressure (SBP) at any measuring time point after standing was considered to demonstrate orthostatic hypotension. Twenty-four hour ambulatory blood pressure monitoring (ABPM) was performed in 31 patients; two patients in Hoehn-Yahr's stage I, seven in stage II, 12 in stage III, four in stage IV, and one in stage V, and in all five patients with MSA. During the procedure, all patients recorded their sleeping time in a diary. Automatic measurements were collected at 30-minute intervals from 6:00 a.m. to 10:00 p.m. (day-time), and at 60-minute intervals from 10:00 p.m. to 6:00 a.m. (night-time). The nocturnal fall in SBP was calculated as (Day SBP-Night SBP)/Day SBP. On the basis of these ABPM data, patients were considered normal "dipper" subjects if the nocturnal BP fall exceeded 10% of the day-time mean SBP. When patients had a blunted nocturnal blood pressure fall (<10% of the mean SBP during the diurnal period), they were judged as having abnormal circadian patterns of variation in blood pressure.

3. RESULTS AND INTERPRETATION

Most of the PD patients showed a marked reduction in cardiac ^{123}I-MIBG accumulation(Fig.1).

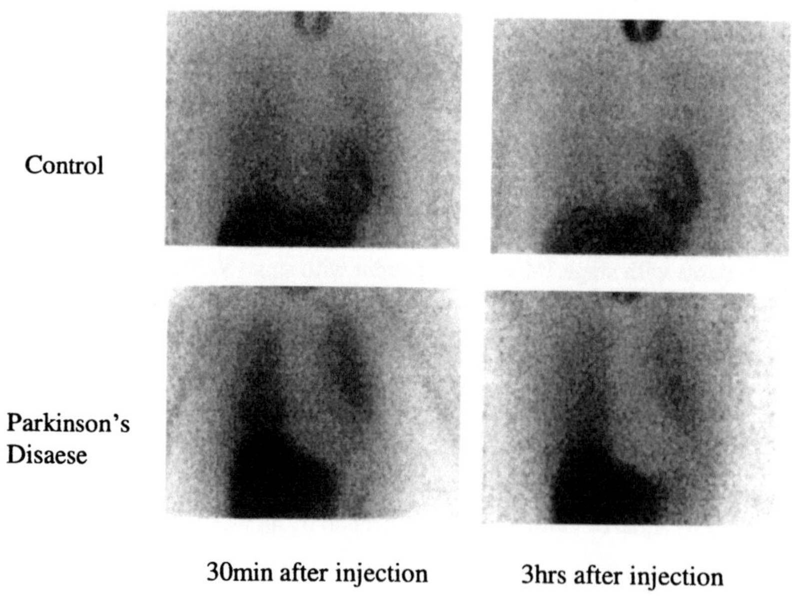

Figure 1. Cardiac Imaging of Sympathetic Nerves with I-123 MIBG in Control and Patint with Parkinson's disease. Significant reduction of I-123 MIBG accumulation in both early and delayed imagings in a patient with Parkinson's disease as compared with control subject.

The mean H/M ratio for all PD patients was 1.58±0.37 and 1.33±0.28 in early and delayed images, respectively, while the corresponding ratios were 2.42±0.27 and 2.60±0.15 in the 7 control subjects. From semi-quantitative assessment with the H/M ratio, all 32 PD patients showed a lower H/M ratio than the lower limit of normal controls (mean minus 2SD of normal controls). A marked reduction in the cardiac ^{123}I-MIBG accumulation was seen in the early-stage (Hoehn-Yahr stage I or II) group, although the reduction was most significant in the advanced PD stage (stage IV or V) groups. Cardiac ^{123}I-MIBG accumulation was also compared with the disease duration; no significant correlation was found between the H/M ratio and the number of years of PD. PD and control patients' H/M ratios were compared with the data of blood pressure monitoring. Among the 26 patients who underwent both an orthostatic test and ABPM, the ^{123}I-MIBG accumulation was compared between the orthostatic hypotension (-) and nocturnal blood pressure fall (+) group, i.e. the normal blood pressure response group, and the orthostatic hypotension (+) or nocturnal blood pressure fall (-) group, i.e., the abnormal blood pressure response group. Interestingly, the cardiac ^{123}I-MIBG accumulation was significantly low even in the group of patients who did not have an abnormal blood pressure response to the orthostatic challenge or abnormal circadian patterns of variation in blood pressure. In addition, there were significant differences in ^{123}I-MIBG accumulation not only between the control and PD patients but also between PD patients and MSA patients.

4. DISCUSSION

The findings of the present study demonstrate that most of the Parkinsonian patients studied, showed a marked reduction of cardiac ^{123}I-MIBG accumulation, indicating a severe impairment of the cardiac sympathetic nervous system. Interestingly, the reduction of ^{123}I-MIBG accumulation was markedly reduced even in patients at stage I-II, and in many patients within 5 years after the onset of the symptoms, although the reduction was also significant in patients at an advanced stage. The reduction appeared to be independent of clinical stage and disease duration. Furthermore, sympathetic dysfunction was evaluated by the postural change or diurnal deviation of blood pressure. However, this reduction of cardiac ^{123}I-MIBG accumulation was also independent of the sympathetic dysfunction evaluated by these two methods. Consequently, the marked reduction of cardiac ^{123}I-MIBG accumulation from the earlier stage could be a special feature of PD, although the reduction may worsen as the sympathetic dysfunction progresses.

To confirm this hypothesis, we examined cardiac ^{123}I-MIBG accumulation in MSA, which is related to Parkinsonism, but is classified into a different disease category from PD. The ^{123}I-MIBG accumulation was significantly reduced in PD patients compared with control and MSA patients, although all of these patients were selected based on their lack of orthostatic blood pressure change and diurnal blood pressure variation. These findings indicate that the cardiac sympathetic denervation in its earlier stage might be specific for PD among the akinetic-rigid syndromes.

5. CONCLUSIONS

In the patients with Parkinson's disease, cardiac ^{123}I-MIBG accumulation was reduced from the early stages of PD. In addition, this reduction is also independent of sympathetic dysfunction evaluated by other methods. Since the reduction is not so significant in patients with multiple system atrophy, this phenomenon of PD may be disease-specific. Then, cardiac MIBG scintigraphic study may help us differentiate PD from MSA.

REFERRENCE

1. Takatsu H, et al. Cardiac Sympathetic Denervation from the Early Stage of Parkinson's Disease -Clinical and Experimental Studies with Radiolabeled MIBG. *J Nucl Med* 2000; **41**: 71-77.
2. Takatsu H, et al. Differentiating Parkinson Disease From Multiple-System Atrophy by Measuring Cardiac Iodine-123 Metaiodobenzylguanidine Accumulation. *JAMA* 2000 **284**: 44-45.

MIBG SCINTIGRAPHY IN PARKINSONIAN SYNDROMES

Stefan Braune*

1. SUMMARY

Active uptake of [^{123}I] metaiodobenzylguanidine (MIBG) reflects localisation and functional integrity of postganglionic adrenergic neurons. There is accumulating evidence, that in patients with Parkinson's disease MIBG uptake into cardiac sympathetic neurons is selectively and grossly reduced. Currently there are studies available on 239 patients with idiopathic Parkinson's disease (IPD), 73 patients with MSA and 14 with progressive supranuclear palsy (PSP) in comparison to 78 healthy controls. The overall mean heart/mediastinum ratio of MIBG uptake was 1.29 in IPD, 2.13 in MSA, 1.89 in PSP and 2.24 in controls. The sensitivity to identify IPD patients correctly by their very low cardiac MIBG uptake appears to be higher than 95% and the specificity is 100%, if Hoehn and Yahr stage is II or higher, if no interacting medication is given and if there is no additional polyneuropathy. This significant reduction of uptake is independent of duration of Parkinsonian or autonomic symptoms and the degree of abnormality in cardiovascular function tests. Although it remains to be cleared whether the findings in IPD represent true postganglionic damage or a functional neuronal deficit, in clinical practice cardiac MIBG scintigraphy allows an early discrimination of IPD in relation to MSA and PSP, which appears to be more reliable and less expensive compared to PET and SPECT imaging of the brain.

2. INTRODUCTION AND METHOD

In the past few years it has become possible to visualize and quantify cardiac sympathetic innervation *in vivo* with [^{123}I] metaiodobenzylguanidine (MIBG) scintigraphy and single-photon emission-computed tomography. MIBG is a derivative of urea, with a similar biochemical structure like noradrenaline, but without pharmacological activity. Both substances compete for the same cellular transport mechanisms on postganglionic

* Department of Neurology, University of Freiburg, Breisacherstr. 64, D-79106 Freiburg, Germany. Braune@nz11.ukl.uni-freiburg.de

adrenergic neurons. MIBG was shown to be incorporated into adrenergic cells by the same active uptake mechanism like noradrenaline, is stored in granules and is secreted after stimulation by acetylcholine. Therefore the concentration of [^{123}I]MIBG reflects not only the localization of postganglionic adrenergic neurons in organs but also serves as an index of their integrity and function. In animal experiments myocardial MIBG uptake is reduced after destruction of the stellate ganglia or of the sympathetic efferents locally at the epicardium. This underlines the specificity of this technique in relation to postganglionic lesions of sympathetic efferents.

3. RESULTS AND DISCUSSION

In patients with Parkinson´s disease uptake of MIBG in cardiac neurons was found to be grossly reduced (Braune et al. 1998), indicating involvement of postganglionic sympathetic neurons in this disorder. Recent studies provided accumulating evidence that this is a typical finding in almost all patients with Parkinson´s disease. After clearing for repeatedly published patients up to now data on 239 patients with idiopathic Parkinson´s disease are available, which showed a mean heart/mediastinum ratio of MIBG uptake of 1.29 (figure 1).

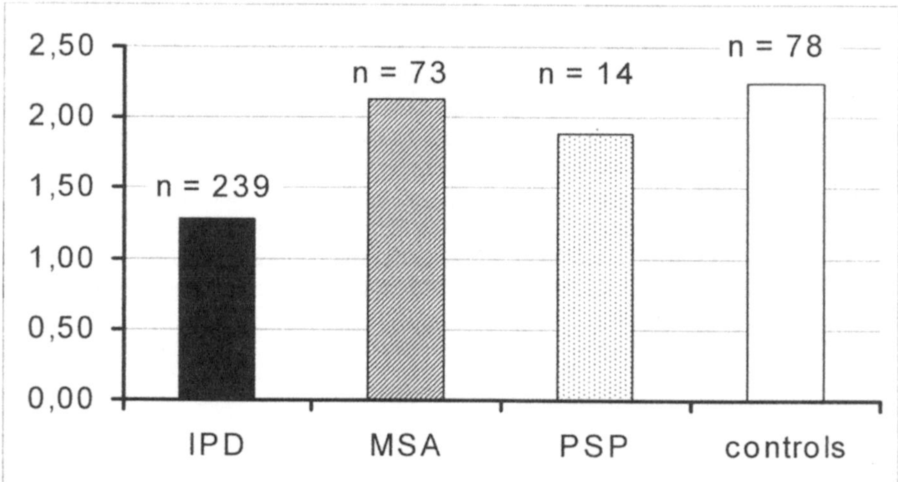

Figure 1. Overall mean heart/mediastinum ratio from all studies (see figure 1) in patients with Parkinson´s disease (IPD), multiple system atrophy (MSA), healthy controls and patients with progressive supranuclear palsy (PSP).

This was significantly lower in all studies (figure 2) and also compared to the overall mean value from a total of 78 healthy controls of 2.24. The sensitivity to identify IPD patients correctly by their very low cardiac MIBG uptake appears to be higher than 95% and the specificity is 100%, if Hoehn and Yahr stage is II or higher. This significant reduction of uptake is independent of duration of Parkinsonian or autonomic symptoms and the degree of abnormality in cardiovascular function tests. It is mandatory to avoid concomittant medication which can interfere with the active process of MIBG uptake into the neurons (Solanki et al. 1992, Yoshita & Braune 2000). It remains to be cleared whether the findings in IPD represent true postganglionic damage to neurons or a specific functional deficit i.e. of the transmembrane human norepinephrine transport mechanism.

Besides the pathophysiologically important identification of involvement of postganglionic sympathetic neurons in Parkinson's disease, this approach enables also the improvement of the clinically relevant differential diagnosis between several neurodegenerative disorders. Even with longer duration of symptoms about one quarter of patients with a parkinsonian syndrome of other etiology are misdiagnosed as Parkinson's disease. Recent MIBG studies reported data on 73 patients with multiple system atrophy and 14 with progressive supranuclear palsy. The overall mean heart/mediastinum ratio was 2.13 and 1.89, respectively, which has to be regarded as normal. This provides evidence that quantification of cardiac MIBG uptake can separate Parkinson's disease from other neurodegenerative disorders already at an early stage of disease and independent of clincial symptoms of autonomic failure. Interestingly PET studies of cardiac [18F]fluorodopamine uptake did not allow to identify all patients with IPD (Goldstein et al. 2000, Druschky et al. 2000), which point to a higher sensitivity of MIBG scintigraphy as a diagnostic tool. In addition scintigraphy with MIBG is a standard method of nuclear medicine, is widely available, needs less-sophisticated equipment and analytical procedures and is cheaper than PET studies with [18F]fluorodeoxyglucose and [18F]fluorodopa to distinguish IPD from MSA.

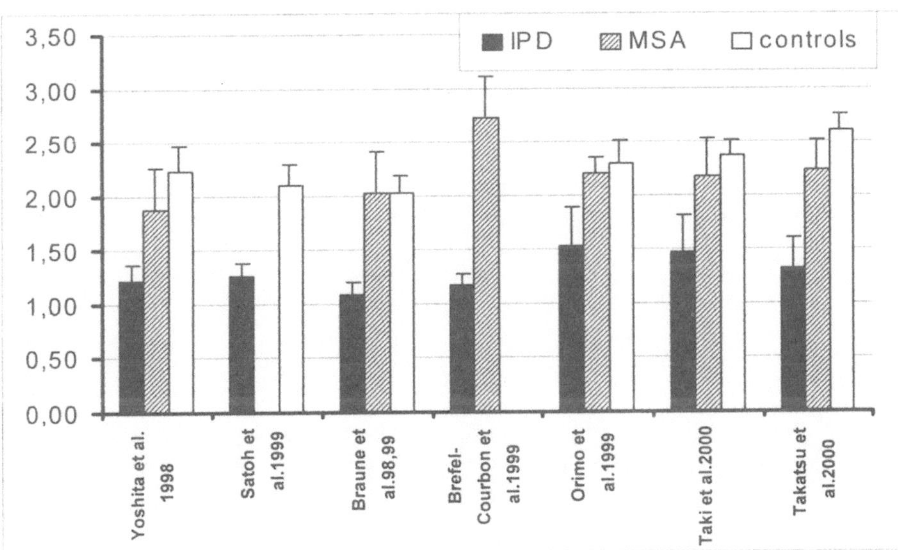

Figure 2. Mean heart/mediastinum ratio in patients with Parkinson's disease (IPD), multiple system atrophy (MSA) and healthy controls.

REFERENCES

Brefel-Courbon C, Courbon F, Thalamas C et al. Myocardial meta-[123I] iodobenzylguanidine (MIBG) uptake is different in Parkinson's disease and multiple system atrophy with autonomic failure. *Clin Autonom Res* **9**, 285 (1999)

Braune S, Reinhardt M, Bathmann J, et al. . Impaired cardiac uptake of meta-[123I]iodobenzylguanidine in Parkinson's disease with autonomic failure. *Acta Neurol Scand* **97**,307-314 (1998)

Braune S, Reinhardt M, Schnitzer R, et al. . Cardiac uptake of [^{123}I] MIBG separates Parkinson's disease from multiple system atrophy. *Neurology*, **53**:1020-1025 (1999)

Druschky A, Hilz MJ, Platsch G, et al. Differentiation of Parkinson's disease and multiple system atrophy in early disease stages by means of I-123-MIBG-SPECT. *J Neurol Sci* **175**,3-12 (2000)

Goldstein DS, Holmes C, Li ST, et al. Cardiac sympathetic denervation in Parkinson disease. *Ann Intern Med* **133**, 338-47 (2000)

Orimo S, Ozawa E, Nakade S, et al. (123)I-metaiodobenzyl-guanidine myocardial scintigraphy in Parkinson's disease. *J Neurol Neurosurg Psychiatry* **67**,189-194 (1999)

Satoh A, Serita T, Seto M et al. Loss of ^{123}I-MIBG uptake by the heart in Parkinson's disease: assessment of cardiac denervation and diagnostic value. *J Nucl Med* **40**,371-375 (1999)

Solanki KK, Bomanji J, Moyes J et al.. A pharmacological guide to medicines which interfere with the biodistribution of radiolabelled meta-iodobenzylguanidine (MIBG). *Nucl Med Commun* **13**, 513-521 (1992)

Takatsu H, Nagashima K, Murase M, et al. Differentiating Parkinson disease from multiple-system atrophy by measuring cardiac iodine-123 metaiodobenzylguanidine accumulation. *JAMA* **284**,44-45 (2000)

Taki J, Nakajima K, Hwang EH, et al. Peripheral sympathetic dysfunction in patients with Parkinson's disease without autonomic failure is heart selective and disease specific. *Eur J Nucl Med* **27**,566-573 (2000)

Yoshita M. Differentiation of idiopathic Parkinson's disease from striatonigral degeneration and progressive supranuclear palsy using iodine 123 metaiodobenzyl-guanidine myocardial scintigraphy. *J Neurol Sci* **155**,60-67 (1998)

Yoshita M, Braune S. Cardiac uptake of [123I]MIBG separates PD from multiple system atrophy. *Neurology* **54**,1877-1878 (2000)

EVALUATION OF SYMPATHETIC NERVE FUNCTION IN PATIENTS WITH HEART FAILURE BY MIBG MYOCARDIAL SCINTIGRAPHY

Junichi Yamazaki and Shohei Yamashina*

1. INTRODUCTION

Patients with heart failure show signs of sympathetic dysfunction, including an elevation in blood norepinephrine levels, and down-regulation of beta-receptors. The reported changes in these patients include a reduction in the number of sympathetic nerves, and a decrease in myocardial norepinephrine (NE) content (1). ^{123}I-Metaiodobenzylguanidine (MIBG) has been developed for imaging of the kinetics of NE in vivo. MIBG uptake is reduced, and washout is increased in the failing heart due to increased turnover or spillover of NE plasma, poor retention, and NE re-uptake disorder at sympathetic nerve endings (2,3). We performed this study to determine which parameter of ^{123}I-MIBG image is an accurate predictor of the response to beta-blocker therapy, and to evaluate differences in therapeutic effects among various beta-blockers in patients with dilated cardiomyopathy (DCM).

2. SUBJECTS

Fifty patients (41 males and 9 females; mean age of 56 years) with DCM were diagnosed from clinical findings, cardiac catheterization and myocardial biopsy. The control group consisted of 17 healthy volunteers (15 males and 2 females; mean age of 35 years). ^{123}I-MIBG myocardial SPECT was performed before and after beta-blocker therapy, when the symptoms of acute heart failure had improved and had been confirmed to be stable. The patients were given a beta-blocker, either Carvedilol, Metoprolol or Nipradilol. Fifty patients were classified into two groups, one with left ventricular ejection fraction (LVEF) improved by 10% or more within 6 months after the administration of beta- blocker, and the other with less than 10% improvement in LVEF.

* Junichi Yamazaki, The First Department of Internal Medicine, Toho University School of Medicine, 6-11-1 Ohmori-Nishi, Ohta-ku, Tokyo, 143-8541 Japan

Catecholamine Research, Edited by Nagatsu *et al.*
Kluwer Academic/Plenum Publishers, 2002

3. METHODS

With the patient resting, 3 mCi ^{123}I-MIBG were injected intravenously, and MIBG-myocardial SPECT images were obtained. The early image was 20 min. after, and the delayed image was 4 hours after injection. LVEF calculated from echo-cardiogram was obtained within 2 weeks of the MIBG myocardial SPECT procedure. A washout rate (WR) for the entire left ventricle was calculated from the early and delayed images based on the polar map. A polar map was prepared from the data acquired from 17 healthy volunteers to obtain a normal range (mean$\pm$2 SD) of myocardial MIBG uptake. For further objective evaluation, a polar map was constructed from short axis images, from the apex to the base of the left ventricle in both the early and delayed images. The extent score (ES), representing the area of reduced MIBG uptake, the severity score (SS) representing the severity of defect, were calculated in the patients.

4. RESULTS

4.1. Comparison between LVEF and ES / SS of MIBG

LVEF was significantly correlated with both ES (r=-0.401, p<0.001), and SS (r=-0.314, p<0.0001) among the 185 early MIBG images. LVEF was significantly correlated with both ES (r=-0.346, p<0.0001), and SS (r=-0.377, p<0.0001) in the delayed images.

4.2. Comparison between LVEF and WR of MIBG

There was a good correlation between LVEF and WR (r=-0.521, p<0.0001) in all 185 studies. In the comparison of changes of WR and LVEF before and 6 months after beta-blocker therapy, there was a good correlation between WR and LVEF before and after beta-blocker therapy (r=-0.735, p<0.0001).

4.3. Comparison of WR, LVEF, ES and SS between the improved and the unimproved group

There were no significant differences in WR and LVEF between both groups, before beta-blocker therapy. ES and SS in the early image were significantly lower for the improved group before beta-blocker therapy.

4.4. Comparison of the Efficacy of Various Beta-Blockers

There was no significant difference in the improvement of LVEF or MIBG findings among the three kinds of beta-blocker therapy.

4.5. Serial Changes in LVEF and WR for Long Term Follow-up in Two Groups after Beta-Blocker Therapy

After beta-blocker therapy, LVEF significantly improved from 36% to 55% after 6 months in the improved group, then to 63% after 3 years. And almost all patients exhibited a stable cardiac function. On the other-hand, no significant change occurred in

LVEF from 30% to 34% in the unimproved group. But, LVEF slightly improved to 45% with improvement of the symptoms, 3 years after beta-blocker therapy (Fig. 1).

After beta-blocker therapy, WR significantly improved from 51% to 35% after 6 months in the improved group, then to 33% after 3 years. On the other-hand, no significant change occurred in WR from 51% to 48% in the unimproved group. But, WR slightly improved to 40%, 3 years after beta-blocker therapy (Fig.2).

4.6. Relationship between the Cardiac Events and the Serial Changes in WR or LVEF for Long Term Follow-up

There was a tendency for an increase of WR in patients with a cardiac event during the follow-up period (Fig.3). But, there was no change in LVEF, when a cardiac event occurred (Fig.4).

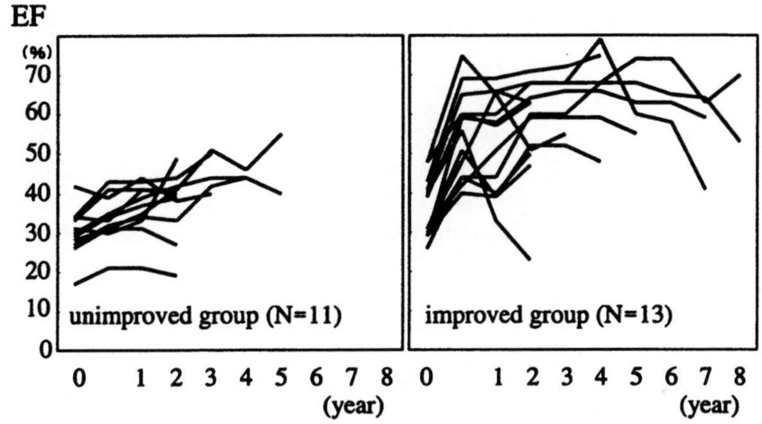

Fig.1: Changes of LVEF in long term follow-up

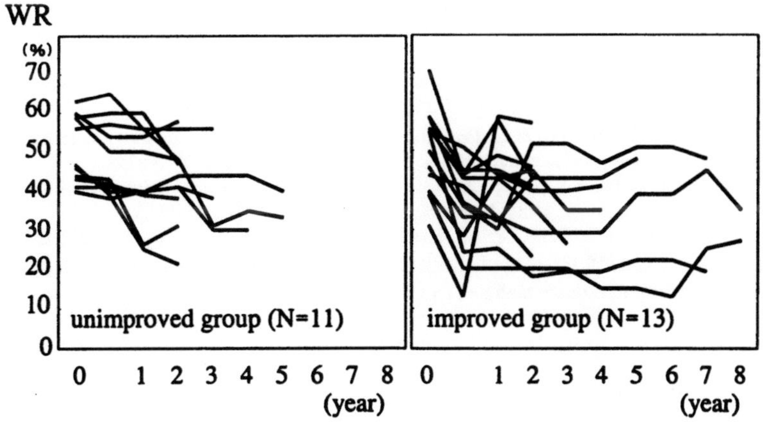

Fig.2: Changes of WR in long term follow-up

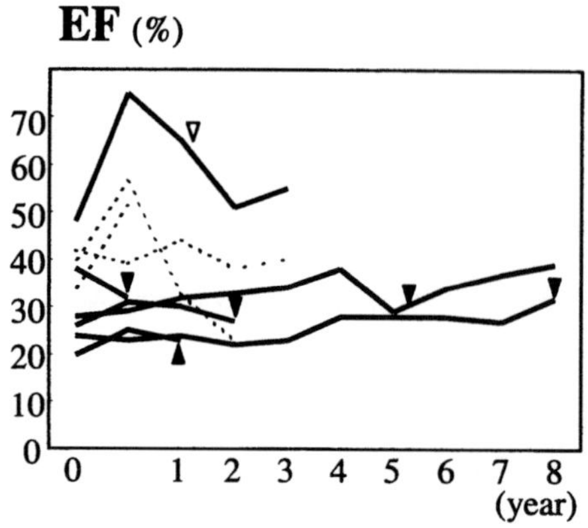

Fig.3: Changes in WR in patients having cardiac events

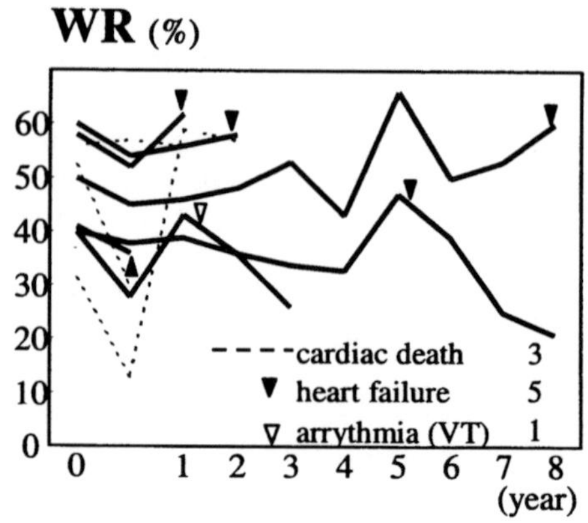

Fig.4: Changes in LVEF in patients having cardiac events

5. CONCLUSION

We performed MIBG myocardial SPECT at successive intervals during beta-blocker therapy in patients with DCM, as well as before and after the therapy. As a significant correlation was observed between MIBG findings and LVEF, WR of MIBG seemed to reflect sympathetic nervous function regardless of the existence of beta-blocker therapy. Prediction of therapeutic effects of the beta-blocker is possible by conducting MIBG before treatment. While WR is useful for evaluation of therapeutic effects, it should be noted that WR and LVEF change in some cases during long term follow-up. It was an interesting suggestion that cardiac accidents could be predicted by changes in WR. In the present study, there was no significant difference in improvement of cardiac function or MIBG findings among various beta-blockers.

From the above findings, it was suggested that MIBG was useful in examining the severity of DCM, determining the applicability of β-blocker therapy, estimating the maintenance dosage, and evaluating the prognosis.

REFERENCES

1. Rector TS, Olivari MT, Levine TB, et al, Predicting survival for an individual with congestive heart failure using the plasma norepinephrine concentration, Am. Heart J, **113**, 148-152 (1987).
2. Yamazaki J, Muto H, Ishiguro K, et al, Quantitative scintigraphic analysis of ^{123}I-MIBG by polar map in patients with dilated cardiomyopathy, Nucl. Med. Commun, **18**, 219-229 (1997).
3. Yamazaki J, H Muto, T Kabano, et al, Evaluation of ß-blocker therapy in patients with dilated cardiomyopathy —Clinical meaning of iodine 123- MIBG metaiodobenzylguanidine myocardial single-photon emission computed tomography—, Am. Heart J, **141**, 645-652 (2001).

FAMILIAL ORTHOSTATIC TACHYCARDIA DUE TO NOREPINEPHRINE TRANSPORTER (SLC6A2) DEFICIENCY

Maureen K. Hahn, Michelle Mazei, Nancy Flattem, John R. Shannon, Randy D. Blakely, and David Robertson[1]*

The availability of extracellular NE is limited by a presynaptically localized NE transporter, NET. NET retrieves released NE, thus limiting the spread and duration of synaptic excitability and allowing neurotransmitter to be repackaged into synaptic vesicles. NET is a target for both antidepressants, such as the tricyclic agents and for the psychostimulants, amphetamine and cocaine.[2] NET is a member of a gene family of 12 transmembrane domain (TMD)-containing, Na^+- and Cl^--coupled cotransporters which also includes dopamine, serotonin, GABA, glycine, proline and taurine transporters. hNET is a single copy gene localized to chromosome 16 at 16q12.2 (SLC6A2).[5,10] The hNET gene was cloned from a human lung fibroblast library and the intron-exon borders were determined, reporting a gene of ~45 kb with 16 exons and 15 introns.[17,23,27] The hNET gene codes for a 617 amino acid protein with a predicted molecular weight of 69,000 Da.[26]

Genetic variability in the NET gene that impacts transporter expression could contribute to complex genetic disorders. Single nucleotide polymorphisms (SNPs) throughout the genome provide the genetic variability responsible for most behavioral and physiological traits and common diseases. Polymorphisms in promoter or intron regions could alter expression of the hNET gene and noradrenergic transmission through effects on transcriptional regulation and alternative splicing events.[17] Alternative splicing occurs in the 3' end of hNET resulting in proteins that differ in the COOH-terminus sequence,[27] and evidence supports that the COOH-terminus mediates interactions with proteins involved in cell surface targeting.[28,37] Synonymous coding SNPs (cSNPs) that don't alter amino acid sequence may contribute to disease through effects on gene expression or may be markers linked to more functional polymorphisms located at a distinct site in the gene or nearby

[1]Maureen K. Hahn, Center for Molecular Neuroscience, Department of Pharmacology; Michelle Mazei, Department of Pharmacology; Nancy Flattem, Department of Pharmacology; John R. Shannon, Autonomic Dysfunction Center; Randy D. Blakely, Center for Molecular Neuroscience, Department of Pharmacology; David Robertson, Autonomic Dysfunction Center, Vanderbilt University, Nashville, TN 37232.

genes. Nonsynonymous cSNPs alter the amino acid sequence of a protein, potentially modifying protein structure and function. Therefore, nonsynonymous SNPs have a greater potential both to cause functional disturbances to NE transport *in vivo*, and contribute to genetic risk for disease.

Some hNET SNPs have been identified but are not as yet linked with a disease phenotype. For example, a Taq1 restriction fragment length polymorphism has been identified, although its functional role remains unknown.[9] Stöber and coworkers identified SNPs in the coding region of the hNET gene, including several nonsynonymous cSNPs.[36] They identified thirteen variants; five nonsynonymous, three synonymous and five in introns. The nonsynonymous polymorphisms were present in the population with frequencies of 0.01-0.02, while synonymous and noncoding SNPs were present at frequencies up to 0.40. An association study on new groups of 374 controls, 206 bipolar patients and 456 schizophrenics revealed no association of any of the nonsynonymous cSNPs with schizophrenia or bipolar disorder.[36] Subsequently, this laboratory also demonstrated a lack of association of these SNPs with Tourette's Syndrome (TS).[35] A lack of association for the common synonymous G1287A polymorphism with major depression or with suicidal ideation has also been reported.[25] The functional impact of the five identified nonsynonymous SNPs that cause amino acid changes in hNET protein have been assayed in *in vitro* heterologous expression systems.[31]

The contribution of hNET to the pathology of cardiovascular disorders, as described above, suggests that these populations may be enriched for hNET SNPs. A recent study to determine the prevalence of SNPs in the human genome examined a group of genes that encode proteins involved in cardiovascular function. Seven nonsynonymous, three synonymous and three noncoding SNPs were identified in the hNET gene, with reported frequencies up to 0.1 were revealed.[14] Our laboratory has studied the effects of the nonsynonymous cSNPs on hNET expression and transport through transfection into heterologous cell culture systems.[12] These studies reveal both loss-of-function as well as gain-of-function hNET alleles that now must be explored in the context of individual patient phenotypes to determine the extent to which these hNET SNPs contribute to symptoms. In summary, while there has not yet been evidence of a relationship between identified hNET SNPs and a disease state, the identification of nonsynonymous cSNPs with functional consequences in a study of blood pressure extremes suggests that cardiovascular disorders might be likely to involve hNET SNPs.

Noradrenergic systems and NET have been implicated in diseases of the autonomic nervous system. NE is the major neurotransmitter at postganglionic sympathetic synapses and it is estimated that NET can recapture approximately 90% of NE released from sympathetic nerves terminals in the heart.[6,7] Additionally, as NET is dependent on the Na^+ and Cl^- gradients, efflux of NE through the transporter can be achieved if ion gradients are reversed or membranes depolarized, which can occur during cardiac ischemia.[15,19] Decreases in NE uptake sites and activity are observed in cardiomyopathy and heart failure.[4,21,22,32] NE uptake is also impaired in a proportion of hypertensive patients.[8]

Orthostatic Intolerance (OI) is a disease of the autonomic nervous system with symptoms that become manifest upon assuming upright posture and include cerebral hyperperfusion, altered cognitive function, tachycardia, anxiety, visual changes, nausea, chest discomfort, lightheadedness, dizziness, presyncope and syncope.[16,30] The etiology of OI is not well-understood, but it appears to be a heterogenous disorder with multiple causes. For example, evidence of pooling of blood in the legs and increased sympathetic nerve conduction velocities suggest that sympathetic denervation of the lower extremities is a primary feature of some OI cases, whereas a primary hyperadrenergic state is suggested by evidence of increased supine vascular resistance.[16] In order to simplify such a complex cluster of pathologies, our laboratory defined a select group of patients exhibiting orthostatic tachycardia of greater than 30 beats per minute with no concomitant hypotension.[16] These patients exhibit a pronounced hypernoradrenergic state particulary upon standing.[16] Plasma NE levels in these OI patients are elevated while supine, demonstrate a standing-induced increase that is greater than that seen in healthy controls and NE levels are further elevated if patients remain standing for extended periods of time, the latter effect not observed in healthy controls.[16] These patients also demonstrate a decreased disappearance of NE from plasma, or clearance, upon standing.[16] These patients were 2 times more sensitive to both the effects of isoproterenol to increase heart rate and phenylephrine to increase systolic blood pressure.[16] Finally, patients show a decreased sensitivity to the drug tyramine with smaller changes in NE spillover into plasma than controls and decreased clearance.[16] Tyramine is a compound that normally causes an increase in plasma NE following uptake through NET (Figure 1). These findings supported enhanced activation of sympathetic noradrenergic pathways and indicate the involvement of altered hNET function.

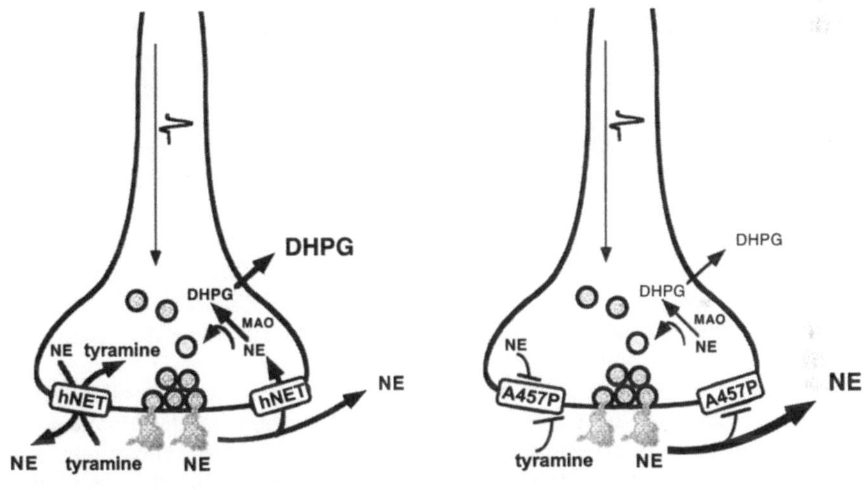

Figure 1. Schematic illustration of the function of wild-type-hNET (A) and A457P (B) in the presynaptic terminal of a neuron. A) Normally hNET functions to take up released NE that once in the cytoplasm can be recycled into exocytotic vesicles or metabolized by monoamine oxidase (MAO) to dihydroxy-phenylglycol (DHPG). Some NE escapes the reuptake process and appears in plasma. DHPG can diffuse out of the neuron and also be measured in plasma. Tyramine is a substrate for hNET and once inside the cell is also a substrate for the vesicular monoamine transporter. Tyramine displaces NE from vesicles, thereby increasing cytoplasmic NE levels and resulting in the exit of NE from the neuron through hNET via a reverse transport mechanism. B) The failure of A457P to transport NE results in a larger than normal spillover of NE and a diminished clearance of NE from plasma. This lack of uptake diminishes the flux of NE through the MAO degradation pathway and plasma DHPG levels following NE release do not increase as compared to normal. Additionally, tyramine is not able to induce release of NE through reverse transport. Thus, whereas A457P enhances basal and stimulated plasma NE levels, there is a blunted NE response to tyramine compared to normal.

The suggestion of hNET dysfunction in autonomic nervous system disorders and the demonstrated hyperadrenergic state of many OI patients led our laboratory to search for hNET SNPs in OI patients. A proband with OI was identified who displayed symptoms highly suggestive of diminished hNET activity. The proband demonstrated standing-induced increased NE appearance in plasma (spillover) and decreased NE clearance.[34] In addition, the proband demonstrated a blunted response to tyramine similat to previous OI patients examined.[34] Furthermore, the proband exhibited decreased intraneuronal metabolism of NE as measured by decreased DHPG/NE ratios.[34] DHPG is the metabolite formed by monoamine oxidase in the cytoplasm of cells and therefore is considered a correlative measure for NE that has been taken up by NET following release (Figure 1).[7] Taken together, these findings suggested a selective involvement of an hNET dysfunction in this OI patient.

Direct sequencing of the hNET gene revealed the proband to be a heterozygote for a G to C substitution at nucleotide 237 in exon 10 that results in a proline substitution for alanine at amino acid position 457.[34] This amino acid lies in a highly conserved region of transmembrane domain 9. Family members of the proband were genotyped identifying several heterozygotes for A457P. A high correlation was found between presence of the A457P polymorphism and standing-induced heart rate, plasma NE and the plasma DHPG to NE ratio.[34] A457P was not identified in 254 unrelated normal subjects, patients with hypertension or other OI patients.[34] The location of a non-conservative amino acid substitution in a highly conserved TMD of hNET suggested that the A457P mutation could be highly disruptive to hNET structure and function. A proline substitution may cause a conformational change in the alpha-helical structure of TMD 9. Furthermore, a series of charged residues in the intracellular loop immediately preceding TMD 9 suggests a potential role in positioning the protein in the membrane and therefore TMD 9 may be an important determinant of hNET structure and folding.[3]

Examination of the transport properties of A457P confirmed that the proline substitution resulted in a nonfunctional transporter. Transient transfection of A457P into heterologous expression systems revealed a protein that is devoid of transport activity.[34] In addition, A457P is not expressed in the mature, fully glycosylated form and exhibits greatly diminished surface expression.[13] Furthermore, A457P exerts a dominant-negative effect on wt hNET uptake activity.[13] This suggests that transporters form multimeric complexes and that individuals heterozygous for A457P may be affected to a greater extent than predicted for harboring one mutant allele.

This is the first demonstration of a coding mutation in a transporter gene with consequences for both *in vitro* transporter activity and neurotransmitter biochemistry and behavioral phenotype in an affected proband and family. Designating alterations in DHPG levels and tyramine responsiveness as indicative of hNET dysfunction, the identification of the nonfunctional polymorphism, A457P was achieved. Elevated standing heart rate may be a highly predictive endophenotype, particularly for those polymorphisms highly disruptive to hNET function. The contribution of an hNET mutation to the tachycardia observed in OI may relate to the prominence of the reuptake mechanism in the heart, where synapses are 3 times narrower than those in the vasculature.[11,24] Additionally, NE is the major neurotransmitter mediator of sympathetic nervous system regulation of heart rate. In contrast, the actions of NE in the brain occur within complex pathways utilizing multiple neurotransmitters. An hNET mutation in such an environment may have more subtle effects or greater opportunity for compensation.

Elevated standing heart rate may therefore serve as an endophenotype indicative of hNET dysfunction that may successfully identify subgroups of subjects with hNET polymorphisms within populations with broader psychiatric diagnoses. For example, the role of the noradrenergic system in cognition and attention through actions in the prefrontal cortex suggests an involvement in Attention Deficit/Hyperactivity Disorder (ADHD).[3] TCAs such as desipramine and the experimental selective norepinephrine reuptake inhibitor, tamoxetine, have been effective in ADHD.[3] Perhaps a subset of ADHD involves the NE system to a greater extent than the DA system, and a subset of these patients have hNET polymorphisms. Selecting a subset of ADHD patents with elevated standing heart rate may greatly increase the chances of finding an association of hNET with this disorder. The cardiovascular phenotype may also provide a means of identifying subgroups of depression likely to harbor hNET SNPs. It has long been hypothesized that depression involves the noradrenergic system as evidenced by the mood altering effects of reserpine, monoamine oxidase inhibitors and tricyclic antidepressants, all compounds that impact catecholamine levels, and by studies of noradrenergic metabolite levels and receptor binding sites in depression.[20,29,33] Furthermore, hNET levels measured by ^{3}H-nisoxetine binding are decreased in the brains of patients with major depression, the majority of whom died by suicide.[18] We are currently assessing DNA from these suicide victims for hNET SNPs. Additionally, we are identifying depressed subjects with elevated standing heart rate as likely candidates for hNET SNPs.

REFERENCES

1. S. G. Amara and M. J. Kuhar. Neurotransmitter transporters: recent progress. Annu Rev Neurosci **16** (1993) 73-93.
2. E. L. Barker and R. D. Blakely. Norepinephrine and serotonin transporters: molecular targets of antidepressant drugs. In F. E. Bloom and D. J. Kupfer (Eds.), Psychopharmacology: the fourth generation of progress, Raven Press, Ltd., New York, 1995.
3. J. Biederman and T. Spencer. Attention-deficit/hyperactivity disorder (ADHD) as a noradrenergic disorder. Biol Psychiatry **46** (1999) 1234-1242.
4. M. Bohm, K. La Rosee, R. H. Schwinger and E. Erdmann. Evidence for reduction of norepinephrine uptake sites in the failing human heart. J Am Coll Cardiol **25** (1995) 146-153.
5. M. Brüss, J. Kunz, B. Lingen and H. Bönisch. Chromosomal mapping of the human gene for the tricyclic antidepressant-sensitive noradrenaline transporter. Hum Genet **91** (1993) 278-280.
6. G. Eisenhofer, H. S. Cox and M. D. Esler. Parallel increases in noradrenaline reuptake and release into plasma during activation of the sympathetic nervous system in rabbits. Naunyn Schmiedebergs Arch Pharmacol **342** (1990) 328-335.
7. G. Eisenhofer, M. D. Esler, I. T. Meredith, A. Dart, R. O. d. Cannon, A. A. Quyyumi, G. Lambert, J. Chin, G. L. Jennings and D. S. Goldstein. Sympathetic nervous function in human heart as assessed by cardiac spillovers of dihydroxyphenylglycol and norepinephrine. Circulation **85** (1992) 1775-1785.
8. M. Esler, G. Jackman, A. Bobik, P. Leonard, D. Kelleher, H. Skews, G. Jennings and P. Korner. Norepinephrine kinetics in essential hypertension. Defective neuronal uptake of norepinephrine in some patients. Hypertension **3** (1981) 149-156.
9. J. Gelernter, S. Kruger, K. K. Kidd and S. Amara. TaqI RFLP at norepinephrine transporter protein (NET) locus. Hum Mol Genet **2** (1993) 820.
10. J. Gelernter, S. Kruger, A. J. Pakstis, T. Pacholczyk, R. S. Sparkes, K. K. Kidd and S. Amara. Assignment of the norepinephrine transporter protein (NET1) locus to chromosome 16. Genomics **18** (1993) 690-692.
11. D. S. Goldstein, J. E. Brush, Jr., G. Eisenhofer, R. Stull and M. Esler. In vivo measurement of neuronal uptake of norepinephrine in the human heart. Circulation **78** (1988) 41-48.
12. M. K. Hahn, M. S. Mazei, D. Robertson and R. D. Blakely. Role of human norepinephrine transporter gene single nucleotide polymorphisms in cardiovascular disease. Am J Med Genet **67** (2000) 369.
13. M. K. Hahn, M. S. Mazei, D. Robertson and R. D. Blakely. Single nucleotide polymorphisms in the human norepinephrine transporter gene alter *in vitro* norepinephrine transport. Society for Neuroscience Abstracts **26** (2000) 25.
14. M. K. Halushka, J. B. Fan, K. Bentley, L. Hsie, N. Shen, A. Weder, R. Cooper, R. Lipshutz and A. Chakravarti. Patterns of single-nucleotide polymorphisms in candidate genes for blood-pressure homeostasis. Nat Genet **22** (1999) 239-247.
15. M. Imamura, H. M. Lander and R. Levi. Activation of histamine H3-receptors inhibits carrier-mediated norepinephrine release during protracted myocardial ischemia. Comparison with adenosine A1-receptors and alpha2-adrenoceptors. Circ Res **78** (1996) 475-481.
16. G. Jacob, J. R. Shannon, F. Costa, R. Furlan, I. Biaggioni, R. Mosqueda-Garcia, R. M. Robertson and D. Robertson. Abnormal norepinephrine clearance and adrenergic receptor sensitivity in idiopathic orthostatic intolerance. Circulation **99** (1999) 1706-1712.

17. C. H. Kim, H. S. Kim, J. F. Cubells and K. S. Kim. A previously undescribed intron and extensive 5' upstream sequence, but not Phox2a-mediated transactivation, are necessary for high level cell type-specific expression of the human norepinephrine transporter gene. Journal of Biological Chemistry **274** (1999) 6507-18.
18. V. Klimek, C. Stockmeier, J. Overholser, H. Y. Meltzer, S. Kalka, G. Dilley and G. A. Ordway. Reduced levels of norepinephrine transporters in the locus coeruleus in major depression. J Neurosci **17** (1997) 8451-8458.
19. T. Kurz, G. Richardt, S. Hagl, M. Seyfarth and A. Schomig. Two different mechanisms of noradrenaline release during normoxia and simulated ischemia in human cardiac tissue. J Mol Cell Cardiol **27** (1995) 1161-1172.
20. B. E. Leonard. The role of noradrenaline in depression: a review. J Psychopharmacol **11** (1997) S39-S47.
21. C. S. Liang, T. H. Fan, J. T. Sullebarger and S. Sakamoto. Decreased adrenergic neuronal uptake activity in experimental right heart failure. A chamber-specific contributor to beta-adrenoceptor downregulation. J Clin Invest **84** (1989) 1267-1275.
22. P. Merlet, J. L. Dubois-Rande, S. Adnot, M. H. Bourguignon, C. Benvenuti, D. Loisance, H. Valette, A. Castaigne and A. Syrota. Myocardial beta-adrenergic desensitization and neuronal norepinephrine uptake function in idiopathic dilated cardiomyopathy. J Cardiovasc Pharmacol **19** (1992) 10-16.
23. J. Meyer, P. Wiedemann, O. Okladnova, M. Bruss, T. Staab, G. Stober, P. Riederer, H. Bonisch and K. P. Lesch. Cloning and functional characterization of the human norepinephrine transporter gene promoter. J Neural Transm **105** (1998) 1341-1350.
24. A. M. Novi. An electron microscopic study of the innervation of papillary muscles in the rat. Anat Rec **160** (1968) 123-141.
25. D. Owen, L. Du, D. Bakish, Y. D. Lapierre and P. D. Hrdina. Norepinephrine transporter gene polymorphism is not associated with susceptibility to major depression. Psychiatry Res **87** (1999) 1-5.
26. T. Pacholczyk, R. D. Blakely and S. G. Amara. Expression cloning of a cocaine- and antidepressant-sensitive human noradrenaline transporter. Nature **350** (1991) 350-354.
27. P. Pörzgen, H. Bönisch and M. Brüss. Molecular cloning and organization of the coding region of the human norepinephrine transporter gene [published erratum appears in Biochem Biophys Res Commun 1996 Oct 14;227(2):642-3]. Biochem Biophys Res Commun **215** (1995) 1145-1150.
28. P. Pörzgen, H. Bönisch, R. Hammermann and M. Brüss. The human noradrenaline transporter gene contains multiple polyadenylation sites and two alternatively spliced C-terminal exons. Biochim Biophys Acta **1398** (1998) 365-370.
29. K. J. Ressler and C. B. Nemeroff. Role of norepinephrine in the pathophysiology and treatment of mood disorders. Biol Psychiatry **46** (1999) 1219-1233.
30. D. Robertson. The epidemic of orthostatic tachycardia and orthostatic intolerance. Am J Med Sci **317** (1999) 75-77.
31. F. Runkel, M. Bruss, M. M. Nothen, G. Stober, P. Propping and H. Bonisch. Pharmacological properties of naturally occurring variants of the human norepinephrine transporter [In Process Citation]. Pharmacogenetics **10** (2000) 397-405.
32. M. Schafers, D. Dutka, C. G. Rhodes, A. A. Lammertsma, F. Hermansen, O. Schober and P. G. Camici. Myocardial presynaptic and postsynaptic autonomic dysfunction in hypertrophic cardiomyopathy. Circ Res **82** (1998) 57-62.
33. J. J. Schildkraut. The catecholamine hypothesis of affective disorders: a review of supporting evidence. American Journal of Psychiatry **122** (1965) 509-522.
34. J. R. Shannon, N. L. Flattem, J. Jordan, G. Jacob, B. K. Black, I. Biaggioni, R. D. Blakely and D. Robertson. Clues to the origin of orthostatic intolerance: a genetic defect in the cocaine- and antidepressant sensitive norepinephrine transporter. New England Journal of Medicine **342** (2000) 541-549.
35. G. Stöber, J. Hebebrand, S. Cichon, M. Bruss, H. Bonisch, G. Lehmkuhl, F. Poustka, M. Schmidt, H. Remschmidt, P. Propping and M. M. Nothen. Tourette syndrome and the norepinephrine transporter gene: results of a systematic mutation screening. Am J Med Genet **88** (1999) 158-163.
36. G. Stöber, M. M. Nothen, P. Porzgen, M. Bruss, H. Bonisch, M. Knapp, H. Beckmann and P. Propping. Systematic search for variation in the human norepinephrine transporter gene: identification of five naturally occurring missense mutations and study of association with major psychiatric disorders. Am J Med Genet **67** (1996) 523-532.
37. G. E. Torres, W. Yao, A. R. Mohn, H. Quan, K. Kim, A. I. Levey, J. Staudinger and M. G. Caron. Functional Interaction between Monoamine Plasma Membrane Transporters and the Synaptic PDZ Domain-Containing Protein PICK1. Neuron **30** (2001) 121-134.

SYMPATHETIC INNERVATION AND BLOOD FLOW IN THE DIABETEC HEART

Oliver Schnell*

1. INTRODUCTION

The balance of the activity of the cardiac sympathetic and parasympathetic nervous system plays a key role for the functional and structural performance of the diabetic heart.[1] Metabolic, vascular/hemorrheological and, in type 1 diabetes mellitus, immunological factors are currently hypothezised to contribute to the pathogenesis, which, however, has not been fully elucidated.[1] The heterogeneous pattern and the high extent of dysfunction of the cardiac sympathetic nervous system in diabetes mellitus had remained largely unknown until advanced scintigraphic techniques for direct characterization of dysfunction of sympathetic neurons in the living heart emerged in the 1990's. Previously, diagnosis of cardiac sympathetic abnormalities had been mainly based on conventional non-invasive cardiac reflex tests, including measurements of heart rate variation and blood pressure changes to standing. These tests, however, only indirectly assess diabetic cardiac autonomic dysfunction, they do not differentiate absolutely between sympathetic and parasympathetic function and they do not assess the heart regionally. Imbalance of the activity of the cardiac sympathetic nervous system is suggested to lead to multiple functional disturbances and to increase the risk for arrythmogenicity of the diabetic heart.[2]

2. SCINTIGRAPHICALLY ASSESSED CARDIAC SYMPATHETIC DYSINNERVATION

Advanced single-photon emission computed tomography (SPECT) and positron emission tomography (PET) using radiopharmaceuticals I-123-metaiodobenzylguanidine (I-123-MIBG) and 11C-hydroxyephedrine (11C-HED) enable the characterization of the pattern and extent of cardiac sympathetic dysinnervation.[1,2] The applied radiolabeled norepinephrine analogues share the same active uptake mechanisms as norepinephrine.[1,2]

* Diabetes Reseach Institute Munich, Munich, Germany

In 1992, Mäntisaari et al. reported the presence of cardiac sympathetic dysinnervation, as assessed by I-123-MIBG scintigraphy, in diabetic patients with autonomic neuropathy[3]. Succeeding scintigraphic investigations of the human diabetic heart demonstrated that, both in type 1 and type 2 diabetes, cardiac sympathetic dysinnervation is present more frequently than previously detected with conventional cardiac reflex tests.[1,2,4,5]

It is important to realize that sympathetic dysinnervation may be present to a large extent both in Type 1 and Type 2 diabetes. Scintigraphically, cardiac sympathetic dysinnervation was detected in 77% of newly diagnosed metabolically stabilized type 1 diabetic patients.[4] 9% of these patients have been reported to exhibit advanced cardiac sympathetic dysinnervation, as evidenced by a severe reduction of global I-123-MIBG uptake or no uptake of the radiotracer.[5] Supporting the view that neuronal abnormalities progress with duration of diabetes,[6] defects of cardiac sympathetic innervation are more enhanced in long-term type 1 diabetic patients than in newly diagnosed type 1 diabetic patients. In long-term type 1 diabetes mellitus 80-90 % of patients scintigraphically exhibit cardiac sympathetic abnormalitites.[1,2] The pattern of cardiac sympathetic dysinnervation of both type 1 and type 2 diabetic patients is heterogeneous with a predominant affection of the posterior myocardial region.[1,2] Improvement of metabolic control in newly-diagnosed metabolically-stabilized type 1 diabetic patients has been shown to partially reduce scintigraphically assessed cardiac sympathetic abnormalities at one-year follow-up.[7] Poorly controlled long-term type 1 diabetic patients (HbA1c < 10%) also demonstrated a propensity of cardiac sympathetic abnormalities to restoration with amelioration of metabolic control to the near normal range.[8] In a group of long-term Type 1 diabetic patients, scintigraphically assessed cardiac sympathetic dysinnervation did neither regress nor progress on the average during a three-year follow-up. In the study, HbA1c was 8.0% at entry compared with 7.9% at three-year follow-up.[9]

The superior sensitivity of I-123-MIBG scintigraphy to conventional cardiac reflex tests is underlined by the observation that the majority of newly diagnosed and long-term type 1 diabetic patients without ECG-detectable cardiac autonomic neuropathy demonstrate cardiac sympathetic dysfunction at scintigraphic investigation.[1,2,5] Assessment of the heart by PET with 11C-HED in type 1 diabetes mellitus confirms the frequent sympathetic dysinnervation and the predominance in the posterior myocardial region.[10] Complicating the heterogeneous cardiac sympathetic dysinnervation, it has been observed with PET that diabetic patients exhibit proximal sympathetic hyperinnervation of the heart.[11]

In long-term diabetic patients, scintigraphically assessed cardiac sympathetic dysinnervation has been shown to correlate with the presence of complement-fixing sympathetic ganglia (CF-SG) autoantibodies, suggesting the view that autoimmune mechanisms play a role in the pathogenesis of cardiac sympathetic neuropathy in type 1 diabetes mellitus.[12] Autoantibodies against sympathetic ganglia seem to be rather specific for cardiac autonomic dysfunction of type 1 diabetic patients and occur independently of ICA, GADA or IA2A.[13] Autoantibodies against autonomic nervous tissues are not detected to a significant extent in type 2 diabetes mellitus and in patients with neuropathies in non-diabetic neurological diseases.[14] The specificity of autoantibodies against sympathetic ganglia in type 1 diabetes mellitus support the view that autoantibodies targeted against sympathetic postganglionic neurons cause local damage and subsequently deteriorate neuronal function in the patients. Hypothetically they contribute to the more advanced scintigraphically assessed cardiac sympathetic dysinnervation in type 1 than in type 2 diabetes mellitus.

3. CORONARY BLOOD FLOW RESERVE

PET with 13N-ammonia is an established measure to quantify hemodynamic abnormalities and to assess both endothelial-dependent and endothelial-independent coronary vasodilatatory reserve.[1] The technique is based on the use of 13-N-ammonia, which is retained in myocardial tissue in proportion to blood flow. Relative and absolute measures of regional myocardial blood flow can be obtained.[15] PET has been reported to detect coronary artery disease with a sensitivity of 92% and a specificity of 89 %.[15] Preferably, resting myocardial perfusion scintigraphy is combined with exercise testing to unmask reduced myocardial blood flow reserve. The cold pressor test assesses myocardial blood flow in response to sympathetic-nerve stimulation.[1, 15] Applying the technology to determine endothelial-dependent blood flow reserve in type 1 diabetes mellitus, a reduction of coronary blood flow reserve has been detected. Additional assessment of cardiac sympathetic dysfunction with SPECT showed a trend towards an advanced decrease in flow reserve with progression of cardiac sympathetic dysfunction.[1, 16]

The observations are in line with an increasing number of observations demonstrating reduced coronary blood flow reserve early on in diabetes mellitus. Recent observations show that cardiac sympathetic dysfunction affecting cardiac sympathetic efferent sympathetic signals is directly related to coronary blood flow in diabetes mellitus.[1, 16] The magnitude of impairment of sympathetically mediated coronary flow reserve has been shown to be related to the degree of diabetic cardiac sympathetic dysfunction. Therefore, a link between sympathetic nervous defects and alterations of coronary perfusion is suggested. The hypothesis is enhanced by the observation that flow reserve in humans is largely independent of changes in hemodynamics and circulating catecholamines.[16] Also beta-adrenergic blockade does not diminish increases in blood flow in normal coronary arteries in response to cold pressure test.[17]

Hypothetically, a reduction of sympathetically mediated coronary blood flow in diabetes mellitus with cardiac sympathetic dysfunction may directly result from a decrease in endothelium-derived NO activity (1). Supporting the view, norepinephrine and BHT 920, a selective alpha2 adrenergic receptor agonist, have been reported to increase dose-dependently nitrate release from human coronary microvessels.[18]

It is hypothetsized that the frequently detectable cardiac sympathetic dysinnervation in the course of diabetes mellitus is associated with subsequent reduction of coronary blood flow reserve leading to an impaired vascular performance of the diabetic heart.

4. CONCLUSION

Scintigraphic techniques have largely contributed to characterize cardiac sympathetic innervation and blood flow regulation of the human diabetic heart. It is important to realize that, even in long-term diabetes, cardiac autonomic disturbances can be partially reversed. Substantial amelioration of metabolic control has been shown to be effective in both primary and secondary prevention. Rapid recompensation of hyperglycemia and long-term optimisation of glucose control are important factors of a successful approach for treatment of neuronal and vascular impairment of cardiac performance in diabetes mellitus.

REFERENCES

1. E. Standl, O. Schnell. A new look upon the heart in diabetes mellitus: from ailing to failing. Diabetologia **43**: 1455-1469 (2000)
2. O. Schnell, C.-M. Kirsch, J. Stemplinger, M. Haslbeck, E. Standl. Scintigraphic evidence for cardiac sympathetic dysinnervation in long-term Type 1 diabetic patients with and without ECG-based autonomic neuropathy. Diabetologia **38**:1345-1352 (1995)
3. M. Mäntysaari, J. Kuikka, J. Mustonen, K. Tahvanainen, E. Vanninen, E. Länsimies, M. Uusitupa: Noninvasive detection of cardiac sympathetic nervous dysfunction in diabetic patients using (123I) metaiodobenzylguanidine. Diabetes **41**: 1069-1075 (1992)
4. A. Langer, M.R. Freeman, R.G. Josse, P.W. Armstrong: Metaiodobenzylguanidine imaging in diabetes mellitus: Assessment of cardiac sympathetic dysinnervation and its relation to autonomic dysfunction and silent myocardial ischemia. J Am Coll Cardiol **25**: 610-618 (1995)
5. O. Schnell, D. Muhr, M. Weiss, S. Dresel, M. Haslbeck, E. Standl. Reduced myocardial ^{123}I-metaiodobenzylguanidine uptake in newly diagnosed IDDM patients. Diabetes **45**:801-805 (1996)
6. M.A. Pfeifer, M.P. Schumer: Perspectives in Diabetes: Clinical trials of diabetic neuropathy: past, present, and future: Diabetes **44**: 1355-1361 (1995)
7. O. Schnell, D. Muhr, S. Dresel, M. Weiss, M. Haslbeck, E. Standl. Partial restoration of scintigraphically assessed cardiac sympathetic denervation in newly diagnosed patients with insulin dependent (Type 1) diabetes mellitus at one-year follow-up. Diabetic Medicine **14**:57-62 (1997)
8. D. Muhr-Becker, M. Weiss, K. Tatsch, G. Wolfram, E. Standl, O. Schnell. Scintigraphically assessed cardiac sympathetic dysinnervation in poorly controlled type 1 diabetes mellitus: One-year follow-up with improved metabolic control. Exp Clin Endocrinol Diabetes **107**:306-312 (1999)
9. O. Schnell, D. Muhr, M. Weiss, C.M. Kirsch, M. Haslbeck, E. Standl. Three-year follow-up on scintigraphically assessed cardiac sympathetic denervation in patients with long-term Insulin Dependent (Type 1) Diabetes mellitus. J Diab Compl, **11**:307-313 (1997)
10. K.C. Allman, M.J. Stevens, D.M. Wieland, G.D. Hutchins, E.R. Wolfe, D.A. Greene, M. Schwaiger. Noninvasive assessment of cardiac diabetic neuropathy by C11 hydroxyephedrine and positron emission tomography. J Am Coll Cardiol **22**: 1425-1432 (1993)
11. M.J. Stevens, D.M. Raffel, K.C. Allman, F. Dayanikli, E. Ficaro, T. Sandford, D.M. Wieland, M.A. Pfeifer, M. Schwaiger et al. Cardiac sympathetic dysinnervation in diabetes. Circulation **98**: 961-968 (1998)
12. O. Schnell, D. Muhr, S. Dresel, K. Tatsch, A.G. Ziegler, M. Haslbeck, E Standl. Autoantibodies against sympathetic ganglia and evidence of cardiac sympathetic dysinnervation in newly diagnosed and long-term IDDM patients. Diabetologia **39**:970-975 (1996)
13 D. Muhr, U. Mollenhauer, A. Ziegler, M. Haslbeck, E. Standl, O. Schnell. Autoantibodies to sympathetic ganglia, GAD, or tyrosine phosphatase in long-term IDDM with and without ECG-based cardiac autonomic neuropathy. Diabetes Care **20**:1009-1012 (1997)
14. O. Schnell, A. Schwarz, D. Muhr-Becker, E. Standl: Cardiac autonomic dysfunction in Type 2 Diabetes mellitus: No association with autoantibodies against autonomic nervous tissues. Exp Clin Endocrinol Diabetes **108**: 181-186 (2000)
15. M. Schwaiger, J. Melin. Cardiological applications of nuclear medicine. Lancet 354: 661-666 (1999)
16. M.F. Di Carli, D. Bianco-Batlles, M.E. Landa, A. Kazmers, H. Groehn, O. Muzik, G. Grunberger. Effects of autonomic neuropathy on coronary blood flow in patients with diabetes mellitus. Circulation **100**: 813-819 (1999)
17. A.M. Zeiher, H. Drexler, H. Wollschlager, B. Saurbier, H Just. Coronary vasomotion in response to sympathetic stimulation in humans: importance of the functional integrity of the endothelium. J Am Coll Cardiol **14**: 1181-1190 (1989)
18. M.R. Kichuk, N. Seyedi, X Zhang et al. Regulation of nitric oxide production in human coronary microvessels and the contribution of local kinin formation. Circulation **94**:44-51 (1996)

DYNAMIC MONITORING OF NEUROLOGICAL DISEASES BY CSF CLINICAL MARKERS

Hasan Parvez,[1] Shahid Baig,[2] Catherine Collin,[1] Ali Qureshi,[3] Simone Parvez,[4] and Claude Reiss[1*]

INTRODUCTION

Neurological symptoms and neurodegeneration are silent features of many neuropathological conditions. These include acute metabolic and physiological insult, chronic motor and dementia disorders as well as inflammatory diseases represting important causes of morbidity and mortality in humans. Studies during the last decade have provided convincing evidence that numerous neuroactive molecules play a vital role in chemical signaling and neurodegenerative processes of the nervous system.[1] These substances before mainly represented putative transmitters and neuropeptides. Investigations on oxidative stress involving reactive oxygen species (ROS) and reactive nitrogen species (NOS) are now largely believed to be in part responsible for the induction of neurogenic lesions not only by producing neurotoxins but also leading to DNA adduct formation.[2] The interdependence of neuropeptide/neurotransmitter pathways and their co-existence in the same neuron is a tremendous advance to understand how neurosecretion can control the processes of transmitter synthesis, release and metabolism. Receptor active opioid peptides in human brain were demonstrated more than 20 years ago.[3] Regardless of our advance knowledge on the neurochemitsry of neuropeptide/neurotransmitter and free radicals, it still remains extremely difficult to have a precise diagnosis of the early onset of neurodegenerative processes and only the post-mortem histopathological and laboratory tests can provide the true neuropathological picture. Despite the existence of several barriers between the blood and the brain, cerebrospinal fluid (CSF) can still be considered as an ultrafiltrate of plasma. Involvement of neuropathological processes can be reflected by changes in CSF celular and chemical composition. CSF obtained by lumber puncture is normally employed to assess the cell counts, protein, IgG levels and among all these parameters,

*Corresponding authors: [1]Hasan Parvez, Catherine Collin and Claude Reiss, Neuroendocrinologie & Neuropharmacologie du Développement, CNRS-Institut Alfred Fessard de Neurosciences, IFR 2118, Bât 5, 91190 Gif Sur Yvette, France; [2]Shahid Baig, Department of Neurology, Agha Khan Medical College, Karachi, Pakistan; [3]Ali Qureshi, Department of Clinical Neuroscience, Huddinge University Hospital, Novum,S-141-57, Sweden; [4]Simone Parvez, Laboratoire de Neuroendocrinologie du Développement, Université de Reims, UFR Sciences, B.P. 1039, 51687 Reims, France.

Catecholamine Research, Edited by Nagatsu *et al.*
Kluwer Academic/Plenum Publishers, 2002

pleocytosis or increase of cells in CSF is the hall mark of all types of neuropathologies.[4] Even more important is the possibility to follow each disease during its different stages, and to understand the metabolic and pathogen changes of the disease identified *by neurochemical markers such as neuropeptides, neurotransmitters and free radicals.*

The present investigations summarized in this chapter provide a brief reflection of the invited conference at Kyoto Joint Conference on natural variations in human CSF content of the neuropeptides, catechol and indole transmitters, excitatory and inhibitory amino acids and free radical NO. We shall discuss in detail the true impact of neuropeptide Y (NPY) in different neurological diseases reflecting chronic and acute degenerative processes. Some clinical data on neuroprotective role of vitamin B-12 to reverse neurodegenerative processes has also been provided.

Our clinical studies were oriented to differentiate the effects of chronic long term neurodegenerative diseases such as multiple sclerosis, cerebro vascular stroke and Parkinson's disease from acute clinical situations such as aseptic meningitis and tuberculosis meningities.

CSF NEUROPEPTIDE-Y IN NEUROLOGICAL DISORDERS

The very first peptide that we analyzed represents Neuropeptide Y (NPY) consisting of 36 amino acid residues and has been shown to modulate a number of CNS functions.[5,6] As compared to healthy subjects, NPY levels were significantly decreased from 29 ± 3 pmol/l to 24 ± 2 in patients suffering from cerebro vascular stroke ($p<0.05$). The most important decrease in NPY content of CSF was observed in patients with multiple sclerosis and the value as compared to healthy subjects was decreased to 12 ± 1 pmol/l ($p<0.01$). Patients suffering from Parkinson's disease manifested significant increase ($p<0.01$) in their CSF content as compared to healthy controls. The effect of acute neurodegenerative processes caused by aseptic meningitis also showed decrease in NPY CSF content. Tuberculosis meningitis patients had a similar spectrum of decreased NPY levels of CSF. These clinical effects on NPY concentration reflect a significant modulation of cerebral function during the progress of the disease. The most potent effect of NPY causes increase in vasoconstriction and an increase in food intake[5] and this neuropeptide is described as the most active stimulator of appetite. NPY is localized in high concentrations in several regions of the brain including nuclei of the brain stem and nerve fibres surrounding cerebral vessels which can have an important impact on regulation of cerebral blood flow and systemic vegetative functions. Vascular inflammation resulting in spasm and thrombosis has been described during acute neurodegeneration caused by meningitis. These clinical observations can further document evidence that autoregulation of cerebral blood flow is constantly decreasing during pathogenesis of degenerative nervous diseases and the ultimate proof for its complete loss can be seen in meningitis patients.[5,7] Therefore, NPY in central nervous system can modulate cerebral perfusion pressure and make cerebral blood flow completely dependent to lost autoregulation. The co-existence of NPY with noradrenaline was elegantly demonstrated by Hökfelt[6] and this observation could explain to some extent the bahavioral effects of NPY in patients. The onset of most of neurological diseases are accompanied by important variations in anxiety and depressive states besides eating disorders. Depression is a multifactoral process, therefore, it is of interest to study the coexistence of NPY with noradrenergic, serotonergic and dopaminergic systems in relevance to depressive states as commonly observed in neurological patients. Our observations reflecting low CSF levels of NPY in vascular stroke, multiple sclerosis and meningitis patients can be of some value to explain depressive state of patients. Among our patients, all groups except Parkinson's patients showed decreased CSF levels of NPY and are of special interest since the infection of the CNS due to meningitis and their potential relationship to the local and systematic consequences could be the reason for this change in infection disorders such as cerebral stroke andf mutiple sclerosis could be eventually attributed to reduction in cerebral blood flow. NPY has been shown to exert powerful vasoconstrictive properties in vivo as well as in vitro, a finding that leads to suggest the hypothesis that NPY may contribute to the vasospasm associated with subarachnoid hemorrhage. These clinical determinations of dynamic decreases in CSF NPY can explain to some extent the basis of neurodegenerative processes linked to low blood flow.

Nitric Oxide, Homocysteine Neurotoxicity and Neurodegenerative Diseases: Neuroprotection by Vitamin B-12

The relation between vitamin B-12 deficiency and neurotoxicity of homocysteine with nitric oxide is defined on the basis of accumulated results obtained in CSF from 20 Parkinson's patients, 16 patients with cerebrovascular disorders, 20 patients with multiple sclerosis, 18 patients with meningitis and 14 patients with tuberculosis meningitis. These results were compared with 14 healthy subjects and demonstrated significant increase of nitrite (a metabolite of nitric oxide), arginine and homocysteine. Simultaneously, we observed significant decreases in vitamin B-12 concentration of CSF in all patients except those with aseptic meningitis. A linear relationship between CSF levels of nitrite and homocysteine was found, suggesting that the production of the free radical nitric oxide is interrelated with neurotoxic levels of homocysteine. Our results show that in comparison to healthy subjects, CSF levels of arginine (ARG) were significantly increased in Parkinson's patients, patients with cerebrovascular stroke, multiple sclerosis and tuberculosis meningitis (significance from $p<0.01$ to $p<0.001$)) Aseptic meningitis did not manifest any important variation of ARG in the CSF. At the same time nitrite levels in the CSF were also increased significantly ($p<0.01$ to $p<0.001$) in Parkinson's, cerebrovascular stroke, multiple sclerosis and tuberculosis meningitis patients. On the contrary, vitamin B-12 levels in the CSF decresaed significantly ($p<0.01$ to p0.001) in the above 4 degenerative diseases. Homocysteine CSF level showed significant increase in Parkinson's patients, cerebrovascular stroke patients, multiple sclerosis and tuberculosis meningitis patients. The correlation between CSF level of nitrite and homocysteine in various neuronal diseases studied was linear reflecting a coefficient of variance (R2) between 0.977 to 0.984 for healthy subjects and all types of patients studied.

Our data on decreased levels of vitamin B-12 in the CSF are consistent with published observations that this vitamin plays a vital role for brain aging and intellectual activity. Experimental studies done in close collaboration with Professor Hideya Saito utilizing stroke-prone hypertensive rats (SHR-SP) clearly demonstrate that the hypothalamin acetylcholine (ACh) during post-natal development of SHR-SP decreases tremendously with direct impact on Passive Avoidance Response.[8] Simultaneously, the SHR-SP also show decreased levels of vitamin B-12 in their brain nuclei. Administration of vitamin B-12 to these rats for 10 days can induce neuroprotection and completely establish Passive Avoidance Response and Ach levels to normal levels. The patients with vitamin B-12 deficiency manifest pernicious anemia and macrocytosis of red cells. Mild macrocytosis is relatively common in multiple sclerosis patients. In fact, clinical detection of vitamin B-12 in humans still remains a problem to solve since blood data do not always correlate with CSF levels.

However, now there is a good evidence that more than 90% of neurologic patients who are deficient in vitamin-B-12 accumulate in their blood and CSF methylmalonic acid and homocysteine, substances which depend on vitamin B-12 levels and therefore, this alternative is considered as a better marker for clinical deficiency. A disturbance in the integrated mechanism between these vitamin B-12 cofactors is expected to cause significant changes in the synthesis and bioavailability of a number of compounds directly or indurectly linked to B12/folate metabolism. Nitric oxide toxicity due to its interaction with vitamin B-12 dependent enzyme *i.e.,* methionine synthase produces haematological and less frequently neurological symptoms resembling those of pernicious anemia. NO interaction with cobalamin results in the liberation of hydroxy free radicals responsible for inactivation of methionine synthase. Our results showing reduced levels of vitamin B-12 and elevated levels of homocysteine in all patients and a linear relationship between CSF nitrite and homocysteine are interconnected to the processes leading to neurodegeneration. This does also suggest the involvement of NO, low vitamin B-12 and increased homocysteine could add up to multifactoral processes of neuronal cell death in some of the patients studied.

Therefore, from the observed results, a variety of therapeutic approaches are possible. These may include NMDA receptor antagonists, agents to improve mitochondrial function, free radical scavengers and neurotropic factors. Combinations of therapies with vitamin B-12 acting at sequential steps in neurodegenerative processes may be improved for better efficacy.

ACKNOWLEDGMENTS

The authors express their sincere gratitude to Prof. Toshiharu Nagatsu for his kind advice and moral support over the years. The generous support by Dr. Dolbeau of Perkin Elmer France for microanalytical equipment is kindly acknowledged.

REFERENCES

1. G.A. Qureshi, Sh. Baig, I. Bednar, A. Halawa and S. H. Parvez, The neurochemical markers in cerebrospinal fluid to differentiate between aseptic and tuberculosis meningitis, *Neurochem. Int.* **32**, 197-203 (1998).
2. I. N. Acworth, M. Naoi, H. Parvez and S. Parvez, *Coulometric Electrode Array Detectors for HPLC* (VSP Press, Zeist, The Netherlands1997), pp. 97-126.
3. L. Terenius and A. Wahlström, Morphine-like ligand for opioid receptors in human CSF, *Life Sci.* **13**, 1759–1764 (1975).
4. C. Reiss, S. Parvez, G. Labbe and H. Parvez, *Advances in Molecular Toxicology* (VSP Press, Zeist, The Netherlands 1998), pp.293–312.
5. G. A. Qureshi, Sh. Baig, and S.H. Parvez, Biochemical markers in tuberculosis meningitis, *Biogenic Amines*, **12**, 499–519 (1996).
6. T. Hökfelt, J. Lundberg, H. Lagercrantz, K. Tatemoto, V. Mutt, J. Lindberg, L. Terenius, B. Everitt, K. Fuxe, L. Agnati and M. Goldstein, Occurrence of neuropeptide Y (NPY)-like immunoreactivity in catecholamine neurons in human medulla oblongata, *Neurosci. Lett.* **36**, 217–222 (1983).
7. G. A. Qureshi, Sh. Baig, C. Collin and S. H. Parvez, Variations in cerebrospinal fluid levels of neuropeptide Y, cholecystokinin and substance-P in patients with neurological disorders. *Neuroendocrinology Lett.* **21**, 409–416 (2000)
8. H. Saito, H. Togashi, M. Yoshioka, N. Nakamura, M. Minami and S.H. Parvez, Animal models of vascular dementia with emphasis on stroke-prone spontaneously hypertensive rats. *Clin. Exp. Pharmacol. Physiol.* **Suppl.1**, S-257-S-259 (1995).

COMPARATIVE PHYSIOLOGY AND MORPHOLOGY OF CATECHOLAMINE SYSTEMS

Beta adrenergic receptor-mediated attenuation of TNFα-stimulated ICAM-1 expression on human brain microvascular endothelial cells

Yukoh Ohara[1], Richard M. McCarron[1], Susanne Golech[1], Joliet Bembry[2], Fred A. Lenz[3], and Maria Spatz[2]

1. INTRODUCTION

The involvement of the adrenergic system derived from locus ceruleus (LC) in the regulation of blood-brain barrier has been recognized to be mediated to a large extent by the endothelium.[1-6] These cells, which produce many vasoactive and inflammatory substances, have been shown to participate in the interaction between immune cells and the central nervous system (CNS). For example, the infiltration of inflammatory cells into the CNS, a critical step in the pathogenesis of many CNS diseases, is limited in normal brain. In inflammation, the endothelium has been recognized as an active participant in pathogenic mechanisms by its capacity to express adhesion molecules. These molecules facilitate binding with circulatory leukocytes, which may result in their activation and ultimate passage across the endothelium. In addition, tumor necrosis factor-α (TNFα) is one of the most important endogenous cytokines that mediates local inflammation. Its proinflammatory activities toward vascular endothelium have also been described previously.[2,7] Many reports also suggest that the changes in BBB properties in neurological disorders may be associated with disturbances of adrenergic input.[3,4] Thus, we examined the influence of adrenergic agents on TNFα-induced intercellular adhesion molecule-1 (ICAM-1) on human brain microvascular endothelial cells (HBEC).

[1] Resuscitative Medicine Department, Naval Medical Research Center, Silver Spring, MD, [2] Stroke Branch, NINDS, National Institutes of Health, Bethesda, MD, and [3]Department of Neurosurgery, JHU Medical School, Baltimore, MD

2. MATERIALS AND METHODS

2.1 HBEC Cultures, Determination of ICAM-1 expression, cAMP production

HBEC were isolated and cultivated as reported previously.[8] Confluent EC cultures (purity>95%) grown in 96-well microtiter plates were treated with indicated concentrations of recombinant TNFα in the presence or absence of isoproterenol (added 15 min prior to TNFα). Where indicated 10-50 μM atenolol, 10-50 μM butoxamine or 1 or 10 μM propranolol were added 20 min prior to TNFα and cultures incubated for 6-48 h at 37º C. Some of the HBEC cultures were exposed to hypoxia/reoxygenation [(H/R) 95% N_2/ 5% CO_2 (6h)/95% O_2/5% CO_2 (24h)]. Enzyme-linked Immunosorbant Assay (ELISA) was used to quantitate the level of ICAM-1 expression on HBEC.[9] The formation of cAMP was assayed as described by Karnushina *et al.*[10]

3. RESULTS

3.1 Effect of β-adrenergic agents, db-cAMP or forskolin on constitutive and TNFα-induced ICAM-1 expression; cAMP formation

HBEC expressed relatively low levels of ICAM-1, which were not significantly changed by exposure to various concentrations of adrenergic receptor antagonists [*i.e.*, propranolol (non-specific β1/β2-AR antagonist), butoxamine (β2-AR antagonist) and atenolol (β1-AR antagonist)] (Table 1). The EC were incubated with TNFα for 6, 24 and 48 h; the highest level of ICAM-1 expression was seen at 24 h (2-4-fold increase over controls). This time period was chosen for all subsequent experiments. TNFα stimulated ICAM-1 expression in a dose-dependent manner (not shown). Isoproterenol, the β-adrenergic agonist, reduced the TNFα-stimulated expression of ICAM-1 (Table 1). Propranolol or butoxamine prevented this effect, whereas atenolol was ineffective (Table 1). This inhibition was observed at all TNFα concentrations studied (not shown).

Neither db-cAMP nor forskolin had any significant effect on the constitutive expression of ICAM-1. However, db-cAMP (1 mM) or forskolin (10 μM) had a similar effect as Iso in decreasing (46%, 47% and 31%, respectively) the level of ICAM-1 expression induced by TNFα. Pretreatment with propranolol was ineffective.

Table 1. Effect of adrenergic mediators on ICAM-1 expression on HBEC

	Pretreatment			
Addition	None	Pro	But	Aten
None	0.267 ± 0.023	0.275 ± 0.019	0.258 ± 0.020	0.249 ± 0.031
TNFα	1.479 ± 0.054	1.499 ± 0.062	1.507 ± 0.143	1.398 ± 0.101
Iso	0.294 ± 0.023	-	-	-
TNFα + Iso	1.093 ± 0.017	1.594 ± 0.092	1.488 ± 0.103	0.997 ± 0.083

HBEC were incubated for 24 h in media alone (None) or in the presence of 1 μM isoproterenol (Iso) ± 25 U/ml TNFα. Pretreatment consisted of propranolol (Pro), butoxamine (But), or atenolol (Aten) administered 20 min prior to TNFα as indicated in the Methods section. Data represent ICAM-1 expression (mean ± S.E.M., O.D. at 490 nm; of 3-8 total experiments performed in triplicate).

HBEC produced variable amounts of cAMP that were not affected by propranolol. A 2-3-fold increase of cAMP formation was seen after the cells were exposed to isoproterenol (1 μM) alone or in the presence of TNFα. Propranolol completely abolished the increased production of cAMP induced by isoproterenol ± TNFα.

3.2 Effect of hypoxia/reoxygenation and β-adrenergic agents on constitutive and TNFα-induced ICAM-1 expression and TNFα production by HBEC

Hypoxia/reoxygenation enhanced (57%) constitutive expression of ICAM-1 on HBEC as compared to normoxia (mean ± S.E.M.; 0.908 ± 0.031 and 0.578 ± 0.082, respectively). The addition of TNFα to HBEC incubated under H/R further increased the level of ICAM-1 expression (69%); this level was similar to that seen induced by TNFα under normoxia (mean ± S.E.M.; 1.538 ± 0.09 and 1.510 ± 0.09, respectively). Interestingly, as opposed to observations under normoxia, pretreatment with isoproterenol had no effect on either the constitutive or TNFα-induced level of ICAM-1 expression of HBEC incubated under H/R (not shown).

HBEC cultured in media alone were shown to produce low, but significant levels of TNFα under normoxia (0.53 pg/ml). Incubation of HBEC under H/R increased (36% over control) the level of TNFα produced (0.73 pg/ml). The addition of isoproterenol to HBEC cultured under H/R significantly increased (75%) the level of TNFα produced (1.269 pg/ml). A smaller increase (17%) in TNFα production was observed when isoproterenol was added to HBEC incubated under normoxia (0.623 pg/ml). As opposed to the effects of isoproterenol, pretreatment of HBEC incubated under both normoxia and H/R with propranolol resulted in decreased production of TNFα (47% and 22%, respectively). The concomitant treatment of HBEC with both isoproterenol and propranolol resulted in levels of TNFα production similar to those observed under basal normoxia and H/R (0.55 and 0.80 pg/ml, respectively). It thus appears that propranolol inhibits the isoproterenol-induced secretion of TNFα observed in H/R.

4. DISCUSSION

The results of this study clearly indicate that TNFα-induced up-regulation of ICAM-1 expression is diminished by β-adrenergic agents such as isoproterenol. The data also demonstrate that this effect is specifically modulated by β_2-AR antagonists (propranolol, a non-specific β_1/β_2-AR antagonist and butoxamine, a β_2-AR antagonist) but not atenolol, a β_1-AR antagonist. These findings are in agreement with previous *in vivo* and *in vitro* studies describing the participation of the adrenergic system in BBB and transendothelial permeability.[3, 4] It is known that the isoproterenol activation of β-adrenergic receptors linked to Gs protein involved stimulation of cAMP. The observed decreases of TNFα-stimulated ICAM-1 expression by db-cAMP or forskolin confirm this supposition. The inability of β-AR antagonists to reverse the forskolin effects on TNFα-induced ICAM-1 expression suggests that the inhibitory effect by these agents occurs distal to the receptor.

The involvement of cAMP in modulating both peripheral and cerebral vascular EC adhesion molecule expression induced by various cytokines has been previously shown.[7, 11] However, the findings presented here represent a first demonstration regarding the effect of adrenergic agents on this phenomenon. The precise mechanism(s) involved in this process remain to be clarified. A few studies implicate the potential involvement of protein kinase A (PKA) and/or protein kinase C (PKC) in expression of adhesion molecules induced by TNFα as well as other cytokines such as IL-1[12]; additional possible participants include NF-kB and/or NO.[11] Our preliminary results indicate that CREB might be also involved in this process.

The up-regulation of TNFα-induced ICAM-1 expression or TNFα production was previously observed in a variety of cells under normoxia or hypoxia.[3, 7, 13-15] The data here represent the first demonstration of adrenergic stimulation of ICAM-1 expression and TNFα secretion by cerebral endothelial cells under normoxia and H/R. In addition, the results indicate that isoproterenol is ineffective in down-regulating the TNFα-induced ICAM-1 expression seen under H/R. This implies that the mechanisms responsible for TNFα-induced ICAM expression on H/R cells differs from that utilized by normoxic cells. Furthermore these findings indicate a possible role for the disturbances in adrenergic effectors seen under various ischemic conditions.

REFERENCES

1. Cohen Z, Molinatti G, Hamel E (1997) Astroglial and vascular interactions of noradrenaline terminals in the rat cerebral cortex. J Cereb Blood Flow Metab **17**: 894-904.
2. McCarron RM, Wang L, Stanimirovic DB, Spatz M (1995) Differential regulation of adhesion molecule expression by human cerebrovascular and umbilical vein endothelial cells. Endothelium **2**:339-346.
3. Raichle M, Hartman BK, Echling JO, Sharpe LG (1975) Central noradrenergic regulation of cerebral blood flow and vascular permeability. Proc Nat Acad Sci (USA) **72**:3726-3730.
4. Spatz M, McCarron RM (1998) Blood-brain barrier and monoamines, revisited. In: Partridge WM (Ed.): Introduction to the Blood-Brain Barrier, Cambridge University Press, Cambridge, pp. 362-376.
5. Stanimirovic D, Satoh (2000) Inflammatory mediators of cerebral endothelium: a role in ischemic brain inflammation. Brain Pathol **10**:113-26.
6. Webb AA, Muir GD (2000) The blood-brain barrier and its role in inflammation. J Vet Intern Med **14**:399-411.
7. Pober JS, Slowik MR, De Luca LG, Ritchie AJ (1993) Elevated cyclic AMP inhibits endothelial cell synthesis and expression of TNF-induced endothelial leukocyte adhesion molecule-1, and vascular cell adhesion molecule-1, but not intercellular adhesion molecule-1. J Immunol **150**:5114-5123.
8. Spatz M, Kawai N, Merkel N, Bembry J, McCarron RM (1997) Functional properties of cultured endothelial cells derived from large microvessels of human brain. Am J Physiol **272**:C231-239.
9. Tanaka M, McCarron RM (1990) The inhibitory effect of tumor necrosis factor and interleukin-1 on Ia induction by interferon-gamma on endothelial cells from murine central nervous system. J Neuroimmunol **27**:209-215.
10. Karnushina I, Spatz M, Bembry J (1983) Cerebral endothelial cell culture II. Adenylate cyclase response to prostaglandins and their interaction with the adrenergic system. Life Sci **32**:1427-1435.
11. Balyasnikova IV, Pelligrino DA, Greenwood J, Adamson P, Dragon S, Raza H, Galea E (2000) Cyclic adenosine monophosphate regulates the expression of the intracellular adhesion molecule and the inducible nitric oxide synthase in brain endothelial cells. J Cerebral Blood Flow Metab **20**:688-699.
12. Turunen JP, Ustinov J, Renkonen R (1993) Adhesion molecules involved in protein kinase A- and C-dependent lymphocyte adherence to microvascular endothelial cells. Scand J Immunol **37**:282-288.
13. Ghezzi P, Dinarello CA, Bianchi M, Rosandich ME, Repine JE, White CW (1991) Hypoxia increases production of interleukin-1 and regulation of tumor necrosis factor production by adrenaline and tumor necrosis factor by human mononuclear cells. Cytokine **3**:189-194.
14. Severn A, Rapson NT, Hunter CA, Liew FY (1992) Regulation of tumor necrosis factor production by adrenaline and beta-adrenergic agonists. J Immunol **148**:3441-3445.
15. Kokura S, Wolf RE, Yoshikawa T, Granger DN, Aw TY (2000) Postanoxic T lymphocyte-endothelial cell interactions induce tumor necrosis factor-alpha production and neutrophil adhesion: role of very late antigen-4/vascular cell adhesion molecule-1. Circ Res **86**:1237-1244.

CLONING OF NOVEL GENES FROM THE HUMAN SPINAL CORD BY MOLECULAR PROFILING

Jun-ichi Niwa, Shinsuke Ishigaki, Manabu Doyu, and Gen Sobue*

1. BACKGROUND

Amyotrophic lateral sclerosis (ALS) is a neurodegenerative disease which etiology is unknown. In ALS, large motoneurons in the brain stem and the anterior horn of the spinal cord degenerate selectively and progressively. To uncover the pathogenic mechanism of ALS, it is crucial to identify molecules maintaining motor neuronal integrity or leading to eventual cell death. To this aim, it is necessary to explore the molecules expressed in the spinal cord, and to analyze its function and the interaction they have. Several high throughput methods have been developed to screen many gene expression simultaneously, such as differential displays and cDNA microarrays. Molecular indexing is a method developed by Kato[1] to explore the wide range of expressed genes effectively and enables comparison of profiles of gene expression among two or more samples. In the previous study, we have got 70 genes differentially expressed in autopsied Alzheimer's disease brain compared with normal brain[2]. In this study, we screened genes expressed in the human spinal cord anterior horn by molecular indexing. Using this technique, we cloned two novel genes, *FI58G* and *FI39A*.

2. METHODS

Human spinal cord was obtained at autopsy. Autopsy was performed within 6 hour post mortem. At autopsy, the anterior horn of the lumbar spinal cord was incised and immediately frozen in liquid nitrogen. Total RNA was extracted from each spinal cord sample using TRIsol reagent (Gibco), and double-stranded cDNA was produced using SUPERSCRIPT II (Gibco). cDNAs were cut with *Bsm*FI and ligated to those from a pool of 64 biotinylated adapters cohesive to all possible overhangs. Ligated molecules were

* Jun-ichi Niwa, Shinsuke Ishigaki, Manabu Doyu, and Sobue Gen, Nagoya University Graduate School & Faculty of Medicine, Nagoya, Japan 466-8550.

digested by *BsmAI* and *FokI*, and were recovered with streptavidin-coated paramagnetic beads. PCR was performed with the adapter-primer and an anchored oligo-dT primer. By repeating the experiment with 64 adapters and three anchored oligo-dT primers, 192 cDNA libraries were established. This procedure has been described in detail in previous reports[1]. By random sequencing, two 3'-fragments of new genes were found in the library and were named *FI58G* and *FI39A*. The full length of the cDNAs were obtained screening a human fetal brain cDNA library panel (Origene) by PCR and 5'-RACE using Marathon cDNA Amplification Kit (Clontech). Northern blot analysis was performed using Human MTN Blot and Human Brain II Blot (Clontech). Mouse Embryo MTN Blot (Clontech) and Hybond N^+ membrane (Amersham) blotted with 2μg poly (A^+) RNA from cultured neuro2a or SH-SY5Y cells were also used. DNA probes were radiolabeled with [α-^{32}P] dCTP using the Rediprime DNA labeling system (Amersham). Northern hybridization was performed using an ExpressHyb Hybridization Solution (Clontech), according to the supplier's protocol. *In situ* hybridization was performed with RNA probes labeled by digoxigenin using the DIG-RNA labeling kit (Boehringer). Sense riboprobes were used as controls. Antisense riboprobes were obtained from the T7 promoter of pGEM-Teasy-*FI39A* or -*FI58G*. Sense riboprobes were synthesized from the SP6 promoter of the same plasmid. Six mm-thick sections were treated as previously reported[3] and observed microscopically.

Neuro2a and SH-SY5Y cell lines were maintained in Dulbecco's modified Eagle's medium (DMEM) supplemented with 10% fetal calf serum (FCS), streptomycin and penicillin. SH-SY5Y cells were treated in DMEM with 5% FCS and 10μM retinoic acid (RA) for 12, 24, 48, and 72 hours for differentiation. Differentiation of neuro2a cells was induced in DMEM with 2% FCS and 20μM RA or DMEM with 10% FCS and 0.5μM N-carbobenzoxyl-Leu-Leu-leucinal (MG132), a proteasome inhibitor. Cells were harvested after incubation in medium with each agent for 12, 24, 48 and 72 hours.

3. RESULTS AND INTERPRETATION

FI39A and *FI58G* cDNA were cloned from the anterior horn of the human lumbar spinal cord[4, 5]. The full-length *FI39A* transcript was 4.4kb and had a 2.5kb open reading frame (ORF). The ORF of *FI39A* encoded 838 amino acids. Amino acids sequence of FI39A displays several notable primary structural characteristics (Fig. 1a); three Cys/His-rich motifs located in the N-terminal region of the protein. Two of the Cys/His-rich motifs, encompassing amino acids 132-179 and 301-332, respectively, correspond to the consensus sequence of the C3HC4-type RING-finger motif. The central Cys/His-rich region has recently been termed the "in between RING-fingers" (IBR) motif (residues 219-264)[6]. Two hydrophobic regions exist in the C-terminal of FI39A (365-395 and 414-454), and there are three putative nuclear localization sequences (NLSs), one monopartite NLS (357-360) and two bipartite basic NLSs (61-75 and 396-412). In northern blot, a single 4.4kb *FI39A* transcript was observed in all tissues, although relatively higher expression was observed in the heart. In the central nervous system, *FI39A* was ubiquitously expressed, including the spinal cord. FI39A protein may participate in the ubiquitin-proteasome passway because it has the RING-finger/IBR motif, which was recently shown to be the recruiting motif of specific ubiquitin-conjugating enzymes[7]. Recently, *parkin*, a gene responsible for one of the most common forms of the familial

Parkinson's disease, was shown to have a RING-finger/IBR motif[6], to bind with specific E2s (UbcH7 and UbcH8), and to have E3 ubiquitin ligase activities[8].

The full-length *FI58G* transcript was 1.8kb in size and had a 660bp ORF (GenBank Accession Number AB029315). An NLS-like domain 'RRRK', encompassing amino acids 133-136, locates in the middle of FI58G protein (Fig. 1b). No other previously known motif was found. Northern blot analysis revealed that *FI58G* was expressed mainly in the brain, heart and skeletal muscle. We also cloned a mouse homologue of *FI58G* gene (GenBank Accession Number AB047544). The expression level of mouse *FI58G* mRNA increased with embryonal development. *In situ* hybridization of the adult human spinal cord revealed that the *FI58G* transcript was mainly expressed in the motor neurons. In cultured cell line, nuclear localization of FI58G -GFP fusion protein was compatible with the existence of an NLS-like signal.

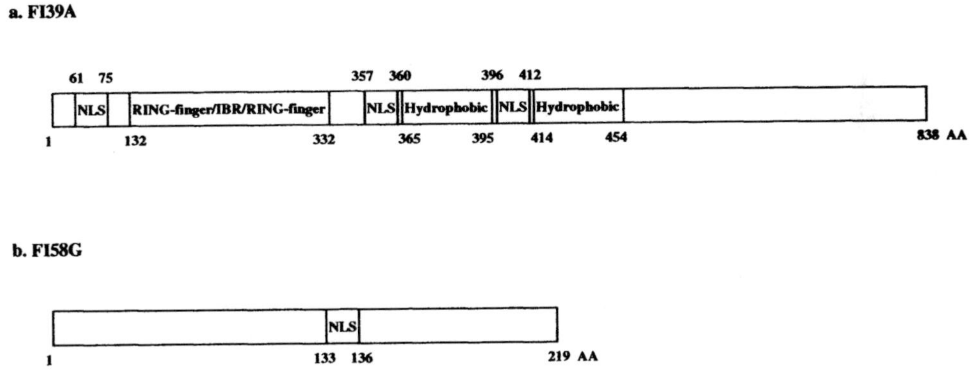

Figure 1. Schematic representation of structures of FI39A and FI58G proteins. NLS: putative nuclear localization sequence.

SH-SY5Y, a human neuroblastoma cell line, was differentiated with RA treatment. Most of the cells differentiated to neuronal phenotypes with long neurite processes by 72 hours of treatment. Neuro2a, a mouse neuroblastoma cell line, was differentiated with RA or MG132. In the presence of RA, cells extended bipolar neurites, and MG132 treatment also caused the cells to differentiate into spindle-like bipolar forms. A semi-quantitative RT-PCR showed that the RA treatment of SH-SY5Y cells upregulated the expression of the *FI58G* mRNA with time. The mouse *FI58G* expression also increased by the RA and MG132 treatment in neuro2a cells. The RA- and MG132-induced upregulation of *FI58G* seemed to parallel with the degree of neuronal differentiation. RA has profound effects in regulating a broad range of biological process including cell differentiation. Some genes like *NF-κB* or *Bcl-2* have been directly implicated in neuronal differentiation after RA treatment[9, 10]. RA may work downstream of the NGF signal pathways in mouse sensory neurons[11]. Despite these findings, the precise molecular mechanism of RA-promoted differentiation still remains unclear. It is suggested that proteasome inhibitors induce cell differentiation through cell cycle arrest[12]. MG132 has been shown to induce neurite outgrowth by inhibiting ubiquitin-dependent proteolysis associated with an increase in ubiquitin-protein conjugates[13]. Molecules proteolysed through ubiquitin-dependent passway have been supposed to be involved in neuronal differentiation. RA and MG132 may induce neuronal differentiation by different molecular mechanism. However, common upregulation of *FI58G* expression with neurite outgrowth by both RA and MG132, parallel increase of the mRNA level with development in mouse embryo and its high expression in neurons of the central nervous system suggest that *FI58G* may conceivably play an important role in neuronal cell differentiation or the maintenance of neuronal function in the nervous system.

4. CONCLUSION

We have identified two novel human genes, *FI39A* and *FI58G*, expressed in the adult human spinal cord. The *FI39A* gene product has RING-finger/IBR motif and may be involved in the ubiquitin-proteasome passway. *FI58G* is upregulated with neuronal differentiation and may have an important role in neurite extension. Further investigation and more evidence are needed to elucidate their function in the spinal cord and their relation to the pathogenic process in motor neuronal cell death.

REFERENCES

1. K. Kato, Description of the entire mRNA population by a 3' end cDNA fragment generated by class IIS restriction enzymes, *Nucleic Acids Res.* **23**, 3685-3690 (1995).
2. M. Doyu, K. Sawada, N. Mitsuma, J. Niwa, M. Yoshimoto, Y. Fujii, G. Sobue, and K. Kato, Gene expression profile in Alzheimer's brain screened by molecular indexing, *Brain Res. Mol. Brain Res.* **87**, 1-11 (2001).
3. U. Strähle, P. Blader, J. Adam, and P.W. Ingham, A simple and efficient procedure for non-isotopic *in situ* hybridization to sectioned material, *Trends Genet.* **10**, 69-108 (1994).
4. S. Ishigaki, J. Niwa, T. Yoshihara, N. Mitsuma, M. Doyu, and G. Sobue, Two novel genes, human neugrin and mouse m-neugrin, are upregulated with neuronal differentiation in neuroblastoma cells, *Biochem. Biophys. Res. Commun.* **279**, 526-533 (2000).
5. J. Niwa, S. Ishigaki, M. Doyu, T. Suzuki, K. Tanaka, and G. Sobue, A novel centrosomal ring-finger protein, dorfin, mediates ubiquitin ligase activity, *Biochem. Biophys. Res. Commun.* **281**, 706-713 (2001).
6. E. Morett, and P. Bork, A novel transactivation domain in parkin, *Trends Biochem. Sci.* **24**, 229-231 (1999).
7. T.P. Moynihan, H.C. Ardley, U. Nuber, S.A. Rose, P.F. Jones, A.F. Markham, M. Scheffner, and P.A. Robinson, The ubiquitin-conjugating enzymes UbcH7 and UbcH8 interact with RING finger/IBR motif-containing domains of HHARI and H7-AP1, *J. Biol. Chem.* **274**, 30963-30968 (1999)
8. H. Shimura, N. Hattori, S. Kubo, Y. Mizuno, S. Asakawa, S. Minoshima, N. Shimizu, K. Iwai, T. Chiba, K. Tanaka, and T. Suzuki, Familial Parkinson disease gene product, parkin, is a ubiquitin-protein ligase, *Nat. Genet.* **25**, 302-305 (2000).
9. Z. Feng, and A.G. Porter, NF-kappaB/Rel proteins are required for neuronal differentiation of SH-SY5Y neuroblastoma cells, *J. Biol. Chem.* **274**, 30341-30344 (1999).
10. M. Hanada, S. Krajewski, S. Tanaka, D. Cazals-Hatem, B.A. Spengler, R.A. Ross, J.L. Biedler, and J.C. Reed, Regulation of Bcl-2 oncoprotein levels with differentiation of human neuroblastoma cells, *Cancer Res.* **53**, 4978-4986 (1993).
11. J. Corcoran, and M. Maden, Nerve growth factor acts via retinoic acid synthesis to stimulate neurite outgrowth, *Nat Neurosci.* **2**, 307-308 (1999).
12. G. Fenteany, and S.L. Schreiber, Lactacystin, proteasome function, and cell fate, *J. Biol. Chem.* **273**, 8545-8548 (1998).
13. M. Obin, E. Mesco, X. Gong, A.L. Haas, J. Joseph, and A. Taylor, Neurite outgrowth in PC12 cells. Distinguishing the roles of ubiquitylation and ubiquitin-dependent proteolysis, *J. Biol. Chem.* **274**, 11789-11795 (1999).

MOLECULAR BIOLOGY OF CONFORMATIONAL DISEASES

Claude Reiss, Thierry Lesnik, Hasan Parvez, Simone Parvez and Ricardo Ehrlich*

1. INTRODUCTION

An important family of diseases is characterized by the presence of a disease-specific protein that accumulates as "amyloïd" fibers and thereby becomes toxic to the cell. Although the protein has the authentic primary sequence (except in familial cases), it adopts a non-native conformation which resists proteolysis, allowing the protein to accumulate and form fibers. Non-familial amyloid conditions are clearly epigenetic.

The most prevalent of these so-called "conformational" diseases[1] are of the neurodegenerative type, in particular Alzheimer's disease and prion diseases in cattle, which threaten to give rise through the food chain to a major human epidemic.

Conformational diseases pose two provocative questions: can the cell produce misfolded versions of a protein, and how? can the latter escape proteolysis, and how?

2. MOLECULAR MECHANISMS LEADING TO PROTEIN MISFOLDING

We are used to seeing the cell producing proteins in their native conformation. It is indeed generally accepted that the native conformation of a protein is fully determined by its primary structure, which would argue against the existence in the cell of some alternative conformation. Lately however, it was reported at several places that a significant part of the proteins synthesized in the cell -up to one out of three or even two- has a faulty conformation.[2] Most of the faulty proteins belong to the class of Defective Ribosomal Products (DRiPs). The majority of DRiPs are nascent proteins having the correct primary sequence, ubiquitin-tagged for degradation while still attached to the ribosome. Misconformers are also found in the endoplasmic reticulum. Following

* Claude Reiss and Thierry Lesnik, Alzheim'R&D, Gif F91190, Hasan Parvez CNRS IAF Gif F91190, Simone Parvez, Reims University, Reims F51000, Ricardo Ehrlich , University of the Republic, Uruguay 11400 Montevideo.

Catecholamine Research, Edited by Nagatsu *et al.*
Kluwer Academic/Plenum Publishers, 2002

detection upon "quality control", they are "dislocated" to the cytosolic side of the ER and ubiquitinated there.[3] In general, misfolded proteins are rapidly processed by the proteasome.

How can misconformers like DRiPs be produced? Most protein folding is co-translational, that is vectorial, and in general does not occur under thermodynamic equilibrium conditions. Indeed, in addition to taking place in a crowded environment, which excludes part of the possible folding space, the rate of translation elongation controls the time left to nascent protein segments for folding, exposure to other segments (formation of secondary structures), or to folding factors (foldases) needed for tertiary structure formation. Hence the local translation rate is an important contributor to protein folding, in addition to the primary sequence of the protein.

What determines the translation rate of a codon? It is generally assumed that the rate-limiting factor is the availability of the cognate ternary complex. In all species investigated so far, most amino-acids (aa) can be transferred to ribosomes by two or more transfer RNAs (tRNA), most of them cognate for synonymous codons. Isoamino-acylated tRNAs are usually present in the cell in rather different, for some aas very different, amounts. Furthermore, there is a good correlation between the codon demand and cognate tRNA supply. It follows that the rate at which the messenger RNA is translated, i.e. the rate at which the nascent protein is produced, is controlled to a large extend by synonymous codon selection along the mRNA.[4]

Alpha helices tend to fold rather fast, as mainly neighbouring aas participate. The corresponding messenger sequences tend to bear frequent synonymous codons, corresponding to abundant cognate tRNAs and are therefore translated rapidly.[5, 6] Had this sequence be translated slowly, its prolonged exposure could have favored interaction with some distant sequence, thereby precluding helix formation. Beta structures in contrast form much more slowly. Slow translation of the corresponding sequence grants enough time for retrieval and adequate chaperoning of the distant structural elements involved. Rare codons in the corresponding messenger sequence favor the beta fold, frequent codons would not.[5, 6]

The folding of many proteins needs assistance or catalysis from a variety of factors (foldases), crucial for many proteins to attain their native conformation. Members of the Hsp70 family for instance can transiently bind to sequences of 10-15 amino-acids with a hydrophobic core, and discriminate between native and non-native conformers.[7] Proline cis/trans isomerization is catalyzed by specific enzymes, and so is disulfide bond formation and isomerization, etc. The "reaction" of a particular foldase with its specific target in the nascent protein, or during post-translational folding steps, requires that the target remains available to the foldase for sufficient time before becoming buried inside the protein.

The contribution of the local translation kinetics to folding is constitutive, as it is controlled by the (permanent) selection of synonymous codons. Nevertheless, the cell keeps control over the local translation rate, primarily through its pool of ternary complexes. For instance, in response to a change in its growth status, the cell can deeply modify the composition of this pool or the efficiency or accuracy in codon-anticodon recognition (controlled by the base modifications of tRNAs), in synthase expression levels or in amino-acylating tRNAs etc. As we will see, changes in the cell status correlate indeed with the rate of protein misfolding and yield DRiPs.

There is experimental evidence showing that translation rate, as specified by the synonymous codon selection, affects protein conformation. For instance, the translation, under identical conditions, of two mRNAs coding for the same protein and differing only by a short sequence of synonymous codons, produce two enzymes having different specific activities despite sharing the same primary structure.[8] Conversely, translation of the same prion mRNA in two different translation systems, differing in particular by their tRNA and foldase contents, produce proteins of the same primary structure, but having respectively the native and pathologic prion conformations, as characterized by their large difference in resistance to proteolysis.[9]

The load of the translation apparatus can change within large limits, depending on the cell status. Modifications in the growth rate of the cell, its secretory activity or its position in the cell cycle can correspond to robust changes of its translation activity and demand for the associated resources (ribosomes, amino-acids, tRNAs, elongation factors, synthases, foldases etc). The production of these resources in due quantities does not necessarily meet that demanded by the new cell status. For instance, as the growth rate of *E.coli* is shifted from .4 to 2.5 doublings/hr, the total tRNA population increases by 80%, the ribosome population by 250%; hence on average, the total tRNA to ribosome ratio drops by 50% and the drop affects primarily rare tRNAs.[10] The change in growth rate modifies the absolute and relative codon translation rate, possibly leading to protein misfolding. Similar effects would result from the insufficient production of foldases, synthases, elongation factor, aas, etc

Unscheduled environmental, para- or endocrine signals, stressors, metabolic toxicants etc, and cellular senescence, could all unbalance the cell's translation resources and produce misconformers. These would in turn require and sequester more foldases, a process that could enter an avalanche regime leading to massive production of misfolded proteins. Inclusion bodies in microorganisms expressing heterologous genes examplify this process.

We notice that the limited ternary complex availability reduces the codon translation rate, which favors the formation of beta structures (see above). Amyloid "plaques" or "fibrils" have indeed a high content of beta structures, indicating that their formation may well come from an imbalance of the translation load with the ressources needed.

3. RESISTANCE OF MISFOLDED PROTEINS TO PROTEOLYSIS

Misfolded proteins (DRiPs etc.) would rapidly clutter up the cell and possibly induce apoptosis, if they were not immediately proteolyzed. Proteolysis of a majority of cytosolic proteins follows the proteasome pathway.[11] Proteins to be degraded are covalently marked by a chain of several units of a small peptide (Ubiquitin (Ub) in eukaryotic cells), with the help of Ub-activator E1 (ATP dependent), Ub-carrier E2, and Ub-protein ligase enzymes, E3. The cell has over one hundred species of these enzymes, each tagging preferentially a group of proteins. Once tagged, the protein has access to a compartmentalized, high molecular weight, ATP-consuming proteolytic complex, the 26S proteasome. A capping regulator (ATP-dependent), recognizes and unfolds tagged proteins and links them to the proteolytic core of the proteasome.

Secreted and membrane proteins have their tertiary and quaternary structures "quality controlled" in the endoplasmic reticulum (ER) before entering the vacuolar system. If defective, they are retro-translocated into the cytoplasm with the help of members of the ER membrane, and processed there by the proteasome pathway.

For several conformational diseases, it was observed that the specific protein deposit contains mostly ubiquitinated species. Thus, the proteolysis pathways works well for all other protein species of the cell, hence the resistance of the specific protein is due to its misconformation or aggregation, not to some proteolysis defect. Protein misfolding or aggregation could sterically hinder the required ubiquitin ligation by E3, or processing by the proteasome cap in charge of the protein denaturation. Alternatively, a massive production of misfolded proteins could saturate the ubiquitination resources of the cell (in particular substrate-specific E3 enzymes) and thereby favor aggregation. Although failure of the enzymatic action of the substrate-specific E3 enzyme is probably a major cause of resistance to proteolysis, little is known about these ubiquitin ligases,[12] and those specific for proteins involved in conformational diseases have not been identified so far. Even less is known about cytosolic proteolysis not involving the 26S proteasome.

4. CONCLUSION

Conformational diseases (others than familial) involve two epigenetic steps: misconformation (probably co- or post-translational) of a specific protein and its resistance to proteolysis, leading to its toxic accumulation in and around the cell. The various mechanisms leading to *de novo* protein misconformation begin to be understood. Monitoring of the cellular level of misconformers, of foldases, ternary translation complexes etc could on one the hand provide an early diagnostic of the condition -long before neural loss for instance would give rise to clinical signs-, and also allow one to design therapies -gene therapies in particular- which could possibly stabilize the condition. The other route to curing conformational diseases, namely specific proteolysis of amyloid fibers, seems at present less practicable.

REFERENCES

1.Kelly J.W. (1998) The alternative conformations of amyloidogenic proteins and their multi-step assembly pathways. Curr. Opin. Struct.. Biol. **8** 101-106.

2. Schubert U., Anton L.C., Gibbs J., Norbury C.C.,Yewdell J.W. and Bennink J.R. (2000) Rapid degradation of a large fraction of newly synthesized proteins by proteasomes, Nature **404** 770-773.

3. Kopito R.R.(1997) ER quality control: the cytoplasmic connection. Cell **88** 427-430.

4. Solomovici J., Lesnik T. and Reiss C. (1997) Does *E.coli* optimize the economics of the translation process. J. Theor. Biol. **185**.511-521.

5. Thanaraj T.A. and Argos P. (1996) Protein secondary structural types are differentially encoded in mRNA. Prot. Sci. **5** 1973-1983.

6. Lesnik T., Solomovici J., Deana A., Ehrlich R. and Reiss C. (2000) Ribosome traffic in *E.coli* and regulation of gene expression. J. Theor. Biol. **202** 175-185.

7. Rüdiger S., Germeroth L., Schneider-Mergener J. and Bukau B. (1997) Substrate specificity of the DnaK chaperone determined by screening cellulose-bound peptide libraries EMBO J. **16** 1501-1507.

8. Komar A.A., Lesnik T. and Reiss C. (1999) Synonymous codon substitution affect ribosome traffic and protein folding during *in vitro* tarnslation FEBS Lett. (1999) **462** p387-391.

9. Komar A.A., Lesnik T., Cullin C., Guillemet E., Ehrlich R. and Reiss C. (1997) Differential resistance to proteinase K digestion of the yeast prion-like (Ure2p) protein synthesized in vitro in wheat germ extract and rabbit reticulocyte cell-free translation systems. FEBS Lett. **415** 6-10.

10. Emilsson V. and Kurland C.G. (1990) Growth rate dependence of transfer RNA abundance in *E.coli*? EMBO J. **9**4359-4366.

11. Baumeister W., Walz J., Zühl F., and Seemüller E. (1998) The Proteasome: paradigm of a self-compartmentalizing protease Cell **92** 367-380.

12. Hershko A. and Ciechanover A. (1998) The ubiquitin system In Ann Rev. Biochem, Richardson C.C. editor **67** 425-480.

AN APPROACH TO ESTABLISH ANTEMORTEM DIAGNOSIS OF DEMENTIA WITH LEWY BODIES

Hiroyuki Arai*, Makoto Higuchi, Nobuyuki Okamura, Xia Sheng Hu, Toshifumi Matsui, and Hidetada Sasaki

1. INTRODUCTION

Dementia with Lewy bodies (DLB) is a neurodegenerative disorder that is characterized by a progressive cognitive decline and the presence of numerous Lewy bodies (LBs) in the cortical and subcortical brain regions with a variable degree of co-existing Alzheimer-type pathology. DLB is likely to be the second most frequent cause of degenerative dementia after Alzheimer's disease (AD) in the elderly. In 1996, the consortium on dementia with Lewy bodies (CDLB) proposed clinical and pathological criteria of DLB. However, some groups indicated a relatively poor sensitivity of the CDLB criteria to accurately detect DLB. Early and reliable antemortem detection of DLB as well as distinguishing DLB from AD are critically important for several reasons. First, extrapyramidal features of DLB patients are successfully treated by L-dopa in most cases. Further, DLB patients seem more likely to respond to cholinesterase inhibitors than AD patients. Second, it is well documented that there is a great need to pay particular attention to life-threatening neuroleptic adverse effects in DLB. Finally, DLB may have a different rate of disease progression than AD. Here, we summarize our recent studies of functional brain imaging using positron emission tomography (PET) in an attempt to establish antemortem diagnosis of DLB.

2. CEREBRAL GLUCOSE METABOLISM USING PET WITH ^{18}F-FLUORODEOXYGLUCOSE[1, 2]

Cerebral glucose metabolism using PET with ^{18}F-fluorodeoxyglucose (FDG) was examined in 11 patients with probable AD, 6 patients with probable and 1 patient with autopsy-confirmed DLB as well as in 10 age-matched normal control subjects. Among

* Hiroyuki Arai, Department of Geriatric Medicine, Tohoku University School of Medicine, Sendai, Miyagi, 980-8574, Japan

Catecholamine Research, Edited by Nagatsu *et al.*
Kluwer Academic/Plenum Publishers, 2002

widespread cortical regions showing glucose hypometabolism in the DLB group, the metabolic reduction was most pronounced in the visual association cortex compared to that in the AD group. Using a metabolic ratio of 0.92 in the visual association cortex as a cut-off (mean-2 S.D. of normal control subjects), DLB could be distinguished from AD with a sensitivity of 86% and a specificity of 91%. In contrast, apolipoprotein E4 allele frequency and cerebrospinal fluid tau levels did not differ significantly between the 2 groups.

3. NEUROPATHOLOGICAL CORRELATES OF OCCIPITAL DYSFUNCTION[1]

In order to further dissect out neuropathological correlates of the dysfunctional occipital lobe, postmortem brains from 19 patients with AD and 17 with DLB as well as 11 brains from normal controls were examined. A distinct and extensive spongiform change with co-existing gliosis was variably noted throughout cerebral white matter with relative sparing of gray matter in DLB. Notably, the white matter spongiform change and gliosis was most prominently and consistently found in the occipital region of DLB, and the severity of the spongiform change in each brain region generally paralleled to the regional difference in reduced glucose metabolism between the living AD and DLB patients. These findings suggest that: 1) Among several potential antemortem bio-markers in the diagnosis of DLB, measures of the glucose metabolism in the occipital cortex may be an informative diagnostic aid to distinguish DLB from AD; 2) A pathological process that generates widespread spongiform change and gliosis in long projection fibers may contribute, at least in part, to the characteristic imaging features of DLB.

4. STRIATAL DOPAMINE UPTAKE USING PET WITH ^{18}F-FLUORODOPA[3]

The study consisted of six patients with probable DLB and one patient with autopsy-confirmed DLB plus AD (66.9±10.0 years, range: 51-79) as well as 10 patients with probable AD (70.5±8.4 years, range: 58-86) and eight age-matched normal controls (67.5±10.2 years, range: 52-85). There was no significant difference in age between the three groups. Age-matched normal subjects were community-dwelling healthy volunteers without any confirmed neuropsychiatric disorders. The severity of dementia as assessed by Mini-Mental State Examination was not significantly different between DLB (15.7± 6.7 points) and AD (16.3±5.0 points). Among three core features of the CDLB criteria, parkinsonism was noted in all of the DLB patients in addition to fluctuating cognition or visual hallucination, or both. Six DLB patients variably responded to L-dopa therapy. Supportive features for the diagnosis of DLB were occasionally noted. All subjects underwent PET scans using a PT931 PET scanner (CTI, USA) with 7mm axial and transaxial resolution. Following an intravenous bolus injection of 2.5-8.3 mCi of FDOPA, a series of 5 min emission scans was carried out for 60 minutes and emission data were simultaneously collected from seven contiguous axial sections. The tissue FDOPA concentration was measured by defining regions of interest (ROIs) on three image planes that included the caudate nucleus and putamen. ROIs in the caudate nucleus and putamen were defined on a summed image of data collected during the PET scan. An influx rate constant (Ki) of FDOPA into the selected regions was then calculated with radioactivity of the cerebellar hemisphere as an input function. In the DLB group, Ki values were reduced in the caudate nucleus (0.0064 ± 0.0017, $p<0.001$) and in the putamen (0.0051 ± 0.0019, $p<0.005$) compared to the AD group (caudate nucleus: 0.0119 ± 0.0021; putamen: 0.0092 ± 0.0014). However, there was no significant difference in the Ki values between the AD and the normal group either in the caudate nucleus or the putamen. Using a cut-off value of 0.0062 (mean-2SD of the normal group) in the putamen, DLB could be distinguished from AD with a sensitivity of 86% and a specificity of 100%, while a cut-off value of 0.0071 (mean-2SD of the normal group) in the caudate nucleus yielded a sensitivity of 71% and a specificity of 100%.

5. CASE PRESENTATION

Patient had been well until 55 years old when depression developed with prominent headache, anxiety and loss of interest. Tricyclic anti-depressants were tried without benefit and discontinued. At age 58, she became progressively disorientated and had memory deficits. She also developed difficulty with performing fine movements such as using chopsticks and buttoning her clothing. However, computed tomography of the brain was unremarkable. On neurological examination performed at age 59, she was emotionless and hypophonic, lost her facial expression, was slowed down and shuffled on walking with reduced arm swing. Cog-wheel rigidity was present at both wrists and elbows. Bilateral hand tremor was noted as well. She drew a clock poorly, and she scored 12 out of 30 on Mini-Mental State Examination. Deep tendon reflexes were normally active. Brain CT and MRI revealed diffuse cortical atrophy, while 7-8 Hz slow background activity was noted on EEG. On FDG-PET images obtained at age 60 showing widespread cortical hypometabolism, the occipital metabolic deficit was particularly noticeable. The glucose metabolic ratio in the visual association cortex (normalized by the radioactivity in cerebellar vermis) was 0.78, which was lower than an average value (1.02) in AD patients. Levo-dopa with carbidopa produced a transient improvement in the motor features of parkinsonism, while cognitive impairment continued to deteriorate with a persistent visual hallucination and delusion that was unrelated to the levo-dopa therapy. At age 63, she became wheelchair bound and had difficulty in verbal communication. At age 64, she was unable to feed herself because of progressive dysphasia, and died of aspiration pneumonia and congestive heart failure 9 years after the onset of the disease. Postmortem brain examination revealed numerous LBs and LB-neurites that were immunoreactive to anti-α-synuclein and anti-ubiquitin antibodies as well as abundant senile plaques and neurofibrillary changes. Hence, a final pathological diagnosis was made as "DLB plus AD". In addition to such well-documented hallmarks, there was an extensive spongiform change and gliosis throughout the white matter. The most extensive white matter change was observed in the occipital region. An extensive gliosis also was present in the white matter with relative sparing of the gray matter. No convincing α-synuclein positive structures were found in and around the white matter changes. Apolipoprotein E genotype was 3/4. There was no mutation in the α-synuclein gene.

REFERENCES

1. Higuchi M, Tashiro M, Arai H, Okamura N, Hara S, Higuchi S, Itoh M, Shin R-W, Trojanowski JQ, Sasaki H. Glucose hypometabolism and neuropathological correlates in brains of Dementia with Lewy bodies. Exp. Neurol. **162**, 247-256, (2000).
2. Okamura N, Arai H, Higuchi M, Tashiro M, Matsui T, Hu X-S, Takeda A, Itoh M, Sasaki H. [^{18}F]FDG-PET study in dementia with Lewy bodies and Alzheimer's disease. Prog. Neuro-psychopharmacology Biol. Psychiat **25**, 447-456, (2001).
3. Hu XS, Okamura N, Arai H, Higuchi M, Matsui T, Shinkawa M, Itoh M, Ido T, Sasaki H. ^{18}F-DOPA PET study of striatal dopamine uptake in the diagnosis of dementia with Lewy bodies. Neurology **55**, 1575-1576, (2000).

SYMPATHOADRENAL ACTIVITY AND CARDIOVASCULAR RESPONSES DURING SPACE FLIGHT

Niels Juel Christensen and Peter Norsk*

1. INTRODUCTION

Several ground-based simulation studies have been performed during which the gravitational stress has been reduced for shorter or longer periods of time. Simulated microgravity induces an increase in central venous pressure and a decrease in plasma norepinephrine (NE). In one experiment lasting 42 days of head-down tilted bed rest plasma NE did not change. These studies have therefore suggested that sympathoadrenal activity remains unchanged or may even be reduced during a space mission. This hypothesis has, however, been refuted.[1]

2. STUDIES OF SYMPATHOADRENAL ACTIVITY INFLIGHT

Studies during microgravity have shown that contrary to expectation, sympathoadrenal activity was increased during space flights as compared to ground based observations in the supine position.[1,2] We investigated endocrine and renal responses to isotonic saline infusion during a flight mission. We observed that forearm venous plasma NE, which mainly reflects muscle sympathetic nerve activity, was increased during the 5th to 6th days of flight. The microgravity adapted renal excretion rate of sodium in response to saline infusion was also reduced as compared to results obtained during ground-based experiments, when subjects rested in the supine position. Plasma aldosterone and renin concentrations inflight were similar to values obtained in ground-based experiments in the seated position. Consistently, all three types of measurements showed that sympathetic, endocrine and renal responses to a manned space flight were approximately similar to that observed in the seated position in ground-based

* Niels Juel Christensen, Division of Endocrinology, Herlev Hospital, University of Copenhagen, 2730 Herlev, Denmark. Peter Norsk, Danish Aerospace Medical Centre of Research, National Hospital, University of Copenhagen, Denmark

Catecholamine Research, Edited by Nagatsu *et al.*
Kluwer Academic/Plenum Publishers, 2002

experiments. Plasma NE concentration was, however, even above seated values at the end of the saline infusion experiments. Thus the hypothesis that microgravity as observed during a manned space flight can be simulated by long-term head-down tilted bed rest in ground-based experiments was refuted.

We did not measure the clearance of NE, because the clearance concept cannot readily be applied to NE.[3,4] For the plasma appearance rate to be valid and a relative index of NE release, the tracer must mix with a constant fraction of NE released at nerve endings. There is no evidence to suggest that this assumption is valid, and it is unlikely to be so during experiments, where the blood flow and the capillary surface exchange area are changed. Discrepancies between results obtained by microneurography and the appearance rate of NE during moderately lower body negative pressure are most likely due to inadequate mixing between NE released at nerve endings and tracer NE.[4] Furthermore, it has been shown in many studies that there is a close correlation between forearm venous plasma NE and muscle sympathetic nerve activity as measured by microneurography.[3]

The explanation of the increased plasma NE in space remains unclear. Cardiovascular reflexes appear to be normal during space flight. The present data are compatible with the observation that blood volume is reduced during a space flight.[5] Previous studies have indicated that there is a close negative correlation between sympathetic activity at rest and the plasma volume.

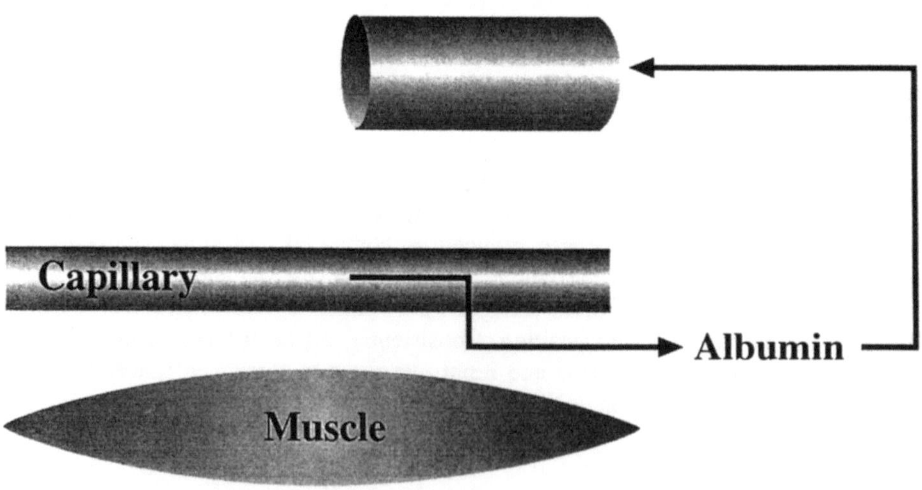

Fig. 1. Transcapillary escape rate of albumin. Albumin is returned to the vascular space through the lymphatics. This process may be altered during microgravity.

The mechanism of the decreased plasma volume observed during space flight remains to be clarified. This response is unlikely to be mediated by redistribution of blood from the lower part of the body to the cardiopulmonary area during microgravity, because the capillary surface exchange area is unlikely to be much greater above the heart than below. Furthermore, there appears to be no indication of an early absolute fluid loss at the beginning of the space flight as suggested by ground-based simulation models. Fasting decreases plasma volume and increases plasma NE. Astronauts tend to lose weight during a space flight probably due to hypocaloric nutrition.[6] The decrease in body weight is unlikely, however, to explain the decrease in plasma volume, because the reduction in plasma volume already is present within the first 24-hours after launch. Other mechanisms are therefore likely to play a role. The plasma volume is dependent on the muscle activity and the degree of physical training.[7,8] Albumin and water spill over to the interstitial fluid. The albumin and water are returned to the vascular compartment, and the transport is in part mediated by the massaging action of light contracting muscles, which increases lymph flow from the area of activity (Fig. 1). We speculate that the abrupt cessation of muscle activity during microgravity of the large muscle groups, which normally control body movement in response to gravitation may contribute significantly to the decrease in plasma volume. Decreased activity leads to accumulation of albumin and fluid in the interstitial space in these muscles. Furthermore, these muscle groups are probably only trained to a limited extent, when the astronauts perform physical training inflight. Further studies are warranted to test this hypothesis.

We have also had the opportunity to study urinary excretion rates of NE and epinephrine (E) in two astronauts during the MIR mission.[2] Urine NE and E concentrations were measured during a 300 min period at several time intervals from day 5 to day 164 during the mission. Urinary NE and E concentrations inflight were above corresponding values obtained in the supine position in ground-based experiments. The elevated excretion rates of NE and E in the astronauts were not correlated to the length of the mission.

It should be noted that urinary excretion rate of catecholamines is dependent not only on plasma concentrations of NE and E but also on the renal handling of the catecholamines. Furthermore this technique does not detect changes in NE released in the gastrointestinal tract, because NE and E in the portal vein are largely removed by the liver.

Measurements of thrombocyte NE and E may be a better index of mean 24-hours plasma NE and E concentrations. Thrombocytes travel through all parts of the organism and take up catecholamines from plasma. Thrombocyte NE and E concentrations do not change in reponse to acute changes in sympathoadrenal activity but reflect the average plasma NE and E level in the preceding one to two days. Furthermore, the ratios between plasma NE and thrombocyte NE and between plasma E and thrombocyte E appear to be constant.

3. CONCLUSION

Our results indicate that sympathoadrenal activity is increased during microgravity. Thus, the head-down tilted bed rest model applied in ground-based experiments to simulate microgravity may not mimic changes in cardiovascular function observed during a long-term space flight. The increase in sympathetic activity is probably due to a decrease in plasma volume, which is already present during the first day inflight. We suggest that the decrease in plasma volume and the subsequent increase in sympathetic activity is at least in part due to the abrupt cessation of muscle activity during microgravity of the large muscle groups, which normally counteract gravitational effects on body positions. This leads to accumulation of albumin and fluid in the interstitial space.

REFERENCES

1. P. Norsk, C. Drummer, L. Röcker, F. Strollo, N.J. Christensen, J. Warberg, P. Bie, C. Stadeager, L. B. Johansen, M. Heer, H-C. Gunga and R. Gerzer, Renal and endocrine reponses in humans to isotonic saline infusion during microgravity *J. Appl. Physiol.* **78**, 2253-2259 (1995).
2. P. Norsk, N. J. Christensen, P. Bie, A. Gabrielsen, M. Heer, C. Drummer, Unexpected renal responses in space. *Lancet* **356** 1577-1578 (2000).
3. N. J. Christensen, The biochemical assessment of sympathoadrenal activity in man. *Clinical Autonomic Research* **1** 167-172 (1991).
4. N. J. Christensen, J. H. Knudsen, Peripheral catecholaminergic function evaluated by norepinephrine measurements in plasma, extracellular fluid, and lymphocytes, from nerve recordings and cellular responses. *Adv. Pharmacol.* **42** 540-544 (1998).
5. C. P. Alfrey, M. M. Udden, C. Leach-Huntoon, T. Driscoll, M. H. Pickett, Control of red blood cell mass in spaceflight. *J. Appl. Physiol.* **81** 98-104 (1996)
6. C. Drummer, R. Gerzer, F. Baisch, M. Heer, Body fluid regulation in μ-gravity differs from that on Earth: an overview. *Pflügers Arch.* **441** (2-3 Suppl) R66-72 (2000).
7. L. C. Senay Jr, Changes in plasma volume and protein content during exposures of working men to various temperatures before and after acclimatization to heat: separation of the roles of cutaneous and skeletal muscle circulation. *J. Physiol.* **224** 61-81 (1972).
8. B. Nielsen, S. Strange, N. J. Christensen, J. Warberg, B. Saltin, Acute and adaptive responses in humans to exercise in a warm, humid environment. *Pflügers Arch. – Eur. J. Physiol.* **434** 49-56 (1997).

ESTROGEN METABOLISM: DOES THE FORMATION OF ESTROGEN QUINONE PROVIDE A POTENTIAL PATHWAY TO BREAST CARCINOGENESIS?

Cyrus R. Creveling*

Estrogen has long been implicated in the initiation of carcinogenesis in estrogen-sensitive tissues through receptor-mediated events (Girder and Brotherick, 2000). However recent studies indicate that estrogens may cause DNA damage directly and independent of receptor-mediated mechanisms through the oxidation products of estrogen metabolism. The major pathway for estrogen metabolism proceeds by two steps: First the aromatic hydroxylation of the primary estrogens to form 2- and 4-hydroxycatechol estrogens. Under normal conditions these catechols rapidly undergo O-methylation, catalyzed by catechol-O-methyltransferase (COMT). The evidence for the rapid O-methylation is several fold. While virtually all catechols are substrates for COMT the highest affinities for the enzyme are exhibited by the catechol estrogens (Ball and Knuppen, 1985). Further the O-methylated products are the major urinary estrogen metabolites. Importantly, the efficiency of the O-methylation pathway is reflected in the extremely short half-life of catechol estrogens (Lipsett et. al. 1983).

Under conditions where the capacity for O-methylation is reduced, inhibited by an excess catechol load or the presence of xenobiotic inhibitors, the half-life of the catechol quinones may be significantly increased (Weisz et al 2000). It should be noted that the activity of COMT in some individuals may be inherently compromised by a phenotypic trait of low activity. The work of Weinshilboum and coworkers clearly demonstrated that among the human population there is a bimodal distribution of COMT into genetically low activity and high activity forms. Molecular studies have identified a single nucleotide polymorphism in the COMT gene that results in a valine108methionine amino acid substitution (Vidgren, J. et al., 1994). The methionine substitution results in a protein with low enzyme activity. The polymorphic allele frequencies within a population can differ widely depending on ethnicity. While still an early approach an increasing number of genetic epidemiological studies have indicated persons with the low activity form of COMT may exhibit a significant risk factor for breast cancer (Lavigne et al., 1997; Thompson et al. 1998; Huang et al. 1999). A similar observation for the role of COMT

* National Institute of Diabetes and Digestive and Kidney Diseases, National Institutes of Health, Bethesda, MD, 20892, USA.

Catecholamine Research, Edited by Nagatsu *et al.*
Kluwer Academic/Plenum Publishers, 2002

has been made with regard to the detoxifying potential of COMT in malignant melanoma (Smit and Pavel, 1995).

With an increased lifetime the catecholestrogens can readily undergo oxidation catalyzed by peroxidases and cytochrome P450 to form their semiquinone and quinone forms. In the presence of quinone reductase a redox-cycle between the quinone and catechol forms ensues. This cycle generates cytotoxic reactive oxygen species and oxidative damage to lipids (Cavalieri and Rogan, 2000). There is some evidence that the catechol quinones can be inactivated by conjugation with a steroid specific glutathione-S-transferase. However little is known about the effectiveness of this pathway. Perhaps that most critical action of the catechol quinones is their reaction with the most reactive of the purine bases in DNA, guanidine and adenine. The reaction with guanine forms a stable adduct which is subject to DNA repair mechanisms. However the catechol quinone derived from 4-hydroxycatechol reacts with the N-3 of adenine with destabilization of the glycosidic bond and the formation of a permanent apurinic site in the DNA. It is suggested that the apurinic site is an important mutagenic site leading to breast carcinogenesis (Balu et al., 1999, Cavalieri et al., 2000). Direct evidence of the presence of a cancer phenotype in DNA from breast tissue is present in many American women (Malins et al. 1995). It should be noted that the components for the biosynthesis and metabolism of estrogen are present locally in the stromal and epithelial cells of breast tissue (Jefcoate et al. 2000).

One of the continuing speculations is that a compound or compounds in the environment is a causative factor in the initiation of carcinogenesis. It has been proposed that in the United States there are no women with "normal" breast tissues. The causative factors are believed to be so ubiquitous that all women in the United States are effected (Weisz et al., 2000; Malins et al. 1995). Some environmental pollutants have been shown to be effective inhibitors of COMT and thus may reduce the protection afforded by COMT (Garner et al. 1999; 2000). The remarkable lower incidence of breast cancer in Japanese women compared to the incidence in the United States has been attributed to a much lower environmental pollution in Japan. However the apparent pollution of the environment in Japan is just as great as that in the United States. If this is true some other factors may be the significant difference. Several epidemiologic studies have shown a low mortality in breast cancer in Japanese women on a traditional Japanese diet (Adlercreutz et al. 1991). Certain diphenolic food components may be converted to biologically active substances be intestinal micro flora which may be cancer protective-protective agents. The Japanese diet is typically low in fat, with much rice, fish, vegetables and most importantly soy products. The excretion of isoflavonoids is correlated with soybean-product intake (Davis et al 1999). The presence of soy products are a constant component of the Japanese diet, for example in miso soup. The possible anticancer effects of the phytoestrogens present in Green tea are known (Creveling, 1994). A major constituent of green tea is epigallocatechin gallate and theaflavins in black tea have been shown to have anti-cancer activity (Isemura, M. et al. 2000).

It has been established that COMT is an inducible enzyme (Creveling and Inoue, 1995; Zhu et al., 1984). The induction of COMT may be an important factor under conditions which might lead to less than optimal O-methylation in breast tissue. COMT plays a protective role which prevents the hazardous increase in the lifetime of the catecholestrogens and the subsequent oxidation of catecholestrogens to their semiquinone and quinone forms. It may be that one or more components of the Japanese diet are responsible for an induction in COMT. The induction of COMT may provide a necessary margin of protection from the genotoxic formation of the estrogen catechol quinones.

REFERENCES

Adlercreutz, H., Honjo, H., Higashi, A. Fotsis, T., Hamalainen, E., Hasegawa, T., and Okada, H., 1991, Urinary-excretion of lignanas and isoflavonoid phytoestrogens in Japanese men and women consuming a traditional Japanese diet., *American Journal of clinical Nutrition* **54**, (6), 1093-1100.

Ball, P., and Knuppen, R. 1985 Catecholestrogens (2- and 4-hydroxyestrogens): chemistry, biogenesis, metabolism, occurrence and physiological significance. *Acta Endocrinol. Suppl (Copenh)* **232**, 1-127.

Balu, N., Li, E-M, Rogan, E., and Cavalieri, E. 1999 A unifying mechanism in the reaction of catechol-, catecholamines- and catecholestrogen quinones with nucleophilic sites in DNA. *Proc. American Assoc. Cancer Res.* **40**, 46.

Cavalieri, E., and Rogan, E., 2000,The key role of catecholestrogen-3, 4-quinones in tumor initiation, in: *Role of Catechol Quinone Species in Cellular Toxicity*, C. R., Creveling, ed., F. P. Graham Publishing, Johnson City, TN pp 247-260.

Cavalieri, E., Frenkel, K., Liehr, J. G., Rogan, E., Roy, D. 2000 Estrogens as endogenous genotoxic agents—DNA adducts and mutations, *Journal of the National Cancer Institute*, USA, **27**, 75-93.

Creveling, C. R. and Inoue, K, 1995 Induction of catechol-O-methyltransferase in the luminal epithelium of rat uterus by progesterone *Drug Metabolism and Disposition* **23**, 430-432.

Creveling, C. R. 1994 Catechol-O-methyltransferase: Role in Estrogen-Induced Carcinogenesis. "Dietary Phytoestrogens: Cancer Cause of Prevention?" National Cancer Institute, Division of Cancer Etiology, Chemical and Physical Carcinogenesis Branch, The Hyatt Dulles Hotel at Washington Dulles International Airport, Herndon VA. September 21-23.

Davis, S.R., Dalsis, F. S., and Simpson, E.R. 1999 Phytoestrogens in Health and Disease, *Recent Progress in Hormone Research* **54**, 185-211.

Garner, C. E., Burka, L. T., Etheridge, A. E., and Matthews, H., B. 2000 Catechol metabolites of polychlorinated biphenyls inhibit the catechol-O-methyltransferase-mediated metabolism of catechol estrogens. *Toxicology and Applied Pharmacology* **162**, 115-123.

Garner, C. E., Matthews, H.B., and Burka, L.T. 1999 Phenolphthalein metabolites inhibit catechol-O-methyltransferased-mediates metabolism of catechol estrogens: a possible mechanism for carcinogenicity. (2000) *Toxicology and Applied Pharmacology* **162**, 124-131

Girdler, F., and Brotheric, I. 2000 The oestrogen receptors and their role in breast cancer: a review. *The Breast* **9**, 194-200.

Huang, C. S., Chern, H.D, Chang, K.J. Hsu, S.M., and Shen, C.Y. 1999 Breast cancer risk associated with genotype polymorphism of the estrogen metabolizing enzymes genes CYP17, CYP1A1, and COMT. *Cancer Researh* **59**, 4870-4875.

Inoue, M., Tajima, K., Mizutani, M., 2001 Regular consumption of green tea and the risk of breast cancer recurrence: follow-up study from the Hospital based Epidemiologic Research Program at Aichi Cancer Center(HERPACC) Japan. *Cancer Letter* **167**, 175-182.

Isemura, M., Saeki, K., and Kimura. T. (2000) Tea catechins and related polyphenols as anti-cancer agents. *Biofactors* **13**, 81-85.

Jefcoate, C. R., Liehr, J. G., Santan, R.J., Sutter, T. R., Yager, J. D., Wei, Y., Santner, S. J., Tekmal, R., Demers, L., Pauley, R., Naftolin, F., Mor, G., and Berstein, L. 2000 Tissue-specific synthesis and oxidative metabolism of estrogens. J National Cancer Institute, monograph **27**, 95-112.

Lavigne, J. A., Helzlsouer, K. J., Huang, H. Y., Strictland P. T., Bell D. A., Selmin O. An association between the allele coding for a low activity varient of catechol-O-methyltransferase and the risk of breast cancer 1997, *Cancer Research* **57**, 5493-5497.

Lipsett, M., Merriam G., Kono, S. Brandon, D., Pheiffer, D., Loriaux, D. Metabolic clearance of catecholestrogens in: *Catechol Estrogens*, Merriam, G. R. and Lipsett, M. D., eds. Raven Press, New York, pp 105-114, 1983.

Malins, D.C., Polissar, N.L., Nishikida, K., Holmes, E.H., Gardner, H.S., and Gunselman S.J. 1995 The etiology and prediction of breast cancer. Fourier transform-infrared spectroscopy reveals progressive alteration in breast DNA leading to a cancer-like phenotype in a high proportion of normal women (1995) *Cancer* **75**, 503-517.

Smit, N. P. M., and Pavel, S. 1995 Induction of cytotoxicity in melanoma cells through inhibition of catechol-O-methyltransferase *Biochemical Pharmacology* **50**, 1955-1962.

Thompson, P. A., Shields, P. G., Freudenheim J. L., Stone A., Vena J. E., Marshall J. R., 1998, Genetic polymorphisms in Catechol-O-methyl transferase, menopausal status and breast cancer risk. *Cancer Research*, **58**, 2107-2110.

Vidgren J., Svenson, L. A., and Lijas, A. 1994, Crystal structure of catechol-O-methyltransferase 1994, *Nature* **368**, 354-358.
Weisz, J., Fritz-Wolz, G., Gestl, S., Clawson, G. A., Creveling, C. R., Liehr, J. G., and Dabbs, D., 2000, Nuclear localization of catechol-O-methyltransferase in neoplastic and nonneoplastic mammary epithelial cells, *American Journal of Pathology*, **156**, 1841-1848.
Zhu B. T., and Liehr, J. G., 1996, Inhibition of catechol-O-methyltransferase-catalyzed O-methylation of 2- and 4-hydroxyestradiol by quercetin, *Journal of Biological Chemistry*. **271**, 149-159.

INDEX

MIX
Papier aus verantwortungsvollen Quellen
Paper from responsible sources
FSC® C105338

If you have any concerns about our products,
you can contact us on
ProductSafety@springernature.com

In case Publisher is established outside the EU,
the EU authorized representative is:
Springer Nature Customer Service Center GmbH
Europaplatz 3, 69115 Heidelberg, Germany

Printed by Libri Plureos GmbH
in Hamburg, Germany